MW01380026

STATISTICS

Tools for Understanding Data in the Behavioral Sciences

Eva D. Vaughan
University of Pittsburgh

Prentice Hall
Upper Saddle River, New Jersey 07458

Library of Congress Cataloging-in-Publication Data

Vaughan, Eva D.
Statistics : tools for understanding data in the behavioral sciences / Eva D. Vaughan.
p. cm.
Includes bibliographical references and index.
ISBN 0-02-422733-1
1. Social sciences—Statistical methods. I. Title.
HA29.V33 1998 97-23655
300'.1'5195—dc21 CIP

Editor-in-Chief: Nancy Roberts
Acquisitions Editor: Jennifer Gilliland
Editorial Assistant: Tamsen Adams
Director of Production and Manufacturing: Barbara Kittle
Managing Editor: Bonnie Biller
Production Liaison: Fran Russello
Project Manager: Linda B. Pawelchak
Manufacturing Manager: Nick Sklitsis
Prepress and Manufacturing Buyer: Tricia Kenny
Creative Design Director: Leslie Osher
Interior and Cover Designs: Levavi & Levavi
Cover Art: Shadows/© Peter B. Kaplan, 1982 NYC.
Electronic Art Creation: Levavi & Levavi
Marketing Manager: Mike Alread
Copy Editing: Mary Louise Byrd
Proofreading: Maine Proofreading Services

This book was set in 10/12 Times Roman by Pub-Set, Inc., and was printed and bound by Courier Companies, Inc. The cover was printed by The Lehigh Press, Inc.

Printed in the United States of America
10 9 8 7 6 5 4 3 2 1

ISBN 0-02-422733-1

Prentice-Hall International (UK), *London*
Prentice-Hall of Australia Pty. Limited, *Sydney*
Prentice-Hall Canada Inc., *Toronto*
Prentice-Hall Hispanoamericana, S. A., *Mexico*
Prentice-Hall of India Private Limited, *New Delhi*
Prentice-Hall of Japan, Inc., *Tokyo*
Simon & Schuster Asia Pte. Ltd., *Singapore*
Editora Prentice-Hall do Brasil, Ltda., *Rio de Janeiro*

Dedicated to my granddaughter,

Yuriko

May she always be as eager to learn new things as she is now

Contents

6 Relations Between Sample and Population Means and Variances 152

7 The Normal Curve 171

8 Using the *t* Distribution to Make Inferences About a Population Mean 202

9 Correlation: Measuring Relationships Between Variables 233

11 Making Inferences About Differences Between Population Means 312

12 Simple or One-Way Analysis of Variance 352

List of Figures

Chapter 13

Chapter 15

List of Tables

Chapter 13

Chapter 14

Chapter 15

Preface

Some years ago, I was asked to teach an introductory statistics class for a distance-education program. Most students in the class would have no face-to-face contact with me or other students. They would work with the course materials at their own pace, mail required exercises or homework assignments to me, and take exams, when ready to do so, at designated "testing centers." Obviously, the students would need to rely entirely on their textbook and adjunct printed materials. As I examined possible textbooks, I realized that few, if any, could be used as stand-alone learning materials. This is understandable since statistics courses typically include lectures and other class activities to augment and reinforce the text. I had two alternatives: Choose a published textbook and prepare an extensive set of supplementary learning aids or write a stand-alone book. I chose the latter course.

Earlier versions of the material in this book have been used not only in distance-education settings but also in traditional courses that meet with an instructor for 3 to 4 hours a week. The responses in both settings have been highly favorable, and both students and instructors have asked me, "Why don't you publish this as a textbook?" I finally took their advice, and the book you have in front of you is the result.

Features of this book that support student learning, whether alone or in a class, include the following:

- Objectives at the start of each chapter so that students know what they will be learning and why it is important.
- A conversational writing style to make students feel as though I am speaking to them (many of my students have commented positively on this feature).
- Repeated explanations, stated in several different ways whenever possible, of difficult concepts.
- Many examples illustrating new ideas and procedures.
- Within-chapter exercises, usually seven or more per chapter, that allow students to check their comprehension of relatively small chunks of new information before going on to the next section.
- An appendix containing worked-out answers to all within-chapter exercises. Students are able to determine whether their answers are correct, and, for incorrect answers, what they did wrong.
- End-of-chapter summaries.

In many ways, the sequence of topics in this book is like that of most introductory statistics textbooks. Perhaps the most marked departure is that the theory and practice of hypothesis testing are introduced earlier than usual (in Chapter 5), using the chi square test rather than the z test for a population mean. The reasons for this choice are that (1) frequencies in categories are more familiar to students than means and standard deviations; (2) the rationale of the test is relatively easy to understand; (3) chi square tests always use one tail of the distribution (the rather difficult concept of one-tailed versus two-tailed tests can be deferred until later, after students have mastered the logic of hypothesis testing); and (4) in practice, chi square tests are used far more often than z tests.

I have found this sequence to be very successful. However, instructors who prefer a more traditional sequence can proceed directly from Chapter 4 to Chapter 6. The z test is introduced near the end of Chapter 6. Students can be asked to read the first part of Chapter 5 just before reading about the z test. The rest of Chapter 5 can be left for later in the course.

Another feature that is, I believe, unique in this book is the inclusion of a final chapter, Chapter 15, Connections, which attempts to integrate material taught throughout the previous 14 chapters and to relate that material to other topics in statistics (e.g., meta-analysis) and beyond statistics (e.g., steps in the research process). I hope that instructors will have time to assign Chapter 15 and will find it useful. I would be especially interested in feedback from both instructors and students regarding this chapter.

Throughout the book, I have attempted to follow a piece of advice attributed to Albert Einstein: "Everything should be made as simple as possible, but not simpler." Yes, material should be presented in ways that make learning as easy as possible. But difficult concepts—and there are many of them in statistics—should not be "dumbed down." I believe that most students *can* learn statistics, even the hard stuff, if they put forth the necessary effort and if that effort is supported by a user-friendly textbook. And that is what this book is intended to be.

Many people have contributed to the writing of this textbook. I am especially grateful to the many students who have used these materials, most of them in a distance-learning setting. They have provided me not only with useful suggestions but also the motivation to persist in the task of reworking the materials into a publishable textbook. I owe a great debt, too, to my family for their patience and support. I am sure there were days when my husband wondered whether I still lived in the same house as he did. My friends, too, must have gotten tired of hearing me say, "Sorry, I can't do so-and-so. I have to work on the book." Many thanks to them for understanding.

Thanks, also, to the reviewers who carefully scrutinized early drafts: Barney Beins, Ithaca College; James Chumbley, University of Massachusetts; Susan Donaldson, University of Southern Indiana; Jack Kirshenbaum, Fullerton Community College; Linda Noble, Kennesaw State College; Carol Pandey, L.A. Pierce College; Toni Wegner, University of Virgina; and Patrick Williams, University of Houston. Their comments and criticisms have made this a better book. Any errors that remain are my own.

Lynn Cooper of the University External Studies Program at the University of Pittsburgh has done a great job preparing the manuscript to send to the publisher. Without her skills the job would have been impossible. Finally, let me acknowledge the assistance of Bill Webber, Jennifer Gilliland, and Linda Pawelchak of Prentice Hall. It has been a pleasure working with all of them.

Eva Vaughan

Basic Concepts: Variables, Measurement, and Research

INTRODUCTION

Researchers in the social and behavioral sciences use statistics to answer questions, test hypotheses, and develop theories about variables in their field of study. But what are variables, and how do researchers study them? And, above all, what is statistics? We begin to answer these questions by discussing the role of statistics in research, the concept and importance of variability, variables and their measurement, and types of research questions.

Objectives

When you have completed this chapter you will be able to

- Define statistics.
- Describe the uses and the limitations of calculators and computers in statistical analysis.
- Define and apply the following concepts: variability, variable, descriptive and inferential statistics.
- Define and distinguish between qualitative and quantitative variables and between discrete and continuous variables. Compare the precision of measurement of discrete and continuous variables.
- Differentiate between a variable and measurement of the variable. Describe and identify examples of measurement by classification into categories, measurement by ranking, and measurements of amount.
- Given a research question, decide whether it involves one variable in one population, the relation between two or more variables in one population, or differences between two or more populations.
- Given an example of a research study comparing two or more populations, decide whether the study is or is not an experiment; if it is an experiment, identify the independent and dependent variables.

What Is Statistics?

You are about to embark on a course of study that is probably quite different from any course you have taken before. Unlike most other courses in the social and behavioral sciences, it emphasizes method rather than content. "Oh," you may say, "I've taken math courses before!" And many of you may add, "Long ago, thank goodness!" or "Ugh! I hate math!" That is too bad, because math can be a lot of fun. But whether math is fun or drudgery, something to be enjoyed or something to be feared, is beside the point. *Statistics is not math.*

"Quit kidding me," you may reply, "Of course statistics is math. People who use statistics are always manipulating numbers." Yes, users of statistics perform certain mathematical operations involving primarily arithmetic and simple algebra. But, mostly, *statistics is a way (or set of ways) for extracting meaning from data.* When the data are quantitative, such as test scores or frequency counts, then certain mathematical manipulations are required. But the math is not an end in itself; rather, it helps us to reason about what the data mean and what we can conclude from them.

Suppose, for example, that an educator has collected reading achievement test scores of 1000 urban fourth-grade children. As an unorganized collection of numbers, the scores tell her nothing. She must organize the scores and perform certain operations on them to answer questions such as: What is the average performance of these children? Is there a difference between boys and girls? How do these scores compare with the children's reading scores a year ago? How do these scores compare with reading scores of fourth graders last year? Five years ago? Are there differences among schools? Are reading scores related to math scores? To IQ test scores? To socioeconomic level? The educator's questions influence the statistical techniques that will be applied to the data, and statistics helps her to answer her questions.

Or suppose a social psychologist finds that a group of 20 adolescents who have been convicted of at least one juvenile offense watch an average of 4.3 hours of TV a day, while a comparable group of 20 adolescents from the same community who have never been convicted of an offense watch an average of 2.6 hours of TV a day. Can the psychologist conclude that juvenile offenders watch more TV? It should be obvious to you that because he has studied only 40 adolescents, he can't be sure that all or even most juvenile offenders watch more TV. Statistics cannot make him sure, but it can help him decide whether he can state, with some degree of confidence, that a true difference exists between the TV-watching habits of offenders and nonoffenders.

What is statistics, then? ***Statistics*** is a set of concepts, principles, and procedures for organizing and manipulating data in order to answer the questions that led to the data collection in the first place. More than anything else, statistics involves logic and critical thinking. Statistics is mainly a set of ways that help us to *reason* about data in precise, quantitative ways. It is the "reasoning," not the "quantity," that is the most important thing.

I hope we have established that statistics is not primarily math. Nevertheless, certain numerical manipulations are necessary in statistics as a means of extracting answers to our questions from data. Although the manipulations are secondary to the reasoning, inaccurate manipulations result in wrong answers and false reasoning. The type of mathematics required includes basic arithmetic: adding, subtracting, multiplying, dividing; finding squares and square roots; and working with whole numbers, decimals, and fractions. Some elementary algebra is needed, too, for operations with positive and negative numbers, equalities and inequalities, putting numbers into formulas and doing the calculations required, and simple manipulations of formulas and equations. If you are uncertain about your skills in these areas, you can use the pretest in Appendix A to test yourself. If you need to brush up on some rusty skills, you can use the self-instructional material in Appendix B to do so.

The Use of Calculators and Computers in Statistical Analysis

When I took an introductory statistics course, many, many years ago, my fellow students and I—as well as our instructors and researchers conducting statistical analyses—had to perform calculations by hand or by using big, bulky, noisy mechanical calculators. In order to make the computations easier, we used a variety of arithmetic shortcuts. This is no longer necessary. Electronic calculators and computers have taken the drudgery out of doing statistics.

Hand-held electronic calculators range widely in the operations they can perform. The simplest do nothing but addition, subtraction, multiplication, and division. But many calculators perform complex mathematical and scientific operations, such as logarithms and trigonometric functions. There are even statistical calculators that automatically calculate means, standard deviations, correlation coefficients, and other statistical functions you will learn about in this book.

If you plan to use a calculator in conjunction with this book, you may wonder what kind you should purchase. At the least, a calculator to be used for introductory statistics should perform two kinds of operations besides basic arithmetic: automatic square root and memory storage. The square root function is important because one of the steps in many statistical procedures is the extraction of square roots. Memory is important because it makes certain sequential operations easier. Of course, a statistical calculator can save you some computational labor, and your instructor may recommend or require that you buy one. I strongly recommend, however, that when you first encounter a new procedure, you do it step by step several times; you will understand it better as a result.

Most of the statistical analyses discussed in this book are simple enough, and the data sets are small enough that a computer is not necessary. Modern researchers, however, are conducting increasingly complex statistical analyses, with increasingly large sets of data. For such analyses, computers are indispensable. A computer can complete in seconds an analysis that would require weeks or even months using a calculator.

There are many statistical software packages for computers. Major ones, for example, are the Statistical Package for the Social Sciences (SPSS), Biomedical Data Analysis Package (BMDP), and Statistical Analysis System (SAS), as well as Minitab, a package designed specifically for students. All of these are available for both mainframes and microcomputers. There are also numerous smaller packages for microcomputers. Your course may be using one or another computer package in conjunction with this book. Learning to use a statistical software package for data analysis can be very useful, especially if you are going on to more advanced statistics courses or to research settings in which you will need to conduct statistical analyses of large data sets. In this book we occasionally mention the use of computers for particular procedures and/or give examples of the results of computer analyses.

You need to remember one thing about computers, however. Although computers can perform many, many complex operations and can do so much faster than you can, they cannot decide which analysis to perform or what the results mean in terms of answering the research question. Computers have no mind of their own (at least not yet) and can do only what you tell them to do. If you tell a computer to conduct an analysis that is not appropriate for the question being asked, it will cheerfully (and speedily) do so; if instructed, it will present you with a beautiful printout of the results. The only problem is that if the wrong analysis was selected, the results are meaningless. You, the human researcher, must decide what type of statistical analysis is needed to answer the question you are asking.

Even if you have selected an appropriate analysis for a set of data, it is *you* who must interpret the results, whether they have been generated by a computer or by hand, and decide what they mean. Suppose, for example, that you have used a computer program to analyze data comparing the TV-watching habits of children and teenagers. The computer printout tells you that "$t = 4.836$, $p = .002$." What does this mean? Can you conclude that the TV-watching habits of children and teenagers are different? You, the user, must interpret the information provided by the computer to make decisions like this.

As these examples show, conducting a statistical analysis involves three stages:

1. Selecting an appropriate statistical procedure or procedures to answer the question
2. Applying the procedure (number-crunching)
3. Interpreting the results and drawing conclusions

All three stages are important. Errors at any stage can result in incorrect conclusions. The computer can help you immensely by doing the work at Stage 2, often the lengthiest and most tedious stage. But you, the user, must do the work at Stages 1 and 3. So think of the computer as a helpful and sometimes indispensable tool. A tool in the hands of a knowledgeable worker can do wondrous things. But without the hands and mind of a qualified user to guide it, a tool is useless. That is why the emphasis in this book is on the rationale of various statistical procedures and on the interpretation of the results—in other words, on statistical reasoning. The purpose is to make you a wise and thoughtful user of statistical tools, whether you perform the operations themselves by hand, with a calculator, or with a computer.

This textbook can be used with or without statistical analysis software. There are no data sets or problems so large or unwieldy as to require computer analysis. Nevertheless, it is useful for you to know a little about how computers perform the operations you are learning to do. Thus, many of the chapters include brief discussions of computer analyses and examples of the results. But keep in mind that these examples apply only to Stage 2 of statistical analysis. The computer analyses that are shown have, in all cases, been preceded by an informed choice of these particular procedures (Stage 1) and must be followed by well-reasoned interpretation (Stage 3).

EXERCISE 1–1

How would you respond to this argument: "You don't need to learn statistics. Nowadays, computers can do all the statistics for you."

Some Basic Concepts

Before we begin to "do" statistics, some basic concepts must be defined and discussed. These include variability, descriptive and inferential statistics, types of variables, and methods of measurement.

Variability

Imagine a world in which everything that happens to people, animals, and institutions always has a precise, constant effect. In such a world every additional pellet given to a rat (any rat) raises its bar-pressing rate by exactly 10%. And every smile given to a child (any child) raises its self-esteem by precisely 2 points on a self-esteem inventory. At exactly 11 months and 16 days of age, every baby says its first word (which is always "Mama"), and the addition of each dissenter to a discussion group lowers the conformity index by precisely .08.

Such a world would still have a need for social scientists, whose job it would be to determine the precise quantitative laws of their field. But statisticians would be out of work, and you would be spared the agony of taking an introductory statistics course, for *statistics exists for the purpose of describing and explaining variability*. Without variability there is no need for statistics.

What is variability? *Variability* is the opposite of constancy. Variability is the fact that, even given equal numbers of pellets and identical treatment, different rats show different gains in bar-pressing rate. Variability is the fact that a smile has different effects on different children; that all babies don't say their first word at the same age (nor is the first word always "Mama"); and that the addition of one dissenter to a discussion group has different effects on different groups, even under highly similar conditions.

Life in a variable world is undoubtedly more interesting than in a constant world. But variability complicates the task of determining the laws that govern the events in question. We can't talk about the precise effects of each additional pellet of food or dissenter to a group but only the average effects. The picture gets even more complex when we try to make comparisons. Suppose, for example, that we want to compare the effects of smiles and verbal praise on children's self-esteem. In a world without variability, we need smile at and praise only one child, for each child is exactly the same as each other child. If a smile increases the child's self-esteem 2 points and praise only 1, we can confidently state that smiles have a greater effect. But in our variable world, a smile may raise one child's score by 3 points, another's by 1, still another's by 5, and so on. Similarly, the effect of praise will vary from one child to another.

Therefore, we must study many children, smiling at some and praising others, before we can conclude that a difference does or does not exist. And even if we do find a difference in the average self-esteem of the two groups of children we test, we must question whether the difference would have been the same for other groups or for children in general. Statistics helps us both to describe the behavior of the children we test and to generalize to other children as well.

Descriptive and Inferential Statistics

Some statistical procedures are applied to data for the purpose of describing the data at hand. The body of statistical operations designed to describe available data is called ***descriptive statistics***. Thus, if you want to describe the self-esteem scores of 50 children you have tested, you will use descriptive statistics to do so.

Usually, however, we seek to generalize beyond the data we have collected, as when we use the data collected on 50 children to draw conclusions about the effect of smiles on the self-esteem of children in general. The entire group of people, objects, or events we seek information about is called a ***population***. Thus, children in general is a population. The subset of the population that we actually study (the 50 children in this case) is a ***sample***. The body of statistical operations designed for drawing conclusions about populations from samples is called ***inferential statistics***, for it deals with making inferences about populations from sample data.

Descriptive and inferential statistical procedures are distinct, and they are generally described and discussed separately. It is well to keep in mind, however, that statistical analyses almost always include elements of both. Answers to research questions almost always require, first, good description of the available data, and second, accurate generalizations and conclusions drawn from the data.

As an example, consider a cognitive psychologist who has conducted a study to compare men's and women's problem-solving styles. He has collected data from samples of 50 men and 50 women. He first needs to examine the data themselves. For this purpose, he constructs tables and graphs, calculates averages, and the like (descriptive statistics); then he needs to apply inferential statistical procedures in order to generalize the results beyond the 100 members of his samples to the populations of men and women in general.

Types of Variables

Variables are the building blocks of science. Each science studies certain classes of variables and their interrelationships. For example, medical scientists study variables that influence health and disease. Psychologists study behavioral and mental process variables and factors that influence behavior and mental processes. Political scientists study variables relating to the behavior of political institutions and events.

What is a variable? Basically, a ***variable*** is any property or characteristic of individuals, objects, or events that can take on different values. Height is a variable because people and objects have different heights. Intelligence is a variable because people differ in intelligence. Gender is a variable because people and animals differ in gender. Political affiliation is a variable because people have different political affiliations. As you can see from these examples, a variable may take on a large number of values (e.g., height, intelligence) or as few as two (e.g., gender, with two values, male and female).

Qualitative and Quantitative Variables

Some variables of interest to scientists are qualitative, whereas others are quantitative. ***Qualitative variables*** *differ in kind, not amount.* Gender is an example of a qualitative variable. Although males and females are undoubtedly different, neither group has more or less gender than the other. Other examples of qualitative variables are political affiliation, marital status, and psychiatric diagnosis.

Quantitative variables, on the other hand, *differ in amount.* Height and intelligence are examples of quantitative variables. A person who is 67 inches tall has more height than a person who is 64 inches tall. A person with an IQ score of 130 has more intelligence than a person with an IQ score of 85. Other examples of quantitative variables include number of children per family, anxiety, and amount of knowledge of statistics.

Discrete and Continuous Variables

Some quantitative variables, by their very nature, can take only certain values, with gaps in between. A family can have 2 children, or 3, but not 2½ or 3¼. A shopper in a department store can buy 6 items, or 7 or 8 items, but not 6½ or 7.999 items. Such variables are called ***discrete variables*** because they take only specific values. As you can see, discrete variables are usually measured by counting.

Most variables of interest to social and behavioral scientists, on the other hand, are not discrete but ***continuous***. Hypothetically, at least, they may take any value between the lowest and highest possible. No matter how close two observations are—of intelligence, aggressiveness, dominance, or group cohesion—there could always be an observation somewhere between; there are no gaps. We can represent the difference between discrete and continuous variables as follows:

lowest . . . . highest	lowest ———— highest
Values of a Discrete Variable	Values of a Continuous Variable

When we measure a discrete variable, we can do so very exactly. Six items purchased in a department store means precisely 6.000 . . . items on the nose—no more, no less. But continuous variables can only be measured to some degree of approximation, depending on the precision of the measuring instrument. Consider the continuous variable length. Suppose that the actual length of an object is 8.736580216 . . . mm. If we measure its length with a ruler accurate to the nearest millimeter, we report its length as 9 mm. If the measuring instrument is accurate to the nearest half millimeter, we report its length as 8.5 mm. With a measuring instrument accurate to the nearest tenth of a millimeter, we report the length as 8.7 mm. And so on.

What these examples show is that a reported measurement of a continuous variable is not an exact value but represents an interval. If we use a ruler accurate to the nearest millimeter, and report that an object is "9 mm long," we are really saying that its actual length is closer to 9 mm than to either 8 mm or 10 mm; that is, its actual length is somewhere in the interval 8.5 mm to 9.5 mm. Similarly, an object that is reported to be "13 mm long" has an actual length somewhere in the interval 12.5 to 13.5 mm.

The limits of the interval for a reported measurement are called the ***exact limits*** of the interval. Thus, the exact limits of the reported length "9 mm" are 8.5 and 9.5 mm. Similarly, the exact limits of "13 mm" are 12.5 and 13.5 mm, and the exact limits of "10 mm" are 9.5 and 10.5 mm. The relation between reported measurements of length and their exact limits can be shown graphically like this:

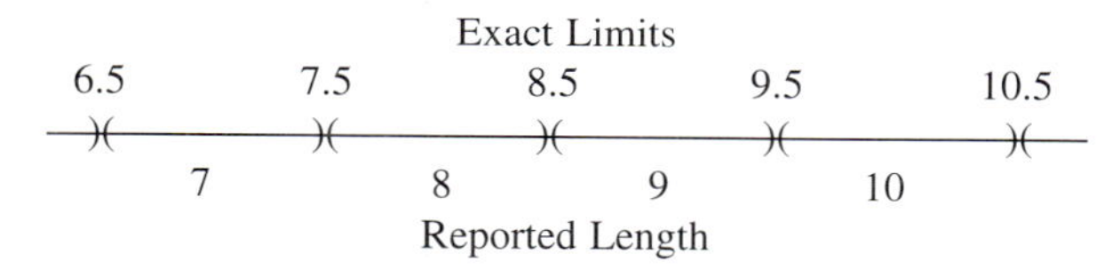

In later chapters of this book, when we work with continuous variables, we will sometimes use reported measurements, and at other times we will use exact limits.

The Measurement of Variables

In order to study variables, we need to measure them. For example, if you want to conduct research on anxiety, you must first find or develop a method of measuring anxiety. The result of a measurement operation is typically a number such as "height = 63 inches" or "IQ = 90" or "anxiety test score = 26." In fact, measurement can be defined as the assignment of numbers to individuals, objects, or events through the application of some systematic procedure.

When you learned arithmetic you did not have to worry about different kinds of numbers; in mathematics, a 9 is a 9 is a 9. But when numbers result from measurement operations, as they do in research, the types of statistical operations that can be applied to the numbers may differ, depending on the method of measurement. So we need to talk about three kinds of measurement operations: measurement by classification of observations into categories, measurement by ranking, and measurement that yields amounts or scores.

Measurement by Classification Into Categories

When we "measure" qualitative variables, we do so by classifying individuals, objects, or events in one of two or more distinct categories. For example, to measure voters' political affiliations, I can classify each voter as Democrat, Republican, Independent, or "Other." I may assign numbers to the categories for coding purposes (such as Democrat = Category 1, Republican = Category 2, Independent = Category 3, Other = Category 4). However, these numbers are simply labels; they do not in any way imply differences in amount. For example, though I have coded Democrat as 1 and Republican as 2, this does not mean that Republicans have more of the variable "political affiliation" than Democrats.

Quantitative variables are sometimes measured by classification, too. But the categories that result are ordered rather than unordered. Consider, for example, the quantitative variable "progress toward college graduation." We usually measure this variable by classifying students as freshmen, sophomores, juniors, and seniors. The four categories are ordered, in that freshmen have made the least progress toward graduation, sophomores a little more, juniors still more, and seniors the most. If we assign numbers to the categories, we could use freshman = 1, sophomore = 2, junior = 3, and senior = 4; or we could start at the other end and use senior = 1, junior = 2, sophomore = 3, and freshman = 4. We would *not*, however, assign numbers such as freshman = 1, junior = 2, sophomore = 3, and senior = 4. We would take care that the order of the numbers preserves the order of the categories.

Certain statistical procedures are specifically designed for analyzing categorical data (both unordered and ordered). One example is the chi square test (discussed in Chapter 5). Many other statistical procedures, however, cannot meaningfully be applied to categorical data. For example, if half the people in a group are Democrats (coded 1) and the other half are Republicans (coded 2), it makes no sense to state that the average political affiliation of the group is 1.5.

Measurement by Ranking

The process of ranking is familiar to most people. In high school, your class rank described your academic standing relative to your classmates'. In sports, the order of the teams in the standings (first, second, third, etc.) is an example of ranking. The measurement operation of ranking is usually easy to apply. For example, if I want to measure the heights of the children in a class by ranking, I can have the children line up in order by height; then I assign the rank "1" to the tallest child, "2" to the next tallest, and so on.

Let us consider some characteristics of ranks. What do ranks mean, and what can we do with them? First, ranks are relative to the particular group in which they are measured. The same rank has different meanings depending on the size of the group. A rank of fourth tallest in a group of 4 children is quite a different thing from a rank of fourth tallest in a group of 40. Moreover, as the composition of a group changes, ranks of individual observations change. If Mary is the fourth tallest child in her class, but a new child who is taller than Mary enters the class, Mary's rank changes to 5. If one of the three tallest children leaves the class, Mary's rank changes to 3.

Besides being relative to the composition of the group, ranks tell us nothing about actual amounts of the variable. For example, is the tallest child (rank 1) 63 inches tall, or 61 inches tall, or 65 inches tall? From the ranks alone, it is impossible to tell. Nor do ranks provide information about differences in amount. For example, the child with rank 1 may be 2 inches taller than the child with rank 2, who in turn is only half an inch taller than the child with rank 3. Conversely, there may be little difference in height between the two tallest children, but a large difference between the second and third.

Because of these limitations, many common statistical procedures are not appropriate for analyzing data in the form of ranks. Some statistical procedures, on the other hand, are specifically designed for use with ranks. Some examples are the Spearman rank-order correlation coefficient (described in Chapter 9) and the nonparametric tests described in Chapter 14.

Measurements of Amount

Usually when we use the word *measurement*, we refer to a third type of operation, that of applying some kind of measuring instrument, calibrated in some unit of measurement, and reporting the amount of the variable in terms of the number of units on the measuring instrument. We give children intelligence tests and report their scores in IQ units, such as 75 IQ points or 112 IQ points. We use a thermometer for measuring temperature and report the results in units of degrees centigrade or Fahrenheit. We measure length with a ruler and report the results in units of inches or centimeters. A great many statistical procedures, including most of those in this book, are designed for use with measures of amount.

Some people have argued that statistical procedures designed for measures of amount should be reserved for data resulting from the application of physical measuring instruments such as thermometers and rulers, whereas data resulting from social or psychological measurement should be treated as little better than ranks. The reason, they argue, is that only physical instruments have units of measurement that are known to be constant throughout the measurement scale. For example, 1 degree Fahrenheit is the same amount of heat, whether we add 1 degree to 20 degrees or to 80 degrees. We cannot be equally confident that 1 point of IQ represents the same amount of intelligence at all levels of intelligence.

This argument has some merit. Nevertheless, it is clear that measurement with a well-constructed social or psychological scale yields more information than just ranks. Suppose, for example, that Ann, Ted, and Mary have IQ scores of 70, 75, and 110, respectively. We can be sure that the difference in intelligence between Ann and Ted is much smaller than the difference between Ted and Mary. If we treat the scores simply as ranks of 1, 2, and 3, we ignore much of the information conveyed by the scores. Because statistical analyses of amounts utilize more of the information conveyed by scores than do analyses of ranks, they are generally the methods of choice for behavioral science data. Fortunately, practical experience and research have shown that the application of these procedures to the results of social and psychological measurement generally yields meaningful results.

In summary, statistical procedures applied to data depend, in part, on the method of measurement. In statistics, a 9 is not a 9 is not a 9. If the 9 refers to one of several categories, techniques designed for analysis of categorical data are appropriate. If the 9 is someone's rank in a group, techniques designed for ranks are appropriate. Finally, if the 9 represents 9 units on a measuring instrument—either a physical instrument or a social or psychological one—techniques designed for measurements of amount are appropriate.

EXERCISE 1–2

1. Decide whether each of the following variables is qualitative or quantitative. If quantitative, state whether it is discrete or continuous.
   a. Neuroticism
   b. College major
   c. Number of correct turns in a maze
   d. Prices of articles in a supermarket
   e. Type of emotion (fear, anger, etc.)
   f. Anxiety
2. What kind of measurement—classification into categories, ranking, or measurement of amount—is involved in each of the following?
   a. Patients admitted to a mental hospital are diagnosed as schizophrenic, depressed, bipolar, or other.
   b. The patients' level of anxiety is measured by administering a carefully constructed 30-item test. Scores on the test range from zero (least anxious) to 30 (most anxious). Patient A scores 5, B scores 27, and C scores 29. The staff psychologist correctly concludes that A is much less anxious than B and C, but the level of anxiety of B and C differs very little.
   c. Hardness of minerals is determined by the scratch test. That mineral that can scratch all the others has hardness = 1, the one that is scratched by #1 but scratches all the others has hardness = 2, and so on.
   d. Army trainees who complete a gunnery course are labeled "proficient," "adequate," or "in need of additional training."
   e. A food critic orders 10 city restaurants from best to worst.

More About Research Questions and How They Are Answered

We have established that statistical operations are applied to data to answer research questions. But what kinds of research questions are there? And what methods do researchers use to answer them?

There are many ways to classify research. For our purposes, it is useful to distinguish among three types: (1) research designed to answer questions about the status of one variable in a single population; (2) research designed to answer questions about the relationship between two or more variables measured in one population; and (3) research designed to answer questions about differences between two or more populations.

Questions About One Variable in One Population

What are the attitudes of Tooterville citizens toward the proposed library construction project? How are mental hospital patients distributed across categories of mental illness? What is the level of reading achievement of American sixth graders? All these are questions about the status of one variable (attitudes, type of mental illness, reading achievement) in one population. Questions like these are usually addressed by selecting a sample from the population of interest (a sample of Tooterville voters, a sample of mental hospital patients, a sample of American sixth graders), measuring the variable in the sample, and performing statistical operations both to describe the sample and to generalize to the population.

The validity of generalizations from a sample to a population obviously depends on the representativeness of the sample, and we will spend some time discussing sampling procedures in Chapter 4. In later chapters, we will see that the particular statistical procedures applied to data depend, to some extent, on the method of sampling.

Most research in the social and behavioral sciences involves two or more variables and/or two or more populations. Nevertheless, procedures for describing and drawing conclusions from studies of one variable in one population lay the foundation for understanding how to deal with more variables or populations. Therefore, most of the material in the early chapters of this book deals with methods for describing and making inferences from measurements of one variable in a single group of individuals, objects, or events.

Questions About the Relations Between Two or More Variables in One Population

How are reading scores of sixth graders related to their math scores? Is there a relation between marijuana use and use of hard drugs in drug users? What are the relationships among the personality traits of aggressiveness, dominance, and anxiety? Can we use the relation between scores on the SAT and college grade point average (GPA) to predict the GPA of college applicants? All of these are questions about the relationship between two or more variables measured in one group. As is the case with one-variable research, we are usually interested not only in describing the relationship between the variables in the group for which we have data (the sample) but also in making inferences about the relationship in the population from which the sample was drawn.

Research designed to determine whether and how two or more measured variables are related is sometimes called ***correlational research***. The analysis of data from correlational research, as well as applications and limitations of such research, is the topic of Chapters 9 and 10.

Questions About Differences Between Two or More Populations

Do male and female college students use different strategies for coping with stress? Are patients with tuberculosis more likely to recover if treated with a new drug than with an old drug? Are there differences in effectiveness of three methods of pilot training? Do children who attended preschool do better in first grade than children who did not attend preschool? All of these are questions about whether and how two (or more) populations differ on some variable of interest to the researcher. Studies of differences between populations are extremely common in the behavioral and social sciences. We refer to them as ***comparative studies***.

Although methods of data collection and analysis differ in correlational and comparative research, comparative studies like correlational studies examine relations between variables. To show this, we reword the research questions of the previous paragraph as follows:

Research Questions Stated in Terms of a Difference Between Populations	Research Questions Stated in Terms of a Relationship Between Variables
Do male and female college students use different strategies for coping with stress?	Is there a relation between gender and strategies for coping with stress?
Are patients with tuberculosis more likely to recover if treated with a new drug than with an old drug?	Is there a relation between drug administered and recovery rate?
Are there differences in effectiveness of three methods of pilot training?	Is there a relationship between method of training and flying ability?
Do children who attended preschool do better in first grade than children who did not attend preschool?	Is there a relation between preschool attendance and achievement in first grade?

The general procedure in conducting a comparative study is to (1) take two (or more) groups that differ on one variable (gender, type of drug, training method, preschool attendance) and (2) compare their performance on the other variable (coping strategies, recovery rate, flying ability, first-grade achievement).

Experimental and Nonexperimental Comparative Research

It is important to distinguish between two types of comparative research, experimental and nonexperimental. In ***experimental comparative research***, the researcher *manipulates* one variable to determine its effect on the other. As an example of an experiment, consider a study comparing an old and a new drug in the treatment of tuberculosis. The researcher manipulates drug treatment by administering the old drug to one group of tuberculosis patients and the new drug to another group; after a suitable period of time, the recovery rates of the two groups are measured and compared. Similarly, in a study comparing three methods of pilot training, the researcher manipulates the training method by assigning some trainees to Method A, others to Method B, and still others to Method C; after training is completed, the flying ability of the three groups is measured and compared. In experiments, the variable that is manipulated by the experimenter (type of drug, method of training) is called the ***independent variable***, and the other variable (recovery rate, flying ability) is called the ***dependent variable*** because it is presumably dependent on the independent variable.

Thus, the procedure in an experiment is roughly as follows:

1. Divide a pool of available research participants (subjects) into subgroups.
2. Administer a different level, amount, or treatment of the independent variable to each subgroup.
3. Measure the dependent variable and compare the groups.

In ***nonexperimental comparative research***, on the other hand, subjects already differ on both variables before the study begins; neither variable is manipulated by the researcher. Consider a study of the relation between gender of college students and their strategies for coping with stress. The researcher obviously cannot take a genderless group of college students and assign half to the gender "male" and the other half to the gender "female"; college students have differed in gender since long before the study was undertaken. Instead, the researcher identifies or selects a group of male college students and a group of female college students and then measures and compares their coping strategies. Similarly, a researcher who is studying the relation between preschool attendance and first-grade achievement cannot send some 3- and 4-year-old children to preschool and withhold preschool from others. Instead, the researcher identifies a group of first-grade children who attended preschool and a second group who did not and then compares their achievement at the end of first grade. Thus, the procedure in a nonexperimental comparative study is as follows:

1. Select a sample from each of the populations to be compared.
2. Measure the variable of interest and compare the groups.

Some people use the terms *independent variable* and *dependent variable* when talking about nonexperimental as well as experimental research. For example, in the study relating gender to coping strategies, gender may be referred to as the independent variable and coping strategies as the dependent variable; in the comparison of preschool attenders and nonattenders, preschool attendance may be referred to as the independent variable and first-grade achievement as the dependent variable. However, the fact that the "independent variable" in such studies has not been manipulated has certain consequences for the interpretation of the results. As you will see, methods for analyzing data from experimental and nonexperimental comparative research are much the same, but the conclusions that can be drawn differ in important ways. Methods for analyzing and interpreting data from comparative studies are the subject of Chapters 11 to 14.

EXERCISE 1–3

1. Classify each of the following research questions in terms of whether it concerns the status of one variable in one population, the relation between two or more variables measured in one population, or differences between two or more populations:
   a. In small groups, do 3-year-old children interact in different ways than do 4-year-old children?
   b. What percentage of the U.S. population is likely to vote for the Republican candidate in the next presidential election?
   c. Does perception of tones differ under three different levels of background noise?
   d. Does the use of videos in training improve the performance of students in an archery course?
   e. In adults, what is the relation between age and muscular strength?
   f. What percentage of the gizmos rolling off this assembly line is defective?
   g. Does addition of a lab component improve performance in statistics as compared with a course having no lab?
2. For each of the questions that asks about differences between populations, (a) decide whether the study that addresses the question is likely to be an experiment or a nonexperimental comparative study; (b) if it is an experiment, identify the independent and dependent variables.

SUMMARY

Statistics is a body of concepts, principles, and procedures for extracting meaning from data. Researchers collect data to answer questions about ***populations*** of individuals, objects, or events. Because populations are usually too large or inaccessible to study as a whole, researchers obtain information from subsets of populations called ***samples***. Statistical procedures for describing sample data are called ***descriptive statistics***, and procedures used to generalize from samples to populations are called ***inferential statistics***.

Any statistical analysis consists of three stages: (1) deciding which statistical procedure(s) is (are) needed to answer the research question; (2) applying the procedure(s); and (3) interpreting the results and drawing conclusions. The mathematical operations required in the second step can be performed by calculators and computers; however, the selection of procedures and interpretation of results can be done only by people.

Statistical analyses are conducted on measures of variables. A ***variable*** is anything that can take two or more values. Variables that differ in kind are called ***qualitative***; those that differ in amount are ***quantitative***. ***Discrete variables*** take on only certain values within their range and can be measured precisely. ***Continuous variables*** can take any value within their range, and can only be measured approximately.

In order to study variables, they must first be measured. Methods of measurement include classification into unordered or ordered categories, ranking, and application of instruments that measure amounts or scores. The choice of a statistical analysis is influenced by the method of measurement.

Research studies can be classified into three categories, which generally require different kinds of statistical analyses. Some studies involve the status of one variable in one population; other studies, ***correlational studies***, concern the relation between two variables measured in one population; and still other studies, ***comparative studies***, concern differences between two or more populations. There are two subclasses of comparative studies, ***experimental*** and ***nonexperimental***. In experiments, an ***independent variable*** is manipulated to determine its effect on a ***dependent variable***. In nonexperimental studies, samples are selected from the populations concerned and compared on the variable of interest.

QUESTIONS

1. Define the following concepts in your own words: variable, descriptive statistics, inferential statistics.

2. List the three stages of statistical analysis, and discuss the role of reasoning and computation in each stage.

3. You overhear the following conversation between two students:

   Student 1: "I'm taking statistics next term."
   Student 2: "Me too."
   Student 1: "It will be a breeze! I got an *A* in algebra last year."
   Student 2: "Lucky you! I've never been good at math. I'll probably flunk statistics."
   Student 1: (turning to you) "You're taking statistics this term. What do you think?"

   What is your reply?

4. Identify two or three variables in the field you are studying. For each one, tell whether it is qualitative or quantitative and, if quantitative, whether it is discrete or continuous. Tell how the variable is usually measured, and decide whether the measurement procedure is an example of classification into categories, ranking, or measurement of amount.

5. Classify each of the following variables as discrete or continuous:

   a. Interest in sports
   b. Outcome when tossing two dice
   c. Socioeconomic level
   d. Aptitude for learning languages
   e. Number of chairs per classroom

Questions 6 to 8 are based on the following four questions from a voter survey.

1. How many hours of TV do you watch per day? ______

2. What is your marital status?

   (1) Single ______ (2) Married ______
   (3) Divorced ______ (4) Widowed ______

3. What is your level of involvement in local politics? (Check one of the following.)

   (1) I am not involved in local politics ______
   (2) I observe the process closely but am not active ______
   (3) I am actively involved, but I do not hold a political office ______
   (4) I hold a local political office ______

4. Four activities are listed below. Place a (4) beside the activity you enjoy the most, a (3) beside the second most enjoyed activity, a (2) beside the next most enjoyed activity, and a (1) beside the activity you enjoy the least.

   Going to movies ______
   Attending athletic events ______
   Reading the newspaper ______
   Shopping for clothes ______

6. Classify each question in the survey as an example of measurement of amount, ranking, or categorical measurement. In cases of categorical measurement, tell whether the categories are ordered or unordered.

7. For which survey question or questions can you say that an answer of (3) denotes a higher level of the variable than an answer of (2)?

8. For which survey question or questions can you say that the difference in amount of the variable is the same between answers of (1) and (2) as between answers of (2) and (3)?

9. Classify each of the following research questions according to whether it asks about the status of one variable in one population, the relation between two variables in one population, or a difference between populations:

   a. What are the attitudes of college students toward required language courses?
   b. Do male and female attitudes toward required language courses differ, and, if so, how?
   c. How are college students' attitudes toward required language courses related to their grade point averages?
   d. Do workers given three short coffee breaks a day produce more than those given two somewhat longer coffee breaks?
   e. Can we use the scores of job applicants on the XYZ test to predict their productivity on the job?
   f. What is the average reading ability of sixth-grade students in the Perrytown school district?

10. In each of the following comparative studies, (1) state the research question in terms of a relation between variables; (2) classify the example as an experimental or a nonexperimental comparative study; and (3) if it is an experiment, identify the independent and dependent variables:

   a. Are the methods mothers use to discipline their children affected by the size of their (the mothers') families? A psychologist identified three groups of mothers: those with no siblings, those with one sibling, and those with two or more siblings. Next, she observed the interactions of each mother with her children. A statistical analysis was conducted to determine whether mothers with different numbers of siblings used different methods of discipline.
   b. What is the effect of prior knowledge on comprehension of text? Two groups of college students—a group of political science majors and a group who had taken no political science courses—read an article analyzing recent political events. After reading the article, the students were asked questions to measure their comprehension, and the comprehension scores of the two groups were compared to find out whether the political science majors' level of comprehension was higher.
   c. What is the effect of prior knowledge on comprehension of text? One group of college students attended a lecture about the history of an imaginary country named

Bassalonia, whereas a second group did not attend the lecture. Next, both groups read a news article about a recent government upheaval in Bassalonia. The researcher measured recall of the article to find out whether the lectured group recalled more facts from the news article.

d. Is biofeedback effective in reducing blood pressure? Twenty patients with high blood pressure were assigned to either biofeedback training or drug treatment. After 2 months, the average resting blood pressure of the two groups was compared.

Displaying Distributions in Tables and Graphs

INTRODUCTION

The process of research often generates large amounts of data, such as 1000 test scores or color preferences of 100 children and adults. The set of measurements of a variable in a group is called the ***distribution*** of the variable. In order to describe a distribution, we reduce the data set to a more compact, manageable form. As a first step, this usually involves ordering the data in some way and presenting them as a table called a frequency distribution or as a graph. When we have done this, we can answer such questions as "What is the range of the test scores?" "Are there any patterns or trends in the data?" "What are the typical color preferences of children, and how do they differ from those of adults?" In this chapter you will learn how to construct tables and graphs of data and how to use the information in tables and graphs to describe the entire distribution, as well as each individual's performance relative to the group.

Objectives

When you have completed this chapter, you will be able to

- Construct frequency distributions and bar graphs of categorical data.
- Construct and interpret contingency tables.
- Construct ungrouped and grouped frequency distributions of quantitative data.
- Determine the exact limits and the midpoint of a class interval or of a score.
- Construct cumulative frequency and cumulative percentage distributions.
- Construct and interpret histograms, frequency polygons, stem-and-leaf plots, and cumulative frequency and cumulative percentage graphs.
- Describe and identify examples of skewed, symmetrical, bimodal, and rectangular distributions.
- Distinguish between theoretical and obtained distributions.
- Describe the relationship between areas under sections of a curve and proportions of observations.
- Define, calculate, and interpret percentiles and percentile ranks.

- Define the first, second, and third quartiles of a distribution. Use distances between quartiles to determine whether a distribution is symmetrical, negatively skewed, or positively skewed.
- Use the quartiles of a distribution to construct box-and-whisker plots.

Tabulating and Graphing Categorical Data

One Variable Measured in One Group

Suppose that 20 people have been asked the following question in an opinion poll: "Some people believe that members of Congress should be limited to two terms. Indicate whether you strongly agree, agree, are neutral, disagree, or strongly disagree." As you can see, the response options form a set of ordered categories. The raw data may look like those in Table 2–1a. A mental health center classifies each new client according to type of problem, a set of unordered categories. The diagnoses of 30 new clients are shown in Table 2–1b.

Construction of tables, called ***frequency distributions***, is the first step in ordering and summarizing data. To construct a frequency distribution for categorical data, first make a list of the categories. Then count the number of individuals, objects, or events (observations) in each category. The number of observations in a category is called the ***frequency*** for the category, sometimes abbreviated f. Frequencies are entered in a column to the right of the category list, as in Table 2–2. The total number of observations, which is the sum of all the frequencies, is symbolized N. Thus, for the opinion poll data, $N = 20$; for the mental health center diagnoses, $N = 30$.

Frequencies can be converted to ***relative frequencies*** (***proportions*** or ***percentages***) if desired. Proportions and percentages are calculated as follows:

Table 2–1 Two Categorical Data Sets

(a) Responses of 20 People to a Question on an Opinion Poll		(b) Diagnoses of 30 Consecutive New Clients at a Mental Health Center	
Agree	Strongly agree	Schizophrenia	Drug abuse
Strongly agree	Agree	Psychopathic personality	Schizophrenia
Disagree	Strongly agree	Depression	Neurosis
Neutral	Strongly disagree	Schizophrenia	Depression
Disagree	Agree	Drug abuse	Mania
Strongly agree	Disagree	Neurosis	Schizophrenia
Disagree	Disagree	Depression	Psychopathic personality
Disagree	Neutral	Schizophrenia	Schizophrenia
Neutral	Strongly agree	Mania	Neurosis
Agree	Disagree	Neurosis	Depression
		Schizophrenia	Depression
		Depression	Neurosis
		Psychopathic personality	Drug abuse
		Schizophrenia	Schizophrenia
		Depression	Psychopathic personality

$$\text{Proportion} = \frac{f}{N}$$

$$\text{Percentage} = \frac{f}{N} \times 100$$

For example, 4 out of 20 opinion poll respondents answered "Agree," so the proportion of "Agree" responses is 4/20 = .20, and the percentage is .20 × 100 = 20%. Table 2–2 shows frequency (f) and percentage (P) distributions for the opinion poll and mental hospital diagnoses data. Notice that the sum of the frequencies is N, and the sum of the percentages is 100 in both data sets.

The information in frequency distributions can also be displayed in graphs. Although graphs provide no additional information, they often show important features of the data particularly vividly. Graphs are visual representations drawn on a pair of axes. The horizontal axis is called the ***abscissa***, and the vertical axis is called the ***ordinate***. Different types of graphs are drawn for different types of data and/or to display different features of the data.

A common type of graph for displaying categorical data is the ***bar graph***. Bar graphs for the opinion poll and mental hospital diagnoses data are shown in Figure 2–1. As the figure shows, category names are displayed on the abscissa, and frequencies (or relative frequencies) are shown on the ordinate.

Use the following rules for drawing bar graphs:

1. Locate the successive categories at equal distances along the abscissa. If the categories are ordered, they must be listed in their natural order. Otherwise, the order of the categories is arbitrary. In the graph of mental health center diagnoses (a set of unordered categories), the categories have been ordered by frequency, but other orders could be used. Label the abscissa by the variable name.
2. The height of a graph should be about two-thirds to three-fourths its width. Use this rule to select a scale for frequencies on the ordinate. In the opinion poll distribution, for example, the largest frequency of responses is 7 (in the category "Disagree"). Therefore, a frequency of 7 is located at a point on the ordinate that is about three-fourths as far from the origin as "Strongly agree" is from the origin on the abscissa. Locate the remaining frequencies at equal intervals down to zero at the origin. Be sure to label the ordinate "Frequency."

Table 2–2 Frequency (f) and Percentage (P) Distributions of the Data in Table 2–1

(a) Responses to the Opinion Poll Question (N = 20)

Response	f	P
Strongly agree	5	25
Agree	4	20
Neutral	3	15
Disagree	7	35
Strongly disagree	1	5
Total	20	100

(b) Diagnoses of 30 New Clients

Diagnosis	f	P
Schizophrenia	9	30
Depression	7	23
Mania	2	7
Neurosis	5	17
Psychopathic personality	4	13
Drug abuse	3	10
Total	30	100

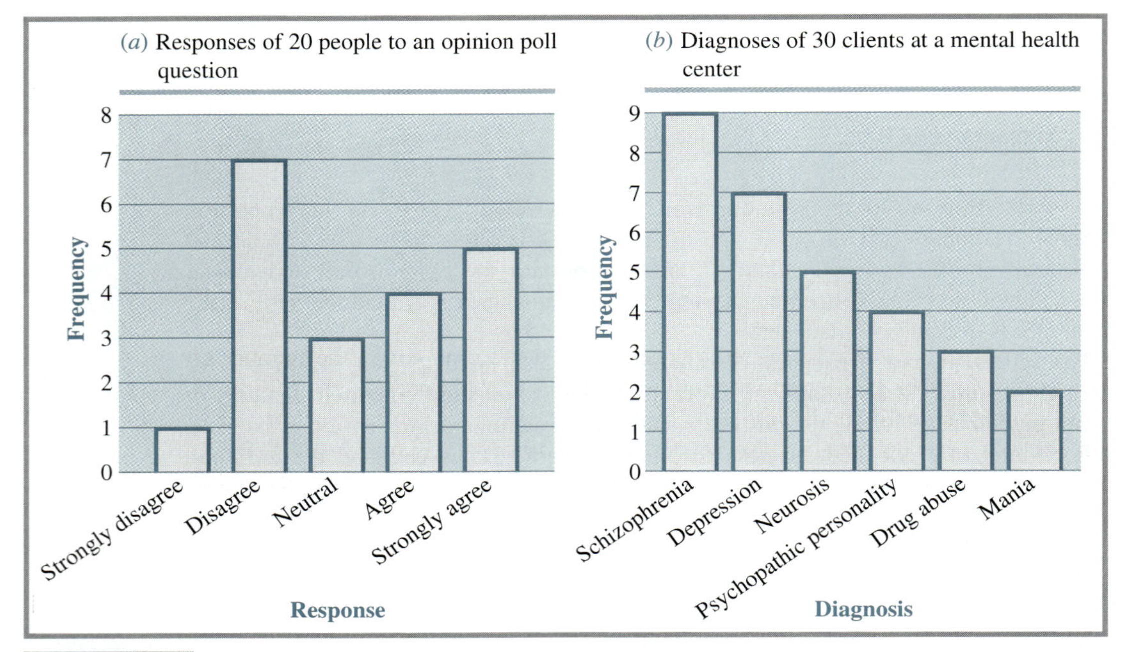

Figure 2–1 **Bar graphs of the frequency distributions in Table 2–2.**

3. Draw a bar for each category. All bars should be equal in width. The height of each bar is the frequency for that category. Bars for adjacent categories are separated by a gap.
4. Write a descriptive title (a caption) above or below the graph.

We could have graphed relative frequencies (percentages or proportions) instead of frequencies. Do you think this would have changed the appearance (shape, configuration) of the graph?

EXERCISE 2–1

The undergraduate majors of 25 students in a graduate social work program are listed below.

Sociology	Psychology	Chemistry	Psychology	Physics
Psychology	Psychology	Business	Elem Educ	Music
Psychology	Sociology	Psychology	Biology	Psychology
Music	Sec Educ	Psychology	Business	Business
English	Philosophy	Painting	Physics	Psychology

1. Construct a frequency distribution.
2. Calculate the percentage of students with each major.
3. Construct two bar graphs of the data, one a graph of the frequencies and the other a graph of the percentages. Compare the graphs.

Comparing Two or More Distributions: Contingency Tables

Suppose that a psychologist studying sex differences in personality development asked 73 women and 56 men, "Were you shy as a child?" with three response options: "Yes," "No," and "Uncertain." Now she wants to display the distribution of responses given by each group in a table. Distributions of the same categorical variable in two or more groups are commonly presented in what are called ***contingency tables***. Table 2–3 is a contingency table that shows the results of the sex-difference study.

As you can see, each column of Table 2–3 represents one group, and each row represents one response option. The entry within each cell is the frequency (number) of persons of that gender who chose that response option. Thus, for example, 39 women answered "Yes," and 21 men answered "Uncertain." The column sums (the number of women and the number of men) and the row sums (the number of persons choosing each response option) are called *marginal sums* or *marginal totals.*

In Table 2–3, group membership (gender) has been assigned to the columns of the table, and the response options have been assigned to the rows. We could just as well have assigned gender to the rows and the response options to the columns, as is done in Table 2–4. Either way of displaying the data is acceptable, and both tables present exactly the same information.

Contingency tables are often referred to in terms of their size, that is, the number of columns and number of rows. Tables 2–3 and 2–4 are 2×3[†] or 3×2 contingency tables, because they include two columns and three rows (or vice versa). A table with six columns and five rows is a 6×5 (or 5×6) contingency table; a table with two columns and two rows is a 2×2 table; and so on.

You may wonder why two-way tables such as these are called "contingency" tables. *Contingent* means "dependent." Thus, the tables are called contingency tables because they show (or allow us to figure out) how one variable is contingent (dependent) upon another. For example, from Table 2–3 or 2–4, we can determine whether childhood shyness is contingent upon gender. Let us examine the tables to see how we can make this determination.

We want to compare men's and women's frequency of choice of the three response options. With a little thought, you will quickly realize that we cannot compare the frequencies in Tables 2–3 and 2–4 directly because of the difference in group size. For example, although the frequency of women and men choosing the response option "No" is the same (22), we cannot conclude that women and men are equally likely to answer "No," because 22 out of 73 women is not the same as 22 out of 56 men. To compare the two groups, we must first convert the frequencies to relative frequencies (proportions or percentages) within each group.

Table 2–3 Contingency Table of the Responses of 73 Women and 56 Men to the Question "Were You Shy as a Child?"

		Women	*Men*	*Sum*
Response	Yes	39	13	52
	No	22	22	44
	Uncertain	12	21	33
	Sum	73	56	

[†] 2×3 is read "2 *by* 3," not "2 times 3."

Table 2–4 Contingency Table of the Same Data in Table 2–3, with Rows and Columns Reversed

	Yes	*No*	*Uncertain*	*Sum*
Women	39	22	12	73
Men	13	22	21	56
Sum	52	44	33	

To calculate the percentage of women choosing each response option, we divide the frequency by 73 and multiply by 100. For example, the percentage of women answering "No" is $(22/73) \times 100 = 30.1$. To calculate the percentage of men choosing each option, we divide the frequency by 56 and multiply by 100. Thus, the percentage of men answering "No" is $(22/56) \times 100 = 39.2$. The percentages in all the cells are shown in Table 2–5. From the table, we can readily see that women were more likely to answer "Yes" and men were more likely to answer "No." It appears that, at least in this group of men and women, childhood shyness is contingent upon gender.

A common method of graphing comparative categorical data is shown in Figure 2–2, a bar graph of the percentage of women and men choosing each response option. Notice that the bars for men and women are coded with distinctive patterns to make them easily distinguishable, and that a key or legend is provided to identify the group associated with each pattern. Placing the two bars for each response option side by side makes it easy to compare the two groups' responses. Notice that, except for the inclusion of two groups rather than one, the rules for constructing bar graphs have been followed.

Figure 2–2

Percentage of 56 men and 73 women responding "yes," "no," and "uncertain" to the question "Were you shy as a child?"

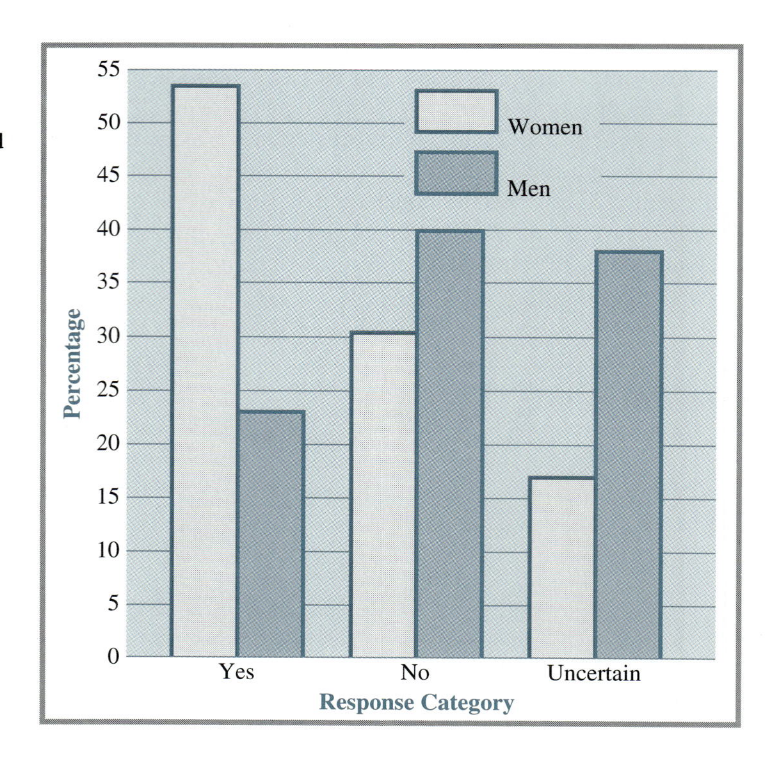

Table 2–5 Percentage of Women and Men in Each Response Category

Response	Women	Men
Yes	53.4	23.2
No	30.1	39.3
Uncertain	16.4	37.5
Sum	99.9[a]	100

[a]The sum is not equal to 100 because of rounding.

EXERCISE 2–2

Here is a 3 × 3 contingency table showing the number of liberal arts, business, and preprofessional students at each of three colleges.

	Liberal Arts	*Business*	*Preprofessional*
College A	450	375	675
College B	480	80	240
College C	1250	3000	750

1. How many students are at College A?
2. What is the total number of business majors?
3. Calculate appropriate percentages to decide whether the distribution of majors differs from one college to another.
4. Graph the data to display the differences among the colleges.

Frequency Distributions for Quantitative Data

Dr. Marrowbone gave a short, 20-point quiz to his 44 statistics students. After recording the scores, he decided to look at the entire distribution. First, he listed the scores as shown on the left side of Table 2–6. As you can see, even with only 44 scores, it is hard to derive even the most basic information, such as the highest and lowest scores. And patterns or trends cannot be distinguished. Just think how much more difficult it would be to make sense of the data if 1000 students had been tested instead of 44.

Next, Dr. Marrowbone proceeded to construct the frequency distribution shown on the right side of Table 2–6. The frequency distribution provides a clear picture of the distribution. We see that the scores *range* from 9 to 20. Most students did well on the quiz; the most frequently occurring score—the ***mode***—is 18; and all but one student scored at least 13. (The student who scored 9 has been observed to sleep through every class!) Clearly, the score of 9 is unusual in this group. An unusual observation is called an ***outlier***.

Constructing a frequency distribution has allowed Dr. Marrowbone to identify quickly the range, the mode, and outliers. Moreover, it would be just as easy to read this kind of information from a frequency distribution based on 1000 observations as it is for 44.

To construct a frequency distribution for quantitative data, you must do the following:

1. In a column labeled by the name of the variable, list the score values from highest to lowest (or lowest to highest). It is not necessary to list scores higher or lower than the highest and lowest actually obtained. For example, since no one scored above 20 or

Table 2–6 Scores of 44 Students on a 20-Point Statistics Quiz

Unordered Data Set

18	19	16	13
16	16	20	19
19	13	14	17
18	17	18	20
20	19	18	17
15	20	15	19
17	20	17	20
19	9	19	16
19	18	18	18
16	17	16	18
15	18	16	18

Frequency Distribution

Score	Tally[a]	f
20	~~////~~ /	6
19	~~////~~ ///	8
18	~~////~~ ~~////~~	10
17	~~////~~ /	6
16	~~////~~ //	7
15	///	3
14	/	1
13	//	2
12		
11		
10		
9	/	1
		44

[a]The "Tally" column is used for counting purposes. It is not included in the finished table.

below 9, score values above 20 and below 9 are not listed. However, do not leave any gaps. For example, although no one scored between 9 and 13, 10, 11, and 12 are listed.

2. Transfer the scores from the unordered data set into the table by tallying each observation into the appropriate row. An easy way to accomplish this is by the use of tally marks. For example, the first four scores of the data set are 18, 16, 19, and 18. After you have tallied them, the part of the frequency distribution between scores of 16 and 19 looks like this:

19	/
18	//
17	
16	/

   Continue until all the scores have been tallied.
3. Count the tally marks beside each score and enter the sums in a column labeled "Frequency," or "f." There are six tally marks beside the score 20, so that 6 is listed in the frequency column; there are eight tally marks beside the score 19, so that $f = 8$; and so on.
4. Always tally twice to make sure you haven't made any errors in counting. And always make sure that the sum of the frequency column is equal to N, the total number of observations.

Frequencies can be converted to proportions or to percentages, if desired. As is the case with categorical data, conversion to proportions or percentages is essential if two or more distributions with different numbers of observations are to be compared. As before, the proportion of observations for any given score value is f/N, and the percentage is $(f/N) \times 100$.

Grouped Frequency Distributions

In Dr. Marrowbone's set of statistics quiz scores, all the scores fall between 9 and 20, a range of only 11 points. The ***range*** is obtained by subtracting the lowest score from the highest; thus, range = 20 – 9 = 11. With a range this small, it is reasonable to list each score value from 9 to 20 in the frequency distribution. But when the range is greater than 15 or 20, it is no longer reasonable to do so. Consider the scores of 50 college students on a 75-item anxiety scale, listed in Table 2–7.

The lowest anxiety score is 26, the highest 68, a range of 68 – 26 = 42 points. If you were to list the scores from highest to lowest and tally the frequency of each, you would end up with a very long "Score" column and many frequencies of zero. Worse still, the general "shape" of the distribution would be very hard to see. In cases like this, scores can be grouped into class intervals, each including two or more score values. The result is called a ***grouped frequency distribution***. A grouped frequency distribution of the anxiety scores is shown in Table 2–8.

In Table 2–8, the scores have been grouped into nine class intervals. Five score values are included in each interval. For example, the interval 65–69 includes the score values 65, 66, 67, 68, and 69. The number of score values in each interval, in this case, five, is called the ***class interval size***, and it is constant throughout a given distribution. The highest obtained score (68) is included within the highest interval (65–69), and the lowest obtained score (26) within the lowest interval (25–29).

Use the following procedure for constructing a grouped frequency distribution:

1. First, decide about how many class intervals are desired. As a rule, the larger the number of observations, the larger the number of intervals. With N = 50, we decided to use eight to ten intervals. If N had been 20, we might have preferred fewer than eight intervals. If N were 1000, we might want as many as 20 class intervals.
2. Find the highest and lowest scores and compute the range. In the example, 68 – 26 = 42.
3. Divide the range by the desired number of classes to determine the class interval size. Since we want eight to ten class intervals, we divide 42 by 8 and by 10.

   $42 \div 8 = 5.25$
   $42 \div 10 = 4.2$

   We should include four or five score values in each interval. A class interval size of 5 was decided upon.
4. Start from either end of the distribution and decide which score values are to be included in the highest (or lowest) interval. The highest score must be included in the

Table 2–7 Anxiety Scale Scores of 50 College Students

36	60	51	48	41
35	44	52	46	26
59	51	37	45	39
42	61	57	48	52
38	44	37	60	41
31	59	62	43	30
49	68	53	49	26
44	53	52	57	47
38	26	46	47	46
43	50	56	34	50

Table 2–8 A Grouped Frequency Distribution of Scores of 50 College Students on an Anxiety Scale

Score	Tally	f
65–69	/	1
60–64	////	4
55–59	𝍸	5
50–54	𝍸 ////	9
45–49	𝍸 𝍸	10
40–44	𝍸 ///	8
35–39	𝍸 //	7
30–34	///	3
25–29	///	3
		50

highest interval and the lowest score in the lowest interval. The highest anxiety score is 68; therefore the highest interval must include the score value 68. Since we have decided to use a class interval size of five, there are five possible ranges of scores for the highest interval:

68–72
67–71
66–70
65–69
64–68

We chose 65–69, but we could have chosen one of the other options.†

5. List the remaining intervals, making sure that each interval has the same size and that no score values have been left out.
6. Finally, tally the number of observations falling in each interval and record the results in the "Frequency" column. Frequencies may be converted into proportions or percentages, if needed.

EXERCISE 2–3

1. Here is information about five data sets (a to e). Some of the information for sets c, d, and e is missing. Fill in the missing information.

	Highest Score	Lowest Score	(A) Range	(B) Approximate Number of Classes Desired	(A) Divided by (B)	Class Interval Size
a.	100	19	81	10	8.1	8
b.	38	3	35	12	2.9	3
c.	122	12		15		
d.	57	20		8		
e.	57	2		12		

2. Here is a set of 20 test scores

18	6	3	8
10	6	16	5
7	9	9	21
10	8	23	17
14	24	4	14

a. Construct
   (1) An ungrouped frequency distribution
   (2) A grouped frequency distribution with about 10 class intervals
   (3) A grouped frequency distribution with about seven class intervals
b. In your opinion, which frequency distribution best shows what the distribution as a whole looks like?
c. By grouping the scores into class intervals, what have you gained? What have you lost?

†It is sometimes considered desirable to make the lowest score in each interval a multiple of the class interval size, as we have done. However, this is not necessary.

Score Limits and Exact Limits

The lowest and highest score values in each class are called the ***score limits*** of the class interval. Thus, in the class 45–49 of the anxiety score distribution, 45 is the *lower score limit* and 49 is the *upper score limit.* However, if the variable being measured is continuous, as anxiety is, each score is assumed to represent an interval with exact limits below and above the reported value. For example, the score 45 represents the interval 44.5–45.5, and the score 49 represents the interval 48.5–49.5. Therefore the exact limits (as opposed to the score limits) of the class interval 45–49 are 44.5–49.5.

Scores in an ungrouped frequency distribution also have exact limits. In an ungrouped distribution, each score is a separate class, and the exact limits of the score are the exact limits of the class. For example, in the distribution of statistics quiz scores in Table 2–6, the exact limits of the score 13 are 12.5 and 13.5.

Exact limits of all intervals in the grouped distribution of anxiety scores are shown in Table 2–9. Notice that the lower exact limit of each class is the upper exact limit of the next lower class, and the upper exact limit of each class is the lower exact limit of the next higher class. Also notice that the class interval size is equal to the upper exact limit minus the lower exact limit of each interval.

As you will see later in this chapter, exact limits are important for drawing certain kinds of graphs and for computing certain statistics, such as percentiles.

The Midpoint of a Class Interval

In a grouped frequency distribution each class interval includes two or more score values. For example, the interval 45–49 includes the score values 45, 46, 47, 48, 49. Sometimes a single value is needed to represent all the scores in the interval. The value generally chosen is the ***midpoint*** of the interval. The midpoint is defined as the point exactly halfway between the lower and upper score limits of the interval. In the interval 45–49, this point is 47.0, halfway between 45 and 49. Notice that the midpoint is also exactly halfway between the lower and upper exact limits of the interval, in this case, 44.5 and 49.5. The midpoints of all intervals in the anxiety scale score distribution are shown in Table 2–9.

Usually you can tell what the midpoint of a class interval is without computing. In case you cannot, do the following:

Table 2–9 Exact Limits and Midpoints of Intervals in the Grouped Distribution of Anxiety Scores

Score	f	Exact Limits	Midpoints
65–69	1	64.5–69.5	67.0
60–64	4	59.5–64.5	62.0
55–59	5	54.5–59.5	57.0
50–54	9	49.5–54.5	52.0
45–49	10	44.5–49.5	47.0
40–44	8	39.5–44.5	42.0
35–39	7	34.5–39.5	37.0
30–34	3	29.5–34.5	32.0
25–29	3	24.5–29.5	27.0

1. Add the lower score limit plus the upper score limit.
2. Divide by 2 to get the midpoint.

For example, consider the interval 45–49:

1. 45 + 49 = 94
2. The midpoint is 94 ÷ 2 = 47.0.

EXERCISE 2–4 Each of the following is a class interval of a distribution. For each one, list the scores included in the interval, the class interval size, the exact limits, and the midpoint. The first one has been completed for you.

Score Limits	Scores in Interval	Class Interval Size	Exact Limits	Midpoint
a. 30–35	30, 31, 32, 33, 34, 35	6	29.5–35.5	32.5
b. 30–39	________	________	________	________
c. 30–31	________	________	________	________
d. 30–32	________	________	________	________
e. 30–49	________	________	________	________
f. 30	________	________	________	________

Cumulative Frequencies and Cumulative Percentages

Sometimes we want to know the number of observations or the percentage of observations falling *at* or *below* a score. For example, how many students scored 15 or less on the statistics quiz? What percentage of the college students scored 50 or less on the anxiety inventory? The number of observations at or below a given score is called the ***cumulative frequency*** of the score, and the percentage of observations at or below a score is called the ***cumulative percentage***. In the statistics quiz score distribution, for exam-

Table 2–10 Cumulative Frequencies (Cum *f*) and Cumulative Percentages (Cum *P*) of Scores on a Statistics Quiz

Score	*f*	Cum *f*	Cum *P*
20	6	44	100
19	8	38	86
18	10	30	68
17	6	20	45
16	7	14	32
15	3	7	16
14	1	4	9
13	2	3	7
12		1	2
11		1	2
10		1	2
9	1	1	2

ple, the cumulative frequency for the score 15 is 7, because seven students had scores of 15 or less.

Although cumulative frequencies (or cumulative percentages) can be calculated for individual scores, it is more common to calculate them for the entire distribution. Table 2–10 displays both cumulative frequencies (Cum f) and cumulative percentages (Cum P) for the statistics quiz score distribution. The table shows that 14 students scored 16 or less on the statistics quiz, 38 students scored 19 or less, and so on. Similarly for cumulative percentages: 68% of the group scored 18 or less, only 7% scored 13 or less, and so on.

To calculate cumulative frequencies, begin with the *lowest* score; in the distribution of statistics quiz scores this is a score of 9. Because no one scored lower than 9, its frequency, 1, is also its cumulative frequency, and "1" is entered in the Cum f column. For each higher score, the cumulative frequency is equal to the frequency of the score plus the number of observations at all lower scores. In other words, for any score

$$\text{Cum} f = f + \text{next lower score's Cum} f$$

For example, the score 17 has Cum $f = 6 + 14 = 20$. The cumulative frequency of the highest score in a distribution is always equal to N, because all observations fall at or below the highest score.

Cumulative percentages can be calculated from cumulative frequencies by using the following formula:

$$\text{Cum } P = \frac{\text{Cum} f}{N} \times 100$$

For example, the quiz score 18 has Cum $f = 30$; therefore Cum $P = \frac{30}{44} \times 100 = 68$. Since the entire group falls at or below the highest score, Cum P for the highest score is always equal to 100.

A cumulative frequency has been defined as the number of observations falling at or below a score. Here is another way of interpreting cumulative frequencies and percentages: Because all the observations with a given score fall below the score's upper exact limit, a cumulative frequency for a given score is the number of observations falling below the score's upper exact limit. For example, 7 students scored 15 or less on the quiz. Because 15 has an upper exact limit of 15.5, we can also say that 7 students scored below 15.5. Similarly, 30 students scored below 18.5, 4 students scored below 14.5, and so on. In the same way, 32% of the students scored below 16.5, 45% below 17.5, and so on.

EXERCISE 2–5

Calculate cumulative frequencies and cumulative percentages in the ungrouped distribution of test scores you constructed in Exercise 2–3, number 2. Then answer the following questions:

1. How many students scored below 16.5?
2. What percentage of the group scored 14 or less?
3. Ten students scored below what point?
4. Seventy-five percent of the students scored below what point?

Graphs of Quantitative Data

There are many methods for graphing quantitative data. To some extent, the method used depends on whether the variable is discrete or continuous. For example, bar graphs are appropriate for discrete variables, whereas histograms and frequency polygons are more suitable for continuous variables. The method chosen also depends on the information to be conveyed. For example, if we want to display the number or percentage of cases falling below various score values, we construct cumulative graphs. The types of graphs mentioned earlier, as well as another type called the stem-and-leaf plot, are discussed in this section.

Bar Graphs for Discrete Variables

You have already encountered the bar graph as a way to display categorical data. The rules for constructing bar graphs of discrete quantitative variables are the same as those for constructing bar graphs of categorical data. The only difference is that, instead of category names, values of the variable are displayed (in order) on the abscissa.

The number of TV sets per household is a discrete variable. Figure 2–3 displays a frequency distribution and a graph of the number of TV sets in 60 suburban households. Notice that values of the variable (number of TV sets) are displayed at equal intervals on the abscissa, with low values at the left and high values on the right. Frequencies are displayed on the ordinate. The frequency of households having each number of TV sets is represented by a vertical bar. All the bars are the same width, and adjacent bars are separated by a gap. As with all graphs, the height of the figure is about two-thirds to three-fourths of its width.

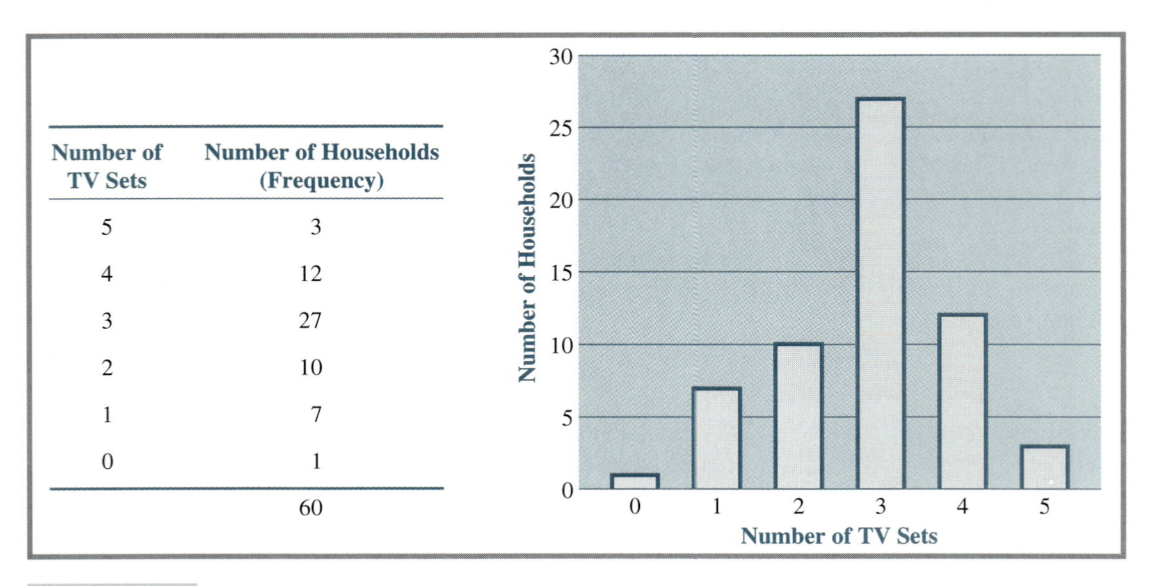

Number of TV Sets	Number of Households (Frequency)
5	3
4	12
3	27
2	10
1	7
0	1
	60

Figure 2–3 **Number of TV sets in 60 suburban households.**

THE HISTOGRAM

A ***histogram*** is a kind of bar graph. Each score (for ungrouped data) or class interval (for grouped data) is represented by a vertical bar whose height corresponds to the frequency of observations in the score or interval. However, because the variable being graphed is continuous, no gaps are left between bars. Histograms of the distributions of statistics quiz scores and anxiety scale scores are shown in Figure 2–4.

In histograms, score values are shown on the abscissa, and frequencies (or relative frequencies) on the ordinate. The width of the bar for each score or interval extends from the lower to the upper exact limits of the score or interval. For example, the bar for the score 15 in the distribution of statistics quiz scores extends from 14.5 to 15.5, and the bar for the interval 50–54 in the distribution of anxiety scores extends from 49.5 to 54.5. The height of each bar corresponds to its frequency.

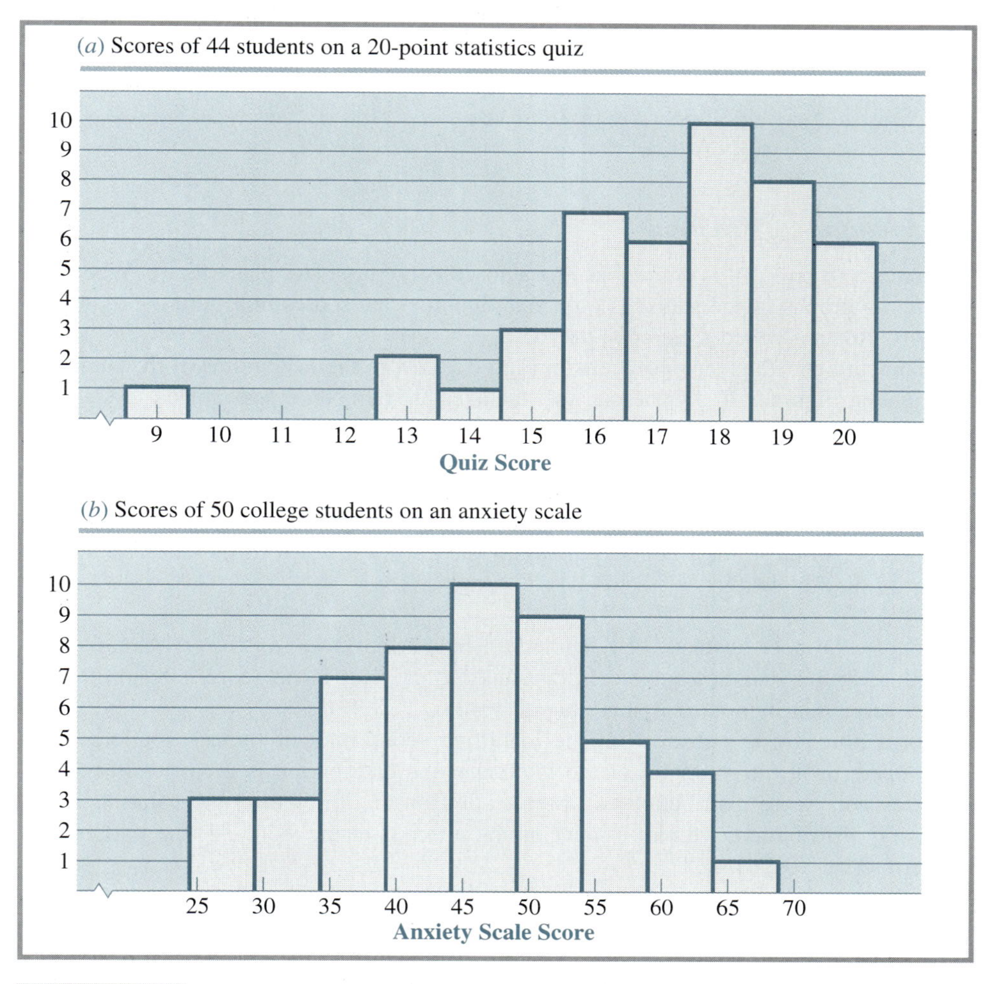

Figure 2–4 **Histograms of statistics quiz and anxiety scale scores.**

Thus the rules for drawing a histogram are as follows:

1. On the horizontal axis (abscissa) locate score values, with lower values at the left and higher to the right. Start, at the left, with a score value somewhat below the lower exact limit of the lowest score or interval in your data. If the lowest score value you have marked on the abscissa is greater than zero, use a break indicator —^v— (as in Figure 2–4) to indicate the omission of values between the lowest displayed score value and zero.
2. Place frequencies, proportions, or percentages on the ordinate, starting from zero at the origin, in such a way that the height of the figure will be two-thirds to three-quarters its width.
3. For each score (or interval) draw a vertical bar. Its width should extend from the lower to the upper exact limit of the score or interval, and its height should be equal to that score's (or interval's) frequency (or proportion or percent).

EXERCISE 2–6

Construct a histogram of one of the grouped frequency distributions you constructed in Exercise 2–3, number 2.

The Frequency Polygon

In geometry, a polygon is defined as an enclosed, straight-sided, plane figure. When you look at the graphs of the statistics quiz and anxiety scale scores in Figure 2–5, you will see why they are called ***frequency polygons***.

Compare the frequency polygons in Figure 2–5 with the histograms of the same distributions in Figure 2–4. As you can see, the axes are numbered and labeled in the same way. However, in the frequency polygons, each score or interval is represented by a point rather than a bar. The point is drawn directly above the score value (ungrouped data) or the midpoint of the interval (grouped data), at a height equal to the frequency. The points are then connected with straight lines.

In order to close the frequency polygon (i.e., bring it down to the abscissa at both ends), do the following:

1. Identify the next lower score (or interval) below the lowest in the frequency distribution. In the statistics quiz score distribution, the next lower score is 8; in the anxiety score distribution, it is the interval 20–24.
2. Plot a point on the abscissa at the identified score value if the distribution is ungrouped, or at the midpoint of the interval if the distribution is grouped. In the statistics quiz score distribution, we plot a point on the abscissa at the value 8; in the anxiety distribution, we plot a point on the abscissa at the value 22 (the midpoint of the interval 20–24).
3. Draw a straight line connecting this point to the nearest point on the right.
4. Repeat the procedure to close the graph at the upper end: Identify the next higher score (or interval) above the highest in the frequency distribution. In the statistics distribution, this is the score 20; in the anxiety score distribution, it is the interval 70–74, with midpoint = 72.

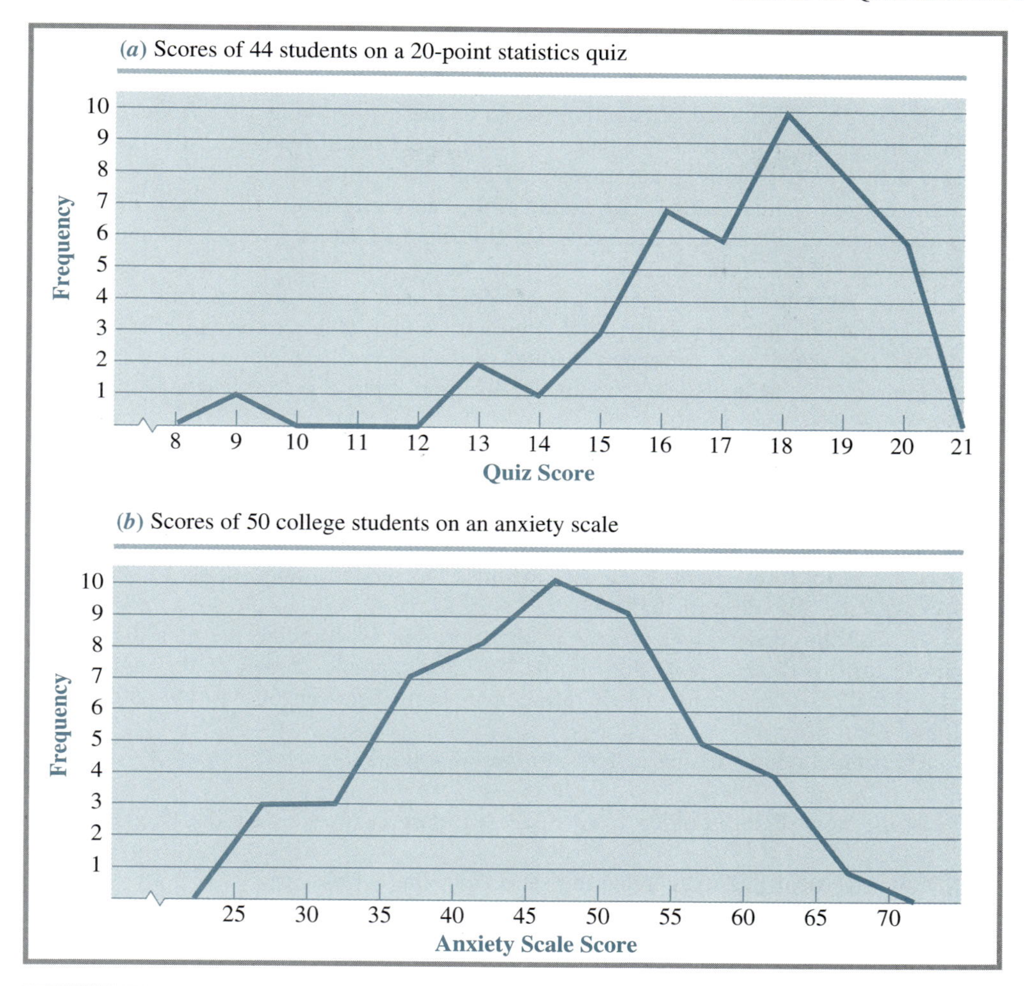

Figure 2–5 **Frequency polygons of statistics quiz and anxiety scale scores.**

5. Plot a point on the abscissa at the score (ungrouped distribution) or at the midpoint of the interval (grouped distribution).
6. Draw a straight line connecting this point to the nearest point on the left.

Notice the similarity of the shapes of histograms and frequency polygons for the same data. A frequency distribution can usually be represented equally well by a histogram or a frequency polygon.

EXERCISE 2–7

Draw a frequency polygon of one of the grouped frequency distributions you constructed in Exercise 2–3, number 2.

Stem-and-Leaf Plots

When the range of scores on a variable is large, we often group the scores into class intervals. Graphs of grouped distributions, such as the histogram and the frequency polygon of the grouped anxiety score data, certainly reveal major features of the distribution better than would graphs of the ungrouped data. But grouping results in the loss of a certain amount of information. For example, we can see from the frequency distribution (or the graph) of the anxiety data that nine students had scores between 50 and 54. But how many of these students scored 50? How many scored 51? How many scored 52? We cannot recover this information without going back to the original data.

A method of data representation that groups without losing information is the ***stem-and-leaf*** plot (***stem plot***, for short). Although a stem plot lists scores, like a frequency distribution, it is, in appearance, more like a graph. Like a graph, it shows major features of the data quite vividly.

A stem-and-leaf plot of the anxiety scale scores is shown in Figure 2–6. The column of numbers to the left of the vertical line is the *stem*; it consists of the first (tens) digits of the anxiety scores. Since the scores range from 26 to 68, the numbers 2, 3, 4, 5, 6 are listed in the stem. The horizontally displayed digits, to the right of each number in the stem, form a *leaf*. Each leaf includes the second (units) digits of scores that begin with the tens digit of that row.

To read the anxiety scale scores in the plot, simply combine the tens digit in the stem, in turn, with each units digit in the leaf. For example, in row 6, the digits in the leaf are 00128. This means that there are five scores between 60 and 69—60, 60, 61, 62, and 68.

The steps in constructing a stem plot are as follows:

1. List the stem digits in order, from smallest to largest (as we have done) or largest to smallest.
2. Draw a vertical line to separate the stem from the leaves.
3. Enter each observation in the row corresponding to its tens digit, and write its units digit to the right of the vertical line. As an example, the first six scores in the anxiety data set (see Table 2–7) are 36, 35, 59, 42, 38, 31. After these scores have been entered, the stem plot looks like this:

2	
3	6 5 8 1
4	2
5	9
6	

Continue until all scores have been entered.

4. The last step is to reorder the leaves within each row from smallest to largest.

Figure 2–6

Stem-and-leaf plot of anxiety scale scores of college students.

2	777
3	0145677889
4	112334445666778899
5	00112223367799
6	00128

The stem-and-leaf plot in Figure 2–6 has only five rows ("class intervals"). We can obtain a finer grouping (smaller class intervals) by splitting each row into two rows, one for scores ending in 0 to 4 and one for scores ending in 5 to 9. This is commonly done by the method shown in Figure 2–7. As you can see in the figure, each stem digit is listed twice, first with an asterisk (*) and then without. Scores ending with the digits 0 to 4 are listed in the asterisked row, and those ending with the digits 5 to 9 are listed in the following row. The result is called a *split-stem plot.*

Notice the similarity between the split-stem plot of the anxiety data in Figure 2–7 and the graphs of the same data in Figures 2–4 and 2–5. This should not be surprising, as all three representations use the same grouping of scores.

If you were preparing a paper for submission to a journal or for presentation at a conference, you would probably use histograms or frequency polygons to display your data. But if you have a relatively small data set with a wide range, and you want a picture of the distribution without losing any of the information in the original data set, a stem-and-leaf plot is a good choice.

EXERCISE 2–8

Construct a stem-and-leaf plot of the data in Exercise 2–3, number 2. The range of scores is only 17, so you will need to split the stem.

Displaying Two or More Distributions in One Graph

If you have measured the same variable in two (or more) groups, you can construct tables or graphs that allow you to compare the two distributions. Suppose that, in addition to the anxiety scale scores of 50 college students, you have scores on the same scale of 35 clients at a mental health center. You want to compare the score distributions of college students and mental health center clients.

Table 2–11 shows the two frequency distributions side by side. Notice that the same grouping has been used for both distributions. Because of the difference in size of the two groups, frequencies cannot be compared directly. The last two columns list the percentage of observations in each class interval for each group. The percentages show clearly that, though there is considerable overlap between the two groups, mental health center clients tended to score higher. (However, notice the low-scoring outlier among the clients.)

Differences between the two groups show up vividly in a frequency polygon displaying both distributions (Figure 2–8). Notice the use of different kinds of points and lines for the two groups and the inclusion of a key to identify each group.

Figure 2–7

Split-stem plot of anxiety scale scores.

Stem	Leaves
2	666
3*	014
3	5677889
4*	11233444
4	5666778899
5*	001122233
5	67799
6*	0012
6	8

Table 2–11 Anxiety Scale Scores of 50 College Students and 35 Mental Health Center Clients

	Frequency		Percentage	
Scores	Students	Clients	Students	Clients
70–74		2		6
65–69	1	6	2	17
60–64	4	8	8	23
55–59	5	5	10	14
50–54	9	6	18	17
45–49	10	3	20	9
40–44	8	4	16	11
35–39	7		14	
30–34	4		6	
25–29	3		6	
20–24		1		3

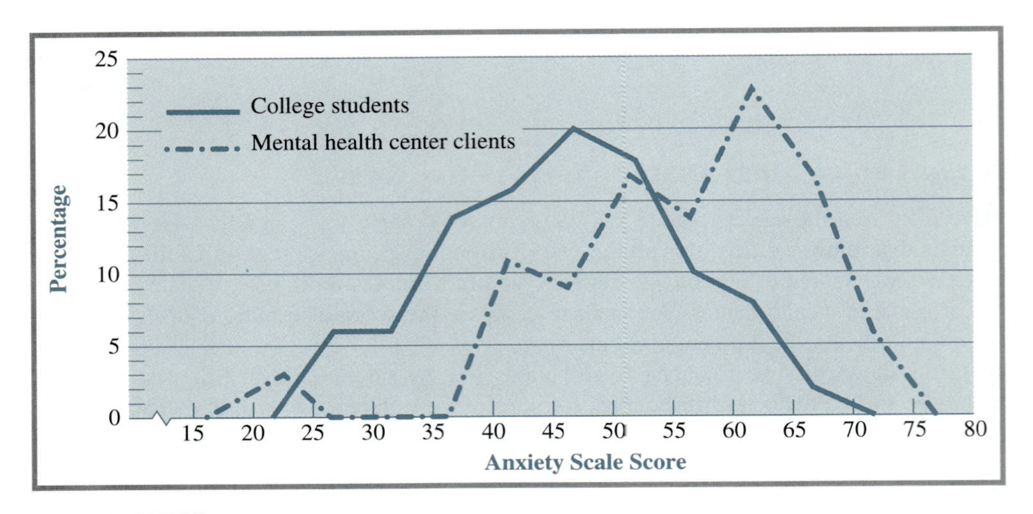

Figure 2–8 **Frequency polygon comparing anxiety scale scores of 50 college students and 35 clients at a mental health center.**

Figure 2–9

Back-to-back stem plots of anxiety scale scores of college students and mental health clients.

	2*	2
666	2	
410	3*	
9887763	3	
44433211	4*	2344
9988776665	4	589
33221100	5*	133344
99776	5	57789
2100	6*	01123334
8	6	556889
	7*	23

To preserve the individual scores that are lost by grouping, back-to-back stem plots can be used. As you can see in Figure 2–9, a single stem is used for both plots and is placed in the center of the figure. Leaves for one group extend to the left of the stem, and leaves for the other group extend to the right.

EXERCISE 2–9

Below are data on the number of defective widgets per batch in 30 consecutive batches produced by each of two different production methods, Method A and Method B. Construct a back-to-back stem plot comparing the two methods. Which method seems to do a better job of producing widgets?

Method A					Method B				
24	9	2	24	24	16	16	8	5	6
14	22	23	13	6	6	23	4	19	14
28	19	22	26	20	20	7	20	18	16
17	28	16	18	19	12	1	1	3	7
15	25	8	15	3	12	16	28	0	15
11	11	8	25	16	24	11	9	27	23

CUMULATIVE GRAPHS

The purpose of a ***cumulative graph*** is to show the frequency, proportion, or percentage of observations falling below each score value in a distribution. Earlier we defined the cumulative frequency of a score as the number of observations falling below the score's upper exact limit. Therefore, in drawing a cumulative graph, points are plotted above the upper exact limit of each score.

The cumulative frequency and cumulative percentage distributions of the 44 statistics quiz scores are graphed in Figure 2–10. As you can see, a point is plotted above the upper exact limit of each score at a height equal to the score's cumulative frequency (or

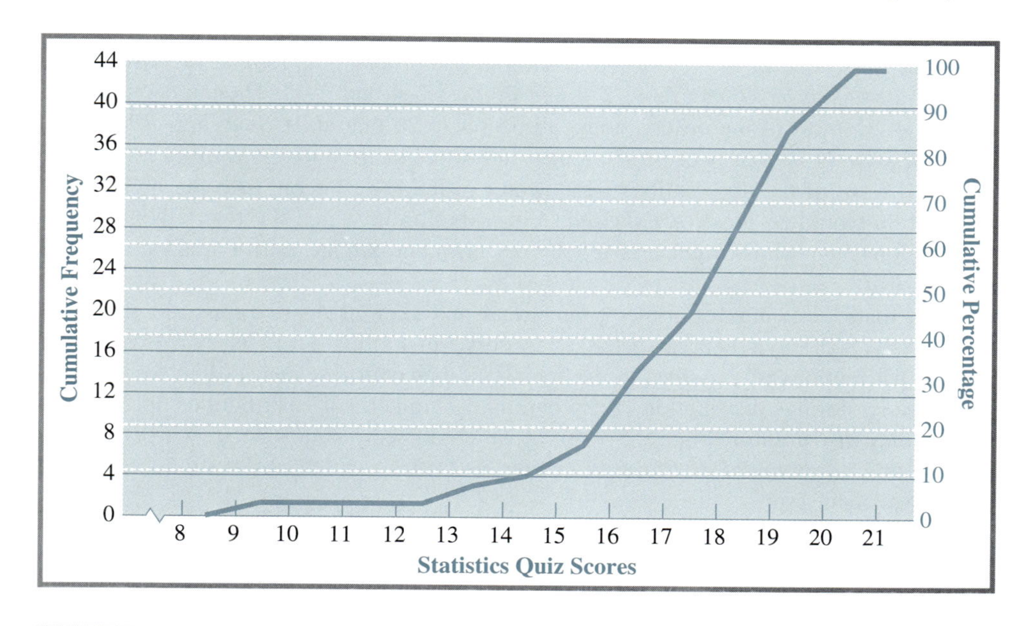

Figure 2–10 **Cumulative frequency and cumulative percentage distributions of scores on a statistics quiz ($N = 44$).**

cumulative percentage). The graph is closed at the left by plotting a cumulative frequency (or percent) of zero at the lower exact limit of the lowest score, in this case, 8.5. Unlike frequency polygons, cumulative graphs are not closed on the right.

Notice the S-shape of the curve. This shape is characteristic of cumulative graphs for data sets with larger frequencies near the center and smaller ones in the tails. Such an S-shaped curve is called an *ogive* (pronounced OH-jive).

In summary, the rules for constructing a graph of a cumulative distribution are as follows:

1. Locate score values along the abscissa, as in a noncumulative graph. On the ordinate, locate cumulative frequencies (or cumulative relative frequencies) in such a way that the height of the figure will be about two-thirds to three-quarters the width.
2. Plot a point precisely above the upper exact limit of each score at a height equal to the cumulative frequency (or cumulative relative frequency) for that score.
3. Plot a point on the abscissa at the lower exact limit of the lowest score.
4. Connect the points with straight lines.

EXERCISE 2–10 Graph the cumulative frequency distribution you constructed in Exercise 2–5.

Different Kinds and Shapes of Distributions

In distributions of social and behavioral science data, most observations tend to fall near the center of the distribution, with frequencies decreasing gradually toward the lower and upper ends of the distribution. Sometimes, however, observations tend to pile up toward either the upper end or the lower end of the range. For example, look at the distribution of the statistics quiz scores (Table 2–6 and Figures 2–4 and 2–5). Though scores range from 9 to 20, most of the observations fall nearer to 20 than to 9. Such a distribution is called ***negatively skewed*** because the longer tail of the distribution goes toward the lower scores. If most of the observations were nearer to the lowest score than the highest, the distribution would be ***positively skewed***. A distribution in which the two tails are about equally long and whose right and left halves are approximate mirror images is called ***symmetrical***.

Schematic drawings of some symmetrical and skewed distributions are shown in Figure 2–11. The symmetrical bell-shaped curve in Figure 2–11a is called a ***normal curve***. Distributions of variables in the social and behavioral sciences often form an approximately normal distribution. Less common symmetrical distributions include the ***rectangular distribution*** (Figure 2–11b), in which all score values occur with equal frequency, and the ***bimodal distribution*** (Figure 2–11c), in which observations pile up at two nonadjacent scores.

Figure 2–11 also shows three skewed distributions. Distribution 2–11d is negatively skewed, and distribution 2–11e is positively skewed. The ***J-shaped distribution*** (Figure 2–11f) is the extreme of a skewed (in this case, positively skewed) distribution.

The graphs in Figure 2–11 are smooth and regular. Most distributions of real data, of course, are rougher and less regular, especially when group sizes are small.

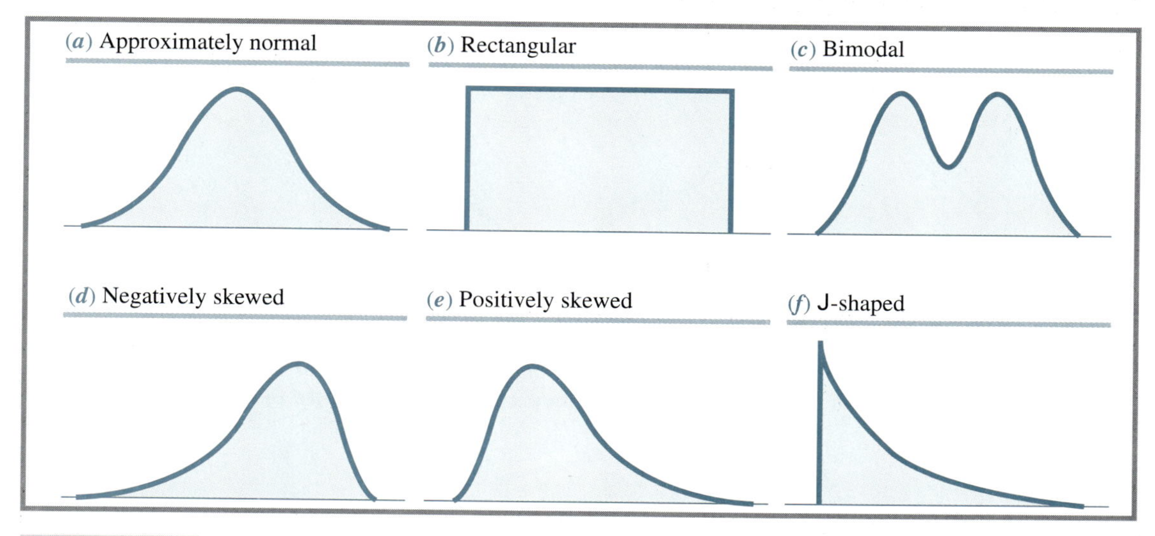

Figure 2–11 **Some common symmetrical (a–c) and skewed (d–f) distributions.**

Theoretical and Obtained Distributions

It may be useful at this point to distinguish between theoretical and obtained distributions. When we measure a variable in a group of individuals, objects, or events, the result is an ***obtained distribution***. We can describe and display this distribution in various ways, some of which have been treated in this chapter. For example, we can construct a frequency distribution and draw a variety of graphs.

Theoretical distributions, on the other hand, are generated by mathematical methods or logical reasoning rather than by measurement. The normal distribution is an example of a theoretical distribution. Normal curves are generated by a mathematical formula; no measured variable is ever exactly normally distributed. However, as you will see in Chapter 7, normal curve procedures are useful for working with distributions that are approximately normally distributed. In addition, theoretical distributions—normal and others—play an important role in statistical inference.

Areas Under a Curve

Suppose we have constructed a graph of a distribution such as the one in Figure 2–12a. Now we draw one or more vertical lines from the curve to the abscissa, anywhere we want as shown. The vertical lines divide the distribution into two or more sections (in this case, three). By using calculus, mathematicians could determine the area of each section as well as the total area under the curve. Then they could express the area of each section as a proportion of the total area. For example, they might find that the leftmost section includes .25 of the total area, the middle section .60, and the rightmost section .15. (The three proportions, of course, add up to 1.00).

Why is this important? It is important because, in graphs of distributions, the proportion of area in any section is equal to the proportion of observations falling in that section. Suppose, for example, that the graph in Figure 2–12a is a distribution of introversion scores in a group of college students. Because .25 of the area falls below 32, .60

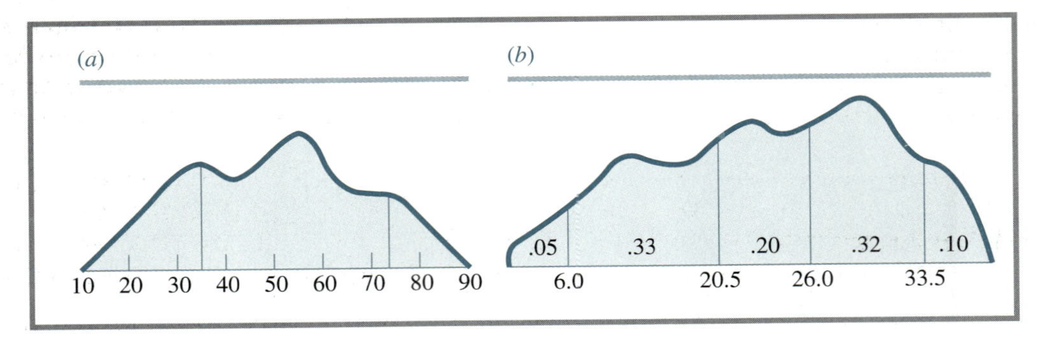

Figure 2–12 **Two continuous distributions divided into three or more sections. The proportion of area within each section is indicated in distribution (b).**

between 32 and 74, and .15 above 74, we can state that .25 (one-fourth) of the students scored below 32, 60% of them between 32 and 74, and the remaining 15% above 74 on the introversion test.

Figure 2–12b provides an example of a distribution divided into five sections. The score values dividing the sections are shown on the abscissa, and proportions of the total area are indicated in each section. Again, note that the areas add to 1.00. What kinds of statements can we make about this distribution? We can say that 10% of the group scored above 33.5, 5% scored below 8.0; 33% scored between 8.0 and 20.5, and so on. We can also say that .53 of the group scored between 8.0 and 26.0 (because .33 + .20 = .53); .42 of the group scored above 26.0; all but 5% of the group scored above 8.0; and the point below which the bottom 38% scored is 20.5.

In later chapters, you will use statistical tables to find areas in sections of distributions, such as the normal distribution, and you will use these areas to make statements about proportions and percentages of a group as well as about probabilities. Probability is the subject of Chapter 4.

EXERCISE 2–11

1. Describe the shape of each of the following frequency distributions.

a.

Score	f
10	2
9	3
8	5
7	3
6	2

b.

Score	f
22	3
21	5
20	12
19	20
18	15

c.

Score	f
5	2
4	5
3	1
2	5
1	1

2. Given the distribution in Figure 2–13, with proportions of the area shown for each section, answer the following questions:
   a. What proportion of the group scored below 6?
   b. What proportion scored between 3 and 8.5?
   c. If half the group scored below 5.5, what proportion scored between 3 and 5.5?
   d. Above what score value does the top 5% of the group fall?

Figure 2–13

Graph of scores on a variable.

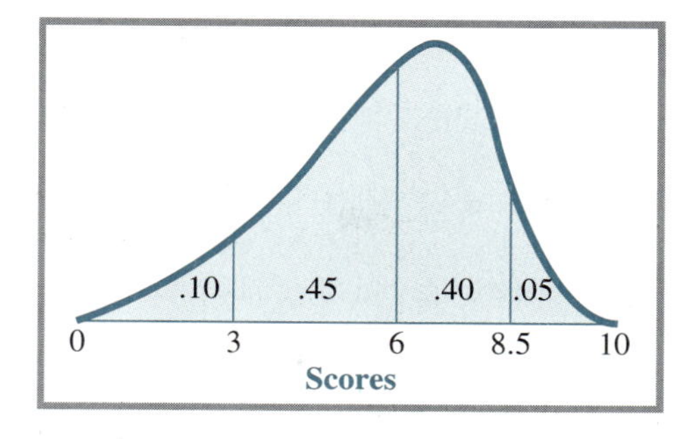

Locating Scores Within a Distribution: Percentiles and Percentile Ranks

Most of you have taken tests such as the Scholastic Assessment Test (SAT) or other standardized achievement tests. On such tests you are usually informed that you scored at such-and-such a percentile, for example, the 26th or the 59th or the 83rd. Suppose you have been told that your score on the SAT-Verbal Test was at the 75th percentile. What does this mean? It means that 75% of applicants who took the SAT-Verbal Test scored lower than you did. In general, *if a score is at the kth percentile, k% of the observations scored below that score.* Thus ***percentiles*** are score values that cut off certain percentages of observations below them.

Percentiles are symbolized by the letter *P*, with a subscript denoting the percentage below the score value. Thus, P_{30} is read "the 30th percentile" and stands for the score value below which 30% of observations fall; P_{83}, the 83rd percentile, is the score value below which 83% of observations fall; and so on. The statement "$P_{65} = 74$" means that 65% of the observations in the distribution fall below the score 74.

The percentage of observations falling below a given score value is called a ***percentile rank***. Thus, if $P_{65} = 74$, the percentile rank for a score of 74 is 65. Similarly, if $P_{20} = 19$, the percentile rank for a score of 19 is 20.

Do not confuse percentiles and percentile ranks. Remember: A percentile is a *score* value below which a certain percentage of observations in a distribution falls; a percentile rank is the *percentage* of observations below a certain score value. Study the following example until you are sure you understand it.

Information Given	Interpretations
$P_{80} = 13$	80% of observations falls below 13.
(The 80th percentile is 13.)	(The percentile rank for a score of 13 is 80.)

Sometimes we want to know the percentile rank for each score in a distribution. At other times, we want to know the exact score value in a distribution below which a certain percentage of observations fall; that is, we want to determine specific percentiles, such as P_{25} or P_{50}. We discuss calculation of percentile ranks first, then of specific percentiles.

Calculating and Interpreting Percentile Ranks

You may have noticed that the definition of a percentile rank is very similar to that of a cumulative percent. Let us look at the two definitions:

Cumulative percent:	Percentage of observations falling below the upper exact limit of a score
Percentile rank:	Percentage of observations falling below a score value

Not surprisingly, percentile ranks are computed in a manner similar to cumulative percentages. There is one difference: Because the midpoint of an interval is considered to be more representative of the interval as a whole than the upper exact limit, the percentile rank is the cumulative percentage of observations to the midpoint, halfway between the lower and upper exact limits of a score.

To calculate percentile ranks for scores in a distribution, we do the following:

1. Construct a cumulative frequency distribution of the scores.
2. Calculate the cumulative frequency to the *midpoint* of each score by the following formula:

$$\text{Midpoint Cum} f = \frac{1}{2} f + \text{next lower score's Cum} f$$

3. To convert the midpoint Cum f to a percentile rank, divide by the group size N and multiply by 100.

$$\textit{Percentile rank} = \frac{\text{midpoint Cum} f}{N} \times 100$$

The calculation of percentile ranks for the distribution of statistics quiz scores is shown in Table 2–12. By consulting this table, the students in Dr. Marrowbone's class can determine what percentage of their classmates scored lower than they did.

Educators and psychologists often use percentile ranks to interpret scores in terms of their location in a distribution. Suppose that Sally has taken a spelling test and reports to her parents that her score was 24. Is this good or bad? From the raw score alone,

Table 2–12 **Calculation of Percentile Ranks on a Statistics Quiz**

Score	f	Cum f	Midpoint Cum f[a]	Percentile Rank[b]
20	6	44	$\frac{1}{2}(6) + 38 = 41.0$	93
19	8	38	$\frac{1}{2}(8) + 30 = 34.0$	77
18	10	30	$\frac{1}{2}(10) + 20 = 25.0$	57
17	6	20	$\frac{1}{2}(6) + 14 = 17.0$	39
16	7	14	$\frac{1}{2}(7) + 7 = 10.5$	24
15	3	7	$\frac{1}{2}(3) + 4 = 5.5$	12
14	1	4	$\frac{1}{2}(1) + 3 = 3.5$	8
13	2	3	$\frac{1}{2}(2) + 1 = 2.0$	5
12		1	$\frac{1}{2}(0) + 1 = 1.0$	2
11		1	$\frac{1}{2}(0) + 1 = 1.0$	2
10		1	$\frac{1}{2}(0) + 1 = 1.0$	2
9	1	1	$\frac{1}{2}(1) + 0 = 0.5$	1

[a]Midpoint Cum $f = \frac{1}{2}(f)$ + next lower score's Cum f.

[b]Percentile rank $= \frac{\text{midpoint Cum} f}{N} \times 100$.

it is impossible to tell. However, if Sally provides the additional information that her percentile rank in the class was 80, her parents know that 80% of her classmates scored lower than she did; therefore, 24 is a relatively high score in the class.

Percentile ranks can also be used to compare the standing of an individual on two or more tests. For example, if Peter's percentile rank on a math test was 30 and his percentile rank on a science test was 10, we know not only that, relative to his peers, Peter performed poorly on both tests but also that he did better on the math test than on the science test.

Before we leave percentile ranks, note this word of caution about their interpretation. Differences between percentile ranks of two or more individuals do not always provide accurate information about differences between their scores. For example, if you are told that, on Dr. Marrowbone's statistics quiz, Mary scored at the 57th percentile and John at the 39th percentile, you may be tempted to conclude that Mary's performance is vastly superior to John's. But if you look at the percentile rank distribution in Table 2–12, you will observe that Mary scored 18 and John 17, a difference of only one point. Now suppose that Alan scored at the 24th percentile and Sue at the 5th percentile. The difference between Alan's and Sue's percentile ranks is almost the same as that between Mary's and John's. Yet Alan's score on the quiz was 16 and Sue's was 13, a score difference three times as great as that between Mary and John. In other words, "equal" differences in percentile ranks do not usually reflect equal differences in raw scores. In the part of a distribution where frequencies are large, very small score differences may result in large percentile rank differences; in other parts where frequencies are small, even large score differences sometimes result in small differences between percentile ranks.

If a transformation of scores from one scale to another does not preserve equality of differences between scores, the transformation is said to be *nonlinear*. Thus, the transformation of raw scores to percentile ranks is a nonlinear transformation. Though such transformations may be useful descriptively, they have little usefulness for further calculations or for inference. Linear transformations, which preserve relationships between scores, are more useful for these purposes, as you will see in later chapters.

EXERCISE 2–12

Using the ungrouped frequency distribution you constructed in Exercise 2–3, number 2, do the following:

1. Calculate the percentile ranks of all scores.
2. The scores 5, 10, 15, and 20 are equidistant from one another. Compare their percentile ranks. Why aren't they equidistant?

Calculating Percentiles

Consider the statistics quiz scores again. Suppose that we wish to know the value of P_{75}, the score value below which 75% of observations fall. $N = 44$, so we want the score value below which 75% of 44, or 33 observations, fall. The Cum f column in Table 2–10 shows that 30 observations (too few) fall below 18.5, and 38 observations (too many) fall below 19.5. Therefore, we know that P_{75} is somewhere between 18.5 and 19.5. But what is its exact value?

To find any percentile, P_k, of a distribution, do the following:

1. Calculate the number of observations falling below P_k. To do so, multiply k by N and divide by 100.

$$\text{Number of observations below } P_k = \frac{kN}{100}$$

For example, the number of observations below P_{75} in the distribution of statistics quiz scores is [(75)(44)]/100 = 33, as we have already seen.

2. Find the lowest score with Cum $f > (kN)/100$. P_k falls between the lower and upper exact limits of this score.

   In the statistics score distribution, the lowest score with Cum $f > 33$ is 19, with Cum $f = 38$. The exact limits of the score 19 are 18.5 and 19.5. Therefore P_{75} falls between 18.5 and 19.5.

3. Use the following formula to calculate P_k:

$$P_k = \text{LEL} + \left[\frac{\frac{kN}{100} - \text{Cum}\,f\,\text{below}}{f\,\text{within}}\right] i$$

where:

LEL = lower exact limit of the interval in which P_k falls
Cum f below = Cum f of next lower score
f within = frequency within the interval in which P_k falls
i = class interval size (with integer data, $i = 1$)

We already know that P_{75} of the statistics quiz score distribution is between 18.5 and 19.5. Let us complete the calculation of P_{75}.

LEL = 18.5
Cum f below = 30 (this is Cum f of the score 18)
f within = 8 (this is the frequency in the interval 18.5–19.5)
i = 1

We have already calculated $\frac{kN}{100} = 33$

Therefore $P_{75} = 18.5 + \left(\frac{33-30}{8}\right) = 18.5 - .375 = 18.875$ (which we round to 18.9)

As another example, let us calculate P_{50}, the point below which 50% (half) of the observations fall.

$k = 50$ and $N = 44$

Number of observations below $P_{50} = \frac{kN}{100} = \frac{(50)(44)}{100} = 22$

The lowest score with Cum $f > 22$ is the score 18 with exact limits 17.5–18.5. Therefore P_{50} falls in the interval 17.5–18.5.

LEL = 17.5
Cum f below = 20
f within = 10
i = 1

Therefore $P_{50} = 17.5 + \left(\frac{22-20}{10}\right)1 = 17.5 + \frac{2}{10} = 17.5 + .2 = 17.7$

Percentiles, Quartiles, and Skew

It is possible to calculate any percentile from P_1 to P_{99}. But P_{25}, P_{50}, and P_{75} are particularly useful because they divide a distribution into quarters: 25% (one-fourth) of observations fall below P_{25}, 25% between P_{25} and P_{50}, 25% between P_{50} and P_{75}, and 25% above P_{75}. Because P_{25}, P_{50}, and P_{75} divide the distribution into quarters, they are sometimes referred to as ***quartiles***, symbolized Q with an appropriate subscript. Thus,

P_{25} = first quartile = Q_1
P_{50} = second quartile = Q_2
P_{75} = third quartile = Q_3

P_{50}, the second quartile, has still another name, the ***median***, because it divides the distribution into equal halves; 50% of observations fall below the median and 50% of observations fall above. As you will see in Chapter 3, the median is one of several measures of central tendency of a distribution.

In the statistics quiz score distribution, we have already calculated P_{50}, the second quartile or median, and P_{75}, the third quartile. Let us calculate P_{25}, the first quartile.

$$\text{Number of observations below } P_{25} = \frac{kN}{100} = \frac{(25)(44)}{100} = 11$$

The next higher Cum f, 14, is that for the score 16. P_{25} falls in the interval 15.5–16.5.

LEL = 15.5, Cum f below = 7, f within = 7, and i = 1.

$$\text{Therefore, } P_{25} = 15.5 + \left(\frac{11-7}{7}\right)1 = 15.5 + \frac{4}{7} = 15.5 + .6 = 16.1$$

We have calculated $Q_1 = 16.1$, $Q_2 = 17.7$, $Q_3 = 18.9$. We now know that, in the distribution of statistics quiz scores, 25% of students scored below 16.1, 25% between 16.1 and 17.7, 25% between 17.7 and 18.9, and 25% above 18.9.

Recall that the percentage of observations in a section of a distribution is equal to the percentage of the area in that section of the graph. Thus, when a distribution is graphed, 25% of the area falls below Q_1, 25% of the area between Q_1 and Q_2, 25% of the area between Q_2 and Q_3, and 25% of the area above Q_3. Areas are distributed differently in distributions having different shapes. Therefore, the relative locations of the three quartiles are different. Figure 2–14 shows the location of Q_1, Q_2, and Q_3 in symmetrical, negatively skewed, and positively skewed distributions.

In a symmetrical distribution, such as that shown in Figure 2–14a, Q_2 is halfway between the lowest and highest scores, directly below the peak of the curve. Since the left and right halves of the graph are mirror images, Q_1 and Q_3 are equidistant from Q_2, as shown. In a negatively skewed distribution (like Figure 2–14b), Q_2 is nearer to the high-

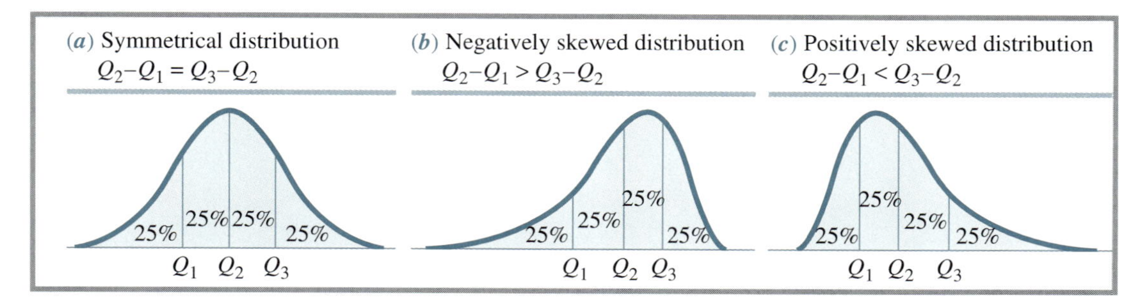

Figure 2–14 **Location of first quartile (Q_1), second quartile (Q_2), and third quartile (Q_3) in symmetrical and skewed distributions.**

est than to the lowest score, because observations tend to pile up at the high scores. Q_1, the point that divides the area in the long lower tail in half, is relatively far below Q_2; but Q_3 is very close to Q_2, as you can see. In a positively skewed distribution (Figure 2–14c), Q_1 and Q_2 are close together, near the low end of the distribution, whereas Q_3 is relatively far from Q_2 in the long upper tail.

In summary:

In a symmetrical distribution, the first, second, and third quartiles are equidistant from one another: $Q_2 - Q_1 = Q_3 - Q_2$.

In a negatively skewed distribution, the first and second quartiles are farther apart than the second and third quartiles: $Q_2 - Q_1 > Q_3 - Q_2$.

In a positively skewed distribution, the first and second quartiles are closer together than the second and third quartiles: $Q_2 - Q_1 < Q_3 - Q_2$.

Consider, once again, the distribution of statistics quiz scores. We have already observed that the distribution is negatively skewed. Let us compare the distance between Q_1 and Q_2 with the distance between Q_2 and Q_3.

$$Q_2 - Q_1 = 17.7 - 16.1 = 1.6$$

$$Q_3 - Q_2 = 18.9 - 17.7 = 1.2$$

As expected in a negatively skewed distribution, $Q_2 - Q_1 > Q_3 - Q_2$.

EXERCISE 2–13

Here is a distribution of ratings, on a scale of 1 to 7, that 25 people gave to a new TV show:

Rating	*f*
7	1
6	
5	2
4	4
3	7
2	8
1	3

1. Calculate P_{10} and P_{90}.
2. Before calculating the first, second, and third quartiles, decide whether the distribution is positively skewed, negatively skewed, or approximately symmetrical.
3. Calculate the first, second, and third quartiles. Then use the results to confirm the skewness of the distribution.
4. Match each of the symbols in column (a) to one or more terms in column (b).

(a)	(b)
P_{50}	25th percentile
P_{25}	2nd quartile
Q_1	3rd quartile
Q_3	first quartile
P_{75}	50th percentile
Q_2	75th percentile median

Box-and-Whisker Plots

The quartiles of a distribution can be used to draw a graph called a ***box-and-whisker plot*** (***box plot***, for short), a useful and simple way to display trends in a distribution. It is easier to draw a box-and-whisker plot than it is to explain how to draw it. Therefore, we start by showing a box plot and then describe the procedure for constructing it.

Box plots may be drawn horizontally or vertically; a horizontal plot of the statistics quiz scores is shown in Figure 2–15. As you can see, score values are marked from lowest on the left to highest on the right. The "box" in the box-and-whisker plot extends from the first quartile ($Q_1 = 16.1$) to the third quartile ($Q_3 = 18.9$). Thus, the box includes the middle 50% of observations. A vertical line crosses the box at the median (17.7). The lines extending from both ends of the box are the "whiskers." The whiskers extend to the lowest and highest scores (9 and 20, in this case).

The box plot in Figure 2–15 effectively reveals the negative skew of the distribution. But it does not reveal the fact that the lowest score, 9, is an outlier. It is often considered useful to indicate the presence of outliers in a box plot. One way to do this is to extend each whisker to a maximum distance equal to 1.5 times ($Q_3 - Q_1$) beyond the box. Scores that fall beyond the end of the whiskers (outliers) are indicated by dots (•).

In the case of the statistics quiz scores, 1.5 ($Q_3 - Q_1$) = 1.5 times (18.9 − 16.1) = 1.5 (2.8) = 4.2. Therefore, the lower whisker should extend no farther than 4.2 points below 16.1, or to a score value of 11.9 (about 12). Therefore, we extend the lower whisker to 13 (the lowest score except for 9), leaving 9 as the single outlier. The resulting modified box plot is shown in Figure 2–16.

Side-by-Side Box Plots

We have already seen that frequency polygons and side-by-side stem plots are useful for comparing two distributions of scores on the same variable. Another way of comparing two or more distributions is the *side-by-side box plot*.

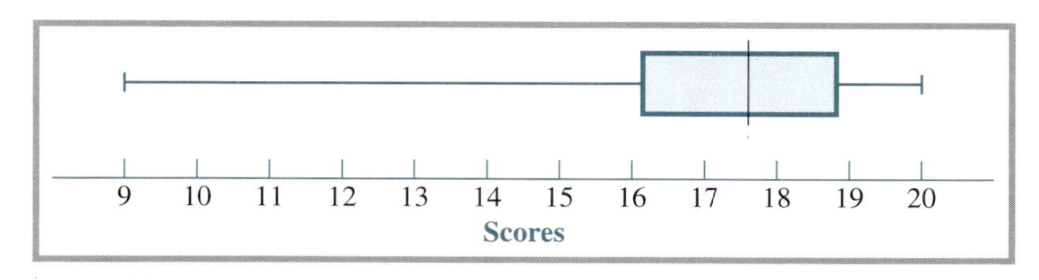

Figure 2–15 **A box-and-whisker plot of the statistics quiz scores.**

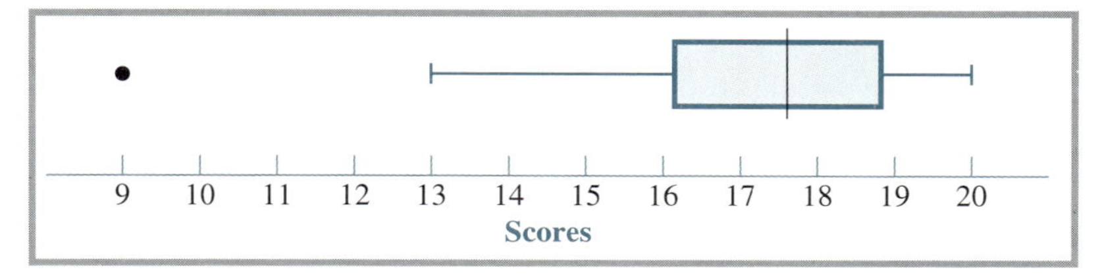

Figure 2–16 **Modified box plot of the statistics quiz scores.**

As an example, let us go back to the anxiety scale scores of college students versus clients at a mental health center. The quartiles of the two distributions are as follows:

	College Students	Clients
Q_1	39.0	50.25
Q_2 (Median)	46.5	58.0
Q_3	52.3	63.75

(If you would like to calculate these quartiles yourself, you can find both sets of scores in Figure 2–9).

Side-by-side box plots are shown in Figure 2–17. Notice how vividly the graph shows the difference between the groups.

The box plots in Figure 2–17 have been drawn vertically; they could, however, be drawn horizontally instead, with one group displayed below the other. In that case, of course, the accompanying scale of score values would go from left to right instead of from the bottom to the top of the figure.

EXERCISE 2–14

1. Draw a box-and-whisker plot of the data in Exercise 2–13. What aspects of the plot indicate the skew of the distribution?
2. Compare the frequency polygons in Figure 2–8, the stem plots in Figure 2–9, and the box plots in Figure 2–17. All three figures are based on the same data. Describe the advantages and disadvantages of each relative to the others. (What features of the two distributions are best displayed by each one? What features are omitted?)

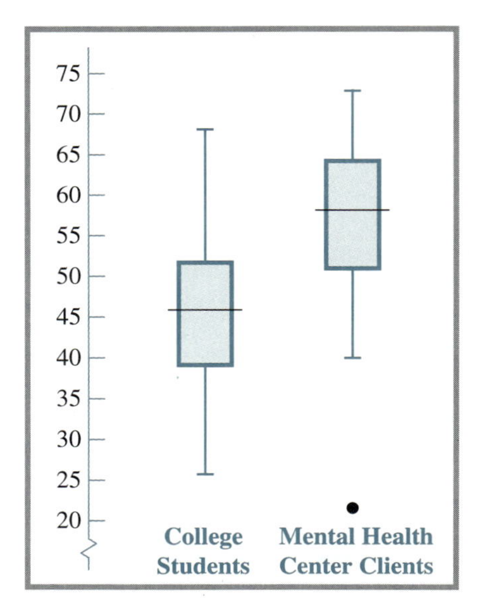

Figure 2–17

Side-by-side box plots comparing anxiety scale scores of 50 college students and 35 mental health center clients.

Data Files and Computer Graphs

In this chapter, we have worked with distributions of a single variable measured in one or more groups, and the groups' sizes have been relatively small. In practice, however, researchers often measure many variables in large groups. Thus, for example, an educational psychologist studying factors that affect the reading ability of fourth-grade children may collect data on sex, age, intelligence, family size, marital status of parents, parents' income, parents' education, number of books in the home, and others, as well as scores on a reading achievement test, of 500 children.

It is common to display data on multiple variables in two-way tables with subjects (cases, records) listed by rows and variables by columns. An example of a relatively small set of data, collected on 28 students in an introductory statistics class, is shown in Table 2–13. As you can see, each student occupies one row of the table. The first col-

Table 2–13 A Data Matrix of Data Obtained from 28 Students Enrolled in an Introductory Statistics Class

Student	Gender	Class[a]	Area of Major[b]	Height (to the nearest inch)	Score on 20-Item Quiz
01	M	Fr	Hum	72	7
02	M	So	Sci	64	16
03	F	Jr	SS	64	15
04	F	Jr	Hum	59	19
05	M	Sr	SS	69	14
06	F	Sr	Sci	65	12
07	M	Sr	SS	70	16
08	F	Jr	Hum	68	10
09	F	So	SS	66	18
10	F	Fr	Hum	64	15
11	M	Jr	SS	67	16
12	F	So	SS	64	18
13	F	Jr	Sci	67	7
14	F	So	Sci	61	6
15	M	Fr	SS	67	12
16	F	So	Hum	62	20
17	M	So	SS	69	17
18	M	So	SS	71	10
19	M	So	Hum	69	17
20	F	So	SS	66	17
21	F	Jr	SS	61	17
22	F	Sr	SS	64	13
23	M	So	Sci	69	16
24	F	Jr	SS	63	14
25	F	So	SS	64	17
26	F	So	SS	64	17
27	F	Jr	Hum	62	18
28	M	So	Sci	64	17

[a]Fr = Freshman; So = Sophomore; Jr = Junior; Sr = Senior
[b]Hum = Humanities; Sci = Science; SS = Social Science

umn lists ID numbers (from 01 to 28); the remaining columns list five variables measured in each student.

The data set in Table 2–13 is small enough to work with manually. (In fact, you will have an opportunity to do this in the end-of-chapter problems.) A researcher who has measured many more variables in many more subjects, and who probably plans multiple analyses, could hardly do all the work by hand. As pointed out in Chapter 1, deciding *which* analyses are needed is the researcher's responsibility. A computer is essential, however, for the actual conduct of the analyses.

Before a computer analysis can be conducted, the data must first be typed into a machine-readable ***data file***. This is a task that requires no intellectual effort but, it does need a high level of care and frequent checking to make sure that all data are entered correctly. In fact, this is the part of computer analysis that takes the most time. But the effort pays off because the data are now available not only for immediate analysis but for additional future analyses as well.

To conduct a particular analysis on data in a computer file, the researcher must tell the computer two things: (1) where to find the data (in which file and in which columns) and (2) what to do with it. Let us take a simple example using data we have worked with in this chapter. Suppose that Dr. Marrowbone is using a statistical package called Minitab to analyze the statistics quiz scores of his 44 students. He has entered the scores into the first column of a data file; Minitab refers to this column as "c1." He has retrieved the file and is ready to work on it. Now Minitab prompts him to enter a ***command***, that is, an instruction to perform an operation, like this:

```
MTB>
```

Dr. Marrowbone has decided that he wants a box plot of the scores, and so he simply types "boxplot c1" at the Minitab prompt, like this:

```
MTB > boxplot c1
```

Within milliseconds, Minitab processes the command and constructs the graph shown in Figure 2–18. As you can see, the computer-drawn box plot is identical with the one shown in Figure 2–17.

Computer packages like Minitab can be instructed to do most of the things we have discussed in this chapter, such as construction of frequency distributions and graphs and calculation of percentiles and percentile ranks, as well as many other kinds of analyses. Other examples of computer use and the printouts that result are discussed in later chapters.

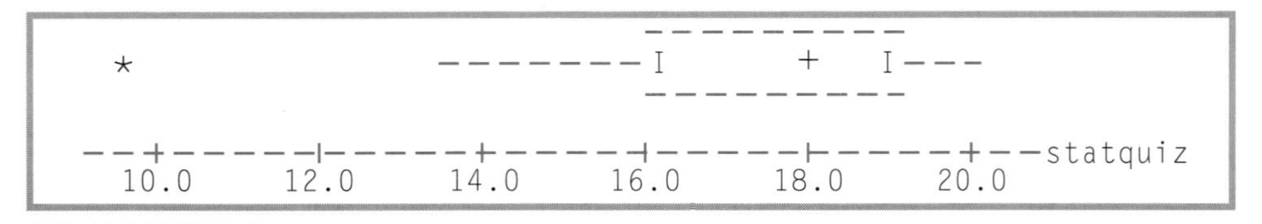

Figure 2–18 A box plot constructed by the Minitab statistical analysis package.

SUMMARY

Tables and graphs of distributions help us to see characteristics of a distribution as a whole. A table that lists categories or scores and the frequency of each one is called a ***frequency distribution***. Two or more distributions of a single categorical variable are typically displayed in a ***contingency table***. If two or more distributions (of either categorical or numerical data) are to be compared, frequencies should be converted to ***relative frequencies*** (proportions or percentages), unless the groups are the same size.

Frequency distributions of quantitative data may be ***ungrouped***, with one score value per interval, or ***grouped*** with two or more score values within each interval. The highest and lowest score values of an interval are the ***score limits***; the lower exact limit of the lowest score and the upper exact limit of the highest score are the ***exact limits*** of the interval. The ***midpoint*** of an interval is the point halfway between the lower and upper limits. Besides frequencies, we may also calculate ***cumulative frequencies*** and/or ***cumulative percentages*** (the number or percent of observations falling at or below each score).

A common graph for displaying categorical data or measurements of a discrete variable is the ***bar graph***. For measurements of a continuous variable, either a ***histogram*** or a ***frequency polygon*** can be constructed. ***Stem-and-leaf plots*** are useful for showing major features of a distribution without losing any details, especially when the range of scores is large. ***Cumulative graphs*** can be drawn to display the number or percentage of observations below any given score point.

Distributions of data may be ***symmetrical*** or ***positively*** or ***negatively skewed***. The ***normal curve*** is an important symmetrical distribution. Other less common distribution shapes include ***bimodal*** and ***rectangular*** distributions. ***Obtained distributions*** are based on actual data, whereas ***theoretical distributions*** are generated mathematically or logically. The normal curve is an example of a theoretical distribution. In a graph of a distribution, the proportion of area within a given section is equal to the proportion of observations in that part of the distribution.

The kth ***percentile*** of a distribution (P_k) is the score value below which k percent of observations fall. The ***percentile rank*** of a score is the percentage of observations below the score. Percentile ranks are useful for locating individuals within a distribution and for comparing individuals' locations in two or more distributions. However, equal differences between percentile ranks do not usually reflect equal differences between raw scores.

The 25th percentile of a distribution is also called the ***first quartile*** (Q_1); the 50th percentile is the ***second quartile*** (Q_2), or ***median***; the 75th percentile is the ***third quartile*** (Q_3). The quartiles divide a distribution into four quarters each containing 25% of the observations. In symmetrical distributions, $Q_2 - Q_1 = Q_3 - Q_2$; in negatively skewed distributions, $Q_2 - Q_1 > Q_3 - Q_2$; in positively skewed distributions, $Q_2 - Q_1 < Q_3 - Q_2$. The ***box-and-whisker plot***, a simple and useful way of displaying a distribution, is based on the quartiles of the distribution.

Computers can do most of the operations discussed in this chapter. If computer analyses are to be performed on data, the data must first be typed into a machine-readable ***data file***. Then ***commands*** can be used to tell the computer what to do with the data. A simple example using Minitab was shown in this chapter.

QUESTIONS

Questions 1–14 are based on the data set displayed in Table 2–13.

1. Construct a frequency distribution and a graph of class.
2. Construct a contingency table showing the distribution of majors for males and females. Then construct a graph comparing the male and female distributions. Does major appear to be contingent on gender?

3. Construct an ungrouped frequency distribution of the quiz score data. Then construct two grouped frequency distributions of the same data, one including two score values within each interval and the second including three score values within each interval. Which of the three distributions do you prefer? Why?

4. What is the shape of the distribution of quiz scores?

5. Construct a split-stem plot of the quiz score data.

6. In each of the three quiz score frequency distributions that you constructed in question 3, list the exact limits and midpoints of all intervals.

7. Calculate cumulative frequencies and cumulative percentages for the ungrouped distribution of quiz scores. Then answer the following questions:

   a. How many students scored 15 or less?
   b. What is the percentage of students who scored 17 or less?
   c. What is the score point below which 7 students fall?

8. Construct a histogram and a frequency polygon of each of the two *grouped* distributions you constructed in question 3.

9. Construct a grouped frequency distribution for the heights of males only and another for the heights of females only. Use the same grouping for both genders. Select a class interval size that will give you about seven class intervals over the entire range of heights.

10. Draw a graph displaying both of the frequency distributions you constructed in question 9. Describe differences between the height distributions of the male and female students.

11. Construct a cumulative percentage graph for the statistics quiz scores. Then use the graph to find the point below which one-fourth of observations fall.

12. Calculate the percentile ranks of all the statistics quiz scores (use the ungrouped distribution).

13. In the distribution of statistics quiz scores, calculate the following:

    a. P_{40}.
    b. P_{80}.
    c. the first, second, and third quartiles. (Does your calculation of the first quartile agree with your answer to question 11?)
    d. Use your answers to part (c) to measure the direction of skew of the distribution. Does your answer agree with your answer to question 4?

14. Construct a box-and-whisker plot of the quiz scores.

The remaining questions are *not* based on the data set in Table 2–13.

15. Describe the shape of each of the following distributions:

(a)		(b)		(c)	
Score	*f*	**Score**	*f*	**Score**	*f*
10	1	10	3	10	4
9	2	9	8	9	4
8	4	8	4	8	4
7	9	7	3	7	4
6	12	6	8	6	4
5	10	5	2	5	4

16. A graph of data collected in a group of 2000 subjects is shown in Figure 2–19. The numbers within the sections are proportions of the total area under the curve.

    a. What proportion of the group scored above 33?
    b. What percentage of the group scored below 29?

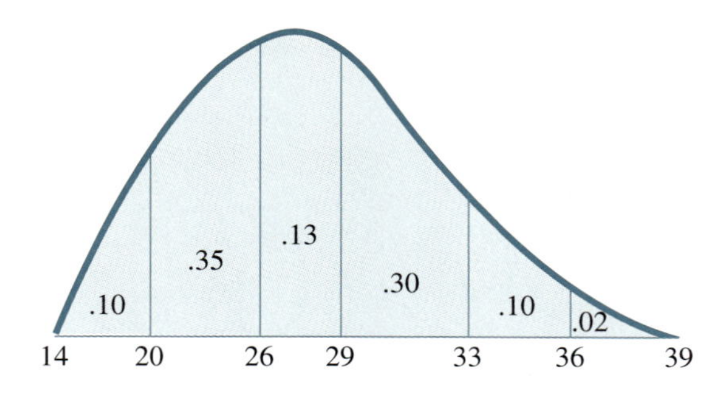

Figure 2–19 **Graph of data collected in a group of 2000 subjects.**

c. What proportion of the group scored between 20 and 29?
d. How many of the 2000 subjects scored between 20 and 29?
e. What is the point above which 90% of the group falls?
f. What is the point below which 45% of the group falls?

17. On a test, 60% of students scored below 40. Which pair of expressions below states this information correctly?

a. $P_{60} = 40$
b. $P_{40} = 60$
c. The percentile rank for a score of 60 is 40.
d. The percentile rank for a score of 40 is 60.

18. In an approximately normal distribution of test scores, Bo scored at the 5th percentile, Flo at the 15th percentile, Joe at the 45th percentile, and Moe at the 55th percentile. Is the difference between the raw scores of Bo and Flo greater or less than the difference between the raw scores of Joe and Moe?

19. Here are Q_1, Q_2, and Q_3 for three distributions. Decide whether each distribution is symmetrical, positively skewed, or negatively skewed.

Distribution A: $Q_1 = 11$, $Q_2 = 13$, $Q_3 = 18$.
Distribution B: $Q_1 = 12$, $Q_2 = 16$, $Q_3 = 20$.
Distribution C: $Q_1 = 2.5$, $Q_2 = 7.0$, $Q_3 = 9.6$.

Measures of Central Tendency and Variability

INTRODUCTION

In Chapter 2 you learned about procedures for examining a distribution as a whole and for locating particular observations within it. This is a good beginning, but we also need a few additional kinds of information to make our descriptions complete. In particular, we need numerical summary measures to answer questions such as "What is the average performance of Group A?" "Did Group A have higher scores, on average, than Group B?" "How much did scores on Variable C vary around the average score?" "Did the experimental treatment increase or decrease differences among subjects?" To answer questions such as these, we calculate measures of central tendency (averages) and measures of variability. In this chapter we consider the calculation, interpretation, and use of several measures of central tendency and variability.

Objectives

After you have completed this chapter, you will be able to

- Explain the need for measures of central tendency and variability.
- Define and calculate the following measures of central tendency: the mode, the median, and the mean (including the weighted mean).
- Define and calculate the following measures of variability: the range, the interquartile and semi-interquartile range, and the variance and standard deviation.
- Name the measure(s) of variability associated with each measure of central tendency.
- For each measure of central tendency and variability, describe its properties, factors that affect it, its strengths and weaknesses, and conditions under which it is appropriate and inappropriate.
- Define linear transformation. Describe the effects of linear transformations on measures of central tendency and variability.
- Calculate and interpret standard scores (*z* scores), and describe the mean, standard deviation, and shape of *z* score distributions.
- Compare standard scores with percentile ranks as a means of locating scores within distributions.

Why Do We Need Measures of Central Tendency and Variability?

Say that you have measured the same variable in three groups of people and have gotten the results shown in Distributions A, B, and C in Figure 3–1. All three distributions are symmetrical, and all have the same shape. Yet they differ from one another in important ways. One of these differences has to do with central tendency and the other with variability.

Suppose someone were to ask you, "What is the single score value in each distribution that is most representative of the entire distribution?" You would probably select the central value 10 in Distributions A and B and 20 in Distribution C. A score value selected as representative of the distribution as a whole is called a measure of ***central tendency***, or ***average***. The central point or average value is one important characteristic of a distribution. In terms of their average values, Distributions A and B are alike, whereas C is different.

Yet, in another sense, Distributions A and C are alike and B is different. Despite their different average values, the scores of A and C spread out equally around the average, whereas the scores of B are more widely spread out or dispersed. The degree to which scores spread out is called the *variability* of a distribution. Distribution B has a greater amount of variability than Distributions A and C.

Differences in central tendency and variability have both theoretical and practical importance. Suppose, for example, that a clinical psychologist has hypothesized that Therapy C is more effective for increasing the self-concept of delinquent boys than Therapy A. To test the hypothesis, the psychologist randomly subdivides a group of delinquent boys and treats half the boys with Therapy A and the other half with Therapy C. When the boys' self-concept is measured a year later, Distributions A and C of Figure 3–1 are the result. The average level of self-esteem is clearly higher under Therapy C. The difference in central tendency supports the psychologist's hypothesis.

Or suppose that you are preparing to teach reading to two classes of beginning first graders, Class A and Class B. The children's scores on a reading readiness test are dis-

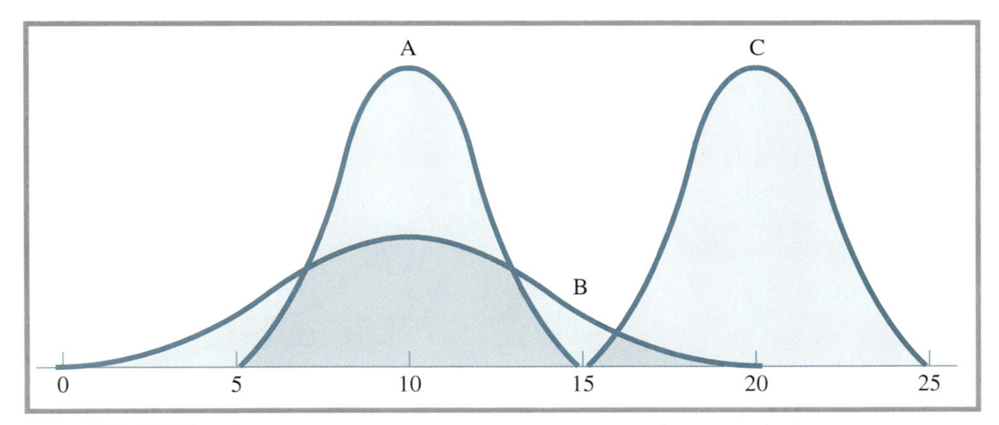

Figure 3–1 **Three distributions illustrating central tendency and variability. Distributions A and B have the same central value but different amounts of variability. Distributions A and C have different central values but the same amount of variability. Distributions B and C differ in both central tendency and variability.**

played in Distributions A and B of Figure 3–1. As you can see, the average readiness of the two classes is the same; but Class B's scores are more variable. This difference in variability has implications for how you will teach. Because Class A children do not differ a great deal in reading readiness, you can probably instruct them as a single group. But in Class B, you may have to divide the children into two or more subgroups, beginning your instruction at different levels for children in different subgroups.

Distributions A, B, and C all have one clearly distinguishable central point. However, this is not always the case. Consider Distributions D and E in Figure 3–2. Both distributions are skewed, but in different directions: Distribution D is negatively skewed and Distribution E is positively skewed. Let us see if we can locate their average values.

In both Distributions D and E, the score value with the highest frequency is 15. The score with the highest frequency is called the *mode*. The mode is one measure of central tendency (or average). Because both distributions have the same mode, in a certain sense, they have the same average value. Yet most of Distribution D falls below 15, whereas most of Distribution E falls above 15. If we locate the score value that divides each distribution in half, such that 50% of observations fall below that value and 50% above (the 50th percentile, or *median*), we find that the median of Distribution D is 13, but the median of Distribution E is 17, as the figure shows. If we use the median as a measure of central tendency, we conclude that the average of Distribution E is greater than the average of Distribution D.

Distributions D and E demonstrate that there is more than one way to measure the central tendency of a distribution. The mode is one measure of central tendency and the

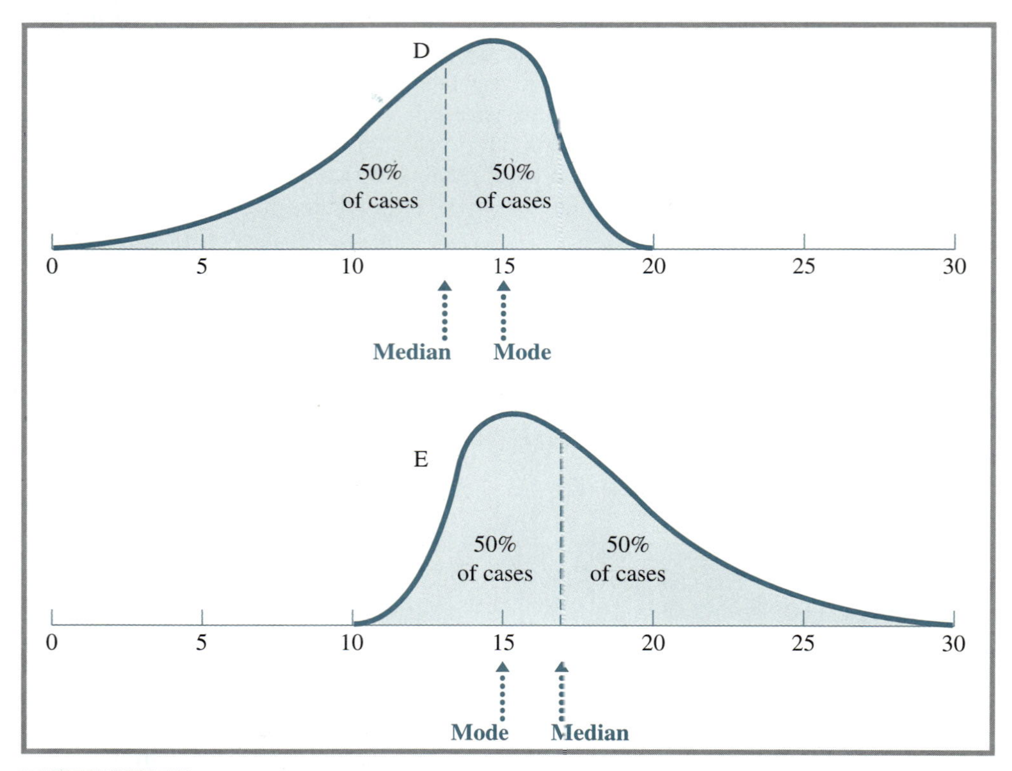

Figure 3–2 **Two distributions with the same mode but different medians. In Distribution D, mode = 15 and median = 13; in Distribution E, mode = 15 and median = 17.**

median is another. Still another measure of central tendency is the *arithmetic mean* (*mean*, for short). Which measure is preferred for a given set of data depends on the variable being measured, the shape of the distribution, and the uses to which the information will be put.

The variability of a distribution can also be measured in more than one way. Consider Distributions F and G displayed in Table 3–1. Notice that scores in both distributions vary from 2 to 10. The difference between the highest and lowest scores of a distribution is called the *range*. Thus, both Distributions F and G have a range of 8. If we use the range as a measure of variability, it would appear that Distributions F and G are equally variable. But observe how differently observations are distributed over the range in the two distributions. In F, observations are evenly distributed over all values between 2 and 10. In G, on the other hand, all but two observations fall between 5 and 7. If we were to calculate the range of the middle 50% of observations, a measure of variability called the *interquartile range*, we would find that the interquartile range is considerably smaller in Distribution G than in Distribution F, and we would conclude that Distribution G is less variable.

The range and interquartile range are two measures of variability. There are still others. Which one best describes the variability of a distribution depends on the nature of the variable, the shape of the distribution, and the uses to which the information will be put. Thus, the factors that influence the choice of a measure of variability are the same as those that influence the choice of a measure of central tendency. In fact, each measure of variability is associated with a particular measure of central tendency. When we report a measure of central tendency to describe the average value of a set of data, we usually report an associated measure of variability to describe the spread or dispersion of scores around the average. The following table lists the measures of central tendency and variability to be considered in this chapter and shows their relationships.

Measure of Central Tendency	Associated Measure(s) of Variability
Mode	Range
Median	Interquartile range
	Semi-interquartile range
Arithmetic mean (mean)	Variance
	Standard deviation

Table 3–1 Two Distributions With the Same Range but Different Distributions of Scores Over the Range

Distribution F		Distribution G	
Scores	*f*	Scores	*f*
10	3	10	1
9	3	9	
8	3	8	
7	3	7	7
6	3	6	11
5	3	5	7
4	3	4	
3	3	3	
2	3	2	1
	27		27

EXERCISE 3–1

Here are four distributions:

(1)		(2)		(3)		(4)	
Score	*f*	Score	*f*	Score	*f*	Score	*f*
18	2	25	2	24	2	6	1
17	4	24	3	23	4	5	2
16	10	23	4	22	10	4	3
15	6	22	8	21	6	3	3
14	3	21	5	20	3	2	5
		20	2			1	8
		19	1			0	3

1. Which two distributions have the same shape and variability but different average values?
2. Which two distributions have the same central value (mode) but different amounts of variability?
3. Without calculating, in which distribution are the mode and the median most different?

Measures of Central Tendency

The Mode

The ***mode*** (***Mo***) of a distribution is defined as *the score with the greatest frequency*. The mode can usually be found by inspection; no calculation is necessary. This simple procedure makes the mode a quick and easy measure of central tendency. Unlike other measures of central tendency, however, the mode is not necessarily a unique value in a distribution. Some distributions have no mode, and some have more than one. Consider the following two sets of scores:

1. Scores: 5, 8, 7, 6, 3, 10, 2
2. Scores: 2, 7, 6, 2, 8, 7, 4

In the first set of scores, each score occurs with equal frequency; there is no mode. In the second set, scores 2 and 7 occur more frequently than other scores; both 2 and 7 are modes, and the distribution is said to be ***bimodal***. When a bimodal distribution is graphed, the figure has two peaks. Figure 3–3a shows a symmetrical bimodal distribution with modes of 25 and 35.

The two peaks of a bimodal distribution need not be equally tall. A distribution is usually considered bimodal if it has two nonadjacent scores with frequencies greater than those of intervening scores, even if one of the two scores has a smaller frequency than the other. Consider the distribution in Figure 3–3b. The single score with the highest frequency is 16. However, since score 12 has a frequency greater than that of scores 13, 14, or 15, the distribution can be considered bimodal with modes of 12 and 16.

The mode has some shortcomings as a measure of central tendency. For one, a small change in a distribution may produce a large change in the mode. Table 3–2 shows two almost identical frequency distributions, yet the mode of Distribution (a) is 16, and that of Distribution (b) is 13. Also, the mode lacks mathematical properties needed for further calculations or for statistical inference. Because of these shortcomings, the mode is rarely the preferred measure of central tendency.

Figure 3–3

Symmetrical (a) and nonsymmetrical (b) bimodal distributions.

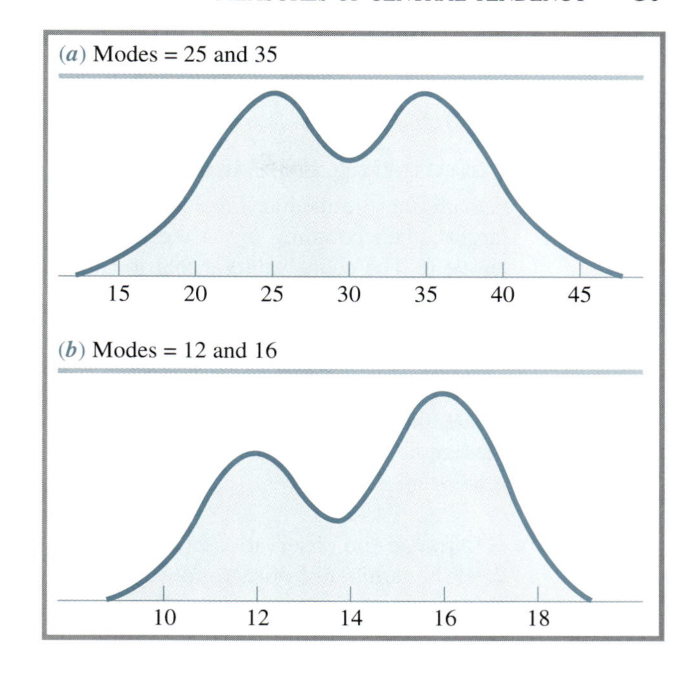

However, the mode is needed to answer certain kinds of questions. First, the mode is the only measure of central tendency applicable to unordered categorical data; it makes sense to talk about the *modal* category—the category having the most members—but not the median category or the mean category. The mode is also the number you need if you want information about the most typical score. Suppose that you need to buy hats for members of a high school marching band, and for some reason, you must order them all the same size. It would probably be best to order the band members' modal size because that size will fit more of the members than any other size.

THE MEDIAN

The ***median (Mdn)*** of a distribution is defined as *the score value or score point below which half the observations fall.* For example, if N is 10, the median is the point below which 5 observations fall; if N is 25, the median is the point below which 12.5 obser-

Table 3–2 Two Similar Distributions of Scores With Different Modes

	(a) Score	(a) f		(b) Score	(b) f
	18	1		18	1
	17	3		17	3
Mode →	16	6		16	5
	15	5		15	5
	14	5		14	5
	13	5	Mode →	13	6
	12	2		12	2
		27			27

vations fall. "The point below which half the observations fall" is also the definition of the 50th percentile (P_{50}) or second quartile (Q_2). Therefore, we can write:

$$Mdn = P_{50} = Q_2$$

Calculating the Median

Calculating the median involves, first, listing the observations in order from smallest to largest, then counting up to the location that has half the observations below and half above it. The score value at that location is the median.

Recall that in distributions of continuous variables, each score value represents an interval. The precise value of the median can fall anywhere within the interval, so that the median can be a number like 17.2 or 63.84. If you need the precise value of the median, use the procedure described in Chapter 2 for interpolating into an interval to calculate the 50th percentile, P_{50}. For most purposes, however, the precise value of the median is not needed. In that case, the median can be found by inspection, using the following method:

1. Arrange the observations (scores) in order from smallest to largest.
2. If the number of observations (N) is odd, locate the middle score. The median is equal to that score.
3. If N is even, there are two middle scores. The median is the number halfway between them.

Here are some examples:

A. Scores: 8, 5, 10, 5, 4, 3, 4 ($N = 7$)

Scores in order: 3, 4, 4, 5, 5, 8, 10

↑ *Mdn* (under the fourth score, 5)

The middle score is 5. The median is 5.

B. Scores: 10, 18, 7, 3, 7, 6, 12, 15 ($N = 8$)

Scores in order: 3, 6, 7, 7, 10, 12, 15, 18

↑ *Mdn* (between 7 and 10)

The middle scores are 7 and 10. The median is halfway between 7 and 10, or 8.5.

If N is large, it may be more efficient to construct a frequency distribution or stem plot at Step 1 rather than listing the scores. In either case, the median is easily located by counting up from the bottom (or down from the top) to the middle score (odd N) or middle two scores (even N). As examples, the frequency distribution of Dr. Marrowbone's statistics quiz scores and the stem plot of college students' anxiety scores are reproduced in Figure 3–4. There are 44 quiz scores, so that the median falls between the 22nd score (18) and the 23rd score (also 18). The median is equal to 18. Similarly, with 50 anxiety scores, the median falls between the 25th score (46) and 26th score (47) and is equal to 46.5.

Properties of the Median

The median has several properties that make it preferable to the mode as a measure of central tendency. First, it is more stable than the mode. We have already noted how a small change in frequencies can produce a large change in the mode of a distribution

(a)

Quiz Scores of 44 Students

Score	Frequency	
20	6	
19	8	
(18)	10	The two middle
17	6	scores (the 22nd
16	7	and 23rd) are
15	3	both 18.
14	1	*Mdn* = 18
13	2	
12		
11		
10		
9	1	

(b)

Anxiety Scale Scores of 50 Students

2	666	
3*	014	
3	5677889	The two middle
4*	11233444	scores (the 25th
4	566⑥⑦78899	and 26th) are
5*	122233	46 and 47.
5	67799	*Mdn* = 46.5
6*	0012	
6	8	

Figure 3–4 **Locating the median in a frequency distribution (a) and stem plot (b).**

(see Table 3–2). The modes of the two distributions in Table 3–2 are 16 and 13, respectively. However, both distributions have the same median, 15.

A second advantage of the median is that it is little affected by unusual extreme scores, that is, outliers. For example, the following three sets of scores all have the same median, despite the fact that the highest and lowest scores of Set 1 are replaced by extreme outliers in Sets 2 and 3.

Set 1	Scores:	8, 8, 10, 11, 13	*Mdn* = 10
Set 2	Scores:	8, 8, 10, 11, 200	*Mdn* = 10
Set 3	Scores:	−100, 8, 10, 11, 13	*Mdn* = 10

The median, in other words, is affected by whether scores fall above or below the middle of the distribution, but not how far above or below. When we want a measure of central tendency that is not unduly influenced by a few extreme scores, as in highly skewed distributions, the median is the preferred measure of central tendency.

Despite its usefulness as a descriptive measure of central tendency, especially in distributions that are skewed or have one or more outliers, the median has certain shortcomings. In particular, the fact that it does not utilize all the observations in its calculation limits its use for further calculations, especially in statistical inference. You will see that when we have questions about averages of populations, we usually seek answers based on sample means rather than medians.

EXERCISE 3–2

1. Find the mode and median of the following distributions:
   a. Scores: 7, 7, 9, 8, 7
   b. Scores: 15, 13, 12, 15, 16, 10, 12, 11
   c. Scores: 18, 15, 19, 21, 9, 24, 8
   d. Scores: 8, 11, 10, 10, 15, 6, 10, 8
2. In Exercise 2–8 in Chapter 2, you constructed a stem plot of 20 test scores. Locate the median in the stem plot.

The Mean

You probably were taught in elementary school that the average of a set of scores is the sum of the scores divided by the number of scores. The average that you learned to calculate was the ***arithmetic mean***. There are other kinds of means, too—the geometric mean and the harmonic mean. But the arithmetic mean is the most widely used in the social and behavioral sciences. We refer to it simply as the ***mean***, symbolized ***M***.†

In order to write a formula for the mean, we let X represent a score. The symbol Σ stands for "add whatever follows." Therefore, to say "Add (Σ) the scores (X) and divide by the number of scores (N) to get the mean," we write

$$M = \frac{\Sigma X}{N} \qquad \textit{Formula for calculating the mean}$$

For example, for the set of five scores 5, 8, 7, 6, 3

$$M = \frac{5+8+7+6+3}{5} = \frac{29}{5} = 5.8$$

Notice that each and every score is used in calculating the mean. Therefore, any change in one or more scores produces a change in the mean. For example, if the last score were changed from 3 to 1, then

$$M = \frac{5+8+7+6+1}{5} = \frac{27}{5} = 5.4$$

Properties of the Mean

If you were to draw a graph of a distribution on plywood or heavy cardboard, cut it out, and then balance it on a fulcrum or wedge △, the distribution would balance only if the point of the wedge is directly below the mean. The reason is that deviations of scores from the mean in the positive direction exactly balance off deviations in the negative direction. That is, *the sum (Σ) of deviations from the mean (X-M) is equal to zero.*

$$\Sigma(X - M) = 0$$

For example, the scores 5, 8, 7, 6, 3 have mean (M) = 5.8, as we have seen. Subtract 5.8 from each score and then add the deviations from the mean.

Score	Deviation from the Mean $(X - M)$
5	5 − 5.8 = −0.8
8	8 − 5.8 = 2.2
7	7 − 5.8 = 1.2
6	6 − 5.8 = 0.2
3	3 − 5.8 = −2.8
	$\Sigma(X - M)$ = 0

As you can see, the sum of the deviations is zero. This is always true for any distribution. In fact, *the mean can be defined as the number from which the sum of deviations equals zero.*

Another property of the mean is that, in any distribution, *the sum of squared deviations from the mean is a minimum.* That is, if you calculate the deviation of each score

†Many textbooks use the symbol $\bar{X}$ for the mean. In journal articles and scientific reports, however, it is common practice to use M, and we choose it for that reason.

from the mean $(X - M)$, square each deviation $(X - M)^2$, and then add all the squared deviations, the sum is smaller than the sum of squared deviations from any other number:

$$\Sigma(X - M)^2 < \Sigma(X - \text{any other number})^2$$

We illustrate using the scores 5, 8, 7, 6, 3. The mean of the scores is 5.8, and the median is 6. Let us compare the sum of squared deviations from 5.8 (the mean) with that from 6 (the median).

X	$X - M$	$(X - M)^2$	$X - Mdn$	$(X - Mdn)^2$
5	−0.8	0.64	−1	1
8	2.2	4.84	2	4
7	1.2	1.44	1	1
6	0.2	0.04	0	0
3	−2.8	7.84	−3	9
		14.80		15
		$\Sigma(X - M)^2$		$\Sigma(X - Mdn)^2$

As you can see, $\Sigma(X - M)^2 < \Sigma(X - Mdn)^2$. You will have to take my word for the fact that subtracting any number other than 5.8 from each score in the distribution above would result in $\Sigma(X - \text{the number})^2 > 14.80$. In fact, *the mean can be defined as the number from which the sum of squared deviations is a minimum.* For this reason, the mean is sometimes referred to as the "least squares" measure of central tendency. Statisticians often seek numbers (or formulas for calculating numbers) that have the property of "least squared deviations." You will see this property invoked again in later chapters.

The term *sum of squared deviations from the mean* is often shortened to ***sum of squares***, or ***SS***. Thus, $SS = \Sigma(X - M)^2$. As you will see, the sum of squares is involved in calculation of the variance and the standard deviation, two measures of variability associated with the mean.

The Weighted Mean

Suppose that you have calculated the mean of each of three groups of unequal size. The first group of 10 members has a mean of 35.0; the second group of 5 members has a mean of 41.2; the third group of 20 members has a mean of 33.7. Now you want to find the mean of the three groups combined. Your first impulse may be to add the three means and divide by 3:

$$\frac{35.0 + 41.2 + 33.7}{3} = 36.6$$

However, this method is not appropriate because it gives equal weight to each group, even though Group 3 is twice the size of Group 1 and four times the size of Group 2. We need a method that gives each group a weight proportional to its size. The combined mean, which takes group size into account, is called the ***weighted mean***. If we let k stand for the number of groups and use subscripts to indicate group membership, the formula for the weighted mean is

$$\textit{Weighted mean} = \frac{N_1M_1 + N_2M_2 + \cdots + N_kM_k}{N_1 + N_2 + \cdots + N_k}$$

In our three-group example, $N_1 = 10$ with $M_1 = 35.0$, $N_2 = 5$ with $M_2 = 41.2$, and $N_3 = 20$ with $M_3 = 33.7$. Therefore,

$$\text{Weighted mean} = \frac{10(35.0)+5(41.2)+20(33.7)}{10+5+20} = \frac{350+206+674}{35} = \frac{1230}{35} = 35.1$$

Notice that the weighted mean is smaller than the unweighted mean of the three group means (36.6). This is because the largest group had the smallest mean.

EXERCISE 3–3

1. Calculate the mean of each of the following sets of scores:
   a. 8, 6, 5, 9, 12, 3, 2
   b. 2, 5, 1, 8
   c. 3.6, 9.4, 6.7, 7.1, 6.5
2. Shaw Elementary School has two kindergarten classes. In one class of 20 students, the mean score on a reading readiness test was 26.9. In the other class of 35 students, the mean score was 20.2. Find the mean reading readiness score of all of Shaw's kindergarten students.
3. A group of size 20 has a mean of 27.5. For each of the following changes, first predict whether the mean will increase, decrease, or remain the same. Then calculate the new mean.

HINT: Since $M = \Sigma X/N$, $\Sigma X = NM$. Therefore, ΣX of the 20 original scores is $(20)(27.5) = 550$. To calculate each mean below, you must first find the new value of ΣX, then divide by N.

   a. A score of 35 is changed to 45.
   b. A score of 30 is changed to 20.
   c. A score of 50 is removed.
   d. A score of 15 is removed.
   e. A score of 40 is added to the distribution.
   f. A score of 25 is added to the distribution.

Comparison of the Mean, Median, and Mode as Measures of Central Tendency

The fact that the mean is based on all the scores in a distribution gives it some important mathematical properties and makes it particularly useful for inference. Therefore, it is the single most important and most used measure of central tendency. However, the mean's dependence on all scores can be a disadvantage as well as an advantage. Specifically

1. The mean is sensitive to outliers and to skew. In fact, extreme outliers can substantially affect the mean, especially if N is small.
2. If any scores are missing or indeterminate, the mean cannot be calculated.

Effect of Outliers and Skew

We noted earlier that the following three sets of scores have the same median (10); however, the three means are markedly different, as shown here:

Set 1 8, 8, 10, 11, 13 $\quad M = \frac{8+8+10+11+13}{5} = \frac{50}{5} = 10.0$

Set 2 8, 8, 10, 11, 200 $\quad M = \frac{8+8+10+11+200}{5} = \frac{237}{5} = 47.4$

Set 3 −100, 8, 10, 11, 13 $\quad M = \frac{-100+8+10+11+13}{5} = \frac{-58}{5} = -11.6$

In Set 1, the mean is at the center of the distribution and seems to be a good measure of central tendency. In Sets 2 and 3, however, the extreme outliers "pull" the mean to a location well beyond most of the scores. Clearly, the median, which is closer to the center of Sets 2 and 3, is a better measure of central tendency in both cases.

Here is another example that illustrates the effect of outliers on the mean of a small group. It also illustrates how different averages can paint very different pictures of a distribution. The example involves the salaries of the five employees of a small company:

President	\$250,000
President's nephew	125,000
Foreman	20,000
Laborers (2)	12,000 each

What is the average salary at the company? The union is likely to say, "The average salary is only \$12,000" (mode), to which management replies, "Ridiculous! The average salary is \$83,800" (mean). The median salary, \$20,000, is probably a better measure of central tendency than either the mode or the mean; however, in a distribution this skewed, it is useful to report all three.

The effect of outliers on the mean decreases as sample size increases. For example, suppose that we have two groups with the same mean, 25, but the first group has 5 members and the second has 50. Let us add a score of 100 to each group and calculate the new mean. To do this, we first calculate the sum of the scores, ΣX, in the original set. Since $M = \Sigma X/N$, $\Sigma X = NM$. Then we add the score of 100 to ΣX and divide by the new sample size to get the new mean.

Group 1 ($N = 5$): Original $\Sigma X = NM = 5(25) = 125$

$$\text{New } M = \frac{125+100}{6} = \frac{225}{6} = 37.5$$

Group 2 ($N = 50$): Original $\Sigma X = NM = 50(25) = 1250$

$$\text{New } M = \frac{1250+100}{51} = \frac{1350}{51} = 26.5$$

The addition of the score 100 changed the mean of the smaller group from 25 to 37.5, but the mean of the larger group was changed only from 25 to 26.5.

Because the mean is "pulled" toward extreme scores, the mean is greater than the median in distributions with high extreme scores (positively skewed distributions), and the mean is less than the median in distributions with low extreme scores (negatively skewed distributions). In symmetrical distributions, the mean and the median are equal or approximately equal. Figure 3–5 shows the relative locations of the mean, the median, and the mode in symmetrical and skewed distributions.

In Chapter 2 you learned to use the relative distances between the first, second, and third quartiles to measure the skew of a distribution. Relationships between the mean and the median give us another way to measure skew. Specifically

If $M = Mdn$ (at least approximately), the distribution is symmetrical.
If $M < Mdn$, the distribution is negatively skewed.
If $M > Mdn$, the distribution is positively skewed.

For example, suppose that a distribution has $Mdn = 24.6$ and $M = 21.5$. Because $M < Mdn$, the distribution is negatively skewed.

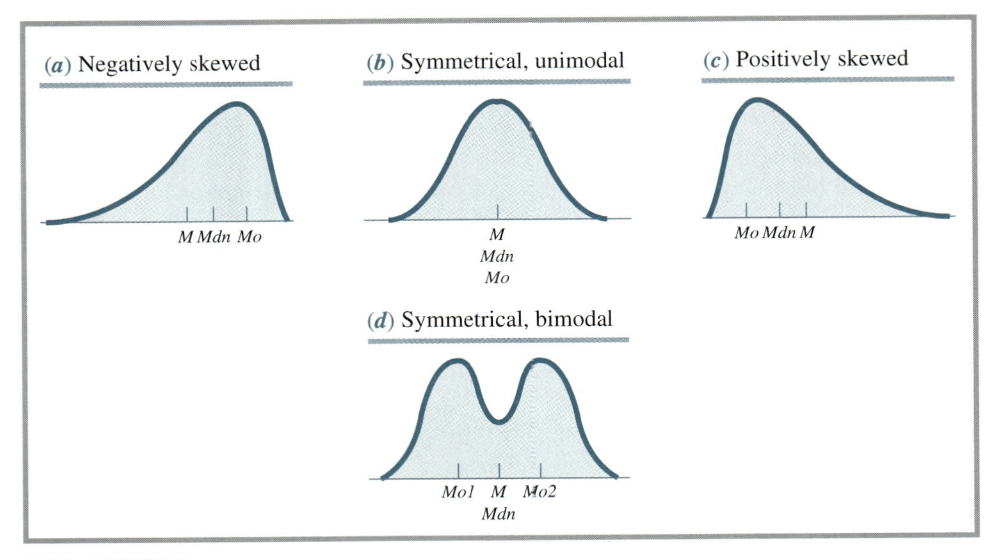

Figure 3–5 **Relative locations of mean *(M)*, median *(Mdn)*, and mode *(Mo)* as a function of the shape of the distribution.**

As an illustration of the relation between the mean and the median in symmetrical and skewed distributions, consider the distributions of anxiety scale scores in 50 college students and 35 clients at a mental health center (see Figure 2–9 in Chapter 2). The students' score distribution is approximately symmetrical, but the clients' distribution is negatively skewed. We can expect, therefore, that the students' mean and median are very similar; for the clients, however, we expect the mean to be smaller than the median. In fact, the students' mean score is 46.2, almost the same as the median, 46.5. But the clients' mean score, 56.7, is less than the median, 58.

Our comparisons of the mean, median, and mode lead us to the following conclusion: For distributions that are reasonably symmetrical, the mean is the most useful measure of central tendency. If, however, a distribution is markedly skewed, the median provides a better description of the central or average value, especially if N is small.

Effect of Missing or Indeterminate Scores

Because of its dependence on every score, the mean cannot be calculated if some score values are missing or indeterminate. Suppose that in a study of problem solving, number of trials to solution was used as a measure of problem-solving efficiency; but subjects were given a maximum of 20 trials. The results for 50 subjects are shown in Table 3–3. We have no scores for the five subjects who did not solve the problem within 20 trials; therefore, we cannot calculate the mean. Because the median is not dependent on every score, it can be calculated in cases like this (unless the missing scores are at the center of the distribution).

Missing data can be a serious problem in research, because the most powerful data analytical tools are based on means and related measures. Faced with missing data, researchers must either eliminate those observations from the analysis or make some kind of adjustment. Neither solution is ideal, and both have attendant difficulties. Methods for coping with missing data are beyond the scope of this book; extended discussions can be found in many textbooks about experimental design.

Table 3–3 Number of Trials to Solution

Number of Trials	Frequency
More than 20	5
20	2
19	1
18	5
17	8
16	12
15	7
14	7
13	3
	50

EXERCISE 3–4

1. A distribution has mean = 53.6, median = 48.7, and mode = 45. What is the shape of the distribution?
2. Here are grade equivalent scores of 15 third graders on a reading test:

5.2	8.0	2.0
2.4	3.1	8.5
9.0	2.0	6.7
2.7	3.5	2.0
1.4	9.5	3.8

   a. Calculate the mode, the median, and the mean.
   b. The teacher reports, "My class is reading right at grade-level." On which average is this statement based?
   c. Pete Buxbaum, running for the school board, says, "The average reading score of our third graders is disgraceful!" Which average does he mean?
   d. The incumbent, Mary Smith, claims, "During my term of office, third graders achieved an average reading score well above grade level." Which average is Mary Smith referring to?
   e. In what sense could one say that all three claims are correct?
3. Each of the following distributions contains 25 scores. First predict the relative sizes of the mean, median, and mode in each distribution, then calculate them.

Note: When calculating the mean of a frequency distribution, be sure to include each score as often as it occurs. For example, in (a) below, there are two scores of 20, three of 19, and so on. Therefore $\Sigma X = (20+20) + (19+19+19) + \cdots$; ΣX is *not* $20 + 19 + 18 + 17 + 16 + 15$.

a.

Score	f
20	2
19	3
18	7
17	8
16	4
15	1

b.

Score	f
20	1
19	2
18	3
17	5
16	10
15	4

c.

Score	f
20	6
19	10
18	5
17	2
16	1
15	1

4. When realtors report the average price of houses in a region, they usually cite the median rather than the mean. Why?

Measures of Variability

In this section we consider the following measures of variability: the range, the interquartile and semi-interquartile ranges, and the variance and standard deviation. The range is the measure of variability most closely associated with the mode; the interquartile and semi-interquartile ranges are associated with the median; and the variance and standard deviation are associated with the mean. As you will see, each measure of variability shares the strengths and weaknesses of its associated measure of central tendency.

The Range

The ***range*** of a set of scores is the difference between the highest and lowest scores. The simplicity of computation is the range's major strength. Its major weakness is its dependence on only two scores in a distribution. A single outlier at either or both ends of the distribution has a marked effect on the range. For example, as we observed earlier, despite the obvious differences between Distributions F and G in Table 3–1, both have the same range. On the other hand, consider Distributions H and I in Table 3–4. Despite their obvious similarities, they have different ranges; the range of Distribution H is 18 − 15 = 3, whereas that of Distribution I is 18 − 10 = 8. Sensitivity to outliers is a major weakness of the range. Although it is common to report the range of any data set, it is rarely the preferred measure of variability. In one situation, however, the range is the appropriate (and the only appropriate) measure of variability; this is for categorical data. Statements such as "Responses to the questionnaire ranged over four categories" make sense. No other measure of variability applies to cases such as this.

The Interquartile and Semi-Interquartile Range

As their names imply, the interquartile and semi-interquartile ranges are based on quartiles of the distribution. Since the median is the second quartile (Q_2), it is not surprising that the interquartile and semi-interquartile ranges are measures of variability associated with the median.

Table 3–4 Two Very Similar Distributions With Different Ranges

H		I	
Score	***f***	**Score**	***f***
18	4	18	4
17	7	17	7
16	3	16	3
15	1	15	
14		14	
13		13	
12		12	
11		11	
10		10	1
Range = 18 − 15 = 3		Range = 18 − 10 = 8	

Note: The outlier of 10 in Distribution I substantially increases the range.

The ***interquartile range*** is defined as *the distance between the first and third quartiles, or $Q_3 - Q_1$*. Recall that the first quartile of a distribution, Q_1, is the score value below which 25% of observations fall, and the third quartile, Q_3, is the score value below which 75% of observations fall. Fifty percent of observations fall between Q_1 and Q_3. Therefore the interquartile range can also be defined as *the range of the middle 50% of the observations*.

The ***semi-interquartile range (Q)***, sometimes called the ***quartile deviation***, is simply half the interquartile range:

$$Q = \frac{Q_3 - Q_1}{2}$$ *Formula for calculating the semi-interquartile range*

For example, if $Q_1 = 12$ and $Q_3 = 73$, the interquartile range is $73 - 12 = 61$, and the semi-interquartile range, Q, is $\frac{61}{2} = 30.5$.

In order to calculate the interquartile or semi-interquartile range, one must first locate the first quartile, Q_1, and the third quartile, Q_3. If the exact values are needed, the interpolation method described in Chapter 2 can be used. Usually, however, the results obtained by the following method are sufficient:

1. Arrange the scores in order by size and locate the median.
2. Divide the distribution into two subsets: scores falling below the median and scores falling above the median. (If N is odd, just drop the middle score.)
3. Find the median of each subset. The median of the lower subset is Q_1; the median of the upper subset is Q_3.

Here are some examples:

Set A Scores: 192, 204, 150, 250, 173, 125, 100, 220 ($N = 8$)

Scores in order: 100, 125, 150, 173, 192, 204, 220, 250

↑

Mdn

The median is halfway between 173 and 192, or 182.5.

Lower subset (scores below *Mdn*): 100, 125, 150, 173

↑

Q_1

The median of the lower subset is halfway between 125 and 150, or 137.5. $Q_1 = 137.5$.

Upper subset (scores above *Mdn*): 192, 204, 220, 250

↑

Q_3

The median of the upper subset is halfway between 204 and 220, or 212. $Q_3 = 212$.

Interquartile range $= Q_3 - Q_1 = 212 - 137.5 = 74.5$

Semi-interquartile range, $Q = 74.5/2 = 37.25$

Set B

Scores: 54, 58, 59, 51, 50, 54, 56, 58, 53 $(N = 9)$

Scores in order: 50, 51, 53, 54, 54, 56, 58, 58, 59

↑

$Mdn = 54$

The middle score, 54, is dropped.

Lower subset: 50, 51, 53, 54

↑

$Q_1 = 52$

Upper subset: 56, 58, 58, 59

↑

$Q_3 = 58$

Interquartile range = 58 − 52 = 6

$Q = 6/2 = 3$

If scores have been arranged in a frequency distribution or stem plot, Q_1 and Q_3 are readily located by using the method just illustrated. We will use the distributions of quiz scores and anxiety scale scores from Chapter 2 to illustrate the process. In Figure 3–6, both distributions have been split at the median. As the figure shows, the median of the lower subset of quiz scores is 16, and this is Q_1; the median of the upper subset of quiz scores is 19, and this is Q_3. Therefore, interquartile range = 19 − 16 = 3 and $Q = 3/2 = 1.5$. In the anxiety scale distribution, the median of the lower subset is 39, so that $Q_1 = 39$; the median of the upper subset is 52, so that $Q_3 = 52$. Therefore, interquartile range = 52 − 39 = 13, and $Q = 13/2 = 6.5$.

(a)

Quiz Scores of 44 Students (The distribution is split at the median into two subsets of 22 scores each.)

Score	Frequency	
20	6	The two middle scores (the 11th and 12th) are both 19.
⑲	8	$Q_3 = 19$
18	8	
Median		
18	2	The two middle scores (the 11th and 12th) are both 16.
17	6	$Q_1 = 16$
⑯	7	
15	3	
14	1	
13	2	
12		
11		
10		
9	1	

(b)

Anxiety Scale Scores of 50 Students (The plot is split at the median into two subsets of 25 scores each.)

2	666	The middle score (the 13th) is 39.
3*	014	$Q_1 = 39$
3	567788⑨	
4*	11233444	
4	5666	
Median		
4	778899	The middle score (the 13th) is 52.
5*	001122②33	$Q_3 = 52$
5	67799	
6*	0012	
6	8	

Figure 3–6 **Locating the first quartile (Q_1) and third quartile (Q_3) in a frequency distribution (a) or stem plot (b).**

Recall from Chapter 2 that the quartiles of a distribution can be used to construct a graph called a *box-and-whisker plot* (*box plot*, for short). Figure 2–15 in Chapter 2 displays a box plot of the statistics quiz scores, and Figure 2–17 shows side-by-side box plots of anxiety scale scores of college students and mental health center clients. Notice that in all cases, the "box" extends from Q_1 to Q_3; that is, it covers the interquartile range. The "whiskers" extend from the box to the lowest and highest scores. The line within each box shows the location of the median.

The interquartile and semi-interquartile ranges have the same strengths and weaknesses as the median. Like the median, the interquartile and semi-interquartile ranges are not dependent on the exact value of each score in a distribution. They are sensitive to whether observations fall above or below the 25% and the 75% points, but not how far above or below. Consequently, they are relatively unaffected by extreme scores.

As an illustration, recall that in Set A of our earlier example, the interquartile range is 74.5. In Set C below, the highest score of Set A, 250, is replaced by the outlier 1750.

Set C Scores in order: 100, 125, 150, 173, 192, 204, 220, 1750

↑ Q_1 (between 125 and 150) ↑ Q_3 (between 204 and 220)

As in Set A, $Q_1 = 137.5$ and $Q_3 = 212.0$. Therefore, the interquartile range is, once again, $212.0 - 137.5 = 74.5$; changing the highest score from 250 to 1750 had no effect on the interquartile or semi-interquartile range. If a measure is needed that is not influenced by one or more extreme scores, as in highly skewed distributions, the interquartile and semi-interquartile ranges are the preferred measures of variability, just as the median is the preferred measure of central tendency.

However, as with the median, the fact that the interquartile and semi-interquartile ranges do not use all scores in their calculation makes them less suitable for further calculations, or for inference, than the variance and the standard deviation, to be considered next.

EXERCISE 3–5

1. Calculate the range and the interquartile range, and construct a box plot of each of the following distributions:
   a. Scores: 105, 110, 97, 115, 117, 110, 109, 107, 113, 110, 95, 113
   b. Scores: 6, 6, 3, 4, 7, 3, 3, 2, 6, 5, 6
   c. Scores: 1.1, 1.3, 0.9, 0.7, 1.2, 2.3, 0.6, 1.2, 1.0
   d. Scores: 30, 26, 33, 32, 27, 24, 35, 28
2. a. One of the sets in problem 1 has an outlier. Which one is it?
   b. Remove the outlier, and calculate the range and the interquartile range of the remaining scores. How does removing the outlier affect the range? How does it affect the interquartile range?
3. Distributions F and G in Table 3–1 have the same range. Calculate and compare their interquartile ranges.
4. You read the following report in a journal article: "Half the children scored below 61 on the test, but the most common score was 58. The lowest score was 40 and the highest was 94. The middle 50% of the children scored between 55 and 68."
   a. What was the median?
   b. What was the mode?
   c. What was the range?
   d. Calculate the interquartile and semi-interquartile ranges.
   e. What was the probable shape of the distribution?

THE VARIANCE AND THE STANDARD DEVIATION

The measures of variability associated with the mean are the variance and its square root, the standard deviation (SD). The variance and standard deviation, like the mean, utilize every score in their calculation and are useful for further calculations and for inference. However, like the mean, they are particularly sensitive to outliers and skew.

The Average Deviation From the Mean

The variance and standard deviation are based on deviations of scores from the mean, that is, on values of $X - M$. The reasoning is that if a distribution has no variability at all, each score is equal to the mean and no score deviates from it. The greater the variability of the distribution, the greater become the deviations of scores from the mean on the average. Therefore, it makes sense to base measures of variability on deviations from the mean.

It is tempting to suggest that we simply add all the deviations from the mean, then divide by N to get the average size of deviations from the mean:

$$\frac{\Sigma(X-M)}{N}$$

The problem with this approach is that the numerator, $\Sigma(X - M)$, is always equal to zero, as you already know. Therefore, $\frac{\Sigma(X-M)}{N}$ always equals zero, too. We need to try another strategy.

The reason that $\Sigma(X - M)$ equals zero, of course, is that negative deviations from the mean always balance off positive deviations. Well, then, you may say, why not ignore signs and just consider the absolute deviations from the mean? In fact, it is possible to do so, and the measure of variability that results is the ***average deviation***. It is defined as the *average distance of scores from the mean*:

$$\textit{Average deviation} = \frac{\Sigma|X-M|}{N}$$

The formula says: Subtract the mean from each score. Treat all the deviations as positive numbers. Add all the absolute deviation values; then divide the sum by N. The calculation is illustrated here for a distribution of five scores with $M = 6$:

X	|X − M|
4	2
6	0
13	7
2	4
5	1
Sum	14

$$\text{Average deviation} = \frac{14}{5} = 2.8$$

Thus, the average distance of the scores from the mean is 2.8.

The average deviation is easy to understand and to calculate. Unfortunately, when we ignore algebraic signs, we wind up with numbers not suitable for further calculations such as those required for statistical inference. Although the average deviation was once a popular measure of variability, it is now little more than a historical curiosity.

The Variance

What we need is a measure of variability that, like the average deviation, is based on deviations from the mean while, at the same time, overcoming the problem that $\Sigma(X - M) = 0$. The variance meets this requirement not by ignoring signs of deviations but by squaring each deviation before adding. Because squaring is a legitimate mathematical operation, and squares of numbers (positive or negative) are always positive, this effectively solves the problem.

The ***variance*** is defined as *the mean of the squared deviations from the mean.* It is equal to the sum of the squared deviations from the mean divided by N. We use Var to represent the variance. Thus,

$$Var = \frac{\Sigma(X - M)^2}{N} \qquad \textit{Formula for the variance}$$

Because the variance is the *mean* of the squared deviations from the mean, it is sometimes called the *mean square,* especially in a method of data analysis called *analysis of variance.* (Analysis of variance is the subject of Chapters 12 and 13.) For now, we will refer to the mean of the squared deviations from the mean by its more common name, the variance.

The formula tells us that to calculate the variance of a set of scores, we must

1. Subtract the mean from each score to get $X - M$, the score's deviation from the mean.
2. Square each deviation from the mean: $(X - M)^2$.
3. Add all the squared deviations: $\Sigma(X - M)^2$.
4. Divide by the number of scores N.

We illustrate by calculating the variance of the five scores 4, 6, 13, 2, 5, whose mean is 6.

	Step 1	**Step 2**
X	$X - M$	$(X - M)^2$
4	–2	4
6	0	0
13	7	49
2	–4	16
5	–1	1
Sum	0	$70 = \Sigma(X - M)^2$ (Step 3)

$$\text{Step 4: Var} = \frac{\Sigma(X - M)^2}{N} = \frac{70}{5} = 14$$

The variance of the scores is 14.

In our discussion of properties of the mean, we pointed out that the numerator term in the variance formula, the sum of squared deviations from the mean, is called the *sum of squares (SS).* Therefore, we can also write the formula for the variance this way:

$$\text{Var} = \frac{SS}{N}$$

One additional comment about the formula we have used for calculating the variance: Many books define the variance of a data set as the sum of squares, *SS*, divided by $N - 1$ rather than by N. As you will see in Chapter 6, there are often very good reasons for doing this, especially when one is concerned with inferring the value of a population variance from the variance of a sample. We prefer to use N for now because, to get the mean of a set of N numbers, we divide the sum of the numbers by N; and the variance is, after all, the *mean* of the N squared deviations from the mean. Later, we will consider conditions under which we need to divide *SS* by $N - 1$ instead.

If you are using a statistical calculator or a computer program to calculate the variance, it may well be using the formula $SS/(N - 1)$. To find out which formula is being used, try this simple test: Enter the numbers 1, 2, 3 and compute the variance. If your calculator or computer reports Var = 1 and/or $SD = 1$, it is dividing by $N - 1$; if it reports Var = .67 (rounded) and/or $SD = .82$ (rounded), it is dividing by N.

If you use a calculator or computer program that is dividing by $N - 1$ to calculate variances and you want to know the value of the variance obtained by dividing *SS* by N, simply multiply the answer reported by your calculator/computer by $(N - 1)/N$. For example, if you enter nine scores and your calculator reports Var $(= SS/(N - 1))$ = 68.72:

$$\text{Var} = \frac{SS}{N} = 68.72\left(\frac{8}{9}\right) = 61.08$$

The Standard Deviation

A problem with the variance as a measure of variability is that it is in squared units. Suppose, for example, that you have measured response time in seconds and calculated Var = 12.25; this means that the variance of response times is 12.25 *square* seconds! Though this may not be a problem in inference, squared units make little sense in describing a distribution.

We can return to the original unit of measurement by taking the square root of the variance. *The square root of the variance* is called the ***standard deviation*** (***SD***, for short) for which we use the symbol ***S***.

$$S = \sqrt{Var} = \sqrt{\frac{SS}{N}} = \sqrt{\frac{\Sigma(X - M)^2}{N}}$$ *Deviation formula for the standard deviation*

In the case of response times with variance (Var) = 12.25 square seconds, the standard deviation, $S = \sqrt{12.25}$ square seconds = 3.5 seconds.

Properties of the Variance and Standard Deviation

The fact that the variance and standard deviation are based on all the scores in a data set gives them mathematical properties that are important for further calculations and for inference. However, they, like their companion measure of central tendency the mean, are considerably affected by outliers and skew. Outliers on either end of a distribution increase both the variance and the standard deviation.

Consider the following two sets of six scores. In Set 1, scores range only from 6 to 9. In Set 2, one score of 8 is replaced by a score of 2, thereby increasing the variability of the distribution. Let us compare the variances and standard deviations of the two sets.

Set 1

Score (X)	X − M	(X − M)²
8	0.5	0.25
7	−0.5	0.25
8	0.5	0.25
6	−1.5	2.25
7	−0.5	0.25
9	1.5	2.25
Sum 45		5.50

$M = \frac{45}{6} = 7.5$

$SS = \Sigma(X - M)^2 = 5.50$

$\text{Var} = \frac{5.50}{6} = 0.917$

$S = \sqrt{0.917} = 0.96$

Set 2

Score (X)	X − M	(X − M)²
8	1.5	2.25
7	0.5	0.25
2	−4.5	20.25
6	−0.5	0.25
7	0.5	0.25
9	2.5	6.25
Sum 39		29.50

$M = \frac{39}{6} = 6.5$

$SS = \Sigma(X - M)^2 = 29.50$

$\text{Var} = \frac{29.50}{6} = 4.917$

$S = \sqrt{4.917} = 2.22$

The small spread of scores in Set 1 results in a small variance, 0.917, and standard deviation, 0.96. In Set 2, the substitution of an outlier for just one of the scores increases the variance by a factor of 5, from 0.917 to 4.917. The effect on the standard deviation is not as great but is still substantial.

As another example, consider Dr. Marrowbone's statistics quiz scores (see Table 2–6), which include one low outlier. The variance of the scores is 4.93 and the standard deviation is $\sqrt{4.93} = 2.22$. If we remove the outlier (the score of 9), we find that the variance of the remaining 43 scores is reduced to 3.41 and the standard deviation to 1.85.

In discussing the interquartile and semi-interquartile ranges, we noted that they, like their companion measure the median, are little affected by outliers and skew. For example, the interquartile range of the statistics quiz scores is 3. If we remove the score 9, the interquartile range is still 3. Therefore, when distributions are markedly skewed, the interquartile and semi-interquartile ranges are generally better descriptive measures of variability than is the variance or standard deviation.

In statistical inference, on the other hand, the variance and the standard deviation are usually the measures of choice. The variance has an important property that makes it, in some ways, the cornerstone of inference: It can be apportioned into components in such a way as to provide information about the sources of variation among scores. The standard deviation is important because it (along with the mean) is a defining characteristic of the normal curve. In addition, certain kinds of standard deviations, called *standard errors*, play a major role in a number of inferential procedures. You will learn about these uses of variances and standard deviations in later chapters.

EXERCISE 3–6

1. The sum of squares of a set of 10 scores is 640. What are the variance and standard deviation?
2. The variance of six scores is 45. What is the sum of squares?
3. The standard deviation of a distribution of 100 scores is 5.2. What is the variance? What is the sum of squares?
4. **a.** Calculate the mean, variance, and standard deviation of the following basic set:

   5, 6, 8, 9

   b. Add a score of 2 to the basic set. First, predict the effect on the mean, variance, and *SD*. Then calculate to see whether your predictions are confirmed.

   c. Add a score of 12 to the basic set. Predict the effect on the mean, variance, and *SD*. Then calculate to see whether your predictions are confirmed.

Raw Score Formulas for Calculating the Sum of Squares, Variance, and Standard Deviation

The formulas we have used so far for calculating the variance and standard deviation are called *deviation formulas* because they utilize deviations from the mean. Deviation formulas are cumbersome for calculation with large data sets because of the need to subtract the mean from each score. Several alternative formulas are available, all of which result in the same values as the deviation formulas. They are called *raw score formulas* because they operate upon the scores directly rather than upon deviations from the mean. They are also called *computational formulas* because they are convenient for computational purposes.

Remember that the variance is equal to the sum of squares (SS) divided by N. The sum of squares can be calculated directly from raw scores by using the following formula:

$$SS = \Sigma X^2 - \frac{(\Sigma X)^2}{N} \qquad \textit{Raw score formula for the sum of squares}$$

The result is the same as that obtained by using the deviation formula: $SS = \Sigma(X - M)^2$.

The raw score formula for the sum of squares includes two terms that are often encountered in statistical formulas: ΣX^2 and $(\Sigma X)^2$. *They are not the same thing.* ΣX^2 says, "Add the squared scores." It means, first, square each score; then, add. $(\Sigma X)^2$ says, "Square the sum of the scores." It means, first, add the scores; then, square the sum. For example, given the scores 1, 2, and 3, compare the following results:

$$\Sigma X^2 = 1^2 + 2^2 + 3^2 = 1 + 4 + 9 = 14$$

$$(\Sigma X)^2 = (1 + 2 + 3)^2 = 6^2 = 36$$

Let us use the raw score formula to calculate the sum of squares of the scores 4, 6, 13, 2, 5. Using the deviation formula, we found $SS = 70$. The scores are listed below in a column labeled X. A second column listing X^2 values is appended.

	X	X^2
	4	16
	6	36
	13	169
	2	4
	5	25
Sum	30	250

$N = 5$, $\Sigma X = 30$, $\Sigma X^2 = 250$.

$$SS = \Sigma X^2 - \frac{(\Sigma X)^2}{N} = 250 - \frac{(30)^2}{5} = 250 - \frac{900}{5} = 250 - 180 = 70, \text{ as before.}$$

As before, $\text{Var} = \frac{SS}{N} = \frac{70}{5} = 14$, and $S = \sqrt{\text{Var}} = \sqrt{14} = 3.74$.

In summary, we have demonstrated the following methods for calculating the sum of squares, variance, and standard deviation.

	Deviation Method	Raw Score Method
Sum of squares (SS)	$\Sigma(X - M)^2$	$\Sigma X^2 - \frac{(\Sigma X)^2}{N}$
Variance (Var)	$\frac{SS}{N}$	
Standard deviation (S)	$\sqrt{\frac{SS}{N}}$	

Occasionally, you may want to calculate the variance directly from raw scores in one step. If so, one or another of the following formulas can be used:

$$\text{Var} = \frac{\Sigma X^2}{N} - \left(\frac{\Sigma X}{N}\right)^2$$ *A one-step computational formula for the variance*

Because $\frac{\Sigma X}{N} = M$, we can substitute and get

$$\text{Var} = \frac{\Sigma X^2}{N} - M^2$$ *Another one-step computational formula for the variance*

For example, given the scores 4, 6, 13, 2, 5, with $\Sigma X = 30$, $\Sigma X^2 = 250$, and $M = 6$

$$\text{Var} = \frac{\Sigma X^2}{N} - \left(\frac{\Sigma X}{N}\right)^2 = \frac{250}{5} - \left(\frac{30}{5}\right)^2 = 14$$

or

$$\text{Var} = \frac{\Sigma X^2}{N} - M^2 = \frac{250}{5} - 6^2 = 14$$

There is nothing conceptually difficult about computing the sum of squares, variance, and standard deviation. If you use a calculator and check your work at each step, you should have little difficulty arriving at a correct answer. (Of course, if you are using a statistical calculator or a computer, you don't need to worry about the calculations at all. All you need to worry about is entering the data correctly, selecting the appropriate program or command, and interpreting the output correctly. For example, "Is this number a variance or a standard deviation?")

Assuming that you are performing the calculations yourself, here are some checks you can use to evaluate your work:

1. Have you obtained a negative value for the sum of squares or variance? If so, you have surely made a mistake. Sums of squares and variances cannot be negative. Go back over your work to find your error.
2. To decide whether your answer is reasonable, use this rule of thumb: The standard deviation of a distribution is usually between one-third and one-sixth of the range. For example, if the range of scores is 30 and you have calculated SD = 5.8, this is quite reasonable. But SD = 2.8 and SD = 15.4 are almost certainly incorrect.
3. Remember the difference between ΣX^2 and $(\Sigma X)^2$. Confusing them is probably the most common source of error in calculating the variance and standard deviation.

EXERCISE 3–7

1. Use raw score formulas to calculate the sum of squares, variance, and standard deviation of the following sets of scores. Then compare the standard deviation with the range to check your answers.
   a. 2, 4, 8, 6, 5, 7, 6
   b. 22, 14, 23, 20, 18, 23, 29, 26
2. Explain why the sum of squares and the variance cannot be negative.

Linear Transformations and Standard Scores

The Effect of Linear Transformations on the Mean, Variance, and Standard Deviation

Sometimes scores in a data set are transformed by adding or subtracting a constant from each score, by multiplying or dividing each score by a constant, or some combination of these operations. This is often done to ease data entry or computation. As an example, suppose that you have a set of 500 SAT Verbal scores that are to be keyed into a computer data file. SAT scores are numbers between 200 and 800 and are always multiples of 10 (e.g., 460 or 680). It will save you a lot of trouble if you simply drop the zero at the end of each score, that is, divide by 10.

Transformations of scores by adding, subtracting, multiplying, or dividing by constants are called ***linear transformations***. The following rules describe the effects of linear transformations on the mean, variance, and standard deviation of a distribution.

1. *If a constant is added to or subtracted from each score in a distribution, the mean changes by the same amount as each score, but the variance and standard deviation do not change.*

For example, if 5 is added to each score in a distribution with mean = 20, and $SD = 4$, the transformed scores have mean = 20 + 5 = 25, but SD is still 4 (and the variance is still $4^2 = 16$). If 3 is subtracted from each score instead, the new mean is 20 − 3 = 17, but SD is still 4.

It is not hard to understand these effects. As you can see in Figure 3–7, adding or subtracting a constant (in this case, 5) simply shifts the whole distribution, including the mean, upward or downward by the amount of the constant. But the spread of the scores is unaffected, so that the variance and standard deviation are unchanged.

2. *If each score in a distribution is multiplied or divided by a constant, both the mean and the standard deviation change by the same amount as each score.*

For example, if each score in a distribution with mean = 20 and $SD = 4$ is multiplied by 3, the transformed distribution has mean = 3(20) = 60 and $SD = 3(4) = 12$. The variance, originally $4^2 = 16$, becomes $12^2 = 144$. Thus, the mean and SD increase by a factor of 3, whereas the variance increases by a factor of 3^2, or 9. Similarly, if each score in the original distribution (mean = 20, $SD = 4$) is divided by 2, the distribution of transformed scores has mean = 20/2 = 10, $SD = 4/2 = 2$, and variance = $16/2^2 = 4$.

To understand the effects of multiplication or division by a constant, consider Figure 3–8. As you can see, if each score is multiplied or divided by 2, the mean increases or decreases by the same factor, 2, as each score. However, multiplication or division by a constant changes the spread of the scores as well. Multiplication (by a number greater than 1) increases the spread, whereas division (by a number greater than 1) decreases it. Because the standard deviation is a measure of spread, it is changed by transformations involving multiplication or division.

Sometimes a linear transformation involves both addition or subtraction and multiplication or division. Suppose, for example, that a distribution has mean = 10 and $SD = 3$. Now you subtract 5 from each score and then divide by 2. What are the mean and SD of the transformed scores? First, consider the mean, originally 10. Subtracting 5 from each

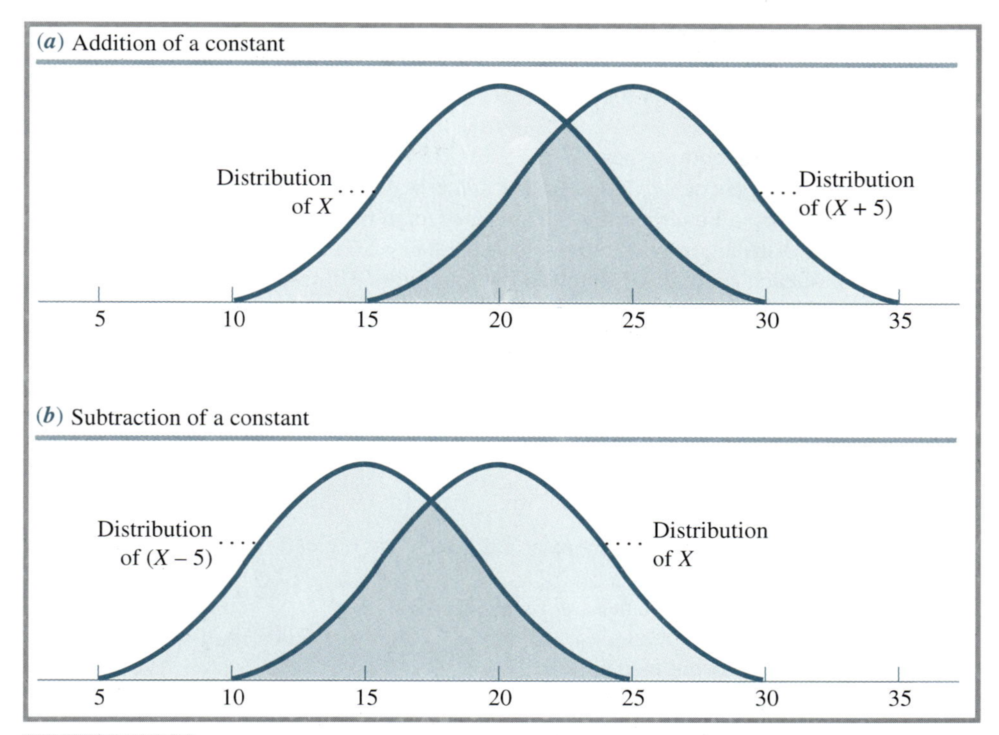

Figure 3–7 Effect of adding or subtracting a constant from each score on the mean and standard deviation.

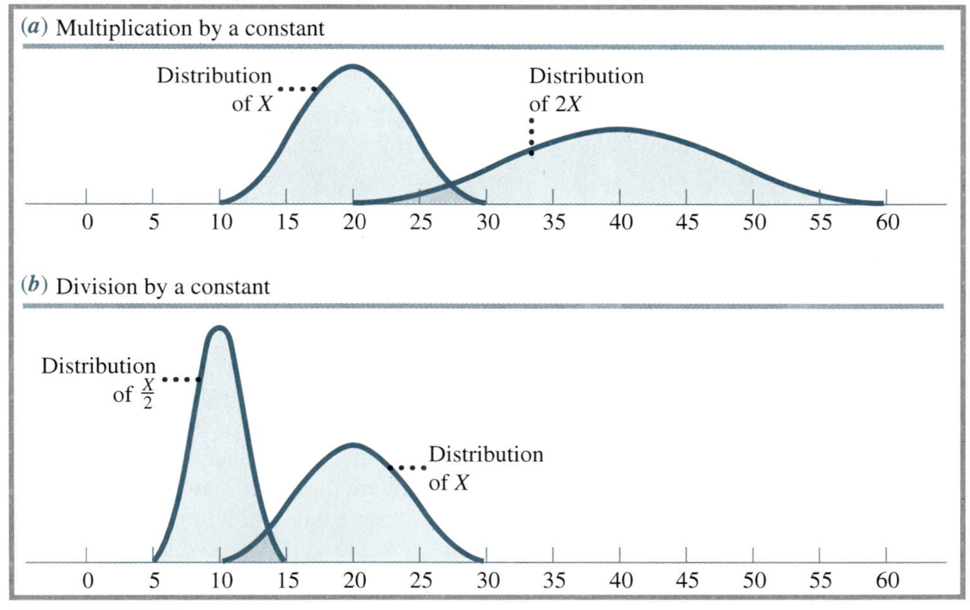

Figure 3–8 Effect of multiplying or dividing each score by a constant on the mean and standard deviation.

score changes the mean from 10 to 5. Then dividing by 2 changes it from 5 to 5/2 = 2.5. The new mean is 2.5. Now consider the standard deviation, originally 3. Subtracting 5 from each score does not affect the *SD*, so *SD* is still 3. But dividing by 2 changes *SD* to 3/2 = 1.5. In sum, after each score has been transformed by subtracting 5 and dividing by 2, the new mean is 2.5 and the new *SD* is 1.5.

The median and mode are affected the same way by linear transformations as the mean, and the range and interquartile range are affected the same way as the standard deviation. Given a distribution with range = 56, median = 37, and interquartile range = 20, multiplication of each score by 2, followed by addition of 10, results in new range = 2(56) = 112, new median = 2(37) + 10 = 84, and new interquartile range = 2(20) = 40.

EXERCISE 3–8

1. A distribution has mean = 60, *SD* = 12. Calculate the new mean, *SD*, and variance if
   a. 13 is subtracted from each score.
   b. each score is multiplied by 4.
   c. 7 is added to each score.
   d. each score is divided by 10.
   e. 60 is subtracted from each score and the result is divided by 12.
2. Scores: 2, 1, 5, 3, 4
   a. Calculate the mean, variance, and standard deviation.
   b. Predict what the new mean, variance, and SD will be if you multiply each score by 2 and add 3 to the result.
   c. Transform the scores by multiplying each one by 2, then adding 3. Calculate the mean, variance, and *SD* of the transformed scores. Do they agree with your predictions?
   d. Draw a graph in which you plot the transformed scores as a function of the original scores. You will see why the kinds of transformations we have discussed in this section are called linear transformations.

The Standard Deviation as a Measure of Distance: Standard Scores

People sometimes make the mistake of trying to locate a measure of variability at a particular place in a distribution. They may say, "All right, I've calculated the standard deviation, and it is equal to ______ . Now, where is it located in the distribution?" Measures of variability are not anywhere. Unlike measures of central tendency, measures of variability are not specific score values or points.

An analogy to cities on a road map may be useful. On a map, cities such as Jonesville and Smithtown are located at particular places. But distances are not. If, for example, Jonesville is 25 miles from Smithtown, the "25 miles" is not at a particular point. Moreover, a distance of 25 miles may be measured off on other parts of the map as well as between Jonesville and Smithtown, for example, between Elmwood and Thornton or east from Lincolnsburg. Measures of central tendency are like Jonesville and Smithtown; they are at particular places in a distribution. But measures of variability are like the 25 miles between the two cities. They are distances that can be measured off between various points in a distribution, but are not located at any particular place.

The standard deviation is often used as a measure of the distance of scores from the mean. Thus, if the mean score on an exam is 50 and the standard deviation is 10, we may describe a score of 40 (10 points below 50) as falling one standard deviation below the mean. On the same exam a score of 52 can be described as falling two-tenths of a standard deviation above the mean. Scores expressed in terms of their standard deviation distance from the mean are called ***standard scores***[†] or ***z scores***. Thus, a score of 40 on the exam has $z = -1$; a score of 52 has $z = 0.2$.

Notice that scores below the mean have negative z scores, whereas scores above the mean have positive z scores. A z score of –2.5 describes a score two and a half standard deviations *below* the mean; a score that is three-fourths of a standard deviation *above* the mean has $z = .75$.

Score values are easily converted into z scores by the following formula:

$$z = \frac{\text{Score} - \text{Mean}}{SD} \qquad \textit{Formula for converting a raw score into a z score}$$

As you can see, z scores are obtained by subtracting one constant (the mean) from each score, then dividing by another constant (the SD). Therefore, z scores are a linear transformation of the original scores.

Here are some examples using the formula:

1. Mean = 10, SD = 2, score = 14: $z = \frac{14-10}{2} = 2.$ A score of 14 is two standard deviations above the mean.
2. Mean = 100, SD = 16, score = 92: $z = \frac{92-100}{16} = -0.5.$ A score of 92 is half a standard deviation below the mean.
3. The mean salary of factory workers in a certain region is \$27,500, with a standard deviation of \$1500. Where does a salary of \$25,000 fall in the distribution?

   $$\text{Answer}: z = \frac{25{,}000 - 27{,}500}{1500} = -1.67 \text{ (rounded to two decimal places).}$$

   A salary of \$25,000 falls one and two-thirds standard deviations below the mean.

Sometimes it is necessary to convert z scores into raw scores. We can obtain a formula for doing so by solving the z score formula for "score." Start with

$$z = \frac{\text{Score} - \text{Mean}}{SD}$$

Now multiply both sides by SD:

$$z(SD) = \text{Score} - \text{Mean}$$

Finally, add mean to both sides:

$$\text{Mean} + z(SD) = \text{Score} \qquad \textit{Formula for converting a z score into a raw score}$$

For example,

1. In a distribution with mean = 12 and SD = 4, what score has $z = -1.50$?
   Answer: Score = 12 + (–1.50)(4) = 12 + (–6) = 12 – 6 = 6.
2. What score is half a standard deviation above the mean in a distribution with mean = 30, SD = 5?
   Answer: Since a score that is half a SD above the mean has $z = 0.5$:
   Score = 30 + (.5)(5) = 30 + 2.5 = 32.5.

[†]In psychological and educational testing and measurement, there are other kinds of standard scores besides z scores, but we do not deal with them in this book.

Properties of Standard Score Distributions

Suppose an entire distribution of raw scores is converted to *z* scores. What are the mean, the variance, and the standard deviation of the distribution of *z* scores? Because *z* scores are linear transformations of the raw scores, we can answer these questions by applying the rules concerning the effects of linear transformations on the mean, variance, and standard deviation.

1. In linear transformations, the mean of the original scores is changed by the same amount as each score. Because *z* scores are obtained by subtracting the mean from each score and then dividing by *SD*:

$$\text{Mean of } z \text{ scores} = \frac{\text{Mean} - \text{Mean}}{SD} = \frac{0}{SD} = 0$$

   The mean of a distribution of z *scores is equal to zero.*

2. In linear transformations, subtraction of a constant does not affect the *SD*. However, dividing each score by a constant changes the original *SD* to the value (*SD*/constant). Because *z* scores are obtained by subtracting the mean (no effect on *SD*) and then dividing by *SD*:

$$SD \text{ of } z \text{ scores} = \frac{SD}{SD} = 1$$

   The standard deviation of a distribution of z *scores is equal to 1.* Because the variance = $(SD)^2$, the variance of a distribution of *z* scores is also equal to 1.

We might also ask: What effect does converting scores to *z* scores have on the shape of the distribution? The answer is: *The conversion of a distribution of raw scores to* z *scores has no effect on the shape of the distribution.* Whatever the shape of the original distribution, that is also the shape of the distribution of *z* scores. For example, if a score distribution is negatively skewed and each score is converted to a *z* score, the distribution of *z* scores is also negatively skewed and to the same degree; if the original distribution was normal, the distribution of *z* scores is also normal; and so on.

Properties of the *z* score distribution and the relation of *z* scores to raw scores are illustrated in Figure 3–9. The distribution in Figure 3–9 is positively skewed; however, the relationships between the original and standard score distributions would be the same in distributions of any shape.

Though distributions of all shapes can be converted to *z* scores, *z* scores are particularly useful in normal distributions, as you will see in Chapter 7.

A Comparison of Standard Scores and Percentile Ranks

If two or more distributions have the same shape, standard scores can be used to compare scores in the distributions. For example, suppose that Alice scored 17 on a spelling test and 58 on a math test, and that both sets of scores were approximately normal. Did she do well on both, poorly on both, or well on one and poorly on the other? Did she do better on the spelling or on the math test? We can't answer these questions based on the scores alone because we don't know where 17 falls in the distribution of spelling scores or where 58 falls among the math scores. However, if we know the means and standard deviations of both distributions, we can convert Alice's scores to *z* scores and compare the *z* scores.

Suppose the means and standard deviations are as follows:

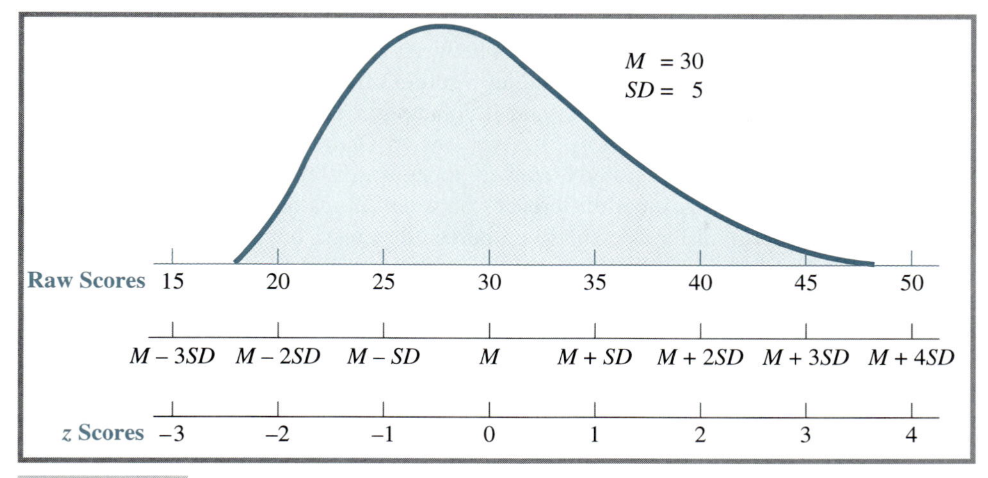

Figure 3–9 **Relation of raw scores to *z* scores.**

	Spelling	Math
M	21	75
SD	4	11

Alice's *z* scores on the two tests are

$$\text{Spelling: } z = \frac{17-21}{4} = \frac{-4}{4} = -1$$

$$\text{Math: } z = \frac{58-75}{11} = \frac{-17}{11} = -1.55$$

Alice scored one standard deviation below the mean in spelling and a little over one and a half standard deviations below the mean in math. You can see that she performed relatively poorly on both tests, but somewhat better on spelling than on math.

In Chapter 2 you learned to calculate and interpret percentile ranks. Percentile ranks are also useful for interpreting and comparing scores in different distributions. For example, if Peter scored at the 76th percentile of his class in reading and the 53rd in math, we know that he reads better than 76% of his classmates but outperforms only 53% of them in math. Relative to his classmates, he is doing better in reading than in math.

Percentile ranks have two advantages over standard scores. First, they are easier for laypersons to understand. For example, statistically unsophisticated parents can understand what it means if their child scored at the 45th percentile on a test: It means that the child scored above 45% of the class (or other comparison group). However, in order to interpret a *z* score, such as –0.13, the parents must know what a mean and a standard deviation are. This is an unreasonable expectation for most people who have not studied statistics.

The second advantage of percentile ranks is that they are meaningful in distributions of any shape. The 45th percentile, for example, is the point below which 45% of observations fall, regardless of whether the distribution is normal, skewed, bimodal, or the like. This is not necessarily true for *z* scores. For example, the percentage of cases falling below a *z* score of +1 may be very different in two distributions, one of which is positively and the other negatively skewed. Therefore, in distributions having different shapes, *z* scores are not comparable.

Standard scores (and other linear transformations), however, have one major advantage over percentile ranks: Equal differences between raw scores are always reflected in equal differences between the corresponding z scores. If John, Ted, and Ann have raw scores of 70, 80, and 90, respectively, and if John's and Ted's z scores are −3 and −2, respectively, then Ann's z score is −1. As we saw in Chapter 2, equal differences between raw scores do not necessarily convert to equal differences between percentile ranks. In most distributions, small differences between scores near the center of the distribution result in large differences between percentile ranks; but at the ends of the distribution, even large score differences tend to result in small percentile rank differences.

When equal differences between raw scores do not convert to equal differences between transformed scores, the transformation is said to be *nonlinear*. The transformation of raw scores to percentile ranks is a nonlinear transformation. If further calculations are to be performed on transformed scores, such as those in certain inferential procedures, linear transformations, such as z scores, are usually more useful than nonlinear ones.

EXERCISE 3–9

1. A distribution has a mean of 25 and standard deviation of 7. Convert the following scores to z scores and interpret each score in terms of its standard deviation distance from the mean.
   a. 10
   b. 25
   c. 27
2. In a distribution of IQ scores with $M = 100$ and $SD = 15$:
   a. What IQ score is two standard deviations below the mean?
   b. What IQ score is two-fifths of a standard deviation above the mean?
   c. What IQ score has a z score of −1.33?
3. Scores: 4, 8, 5, 2, 6
   a. Calculate the mean, variance, and standard deviation.
   b. Convert each score to a z score.
   c. Calculate the mean, variance, and standard deviation of the z scores and confirm that mean = 0 and variance and SD = 1.

Using a Computer to Describe a Data Set

Measures of central tendency and variability and z scores are relatively easy to calculate by hand when data sets are small. However, as the group size increases, manual calculation becomes more and more tedious and time-consuming. If data have been typed into a computer file, all the descriptive measures we have discussed in this chapter, and more, can be quickly obtained by using statistical software. Most statistical packages include commands that permit the user to compute not only individual measures, like a mean or a standard deviation, but also a large number of descriptive measures all at once. In Minitab, for example, the command "describe" produces the mean, median, standard deviation, and first and third quartiles of a data set, as well as a few other measures not discussed in this chapter.

As an example, we use Minitab to obtain descriptive measures of the anxiety scale scores of 50 college students. Let us say that the scores have been typed into the second column (c2) of a computer file. Now we type "describe c2" at the Minitab prompt (MTB>) like this:

```
MTB> describe c2
```

Table 3–5 **Printout From Minitab of Anxiety Scores of 50 College Students**

MTB > Describe c2

	N	MEAN	MEDIAN	TRMEAN	STDEV	SEMEAN
anxiety	50	46.18	46.50	46.36	9.93	1.40
	MIN	MAX	Q1	Q3		
anxiety	26.00	68.00	38.75	52.25		

A printout of the results is shown in Table 3–5. You will recognize most of the statistics in the printout. Two of them, however, were not taught in this chapter: TRMEAN and SEMEAN. TRMEAN stands for "trimmed mean"; it is the mean of the middle 90% of the observations, after the lowest 5% and the highest 5% have been "trimmed" off. The trimmed mean may be useful if there are outliers at one or the other extreme that substantially affect the mean. In the set of anxiety scale scores, which includes no outliers, the mean and the trimmed mean are almost the same. SEMEAN stands for "standard error of the mean." The standard error of the mean, an important concept in inference, is explained in Chapter 6.

The standard deviation given in the printout, 9.93, was obtained using the formula $\sqrt{SS/(N-1)}$. As mentioned earlier in this chapter, division of the sum of squares by $N - 1$ rather than by N is sometimes necessary (the reason why is discussed in Chapter 6).

SUMMARY

In order to describe a distribution, one must know the location of its center, or ***average value***, and the amount of spread, or ***variability***, of scores around the central point. This chapter has dealt with three measures of central tendency—the ***mode***, the ***median***, and the ***(arithmetic) mean***—and with measures of variability associated with each one. The measure of variability associated with the mode is the ***range***; associated with the median are the ***interquartile*** and ***semi-interquartile ranges***; and associated with the mean are the ***variance*** and the ***standard deviation***.

The mode is the most frequently occurring score in a distribution. A distribution may have no mode, one mode, or two or more. The range, the measure of variability associated with the mode, is the difference between the highest and lowest scores. Though simple to compute and understand, the mode and range are dependent on few scores and are relatively unstable measures.

The median, also called the 50th percentile or second quartile, is the score value or score point below which half the observations in the distribution fall. The interquartile range is the difference between the first and third quartiles (the range of the middle 50% of the observations); the semi-interquartile range is half the interquartile range. The median and its associated measures of variability are relatively unaffected by outliers and skew and are the measures of choice for describing highly skewed distributions.

The mean of a distribution is the sum of the scores divided by the number of scores. The sum of the deviations from the mean is zero, and the sum of squared deviations from the mean (the ***sum of squares***) is a minimum. Its mathematical properties make the mean particularly useful for further calculations and for inference. Because of its dependence on every score, however, the mean is affected by outliers and skew. In negatively skewed distributions, the mean is smaller than the median; in positively skewed distributions, the mean is greater than the median. If any scores in the distribution are indeterminate, the mean cannot be calculated.

Measures of variability associated with the mean are the variance and the standard deviation. The variance is the mean of the squared deviations from the mean. It is equal to the sum of squares (SS) divided by N. Because the variance is in squared units, its square root, the standard deviation (*SD*), is more useful descriptively. Like the mean, the variance and standard deviation are sensitive to outliers and skew. They are, however, the most important measures of variability for statistical inference.

Transformations of raw scores that involve addition or subtraction of a constant or multiplication or division by a constant are called ***linear transformations***. Addition or subtraction of a constant changes the mean of a distribution by the same amount as each score but does not affect the standard deviation. Multiplication or division by a constant changes both the mean and the standard deviation by the same amount as each score.

Scores that describe distances from the mean in standard deviation units are called ***standard scores*** or ***z scores***; z scores are a linear transformation of raw scores obtained by subtracting the mean and dividing by the standard deviation. The mean of a distribution of z scores is zero, and the variance and standard deviation are equal to 1. When a raw score distribution is converted to z scores, the shape of the distribution is unchanged.

The z scores, like percentile ranks, can be used to describe the location of scores within a distribution. Percentile ranks are easier to understand and have the same meaning regardless of the distribution shape. However, percentile ranks are nonlinear transformations of raw scores, and equal distances between raw scores do not necessarily convert to equal differences between percentile ranks. The linear transformation of raw scores to z scores, on the other hand, preserves equality of score differences. Consequently, z scores are generally more useful if further calculations are to be performed.

Most statistical software packages for computers include commands for generating a set of descriptive measures.

QUESTIONS

Use the data set from Table 2–13 (p. 49) to answer questions 1–5.

1. Calculate the mode and range, the median and interquartile and semi-interquartile ranges, and the mean, variance, and standard deviation of the distribution of heights.

2. The distribution is bimodal. What do you suppose accounts for the presence of two modes?

3. Which measures of central tendency and variability seem most appropriate for this distribution? Why?

4. Before calculating the median and the mean of the statistics quiz scores in Table 2–13, predict which of the two measures of central tendency will be larger. Then calculate the median and the mean.

5. Calculate the semi-interquartile range and the standard deviation of the quiz scores.

6. What measure of central tendency or variability is needed to answer each of the following questions?

a. What is the range of the middle 50% of the scores?
b. What was the most typical response on this questionnaire?
c. What is the average distance of scores from the mean?
d. From what number is the sum of squared deviations the smallest?
e. If I construct a graph of this distribution out of plywood and try to balance it on a single point, what is the point at which it will balance?
f. This distribution is markedly skewed. What is the best single measure of central tendency?

7. Basic set: 2, 7, 9, 11, 12, 15, 15.

a. Calculate the mean, the median, and the standard deviation.
b. In each of the following, first predict the effect of the change on the mean, the median, and the standard deviation. Then make the change, and calculate the new mean, median, and standard deviation. See whether your predictions are confirmed.
(1) Change the 2 in the basic set to 8.

(2) Change the 2 to 13.
(3) Change one score of 15 to 35.
(4) Eliminate the 2.

8. Here is a set of eight scores: 18, 20, 19, 2, 20, 22, 18, 21.

a. Calculate the mean, the median, the standard deviation, and the semi-interquartile range.
b. Which of the statistics calculated in part (a) best describes the central tendency and variability of this distribution? Why?
c. Assume that the score of 2 was an error in data recording; the score was actually 22. Predict the effect of correcting this error on the mean, median, semi-interquartile range, and standard deviation. Then calculate to determine whether your predictions are confirmed.

9. The five outfielders of a baseball team have a mean batting average of .286. The seven infielders have a mean batting average of .243. The three pitchers have a mean batting average of .155. And the catcher has a batting average of .316. What is the mean batting average of the entire team?

10. Which measure of central tendency and/or measure of variability is best in each of the following situations?

a. A set of test scores is approximately symmetrical. Further calculations are planned.
b. A consumer research organization asked 100 people to name their favorite vegetable and counted the number of people who replied "beans," "corn," and so on.
c. Six out of seven subjects completed a task in 5 minutes or less, but the seventh subject required 35 minutes.
d. A college administrator wants to know the most typical age of nontraditional students.
e. A researcher inadvertently lost the data from 6 of the 50 subjects in an experiment. He does not want to eliminate them from the analysis. Although he does not know the exact scores, he knows that they were the lowest scores of the set.

11. A distribution has mean = 60, SD = 5.

a. Convert the following scores to z scores:
(1) 63
(2) 75
(3) 58
(4) 32
b. Find the raw score equivalents of the following z scores:
(1) −2.50
(2) 0.75
(3) 1.65
(4) zero

12. During final exam week, Mary had exams in psychology, art history, and biology. All three score distributions had the same shape. The mean, standard deviation, and Mary's score on each test are shown below. On which test did Mary do best? On which test did she do worst relative to her classmates?

	Psychology	Art History	Biology
Mean	84	37	152
SD	12	8	13
Mary's score	75	54	168

13. The last section of this chapter defined the "trimmed mean." In Table 3–5, we saw that the mean and the trimmed mean of the anxiety scale scores of 50 college students are almost the same. Table 2–11 (p. 36) shows the anxiety scale scores of 35 clients at a mental health center. Would their mean and trimmed mean be similar, or would they be more different than for the college students? Why?

Sampling and Probability

INTRODUCTION

In Chapters 2 and 3 we were concerned with describing obtained distributions. An obtained distribution, however, is rarely an end in itself. We usually collect data on only a small subset of all the cases we could potentially measure and would like information about. In this chapter we discuss the relationship between the subset we measure, a sample, and the larger set of potential measurements, the population, from which it came. We will talk about how samples are drawn and a little bit about how we use sample data to generalize to the populations from which the samples came. The process of generalizing from samples to populations is called *inference*. In this chapter, then, we are laying the foundation for statistical inference.

We will see that samples drawn from a population are not all alike and, more important, are not all similar to the population. For this reason, we can never be certain that the inferences we make from just one sample are correct. We can only state that we have some degree of confidence in the inferences—that is, that there is a certain probability that they are correct or incorrect. Therefore, we will also need to consider the concept of probability—what it means, how probabilities are computed, and how probability is related to sampling. In doing so, we will consider the construction and use of probability distributions.

Objectives

When you have completed this chapter, you will be able to

- Define the following terms: population, sample, random sample, statistic, parameter, sampling distribution.
- Describe the process of constructing a sampling distribution of a statistic.
- Describe the effect of sample size on sampling distributions of the proportion and the mean, and discuss implications for making inferences from small and large samples.
- State the theoretical and empirical definitions of probability and use them to determine the probability of an event.

- Use the addition and multiplication laws when needed.
- Distinguish between sampling with and without replacement and calculate probabilities of sampling outcomes under both conditions.
- Define probability distribution and construct probability distributions by listing and counting all possible outcomes of an event.
- Describe the conditions for generating a binomial distribution and use binomial distributions to find probabilities of sampling outcomes.
- Compare methods for finding probabilities in discrete and continuous probability distributions.

Sampling and Sampling Distributions

Populations and Samples

You have probably learned to use the term *population* to refer to the inhabitants of some specified region, for example, Pittsburgh, the United States, or the world. In statistics, ***population*** has a different meaning: It refers to *the set of all observations about which one seeks information.* Thus, if we seek information concerning the IQ scores of American schoolchildren, the population is the set of all American schoolchildren, now and, possibly, in the past and future as well. If we seek information about the reaction of depressed persons to a new drug, all depressed persons who could potentially receive the new drug constitute the population. If our interest lies in performance ratings of employees of Company X, all employees of Company X are the population. Sometimes a population is small, such as current employees of a small company. Usually, however, populations of interest tend to be very large.

We are seldom able to obtain data from an entire population. Usually we select a subset of observations and use them to make inferences about the population as a whole. The subset is called a ***sample***. Thus, to draw conclusions about the political behavior and attitudes of all voting Americans, the Gallup poll questions samples of voters. Tea tasters on a tea plantation taste samples of tea leaves and use the results to judge the quality of all the tea. Even though, in these cases and most others, the size of the sample is relatively small compared with the population, inferences can be surprisingly accurate.

Sometimes, of course, inferences from samples to populations are not accurate. This is particularly likely to happen if the sample is biased. A ***biased sample*** is one in which some members of the population have a greater chance of being chosen than others. For example, suppose I want information about attitudes toward statistics among students of elementary statistics; as a sample, I choose 20 math majors who are taking statistics. Because students with many different majors enroll in statistics courses, and because attitudes may vary from one major to another, math majors constitute a biased sample of all statistics students. Inferences made from a biased sample are highly likely to be in error.

One avoids bias in sampling by ensuring that every member of the population has an equal chance of being selected for the sample. Procedures that ensure this are called ***random sampling***. For example, to obtain a random sample of statistics students at the University of Pittsburgh, I would start by getting the names of all students taking elementary statistics this term from the Registrar. I might then write each name on a slip of paper, put the slips in a box, shake the box, and draw 20 slips with my eyes closed. Because each student on the Registrar's list had an equal chance of being chosen, these 20 students are a random sample of all students presently taking elementary statistics at the university. If I measure the attitudes of these 20 students and generalize my findings to the population, my inferences are far more likely to be accurate than those made from a sample of math majors.

Drawing slips from a box is one method of random sampling. Another, more efficient method for drawing a random sample is to use a table of ***random numbers***. An excerpt from a random number table is shown in Table 4–1. The table consists of columns of digits that have been generated in such a way that each digit precedes and follows each other digit (including itself) equally often.

Suppose I use this table to draw a sample of size 20 from the Registrar's list of 600 students currently enrolled in statistics. Here is how I proceed:

1. I number the students on the Registrar's list from 001 to 600.
2. I enter the random number table at some arbitrary point, such as the top of the second five-digit column, Column (2).
3. I examine the last three digits of each five-digit set in Column (2) in turn. If the three-digit number is between 001 and 600, that student is selected for my sample. Otherwise, I go on to the next five-digit set in the column.

   In Column (2) of Table 4–1, the first five-digit set is 72305; the last three digits in the set are 305. Because 305 is between 001 and 600, Student 305 is in my sample. I continue to the next five-digit set, which ends in 460; Student 460 is also selected for my sample, as is Student 380, the last three digits of the next five-digit set. Continuing down the column, I find the number 94710; 710 is not between 001 and 600, so I go to the next five-digit set and select Student 154. I continue in this manner until I have selected 20 students.

The excerpt in Table 4–1 is too short to select any but the smallest samples. Books of statistical tables and textbooks on experimental design usually include multiple-page tables of random numbers. In addition, most statistical software packages and statistical calculators can generate random numbers for sampling purposes. We will illustrate how this is done later in the chapter.

Does random sampling guarantee samples that are representative of the population? No, it does not. One important factor is *sample size*. For example, suppose that I select

Table 4–1 Excerpt From a Table of Random Numbers

(1)	(2)	(3)	(4)
95108	72305	64620	91318
42958	21460	43910	01175
88877	89380	32992	91380
01432	94710	23474	20523
28039	10154	95425	39220
72373	06902	74373	96199
16944	93054	87687	96693
15718	82627	76999	05999
61210	76046	67699	42054
62038	79643	79169	44741

a sample of children enrolled in a certain school and measure their IQ. If I randomly select only two children from the school, I could easily happen to get two children with very low scores or two with very high scores. If I select 10 children, it is less likely that all or most of them will have unusually high or low scores. With 50 children, it is very unlikely indeed. In other words, other things being equal, the larger the random sample, the more likely that it will be representative of the population.

Even large random samples do not guarantee accurate inferences. Whenever one measures only a subset of a population, there is a chance that error will occur. For example, even if I randomly select 100 statistics students from the Registrar's list, just by chance I may draw a disproportionate number of psychology majors, or older students, or females. As a result, the attitudes of the sample may differ to some degree from those of the population as a whole. We can never know whether such an error has occurred in the sample we are working with; but inferential statistical techniques (discussed in later chapters) allow us to determine and, to some degree, control the likelihood that such errors of inference have occurred.

Statistics and Parameters

Statistics (a singular noun ending in *s*) is the subject that you are studying. A ***statistic*** (without the *s*) refers to a numerical characteristic of a sample. In samples of categorical data, the proportion falling in a particular category is a statistic. In samples of quantitative data, measures like the mode, mean, and standard deviation are statistics. For describing a distribution, statistics are important in their own right. However, in inference, we often use statistics computed in samples primarily to estimate (make inferences about) the corresponding values in populations. From the proportion of sample observations in a certain category, we estimate the population proportion in that category; from the mean of a sample, we estimate the mean of the population. The population values, which we rarely measure directly but about which we make inferences, are called ***parameters***. Thus population proportions, modes, means, standard deviations, and the like are parameters.

By convention, Roman letters are used to represent statistics calculated on samples, and Greek letters are used for parameters of populations. For example, the mean of a sample is designated M (or, sometimes, X); the mean of a population is designated μ (mu). S or s may be used to stand for the standard deviation of a sample; σ (sigma) stands for the standard deviation of a population.

EXERCISE

1. Here is a small population of 25 children:

Amy	Alan	Kevin	Bob	William
Catherine	Richard	Josephine	Clark	George
John	Suzy	Barry	Wendy	Mark
Clare	Leonard	Zoe	Carol	Jean
Joseph	Brian	Tim	Mary	Leah

Number the children from 01 to 25. Then use Table 4–1 to select a random sample of size 5. Start with the last two digits of Column (2) and go on to Column (3) if necessary.

2. In each of the following, decide whether or not the sampling process was random.
   a. Linda needed to pick a sample of size 20 from a population of size 40. For each member of the population, she tossed a fair coin. If the coin came up heads, that

member of the population was in her sample; if the coin came up tails, that member was not in her sample.

b. To select 10 of the 100 faculty members at his college, Joe started with the 5th name in the alphabetical faculty directory, then selected every 10th name from that point on. (That is, his sample consisted of the 5th, 15th, 25th, etc. name on the list.)

3. In each of the following, tell which of the two numbers is a parameter and which is a statistic:
   a. A batch of gizmos rolling off a production line during 1 week is examined for defects, and 2.5% of the gizmos are defective. Actually, only 1.8% of all the gizmos produced by the factory are defective.
   b. In Tarrytown 45% of voters plan to vote for Mayor Gordon in the coming election. A poll conducted shortly before the election finds that 48% of a sample of 600 voters plans to vote for Mayor Gordon.
   c. The mean height of young adult males nationwide is 68.5 inches. A group of 50 young men at Rothbart College has a mean height of 68.7 inches.

Sampling Variability and Sampling Distributions

A parameter of a population is assumed to have a particular, fixed value. However, if two or more samples are selected from a population, and the same statistic is computed in each sample, values of the statistic will vary from one sample to another. This will happen even if we have taken care to draw each sample randomly.

This notion of the *variability of a statistic* from one sample to another is extremely important. To illustrate sampling variability, we will conduct some simple sampling experiments. We will draw random samples from two populations and compare certain parameters of the populations with the corresponding sample statistics. In the first case, we will consider a population in which a certain proportion of observations falls in a particular category. We will draw random samples from the population, calculate the proportion of each sample in the category, and compare the sample and population proportions. In the second case, we will consider a population of numerical data having a certain mean. We will draw random samples from the population, calculate the mean of each sample, and compare the population and sample means. Proportions and means have been chosen because they are important for inference and are easy to compute and to understand. However, other population parameters and their corresponding sample statistics could also be used; the conclusions reached would be the same.

Two Sampling Distributions of the Proportion

To examine sampling variability of sample proportions, we consider a population in which exactly half the members are males and the other half females. Thus, the proportion of males in the population is one-half, or .5. If I select a random sample of four members from the population, will exactly half the sample (two members) be males? Not necessarily. The sample may include just one male, or three males, or even, possibly, no males, or four males. But how likely are these various outcomes? If I randomly select not just one sample but many samples of size 4, how many samples will include no males? One male? Two males? Three males? Four males? To answer this question, let us conduct a sampling experiment. Specifically, we will select 50 random samples of size 4 from the population, count the number (and proportion) of males in each sample, and examine the results over all 50 samples.

How can we select 50 random samples from a population in which half the members are males? Here is one way: Take a very large number of slips of paper (say, 1000). Put "M" for male on half of them and "F" for female on the other half. This, then, is the population. Next, put the slips in a bag, shake the bag well, and select four slips without looking. This is the first sample. Count the number of slips with "M" on them. Put the slips back in the bag, and repeat this process 49 more times to get 50 random samples.

Another, simpler method is to use a table of random numbers. Half the digits in a random number table are odd and half even. If we call the odd digits "male" and the even digits "female," the digits in the table represent our population. To select a random sample of size 4, we go down a column of the table and read four digits at a time. Suppose that the first four digits of the column we have chosen are 0, 3, 8, 0 (one odd and three even digits); this means that our first sample has one male and three females. Then we continue to the next four digits in the column: 7, 7, 3, 2 (three odds and one even digit). Our second sample has three males and one female. We continue in this manner until we have examined 50 sets of four digits each (50 samples of size 4).

A table of random numbers was used to select 50 random samples of size 4. Table 4–2 lists the 50 samples and the number of males and proportion of males in each one. (In each sample, the proportion of males is the number of males divided by 4.)

Table 4–2 Fifty Samples of Size 4 From a Population in Which the Proportion of Males Is .5

Sample		Number of Males	Proportion of Males[a]	Sample		Number of Males	Proportion of Males[a]
1	FMFF	1	.25	26	MFFM	2	.50
2	MMMF	3	.75	27	MFMF	2	.50
3	FMFF	1	.25	28	FFFF	0	.00
4	FMFF	1	.25	29	FMFF	1	.25
5	FFFM	1	.25	30	MMMM	4	1.00
6	MMFM	3	.75	31	MMFM	3	.75
7	MFFM	2	.50	32	MFMF	2	.50
8	MFFM	2	.50	33	FMMM	3	.75
9	FFFM	1	.25	34	MMMF	3	.75
10	MFMF	2	.50	35	MMFM	3	.75
11	FMFF	1	.25	36	MMMM	4	1.00
12	MMMF	3	.75	37	MMMF	3	.75
13	FFMM	2	.50	38	FMMM	3	.75
14	MMFF	2	.50	39	FFFM	1	.25
15	FMFM	2	.50	40	MMFF	2	.50
16	MFFF	1	.25	41	FFFM	1	.25
17	FFMF	1	.25	42	MFFF	1	.25
18	FMMF	2	.50	43	MMMF	3	.75
19	MMFF	2	.50	44	MFFF	1	.25
20	MFFM	2	.50	45	FMMM	3	.75
21	FMMF	2	.50	46	MFMF	2	.50
22	FFMM	2	.50	47	FMFF	1	.25
23	FMFM	2	.50	48	FFMF	1	.25
24	FFFM	1	.25	49	FFFF	0	.00
25	FFFM	1	.25	50	FFFM	1	.25

[a]In each sample, the proportion of males is the number of males divided by 4.

When multiple samples are drawn from a population and the same statistic is computed in each sample, it is possible to examine the distribution of the statistic over all the samples. That is, we can construct a distribution showing how many samples had proportion of males = 0, how many had proportion of males = .25, and so on. A distribution of a statistic over successive samples is called a ***sampling distribution*** of the statistic. Thus, the distribution of sample proportions over our 50 samples is a *sampling distribution of the proportion.* The sampling distribution of the proportion is shown in Figure 4–1.

Figure 4–1 shows that although the population proportion of males is .5, only 17 of the 50 samples have a proportion of males exactly equal to .5. Moreover, sample proportions vary from 0 to 1. But notice that more samples have a proportion closer to .5 than farther away: 29 samples have a proportion of males equal to .25 or .75, but only 4 have a proportion equal to 0 or 1. It seems that even in samples as small as size 4, sample proportions are, more often than not, similar to the population proportion.

Earlier in the chapter I pointed out that large samples are more likely to be representative of the population than small samples. Four is a small sample size. Let us see what happens to the sampling distribution of the proportion if we draw 50 samples of size 20 ($N = 20$) from the population, instead of size 4.† The method used to select samples of size 20 was the same as that for samples of size 4. A random number table was used to select 50 sets of 20 digits each, and the number of odd digits in each set (the number of males) was counted. Table 4–3 lists the number of males and proportion of males (the number of males divided by 20) in each sample. The sampling distribution of the proportion is shown in Figure 4–2.

Examination of Figure 4–2 shows that, in some ways, the sampling distribution of the proportion for samples of size 20 is like that for samples of size 4. As with samples of size 4, (1) sample proportions vary over samples, and (2) more samples have

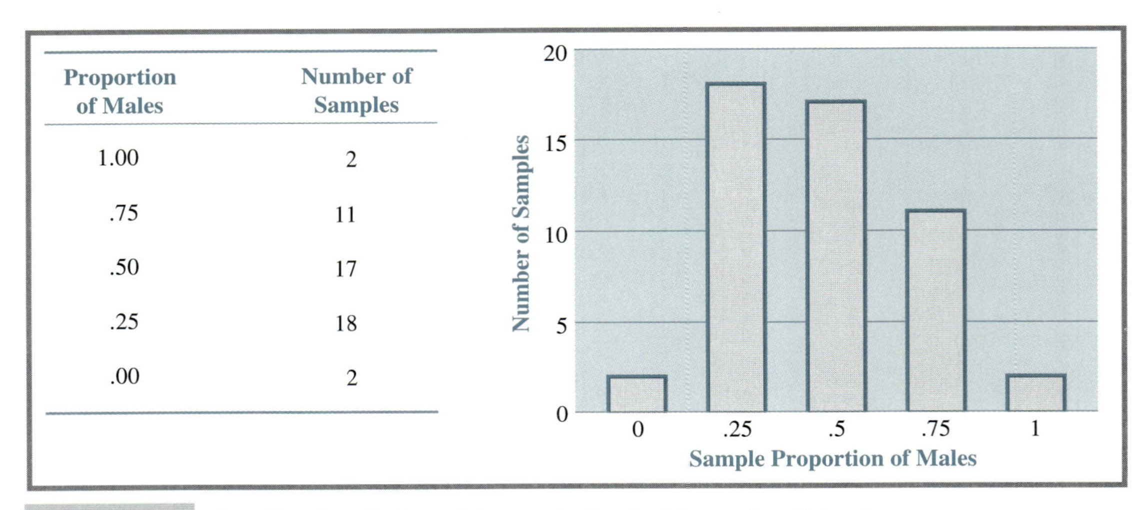

Proportion of Males	Number of Samples
1.00	2
.75	11
.50	17
.25	18
.00	2

Figure 4–1 **Sampling distribution of the proportion for 50 samples of size 4 from a population with proportion of males = .5.**

†Twenty is not really a large sample size, but it is sufficiently larger than 4 to reveal the effect of increasing the sample size.

Table 4–3 Fifty Samples of Size 20 From a Population in Which the Proportion of Males Is .5

Sample	Number of Males	Proportion of Males
1	12	.60
2	11	.55
3	12	.60
4	10	.50
5	13	.65
6	13	.65
7	11	.55
8	6	.30
9	10	.50
10	8	.40
11	9	.45
12	9	.45
13	7	.35
14	9	.45
15	8	.40
16	9	.45
17	10	.50
18	11	.55
19	11	.55
20	12	.60
21	10	.50
22	12	.60
23	9	.45
24	12	.60
25	10	.50
26	13	.65
27	13	.65
28	9	.45
29	9	.45
30	12	.60
31	10	.50
32	9	.45
33	10	.50
34	11	.55
35	10	.50
36	10	.50
37	7	.35
38	8	.40
39	9	.45
40	11	.55
41	8	.40
42	8	.40
43	10	.50
44	8	.40
45	12	.60
46	11	.55
47	12	.60
48	9	.45
49	11	.55
50	12	.60

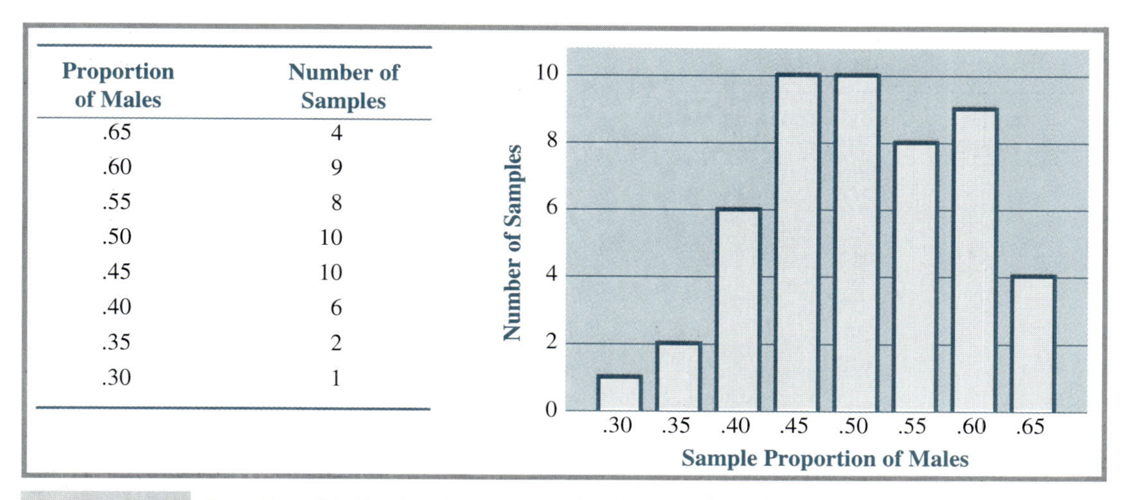

Proportion of Males	Number of Samples
.65	4
.60	9
.55	8
.50	10
.45	10
.40	6
.35	2
.30	1

Figure 4–2 Sampling distribution of the proportion for 50 samples of size 20 from a population with proportion of males = .5.

a proportion closer to .5 than farther away. There is, however, one striking difference between the sampling distributions of the proportion for samples of size 4 and samples of size 20: Whereas proportions of males in samples of size 4 vary from 0 to 1, no sample of size 20 has a proportion of males less than .35 or greater than .65. Thus, the variability of the sample proportions is less over larger samples than over smaller samples.

The difference between the two sampling distributions is clearly shown in Figure 4–3. Note that the spread of sample proportions around the population proportion is less in samples of size 20 than in samples of size 4. Moreover, if we were to select samples larger than size 20 from the population, the variation of sample proportions around the population proportion would be still less than for samples of size 20. This explains what is meant when we say that proportions calculated in large samples are more "representative" of the population proportion than proportions calculated in small samples; they are more often close to the true value of the population proportion.

Two Sampling Distributions of the Mean

In our second sampling experiment, we will consider a population of numerical data having a certain mean. We will draw random samples of size 5 ("small" samples) and of size 20 ("large" samples) from the population, calculate the mean of each sample, and construct sampling distributions of the mean. Then we will see whether the relation between sample and population means is similar to that between sample and population proportions.

The population from which samples are to be drawn is shown in Figure 4–4. As you can see, it is a small population with only 100 members. The population is negatively skewed. However, the shape does not matter; the results of our sampling experiment would be similar with other shapes. Scores in the population range from 3 to 10, and the mean of the population is 8. Recall that the mean of a population is symbolized μ. Thus we can write: $\mu = 8$.

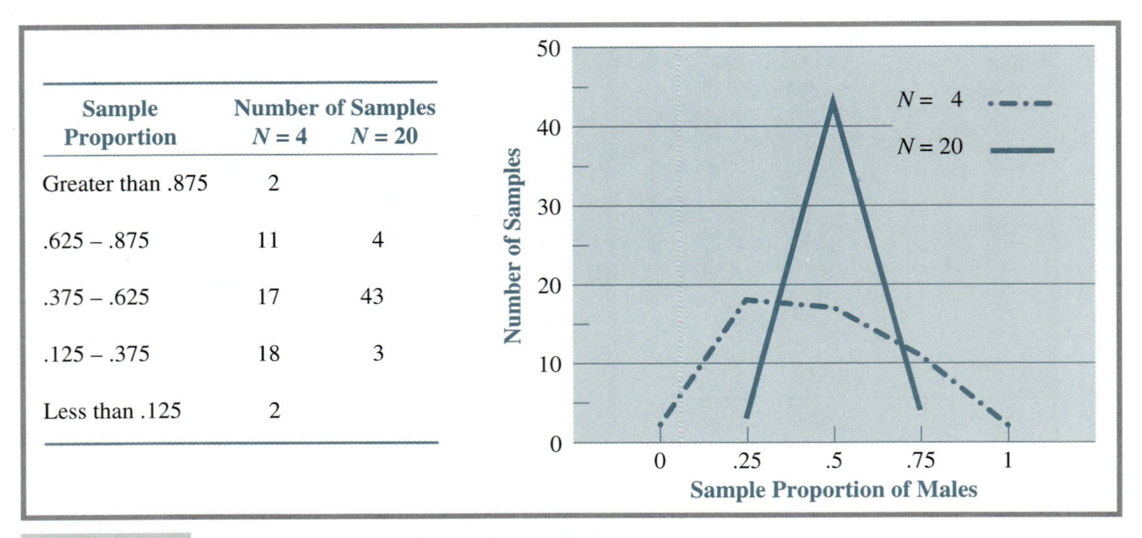

Sample Proportion	Number of Samples $N = 4$	$N = 20$
Greater than .875	2	
.625 – .875	11	4
.375 – .625	17	43
.125 – .375	18	3
Less than .125	2	

Figure 4–3 **Comparison of sampling distributions of the proportion for small ($N = 4$) and large ($N = 20$) samples.**

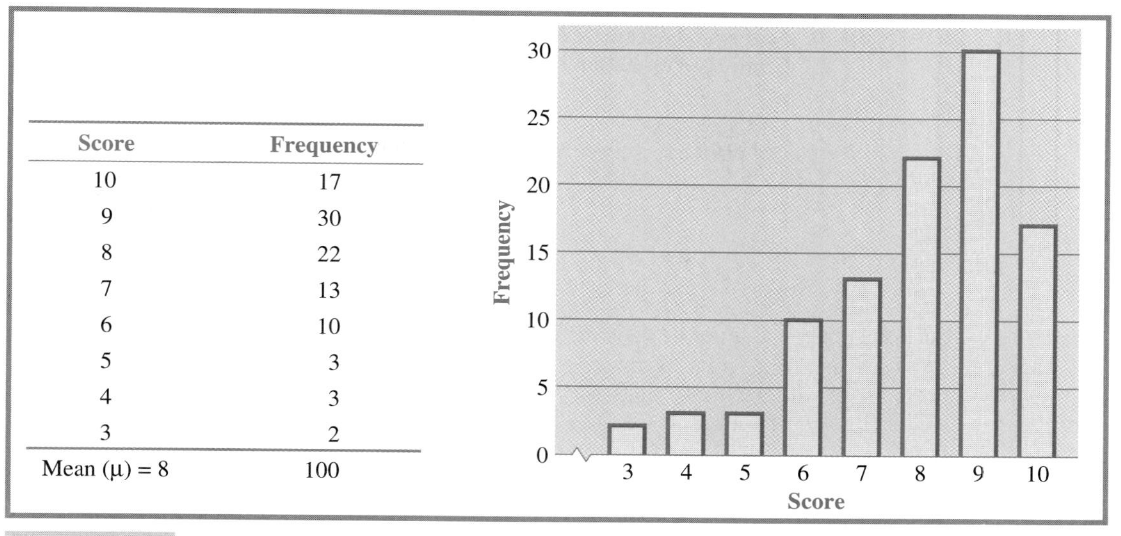

Score	Frequency
10	17
9	30
8	22
7	13
6	10
5	3
4	3
3	2
Mean (μ) = 8	100

Figure 4–4 **A small population of size 100.**

A random number table could be used to select 50 samples of size 5 from the population, but this would be tedious and time-consuming. Instead, we used the following procedure: First, the population distribution was typed into a machine-readable data file. Then the computer was instructed to select 50 random samples of size 5, to calculate the mean of each sample, and to print the results. The 50 samples and their means are listed in Table 4–4. Next, a *sampling distribution of the mean* was constructed by counting the number of samples having each particular mean. The sampling distribution is displayed in Figure 4–5.

When we examine the sampling distribution of the mean for samples of size 5, we notice that, although the population mean is 8, sample means vary from 6.0 to 9.4. However, more samples have a mean close to 8 (e.g., between 7.2 and 8.8) than farther away (less than 7.2 or greater than 8.8). Thus, even in small samples, means are more often close to the population mean than farther away.

Next, we instruct the computer to select 50 random samples of size 20, a larger sample size. Table 4–5 lists the means of the 50 samples.† Again, we construct a sampling distribution of the mean by counting the number of samples having each particular mean. The sampling distribution is displayed in Figure 4–6.

When we examine the sampling distribution of the mean for samples of size 20, we note that (1) the sample means vary around the population mean, 8, and (2) more samples have a mean closer to 8 than farther away. In these respects, the sampling distribution of the mean for samples of size 20 is similar to that for samples of size 5. However, there is one important difference: Whereas the means of samples of size 5 range from 6.0 to 9.4, means of samples of size 20 have a narrower range: from 6.90 to 8.65. Thus, the means of larger samples vary less around the population mean than do the means of smaller samples. This difference in variability can be seen clearly in Figure 4–7, which compares the two sampling distributions of the mean. If samples still larger than size 20 were selected from the population, the sample means would vary even less than in samples of size 20.

†The 20 scores in each sample are not listed in order, to save space.

Table 4–4 Scores and Means of 50 Samples of Size 5 Randomly Selected From the Population in Figure 4–4

Sample	Scores					Sample Mean
1	9	3	6	9	8	7.0
2	8	8	9	10	7	8.4
3	10	8	9	9	8	8.8
4	7	8	8	10	10	8.6
5	5	8	8	6	8	7.0
6	9	7	10	8	7	8.2
7	6	10	8	9	8	8.2
8	7	5	8	10	9	7.8
9	8	9	6	8	10	8.2
10	7	7	3	7	6	6.0
11	7	8	7	9	8	7.8
12	10	8	8	8	9	8.6
13	8	9	7	8	7	7.8
14	9	9	10	9	8	9.0
15	4	9	9	6	10	7.6
16	4	6	7	10	7	6.8
17	9	9	10	9	9	9.2
18	9	8	9	9	10	9.0
19	8	7	10	8	9	8.4
20	6	10	9	9	10	8.8
21	6	8	7	10	5	7.2
22	6	8	9	9	9	8.2
23	9	10	5	6	9	7.8
24	8	9	10	10	10	9.4
25	9	10	8	8	9	8.8
26	8	6	10	8	6	7.6
27	8	9	6	10	8	8.2
28	7	10	8	5	8	7.6
29	3	10	6	8	8	7.0
30	9	8	6	10	8	8.2
31	10	7	8	8	6	7.8
32	8	10	9	7	7	8.2
33	9	8	8	8	8	8.2
34	8	8	3	7	9	7.0
35	4	9	7	6	8	6.8
36	9	9	10	7	6	8.2
37	9	7	10	7	8	8.2
38	6	6	6	10	8	7.2
39	9	10	9	9	9	9.2
40	8	6	8	8	10	8.0
41	9	4	9	8	9	7.8
42	7	7	9	8	10	8.2
43	5	10	9	9	8	8.2
44	9	6	7	9	8	7.8
45	10	9	8	9	10	9.2
46	7	10	7	9	6	7.8
47	10	10	8	5	10	8.6
48	8	6	10	9	4	7.4
49	6	10	9	9	10	8.8
50	10	9	8	6	9	8.4

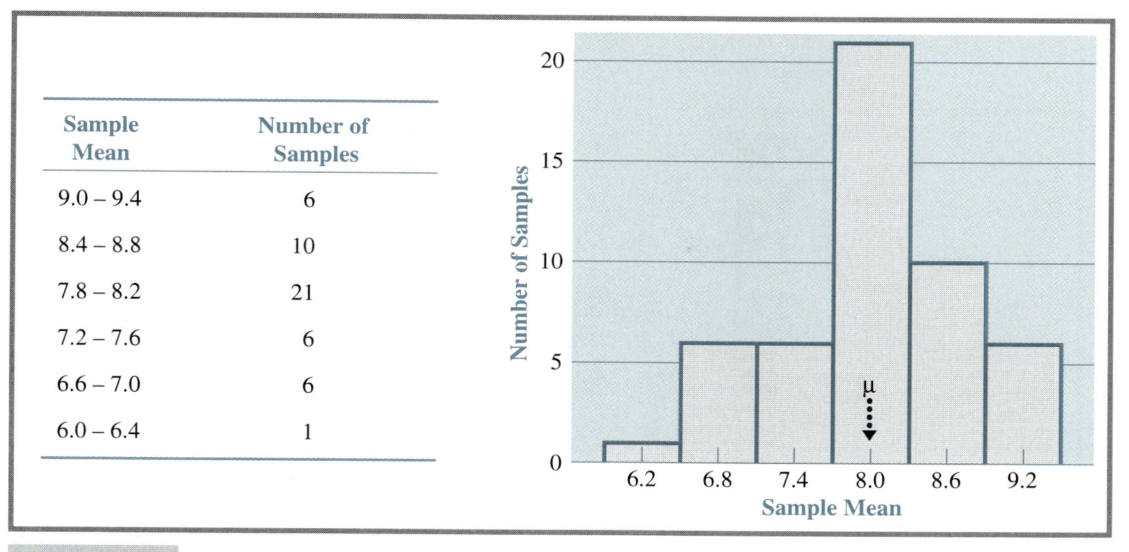

Figure 4–5 Sampling distribution of the mean for 50 samples of size 5 randomly selected from the population in Figure 4–4.

Table 4–5 Means of 50 Samples of Size 20 Randomly Selected From the Population in Figure 4–4

Sample	Sample Mean	Sample	Sample Mean
1	7.90	26	8.35
2	8.25	27	8.25
3	7.85	28	7.55
4	8.05	29	8.65
5	7.20	30	8.50
6	8.55	31	8.60
7	7.70	32	7.95
8	8.05	33	8.05
9	8.65	34	8.30
10	8.10	35	8.00
11	7.85	36	8.40
12	8.05	37	6.95
13	7.60	38	8.60
14	7.95	39	7.25
15	8.60	40	8.10
16	7.95	41	8.15
17	7.85	42	6.90
18	8.40	43	8.05
19	7.35	44	7.65
20	8.35	45	7.75
21	8.10	46	7.70
22	8.55	47	7.95
23	7.55	48	7.75
24	7.70	49	8.00
25	7.65	50	7.85

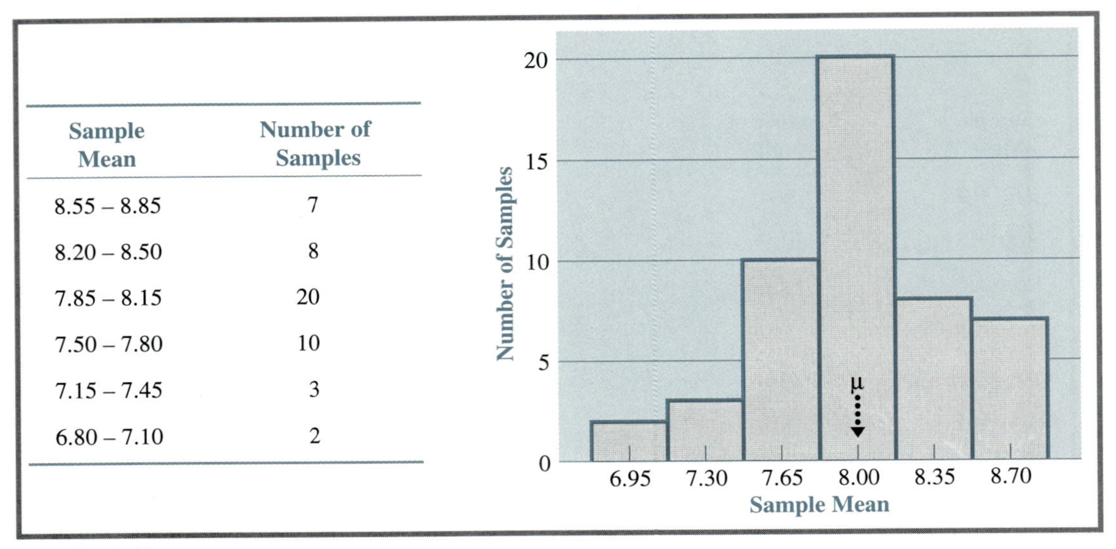

Sample Mean	Number of Samples
8.55 – 8.85	7
8.20 – 8.50	8
7.85 – 8.15	20
7.50 – 7.80	10
7.15 – 7.45	3
6.80 – 7.10	2

Figure 4–6 **Sampling distribution of the mean for 50 random samples of size 20 from the population in Figure 4–4.**

A Comparison of the Sampling Distributions of the Proportion and of the Mean

Now let us compare the sampling distributions of proportions and means that we have constructed. Despite the use of different kinds of variables (categorical vs. quantitative) and different statistics/parameters (proportions vs. means), the sampling distributions of both statistics share the following features:

1. Both sample proportions and sample means vary from one sample to another.

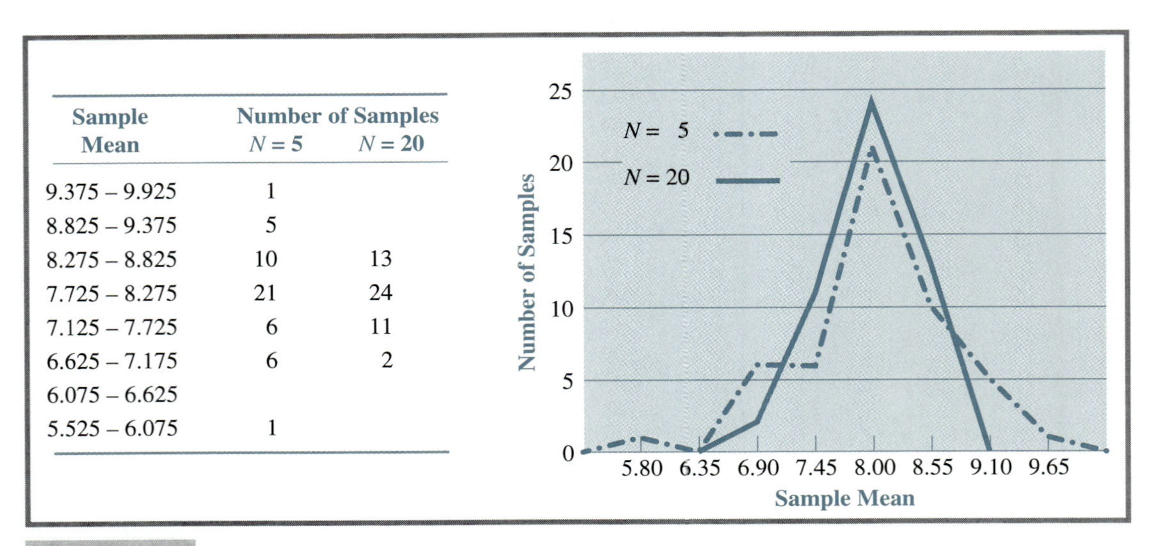

Sample Mean	Number of Samples $N = 5$	$N = 20$
9.375 – 9.925	1	
8.825 – 9.375	5	
8.275 – 8.825	10	13
7.725 – 8.275	21	24
7.125 – 7.725	6	11
6.625 – 7.175	6	2
6.075 – 6.625		
5.525 – 6.075	1	

Figure 4–7 **Comparison of the sampling distributions of the mean for samples of size 5 and size 20.**

2. Sample proportions and sample means vary around the population parameter. That is, samples with means or proportions close to the population value occur more often than samples with means or proportions that are far from the population value.
3. The variability of both sample proportions and sample means decreases as sample size increases. That is, in both cases sample statistics vary less around the population parameter when samples are large than when samples are small.

The tendency for statistics like the proportion and the mean to be closer and closer to the corresponding parameters as sample size increases is called the *law of large numbers*. This law explains why we can be more confident in drawing conclusions about a population from a large sample than from a small one.

Empirical and Theoretical Sampling Distributions

In each of the sampling experiments we have discussed, we drew 50 samples of a certain size. If we were to do the experiments again, different sets of 50 samples would be chosen, and the sampling distributions would be somewhat different. Furthermore, although we chose 50 samples for each sampling distribution, we could have drawn 10 samples, or 100 samples, or 1000 samples. When some arbitrary number of samples is chosen from a population, and a sampling distribution of a statistic is constructed, as we have done, the sampling distribution is called an ***empirical sampling distribution***. As noted earlier, empirical sampling distributions vary somewhat from one sampling experiment to another.

What would happen if we were to draw not 10, 50, 100, or 1,000 samples of a certain size from a population but rather *all possible samples* of that size? For example, what would the sampling distribution of the proportion based on all possible samples of size 20 look like? Of course, we cannot actually draw all possible samples from large populations, but statisticians can use mathematics to derive the properties of the sampling distributions that would result. Such distributions are called ***theoretical sampling distributions*** because they are generated by mathematical methods rather than by actual measurement. Thus, the distribution of proportions of males for all possible samples of size 5 is called the theoretical sampling distribution of the proportion for samples of size 5; the distribution of means for all possible samples of size 20 is the theoretical sampling distribution of the mean for samples of size 20; and so on.

As you will see in later chapters, theoretical sampling distributions provide the foundation for statistical inference. Inference about a particular parameter is based on the sampling distribution of the relevant statistic and on information about the probabilities of various values of the statistic. Before we can proceed to inference, therefore, we need to consider the meaning, calculation, and use of probability.

Addendum

Students are sometimes confused about the difference among population distributions, sampl*e* distributions, and samp*ling* distributions. They are *not* the same thing. Review the following definitions until you are sure you understand them:

Distribution of *observations* in a *population*:	Population distribution
Distribution of *observations* in a *sample*:	Sampl*e* distribution
Distribution of a *statistic* over many (or all possible) samples:	Sampl*ing* distribution of the statistic

EXERCISE

1. Decide whether each of the following distributions is a population distribution, a sample distribution, or a sampling distribution. If it is a sampling distribution, state sampling distribution of *what* statistic.
   a. To determine the typing skills of your firm's secretaries, you give a typing test to all the secretaries and examine the scores.
   b. As inspector on an assembly line, you periodically select 10 gidgets from the moving belt and determine the proportion of defective gidgets in each set. At the end of the day, you examine the distribution of proportions.
   c. You are doing research on affiliation needs of college males. You give 200 male students at your college an affiliation-needs questionnaire and examine the resulting distribution.
2. Why do you suppose that none of the samples from the population with a mean of 8 had a mean equal to 10, even though the population scores ranged from 3 to 10?
3. Perform the following sampling experiment: Assume a population of 50 voters of whom 30 are Democrats and 20 Republicans. On 30 out of 50 slips of paper put the letter "D" and on the remaining 20 the letter "R." Put the slips in a box or bag. Draw 25 random samples of 10 voters each. In each sample, compute the proportion of Republicans by dividing the number of Republicans in the sample by 10. Construct a sampling distribution of the proportion and examine the results. Are they similar to the results of our sampling experiment on proportions?

Probability

Two Definitions of Probability

Most people have an intuitive understanding of "probability." We all know that the probability of winning a million dollars in the lottery is very, very small; that the probability of having a male baby is about one-half; and that the probability of successful completion of a statistics course is relatively large.

A *probability* is a kind of proportion or relative frequency. One way to define probability is to say that it is the observed relative frequency of an event. This is called the *empirical definition of probability*. When the weatherperson says the probability of rain is 70%, this means that on previous occasions when meteorological conditions were like today's, it rained 70% of the time and did not rain 30% of the time. The probability of a male birth, .52, is based on the fact that, over hundreds of thousands of births, about 52% of the babies have been boys and 48% girls.

Usually, however, probabilities are determined logically rather than through experience. When we say that the probability of a head when tossing a fair coin is one-half, we do not base this on the outcomes of thousands of tosses of a coin. In fact, if we were to toss a coin thousands of times it is doubtful that exactly half the tosses would be heads. Instead, we reason that because a coin has two faces, and, if the coin is fair, either face is equally likely to come up, the probability of a head is one-half, and so is the probability of a tail.

This kind of reasoning gives us a second definition of probability: *Providing that all possible outcomes are equally likely, the* probability *of an event is the proportion of the possible outcomes that are examples of the event.* We let $p(A)$ stand for the probability of event A. Thus the theoretical definition of probability states that

$$p(A) = \frac{\text{Number of ways event A can occur}}{\text{Total number of possible equally likely outcomes}}$$

Thus, when all outcomes are equally likely, the probability of an event is the proportion of outcomes that exemplify the event.

Let us use the theoretical definition of probability to determine the probabilities of several events:

1. Geneticists have established that when a mother is a carrier for color blindness and the father has normal color vision, four gene combinations are possible and equally likely. One of these results in color blindness, the other three in normal color vision. Therefore, the probability that such a couple will have a child with normal color vision is

$$\frac{\text{Number of gene combinations resulting in a normal color vision}}{\text{Total number of possible gene combinations}} = \frac{3}{4} = .75$$

2. At the choice point in a T-maze, a rat can turn either left or right. If the rat has no preference, so that either turn is equally likely

$$p(\text{right turn}) = \frac{\text{Number of ways a rat can turn right}}{\text{Total number of possible turns}} = \frac{1}{2} = .5$$

3. A die (singular of "dice") has six faces. If a fair die is thrown, each face is equally likely to come out on top. Therefore, the probability of throwing a 6 with a fair die is

$$p(6) = \frac{\text{Number of faces with the number 6}}{\text{Total number of faces}} = \frac{1}{6}$$

4. In a population of 1000 children, 20 children have IQ scores greater than 130. If one child is randomly chosen (i.e., chosen in such a way that each child is equally likely to be chosen)

$$p(\text{IQ} > 130) = \frac{\text{Number of children with IQ} > 130}{\text{Total number of children in the population}} = \frac{20}{1000} = .02$$

Like all proportions, probabilities range from zero to one. Events that are impossible have probability equal to zero. Consider the probability of drawing a red ace of spades from a standard deck of cards:

$$p(\text{red ace of spades}) = \frac{\text{Number of red aces of spades in the deck}}{\text{Total number of cards in the decks}} = \frac{0}{52} = 0$$

If an event is certain to occur, its probability is one. For example, the probability that a coin will come up heads (H) or tails (T) is

$$p(\text{H or T}) = \frac{\text{Number of faces of a coin that are heads or tails}}{\text{Total number of faces}} = \frac{2}{2} = 1$$

Events that are very unlikely (but not impossible) have a probability close to zero; the more likely an event, the nearer is its probability to one.

If the probability of an event is known, the probability that the event will *not* occur is one minus the probability of the event:

$$p(\text{not A}) = 1 - p(\text{A})$$

For example, if the probability of IQ $>$ 130 is .02, the probability of IQ $\leq$ 130 is $1 - .02 = .98$.

EXERCISE 4–3

1. You are fishing in a lake in which 10% of the fish are over 20 inches long, 50% are between 10 and 20 inches long, and the rest are under 10 inches. Assuming that you catch a fish, and assuming that all the fish are equally likely to bite, what is the probability that your fish is under 10 inches in length? What is the probability that the fish is 10 or more inches long?
2. Here is a population distribution:

Score	*f*
25	3
24	8
23	12
22	7
21	6
20	2

   If you randomly select one member of the population, what is the probability that the person's score is 22 or less?
3. If you randomly select one card from a standard deck, what is the probability of selecting a king? What is the probability of selecting an ace? What is the probability that the card you select is not a spade?
4. The probabilities of three events, A, B, and C, are $p(\text{A}) = .004$; $p(\text{B}) = .0009$; $p(\text{C}) = .01$. Which of the three events is most likely to occur? Which is least likely to occur?

The Addition and Multiplication Laws of Probability

Sometimes we want to know the probability that either of two (or more) events will occur or the probability that both of two (or more) events will occur. For example, what is the probability that a child selected randomly from the U.S. population has IQ < 70 or IQ > 130? What is the probability that a voter is both male and Republican? If we know the probability of each individual event, we can answer questions like these by using the addition and multiplication laws of probability.

The Addition Law

The ***addition law*** states that *if two (or more) events are mutually exclusive, the probability that one or another will occur is the sum of their separate probabilities.* Events are mutually exclusive if they cannot occur together. The addition law is sometimes called the *either-or law* because it is used to determine the probability that *either* event A *or* event B *or* event C . . . will occur.

Let us illustrate. We know that in tossing a die, $p(6) = 1/6$, $p(5) = 1/6$, and so on. We also know that these outcomes are mutually exclusive; for example, a die cannot come up 5 and 6 at the same time. What is the probability of getting either a 4, a 5, or a 6 in tossing one die?

$$p(4 \text{ or } 5 \text{ or } 6) = p(4) + p(5) + p(6) = \frac{1}{6} + \frac{1}{6} + \frac{1}{6} = \frac{1}{2} = .5$$

Notice that the probability of getting either a 1 or a 2 or a 3 or a 4 or a 5 or a 6 in tossing a die is $1/6 + 1/6 + 1/6 + 1/6 + 1/6 + 1/6 = 6/6 = 1.00$. This makes sense: Because 1 or 2 or 3 or 4 or 5 or 6 covers all possible outcomes, one of them is certain to occur, and the probability of an event that is certain to occur is 1.0.

Let us consider another example of the use of the addition law of probability. Suppose that in a population of IQ scores, only 2% of the scores are below 70 and 3% are above 130. If we select one member of the population at random, what is the probability of an IQ score less than 70 or greater than 130? Remember that a relative frequency is a probability. Therefore, $p(\text{IQ} < 70) = .02$ and $p(\text{IQ} > 130) = .03$. Because no one can have an IQ score that is both less than 70 and greater than 130, these two outcomes are mutually exclusive. Then, by the addition law

$$p(\text{IQ} < 70 \textit{ or } \text{IQ} > 130) = p(\text{IQ} < 70) + p(\text{IQ} > 130) = .02 + .03 = .05$$

The Multiplication Law

The ***multiplication law*** states that *if two or more events are independent, the probability that all of them will occur is the product of their separate probabilities.* Two events are independent if the outcome of one in no way affects the outcome of the other. This law is sometimes called the *both-and* law because it is used to determine the probability that *both* event A *and* event B *and* event C . . . will occur.

Here are some examples:

1. In tossing a die, $p(6) = 1/6$ and $p(5) = 1/6$. If I toss two dice, what is the probability of a 6 on the first and a 5 on the second?

   Because the outcome of one die does not affect the outcome of another die, the two outcomes are independent. Therefore, p(6 on the first die and 5 on the second) $= p(6 \text{ on first die}) \times p(5 \text{ on second die}) = (1/6)(1/6) = 1/36$.
2. If we randomly and independently select three children from a population in which $p(\text{IQ} > 130) = .03$, the probability that all three children have $\text{IQ} > 130$ is $(.03)(.03)(.03) = .000027$.
3. In selecting a card from a well-shuffled, standard deck, $p(\text{king}) = 4/52 = 1/13$; $p(\text{spade}) = 13/52 = 1/4$. The suit of a card is independent of its numerical value. Therefore, the probability of drawing a card that is both a king and a spade is $(1/13)(1/4) = 1/52$.

Sometimes we want to find out whether or not two events are independent. If we know the probability of each event, as well as the probability of their joint occurrence, we can use the multiplication law to check for their independence. Here is an example:

> In a population of people over 70 years of age, 35% have had a heart attack, 15% have had a stroke, and 10% have had both a heart attack and a stroke. Are heart attacks and strokes independent in the population?
>
> $p(\text{heart attack}) = .35$, and $p(\text{stroke}) = .15$. If heart attack and stroke were independent, $p(\text{heart attack and stroke})$ would be $(.35)(.15) = .0525$. However, $p(\text{both heart attack and stroke}) = .10$. Therefore, heart attack and stroke are not independent in the population.

EXERCISE

1. Of 2000 students at a college, 700 are freshmen, 550 are sophomores, 350 are juniors, and 400 are seniors. Half the students are male and half are female. Assume that gender and class are independent.
   a. If you select one student at random, what is the probability of selecting a junior or a senior?
   b. What is the probability of selecting a female freshman? (*Hint:* "Female freshman" means both a female and a freshman.)
2. My white rat, Samuel, turned right in a T-maze six times in succession. If he has no left-right preference, what is the probability that he will do this? If I observe him doing this, does it make sense for me to conclude that he has no right-left preference?
3. The partially completed contingency table here shows the marginal totals for two variables measured in 5000 male college students. Find the number of students in each cell if fraternity membership and smoking behavior are independent.

		Smoking Behavior		
		Smoker	*Nonsmoker*	
Fraternity Membership	*Member*			1500
	Nonmember			3500
		1000	4000	

(*A hint to get you started:* Use the marginal totals to find the probabilities of the outcomes on each variable. For example, 1000 of the 5000 students are smokers, so the probability of smoking is 1000/5000 = .2.)

Sampling With and Without Replacement

Samples, whether random or not, may be selected from a population either with replacement or without replacement. In ***sampling with replacement***, after a given observation has been chosen and recorded, it is returned to the population and is available for sampling again. In ***sampling without replacement***, once a given observation has been selected and recorded, it is discarded. It is not returned to the population and is not available for sampling again.

The distinction between sampling with and without replacement is important, because the probabilities of various outcomes may differ under the two methods of sampling. To demonstrate, we consider sampling from a standard deck of cards.

To calculate the probability of drawing any given card at random, we use the theoretical definition of probability. For example

$$p\,(\text{ace}) = \frac{\text{Number of aces}}{52} = \frac{4}{52} = .077$$

$$p\,(\text{heart}) = \frac{\text{Number of hearts}}{52} = \frac{13}{52} = .25$$

Consider what happens if we draw two cards from the deck, one at a time. There are two ways we can do this. Sampling *with* replacement, we draw a card, record its value, put it back, shuffle the deck, and draw another. Sampling *without* replacement, we draw a card, then draw a second without replacing the first. In the first case, the population (the deck) remains unchanged, and the probabilities of various outcomes on the second draw are unaffected by the results of the first draw. In other words, the results of the first and second draws are independent. But in sampling without replacement, the size and composition of the population change after the first draw. As a result, the probabilities of various outcomes on the second draw are different from the first. In other words, the results of the first and second draws are not independent.

Let us calculate the probability of selecting a heart on the second draw, sampling first with and then without replacement. *With* replacement, the card drawn first has been replaced in the deck. The deck once again contains 52 cards, 13 of which are hearts, so that, just as on the first draw, $p(\text{heart on the second draw}) = \frac{13}{52} = .25$. But if the card drawn first has not been replaced, only 51 cards remain in the deck for the second draw. And the number of hearts depends on whether or not a heart was drawn first. If the first card was a heart, only 12 hearts remain in the deck, and

$$p\,(\text{heart on the second draw}) = \frac{12}{51} = .235$$

But if the first card was not a heart, 13 hearts remain in the deck, and

$$p\,(\text{heart on the second draw}) = \frac{13}{51} = .255$$

In either case, the probability of a heart on the second draw is different from the probability of a heart from a full deck.

As you can see, in sampling without replacement, the probability of any outcome on a given draw depends on the results of previous draws, and calculation of probabilities can rapidly become complicated. For example, on the third draw, when only 50 cards are left in the deck:

$$p\,(\text{heart}) \begin{cases} = \dfrac{11}{50} & \textit{if 2 hearts were previously drawn} \\ = \dfrac{12}{50} & \textit{if 1 heart was previously drawn} \\ = \dfrac{13}{50} & \textit{if no heart was previously drawn} \end{cases}$$

Most of the inferential statistical techniques discussed in later chapters assume that samples have been selected with replacement. In practice, though, researchers generally sample without replacement. How, then, do researchers justify the use of inferential procedures that assume sampling with replacement?

Their use is justified by the fact that social and behavioral scientists usually sample from extremely large populations. If one is drawing a sample of 20 or 150 or even 1000 from a population of 4,000,000, it probably makes little difference whether or not selected observations are returned to the population before selecting the next observation. The probability that the same observation will be selected again is minute, even if it has been returned. And the population from which one is sampling is not significantly depleted, even if observations, once selected, are not returned. Despite the fact that sampling is without replacement, probabilities for selections after the first differ very little from what they would have been if sampling had been with replacement.

Therefore, provided that the sample size is small relative to the size of the population, inferential procedures that assume sampling with replacement describe outcomes of sampling without replacement quite accurately. In cases in which samples are selected without replacement from small populations, however, certain adjustments to the standard inferential procedures are needed.

EXERCISE 4–5

1. Suppose that you are sampling from a population of four men and six women.
   a. If you randomly select one member of the population, what is the probability of selecting a man?
   b. If you sample *with* replacement and your first two selections are both women, what is the probability that the third selection is a man?
   c. If you sample *without* replacement and your first two selections are both women, what is the probability that the third selection is a man?
   d. If you use your sample of size 3 to generalize to the population, do you feel that the use of inferential procedures that assume sampling with replacement is justified?
2. A population of size 5000 includes 2000 men and 3000 women.
   a. If you randomly select one member of the population, what is the probability of selecting a man?
   b. If you sample *with* replacement and your first two selections are both women, what is the probability that the third selection is a man?
   c. If you sample *without* replacement, and your first two selections are both women, what is the probability that the third selection is a man?
   d. If you use your sample of size 3 to generalize to the population, do you feel that the use of inferential procedures that assume sampling with replacement is justified?

Probability Distributions

The kinds of events we have been talking about—tossing coins or dice, selecting cards from a deck, running rats in a T-maze, sampling IQs—have two or more possible outcomes. If the outcomes are mutually exclusive, and if we list all the possible outcomes along with the probability of each one, the result is called a ***probability distribution***. Some probability distributions are shown in Table 4–6. Notice that the sum of the probabilities in each probability distribution is 1. This is true for all probability distributions.

Distributions A, C, and F of Table 4–6 are called *empirical probability distributions* because the probabilities are relative frequencies that have been established by experience. For example, the probability distribution of grades in statistics is presumably based on previous experience with classes in statistics, in which 30% of the students received an A, 40% a B, and so on. Distributions B, D, and E are *theoretical probability distributions* because the probabilities are based on logical reasoning rather than actual experience. As you will see in later chapters, theoretical probability distributions are of great importance in inferential statistics.

Let us construct a somewhat more complex probability distribution, that for the outcomes of tossing two fair dice. This will reveal some pitfalls to be avoided in determining probabilities and the importance of listing all possible outcomes when constructing a probability distribution.

You may be tempted to say: "Well, in tossing two dice one can get any number from 2 to 12. There are 11 possible outcomes, so the probability distribution is as follows":

Table 4–6 Some Probability Distributions

A.

IQ	p
IQ ≥ 120	.10
80 < IQ < 120	.80
IQ ≤ 80	.10

B.

Outcome of coin toss	p
Head	.5
Tail	.5

C.

Sex of baby	p
Male	.52
Female	.48

D.

Outcome of card selection	p
Spade	.25
Heart	.25
Diamond	.25
Club	.25

E.

Result of die toss	p
6	1/6
5	1/6
4	1/6
3	1/6
2	1/6
1	1/6

F.

Grade in statistics	p
A	.30
B	.40
C	.20
D	.08
F	.02

Outcome	p
12	1/11
11	1/11
10	1/11
9	1/11
8	1/11
7	1/11
6	1/11
5	1/11
4	1/11
3	1/11
2	1/11

But this is *wrong*, because, as any gambler knows, there are more ways to get some outcomes in tossing two dice (e.g., 7) than others (e.g., 12). We must begin by listing *all* the possible ways two dice may fall. Table 4–7 lists all possible outcomes and the sum of the numbers appearing on the two dice in each case.

As you can see, there are not 11 but 36 possible outcomes in tossing two dice. If the dice are fair, all the outcomes are equally likely. To determine the probability of a given sum over the two dice, we use the theoretical definition of probability:

$$p\,(\text{event}) = \frac{\text{Number of ways this event can occur}}{\text{Total number of possible outcomes}}$$

There is only one way to get a 12 (a 6 on each die—outcome 36 in the table). Therefore

$$p\,(12) = \frac{1}{36} = .0278$$

There are two ways to get 11 when tossing two dice: a 5 on the first die and a 6 on the second, or a 6 on the first die and a 5 on the second (outcomes 30 and 35 in the table). Therefore:

$$p\,(11) = \frac{2}{36} = .0556$$

You can confirm the remaining probabilities yourself:

$$p\,(10) = \frac{3}{36} = .0833$$

$$p\,(9) = \frac{4}{36} = .1111$$

$$p\,(8) = \frac{5}{36} = .1389$$

$$p\,(7) = \frac{6}{36} = .1667$$

$$p\,(6) = \frac{5}{36} = .1389$$

$$p\,(5) = \frac{4}{36} = .1111$$

$$p\,(4) = \frac{3}{36} = .0833$$

$$p\,(3) = \frac{2}{36} = .0556$$

$$p\,(2) = \frac{1}{36} = .0278$$

Note that, as in any probability distribution, the sum of the probabilities is equal to one.

The probability distribution of the dice tosses is graphed in Figure 4–8. Gaps have been left between the bars because the variable being measured is discrete. (One can get 5 when tossing two dice, or 6 or 7, but not 5½ or 6⅔.) This is an example of a ***discrete probability distribution***. Many of the probability distributions used in inferential statistics are continuous probability distributions. We will have more to say about continuous

Table 4–7 All Possible Outcomes When Tossing Two Dice

Outcome	First Die	Second Die	Sum	Outcome	First Die	Second Die	Sum
1	1	1	2	19	4	1	5
2	1	2	3	20	4	2	6
3	1	3	4	21	4	3	7
4	1	4	5	22	4	4	8
5	1	5	6	23	4	5	9
6	1	6	7	24	4	6	10
7	2	1	3	25	5	1	6
8	2	2	4	26	5	2	7
9	2	3	5	27	5	3	8
10	2	4	6	28	5	4	9
11	2	5	7	29	5	5	10
12	2	6	8	30	5	6	11
13	3	1	4	31	6	1	7
14	3	2	5	32	6	2	8
15	3	3	6	33	6	3	9
16	3	4	7	34	6	4	10
17	3	5	8	35	6	5	11
18	3	6	9	36	6	6	12

Figure 4–8

Probability distribution of outcomes of tosses of two fair dice.

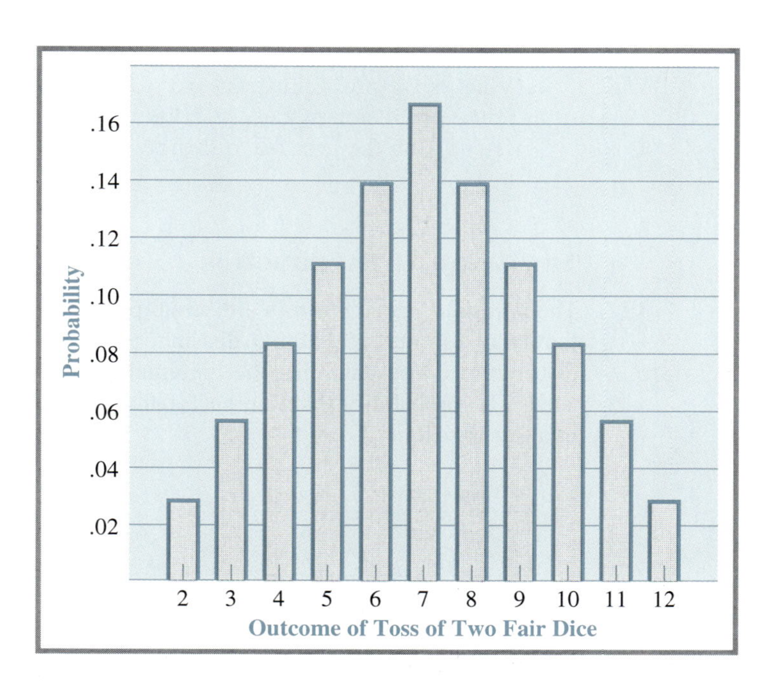

probability distributions later, after we have considered one important discrete probability distribution—the binomial distribution.

EXERCISE

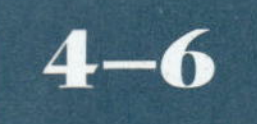

4–6

1. Suppose that you toss four fair coins.
   a. List all the possible outcomes (be sure you don't omit any).
   b. Construct a probability distribution, showing p(4 heads), p(3 heads), and so on.
   c. Find the probability of getting either 3 or 4 heads.
2. Suppose, instead, that you run a rat in a T-maze for four trials. Assuming the rat has no left or right preference, what is the probability that he will turn right on three or four trials?

   HINT: You do not need to construct a new probability distribution. Use the one you constructed for the tosses of four fair coins.

3. The table here shows a sampling distribution of the mode for samples of size 3 from a certain population. (Note that a sampling distribution that lists all possible values of the statistic along with the probability of each is a probability distribution.)

Sample Mode	p
8	.05
7	.12
6	.17
5	.24
4	.20
3	.13
2	.09

   You select one sample of size 3 from the population.
   a. What is the probability that the sample mode is 4, 5, or 6?
   b. If the population mode is 5, what is the probability that the sample mode is different from the population mode?

The Binomial Distribution

The ***binomial distribution*** is a discrete probability distribution. More accurately, it is a family of discrete probability distributions, because there are many specific binomial distributions. You have met the binomial distribution earlier, although the term wasn't used. The probability distribution for the outcomes of tossing one fair coin is a binomial distribution:

Outcome	p
Heads	.5
Tails	.5

So is the distribution of outcomes when tossing four fair coins, which you may have constructed in Exercise 4–6:

Number of Heads in Tossing 4 Fair Coins	*p*
4	.0625
3	.2500
2	.3750
1	.2500
0	.0625

As you may recall, the same distribution describes the probability of 0, 1, . . . , 4 right turns in a T-maze, assuming a rat has no right or left preference.

At this point, you are probably not surprised to learn that a binomial distribution describes the outcomes of any number of coin tosses or T-maze trials. In fact, a binomial distribution is generated whenever the following conditions are met:

1. *An elementary event exists with two possible outcomes.* An elementary event is a single simple action, such as the toss of one coin, one trial in a T-maze, or the like. A coin toss or a T-maze turn has two possible outcomes.

   If more than two outcomes are possible for the elementary event, the binomial distribution does not apply. For example, a single toss of a die has six possible outcomes; therefore the binomial distribution does not apply to dice-tossing experiments.
2. *The probability of the outcome of interest (the elementary probability) is known.* For example, the probability of a head in tossing a fair coin is .5, as is the probability of a right turn in a T-maze for a rat with no preference.

   The elementary probability need not be .5. The binomial distribution applies equally well when the probability is .15, or 90, or anything else, as long as the elementary probability is known.
3. *The elementary event is repeated one or more times.* For example, a fair coin is tossed 3, or 10, or 50 times; a rat in a T-maze has 3, or 20, or 100 trials.

   The repetitions can be over individuals as well as over time. For example, the outcomes and their probabilities when tossing five coins once are the same as those for tossing one coin five times.

Each combination of a specific elementary probability and a specific number of repetitions generates a particular binomial distribution. In earlier sections of this chapter, we generated several binomial distributions by listing all possible outcomes and counting the outcomes of interest. This is impractical or impossible when the elementary probability is not .5 and/or when there are many repetitions of the elementary event. For example, try listing all the possible outcomes of tossing 10 coins. It is not impossible to do so, but it will take you a long time: There are 1024 possible outcomes!

One way to generate binomial distributions for any combination of an elementary probability and number of repetitions is by using an algebraic procedure called the *binomial expansion.* Another way is to use tables prepared by statisticians. The tables are lengthy and are not included in this book. Instead, we will consider just four binomial distributions, those shown in Figure 4–9. (But remember that there are many additional binomial distributions besides the ones shown in the figure.) In each of the illustrated distributions, the letter X is used to represent the outcome of interest (e.g., the number of heads or the number of right turns).

Binomial distribution A is that for elementary probability = .5 and number of repetitions = 5. Thus, if you toss a fair coin five times, Distribution A can be used to find

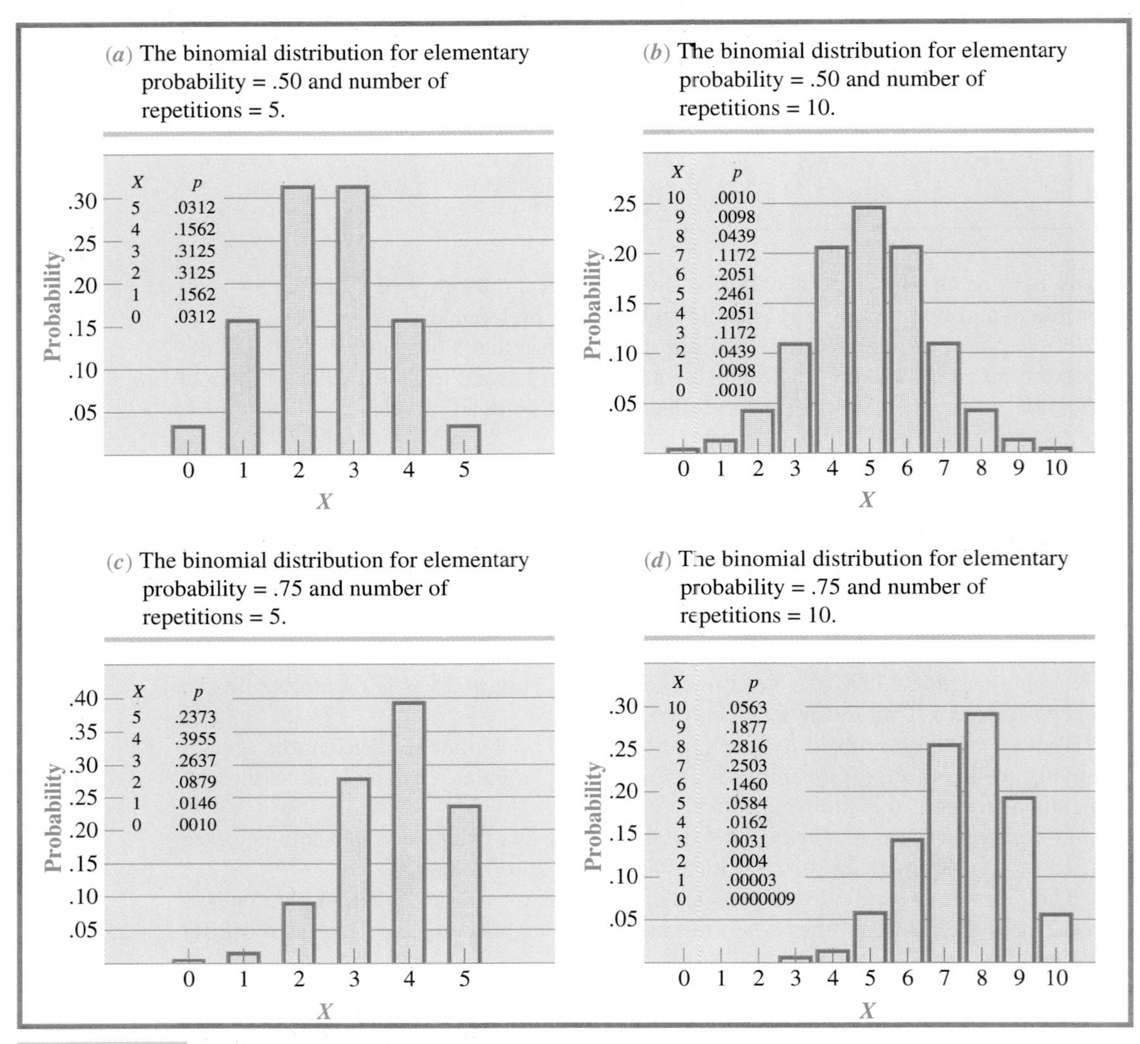

Figure 4–9 **Four binomial distributions.**

the probability of X heads, where X is any whole number between 0 and 5. For example, $p(4 \text{ heads}) = .1562$. As you can see, the events with the highest probabilities are $X = 2$ and 3. Probabilities drop off gradually on either side of $X = 2$ and 3. Notice that the distribution is symmetrical: $p(2) = p(3)$; $p(1) = p(4)$; and $p(0) = p(5)$.

Distribution B is the binomial distribution for elementary probability = .5 and number of repetitions = 10. If, for example, you toss a fair coin 10 times, Distribution B gives the probability of each possible number of heads. As you can see, the most likely outcome is five heads with a probability of .2461. You certainly would not be surprised to get four or six heads; even three or seven heads is not highly improbable. As the number of heads approaches 0 or 10, however, probabilities become very small. As in Distribution A, the distribution is symmetrical about the central value of X, in this case, 5. An elementary probability of .5 always results in a symmetrical binomial distribution, for any number of repetitions.

Distribution C is a binomial distribution in which the elementary probability is not .5 but .75. The number of repetitions is 5. For example, suppose that you are tossing a coin that is biased to come up heads; specifically, on each toss, $p(\text{head}) = .75$ and $p(\text{tail}) = .25$. As you can see, if you toss the coin five times the single most likely outcome is four heads rather than two or three. Though probabilities decrease on either side of $X = 4$, the distribution is not symmetrical but negatively skewed.

Distribution D is the binomial distribution for elementary probability = .75 and number of repetitions = 10. Suppose that you toss a coin with $p(\text{head}) = .75$, ten times instead of five. The most likely outcomes are seven and eight heads. Even as many as nine or ten heads are not too unlikely, but the probability of less than four heads is vanishingly small. As with five repetitions (Distribution C), the distribution is negatively skewed; however, the degree of skew is less for ten repetitions than for 5.

Here are more examples of the use of the binomial distributions in Figure 4–9.

EXAMPLE 1: My rat has a left-turning preference. In fact, his $p(\text{left turn}) = .75$. If he runs five trials in a T-maze, what is the probability that he will turn left three times?

ANSWER: The elementary probability is .75, and there are five repetitions (trials). Therefore Distribution C applies. We need $p(X = 3)$, which is .2637. The probability that he will turn left on three of the five trials is .2637.

EXAMPLE 2: If I toss a fair coin five times, what is the probability that I will get three or more heads?

ANSWER: Since the coin is fair, the elementary probability of a head on one toss is .5. The number of repetitions is five. Distribution A applies, and gives $p(5\text{ heads})$, $p(4\text{ heads})$, and so on. I want to know the probability of *three or more* heads, that is, the probability of three *or* four *or* five heads. I am asking an either-or question, so I must use the addition law:

$$p(3\text{ or }4\text{ or }5\text{ heads}) = p(3\text{ heads}) + p(4\text{ heads}) + p(5\text{ heads})$$
$$= .3125 + .1562 + .0312 = .4999... \text{ or } .5$$

EXAMPLE 3: Assume that children have an equal preference for playing with pinball machines and pianos. If I give 10 children a choice of playing with a pinball machine or a piano, what is the probability that 9 or more of them will choose the pinball machine?

ANSWER: If children have an equal preference, the elementary probability that any one child will choose the pinball machine is .5. Since 10 children are involved, the number of repetitions is 10. Distribution B gives the probability that some number (X) out of 10 children will choose the pinball machine if children have no preference. We need $p(9\text{ or more})$, which means $p(9\text{ or }10)$. According to the table, $p(9) = .0098$ and $p(10) = .0010$. Therefore $p(9\text{ or }10) = p(9) + p(10) = .0098 + .0010 = .0108$. If children have no preference, the probability that 9 or more of 10 children will choose the pinball machine is .0108.

EXAMPLE 4: Again, consider children's preferences for pinball machines versus pianos. Suppose that I do not know whether children have a preference for one or the other. To find out, I give 10 children a choice, and 9 of them choose the pinball machine. Is it reasonable for me to conclude that children have an equal preference for pinball machines and pianos?

ANSWER: In Example 3, we found that if children have an equal preference, the probability that this many children (9 or more) will pick the pinball machine is .0108 (only about one chance in 100). This is a rather unlikely outcome if children have an equal preference. Therefore it seems more reasonable to conclude that children do not have an equal preference for pinball machines and pianos.

Note the reasoning in Example 4. We decided that children do not have an equal preference because, if they do, the outcome we observed would be very unlikely to occur. This is the reasoning that is employed in a method of statistical inference called hypothesis testing. In ***hypothesis testing***, we

1. Hypothesize that a certain state of affairs exists in a population.
2. Select a sample from the population and observe the sample outcome.
3. Find the probability of a sample outcome like this if the hypothesis is true.
4. If the probability is very small, reject the hypothesis.

We discuss hypothesis testing more fully in Chapter 5.

EXERCISE 4–7

1. Your friend has given you a penny that you suspect is biased to turn up tails. You toss it five times, and it comes up tails every time. Do you think it is biased?

   HINT: To answer this, assume it is a fair coin, and find the probability of getting five tails in five tosses if the coin is fair.

2. If children have an equal preference for pinball machines and pianos, what is the probability that, of 10 children, fewer than 3 children will choose to play the piano? What is the probability that fewer than 2 or more than 8 children will choose the piano?

AN INTRODUCTION TO CONTINUOUS PROBABILITY DISTRIBUTIONS

Look at the binomial distribution (B) for elementary probability = .5 and number of repetitions = 10 in Figure 4–9. Imagine a continuous curved line touching the tops of all 11 bars and approaching the abscissa at each end. Doesn't this curve look something like a normal curve? Now imagine what the binomial distribution for elementary probability = .5 and number of repetitions = 50 would look like. It would include 51 closely spaced vertical bars, one for each integer value of X between 0 and 50. A curve connecting the tops of the 51 bars would look even more like a normal curve. In fact, as the number of repetitions increases, the binomial distribution approaches the normal distribution. (Of course, it never actually becomes normal, because the binomial is a discrete distribution whereas the normal is continuous.)

Most of the theoretical probability distributions that underlie statistical inference, including the normal distribution, are ***continuous probability distributions***. Like the binomial, they are used to answer questions concerning the probabilities of various sampling outcomes. Without discussing any one distribution in detail at this point, let us consider how probabilities of various events are found in a continuous probability distribution, and how the method differs from that used with a discrete distribution.

In the probability distribution for a discrete variable, it is possible to list all possible outcomes and find the probability of each one. This is not possible with a continuous variable because the variable can take an infinite number of values between any two points. For example, suppose someone asks, "What is the probability of an outcome between 6 and 8?" If the variable is discrete, we find $p(6)$, $p(7)$, and $p(8)$ and add the probabilities. We need not worry about values between 6 and 7 or between 7 or 8 because there aren't any. On the other hand, if the variable is continuous there are countless outcomes between 6 and 7 and between 7 and 8. We can't list them all, find their probabilities and add; some other method must be used to find the desired probability.

Let us consider the problem graphically. Figure 4–10 displays two hypothetical probability distributions, one discrete and one continuous. In the discrete distribution, the probability of any given outcome, such as between 6 and 8, is the sum of the heights of the respective bars. What represents the probability of a similar outcome in the continuous distribution? Recall that near the end of Chapter 2 you learned that the proportion of the area within any section of a graphed distribution is equal to the proportion of observations in that part of the distribution. A probability is a proportion. Specifically, the proportion of observations in a part of a distribution is equal to the probability that a single, randomly selected observation falls in that part of the distribution. Therefore, we may write: *In a continuous probability distribution, the proportion of the area within any section of the graphed distribution is equal to the probability of an outcome within that section.*

In the continuous probability distribution of Figure 4–10, the area between the outcomes 6 and 8 is shaded. Let us say that this area is .25 of the total area under the curve. This means that the probability of an outcome between 6 and 8 is .25. In the same way it would be possible to find the probability of any other outcome by finding the corresponding area.

The calculation of areas within continuous probability distributions requires calculus. Fortunately, there are relatively few continuous probability distributions that are important for statistical inference. Areas in various parts of these distributions have been computed and put in tables. Thus, there are tables giving probabilities of various sam-

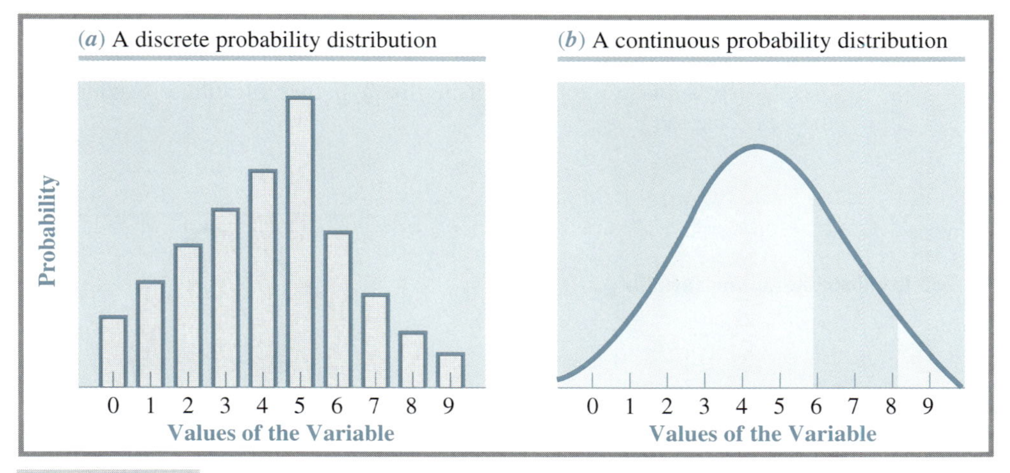

Figure 4–10 **Comparison of discrete (a) and continuous (b) probability distributions.**

pling outcomes in normal distributions as well as in others. In Chapter 5, you will learn to conduct statistical tests called *chi square* (χ^2) tests, because they utilize a continuous probability distribution called the *chi square distribution*, and you will use a chi square table to find probabilities you need for your tests. In later chapters you will learn to use tables of other continuous probability distributions, including the normal distribution.

EXERCISE 4–8

A continuous probability distribution of a variable, X, is graphed in Figure 4–11. The number in each section of the distribution is the proportion of area in that section. Use the graph to answer the following questions.

1. Find the following probabilities:
   a. $p(95 < X < 100)$
   b. $p(X > 110)$
   c. $p(95 < X < 105)$
   d. $p(X < 95)$
   e. $p(X < 80 \text{ or } X > 110)$
2. What is the value of X such that $p(X > \text{that value}) = .15$?

Sampling Distributions and Probability Distributions

We began this chapter by discussing sampling distributions—distributions of a statistic over many (or all possible) samples from a population—and we ended with a discussion of probability distributions. Both sampling distributions and probability distributions give information about various sampling outcomes. What is the relation between them?

Basically, if the probability of each value of a statistic can be determined, the sampling distribution of the statistic becomes a probability distribution. Fortunately, many sample statistics have (or can be converted into) known probability distributions. For example, under certain circumstances, sample means are approximately normally distributed, and their probabilities can be found by referring to a probability distribution called the *standard normal distribution*. How this is done is explained in Chapter 7. Similarly, in samples of categorical data, sample data can often be converted into a statistic called *chi square* (χ^2). The probability distribution of χ^2 is known and can be used to find the probability of various sample distributions of categorical data. The use of the chi square distribution to make inferences, using the hypothesis-testing procedure, is the subject of the next chapter.

Figure 4–11

Probability distribution for variable X.

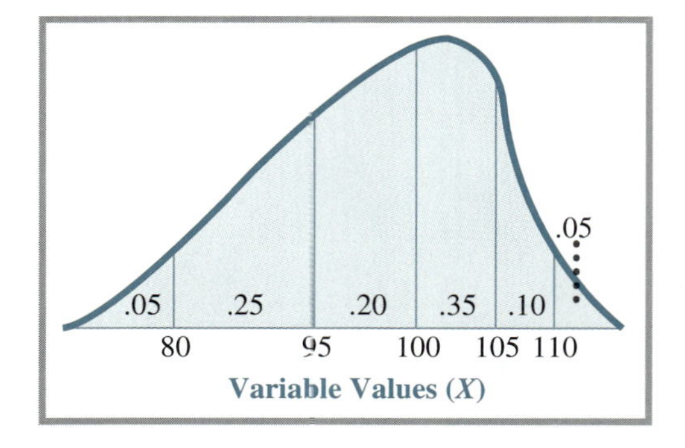

SUMMARY

In order to answer questions about ***populations***, researchers take ***samples*** of the populations and use data collected in the samples to generalize to the populations. If all members of a population are equally likely to be selected for a sample, the sample is ***random***; samples in which not all members have an equal chance of being selected are ***biased***. Inferences from biased samples are likely to be inaccurate.

Numerical characteristics of populations are called ***parameters***. ***Statistics*** calculated in samples are used to make inferences about the corresponding parameters. Values of a statistic vary from one sample to another. It is possible to examine the variation of a statistic over samples by constructing a ***sampling distribution*** of the statistic.

Empirical sampling distributions of the proportion and the mean were constructed, based on small (N = 4 or 5) and large (N = 20) samples from two populations. It was shown that (1) sample proportions vary around the population proportion, and sample means vary around the population mean; and (2) as the sample size (N) increases, the amount of variation of sample statistics decreases. Thus, statistics calculated in large samples are more often close to the actual population parameters.

The ***probability*** of an event is its relative frequency of occurrence (empirical definition) or the proportion of all possible, equally likely outcomes that favor the event (theoretical definition). Probabilities range from 0 for an impossible event to 1 for an event that is certain to occur. The probability that an event will not occur is 1 minus the probability of the event. The probability that either of two (or more) mutually exclusive events will occur is the sum of their probabilities (***addition law***). The probability that both (or all) of several independent events will occur is the product of their probabilities (***multiplication law***).

Samples may be selected from a population either ***with replacement*** or ***without replacement***. In sampling without replacement, probabilities of events after the first change, depending on prior selections. Most procedures used in statistical inference assume sampling with replacement. The procedures can be applied to cases of sampling without replacement, provided that the population size is large relative to the size of the sample.

A distribution of the probabilities of all possible, mutually exclusive outcomes of an event is called a ***probability distribution***. The ***binomial distribution*** is a discrete probability distribution for situations in which an elementary event with two possible outcomes is repeated one or more times. In discrete probability distributions, the probability of an outcome between two values is found by adding the probabilities of all inclusive outcomes. In continuous probability distributions, the probability of an outcome between two values is the proportion of area between those values in the graphed distribution.

The relation between a sampling distribution of a statistic and a probability distribution is this: If the probability of values of the statistic can be determined, the sampling distribution becomes, in effect, a probability distribution. A probability distribution for certain kinds of categorical data is discussed in Chapter 5; probability distributions for means and their use are the subject of later chapters.

QUESTIONS

1. From a population, I select 100 samples of size 25. In each sample, I calculate a statistic called the *blurp*. Then I examine the distribution of sample blurps. What is the name of this distribution?

2. Suppose that, instead of selecting 100 samples of size 25, I select 100 samples of size 10. If blurps behave like proportions and means, how does the distribution of blurps for samples of size 10 differ from that for samples of size 25?

3. In a population of 5000 hospital patients, 2250 have blood type O, 2050 have blood type A, 500 have blood type B, and 200

have blood type AB. If I randomly select one member of the population

a. What is the probability of a patient with blood type A?
b. What is the probability that the patient does *not* have blood type A?
c. What is the probability of a patient with blood type AB?
d. What is the probability of a patient with either blood type A or AB?
e. What is the probability of a patient with either blood type A, B, AB, or O?

4. a. List all the possible outcomes in tossing three fair coins.
b. Construct a probability distribution showing the probability of 0, 1, 2, and 3 tails.
c. Use the probability distribution to find the probability of two or three tails in tossing three fair coins.
d. If you toss three fair coins twice, what is the probability that you will get three tails on both tosses?

5. You have been given a special pair of dice with four faces on each one. The faces are numbered 0, 1, 2, and 3, and each one is equally likely to come up when the die is tossed. Construct a probability distribution showing the probability of getting a sum of 0, 1, 2, . . . , 6 when tossing the two dice.

6. Here are the probabilities of catching five types of fish at Lake Minnikatawalappowwow:

Type of Fish	Probability
Bass	.20
Trout	.15
Catfish	.30
Sunfish	.25
Carp	.10

a. If you catch one fish, what is the probability that it is a carp or a bass?
b. If you catch one fish, what is the probability that it is not a catfish?
c. You catch one fish and toss it back, then catch another. What is the probability that both are trout?
d. You catch a fish and toss it back, then catch another. What is the probability that the first is a sunfish and the second is a catfish?

7. A public affairs club has 12 members. Eight are Democrats and four are Republicans. Three of the members are under 25 years of age, three are between 25 and 35 years of age, and six are over 35. Assume that age and political affiliation are independent.

a. If you select one member at random, what is the probability that she is 25 years old or older?
b. If you select one member at random, what is the probability of a Democrat under 25 years of age?
c. How many members are Republicans over 35 years of age?
d. If you select two members with replacement, what is the probability that both are Republicans?
e. You select two members without replacement. What is the probability that the *second* one will be over 35 if (1) the first one was 35 or under? (2) the first one was over 35?

8. Use the binomial distributions in Figure 4–9 to answer the following questions:

a. A student takes a 10-item, true-false test for which he has not studied. He guesses on every question. What is the probability that he will score 7 or more correct?
b. If the test were a 10-item multiple-choice test, with four alternatives for each item, what is the probability of 7 or more correct by guessing alone? (*Hint:* Rephrase the question: What is the probability of 3 or less *incorrect* by guessing?)
c. My white rat Samuel turned left in a T-maze for five trials in a row. Is it reasonable for me to conclude that he has no preference for turning right or left? (To answer this, find the probability of five left turns for a rat with no preference.)
d. Suppose that, in fact, the probability that Samuel will turn left on any given trial is .75. If so, what is the probability of five consecutive left turns?

Figure 4–12

A probability distribution for a continuous variable.

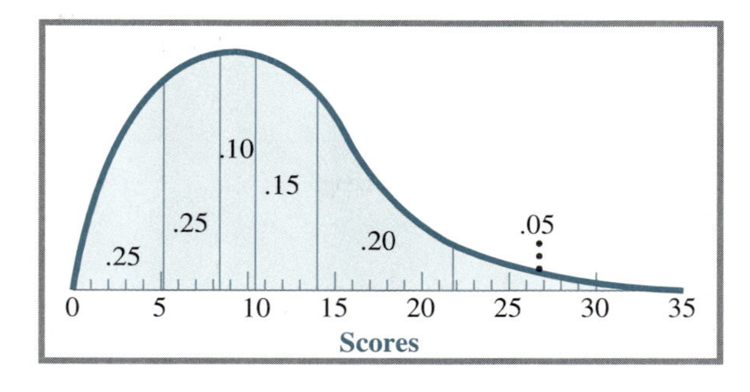

9. Figure 4–12 shows a probability distribution for a continuous variable. Values of the variable (scores) are on the abscissa. The numbers within the graph are proportions of area.

   a. Answer without adding: What is the sum of the numbers within sections of the graph?
   b. What is the probability of
      (1) a score between 5 and 10?
      (2) a score above 22?
      (3) a score below 8?
      (4) a score either below 5 or above 22?
   c. The probability is .40 that a score will fall above what value?

5 An Introduction to Hypothesis Testing: The Chi Square Test

INTRODUCTION

Suppose that Dr. Klingelfinger, a personality theorist, claims that 40% of all mental hospital patients are suffering from thought disorders, 30% from affective disorders, 15% from character disorders, 10% from neuroses, and 5% from "other" or unclassifiable disorders. Of the 200 patients at the Pennsylvania Midwestern Psychiatric Institution (PMPI), 100 (50%) have been classified as thought disorder, 60 (30%) as affective disorder, 20 (10%) as character disorder, 10 (5%) as neuroses, and 10 (5%) unclassified. Do the PMPI data contradict Dr. Klingelfinger's claim?

One week before the mayoral election in Tooterville, both candidates, Smith and Jones, are claiming a decisive victory. A reporter for the *Tooterville Gazette* asks 200 randomly selected citizens for whom they plan to vote; 113 reply Smith and 87, Jones. The reporter tells his editor, "Smith will win the election." Should the editor print this in the newspaper?

A question of interest to both psychologists and parents is "Do boys and girls differ in their preferences for warlike and nonwarlike toys?" In a study, 34 preschool girls and 37 preschool boys are given a choice between playing with a set of toy guns, tanks, soldiers, and other warlike toys or playing with a train and helicopter set. Twenty-five of the girls and 17 of the boys choose the warlike toys. Do the choices of boys differ from those of girls?

These three examples have a number of things in common. All three involve categorical data; that is, each observation falls into one of two or more categories. And, in all cases, obtained distributions of observations across categories are to be compared with certain claims (e.g., Dr. Klingelfinger's claim) or between two or more groups (e.g., between boys and girls). The statistical procedure that is usually applied to test the claims or compare the groups is called the chi square (χ^2) test because it utilizes a probability distribution called the chi square distribution. In this chapter you will learn not only how to use the chi square test to test certain kinds of hypotheses about categorical data but also what hypothesis testing in general is about. In later chapters you will encounter many additional tests, but all use the same basic rationale and procedure that you will learn here.

Objectives

When you have completed this chapter, you will be able to

- List four steps in testing a hypothesis, and use them to test real-life (nonstatistical) hypotheses.
- Compare statistical with real-life hypothesis testing.
- Explain the relevance of the chi square distribution to tests of hypotheses about distributions of observations over categories.
- Describe the chi square distribution and compare it with the binomial distribution.
- Define degrees of freedom (*df*) and describe their relevance to the chi square distribution.
- Use the chi square table (Table D–1 in Appendix D) to find probabilities of various values of χ^2 in distributions with various degrees of freedom.
- Explain the meaning of critical value, level of significance, alpha (α), rejection and nonrejection regions, and significant difference.
- Use the chi square distribution to test hypotheses concerning theoretical distributions of frequencies across categories (the chi square test of goodness of fit).
- Define the null hypothesis (H_0), and explain what is meant by rejection and nonrejection of a null hypothesis.
- Use chi square to test null hypotheses about the distribution of observations over categories in a single population.
- Construct and interpret contingency tables for the relationship between two categorical variables.
- State the null hypothesis for tests of independence in terms of a relation between two variables or a difference between two or more groups.
- Conduct chi square tests of independence.
- List conditions under which chi square tests should not be used and recognize examples of these conditions.
- Define Type I errors and Type II errors. Describe factors that affect each one.

What Is Hypothesis Testing All About?

Why Do We Need to Test Hypotheses?

Let us consider Dr. Klingelfinger's claim concerning the distribution of psychiatric patients. "Look," you may say, "Obviously, Dr. Klingelfinger is wrong. He claims that 40%, 30%, 15%, 10%, and 5% of patients fall in the five categories, respectively; but the percentages of patients in the five categories at PMPI are 50%, 30%, 10%, 5%, and 5%, respectively. The figures aren't the same." That's true. But remember our discussion about sampling and probability in Chapter 4. We showed that sampling outcomes are rarely identical with population values. If we sample from a population with proportion of males = .5, many samples have a proportion different from .5. When tossing a fair coin 10 times, we don't usually get exactly 5 heads and 5 tails. And even a rat with no right-left preference doesn't necessarily turn right and left an equal number of times in a T-maze. In other words, sample results seldom come out precisely the same as expected results. PMPI's present patients are just a sample of all psychiatric patients. Even if Dr. Klingelfinger is right about the population of all psychiatric patients, we expect to find some difference between the distribution of patient classifications in the sample and in the population. The question is not "Is PMPI's present patient distribution identical to Dr. Klingelfinger's claimed distribution?" but rather "If Dr. Klingelfinger is right, how probable (or improbable) is a sample distribution like that at PMPI?"

In the case of Tooterville's mayoral election poll, the need for caution is even more evident. It is true that more than half the 200 voters sampled said they would vote for Smith, but a sample of 200 probably represents only a small fraction of the voting population of Tooterville. Perhaps in another sample of 200, the majority would state an intention to vote for Jones. Or perhaps not. What we need to do is decide how confident we can be that Smith will win the election on the basis of the fact that in a sample of 200 voters, 113 people said they would vote for him.

Now consider the comparison of toy preferences of boys and girls. In the group of children studied, a greater percentage of girls (74%) chose warlike toys than boys (46%). But, again, these boys and these girls are only a sample of the boys and girls about whom we seek information. Is the difference between boys and girls in this sample sufficiently great that we can confidently state that preschool boys and girls in general differ in their choice of warlike and nonwarlike toys?

To answer questions like these, we set up *hypotheses* concerning expected results; then we test the hypotheses by applying appropriate statistical procedures to the sample data. Let us begin by considering the general procedure for testing hypotheses.

General Procedure for Hypothesis Testing

In somewhat simplified form, the general procedure for testing any statistical hypothesis is as follows:[†]

1. State the hypothesis to be tested.
2. Collect the data that will be used to test the hypothesis.
3. Determine the probability of data such as these if the hypothesis is true.
4. Decide whether to reject the hypothesis.
   a. If the probability of data such as these is "reasonably large," do not reject the hypothesis.
   b. If the probability of data such as these is "small," reject the hypothesis.

You have probably used a similar procedure for testing hypotheses in real life, as the following fictitious example shows. Let us say that your friends, Sue and Joe Brown, have always seemed to you a happily married couple. But Mrs. Nosie whispers to you over the garden fence, "Have you heard that Sue and Joe are splitting up?" You are skeptical, so you decide to confirm or disconfirm this rumor yourself.

You start out by hypothesizing: "I think Mrs. Nosie is all wet. Joe and Sue are still happily married." Next you collect data to test the hypothesis. That is, you observe Sue and Joe to obtain information pertaining to your hypothesis. You make the following observations: (1) You see Joe driving down the street with Alice, and they look pretty chummy; (2) your mutual friend John tells you that Joe has moved into an apartment; and (3) at a party you attend, Sue arrives with Ted and is heard to say some pretty nasty things about Joe.

Now you compare the data with your hypothesis. You say to yourself: "I hypothesized that Joe and Sue are still happily married. But when people are happily married, they don't usually date other people, move into apartments, or say nasty things about

[†]Some steps are left out here; they will be added later.

one another in public. Observations like these are pretty improbable if my hypothesis is true. Therefore I was probably wrong and Mrs. Nosie was probably right. I will reject my hypothesis that Sue and Joe are happily married."

In schematic form, what you have done is shown in the following:

1. State the hypothesis: Sue and Joe are happily married.
2. Collect data to test the hypothesis:
   Observation (1): Joe is seen with Alice.
   Observation (2): Joe moves into an apartment.
   Observation (3): Sue, with Ted, says nasty things about Joe.
3. Determine the probability of data like these if the hypothesis is true: If Sue and Joe are happily married, Observations (1), (2), and (3) are improbable.
4. Decide whether to reject the hypothesis: Reject the hypothesis that Sue and Joe are happily married.

Now, suppose that your observations (in Step 2) were as follows instead: You see Joe driving with Alice (Observation 1), but Joe does not move into an apartment, and Joe and Sue come to the party together and show no signs of hostility toward one another. In that case, your reasoning in Steps 3 and 4 would probably proceed as follows: "I hypothesized that Sue and Joe are still happily married. When two people are happily married, observations like mine are not unusual. I certainly don't have sufficient evidence to conclude that Sue and Joe aren't happily married. Therefore, I will not reject my hypothesis that they are still happily married."

Testing statistical hypotheses is a little different from the example of Sue and Joe. However, the differences are primarily differences in degree. The basic logic and reasoning are the same.

EXERCISE 5–1

Hypothesis: Elementary Statistics is an extremely difficult subject.

How would you go about testing this hypothesis (H)?
What kinds of data might lead you to reject H?
What kinds of data might lead you not to reject H?
Make up some fictitious observations and go through Steps 1–4.

Differences Between Statistical and Real-Life Hypothesis Testing

There are several differences between testing hypotheses like "Sue and Joe are still happily married" and testing statistical hypotheses. One difference is that statistical hypotheses are always about quantitative characteristics of populations. Thus, we can test a hypothesis about the distribution of observations across categories in a population, or about a population parameter such as the mean, but not about particular people (like Sue and Joe) or objects or events. Another difference is that, in statistical tests, we decide ahead of time, before collecting data, how improbable the sample outcome must be in order to reject the hypothesis. We do this so that we will have a clear standard for deciding whether or not the probability of our sample result is "small enough" to reject the hypothesis.

The necessity for setting a standard for rejection ahead of time can be illustrated in the case of Sue and Joe. Suppose that you had made the following three observations in your test: (1) Joe is seen driving with Alice; (2) Sue arrives at the party with Joe but is heard to say some pretty nasty things about him; and (3) Joe continues to live with Sue and has not moved into an apartment. Should you reject the hypothesis that Sue and Joe are still happily married? Some people would say that you should reject the hypothesis, because happily married couples don't usually say nasty things about one another in public. But other people would argue that all happily married couples occasionally have disagreements that may lead them to say things that they will later regret. Without some standard for deciding whether your observations are sufficiently unusual in happily married couples, you have no basis for making a decision.

In hypothesis testing, the standard for rejecting a hypothesis is called the ***level of significance***, or ***alpha*** (**α**). Suppose you decide that you will reject the hypothesis if you get a sample outcome whose probability of occurring (if the hypothesis is true) is .05 or less. If so, you are choosing a level of significance of .05 ($\alpha = .05$). The .05 level is common in statistical tests because most researchers agree that a sample outcome that would occur only 5% of the time (or less) if the hypothesis is true is unusual enough to warrant rejection of the hypothesis. However, other levels of significance may be chosen instead.

Another difference between real-life and statistical hypothesis testing has to do with how the sample is selected. In the case of Sue and Joe, it is doubtful that your sample of observations was random. Statistical tests, however, assume that the sample has been randomly selected from the population. (Possible consequences of violating this assumption are discussed later in the chapter.)

The next step in hypothesis testing involves determining whether the probability of selecting a random sample like this one from the hypothesized population is less than alpha (decision: reject hypothesis) or greater than alpha (decision: do not reject hypothesis). This brings us to still another difference between real-life and statistical tests. In the case of Sue and Joe and other real-life hypothesis tests, it would be difficult or impossible to determine the probability of various sample outcomes. In a statistical hypothesis test, on the other hand, the exact probabilities of various sample outcomes can be found by referring to an appropriate probability distribution. In testing hypotheses about categorical data, the relevant probability distribution is the *chi square distribution*. That is why the test is called the *chi square test*.

In most statistical tests, as you will see, finding the probability of a sample outcome entails calculating a value called the *test statistic*. In testing a hypothesis about categorical data, for example, we calculate a value of the test statistic *chi square* (χ^2) from the sample data. Then we can refer to a table of the chi square probability distribution to find out whether the probability of a value of χ^2 like ours is less than or greater than alpha.

Recapitulation: General Procedure in Statistical Hypothesis Testing

Now we can return to the general procedure for hypothesis testing and apply it, specifically, to statistical hypothesis testing. Here are the steps:

1. State the hypothesis to be tested.
   A statistical hypothesis is always about a parameter or other quantitative characteristic of a population.
2. Select a level of significance, alpha (α).
   The hypothesis will be rejected if the probability of the sample outcome is less than alpha when the hypothesis is true.

3. Select a random sample from the population and calculate the test statistic, such as χ^2.
4. Refer to the probability distribution of the test statistic to find out whether the calculated value has probability (p) less than alpha or greater than alpha.
5. Decide whether to reject the hypothesis.
   If $p < \alpha$, reject the hypothesis.
   If $p > \alpha$, do not reject the hypothesis.

EXERCISE 5–2

1. Which of the following could be hypotheses in statistical tests?
   a. The boss is in a bad mood today.
   b. The mean of a population is 50.
   c. The population variance is 95.
   d. The mean of a sample of size 10 is 4.
2. You have chosen $\alpha = .10$ in a hypothesis test. This means that you will reject the hypothesis if
   a. the probability of the hypothesis is less than .10, given the observed sample outcome.
   b. the probability of the observed sample outcome is less than .10, if the hypothesis is true.
   c. you get a sample whose probability is greater than .10 when the hypothesis is true.

Introduction to the Chi Square Test

Rationale: Comparing Observed and Expected Frequencies

The ***chi square test*** is used to test hypotheses about a population distribution of categorical data, based on the distribution of data in a sample from the population. This is done by comparing observed frequencies in the sample with the frequencies that are expected if the hypothesis is true. If the difference is sufficiently large, we reject the hypothesis. To clarify the reasoning involved in the test, let us return to Dr. Klingelfinger.

Suppose that Dr. Klingelfinger is correct: In the population of mental hospital patients, the percentages of patients in the categories thought disorder, affective disorder, character disorder, neuroses, and other are 40%, 30%, 15%, 10%, and 5%, respectively. If so, we expect most samples of patients from the population to have percentages close to these. In a sample of 200 patients, for example, we expect to find about 40% of the 200 patients in the category thought disorder, about 30% of the 200 patients in the category affective disorder, and so on.

The number or frequency of sample observations that is expected to fall in a certain category if the hypothesis is true is called the ***expected frequency*** for the category, symbolized f_e. The expected frequency in a category is equal to the sample size N times the expected proportion, p_e.

$$f_e = Np_e$$

Dr. Klingelfinger's expected proportions for the five categories are .40, .30, .15, .10, and .05, respectively. Therefore, in a sample of size 200, the expected frequencies are

Thought disorders: $f_e = (200)(.40) = 80$
Affective disorders: $f_e = (200)(.30) = 60$
Character disorders: $f_e = (200)(.15) = 30$
Neuroses: $f_e = (200)(.10) = 20$
Other: $f_e = (200)(.05) = 10$

Note that the sum of the expected frequencies, like the sum of the observed frequencies, is equal to the sample size, 200.

The number of sample observations that actually fall in a category is called an ***observed frequency***, symbolized f_o. The observed frequencies in the sample of 200 patients at PMPI are

Thought disorders: $f_o = 100$
Affective disorders: $f_o = 60$
Character disorders: $f_o = 20$
Neuroses: $f_o = 10$
Other: $f_o = 10$

As we have already noted, the observed frequencies and the expected frequencies are different. But are they different enough to make the PMPI sample a very unusual one from the hypothesized population—unusual enough to reject the hypothesis? Or would a sample like this occur relatively often in sampling from the hypothesized population? We need to develop a measure of the difference between the distributions of observed and expected frequencies.

We begin by finding the difference between f_o and f_e, $(f_o - f_e)$, in each category:

	Disorder					
	Thought	Affective	Character	Neuroses	Other	Sum
f_o	100	60	20	10	10	200
f_e	80	60	30	20	10	200
$f_o - f_e$	20	0	−10	−10	0	0

Notice that the sum of the differences between observed and expected frequencies, $\Sigma(f_o - f_e)$, is equal to zero. In fact, no matter how different the observed and expected frequencies of a sample are, $\Sigma(f_o - f_e)$ always equals zero. Therefore we cannot use $\Sigma(f_o - f_e)$ as a measure of the difference between the distributions of observed and expected frequencies.

In Chapter 3 we encountered a similar problem when developing a measure of variability based on deviations from the mean. No matter how much or how little the scores vary around the mean, $\Sigma(X - M)$ is always equal to zero. We solved that problem by squaring each deviation before adding. We use the same strategy here to deal with differences between observed and expected frequencies. That is, we square each value of $(f_o - f_e)$.

	Disorder					
	Thought	Affective	Character	Neuroses	Other	Sum
f_o	100	60	20	10	10	200
f_e	80	60	30	20	10	200
$f_o - f_e$	20	0	−10	−10	0	0
$(f_o - f_e)^2$	400	0	100	100	0	

But now another problem surfaces. In the category "Thought Disorder," $(f_o - f_e)^2$ is much larger than $(f_o - f_e)^2$ in the category "Neuroses." Yet, relative to the number of patients in each category, the difference between f_o and f_e for "Thought Disorder," 100 versus 80, is smaller than that for "Neuroses," 10 versus 20. To correct for this, we divide $(f_o - f_e)^2$ in each category by the expected frequency in the category, f_e.

	Disorder					
	Thought	**Affective**	**Character**	**Neuroses**	**Other**	**Sum**
f_o	100	60	20	10	10	200
f_e	80	60	30	20	10	200
$f_o - f_e$	20	0	−10	−10	0	0
$(f_o - f_e)^2$	400	0	100	100	0	
$\frac{(f_o - f_e)^2}{f_e}$	5.00	0	3.33	5.00	0	

Now we have a corrected (standardized) measure of the difference between the observed and expected frequencies in each category.

To calculate a measure of the overall difference, over all of the categories, we simply add the values of $\frac{(f_o - f_e)^2}{f_e}$. The sum is called ***chi square,*** χ^2.

$$\chi^2 = \Sigma\left[\frac{(f_o - f_e)^2}{f_e}\right] \quad \textit{Formula for calculating } \chi^2$$

As you can see from the formula, the greater the differences between observed and expected frequencies, the larger the value of χ^2.

In the sample of patients at PMPI

$$\chi^2 = 5.00 + 0 + 3.33 + 5.00 + 0 = 13.33$$

Now we must ask: How unusual is a sample of size 200 with $\chi^2 = 13.33$ from the hypothesized population? Is it unusual enough to reject Dr. Klingelfinger's hypothesis? In order to answer this question, we must find the probability of a χ^2 value this big, that is, $p(\chi^2 > 13.33)$. Then if the probability is small enough, we will reject the hypothesis. In order to find $p(\chi^2 > 13.33)$, we must first examine the probability distribution of χ^2, the chi square distribution.

EXERCISE 5–3

Suppose that in a sample of 200 patients at another mental hospital, the number of patients in each category is as follows:

	Disorder				
	Thought	**Affective**	**Character**	**Neuroses**	**Other**
f_o	70	60	35	20	15

1. If Dr. Klingelfinger's hypothesis is correct, is a sample like this one more or less unusual than the one at PMPI?
2. Answer before calculating: Do you expect the sample's χ^2 to be larger or smaller than the χ^2 for the PMPI sample?
3. Calculate χ^2 and see if your prediction is confirmed.

THE CHI SQUARE DISTRIBUTION

The ***chi square distribution***, like the binomial distribution, is not a single distribution but a family of distributions. You may recall that the particular binomial distribution that describes the outcomes when tossing 5 fair coins is not the same as the one that describes the outcomes when tossing 10 fair coins or 5 biased ones. In the same way, different situations call for different chi square distributions.

Degrees of Freedom

The factor that differentiates among various chi square distributions is something called ***degrees of freedom*** (***df***). Degrees of freedom is a very important concept in statistical inference, one that you will encounter again in dealing with other statistical tests.

Basically, degrees of freedom refers to the number of values in a set that are free to vary. Suppose I say to you, "Give me three numbers." You are free to choose any three numbers; you have 3 degrees of freedom. However, if I say, "Give me three numbers whose sum is 16," you can freely choose only two of the three. For example, if you choose the numbers 7 and 3, the third number must be 6. Thus the imposition of one restriction on three numbers reduces the degrees of freedom from 3 to 2.

In statistical tests, the number of degrees of freedom is usually related to the number of observations (N) or the number of categories into which observations fall. In the chi square test, as you will see, degrees of freedom is dependent on the number of categories.

Degrees of freedom is important for chi square because for each number of degrees of freedom, there is a different chi square distribution. Three chi square distributions, those with 3, 9, and 15 degrees of freedom, are shown in Figure 5–1. (Of course, there are also chi square distributions with 1, 2, 4, 5, . . . , degrees of freedom.) As you can see, all three chi square distributions are positively skewed. But as degrees of freedom increases, two things happen to the distribution: It becomes progressively less skewed, and the distribution as a whole moves farther to the right on the abscissa.

The chi square distribution differs from the binomial distribution in being a continuous rather than a discrete distribution. Recall that a variable in a discrete distribution can take only certain specified values with gaps in between. For example, one can get two heads or three heads in tossing five coins, but not two and a half or three and two-thirds. In a discrete probability distribution, it is possible to find the probability of each outcome, for example, the probability of exactly two heads. In a continuous distribution, like the chi square distribution, on the other hand, there are no gaps, and we cannot find

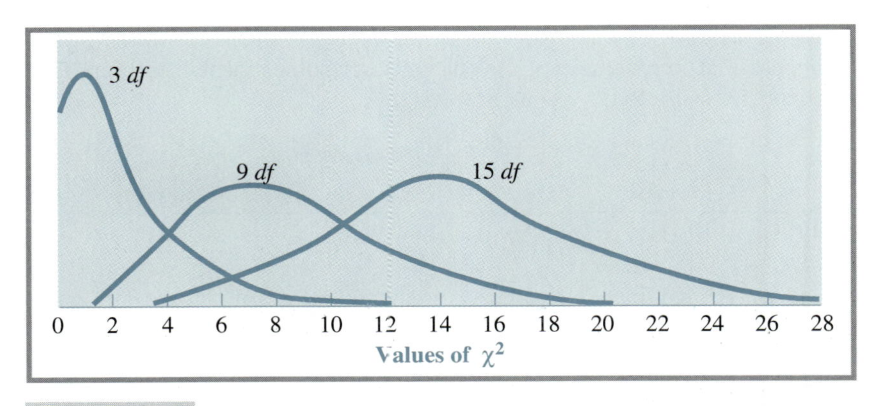

Figure 5–1 **Chi square distributions with 3, 9, and 15 degrees of freedom (*df*).**

the probability of discrete values like $\chi^2 = 2$. Instead, we find the probability that χ^2 is between specified values, such as between 1.5 and 2.5, or greater or less than specified values, such as $\chi^2 < 1.5$ or $\chi^2 > 2.5$.

Suppose we are interested in the probability that χ^2 is greater than 8. Remember that, in the graph of a continuous distribution, the proportion of the area in any section is equal to the probability of values in that section. Therefore, $p(\chi^2 > 8)$ is the proportion of area above the value 8 in the chi square distribution. Now look at Figure 5–1 again. A vertical line drawn at $\chi^2 = 8$ cuts off a very small section in the upper tail of the chi square distribution with 3 degrees of freedom; therefore, with 3 degrees of freedom, $\chi^2 > 8$ has a small probability. In the distribution with 9 degrees of freedom, the same vertical line cuts off a much larger area in the upper tail; and in the distribution with 15 degrees of freedom, most of the distribution falls above $\chi^2 = 8$. Therefore, with 9 degrees of freedom, $\chi^2 > 8$ has quite a large probability, and with 15 degrees of freedom, the probability of $\chi^2 > 8$ is far greater than the probability that $\chi^2 < 8$. Obviously, degrees of freedom need to be taken into account in evaluating the probability of a given value of χ^2.

The Chi Square Table

Areas cut off by various values of χ^2 in distributions with various numbers of degrees of freedom are shown in the chi square table, Table D–1 in Appendix D. You will use Table D–1 whenever you test hypotheses using the chi square test. The rows of the table are degrees of freedom (*df*). Each row provides information about the chi square distribution with that number of degrees of freedom. The column headings are proportions of area in the upper tail of the chi square distribution. Each entry in the table is a value of χ^2 that cuts off a proportion of area in the upper tail equal to the column heading. The values of χ^2 in the table are called ***critical values*** of χ^2.

Let us see how we read the information in the table. Consider the critical value in the row for $df = 6$ and the column headed .05. The critical value is 12.59. This means that in the chi square distribution with 6 *df*, a χ^2 of 12.59 cuts off .05 (5%) of the area in the upper tail. Since the proportion of area in a given part of the distribution is the probability of values within that part of the distribution, we can say that in the chi square distribution with 6 degrees of freedom, the probability of $\chi^2 > 12.56$ is .05. Similarly, the critical value in the row for $df = 3$ and the column headed = .10 is 6.25. Therefore, in the distribution with 3 degrees of freedom, $p(\chi^2 > 6.25) = .10$. Confirm the following examples yourself:

With 8 degrees of freedom, $p(\chi^2 > 15.51) = .05$, and $p(\chi^2 > 20.09) = .01$.
With 1 degree of freedom, $p(\chi^2 > 2.71) = .10$ and $p(\chi^2 > 3.84) = .05$.

Suppose you want to know the probability of a value of χ^2 that is not in the table. For example, with 1 degree of freedom, what is $p(\chi^2 > 3.29)$? In the row with 1 degree of freedom, the two critical values that bracket 3.29 are 2.71, with $p = .10$, and 3.84, with $p = .05$. Since 3.29 is between 2.71 and 3.84, we know that $p(\chi^2 > 3.29)$ is between .10 and .05. That is, $p < .10$ but $p > .05$. Confirm the following examples yourself:

With 3 degrees of freedom, $p(\chi^2 > 11.74)$ is between .01 and .005 ($p < .01$ but $p > .005$).
With 5 degrees of freedom, $p(\chi^2 > 11.74)$ is between .05 and .025 ($p < .05$ but $p > .025$).

Now let us see how the table is used to evaluate the test statistic χ^2 in a chi square test. Recall that, as differences between observed and expected frequencies increase, the value of χ^2 increases. The greater the value of χ^2, the less the probability of selecting a sample like the obtained sample from the hypothesized population. We reject the hypothesis if the value of χ^2 is so large that $p(\chi^2 >$ this value) is less than the level of significance, alpha. Therefore, after calculating χ^2 in a chi square test, you simply need to find $p(\chi^2 >$ this value), using the procedure just illustrated. If $p < \alpha$, you reject the hypothesis; if $p > \alpha$, you do not reject the hypothesis.

Suppose, for example, that you are conducting a test with $\alpha = .05$ and 6 degrees of freedom. You have calculated $\chi^2 = 15.31$. Referring to the row for $df = 6$ in Table D–1, you find that $p(\chi^2 > 15.31)$ is between .025 and .01. Since $p < .05$, you reject the hypothesis. On the other hand, suppose you had calculated $\chi^2 = 10.89$. According to the table, with 6 degrees of freedom, $p(\chi^2 > 10.89)$ is between .10 and .05. Since $p > .05$, you cannot reject the hypothesis.

A little more terminology: In a statistical test, the part of the probability distribution resulting in rejection of the hypothesis is called the ***rejection region***. The rejection region has area = α. The rest of the distribution, with area = $1 - \alpha$, is called the ***nonrejection region***. For example, in a test with 1 degree of freedom and $\alpha = .05$, the rejection region consists of values of χ^2 greater than 3.84, because $p(\chi^2 > 3.84) = .05$. Values of χ^2 less than 3.84 fall in the nonrejection region. The critical value 3.84 is the cutoff point between the rejection and nonrejection regions and can be referred to as the critical value of χ^2 for $\alpha = .05$ and $df = 1$. Figure 5–2 shows the relations between the rejection and nonrejection regions, alpha, and the critical value of χ^2 for that alpha.

Here are two more examples of the use of the chi square table in the chi square test.

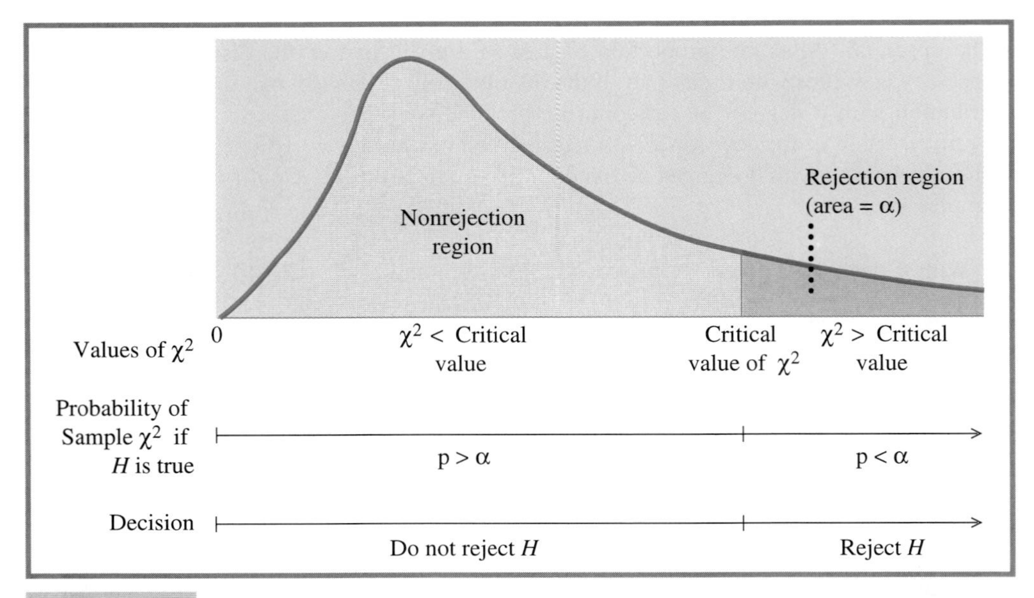

Figure 5–2 **Relation between critical value of χ^2, alpha (α), rejection and nonrejection regions, probability (p) of sample outcome, and decision concerning hypothesis (H) in the chi square test.**

EXAMPLE 1: You have just computed $\chi^2 = 4.55$ in a test with 1 degree of freedom. If you use $\alpha = .05$, will you reject the hypothesis?

ANSWER: According to the chi square table, $p(\chi^2 > 4.55)$ is between .05 and .025. The probability of a sample outcome like yours, if the hypothesis is true, is less than .05. (That is, your sample χ^2 falls in the rejection region for $\alpha = .05$ and $df = 1$.) Therefore, you reject the hypothesis.

EXAMPLE 2: You have calculated $\chi^2 = 17.60$ in a test with 8 degrees of freedom. If $\alpha = .01$ do you reject the hypothesis?

ANSWER: Table D–1 shows that for $df = 8$, $p(\chi^2 > 17.60)$ is between .025 and .01; that is, $p > .01$. The calculated χ^2 falls in the nonrejection region for $\alpha = .01$ and $df = 8$, and the hypothesis is not rejected.

EXERCISE 5–4

1. In testing a hypothesis, you have computed $\chi^2 = 10.73$ with 4 degrees of freedom.
   a. What is $p(\chi^2 > 10.73)$?
   b. If $\alpha = .05$, do you reject the hypothesis?
   c. What if $\alpha = .01$?
   d. What is $p(\chi^2 > 10.73)$ if $df = 5$ instead of 4?
2. a. What is the minimum value of χ^2 required to reject a hypothesis if $df = 10$ and $\alpha = .01$?
   b. If $df = 5$ and $\alpha = .01$?
   c. If $df = 14$ and $\alpha = .025$?
3. You have calculated $\chi^2 = 8.54$ in a test with 2 degrees of freedom.
   a. If the hypothesis is true, is the probability of a sample outcome like yours less than or greater than .05?
   b. If the hypothesis is true, is the probability of a sample outcome like yours less than or greater than .01?

The Chi Square Test of Goodness of Fit

PART 1: TESTING HYPOTHESES ABOUT THEORETICAL DISTRIBUTIONS

In this section and the next, we consider tests of hypotheses about the distribution of one categorical variable in a single population. The test is called the ***chi square test of goodness of fit*** because its purpose is to determine how well a hypothesized distribution fits an obtained one. The discussion in this section deals with situations in which there is a specific hypothesized distribution. The test of Dr. Klingelfinger's hypothesis concerning the distribution of psychiatric patients across various categories is an example.

The hypothesis to be tested is Dr. Klingelfinger's claim that the distribution of psychiatric patients in five diagnostic categories is as follows:

	Thought Disorders	Affective Disorders	Character Disorders	Neuroses	Other or Unclassified
Hypothesized distribution	40%	30%	15%	10%	5%

Based on the hypothesis, we calculated the following expected frequencies in a sample of size 200:

	Thought Disorders	Affective Disorders	Character Disorders	Neuroses	Other or Unclassified
f_e	80	60	30	20	10

The distribution of observed frequencies (f_o) in 200 patients at PMPI is as follows:

	Thought Disorders	Affective Disorders	Character Disorders	Neuroses	Other or Unclassified
f_o	100	60	20	10	10

We calculated $\chi^2 = 13.33$. Now we are almost ready to complete the test. There is just one more question we need to answer first: What is the number of degrees of freedom for the test? The answer is: *In the goodness-of-fit test, the number of degrees of freedom is one less than the number of categories.* This is because the fact that the sum of the observed frequencies always equals the sum of the expected frequencies places one restriction on the data, resulting in a loss of 1 degree of freedom. In the case of Dr. Klingelfinger, with five categories, $df = 5 - 1 = 4$. Therefore, we will compare the calculated χ^2 with the critical values in the row of Table D–1 with $df = 4$.

Conducting the Test

Now we are ready to conduct the goodness-of-fit test of Dr. Klingelfinger's hypothesis. We show all steps of the test.

Step 1. State the hypothesis
The hypothesized distribution of patients over the five categories is thought disorders, 40%; affective disorders, 30%; character disorders, 15%; neuroses, 10%; and other, 5%. In a sample of size 200, the expected frequency distribution is 80, 60, 30, 20, and 10 in the five categories, respectively.

Step 2. Select a level of significance (α)
We will use $\alpha = .05$.

Step 3. Collect data and calculate χ^2 by the formula

$$\chi^2 = \Sigma\left[\frac{(f_o - f_e)^2}{f_e}\right]$$

We have already calculated $\chi^2 = 13.33$.

Step 4. Find the probability of $\chi^2 >$ the calculated value
Referring to the row for 4 degrees of freedom in Table D–1, we find that $p(\chi^2 > 13.33)$ is between .01 and .005. In other words, if Dr. Klingelfinger is right, the probability of obtaining a sample distribution like the one at PMPI is between .005 and .01.

Step 5. Decide whether to reject the hypothesis
Since the probability of a sample like the obtained one is less than the chosen alpha, .05, we reject the hypothesis and conclude that Dr. Klingelfinger's hypothesis is not a good fit to the data at PMPI.

When a hypothesis is rejected at a certain level of alpha, the result is said to be *significant* at that level. Thus, we can state that the difference between the obtained distribution of patient classifications at PMPI and Dr. Klingelfinger's hypothesized distribution is significant at the .05 level.

Recapitulation

In using chi square to test a hypothesis regarding a theoretical distribution, we go through the following steps:

1. State a hypothesis concerning the expected distribution in the population. Convert hypothesized percentages or proportions into expected frequencies by the formula $f_e = Np_e$.
2. Choose a level of significance, α.
3. In a sample of size N from the population, count the number of observations in each category (the observed frequency, f_o). Then calculate χ^2 by the formula

$$\chi^2 = \Sigma\left[\frac{(f_o - f_e)^2}{f_e}\right]$$

4. Find the probability of $\chi^2 >$ the calculated value in the distribution with $df =$ number of categories minus 1.
5. Decide whether to reject the hypothesis:

   If $p < \alpha$, reject the hypothesis. The difference between the obtained and hypothesized distributions is significant at level α.

   If $p > \alpha$, do not reject the hypothesis. The difference between the obtained and hypothesized distributions is not significant at level α.

EXERCISE 5–5

I hypothesize that in introductory statistics courses, 60% of students pass, 20% fail, and 20% withdraw. But in a class of 50 students, 45 pass, 1 fails, and 4 withdraw. Does my hypothesis fit the data in this statistics class? Use $\alpha = .05$.

Part 2: Testing the Null Hypothesis

In the example of Dr. Klingelfinger, a specific hypothetical distribution was postulated, and observed frequencies were compared with the frequencies expected on the basis of the hypothesis. At first glance, the mayoral election in Tooterville looks similar. We have a distribution of observed frequencies on a categorical variable: voters' replies to the question "For whom do you plan to vote?" with two categories, Smith and Jones; and we want to test the hypothesis that Smith will win the election.

However, unlike Dr. Klingelfinger's hypothesis, which stated specific expected percentages, the hypothesis "Smith will win the election" is ambiguous, for Smith could win the election with anywhere from 51% to 100% of the vote. The hypothesis for a statistical test must always be stated in precise quantitative terms. The hypothesis "Smith will win" does not specify whether the expected percentage of votes for Smith is 51, 52, 53, . . . , 99, or 100. It simply says, "Smith will get more votes than Jones," but says nothing about how many more.

Whenever hypotheses are stated in terms of "more" or "less," without specifying the exact amount, statisticians test the converse of the stated hypotheses, called the ***null hypothesis*** (symbolized H_0). The null hypothesis is essentially a hypothesis of no difference. In the present example, the null hypothesis states, "Neither Smith nor Jones will win the election," that is, "50% of the voters will vote for Smith and 50% for Jones." Since the null hypothesis specifies particular expected percentages, it is a precise, quantitative, testable statistical hypothesis.

The procedure for conducting the test is the same as the one we used in the previous section. First, the null hypothesis is stated and expected frequencies are calculated. Second, a level of significance is chosen. Next, the test statistic, χ^2, is calculated, and the probability of a χ^2 this large is found in the distribution with degrees of freedom equal to the number of categories minus one. If $p < \alpha$, the hypothesis is rejected; if $p > \alpha$, the hypothesis is not rejected.

Let us consider how to interpret the results of a test of a null hypothesis (H_0). Suppose the test results in rejection of H_0. Since H_0 states that no difference exists, rejection of H_0 leads to the conclusion that a difference *does* exist. In the case of the mayoral election poll, for example, rejection of H_0 (50% of votes for each candidate) would lead us to conclude that one candidate will get more votes than the other and win the election. But what if the test results in nonrejection of H_0? As you will see, the interpretation of this outcome is a little more complicated. We will defer discussion until after we have tested H_0: "Smith and Jones will each get just 50% of the vote in the Tooterville election."

Computing and Evaluating Chi Square in the Goodness-of-Fit Test of the Null Hypothesis

Let us proceed through all the steps in the statistical test for the mayoral election poll.

1. *State the hypothesis.* The statistical hypothesis is the null hypothesis, H_0: 50% of the voters will vote for Smith and 50% for Jones. Since 200 voters have been sampled, the expected frequency of votes for Smith is 100, as is the expected frequency of votes for Jones.
2. *Select a level of* α. We will use $\alpha = .05$.
3. *Collect data and calculate* χ^2. Of the 200 voters who were sampled, 113 indicated an intention to vote for Smith, and 87 for Jones. These are the observed frequencies. χ^2 is calculated below.

	Smith	Jones	Total
f_o	113	87	200
f_e	100	100	200
$f_o - f_e$	13	−13	
$(f_o - f_e)^2$	169	169	
$\frac{(f_o - f_e)^2}{f_e}$	1.69	1.69	

$$\chi^2 = 1.69 + 1.69 = 3.38$$

4. *Find the probability of* χ^2 *greater than the obtained value.* Since there are two categories, $df = 2 - 1 = 1$. Referring to the row in Table D–1 with $df = 1$, we find that $p(\chi^2 > 3.38)$ is between .10 and .05.
5. *Decide whether to reject* H_0. Since the probability of the calculated χ^2 is greater than .05, we do not reject H_0.

Our test has resulted in nonrejection of the null hypothesis that each candidate will get 50% of the vote. Does this mean that we should conclude, "Neither Smith nor Jones will win the election?" Since elections rarely end in a 50–50 tie, this seems unrealistic. Let us back off a little.

Remember: When we fail to reject a statistical hypothesis, we do so because, if the hypothesis is true, the obtained result is not too improbable. But this is not the same thing as saying that we strongly believe the tested hypothesis to be true.

Consider the present example. H_0 states, "50% of the voters will vote for Smith and 50% for Jones," and our chi square test resulted in nonrejection of H_0. This means that if 50% of the population favors Smith, the results of the poll (113 of 200 sampled voters favoring Smith) are not too unlikely. That is, the results are *consistent* with the null hypothesis of no preference. However, the results of the poll are also consistent with several other hypotheses, such as the hypothesis that 51% of voters favor Smith, or the hypothesis that 52% of voters favor Smith, or 53%, or 54%, or 55%, or 56%. Yet not all these hypotheses can be true at the same time.

What this shows is that nonrejection of a null hypothesis does not imply that we accept H_0 to the exclusion of all other hypotheses. It means not that we are confident that there is no difference but that we do not have enough evidence to conclude that there *is* a difference. In the case of the Tooterville poll, we conclude that we cannot confidently state that either candidate will obtain more votes than the other. The editor of the newspaper would be well advised to tell the readers, "The results of our opinion poll are inconclusive. At present, we are unable to predict who will win the election."

EXERCISE 5–6

The dean of a small college wonders whether prospective freshmen are more likely to select certain majors than others. He telephones 100 randomly selected prospective freshmen and asks which field they plan to major in. Here are the results:

Humanities	32
Natural sciences	19
Social sciences	27
Education	22

Conduct a test to decide whether some majors are more popular than others. Use $\alpha = .05$.

HINT: The null hypothesis states that an equal number of incoming freshmen plans to major in each field.

The Chi Square Test of Independence

The chi square test of goodness of fit is used to test hypotheses about the distribution of one categorical variable in one population. Sometimes, however, we want to know whether two categorical variables are related. For example, is sex (male, female) related to choice of toys (warlike, nonwarlike)? Is type of prior information (relevant, nonrelevant) related to level of comprehension (high, moderate, low) of a difficult reading passage? Is political affiliation (Democrat, Republican, Independent) related to attitudes toward big government (for, against)?

To answer questions like these, we select samples from the relevant populations and measure the variables of interest. We select samples of preschool boys and girls and observe the type of toys they choose to play with. We randomly subdivide a group of experimental subjects in half, give the two subgroups different types of information concerning a particular topic, have the subjects read a difficult passage related to the topic, and then ask them to rate their comprehension of the passage. We ask a sample of voters, "What is your political affiliation?" and "Are you for or against a strong federal government?"

The statistical test for situations like these is the ***chi square test of independence***. It is called the "test of independence" because it tests the null hypothesis that the two variables are independent (unrelated). Thus, in our three examples, the null hypotheses are, respectively, (1) H_0: sex is not related to choice of toys; (2) H_0: type of prior information is not related to rated comprehension; (3) H_0: political affiliation is not related to attitudes toward big government.

Another way of stating null hypotheses for tests of independence is in terms of a difference between groups. For example, in the study of boys' versus girls' choice of toys, we can state the null hypothesis like this: H_0: There is no difference between boys' and girls' choice of toys. Notice that this is just another way of saying that there is no relation between sex and choice of toys; it is not a different null hypothesis.

Contingency Tables

Data obtained in studies such as our three examples are generally displayed in tables called ***contingency tables***. Table 5–1 shows contingency tables that might result from the three studies. The rows of a contingency table represent the categories on one variable (sex, political affiliation, type of prior information), and the columns represent the categories on the other variable (nature of toys chosen, attitude toward big government, comprehension). The numbers within the cells are observed frequencies. Table 5–1A, for example, shows that 25 girls and 17 boys chose warlike toys and that 9 girls and 20 boys chose nonwarlike toys.

Table 5–1 Three Contingency Tables

(A)

		Nature of Toys Chosen	
		Warlike	*Nonwarlike*
Sex	*Girls*	25	9
	Boys	17	20

(B)

		Attitude Toward Big Government		
		In Favor	*Opposed*	*Uncertain*
Political Affiliation	*Republican*	8	27	10
	Democrat	35	8	5
	Independent	10	10	10

(C)

		Rated Comprehension		
		Low	*Moderate*	*High*
Type of Prior Information	*Relevant*	2	11	17
	Irrelevant	15	8	5

All three tables are *two-way* contingency tables because they have two dimensions, or axes. Two-way contingency tables are often referred to by naming the number of rows times the number of columns. Thus the table showing the relation between sex and toys is a 2×2 ("2 by 2") contingency table; the table for political affiliation and attitude toward big government is a 3×3 contingency table; the table for prior information and comprehension is a 2×3 contingency table.

EXERCISE 5–7

Table 2–13 (p. 49) includes data on the sex and majors of 28 students.

1. Construct a contingency table for the relationship between these two variables.
2. Assume that the 28 students are a random sample from a population. If we conduct a test to determine whether sex and major are related in the population, state the null hypothesis to be tested in two ways: in terms of a difference between groups and in terms of a relation between variables.

Procedures in the Chi Square Test of Independence

In many ways, the chi square test of independence is similar to the test of goodness of fit. As in the test of goodness of fit, we determine the expected frequency (f_e) in each cell of the contingency table if the null hypothesis is true, and we calculate χ^2 by the formula $\Sigma\left[\frac{(f_o - f_e)^2}{f_e}\right]$. Then we evaluate χ^2 by referring to the chi square table. However, there are two differences from the chi square test of goodness of fit: (1) the procedure for calculating expected frequencies, and (2) the determination of degrees of freedom (df). We discuss the calculation of expected frequencies and df, using data from the study of boys' and girls' choice of toys. The contingency table is given below. Row sums and column sums (marginal totals) are also shown.

		Nature of Toys Selected		
		Warlike	*Nonwarlike*	*Total*
Sex	*Girls*	25	9	34
	Boys	17	20	37
	Total	42	29	71 = N

Calculation of Expected Frequencies

Remember that an expected frequency is the frequency that is expected if the hypothesis is true. We are testing the null hypothesis that choice of toys does not differ as a function of sex. If the null hypothesis is true, we expect that the proportion of boys selecting warlike and nonwarlike toys will be the same as the proportion of girls selecting each type of toy. Over both groups combined, the proportion of children choosing warlike toys is 42/71. Therefore, if the null hypothesis is true, 42/71 of the girls and 42/71 of the boys are expected to choose warlike toys. Similarly, since 29/71 of all the children chose nonwarlike toys, the expected proportion for each sex, if H_0 is true, is 29/71. Notice that the expected proportion (p_e) for each cell is (column sum/N).

Remember that an expected frequency is equal to group size times expected proportion ($f_e = Np_e$). There are 34 girls; therefore, if H_0 is true, we expect that $34 \times (42/71)$ girls will choose warlike toys and that 34 × (29/71) of the girls will choose nonwarlike toys. Similarly, in the 37 boys, the expected frequency of choice of warlike toys is 37 × (42/71) and of nonwarlike toys is 37 × (29/71). Notice that, in each cell, the expected frequency is (row sum)(column sum/N), which can be written as $\frac{\text{(row sum) (column sum)}}{N}$. Therefore, the general formula for calculating expected frequencies in a two-way contingency table is

$$f_e(\text{cell}) = \frac{\text{(row sum)(column sum)}}{N}$$

Formula for calculating expected frequencies

We complete our calculations of expected frequencies:

$$f_e(\text{girls, warlike}) = \frac{(34)(42)}{71} = 20.1$$

$$f_e(\text{girls, nonwarlike}) = \frac{(34)(29)}{71} = 13.9$$

$$f_e(\text{boys, warlike}) = \frac{(37)(42)}{71} = 21.9$$

$$f_e(\text{boys, nonwarlike}) = \frac{(37)(29)}{71} = 15.1$$

All four expected frequencies are shown in the following table:

		Nature of Toys Selected		
		Warlike	*Nonwarlike*	
Sex	*Girls*	20.1	13.9	34.0
	Boys	21.9	15.1	37.0
		42.0	29.0	

Notice that the row and column sums of expected frequencies are the same as the row and column sums of observed frequencies. This is always the case in the chi square test of independence.

Degrees of Freedom

In a two-way contingency table, the number of degrees of freedom is dependent on both the number of rows and the number of columns and is equal to (number of rows minus one) times (number of columns minus one). In the choice-of-toys case, for example, the number of rows is 2 and the number of columns is 2; therefore, $df = (2 - 1)(2 - 1) = (1)(1) = 1$. Similarly, in a 3 × 3 table, df would be (2)(2) = 4; in a 3 × 4 table, df would be (2)(3) = 6; and so on.

The reason that $df = (\text{rows} - 1)(\text{columns} - 1)$ is that, in a table with j rows and k columns and with fixed row and column sums, only $(j - 1)(k - 1)$ cell entries are free to vary. For example, if you are given an empty table with 4 rows and 5 columns (20 cells), with preset row and column sums, you can fill in (3)(4) = 12 cells with numbers of your choice. But the rest of the cell entries are determined by the requirement that the numbers within each row must add to the row sum, and the numbers within each column must add to the column sum. (You can check this for yourself in Exercise 5–8.)

Calculation and Evaluation of χ^2

The calculation and evaluation of χ^2 in the test of independence are the same as in the test of goodness of fit. The value of χ^2 is calculated by the formula $\chi^2 = \Sigma\left[\frac{(f_o - f_e)^2}{f_e}\right]$; then the probability of χ^2 is found in the row of the table with df = (rows − 1)(columns − 1). If $p < \alpha$, the hypothesis is rejected.

We are now ready to conduct the test for the choice-of-toys study. All steps are shown below.

1. *State the hypothesis.* H_0: There is no difference in choice of toys between preschool girls and boys. (This can also be stated: There is no relationship between sex and choice of toys.) The expected frequencies based on this hypothesis have been computed.
2. *Select a level of* α. We will use $\alpha = .05$.
3. *Collect data and calculate* χ^2. The observed frequencies (f_o) and expected frequencies (f_e) are given in the following table, and χ^2 is computed in the usual manner.

		Warlike	*Nonwarlike*
Girls	f_o	25	9
	f_e	20.1	13.9
Boys	f_o	17	20
	f_e	21.9	15.1

$$\chi^2 = \frac{(25-20.1)^2}{20.1} + \frac{(9-13.9)^2}{13.9} + \frac{(17-21.9)^2}{21.9} + \frac{(20-15.1)^2}{15.1} = 5.61$$

4. *Find the probability of the obtained* χ^2. With 2 rows and 2 columns, $df = (1)(1) = 1$. According to the row for $df = 1$ in the chi square table, $p(\chi^2 > 5.61)$ is between .025 and .01.
5. *Decide whether to reject* H_0. Since $p < .05$, we reject the null hypothesis. The difference between the groups is significant at the .05 level. We conclude that girls and boys tend to select different types of toys. In order to describe the nature of the difference, we refer back to the contingency table and find that girls were more likely than boys to select warlike toys.

Before we leave the choice-of-toys example, let us consider what would happen if we had used $\alpha = .01$ in our test instead of $\alpha = .05$. We found that $p(\chi^2 > 5.61)$ is between .025 and .01. Since $p > .01$, we cannot reject H_0 at the .01 level. The difference between boys' and girls' choice of toys is not significant at the .01 level. This example shows that it is possible for a result to be significant at the .05 level but not at the .01 level.

Recapitulation: Chi Square Test of Independence

The step-by-step procedure in the chi square test of independence is summarized below:

1. *State the null hypothesis* (H_0) either in terms of no relation between variables or no difference between groups.
2. *Select a level of significance, alpha* (α).
3. *Collect data*, prepare a contingency table, and calculate marginal sums. Calculate the expected frequency (f_e) in each cell by the formula

$$f_e = \frac{(\text{row sum})(\text{column sum})}{N}$$

Then *calculate* χ^2.

$$\chi^2 = \Sigma\left[\frac{(f_o - f_e)^2}{f_e}\right]$$

4. *Find the probability of the obtained* χ^2. Determine the number of *df*: $df = (\text{rows} - 1) \times (\text{columns} - 1)$. Then find $p(\chi^2 >$ calculated value) in the table.
5. *Decide whether to reject* H_0. If $p < \alpha$, reject H_0; if $p > \alpha$, do not reject H_0. If H_0 is rejected, the relation between the variables (or difference between groups) is significant at level alpha. If H_0 is not rejected, the relation between variables (or difference between groups) is not significant at level alpha.

EXERCISE 5–8

1. Here is an empty 3 × 4 contingency table with preset row and column sums.

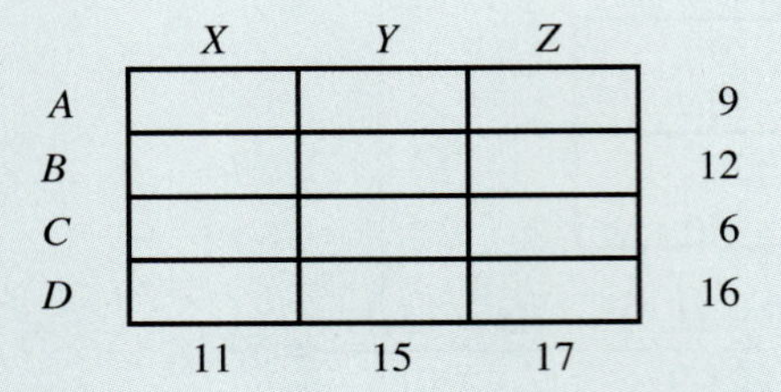

	X	Y	Z	
A				9
B				12
C				6
D				16
	11	15	17	

   a. How many degrees of freedom do you have for filling in the cells of the table?
   b. See how many cells of the table you can fill with numbers of your choice. Remember: The sum of entries in each row must equal the row sum and the sum of entries in each column must equal the column sum. You will find that the number of entries you can freely select is equal to your answer to part (a).
2. Use the data on rated text comprehension as a function of type of prior information in Table 5–1. Conduct a test of independence with $\alpha = .05$ to decide whether the two variables are related. If they are related, describe the nature of the relationship.
3. *Thought question:* In a chi square test, is it possible to obtain a value of χ^2 that is significant at the .01 level but not at the .05 level?

Scope and Limitations of the Chi Square Test

All statistical tests have certain underlying assumptions or conditions. If these conditions are not met, conclusions drawn from the tests are dubious at best. In this section we discuss the limitations of the chi square test. Specifically, the chi square test is appropriate only if the following conditions are met: (1) independence of observations within and between cells, and (2) no expected frequencies smaller than 2 and few smaller than 5. In addition, if the sample used to test a hypothesis has not been randomly selected from the population, great care must be taken in drawing conclusions from the results of a chi square test (or, indeed, any test).

Independence of Observations

The assumption of independence of observations means that each sample observation can fall in one and only one cell. Here are examples of how this assumption might be violated: In a study of the relation between political affiliation and attitudes toward big government, one respondent, who says that he sometimes votes Democratic and sometimes Republican, is entered into the contingency table twice, once in the row for Democrats and once in the row for Republicans; some children, given an opportunity to play with warlike or nonwarlike toys, play with both and are entered in both cells; patients at PMPI who are diagnosed as having a combination of two or more disorders are entered in all the corresponding categories. If a chi square test is to be conducted, care must be taken to ensure that each observation falls in one and only one cell.

A common way to obtain nonindependent observations is to measure each subject two (or more) times. Suppose, for example, that a physical education instructor counts the number of children in her class of 30 children who can do at least 10 push-ups both before and after a fitness training program. Her results look like this:

	Fewer Than 10 Push-ups	*10 or More Push-ups*
Before Training	20	10
After Training	8	22

The table looks very much like a 2 × 2 contingency table. The teacher may be tempted to conduct a chi square test of independence to determine whether provision of instruction is related to ability to perform 10 or more push-ups. However, she cannot do so, because each child is entered in the table twice, once before and once after instruction. Similarly, if an experimenter exposes each of a group of subjects to two or more conditions (such as an experimental and a control condition) and then compares outcomes under the two conditions, a test other than the chi square test must be used. A chi square test can be used only if each subject is exposed to just one condition.

Studies that compare two or more groups of subjects are sometimes referred to as *between-subjects* studies or comparisons, whereas studies that compare a single group under two or more conditions or on two or more trials are called *within-subjects* or *repeated-measures* studies or comparisons. The chi square tests discussed in this chapter are appropriate only for between-subjects comparisons.

Do not confuse the assumption of independence of observations with the chi square test of independence. The chi square test is used to determine whether two variables are independent. But the test should be used only if observations are independent. Independence of observations and independence of variables are different concepts.

Small Expected Frequencies

The chi square test should not be used if expected frequencies (f_e) in one or more cells are very small. A good rule of thumb is to avoid using chi square if there are several cells with $f_e < 5$ or any cells with $f_e < 2$. The reason for this restriction is that the chi square probability distribution, a continuous distribution, does not adequately fit expected distributions with very small (discrete) frequencies. Note that there is no restriction on the *observed* frequencies, which can be any size.

EXAMPLE 1: An investigator has theorized that 5% of his subjects will fall in Category 1, 30% in Category 2, and 65% in Category 3. If he tests 25 subjects, the expected frequencies are the following:

Category 1: $f_e = (25)(.05) = 1.25$
Category 2: $f_e = (25)(.30) = 7.50$
Category 3: $f_e = (25)(.65) = 16.25$

Since f_e for Category 1 is less than 2, χ^2 should not be computed. However, if the investigator were to use 50 subjects, he could use a chi square test, because f_e for Category 1 would then be $(50)(.05) = 2.5$.

EXAMPLE 2: The college majors of two groups of 25 college students are being compared, and a chi square test is to be used to find out whether the distribution of majors differs significantly between the two groups. Here are the results (observed frequencies):

	Major				
	Fine Arts	*English*	*Psychology*	*Physics*	*Chemistry*
Group 1	5	5	12	3	0
Group 2	2	1	8	5	9

The fact that some observed frequencies are very small doesn't matter. However, 8 of the 10 expected frequencies are less than 5, as shown in the following table. [Remember. $f_e = (\text{row sum}) \times (\text{column sum})/N$.]

	Major				
	Fine Arts	*English*	*Psychology*	*Physics*	*Chemistry*
Group 1	3.5	3.0	10.0	4.0	4.5
Group 2	3.5	3.0	10.0	4.0	4.5

With so many expected frequencies smaller than 5, χ^2 should not be computed.

In cases such as Example 2, there are two ways to increase expected frequencies. One is to increase the total number of subjects. With 40 to 50 subjects in each of the two groups, most expected frequencies would probably be 5 or more. A second way is to combine categories with low frequencies, if a logical way exists of doing so. For instance, Fine Arts and English majors can be combined into a category of "Humanities," and Physics and Chemistry can be combined to form the category "Natural Sciences." The results (observed and expected frequencies) are

		Major		
		Humanities	*Psychology*	*Natural Sciences*
Group 1	f_o	10	12	3
	f_e	6.5	10.0	8.5
Group 2	f_o	3	8	14
	f_e	6.5	10.0	8.5

Since all expected frequencies are now greater than 5, χ^2 can be computed.

RANDOM SAMPLING

The chi square test rests on the assumption of random sampling, because the probabilities listed in the chi square table are accurate only under random sampling. What does this mean? Consider the critical value of χ^2 for 1 *df* and $\alpha = .05$: $\chi^2 = 3.84$. This means that if the hypothesis we are testing is really true in the population, and if we draw all possible samples with $df = 1$ from the population and compute χ^2 for each one, the resulting χ^2 will be less than 3.84 in 95 out of every 100 samples and greater than 3.84 in only 5 out of every 100 samples. That is, the probability is 5/100 or .05 that any given sample, *chosen randomly* from all possible samples, will have $\chi^2 > 3.84$. If nonrandom sampling procedures are used, the probability may be quite different. For example, if biased sampling results in samples with unusual frequencies, the probability of $\chi^2 > 3.84$ may be quite large—for example, .25 or .40—even when the hypothesis being tested is true in the population.

Of course, truly random sampling is rare in research. For example, in comparing toy preferences of preschool boys and girls, a researcher could not choose randomly from all the preschool boys and girls all over the world, or even one country, or even one region. Instead, he is likely to use a readily available sample—children attending a nearby preschool, for example. In order to use a sample like this to test hypotheses concerning preschool boys and girls in general, he must at least assume that the boys and girls in the sample are typical preschoolers and that, had he chosen randomly, he would probably have gotten much the same kind of sample. Sometimes this assumption is justified and sometimes it is not. Researchers and readers of research reports must use their own judgment in cases like this.

At any rate, do not attempt to generalize to any population (whether with chi square or any other statistical technique) if you have reason to believe that the sample in a study was biased.

EXERCISE 5–9

1. Thirty boys and 30 girls are given a verbal and a mechanical puzzle to solve. The researcher wants to compare boys' and girls' solutions of the two types of puzzles. Here are the data:

Number Solving Correctly

	Mechanical Puzzle	*Verbal Puzzle*
Boys	25	12
Girls	18	20

Are the observations independent? Can a χ^2 test of independence be used to determine whether there is a difference between boys and girls?

2. Here are data collected on 50 teenagers. The investigator wonders whether urban and rural teenagers tend to prefer different TV programs.

Preferred TV Program

	"Murder She Wrote"	"20/20"	"Designing Women"	"Murphy Brown"	"48 Hours"	"Columbo"
Urban	4	1	2	7	4	2
Rural	3	2	8	5	4	8

Calculate the expected frequencies. Should a chi square test be conducted?

An Introduction to Errors of Inference

Conclusions drawn from statistical tests are always probabilistic. If we reject a hypothesis, it is because the results we obtained are unlikely to occur (but not impossible) if the hypothesis is true. If we do not reject a hypothesis, it is because the results are not unlikely (but not necessarily very likely) if the hypothesis is true. Sometimes, by chance, we reject a hypothesis that is actually true or fail to reject one that is false. Such an occurrence is called an ***error of inference***.

Suppose, for example, that Dr. Klingelfinger is right. The population distribution of psychiatric patients is what he claims it to be. Unknown to us, however, the sample distribution at PMPI happens to be an unusual one, one of those in the outer 5% of all samples. Since Dr. Klingelfinger's hypothesis was true but we rejected it, we have made an error of inference. This type of error of inference—rejecting a true hypothesis—is a ***Type I error***. Also, a test may sometimes result in nonrejection of a hypothesis that is false. This type of error of inference is a ***Type II error***. As an example, suppose that, in fact, the majority of Tooterville voters plans to vote for Candidate Smith. By failing to reject the null hypothesis that only 50% will vote for Smith, we have made a Type II error.

In any given test, we cannot be certain that no error of inference has occurred. If we reject a hypothesis, we may have done so correctly, or we may have made a Type I error. If we do not reject a hypothesis, we may have done so correctly, or we may have made a Type II error. Fortunately, we can take steps to reduce the probability of making Type I and Type II errors.

We control the probability of making a Type I error by the selection of a level of significance, alpha. Remember: When we select a particular alpha, we are saying, "I will reject the hypothesis if I get a sample outcome whose probability of occurring, if the hypothesis is true, is less than alpha." Therefore, alpha is the probability of rejecting the hypothesis when the hypothesis is true; alpha is the probability of making a Type I error.

$$p\,(\text{Type I error}) = \alpha$$

We can reduce the probability of a Type I error by choosing a small level of alpha. For example, the probability of a Type I error is smaller in a test with $\alpha = .01$ than a test with $\alpha = .05$.

The probability of a Type II error is called *beta* (β).

$$p\,(\text{Type II error}) = \beta$$

The probability of a Type II error is influenced by a number of factors, one of which is alpha. Unfortunately, as the probability of a Type I error, alpha, decreases, the probability of a Type II error, beta, increases (and vice versa). Thus, if you take steps to decrease the probability of rejecting a true hypothesis, you are increasing the probability of not rejecting a false one; if you take steps to decrease the probability of not rejecting a false hypothesis, the probability of rejecting a true one increases.

Consider, for example, the study of boys' and girls' choice of toys. Recall that the difference between boys and girls was significant at the .05 but not at the .01 level. Let us consider two scenarios:

Scenario 1: The null hypothesis is true: There is *no* difference between boys' and girls' choice of toys.

Using $\alpha = .05$, we rejected the null hypothesis. Since the null hypothesis is true, we made a Type I error.

Using $\alpha = .01$, we did not reject the null hypothesis, a correct inference since the null hypothesis is true.

Thus, selecting a smaller alpha did reduce the probability of a Type I error.

Scenario 2: The null hypothesis is false: There *is* a difference between boys' and girls' choice of toys.

Using $\alpha = .05$, we rejected the null hypothesis, a correct inference since the null hypothesis is false.

Using $\alpha = .01$, we did not reject the null hypothesis, thereby making a Type II error.

Thus, selecting a smaller alpha increased the probability of a Type II error.

Fortunately, the probability of a Type II error is affected by other factors besides alpha. One of these factors is sample size N. If the size of a sample used in a test is increased, the probability of a Type II error is reduced, even if the level of alpha remains the same.

Type I and Type II errors and related concepts are discussed at greater length in Chapter 8. For now, just keep in mind that drawing conclusions about a population, based on a subset or sample from that population, always entails a certain risk of being wrong. Although you can take steps to reduce the risk, you cannot eliminate it entirely. When discussing the results of statistical tests, always be careful to avoid making statements that sound as though you are absolutely certain that your conclusions are correct.

EXERCISE 5–10

1. In each of the following, decide whether a Type I error, a Type II error, or no error of inference has occurred.
   a. A school administrator tests the hypothesis that two methods of teaching algebra are equally effective and does not reject the hypothesis. Actually, Method B is more effective than Method A.
   b. A psychologist tests the null hypothesis that attitudes toward the death penalty do not differ among ethnic groups and rejects the hypothesis. Actually, attitudes do differ among ethnic groups.
   c. A geneticist tests the hypothesis that 60% of a certain species of plant have blue flowers, 30% have pink flowers, and 10% have yellow flowers. The test results in rejection of the hypothesis. Actually, the percentages of blue, pink, and yellow flowers in plants of this species are 60, 30, and 10, respectively.
2. Recall that in a poll of 200 voters, 113 (56.5%) said they intended to vote for Smith. We tested the null hypothesis: 50% of voters will vote for each candidate, with $\alpha = .05$. We calculated $\chi^2 = 3.38$. Since $p(\chi^2 > 3.38)$ is greater than .05, we did not reject the null hypothesis.

   Suppose the number of voters polled were 400 rather than 200, and that 226 (56.5%) said they planned to vote for Smith. Test the null hypothesis that 50% of voters will vote for each candidate. Use $\alpha = .05$. Do you reach the same conclusion as before? Explain the difference.

Using Statistical Software to Conduct Tests

All the tests discussed in this book, and many more, can be conducted using statistical software. If, for example, the data in Table 2–13 (p. 49) were typed into a machine-readable data file, we could use a package like Minitab or SPSS to conduct a chi square test of the relation between gender and major. Appropriate (relatively simple) commands would produce contingency tables, calculate cell percentages (if desired), and calculate and evaluate χ^2.

One advantage of computer-conducted tests over manually conducted tests (besides the fact that they reduce computational labor) is that the computer not only calculates the test statistic but also reports its probability. This saves the user the trouble of referring to a table; in addition, it provides more precise information about the probability of the sample outcome.

Recall our test comparing the choice of toys of preschool girls and boys, a test with 1 degree of freedom. We calculated $\chi^2 = 5.61$ and found that $p(\chi^2 > 5.61)$ was between .01 and .025. But the chi square table did not enable us to find the exact probability. Had we used a computer to conduct the analysis instead, we would have found out that $p = .0195$. (In other words, in the chi square distribution with 1 degree of freedom, a χ^2 of 5.61 cuts off a proportion of area equal to .0195 in the upper tail.) Thus we can pinpoint the probability of a sample outcome more precisely in computer-conducted tests.

Note, however, that the computer cannot tell us whether or not to reject the hypothesis and whether or not the test result is significant. Although we do not need to compare the calculated test statistic with tabled critical values, we still need to compare the p value reported by the computer with the level of significance (α) chosen before conducting the test. As always, if $p < \alpha$, we reject the hypothesis and state that the difference is significant at that level; if $p > \alpha$, we do not reject the hypothesis, and we state that the difference is not significant at that level.

Nor does the computer tell us the nature of the relationship (if any). For example, if we chose $\alpha = .05$ before conducting the choice-of-toys analysis, the fact that $p = .0195$ is less than .05 leads us to reject the null hypothesis and state that the difference between girls and boys is significant at the .05 level. But in order to describe the nature of the difference, we must refer to the contingency table and compare the percentages of girls versus boys choosing warlike toys. On that basis, we conclude that girls are more likely to choose warlike toys.

EXERCISE

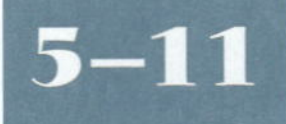

1. You have chosen $\alpha = .05$ in a chi square test of independence. The computer printout reports $p = .0470$. Do you reject the null hypothesis? Do you reject the null hypothesis if the printout reports $p = .0635$?
2. A computer printout reports $p = .00365$. Is the result significant at the .05 level? At the .01 level? At the .005 level? At the .001 level?

SUMMARY

In tests of both real-life and statistical hypotheses, we state a hypothesis, make observations to test it, then reject the hypothesis if the observations have a small probability of occurring when the hypothesis is true. In statistical hypothesis tests, the hypothesis is always about a quantitative characteristic or parameter of a population. A criterion for rejecting the hypothesis (how small the probability of the observations must be to reject the hypothesis), called the ***level of sig-***

nificance, or ***alpha* (α)**, is selected in advance of data collection. A random sample is selected from the population, and the probability of a sample outcome this different from the hypothesized population is determined. If the probability is less than alpha, the hypothesis is rejected.

In this chapter, we considered tests of hypotheses about distributions of categorical data. We use f_e to stand for the ***expected frequency*** in each category or cell if the hypothesis is true. ***Observed frequencies*** are symbolized f_o. The overall difference between observed and expected frequencies for a set of data is measured by calculating $\chi^2 = \Sigma\left[\frac{(f_o - f_e)^2}{f_e}\right]$. By referring to a table of the ***chi square distribution***, we can find out whether the probability of χ^2 greater than the calculated value is less than or greater than alpha.

The chi square distribution is a family of probability distributions, each with a particular number of ***degrees of freedom* (*df*)**. The chi square table gives, for each number of degrees of freedom, the values of χ^2 that cut off given proportions of area in the upper tail of that chi square distribution. The table entries are called ***critical values*** of χ^2. The proportion of area above any critical value is equal to the probability that χ^2 is greater than the critical value. In a chi square test, the sample χ^2 is calculated, and the probability of a χ^2 this large, $p(\chi^2 >$ sample value) is found in the table. If $p < \alpha$, the hypothesis is rejected, and the result is said to be *significant* at level alpha.

When a test is conducted with a particular alpha, the region in the upper tail with area = α is called the ***rejection region***, and the rest of the distribution is called the ***nonrejection region***. The value of χ^2 that is the cutoff point between the rejection and nonrejection regions is the critical value of χ^2 for that particular alpha. (See Figure 5–2 for relations between critical values, alpha, and rejection and nonrejection regions.)

The chi square test of ***goodness of fit*** is used to determine how well an obtained distribution is fitted by a theoretical one, or to test the null hypothesis that an equal proportion of the population falls in each of two or more categories. In a goodness-of-fit test, we calculate the frequency expected in each category if the hypothesis is true and calculate χ^2. To evaluate χ^2, we find the probability of χ^2 for df = the number of categories minus one. If the probability is less than alpha, the hypothesis is rejected.

The ***chi square test of independence*** is used to test the null hypothesis of no difference between the distribution of observations over categories in two or more populations, or the null hypothesis that two categorical variables are unrelated (independent). Data (observed frequencies) are displayed in a two-way ***contingency table***. In the test, the frequency expected in each cell if the null hypothesis is true, f_e, is calculated. Then χ^2 is calculated by the same formula as in the goodness-of-fit test, and its probability is found in the row with df = (number of rows − 1) × (number of columns − 1). If $p < \alpha$, the null hypothesis is rejected, and the difference between groups (or relation between variables) is significant at level alpha.

In both tests of goodness of fit and of independence, nonrejection of a hypothesis is not interpreted as strong evidence that the hypothesis is true but as insufficient evidence to reject it.

Though the chi square test is widely applicable, it should not be used if observations are not independent or if expected frequencies are very small. Great care should be exercised in drawing conclusions from chi square tests, or any other tests, if the samples used in the test were not selected randomly.

Conclusions drawn from statistical tests are always probablistic in nature. Sometimes a test results in rejection of a true hypothesis (a ***Type I error***), and sometimes a test results in nonrejection of a false hypothesis (a ***Type II error***). The probability of a Type I error is equal to alpha. The probability of a Type II error increases as alpha decreases and decreases as N increases.

Computer software can be used to conduct most statistical tests. An advantage of computer analysis is that the exact probability of the test statistic is reported. As always, however, interpretation of the results depends on the knowledge and reasoning of the human analyst.

QUESTIONS

1. Decide whether each of the following situations calls for a test of goodness of fit, a test of independence, or neither.

   a. "Is membership in fraternities or sororities related to community service?" To answer this question, 350 college students are asked whether they belong to fraternities or sororities (yes or no) and whether they perform any community service (yes or no).
   b. A cereal manufacturer wonders whether consumers prefer certain package colors over others. He presents a sample of 100 consumers with blue, green, yellow, and red boxes of the cereal and asks them which color they prefer.
   c. Fifty members of a community are asked their opinions about a proposed library construction project (for, against, undecided) 2 months before a local election and then again 1 month after the election. The question is "Did attitudes change from before to after the election?"
   d. Based on past experience, a college dean expects twice as many students to enroll in French 1 than in French 2. She checks the spring term enrollment at her school to see whether her expectations are confirmed.
   e. A sample of third-grade students takes two tests: a 100-item reading test and a 50-item math test. A school administrator wonders whether scores on the reading and math tests are related.

2. In Table 2–13 (p. 49), data are presented for 28 students enrolled in a statistics course. Assume that these 28 students were randomly selected from the population of all statistics students at a large university. Use the data to answer the following two questions ($\alpha = .05$):

   a. In the population, are there more students from some classes than others?
   b. Is gender related to area of major?

3. A botanist predicts from genetic theory that 50% of flowers of a certain plant will be white, 25% pink, and 25% red. She grows 84 plants. Thirty-seven have white flowers, 27 pink, and 20 red. Are the results in accord with the theory? ($\alpha = .01$.)

4. Here are data on preference for stimulus complexity of 25 two-year-old and 25 three-year-old children.

	Preferred Complexity	
	Low	*High*
2-year-olds	15	10
3-year-olds	8	17

   Use the data to answer the following questions:

   a. Is the relationship between age and preference for stimulus complexity significant at the .01 level?
   b. In 3-year-olds, is the difference in preference for high and low complexity significant at the .05 level?

5. Restate the following null hypothesis in terms of a relationship between two variables: "Experienced and inexperienced teachers do not differ in the types of punishment they use."

6. You have tested the null hypothesis in question 5. You used alpha = .01 and rejected the hypothesis. Which of the following conclusions can you draw?

   a. The difference between experienced and inexperienced teachers is significant at the .01 level.
   b. The difference between experienced and inexperienced teachers is not significant at the .01 level.
   c. The difference between experienced and inexperienced teachers is significant at the .05 level.

d. Experienced and inexperienced teachers use different methods of punishment.
e. Amount of teaching experience and punishment are not related.

7. Suppose that the null hypothesis tested in question 6 is true, but you rejected it.

   a. What kind of error of inference have you made?
   b. Under the conditions of your test, what was the probability of an error of this type?
   c. If you used alpha = .05 instead of .01, would the probability of an error of this type increase, decrease, or remain the same?

8. Restate the following null hypothesis in terms of a difference between groups: "Gender and college major are unrelated."

9. You have tested the null hypothesis in question 8. You used $\alpha = .05$ and did not reject the null hypothesis. Which of the following conclusions are justified?

   a. The college majors of males and females do not differ significantly at the .05 level.
   b. Gender and college major are related.
   c. The distributions of college majors of males and females are exactly the same.
   d. There is not sufficient evidence to conclude that the majors selected by males and females are different.

10. Suppose that the null hypothesis tested in question 9 is false, but you did not reject it.

    a. What kind of error of inference have you made?
    b. If you used alpha = .01 instead of .05, would the probability of an error of this type increase, decrease, or remain the same?
    c. How could you reduce the probability of this type of error without changing the level of significance of your test?

Relations Between Sample and Population Means and Variances

INTRODUCTION

In Chapter 3, we discussed the calculation and use of measures of central tendency and variability as descriptors of a set of available data. As we saw in Chapters 4 and 5, however, available data are often just a sample from a population about which we seek information. The sample has been selected from the population in order to make inferences about the population.

In Chapter 5 you learned to use the chi square test to make inferences about categorical data. When data are quantitative rather than categorical, questions about populations are usually (though not always) questions about the population means. For example, a researcher interested in the effect of an experimental treatment usually wants to know about the mean effect. Similarly, questions about differences between populations are usually questions about differences between means. Because of this, we need to pay particular attention to the relations between sample means and population means, so that we can understand processes for generalizing from the former to latter.

As you will see, in order to draw conclusions about population means from sample data, we need to take variances into account. Therefore, we will also need to examine the relationship between sample variances and population variances. The variance is important in its own right, too. The chapter closes with an introduction to a concept that is at the heart of statistical procedures that attempt to sort out sources of variability in data, the concept of *components of variance*.

Objectives

When you have completed this chapter, you will be able to

- Define the sampling distribution of the mean, and describe procedures for generating empirical and theoretical sampling distributions of the mean.
- Define the standard error of the mean.
- State the central limit theorem; specifically, explain how the shape, the mean, and the variance of the sampling distribution of the mean are related to the shape, mean, and variance of the population and to sample size.

- Distinguish between biased and unbiased estimators of population parameters. Explain why the sample variance (described in Chapter 3) is a biased estimator of the population variance.
- Calculate unbiased estimates of population variances from sample data.
- Explain the concept of components of variance in your own words.

Parameters and Statistics

In Chapter 3 you learned how to calculate and interpret measures of central tendency—the mean, median, and mode—and measures of variability—the range, interquartile and semi-interquartile ranges, and variance and standard deviation. The emphasis in Chapter 3 was on the use of these statistics to describe the data at hand. But, as you know, the data at hand are usually just a sample from a large, often inaccessible population. The sample statistics are calculated not primarily to describe the sample but rather to provide information about parameters of the population. Although any statistic can be an estimator of a population parameter, we will limit our discussion in this chapter to sample means and variances as estimators of population means and variances, respectively. The reason is that means and variances are involved in many widely used and powerful inferential procedures. In this chapter we lay the foundation for methods of inference discussed in the rest of the book.

Recall that it is common practice to use Greek letters to represent parameters of populations (and of theoretical distributions). We use the symbol μ to stand for a population mean. Thus, if the mean of a population is 100, we write "$\mu = 100$." For the standard deviation of a population, we use the symbol σ (sigma). Since the variance is the squared standard deviation, the population variance is denoted σ^2.

Statistics calculated in samples are commonly represented by Roman letters. For the mean of a sample, we continue to use the symbol M. Another symbol that is often used for a sample mean is X ("X bar"). In either case, the calculation is the same:

$$M = \overline{X} = \frac{\Sigma X}{N}$$

In Chapter 3, we used Var to stand for the variance of a data set and calculated Var by the following formula:

$$\text{Var} = \frac{SS}{N}$$

SS stands for "sum of squares," which is short for "sum of squared deviations from the mean." It can be calculated by either of the following formulas:

$$SS = \Sigma(X - M)^2 \qquad \textit{Deviation formula for the sum of squares}$$

$$SS = \Sigma X^2 - \frac{(\Sigma X)^2}{N} \qquad \textit{Computational (raw score) formula for the sum of squares}$$

The sample standard deviation, S, is the square root of the variance. Later in the chapter, we will introduce a slightly different method for calculating sample variances and standard deviations when the purpose is to estimate the variances and standard deviations of the populations from which the samples came.

EXERCISE

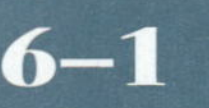

1. You have selected a sample of size 20 from a population. You calculated mean = 36.4 and standard deviation = 6.81. The population has a mean of 40.3 and variance of 64. What are the values of μ, σ^2, σ, M, Var, and S?
2. *Review:* Calculate the mean, sum of squares, variance, and standard deviation of the following data: 6, 5, 8, 9, 6, 4, 4.

The Relation Between Population and Sample Means

Suppose that you are an educational researcher interested in finding out whether a new method of teaching algebra to ninth graders is more effective than the method currently in use. Under the old method, the mean score on an end-of-year algebra test in the population of ninth-grade algebra students has been equal to 75 over the past 10 years. You believe that the new method will raise the mean level of performance in the population. To test your hypothesis, you try the new method with a random sample of 25 students. The 25 students attain a mean end-of-year test score of 80. Since 80 is higher than the old mean, 75, it looks as though the new method is more effective than the old.

But wait a minute. You have used the new method with a sample of only 25 ninth-grade algebra students, not the whole population. Means of samples, like any other statistic, vary from one sample to another. Can you be confident, based on the fact that the sample mean is 80, that the mean of the entire population would be higher than 75 if the new method were used? Perhaps or perhaps not. It depends on the relation between sample means and the mean of the population.

If you can show that samples of size 25 from a population like this one usually have means that are within 2 or 3 points of the population mean, you are justified in concluding that, with the new method, the population mean probably is greater than 75. But if means of samples of size 25 are expected to differ from the population mean by 5 or 6 points or more just by chance, a sample mean of 80 is not convincing evidence that the population mean is greater than 75.

Differences between sample statistics and the corresponding population parameters are called ***sampling errors***. (This does not mean that a mistake has been made in sampling or calculation; in statistics, the term "error" just means a difference that is due to chance.) Thus, the difference between the mean of a sample and the population mean, $M - \mu$, is a sampling error. In order to draw conclusions about a population mean from just one sample, we need to know how big sampling errors are likely to be. That is, we need to know how much sample means vary from one sample to another, and how different they tend to be from the population mean. We need to examine the ***sampling distribution of the mean***.

Recall from Chapter 4 that a sampling distribution of a statistic is the distribution of the statistic over many samples of a given size from the same population. An *empirical sampling distribution* is obtained by randomly selecting a limited number of samples of size N and calculating the statistic in each sample. In Chapter 4 we constructed two empirical sampling distributions of the mean, one based on 50 random samples of size 5, the second based on 50 random samples of size 20, from a population with a mean of 8 ($\mu = 8$). We discovered the following relations between sample and population means:

1. Though sample means varied from one sample to another, they varied around the population mean, 8, and most of the sample means were reasonably close to the population mean.

2. The means of larger samples varied less from one sample to another and tended to be closer to the population mean, on the average, than means of smaller samples.

A Theoretical Sampling Distribution of the Mean

From our two empirical sampling distributions of the mean, we learned something about the relations between sample and population means. But our generalizations were limited by the small number of samples in each sampling distribution. A different set of 50 samples would result in a somewhat different sampling distribution. We would get a more accurate picture of the relation between population and sample means if we could examine the means of all possible samples of size 5 and size 20—the ***theoretical sampling distributions of the mean*** for samples of these sizes. We cannot do so, however, because there are too many possible samples.

In order to make the task of constructing a theoretical sampling distribution of the mean manageable, we will sample from a very small population, one that contains only four members with scores of 3, 5, 7, and 9, respectively. Since each of the four scores occurs equally often, the population distribution is rectangular. The population is displayed in Figure 6–1.

Let us calculate the population mean (μ), variance (σ^2), and standard deviation (σ). We use N_p to represent the size of the population.

$$\mu = \frac{\Sigma X}{N_p} = \frac{3+5+7+9}{4} = 6$$

$$\sigma^2 = \frac{\Sigma(X-\mu)^2}{N_p} = \frac{(3-6)^2+(5-6)^2+(7-6)^2+(9-6)^2}{4} = 5$$

$$\sigma = \sqrt{5} = 2.236$$

Thus, the population mean is 6, the population variance is 5, and the population standard deviation is 2.236.

We will construct the theoretical sampling distribution of the mean for samples of size 2 (N = 2) by listing all possible samples of size 2 and calculating the mean (M) of

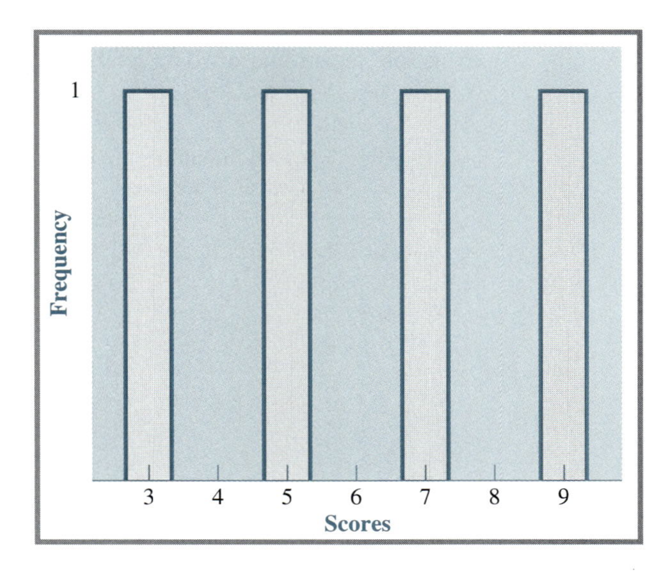

Figure 6–1

A small population consisting of the four scores 3, 5, 7, and 9.

each sample. Before we do this, however, we need to discuss several aspects of our sampling procedure.

1. In Chapter 4 we distinguished between sampling *with replacement* and sampling *without replacement*. In sampling with replacement, each member selected for a sample is replaced in the population before the next member of the sample is drawn. In sampling without replacement, a member selected for a sample is not replaced in the population before the next member is drawn.
   The sampling and probability distributions underlying most inferential procedures assume sampling with replacement. If populations are large, probabilities of various sampling outcomes are much the same when sampling with and without replacement, but with small populations, the probabilities are quite different. Since our population is very small, we must sample with replacement. This means that, for each sample of size 2, we select a member of the population and record the score; then we return the first member to the population before drawing a second member for the sample.
2. In sampling with replacement, it is possible to get a sample consisting of the same member, chosen twice. For example, if 3 is drawn first and then, having been replaced, is drawn again, the sample is 3, 3.
3. Order of selection must be considered in listing the samples. The sample 3,7 (3 drawn first and 7 second) and the sample 7, 3 (7 drawn first and 3 second) are different samples.

Now we list all possible samples of size 2 from our small population. In sampling with replacement, there are 16 possible samples. Table 6–1 lists the 16 samples and the mean (M) of each one (M = sum of the two scores divided by 2). A frequency distribution and a graph of the sample means are shown in Figure 6–2. We will examine three features of this theoretical sampling distribution of the mean: its shape; its average value, the mean of the sample means; and its variability, the variance and the standard deviation of the sample means.

First, notice that the shape of the sampling distribution of the mean is quite different from that of the population. Although the population is rectangular, the sampling distribution of the mean is triangular. The most frequently occurring sample mean is 6 (the population mean), and frequencies decrease at a constant rate both below and above 6.

Next, we calculate the mean of the sampling distribution of the mean by adding the 16 sample means and dividing by 16:

Table 6–1 All Possible Samples of Size 2 From the Population 3, 5, 7, 9

Sample	Mean (M)	Sample	Mean (M)
3, 3	3	7, 3	5
3, 5	4	7, 5	6
3, 7	5	7, 7	7
3, 9	6	7, 9	8
5, 3	4	9, 3	6
5, 5	5	9, 5	7
5, 7	6	9, 7	8
5, 9	7	9, 9	9

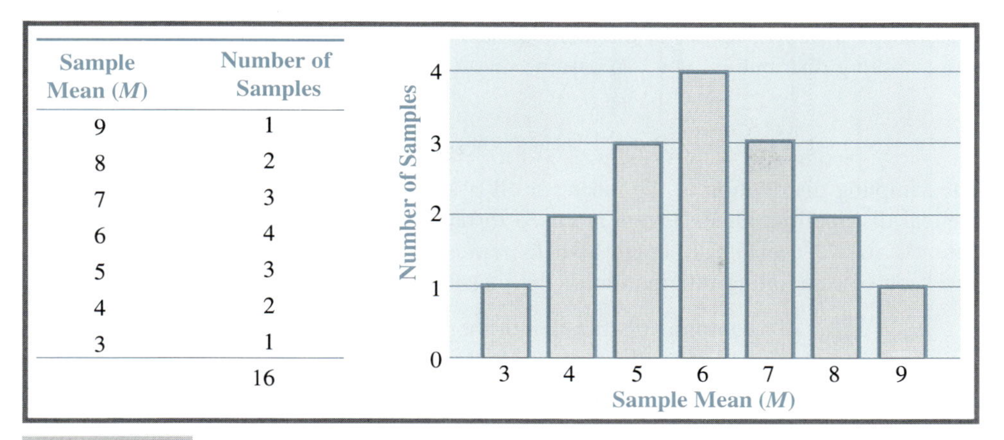

Sample Mean (M)	Number of Samples
9	1
8	2
7	3
6	4
5	3
4	2
3	1
	16

Figure 6–2 **The theoretical sampling distribution of the mean for samples of size 2 from the population 3, 5, 7, 9.**

$$\text{Mean of sample means} = \frac{3+4+5+6+4+\cdots+8+9}{16} = 6$$

Thus, even though sample means vary from 3 to 9, the mean of the sample means is equal to the population mean, 6.

Finally, we examine the variability of the sample means. Though the range of sample means is the same as the range of scores in the population (from 3 to 9 in both cases), the sample means are more concentrated about the population mean than are the population raw scores. Therefore, the variance of the distribution of means is smaller than the population variance. To calculate the variance of the 16 sample means, we subtract the mean of the sample means, 6, from each sample mean, square each deviation, add the squared deviations, and divide by 16.

$$\text{Variance of sample means} = \frac{(3-6)^2+(4-6)^2+\cdots+(8-6)^2+(9-6)^2}{16} = 2.5$$

The variance of the sample means is half the population variance (5). The standard deviation of the sample means is the square root of the variance, $\sqrt{2.5} = 1.58$.

In sum, we have observed the following relationships between our small population and the theoretical sampling distribution of the mean for samples of size 2:

1. Though the population is rectangular in shape, the distribution of sample means is triangular with a peak at the population mean.
2. The mean of the distribution of sample means is equal to the population mean, μ.
3. The variance of the sample means is smaller than the population variance. In fact, it is smaller by a factor equal to the sample size, 2.

But ours is only one very small population. And we have considered samples of only one size. Would the same kinds of relationships hold for samples of various sizes from various populations? Do we need to construct hundreds and hundreds of sampling distributions to find out? Fortunately not, as you will soon see. However, first, we introduce some new terms and symbols.

A brief note: Up to now, we have considered both empirical and theoretical sampling distributions of the mean. In the rest of the book, however, discussions of sampling distributions will always be about theoretical ones. Therefore, we will drop the word "theoretical" from the names of the distributions. From now on, when you see a reference to a sampling distribution, you can assume that it is a theoretical sampling distribution.

Symbols and Terms for the Sampling Distribution of the Mean

The sampling distribution of the mean for all possible samples of a certain size is a theoretical distribution. Greek letters are used to represent parameters of theoretical distributions, so we use Greek letters also to represent the mean, variance, and standard deviation of a sampling distribution of the mean. For the mean, we use the symbol μ_M.

μ_M = Mean of a sampling distribution of the mean; the average value of sample means over all possible samples of size N

For example, in the theoretical sampling distribution that we constructed in the previous section, the mean of the 16 sample means is 6. Therefore, $\mu_M = 6$.

The variance of a sampling distribution of the mean is symbolized σ_M^2.

σ_M^2 = Variance of a sampling distribution of the mean; the variance of sample means over all possible samples of size N

For example, in the theoretical sampling distribution that we constructed, the variance of the sample means is 2.5. Therefore $\sigma_M^2 = 2.5$.

The standard deviation of a sampling distribution of the mean is symbolized σ_M, and it has a special name. It is called the ***standard error of the mean***.

σ_M = Standard error of the mean; the standard deviation of sample means over all possible samples of size N

In the sampling distribution constructed earlier, the standard error of the mean, σ_M is $\sqrt{2.5} = 1.58$.

Reminder: The word *error* in the term *standard error* does not mean that any kind of mistake has occurred in sampling or in calculation. In statistics, the term *error* simply means "chance difference." σ_M is called the standard error of the mean because it measures the amount by which sample means differ from the population mean by chance.

EXERCISE 6–2

1. Suppose you were to construct the sampling distribution of the mean for samples of size 3 instead of 2, sampling with replacement from the population 3, 5, 7, 9. (*Note:* You are not expected to actually construct it, since the number of possible samples is large, 64.) Do you expect the variance of the sample means (σ_M^2) to be greater than, less than, or equal to the variance of the means of samples of size 2?
2. Here is a very small population: 2, 8, 14.
   a. Calculate the mean (μ), variance (σ^2), and standard deviation (σ) of the population.
   b. List all possible samples of size 2 ($N = 2$), sampling with replacement (there are nine possible samples).
   c. Calculate the mean (M) of each sample, and graph the sampling distribution of the mean.
   d. Calculate the mean of the sampling distribution of the mean, μ_M. How is it related to μ?
   e. Calculate the variance of the sampling distribution of the mean (σ_M^2) and the standard error of the mean (σ_M). How is σ_M^2 related to σ^2 and N?

3. Fill in the blanks:

Ms. Armitage, a human resources manager, is interested in the typing speed (words per minute) of the 250 secretaries at her company. She decides to measure the typing speed of a random sample of 30 secretaries. In this case, M stands for ____________, and μ stands for ____________. The variance of the typing speed scores of all the secretaries is symbolized ____________, whereas the variance of her sample of 30 secretaries is ____________.

If Ms. Armitage were to list the means of all possible samples of size 30 (a gargantuan task!), the distribution of sample means would be called ____________. The mean of the distribution would be symbolized ______. The standard deviation of the distribution would be symbolized ______ and would be called ____________.

The Central Limit Theorem

Now we are ready to consider the properties of the sampling distribution of the mean for samples of all sizes from all populations. Fortunately, we do not need to construct hundreds and hundreds of sampling distributions of the mean in order to find out what these properties are. The shape, the mean, and the variance of the sampling distribution of the mean for samples of any size drawn from any population are described by a very important theorem called the ***central limit theorem***. The central limit theorem states

If samples of size N are drawn from a population with mean, μ, and variance, σ^2,

1. (a) If the population is normal, the sampling distribution of the mean is normal for all sample sizes.
   (b) If the population is not normal, the sampling distribution of the mean approaches a normal distribution as sample size increases.
2. The mean of the sampling distribution of the mean is equal to the population mean. That is

$$\mu_M = \mu$$

3. The variance of the sampling distribution of the mean is equal to the population variance divided by the sample size:

$$\sigma_M^2 = \frac{\sigma^2}{N}$$

Let us examine the central limit theorem more closely, one part at a time.

The Shape of the Sampling Distribution of the Mean

Sometimes we sample from populations that are known or assumed to be normal (e.g., IQ test scores of all children in the United States). The central limit theorem tells us that, if we were to select or list all possible samples of size 2, or size 10, or size 50—or any size whatsoever—from a normal population and calculate the mean of each sample, the distribution of the sample means would be normal.

Suppose, however, that the population from which we are sampling is not normal. According to the central limit theorem, if we select all possible samples of a small size, such as $N = 2$ or $N = 5$, the shape of the distribution of sample means may be quite different from normal. However, if the size of the samples drawn from the very same nonnormal population is large, for example, $N = 30$, the distribution of sample means is very similar to a normal distribution, despite the non-normality of the population. The larger the sample size, the more nearly normal is the sampling distribution of the mean.

In an earlier section, we constructed the sampling distribution of the mean for samples of size 2 from the rectangular population 3, 5, 7, 9. The population was markedly non-normal. Yet, even with $N = 2$, the distribution of sample means, though clearly not normal, was closer to a normal distribution than the population distribution. If we were to list all possible samples of size 20 or 30 from the same population and calculate the mean of each sample, the sampling distribution of the mean would be very close to a normal distribution.†

The fact that the sampling distribution of the mean is approximately normal provided that N is reasonably large, regardless of the shape of the population, is extremely important, because it enables us to use normal curve procedures to find probabilities of various sample means, as you will see in the next chapter.

The Mean of the Sampling Distribution of the Mean

In the theoretical sampling distribution of the mean that we constructed in an earlier section, we found that the mean of the distribution of sample means was equal to the population mean, 6. The central limit theorem tells us that this is true for all sampling distributions of the mean. If we had listed all possible samples of size 3, or 10, or 50, instead of size 2, the mean of the sampling distribution of the mean would have been 6. Similarly, if you sample from a population with mean = 50, the mean of all possible sample means is 50; in sampling from a population with mean = –63.2, the mean of all possible sample means is –63.2; and so on.

Be sure that you understand that we are *not* saying that every sample from a population has a mean equal to the population mean. Means of samples, like any other statistic, vary from one sample to another. What the central limit theorem tells us is that *sample means vary around the population mean,* μ. Though some samples have a mean smaller than μ, and others have a mean greater than μ, *the average value of M over all samples is* μ.

The Variance of the Sampling Distribution of the Mean

The central limit theorem states that the variance of the sampling distribution of the mean, σ_M^2, is equal to the population variance divided by the sample size:

$$\sigma_M^2 = \frac{\sigma^2}{N}$$

For example, the population 3, 5, 7, 9 has variance (σ^2) = 5. We found that the variance of the means of samples of size 2 ($N = 2$) was 2.5, which is equal to the population variance 5, divided by the sample size, 2. Similarly, if you were to examine the sampling distribution of the mean for samples of size 25 from a population with $\sigma^2 = 100$, the variance of the sample means, σ_M^2, would be equal to 100/25 = 4.

The formula relating σ_M^2 to σ^2 and N shows that, *as* σ^2 *increases,* σ_M^2 *increases*. That is, the greater the variation of scores in the population, the greater will be the variation of sample means. For example, if samples of size 10 are selected from a population with variance (σ^2) = 100, the variance of the sample means is 100/10 = 10. But if the population variance is 500, the variance of the sample means is 500/10 = 50.

†You may wonder how we can take samples of size 30 from a population of size 4. When sampling with replacement, there is no limit to sample size, no matter how small the population.

However, *as N increases,* σ_M^2 *decreases.* For example, if the population variance is 100, the variance of means of samples of size 10 is 100/10 = 10; but the variance of means of samples of size 50 is only 100/50 = 2. In other words, as we found in our empirical sampling distributions of the mean, means of large samples vary less from one sample to another than do means of small samples.

The effect of the population variance σ^2 and sample size N on the variance of the sampling distribution of the mean is illustrated in Figure 6–3. The figure shows four sampling distributions of the mean In all four cases, the population from which the samples came is normal with $\mu = 70$, but σ^2 and N differ from one case to another. In case (a), $\sigma^2 = 25$, (a relatively small population variance) and $N = 5$ (a small sample size). In case (b) σ^2 is again 25, but $N = 50$ (a larger sample size). In case (c), $\sigma^2 = 100$ (a larger population variance) and N is small (5); in part (d), a large population variance ($\sigma^2 = 100$) is combined with a large sample size ($N = 50$). Note that means of samples vary least when large samples are selected from a population with a small variance, as in case (b); and means of samples vary most when small samples are selected from a population with a large variance, as in case (c).

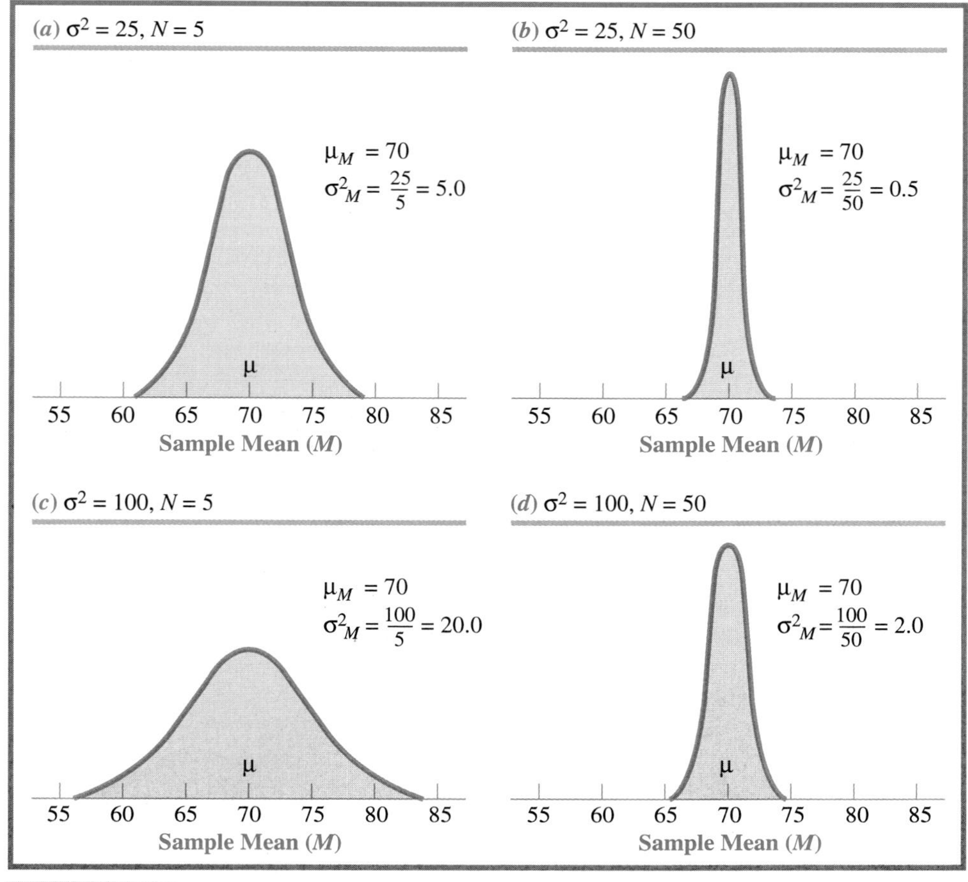

Figure 6–3 **Four sampling distributions of the mean of samples from normal populations with mean (μ) = 70. In each case, the mean of the sample means (μ_M) = μ = 70, and the variance of the sample means (σ_M^2) = σ^2/N.**

Recall that the standard deviation of the sampling distribution of the mean, σ_M, is called the standard error of the mean. Since a standard deviation is the square root of a variance, σ_M is the square root of σ_M^2:

$$\sigma_M = \sqrt{\sigma_M^2} = \sqrt{\frac{\sigma^2}{N}} = \frac{\sigma}{\sqrt{N}}$$

Thus, the standard error of the mean is equal to the population standard deviation divided by the square root of the sample size. In the sampling distribution of the mean for samples of size 2 from the population 3, 5, 7, 9, we calculated $\sigma_M = 1.58$, which is equal to the population standard deviation, 2.236, divided by $\sqrt{2}$.

In summary, if we know the mean and variance (or standard deviation) of a population, we can calculate the mean, variance, and standard deviation (standard error) of the sampling distribution of the mean for samples of any size by using the following formulas:

Mean of the sampling distribution of the mean: $\mu_M = \mu$

Variance of the sampling distribution of the mean: $\sigma_M^2 = \dfrac{\sigma^2}{N}$

Standard error of the mean: $\sigma_M = \dfrac{\sigma}{\sqrt{N}}$

For example, if we drew all possible samples of size 25 from a population with a mean (μ) of 80 and a standard deviation (σ) of 15, and if we calculated the mean of each sample, we would find that the mean of the sample means is 80 and the standard deviation of the sample means (the standard error of the mean) is $15/\sqrt{2.5} = 3$.

Moreover, if the population is normal or the sample size is sufficiently large, the sampling distribution of the mean is approximately normal. In Chapter 7 we will use these facts to find probabilities of various sample means and to test hypotheses about population means.

EXERCISE

6–3

1. You draw all possible samples of size 9 from a normal population with mean = 30 and standard deviation = 5. Next, you calculate the mean of each sample and look at the distribution of sample means.
   a. What is the distribution of sample means called?
   b. What is its shape?
   c. What is the symbol for and value of its mean?
   d. What is its standard deviation called, and what symbol is used to represent it?
   e. What is the value of its standard deviation?
   f. If you drew all possible samples of size 16 instead, which of your answers above would change? Calculate the changed answers.
   g. If the shape of the population were markedly non-normal, which of your answers above would change?
2. A markedly non-normal distribution has mean = 100 and *SD* = 20. If you examine the sampling distribution of the mean for samples of size 10 and size 25, which of the two distributions is more nearly normal in shape?

Estimating the Variance of a Population

As you have seen, the relationship between the mean of a population and the means of samples of size N from the population depends not only on the population mean and sample size but also on the population variance, σ^2. Therefore, inferences about population means must take variances into account. Population variances, however, can rarely be measured directly. Just as we rely on sample data to provide information about the population mean, we must also use sample data to estimate the population variance. We need to consider the relation between the population variance and the variances of samples from the population.

In Chapter 3 you learned to calculate the variance of a data set by using the formula

$$\text{Var} = \frac{SS}{N}$$

where SS (the sum of squares) is equal to $\Sigma(X - M)^2$ or $\Sigma X^2 - \frac{(\Sigma X)^2}{N}$. Now we need to ask: Is Var a good estimator of the population variance σ^2? Unfortunately, the answer is no. Although Var is a good measure of the variability of a sample, it is not a good estimator of the population variance.

In order to explain this statement we must stop to consider what is meant by a "good" estimator of a population parameter and what makes one statistic a "better" estimator than another. There are several ways in which a statistic may be a good or poor estimator. In one sense, a statistic is a good estimator if it shows relatively little variation from one sample to another. In this sense, the sample mean is a better estimator of the average of a population than the sample mode because the mean varies less from one sample to another. Similarly, the sample variance and standard deviation are better estimators of population variability than the sample range because they vary less over different samples.

In another sense, a statistic is considered a good estimator of a parameter if it does not tend to overestimate or to underestimate the parameter, in other words, if its mean value over all possible samples is equal to the parameter. A statistic for which this is true is called an ***unbiased estimator*** of the parameter. The sample mean, M, for example, is an unbiased estimator of the population mean, μ, because, as the central limit theorem states, the average value of M over all possible samples is equal to μ.

Now let us consider the sample variance, Var, as an estimator of the population variance, σ^2. Var varies from one sample to another, as does any statistic. In some samples, Var is less than σ^2, and in some it is greater. Unfortunately, however, Var is more often smaller rather than larger than σ^2, and the average value of Var over all possible samples is less than σ^2. Suppose, for example, that you are sampling from a population with variance $\sigma^2 = 100$. Most of your samples will have a variance (Var) that is less than 100, and the average value of Var over all your samples will be less than 100 rather than equal to it. Since Var tends to underestimate σ^2, it is a biased estimator of σ^2.

As a demonstration, we return to the 16 samples of size 2 from the population 3, 5, 7, 9. The variance of the population is 5. Let us calculate Var in each sample and compare the average value of Var over all samples with the population variance, 5. The samples and their means are listed in Table 6–1.

The first sample has scores 3, 3, with a mean (M) of 3. Therefore, the sample variance is

$$\text{Var} = \frac{\Sigma(X-M)^2}{N} = \frac{(3-3)^2+(3-3)^2}{2} = \frac{0+0}{2} = 0$$

The second sample, with scores 3, 5, has $M = 4$. Therefore,

$$\text{Var} = \frac{(3-4)^2+(5-4)^2}{2} = \frac{(-1)^2+1^2}{2} = \frac{1+1}{2} = \frac{2}{2} = 1$$

Var is calculated in the same way for the remaining samples. Table 6–2 lists the 16 sample variances.

Now examine the distribution of Var over the 16 samples. First, notice that in 14 of the 16 samples, Var is smaller than the population variance, 5; only two samples have Var greater than 5. The mean value of Var over all samples is also smaller than 5:

$$\text{Mean value of Var} = \frac{0+1+4+\cdots+1+0}{16} = \frac{40}{16} = 2.5$$

It can be shown that the mean value of Var over all possible samples is always less than the population variance, not only in this small sampling distribution but in all sampling distributions of Var, for samples of all sizes from all populations. For example, if you were to take all possible samples of size 20 (or any other size) from a population with variance (σ^2) = 60, and if you calculated Var in each sample, the average value of Var over all the samples would be less than 60. Since the mean value of Var is not equal to σ^2, Var is a biased estimator of σ^2.

How can we obtain a better, unbiased, estimator of the population variance from sample data? Consider the formula for Var = SS/N. As just indicated, this formula yields a number that tends to be smaller than σ^2. Suppose we divide SS by $N - 1$ instead of N. This will result in a larger value. For example, if SS = 450 and N = 10

$$\text{Var} = \frac{SS}{N} = \frac{450}{10} = 45, \quad \text{but} \quad \frac{SS}{N-1} = \frac{450}{9} = 50$$

It can be shown that SS divided by ($N - 1$) is an unbiased estimator of σ^2. That is, if we were to draw or list all possible samples of size N from a population having a certain variance σ^2, and if we computed $SS/(N - 1)$ in each sample, the average value of

Table 6–2 Variances (Var) of All Possible Samples of Size 2 From the Population 3, 5, 7, 9

Sample	Var	Sample	Var
3, 3	0	7, 3	4
3, 5	1	7, 5	1
3, 7	4	7, 7	0
3, 9	9	7, 9	1
5, 3	1	9, 3	9
5, 5	0	9, 5	4
5, 7	1	9, 7	1
5, 9	4	9, 9	0

Note: In each sample, Var = $\frac{\Sigma(X-M)^2}{N}$.

$SS/(N-1)$ over all the samples would be equal to σ^2. Using s^2 to symbolize the unbiased estimator of σ^2, we can write

$$\text{Unbiased estimator of } \sigma^2: s^2 = \frac{SS}{N-1}$$

To illustrate that s^2 is an unbiased estimator of σ^2, let us calculate s^2 for each of the 16 samples of size 2 from the population 3, 5, 7, 9. In the first sample, with scores 3, 3 and $M = 3$:

$$s^2 = \frac{\Sigma(X-M)^2}{N-1} = \frac{(3-3)^2+(3-3)^2}{2-1} = \frac{0+0}{1} = 0$$

In the second sample, with scores 3, 5 and $M = 4$

$$s^2 = \frac{(3-4)^2+(5-4)^2}{2-1} = \frac{1+1}{1} = 2$$

The values of s^2 for all 16 samples are listed in Table 6–3.

Now we calculate the mean value of s^2 over the 16 samples.

$$\text{Mean of } s^2 = \frac{0+1+\cdots+2+0}{16} = \frac{80}{16} = 5 = \sigma^2$$

Although s^2 varies from one sample to another, the mean value of s^2 over all samples is equal to the population variance, 5. Thus, s^2 is an unbiased estimator of σ^2.

Why is $s^2 = SS/(N-1)$ a more accurate estimate of the population variance than Var = SS/N? One reason is that in drawing samples, especially small ones, from a population, the probability of drawing observations from either extreme of the population is small. Most samples, then, are less variable than the population from which they came; this is one reason that Var is usually smaller than σ^2.

A more technical reason has to do with degrees of freedom *(df)*. Recall from Chapter 5 that "degrees of freedom" refers to the number of observations in a set that are free to vary. Each restriction on a data set reduces the degrees of freedom by one. Our estimation of the population variance from sample data is restricted by the fact that deviations are measured from the sample mean, M, rather than from the true population mean. This restriction reduces the number of degrees of freedom for estimating the variance to

Table 6–3 Variance Estimates (s^2) of All Possible Samples of Size 2 From the Population 3, 5, 7, 9

Sample	s^2	Sample	s^2
3, 3	0	7, 3	8
3, 5	2	7, 5	2
3, 7	8	7, 7	0
3, 9	18	7, 9	2
5, 3	2	9, 3	18
5, 5	0	9, 5	8
5, 7	2	9, 7	2
5, 9	8	9, 9	0

Note: In each sample, $s^2 = \dfrac{\Sigma(X-M)^2}{N-1}$.

$N - 1$ rather than N. In order to obtain a good estimate of the population variance around the *population* mean, we must divide the sum of squared deviations around the *sample* mean by the degrees of freedom, $N - 1$.

In summary, we now have two ways to measure the variance of a sample:

$$\text{Var} = \frac{SS}{N}$$ *Descriptive variance; biased estimator of population variance, σ^2*

$$s^2 = \frac{SS}{df} = \frac{SS}{N-1}$$ *Unbiased estimator of population variance, σ^2*

From now on, when we are concerned with inference rather than description, we will use s^2 rather than Var as a measure of the variance, and we will use its square root s as a measure of the standard deviation.

Incidentally, when a sample is very large it makes little difference whether the sum of squares, SS, is divided by N or by $N - 1$. The values of Var and s^2 will be very nearly the same. For example, if $SS = 2000$ and $N = 100$

$$\text{Var} = \frac{2000}{100} = 20.00$$

$$s^2 = \frac{2000}{99} = 20.20$$

However, if N is small, Var and s^2 may be quite different. With $SS = 200$ and $N = 10$, we have

$$\text{Var} = \frac{200}{10} = 20.00$$

$$s^2 = \frac{200}{9} = 22.22$$

Therefore, when population variances or standard deviations are to be estimated from small samples, it is *essential* to divide the sum of squares by $N - 1$ rather than by N.

EXERCISE 6–4

1. From the following sample, compute estimates of the population variance and standard deviation. Scores: 6, 3, 4, 6, 5, 8, 5, 9, 8.
2. The square root of s^2, $s = \sqrt{\frac{SS}{N-1}}$, is our best estimate of σ, the population standard deviation. Yet, although s^2 is an unbiased estimator of σ^2, s is *not* an unbiased estimator of σ. To show that this is so, calculate s for each of the 16 samples of size 2 from the population 3, 5, 7, 9; then calculate the mean value of s. You will find that it is not equal to the population standard deviation, 2.236.

An Introduction to Components of Variance

As mentioned in Chapter 1, statistics is necessary because of variability in data. To a large extent, the purpose of statistical analysis is to explain variability. An experimenter who has applied different experimental treatments to two groups of subjects conducts a statistical test to determine the degree to which differences (variability) among the subjects are due to the treatment variable and to what degree they are due to other factors. A political scientist who has measured voting patterns of American adults analyzes the data to find out to what extent differences in voting patterns can be attributed to age and to socioeconomic status. An educator, finding a great deal of variability in readiness test scores of first graders, wonders what kinds of factors account for the differences among the children. Though the specific statistical techniques employed by these three investigators may differ, their ultimate goal is the same: *to apportion the total variation of their data into components reflecting the sources of variation.*

Let us consider the example of children's readiness scores. Suppose that the educator is able to determine that two factors, verbal ability and quantitative ability, contribute to differences among children's readiness scores; that is, there are two components of readiness variance. But which is more important? Do differences in verbal ability produce greater differences in readiness scores, or are the score differences mostly due to differences in quantitative ability? We rephrase the question: "What proportion of the variation of the readiness scores is accounted for by the differences in verbal ability and what proportion by differences in quantitative ability?"

The educator can answer this question by measuring each child's verbal ability and quantitative ability, and then applying certain statistical procedures. Let us suppose that he has done this, and that the statistical analysis shows that 75% of the variation of readiness scores is due to differences in verbal ability and 25% to differences in quantitative ability. At this point, of course, you do not know how figures such as 75% and 25% are arrived at. The important thing for you to note is that (1) such apportionment yields vital information about the relative importance of various factors in contributing to individual differences; and (2) *only measures of variation can be statistically apportioned into components* in this way. We cannot conclude, for example, that each child's readiness score is 75% verbal and 25% quantitative, or that the mean of the readiness scores is 75% verbal and 25% quantitative.

Even when statistics are computed that seemingly bear little relation to variances, it is often possible to interpret the results in terms of ***components of variance***. In the readiness example, the educator probably calculated statistics called *correlation coefficients*. On the other hand, a researcher comparing the effects of two experimental treatments would probably calculate means and conduct a test for the difference between the means. But regardless of the particular statistical analysis, the results can be used to decide what proportion of the variation of the data is due to one factor and what proportion to another.

To illustrate the concept of components of variance graphically, we can depict the total variance of a set of scores as a circle. By using certain statistical analyses, we are able to divide the circle into two or more sectors. The area of each sector represents the contribution of one factor to the total variance. For example, the educational psychologist's conclusion concerning the relative contributions of verbal and quantitative abilities to the variance of readiness scores can be depicted as shown in Figure 6–4a.

Here is another example: In a study of voting patterns of American adults, a political scientist finds that age accounts for 20% of the variance in voting behavior and socioeconomic status (SES) for another 30%. That is, 20% of the differences among voting patterns are due to voters' age, 30% of the differences are due to SES, and the remaining 50% of the differences are due to other factors. These results can be illustrated as shown in Figure 6–4b.

The concept of components of variance allows scientists to attribute differences among people, animals, objects, or events to various sources and to determine how much each source contributes to the differences. In later chapters you will see how this concept is applied in a variety of statistical analyses.

Figure 6–4

Components of variance of (a) reading readiness scores and (b) voting behaviors. In each case, the area of the circle represents the total data variance.

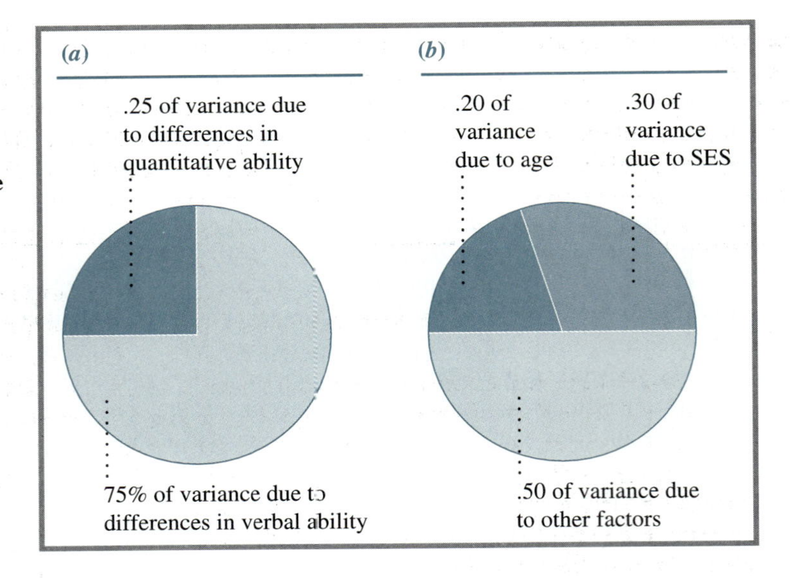

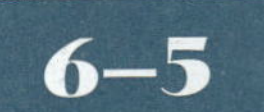

EXERCISE 6–5 Batting averages of baseball players on a team vary greatly from one player to another. This variation is due to many factors, including differences among players in physical strength, eye-hand coordination, amount of practice, and others. Make up a components of variance explanation of differences in batting average among the players on a baseball team, using the factors discussed and any others you may want to add. Draw a diagram like Figure 6–4 to illustrate your analysis.

SUMMARY

This chapter, though short, has introduced concepts that lay the groundwork for statistical inference about population means and variances from sample data. The symbol for the mean of a population is μ; σ^2 and σ are the symbols for a population variance and standard deviation, respectively. A sample mean is symbolized M (or, sometimes, $\overline{X}$), and the descriptive variance and standard deviation of a data set are Var and S, respectively.

A distribution of means of samples of size N over all possible samples is called a ***sampling distribution of the mean***. The ***central limit theorem*** describes the following characteristics of the sampling distribution of the mean: (1) The shape of the sampling distribution of the mean is normal if the population is normal; if the population is not normal, the sampling distribution of the mean approaches normality as N increases. (2) The mean of the sampling distribution of the mean, μ_M, is equal to the population mean, μ. (3) The variance of the sampling distribution of the mean, σ^2_M, is equal to the population variance, σ^2, divided by the sample size, N. The standard deviation of the sampling distribution of the mean, σ_M, is called the ***standard error of the mean*** and is equal to $\sigma/\sqrt{N}$.

If the average value of a statistic over all possible samples is equal to a population parameter, the statistic is an ***unbiased estimator*** of the parameter. Since $\mu_M = \mu$, the sample mean, M, is an unbiased estimator of the population mean, μ. However, the sample descriptive variance, Var, is a biased estimator of the population variance, σ^2 because the average value of Var over all samples is smaller than σ^2. An unbiased estimator of σ^2, symbolized s^2, is obtained by dividing the sample sum of squares, SS, by $N - 1$.

$N - 1$ is the number of degrees of freedom for estimating σ^2 from sample data.

The variance is important because it plays a role in inference about means, but it is important in its own right, too. The purpose of statistical analysis is to explain variability of data, that is, to determine the proportion of the variation that is attributable to each of two or more factors. A variety of statistical procedures enable us to apportion the total variation of a set of data into components. Though the particular statistical techniques vary, depending on the research question and type of data, the description of ***components of variance*** is the end goal of many statistical analyses.

QUESTIONS

1. Give the name of each of the following symbols:

a. μ_M
b. σ^2
c. σ_M
d. Var
e. M
f. s^2
g. μ
h. σ

2. A normal population has mean = 95 and SD = 30. You draw all possible samples of size 10 from the population and calculate the mean of each sample.

a. What is the distribution of sample means called?
b. What is the shape of the distribution?
c. What is the mean of the distribution of sample means?
d. What is the standard deviation of the distribution of sample means called?
e. Calculate the value of the standard deviation of the distribution of sample means.
f. If the population is not normal, which of the answers above will change?
g. If the population SD is 20 instead of 30, which of the answers above will change? Calculate the changed answer.
h. If the population mean = 95 and SD = 30 (as before), but you draw samples of size 50 instead of size 10, which of the answers above will change? Calculate the changed answer.

3. If you draw all possible samples of size 10 and all possible samples of size 50 from the same non-normal population, which of the two distributions of sample means is more nearly normal in shape?

4. First, state which of the following will have the smallest and which the largest standard error of the mean. Then calculate the standard error of the mean for each one.

a. Samples of size 10 from a population with variance = 100.
b. Samples of size 25 from a population with variance = 100.
c. Samples of size 10 from a population with variance = 400.
d. Samples of size 25 from a population with variance = 400.

5. The variance of a population is 250. What sample size must you use to obtain a standard error of the mean equal to 5? What sample size is needed to obtain a standard error of the mean equal to 3?

6. Why is Var = SS/N called a "biased" estimator of σ^2?

7. Calculate the estimated population variance from the following sample: 8, 4, 3, 4, 7, 7, 5.

8. A random sample has scores of 25, 23, 27, 27, 19, 19, and 25. Calculate an estimate of the standard deviation of the population from which the sample was drawn.

9. Use the data set in Table 2–13 (p. 49) to answer the following questions:

a. Calculate the descriptive variance of the quiz scores of the 28 statistics students.

b. Assume that the 28 students are a random sample of all statistics students at the university. Estimate the variance of the population of all statistics students on the quiz. Compare the result with the descriptive variance.

c. If there were 15 students instead of 28 in the class, would the descriptive variance and the unbiased variance estimate be more similar or more different than with 28 students?

10. Assume that there is a parameter/statistic called a guggle. I have a population whose guggle (γ) is 15. Also assume that there are only 10 possible samples of a certain size from this population. The distribution of sample guggles (G) is shown below.

G	Number of Samples
19	1
18	1
17	3
16	2
15	1
14	1
13	1

Is G a biased or an unbiased estimator of γ? Explain.

11. A social scientist has conducted a study to see how attitude change is affected by the feeling tone of messages and by subjects' prior attitudes. She finds that 10% of subjects' attitude change variance is due to feeling tone, 32% to prior attitudes, and the rest to other factors. Draw a diagram portraying these results.

7 The Normal Curve

INTRODUCTION

The theoretical distribution called the *normal curve* is the most important distribution in statistics, for several reasons. First, although no obtained distribution can truly be normal, many variables are approximately normally distributed; therefore, normal curve statistics can be used to answer questions about the variables. Second, many inferential procedures, used for drawing conclusions about populations from sample data, rest on the assumption that the populations are normally distributed. And, third, as a probability distribution, the normal curve can be used to answer questions about the likelihood of obtaining various particular outcomes when sampling from a population. In this chapter, you will learn about the properties of the normal curve and some of its important applications.

Objectives

When you have completed this chapter, you will be able to

- Describe major features of the normal curve.
- Define the standard normal curve, and use the table of the standard normal curve to find areas cut off by z scores or z scores that cut off given areas.
- Given the mean and standard deviation of a normal population, answer questions about proportions, percentages, probabilities, and expected frequencies of observations in various parts of the distribution.
- Use the normal curve to find approximate probabilities in binomial distributions.
- Given the mean and standard deviation of a population and the size of samples drawn from it, answer questions concerning proportions, percentages, and probabilities of sample means.
- Explain the rationale and the steps of the normal curve (z) test to test hypotheses about population means.
- Explain the difference between one-tailed and two-tailed tests.
- Conduct two-tailed and one-tailed z tests, and draw appropriate conclusions based on the outcomes.

Introduction to the Normal Curve

Properties of the Normal Curve

Although we commonly say *the* normal curve, the normal curve is actually a family of curves generated by a mathematical equation. Each curve in the family is defined in terms of its mean and standard deviation. There is a normal curve for every possible combination of a mean and a standard deviation. For example, there is one (and only one) normal curve with mean (μ) = 10 and standard deviation (σ) = 2. There is also one normal curve with $\mu = 10$ and $\sigma = 20$, one with $\mu = 50$ and $\sigma = 20$, one with $\mu = 1056$ and $\sigma = 19.7$, and so on. If two normal distributions have the same mean and standard deviation they are identical distributions.

Despite the infinite number of normal curves, all normal curves share certain common features, illustrated in Figure 7–1. These features include the following:

1. The normal curve is symmetrical and unimodal.
2. The mean lies at the center, directly below the peak.
3. At its center, the curve is concave downward; but the curvature changes to concave upward both above and below the center. The location of the changes in curvature (the *points of inflection*) are above the score values that are one standard deviation below the mean ($\mu - \sigma$) and one standard deviation above the mean ($\mu + \sigma$). For example, in the normal curve with $\mu = 50$ and $\sigma = 10$, the points of inflection are above the scores of $50 - 10 = 40$ and $50 + 10 = 60$.
4. As the curve extends still farther from the mean in both directions, it gradually approaches the abscissa, but never quite reaches it. (In mathematical terms, the normal curve is *asymptotic* to the *x*-axis.) For practical purposes, however, almost all the area under the curve is within three standard deviations of the mean in both directions, that is, between $\mu - 3\sigma$ and $\mu + 3\sigma$.

Figure 7–1 also shows the proportion of area in selected portions of the normal curve. Note that the proportion of area between the mean and a score that is one standard deviation above the mean (i.e., between μ and $\mu + \sigma$) is .3413. The proportion of area between μ and $\mu - \sigma$ is also .3413. Therefore, .3413 + .3413, or .6826 (about 68%), of the area is within one standard deviation of the mean in either direction. Similarly, the proportion of area within two standard deviations of the mean (between $\mu \pm 2\sigma$) is .1359 + .3413 + .3413 + .1359 = .9544 (about 95½%). Almost all the area is within three standard deviations of the mean; more precisely, .9974 (99¾%) of the area is between $\mu \pm 3\sigma$. For example, in the normal distribution with $\mu = 25$ and $\sigma = 5$, about 68% of the area is between 25 ± 5, or 20 and 30; 95½ % of the area is between 25 ± 10, or 15 and 35, and almost all of the area is between 25 ± 15 or 10 and 40.

Recall that in any distribution, the proportion of area in a part of a distribution is equal to the proportion or the probability of observations within that part. For example, in the normal distribution with $\mu = 25$ and $\sigma = 5$, since 68% of the area is between 20 and 30, 68% of the observations are between 20 and 30, and the probability is .68 that a randomly chosen observation will fall between 20 and 30. Thus, we can use areas in various parts of a normal distribution to answer a variety of sampling and probability questions.

In Chapter 5, you used a chi square table to find proportions of area in various chi square distributions. Similarly, you will use a normal curve table to find areas in a normal distribution. Before we discuss the use of the normal curve table, however, we need to consider the following problem: There is an infinite number of normal curves,

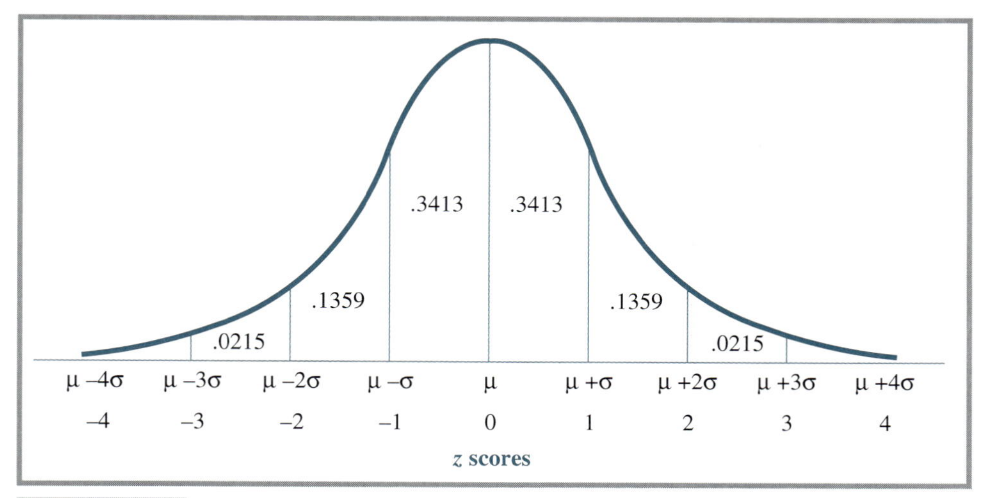

Figure 7–1 **The normal curve.**

each having a different combination of μ and σ. How can we provide information about all these curves in just one table? We do so by converting all normal distributions into a single one, the standard normal curve. Then we only need one table—a table of the standard normal curve—to find areas and answer questions about any normal curve.

The Standard Normal Curve

In Chapter 3 you learned that score values in any distribution can be converted to standard scores or z scores by the formula

$$z = \frac{Score - Mean}{SD}$$ *General formula for converting a score into a z score*

A z score expresses a score's distance from the mean in standard deviation units. For example, $z = -1.5$ means a score that is one and a half standard deviations below the mean.

If all the scores in a distribution are converted to z scores, the distribution of z scores has mean = 0 and standard deviation = 1, and the shape of the z score distribution is the same as that of the original distribution. Therefore, if a distribution is normal, and if all the scores are converted into z scores, the distribution of z scores is normal, with mean = 0 and SD = 1. The normal distribution of z scores is called the ***standard normal distribution*** or ***standard normal curve***. Any normal distribution, with any mean and standard deviation, can be converted into a standard normal distribution by converting the scores into z scores.

Just one word of caution: People sometimes believe that if raw scores are converted to z scores, the result is a standard normal distribution, even if the original distribution was not normal. *This is not true.* A distribution of z scores always has exactly the same shape as the original distribution. If the original distribution is normal, the distribution of z scores forms a standard normal curve. If, however, the original distribution is skewed, or rectangular, or bimodal, or any shape other than normal, the transformed distribution of z scores is also skewed, or rectangular, or whatever; it is not normal.

Using the Table of the Standard Normal Curve

Table D–2 in Appendix D provides information about proportions of area within various parts of the standard normal curve. The first column, Column (A), lists z scores ranging from 0.00 to 4.00. Column (B) lists, for each z score, the proportion of area between the z score and the mean; and Column (C) lists the proportion of area in the tail beyond the z score. For example, for the z score 1.26, the table shows

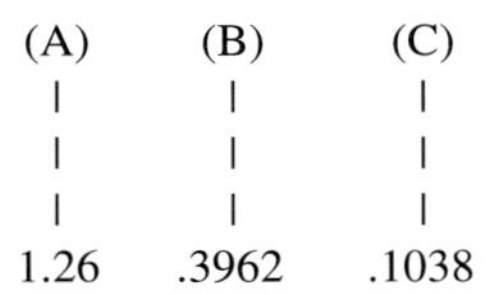

This means that in the standard normal curve, .3962 of the area is between the mean and $z = 1.26$, and .1038 of the area is beyond (above) $z = 1.26$, as Figure 7–2 shows.

In any normal distribution, half the scores are below the mean and have negative z scores. Yet the normal curve table does not list negative z scores. Is this a problem? Not at all; a listing of negative z scores is unnecessary. Because the normal curve is symmetrical, a negative z score cuts off the same proportions of area as the positive z score with the same absolute value. Thus, for $z = -1.26$, as for $z = +1.26$, the proportion of area between the mean and z is .3962, and the proportion of area beyond (below) z is .1038, as Figure 7–3 shows. (Note that although z scores may be positive or negative, *areas are always positive*.)

Sometimes we need to know areas other than those listed in the table, such as the area above $z = -1.26$ or the area between $z = .50$ and $z = 1.50$. In order to find such

Figure 7–2

Proportions of area cut off by $z = 1.26$.

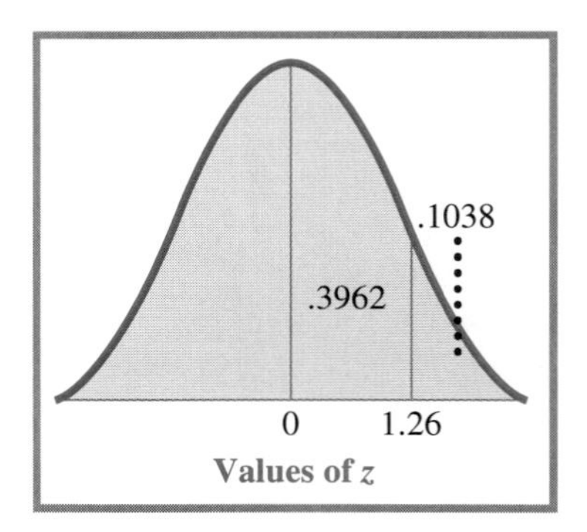

Figure 7–3

Proportions of area cut off by $z = -1.26$.

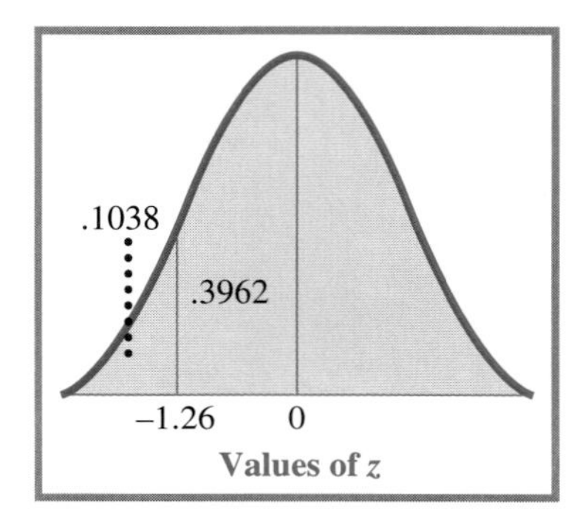

areas, we combine areas that are listed in the table in various ways. A few general rules will help you to find the area within any part of the standard normal curve with ease:

1. The total area under the standard normal curve is 1.
2. Since the normal curve is symmetrical, half the area (.5) lies on either side of the mean.
3. Given a normal curve problem, always draw a picture of a normal curve and shade in the area you wish to find. The picture will help you figure out what you need to do.
4. In drawing your picture, be sure to locate positive z scores above (to the right of) the center and negative z scores below (to the left of) the center of the distribution.

See Examples 7–1 and 7–2.

EXAMPLE 7–1

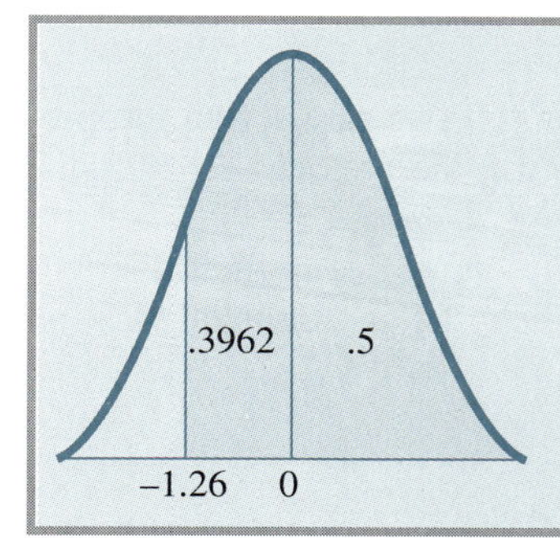

What is the proportion of area above $z = -1.26$?

The desired area is shaded in the diagram to the left. The shaded area consists of two portions: a portion between the z score and the mean and a portion above the mean. According to the normal curve table, the area between the mean and z is .3962. The area in the upper half of the curve, above the mean, is .5. The area we seek is the sum of the two areas, or .3962 + .5 = .8962.

EXAMPLE 7–2

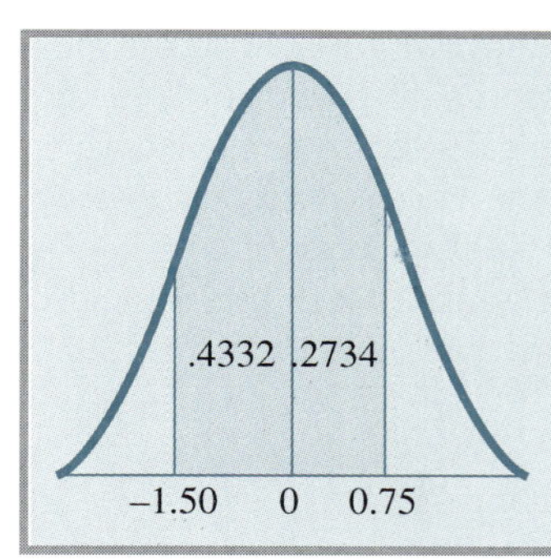

What is the area between z scores of −1.50 and +0.75?

The desired, shaded area consists of two parts: the area between the mean and $z = -1.50$ and the area between the mean and $z = 0.75$. From the table we find that the area between the mean and $z = -1.50$ is .4332, and the area between the mean and $z = 0.75$ is .2734. The picture shows that the area we seek is the sum of these two areas, or .4332 + .2734 = .7066.

As these two examples illustrate, in general, to find the area between two z scores on *opposite* sides of the mean, find the area between each z score and the mean and *add* the two areas.

EXAMPLE 7–3

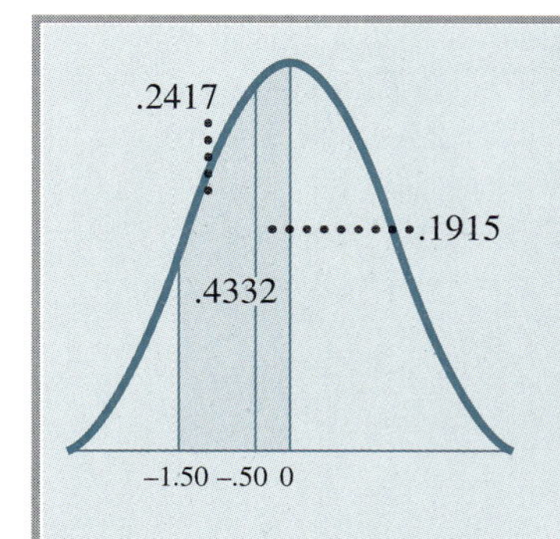

What is the area between z scores of −1.50 and −0.50?

The picture shows that, to find the desired shaded area, we must find the area between each z score and the mean, then subtract the smaller area from the larger one. According to the table, the area between the mean and $z = -1.50$ is .4332; the area between the mean and $z = -0.50$ is .1915. Therefore, the desired area is .4332 − .1915 = .2417.

As Example 7–3 illustrates, in general, to find the area between two z scores on the *same* side of the mean, find the area between each z score and the mean and *subtract* the smaller area from the larger.

In the three examples we solved for areas above, below, or between z scores. We first located the z scores in Column (A) of the table, then found the needed areas in Column (B) or (C). That is, we started with z scores and used the table to find areas. Sometimes we need to work in the other direction and find z scores that cut off certain areas. For example, what is the z score above which only 10% of the area falls? To answer questions such as this, we must first locate the given area in Column (B) or (C), then find, in Column (A), the z score that cuts off that area. The general procedure is like this:

(A)		(B) or (C)
		↓
z	←	given area

When we look for a particular area in Column (B) or (C), we rarely find the precise value we want in the table. It is usually sufficient to use the area closest in value to the one sought, as illustrated in the examples that follow.

As in previous examples, always begin by drawing a picture of a normal distribution. Shade and label the area you are working with, and show the approximate location of the z scores to be found. See Examples 7–4 and 7–5.

EXAMPLE 7–4

What is the z score that cuts off only 10% of the area above it?

.1000

0 z

As the picture shows, we want the z score that cuts off above it 10%, or .1, of the area. Since .1 is the area *beyond* the z score, we go down Column (C), the column of areas beyond z, looking for the value .1000. The area .1000 is not listed in Column (C); the nearest two entries are .1003 and .0985. Since .1003 is closer to .1000, we use it.

Next, we look in Column (A) of the same row to see which z score cuts off an area of .1003 beyond it and find that z = 1.28. Therefore, the z score that cuts off only 10% of the area above it is 1.28.

EXAMPLE 7–5

Only one-third of the standard normal distribution has z scores below what value?

.3333

z 0

In decimal form, one-third is equal to .3333. The picture shows that we need the z score that cuts off an area of .3333 below (beyond) it. We go down Column (C) to find the entry closest to .3333. The nearest entry is .3336; the z score in that row of the table is 0.43.

However, the answer is *not* +0.43. Since the z score we are seeking is below the mean, the answer is z = −0.43.

When finding z scores that you know are below the mean, remember always to put a minus sign in front of the z that you find in the table.

EXAMPLE 7–6

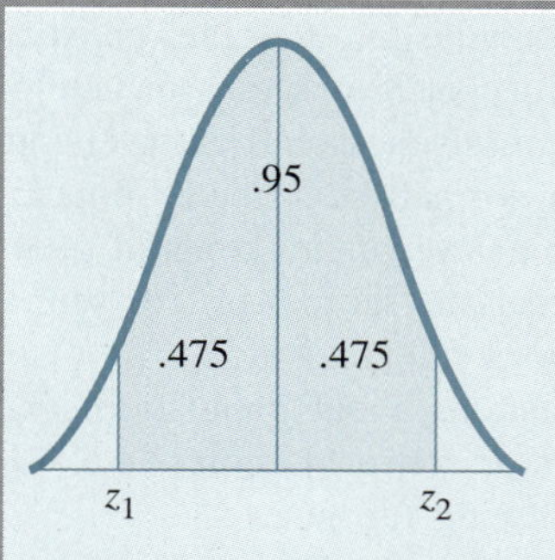

What two z scores cut off the middle 95% of the distribution?

In the picture to the left, the two z scores we seek are labeled z_1 and z_2. The area between the two z scores is .95.

However, we cannot enter the table with an area between two z scores; we need either the area between each z and the mean or the area beyond each z. We reason as follows: If the two z scores cut off the middle .95 of the distribution, half of that .95 (or .475) must be above the mean and half (.475) below. Therefore, the area between each z and the mean is .475, as the picture shows. We go down Column (B) until we find the entry .4750. In Column (A) of the same row, we read $z = 1.96$. Therefore, the two z scores that cut off the middle .95 of the area are $z_1 = -1.96$ and $z_2 = 1.96$.

EXERCISE 7–1

1. Find the following proportions of area in the standard normal curve. In each case, first draw a picture and shade the portion of the curve whose area you wish to find.
   a. above $z = 2.25$
   b. above $z - 1.32$
   c. below $z = 0.83$
   d. below $z = -1.35$.
   e. between $z = -1.11$ and $z = -1.53$
   f. between $z = -1.11$ and $z = 0.68$
   g. between $z = 0.68$ and $z = 1.22$.
2. Find the z scores that meet the following criteria. In each case, first draw a picture, shade the portion of the curve with the indicated area, and show the location of the z score(s) you wish to find.
   a. the z score below which 80% of the area falls
   b. the z score below which .15 of the area falls
   c. the z scores between which the middle half of the area falls
3. Why does the normal curve table not list proportions of area for z scores greater than 4.00?
4. Use the table of the standard normal curve to confirm the fact that approximately 68% of the area falls within one standard deviation of the mean, approximately 95½% within two standard deviations of the mean, and well over 99% within three standard deviations of the mean.

Some Applications of the Normal Curve

The table of the standard normal curve enables us to answer a variety of questions. We consider three types of questions in this section: questions about normal populations, questions about binomial distributions, and questions about distributions of sample means.

Using the Normal Curve Table to Answer Questions About Normal Populations

Many populations of interest to social and behavioral scientists are either known or assumed to be approximately normally distributed. These include, for example, physical characteristics such as height and weight, measures of ability such as scores on intelligence tests, and measures of personality traits such as aggressiveness and extraversion. If we know the mean (μ) and standard deviation (σ) of a normally distributed population, we can use the table of the standard normal curve to answer many kinds of questions about the distribution, such as the probability of certain scores, or the scores between which certain proportions of population observations fall.

In order to use the normal curve table to answer questions about populations, we must either convert population raw scores (X) into z scores or convert z scores into values of X. Raw scores are converted to z scores by the following formula:

$$z = \frac{X - \mu}{\sigma}$$ *Formula for converting population raw scores to z scores*

For example, in a population with mean (μ) = 60 and *SD* (σ) = 5, a score of 53 has

$$z = \frac{53 - 60}{5} = -1.4.$$

To convert z scores to raw scores, we use the following formula:

$$X = \mu + z(\sigma)$$ *Formula for converting z scores to population raw scores*

Thus, in the population with $\mu = 60$ and $\sigma = 5$, the score that is three-fourths of a standard deviation below the mean ($z = -.75$) is $60 + (-.75)(5) = 56.25$.

Keep the following facts in mind as you work with questions about normally distributed populations:

1. The proportion of area within a given part of a distribution is the proportion of observations in that part or the probability that a single observation, selected randomly, falls in that part.
2. The percentage of observations in part of a distribution is equal to the proportion multiplied by 100.
3. The expected frequency of observations in part of a distribution is equal to the proportion multiplied by the group size (N).

For example, if the proportion of area between two scores is .48, we know that (1) the proportion of observations between the scores is .48; (2) the probability of randomly selecting an observation between the two scores is .48; (3) 48% of observations fall between the scores; and (4) in a group of size 1000, the number of observations expected to fall between the two scores is 1000 (.48) = 480.

Examples 7–7 and 7–8 use the normal curve table to answer questions about normal populations.

EXAMPLE 7–7

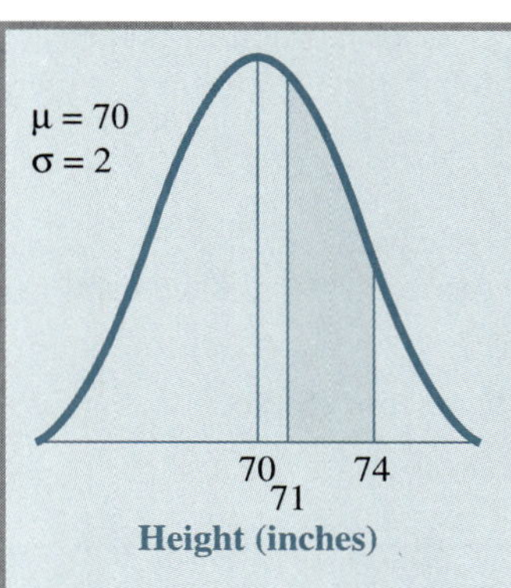

If heights of adult males are normally distributed, with mean = 70 inches and SD = 2 inches, what is the probability that a randomly selected adult male will be between 71 and 74 inches tall?

The picture to the left shows that the needed area is that between scores of 71 and 74 in the normal distribution, with mean (μ) = 70 and SD (σ) = 2. The first step is to convert the heights 71 and 74 inches to z scores:

$$z_{71} = \frac{71 - 70}{2} = 0.50$$

$$z_{74} = \frac{74 - 70}{2} = 2.00$$

The probability of a height between 71 and 74 inches is equal to the proportion of area between z = 0.50 and z = 2.00 in the standard normal curve.

From the table, the area between the mean and z = 0.50 is .1915, and the area between the mean and z = 2.00 is .4772. The area between the z scores is .4772 − .1915 = .2857.

The probability of a randomly selected adult male between 71 and 74 inches tall is .2857.

EXAMPLE 7–8

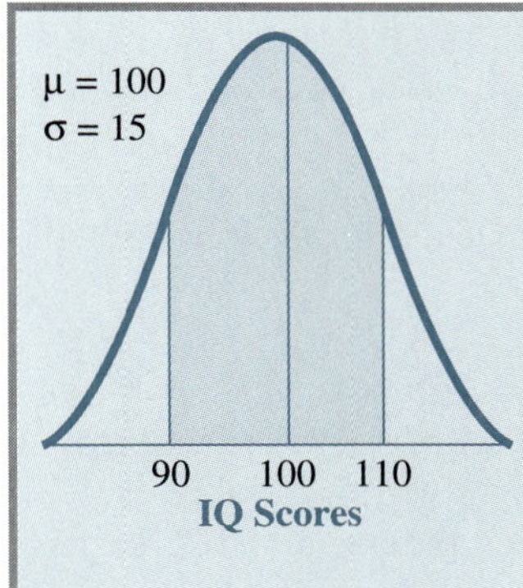

IQ scores are approximately normally distributed, with mean = 100 and SD = 15. In a group of 500 individuals, how many are expected to have IQ scores between 90 and 110?

First, we convert 90 and 110 to z scores:

$$z_{90} = \frac{90 - 100}{15} = -0.67$$

$$z_{110} = \frac{110 - 100}{15} = 0.67$$

The area between each of the z scores and the mean is .2486. Therefore, the area between the two z scores is .2486 + .2486 = .4972.

In a group of 500 individuals, the expected frequency of individuals between the two z scores is 500 (.4972) = 248.6. Therefore, in a group of 500 individuals, 248.6 of them (about half) are expected to score between 90 and 110.

In Examples 7–7 and 7–8, the sequence of steps to solution was as follows:

	(1)		(2)		(3)	
Raw scores (X)	→	z scores	→	Proportion of area	→	Probability, percent, or expected frequency of observations

Sometimes we are given information about a probability or proportion or percent of observations and need to find the scores that cut off that probability, proportion, or percent. If so, we need to go in the opposite direction:

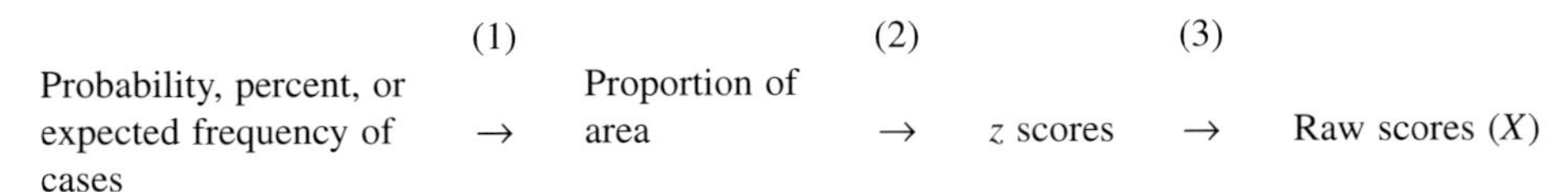

Remember that the formula for converting a z score to a raw score (Step 3) is $X = \mu + z\,(\sigma)$. See Examples 7–9 and 7–10.

EXAMPLE 7–9

In a normally distributed population of IQ scores, with mean = 100 and SD = 15, below what IQ score does only 1% of the population fall?

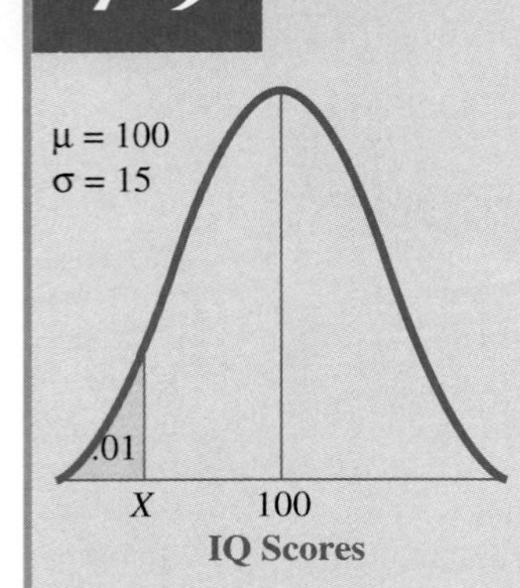

Since 1% is a proportion equal to .01, we need to find the z score with area below = .01. Then we must convert the z score into an IQ score. In the figure at the left, the IQ score we seek is indicated by X.

According to the normal curve table, the z score with area beyond = .01 is 2.33. Because the z score we seek is below the mean, $z = -2.33$.

Finally, we convert z into a raw (IQ) score in the population. Since $\mu = 100$ and $\sigma = 15$, the IQ score we seek = $100 + (-2.33)(15) = 65.05$. Only 1% of the population scores below 65.05.

EXAMPLE 7–10

Scores on a reading test are approximately normally distributed with mean = 25 and SD = 5. What is the 70th percentile?

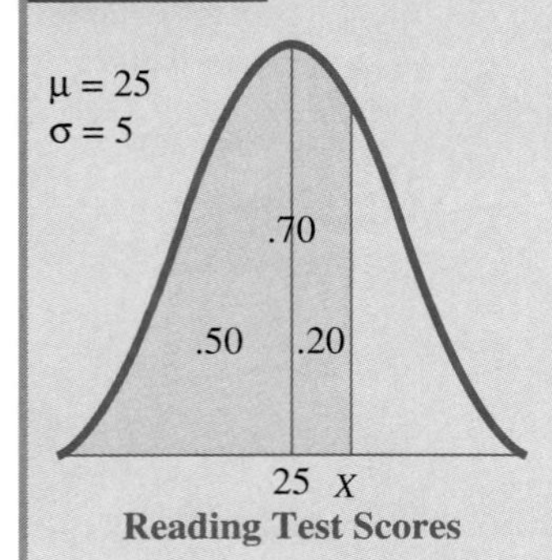

The "70th percentile" means the score value below which 70% of observations fall. We need to find the test score, X, with area below = .70, as the picture shows.

Of the .70 area below X, .20 is between the mean and X. In the normal curve table, we find that the z score that cuts off an area of approximately .20 between itself and the mean is 0.52.

On the reading test, $\mu = 25$ and $\sigma = 5$. Therefore, $X = 25 + (0.52)(5) = 27.6$. The 70th percentile on the test is 27.6.

EXERCISE 7–2

1. Draw a picture before answering each of the following questions. In the distribution of IQ scores with mean = 100 and SD = 15
   - **a.** What percentage of IQ scores is between 70 and 90?
   - **b.** What is the probability of an IQ score either less than 75 or greater than 125?

c. In a group of 1000 children, how many are expected to score between 80 and 120?
d. Between what two IQ scores does the middle three-fourths of the population fall?
e. The probability of scoring lower than a certain IQ score is only .05. What is the IQ score?

2. A distribution of exam scores is negatively skewed with a mean of 40 and *SD* of 4. The instructor argues that if he converts the exam scores to *z* scores, he can use the normal curve table to find expected frequencies in various parts of the distribution. What do you think of this argument?

The Normal Approximation to the Binomial Distribution

The ***binomial distribution*** is a family of discrete probability distributions that can be used to find probabilities of two-outcome events, such as tosses of a coin or turns at a choice point in a maze. Each possible combination of an elementary probability and a number of repetitions generates one distribution in the family. In Chapter 4 we used binomial distributions to answer such questions as, "In tossing five coins, what is the probability of getting three or more heads?"

We observed, in Chapter 4, that the binomial distribution approaches the normal distribution as the number of repetitions increases. Because of this, the normal distribution can be used to find approximate probabilities of binomial events, provided that the number of repetitions is sufficiently large. In fact, even when the number of repetitions is relatively small (e.g., 10), normal curve procedures give quite accurate solutions to binomial problems.

Consider the problem of finding the probability of getting 8 or more heads in tossing 10 coins. Assuming that the coins are fair, the elementary probability (the probability of a head on one toss of one coin) is .5. The number of repetitions is 10. The relevant binomial distribution is shown as a bar graph in Figure 7–4. The probabilities of various outcomes (0 heads, 1 head, 2 heads, etc.) are printed above the bars. To find the probability (p) of 8 or more heads, we use the addition law of probability; $p(8 \text{ or more}) = p(8) + p(9) + p(10) = .0439 + .0098 + .0010 = .0547$. The exact probability of 8 or more heads in tossing 10 fair coins is .0547.

Superimposed on the binomial distribution in Figure 7–4 is its normal approximation. We will use this normal curve to find the approximate probability of 8 or more heads. To work with a normal curve, however, we need to know its mean and standard deviation. If we let N stand for the number of repetitions and p_E for the elementary probability, the formulas for the mean and standard deviation of a binomial distribution are

$$Mean = Np_E \qquad SD = \sqrt{Np_E(1-p_E)}$$

Mean and standard deviation of a binomial distribution

In this particular binomial distribution, $N = 10$ and $p_E = .5$. Therefore,

$$Mean = 10(.5) = 5$$
$$SD = \sqrt{10(.5)(1-.5)} = 1.58$$

The problem can now be stated: What is $p(X = 8 \text{ or more})$ in the normal distribution with mean = 5 and $SD = 1.58$?

Before we can apply normal curve procedures, however, we must make an adjustment called a ***correction for continuity***.† The correction is necessary because the bino-

†If N is large ($N \geq 20$), the correction for continuity can be omitted.

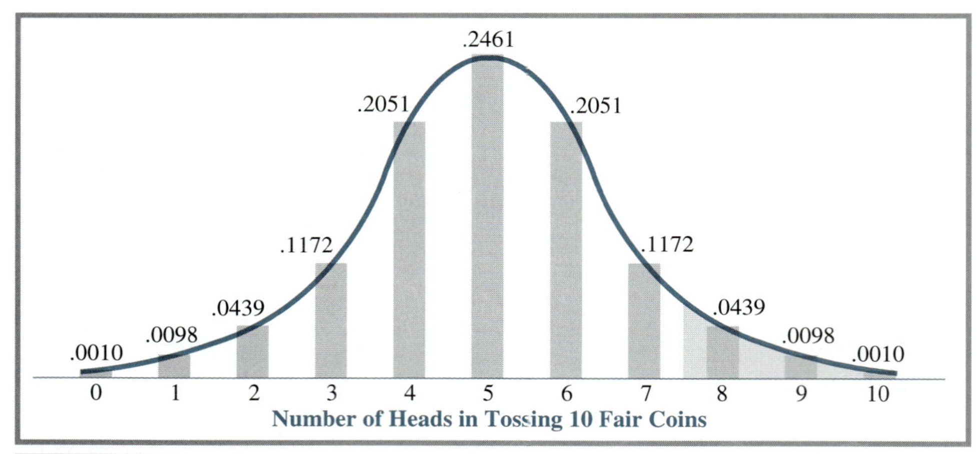

Figure 7–4 **Binomial distribution for elementary probability = .5, number of repetitions = 10, with superimposed normal curve approximation. The shaded area represents the probability of 8 or more heads in the normal distribution.**

mial is a discrete distribution, whereas the normal distribution is continuous. If we want to use a continuous distribution to answer questions about events having discrete numerical values, we must treat the numbers as though they represented measurements of a continuous variable; we must use the exact limits of the numbers. For example, the value 8 in the binomial distribution becomes 7.5 – 8.5 in the normal approximation, 9 becomes 8.5 – 9.5, and so on. Moreover, because the normal curve extends indefinitely far in both directions, the discrete value 10 (the largest value of X in the binomial distribution for $N = 10$) becomes "greater than 9.5" in the normal distribution, and the discrete value 0 becomes "less than 0.5." The number line below shows how discrete values in the binomial distribution with $N = 10$ are converted to intervals in the normal distribution.

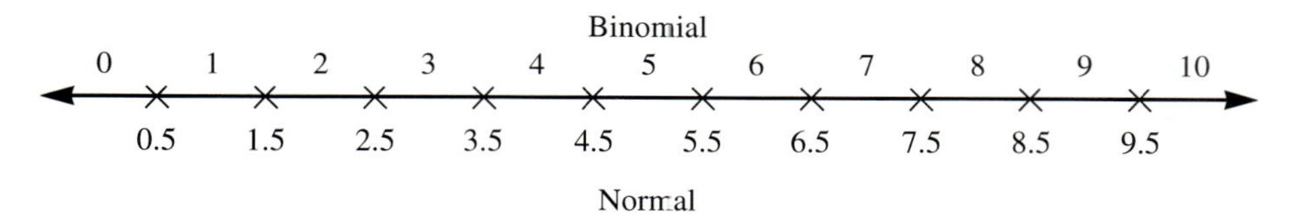

We want to find the probability of 8 or more heads. The event "8 or more" in the binomial distribution becomes "greater than 7.5" in the normal approximation. Therefore, in the normal approximation, the probability of 8 or more heads is the probability of $X > 7.5$. The needed area is shaded in Figure 7–4.

Now we are ready to solve the problem. First, we convert 7.5 to a z score in the normal distribution with mean = 5 and SD = 1.58.

$$z = \frac{7.5 - 5}{1.58} = 1.58$$

Then we go to the normal curve table to find the area above (beyond) a z score of 1.58. According to the table, the area is .0571. The answer is: The approximate probability of 8 or more heads in tossing 10 fair coins is .0571.

Compare this answer ($p = .0571$) with the answer obtained using the binomial distribution ($p = .0547$). As you can see, the difference is small. Thus, when the elementary probability is .5, normal curve procedures give reasonably accurate solutions to binomial problems, even when the number of repetitions is only 10. Accuracy increases as the number of repetitions (N) increases. (In fact, the normal approximation yields reasonably accurate probabilities even when p_E is not .5, though a larger N is needed in such cases.)

EXERCISE 7–3

1. Assuming no right or left preference, find the probability that 3 or fewer of 10 rats will turn right using (a) the binomial distribution and (b) the normal approximation. Don't forget to use the correction for continuity to express $p(X = 3 \text{ or less})$ in the normal distribution.
2. Figure 4–9d (p. 114) shows the binomial distribution for $N = 10$ and $p_E = .75$. Find the exact value of $p(X = 8 \text{ or more})$ in this distribution. Then find the approximate probability, using the normal approximation. Compare the accuracy of the normal approximation when $p_E = .75$ with the accuracy when $p_E = .5$.

Using the Normal Curve to Answer Sampling and Probability Questions About Sample Means

Suppose that you plan to select a random sample of size 50 from a population with mean = 70 and SD = 10. You want to know the probability that the mean of your sample will be greater than 71.0. You are asking a question not about the probability of a raw score in a distribution of raw scores but about the probability of a sample mean in a distribution of sample means. In effect, you are asking: "If I were to select (or list) all possible samples of size 50 from this population, what proportion of the samples would have a mean greater than 71.0?" The distribution of means of all possible samples of size N from a population is called a ***sampling distribution of the mean***. Therefore, you are asking about the probability of particular sample means in a sampling distribution of the mean.

In Chapter 6 we discussed relationships between the sampling distribution of the mean and properties of the population from which samples are drawn. We found that the sampling distribution of the mean is approximately normal, provided that the population is normal or the sample size N is reasonably large, and that the mean (μ_M) and standard deviation (standard error of the mean, σ_M) of the sampling distribution of the mean are related to the population mean (μ) and standard deviation (σ) by the following formulas:

$$\mu_M = \mu$$

$$\sigma_M = \frac{\sigma}{\sqrt{N}}$$

Therefore, if we know the mean and standard deviation of a population and the size of samples drawn from it, we can calculate the mean and standard deviation of the sampling distribution of the mean and use normal curve procedures to answer questions about the probabilities of various sample means.

Before we can use the table of the standard normal curve to answer sampling and probability questions about sample means, we must convert either sample means to z scores or z scores to values of sample means. Remember that the general formula for converting a score in a distribution into a z score is z = (score − mean)/SD. In a sampling distribution of the mean, the mean (μ_M) = μ; $SD = \sigma_M$; and the "scores" in the distribution are sample means (M). Therefore the formula for converting a sample mean, M, into a z score is

$$z = \frac{M - \mu}{\sigma_M}$$ *Formula for converting a sample mean into a z score*

Suppose, for example, that samples of size 100 are drawn from a population with mean (μ) = 35 and *SD* (σ) = 6. What is the z score for a sample mean of 33.5? We first compute σ_M.

$$\sigma_M = \frac{\sigma}{\sqrt{N}} = \frac{6}{\sqrt{100}} = 0.6$$

Then

$$z = \frac{M - \mu}{\sigma_M} = \frac{33.5 - 35}{0.6} = -2.5$$

Recall that a z score tells the number of standard deviations of a score value above or below the mean of the distribution. Therefore, this sample mean, 33.5, is two and one-half standard errors below the population mean.

To convert z scores into sample means, the following formula is used:

$$M = \mu + z(\sigma_M)$$ *Formula for converting a z score into a sample mean*

Suppose we ask: If samples of size 16 are drawn from a population with μ = 40 and σ = 10, what sample mean has z = 1.5? First, we calculate σ_M:

$$\sigma_M = \frac{10}{\sqrt{16}} = 2.5$$

The sample mean (*M*) is

$$M = 40 + (1.5)(2.5) = 43.75$$

After sample means have been converted into z scores the procedure for using the normal curve table to answer sampling and probability questions is the same as for raw scores. See Examples 7–11 and 7–12.

EXAMPLE 7–11 If you randomly select a sample of size 50 from a population with a mean of 70 and *SD* = 10, what is the probability that the sample mean will be greater than 71.0?

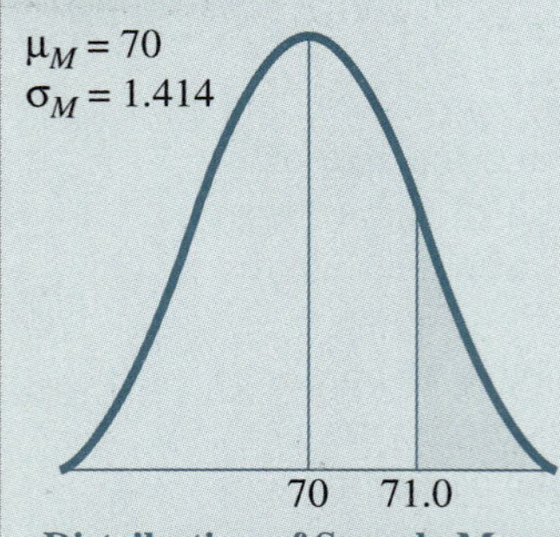

Distribution of Sample Means

Since the sample size is large, the sampling distribution of the mean can be assumed to be normal. The sampling distribution of the mean is shown at the left.

First, we calculate σ_M:

$$\sigma_M = \frac{\sigma}{\sqrt{N}} = \frac{10}{\sqrt{50}} = 1.414$$

Next, we convert the sample mean 71.0 into a z score:

$$z = \frac{71.0 - 70}{1.414} = 0.71$$

According to the normal curve table, the proportion of area above (beyond) a z score of 0.71 is .2389. Therefore, the probability of a sample mean greater than 71.0 is .2389.

EXAMPLE

7–12

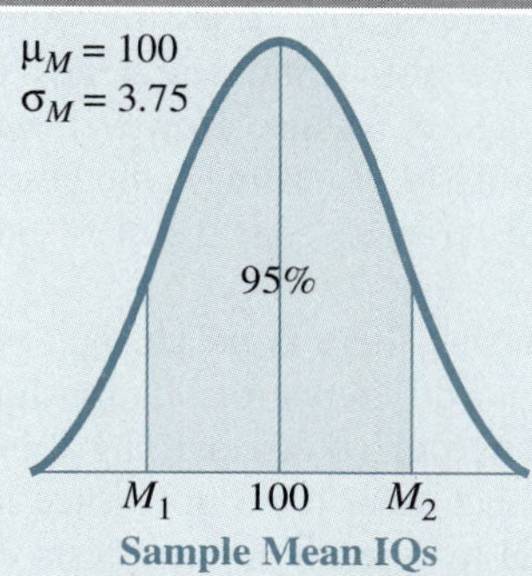

If I draw samples of size 16 from a population of IQ scores with mean = 100 and *SD* = 15, between what two values will the middle 95% of sample means fall?

First, we calculate σ_M:

$$\sigma_M = \frac{15}{\sqrt{16}} = 3.75$$

Since half of the middle 95%, or 47.5%, falls above the mean and half below, we use the normal curve table to find the two z scores, one positive and one negative, which have area = .475 between themselves and the mean. The table gives us $z = \pm 1.96$.

Next, we convert each z score to a value of M using the formula $M = \mu + z\,(\sigma_M)$:

$$M_1 = 100 + (-1.96)(3.75) = 92.65$$

$$M_2 = 100 + (1.96)(3.75) = 107.35$$

The middle 95% of sample means fall between 92.65 and 107.35.

EXERCISE

7–4

1. If you draw samples of size 25 (instead of size 16) from the population of IQ scores, between what two values will the middle 95% of sample means fall? Compare your answer with the answer in Example 7–12. Explain the difference.
2. If you select samples of size 20 from a normal population with a mean of 10 and a standard deviation of 2, below what value will only 5% of sample means fall?
3. A normal population has mean = 50, *SD* = 10.
   a. If you randomly select a sample of size 10, what is the probability that the sample mean is less than 47?
   b. If you randomly select a sample of size 50, first predict whether the probability of a sample mean less than 47 is greater or smaller than your answer in part (a). Then find the probability.
   c. If you randomly select a single *member* of the population, what is the probability that the member's score is less than 47?

A Return to Hypothesis Testing: The Normal Curve Test

Review of Steps in Hypothesis Testing

In this section we introduce methods for testing hypotheses about population means. The test in question is the *normal curve test*, or *z test*. The steps in conducting a normal curve test are the same as in any hypothesis test. We begin with a brief review of the logic and steps in testing a statistical hypothesis. (See Chapter 5 for an extended discussion.)

Step 1. State the Hypothesis to Be Tested

A statistical hypothesis is always about some characteristic or parameter of a population, such as the proportion of members in certain categories, or the population mean. Moreover, a statistical hypothesis always specifies a particular value (or set of values), not a range of values. For example, in Chapter 5 we found that we could not test a hypothesis like "The percentage of voters favoring Candidate Smith over Candidate Jones is greater than 50%," since "greater than 50%" includes a range of values. However, we could and did test hypotheses like "Percentage = 50%." Similarly, in testing hypotheses about population means (μ), we cannot test hypotheses like "$\mu < 35$" or "$\mu > 10$," but we can test hypotheses like "$\mu = 35$" or "$\mu = 10$."

In statistical tests, the hypothesis that is tested is often that there is *no* change, *no* difference, *no* effect, or *no* relationship. The hypothesis of no difference or relationship is called the ***null hypothesis*** (H_o). The hypothesis "Percentage of voters favoring Candidate Smith is 50%" is a null hypothesis because it states that there is no difference in preference for two candidates. As you will see, the statistical hypothesis in most tests of hypotheses about population means is also a null hypothesis.

Step 2. Select a Level of Significance, Alpha (α)

A statistical test results in either rejection or nonrejection of the hypothesis. The hypothesis is rejected if a sample is obtained that has a small probability of occurring when the hypothesis is true. But how small is "small"? It is necessary to specify, before collecting data, how unlikely a sample outcome must be in order to reject the hypothesis. The level of probability selected is called the ***level of significance***, or ***alpha*** (α). For example, if you select $\alpha = .05$, you will reject the hypothesis if your sample outcome has less than 5 chances in 100 of occurring when the hypothesis is true.

In the biological, social, and behavioral sciences, researchers most often use $\alpha = .05$, because a sample outcome with probability less than .05 is considered sufficiently unusual to warrant rejection of the hypothesis. Occasionally, researchers feel that .05 is not unusual enough and select $\alpha = .01$ instead. Though other levels of significance could be chosen (e.g., .10 or .02 or .005), .05 and .01 are by far the most common.

Factors to be considered in selecting a level of significance are discussed in Chapter 8. For now, we will use, more or less arbitrarily, either $\alpha = .05$ or $\alpha = .01$ in most tests.

Step 3. Collect Data to Test the Hypothesis and Calculate the Relevant Test Statistic

A representative (ideally, random) sample is selected from the population and relevant sample statistics are calculated. For example, to test a hypothesis about voter preferences for two candidates, a random sample of voters is asked which candidate they plan to vote for, and the number of voters favoring each candidate is tallied. Similarly, to test a hypothesis about a population mean, we select a random sample from the population and calculate the sample mean.

In order to find out whether the sample outcome has a probability less than alpha, we need to refer to a relevant ***probability distribution***, a distribution that provides information about probabilities of various sampling outcomes. Different tests use different probability distributions. In the chi square test, for example, the relevant probability distribution is the chi square distribution. In the normal curve test, the relevant probability distribution is the standard normal curve. Information about the sample and about the hypothesis being tested is used to convert the sample outcome into a value in the probability distribution (a value of the ***test statistic***). Thus, in the chi square test, observed and expected frequencies are used to calculate a value of χ^2. Similarly, in the normal curve test, the sample mean and sample size, as well as the hypothesized population mean and the population standard deviation, are used to calculate a value of the test statistic z.

Step 4. Find the Probability of the Obtained Test Statistic if the Hypothesis Is True

More specifically, we determine whether the probability is greater or less than alpha. This is usually done by consulting a table that gives probabilities of the test statistic. For example, when we conducted chi square tests in Chapter 5, we consulted a chi square table. Similarly, in normal curve tests, we consult a table of the standard normal curve to find out whether the value of z that we have calculated has probability less than alpha ($p < \alpha$) or greater than alpha ($p > \alpha$).

Step 5. Decide Whether to Reject the Hypothesis

If the table shows that the probability of the test statistic is less than alpha, the hypothesis is rejected; if the probability is greater than alpha, the hypothesis is not rejected.

When a statistical hypothesis is rejected, the result is said to be *significant* at level alpha. Thus, if a chi square test of voter preferences with alpha = .05 results in rejection of the null hypothesis of *no* preference, we say that the difference in preference for the two candidates is significant at the .05 level. Similarly, if a normal curve test results in rejection of a hypothesis about a population mean, we say that the difference between the sample mean and the hypothesized population mean is significant at the level of alpha used in the test. If, on the other hand, the hypothesis is not rejected, the difference between the sample mean and the hypothesized population mean is not significant at level alpha.

Rejection Region, Nonrejection Region, and Critical Values

The selection of a level of significance, alpha, in Step 2 of a hypothesis test, divides the relevant probability distribution into two regions, one containing values of the test statistic with probability (area) less than alpha (the ***rejection region***) and one containing values of the test statistic with probability (area) greater than alpha (the ***nonrejection region***). The value of the test statistic that is the cutoff point between the rejection and nonrejection regions is called the ***critical value*** of the test statistic for that level of significance.

For example, suppose that the chi square table shows that probability of χ^2 greater than 3.84 is .05 [$p(\chi^2 > 3.84) = .05$]. In that case, if $\alpha = .05$, the rejection region consists of all values of χ^2 greater than 3.84, and the nonrejection region consists of all values of χ^2 less than 3.84, and 3.84 is the critical value of χ^2 for $\alpha = .05$. For a pictorial representation of the relation between alpha, rejection and nonrejection regions, and the critical value in the chi square test, see Figure 5–2 (p. 132).

Rationale and Assumptions of the Normal Curve Test

The ***normal curve test***, or ***z test***, is used to test the hypothesis that a population mean has a particular value, which we will symbolize μ_H. The rationale for the normal curve test is as follows: If the population mean is μ_H, and if we were to list or select all possible samples of size N from the population, the sample means would vary around μ_H, with most sample means falling relatively near μ_H and few falling very far away. If we converted each sample mean into a z score by the formula

$$z = \frac{M - \mu_H}{\sigma_M}$$

most of the sample means would have z scores close to zero, with fewer and fewer sample means having z increasingly different from zero. For example, only 5% of the sample means would have values of z beyond $\pm$ 1.96.

In real life, we don't choose all possible samples, or even 100, or 10, or 2; we usually select just one. However, we can use the formula to compute what its z score *would* be if the sample were selected from the hypothesized population. If we get a value of z close to zero, we know that the probability of a sample like ours from the hypothesized population is quite large. Therefore, our sample could well come from the hypothesized population, and we do not reject the hypothesis. On the other hand, if we get a value of z far from zero we reason that, in sampling from the hypothesized population, a sample like ours has a small probability. It is more likely that we are not sampling from the hypothesized population. Therefore, we reject the hypothesis.

Three conditions (assumptions) underlie the z test of hypotheses about population means:

1. Either the population is normal or the sample chosen to test the hypothesis is reasonably large.
2. The sample has been chosen randomly from the population (or at least is likely to be representative of the population).
3. The population standard deviation, σ, is known.

The first condition is necessary because the sampling distribution of the mean is normal only if either the population is normal or the sample size is large. The second condition is necessary because, if samples are biased, the probabilities of various outcomes may be different from those given in the normal curve table. The third condition is necessary because the computation of σ_M requires that σ be known, and σ_M in turn, is necessary to compute z.

The z test is particularly applicable to situations in which we wonder whether a sample has been selected from a known population with known mean μ and standard deviation σ. For example, we know that in the U.S. population, IQ scores on certain tests are approximately normally distributed with mean = 100 and standard deviation = 15. Therefore, we can use the z test to test the hypothesis that a sample of persons whose IQs we have measured came from this population. Since this amounts to testing a hypothesis of "no difference from the known population," the hypothesis is a kind of null hypothesis and is symbolized H_0.

Assuming that the necessary conditions have been met, the normal curve test, or z test, consists of the following steps:

1. State the hypothesis H_0: μ = some value μ_H.
2. Choose a level of α.
3. Select a sample of size N and perform the following calculations:
   a. Compute M, the sample mean.

$$M = \frac{\Sigma X}{N}$$

   b. Compute $\sigma_M = \frac{\sigma}{\sqrt{N}}$, where σ is the standard deviation of the population.

   c. Convert M into a value of the test statistic, z.

$$z = \frac{M - \mu_H}{\sigma_M}$$

4. Consult the normal curve table to determine whether the probability of the obtained z is less than or greater than alpha.
5. If the probability is less than alpha ($p < \alpha$), reject H_0. Otherwise, do not reject H_0.

One–Tailed and Two–Tailed Tests

Before we can illustrate the normal curve test, there is one additional factor we need to consider that is not relevant in the chi square test. In the chi square test, the rejection region always falls in the upper tail of the chi square distribution. However, in the z test, which utilizes the standard normal distribution, the rejection region may be in just the upper tail (*upper one-tailed test*), or just the lower tail (*lower one-tailed test*), or it may be equally divided between the two tails (*two-tailed test*).

Whether a z test is two-tailed or one-tailed and, if the latter, which tail, depends on the nature of the research question or hypothesis that the test is addressing. Therefore, we need to consider the nature of research hypotheses and their relation to statistical hypotheses.

Research Hypotheses and Statistical Hypotheses

Researchers often hypothesize that the mean of a population is greater than some value, or that it is less than some value. For example, a medical researcher may hypothesize that with a new drug treatment, patients will require, on the average, *fewer* than the standard 5 days of hospital treatment; the research hypothesis is "$\mu < 5$." Or a college professor, trying out a new instructional method, may hypothesize that the mean score on a final exam will be *higher* than the past mean of 60 points ("$\mu > 60$"). Hypotheses like these are called ***directional hypotheses*** because they specify the direction of an expected difference.

Researchers who hold directional hypotheses are interested in detecting differences only in the expected direction. The medical researcher, for example, is not interested in new drugs which increase treatment time or leave it unchanged; only a shortened treatment time is of interest. Similarly, the college professor wants to identify new instructional methods that improve student performance, not those that worsen it or leave it unchanged.

Sometimes, on the other hand, researchers are interested in detecting any kind of difference. A consumer psychologist working for a cereal producer may hypothesize that a new method of packaging will change customers' attitudes toward the product, but the direction of the change is unknown and could be positive or negative. If the mean attitude score in the past was 25, the psychologist's hypothesis is that, with new packaging, $\mu \neq 25$. Hypotheses that do not specify a particular direction of expected difference are called ***nondirectional hypotheses***. Researchers who pose nondirectional hypotheses are interested in detecting any kind of difference, in either direction.

Research hypotheses such as $\mu < 5$ or $\mu > 60$ or $\mu \neq 25$ include a range of possible values of μ. For example, $\mu < 5$ includes all values of μ that are less than 5. Since hypotheses about ranges of population values cannot be tested directly, the researcher restates the hypothesis as a null hypothesis (H_0), H_0: $\mu = 5$ or H_0: $\mu = 60$ or H_0: $\mu = 25$. The null hypothesis specifies a particular value of μ; therefore it is a testable hypothesis. Note, however, that the null hypothesis states that there is *no* change or difference, quite the opposite of the research hypothesis. The researcher will test the null hypothesis and reject it in favor of the research hypothesis, provided that the sample mean differs sufficiently from the value specified by the null hypothesis *in a direction consistent with the research hypothesis*.

For example, with research hypothesis $\mu < 5$, the null hypothesis (H_0) that $\mu = 5$ will be rejected only if the sample mean M is so far *below* 5 that its z score falls well out in the *lower tail* of the standard normal curve. The test is called a ***lower one-tailed test*** because the entire rejection region, with area = α, falls in the lower tail of the curve. Figure 7–5a shows the relationship between the rejection region, the nonrejection region, and alpha in a lower one-tailed test.

If the research hypothesis is $\mu > \mu_H$, such as $\mu > 60$, the null hypothesis (H_0) that $\mu = \mu_H$ will be rejected only if M is so far *above* μ_H that it converts to a z score far out in the upper tail of the standard normal curve. The test is called an ***upper one-tailed test*** because the entire rejection region with area = α falls in the upper tail of the curve. Figure 7–5b shows the relationship between the rejection region, the nonrejection region, and alpha in an upper one-tailed test.

If the research hypothesis is $\mu \neq \mu_H$, such as $\mu \neq 25$, the null hypothesis (H_0) that $\mu = \mu_H$ will be rejected if the sample mean is so far *above or below* μ_H that it converts to a z score well out in either the lower or upper tail of the standard normal curve. The test is called a ***two-tailed test*** because the rejection region is split evenly between the two tails with area = ½α in each tail. Figure 7–5c shows the relationship between the rejection region, the nonrejection region, and alpha in a two-tailed test.

Figure 7–5

Location of rejection region and critical value of z in one-tailed and two-tailed normal curve (z) tests.

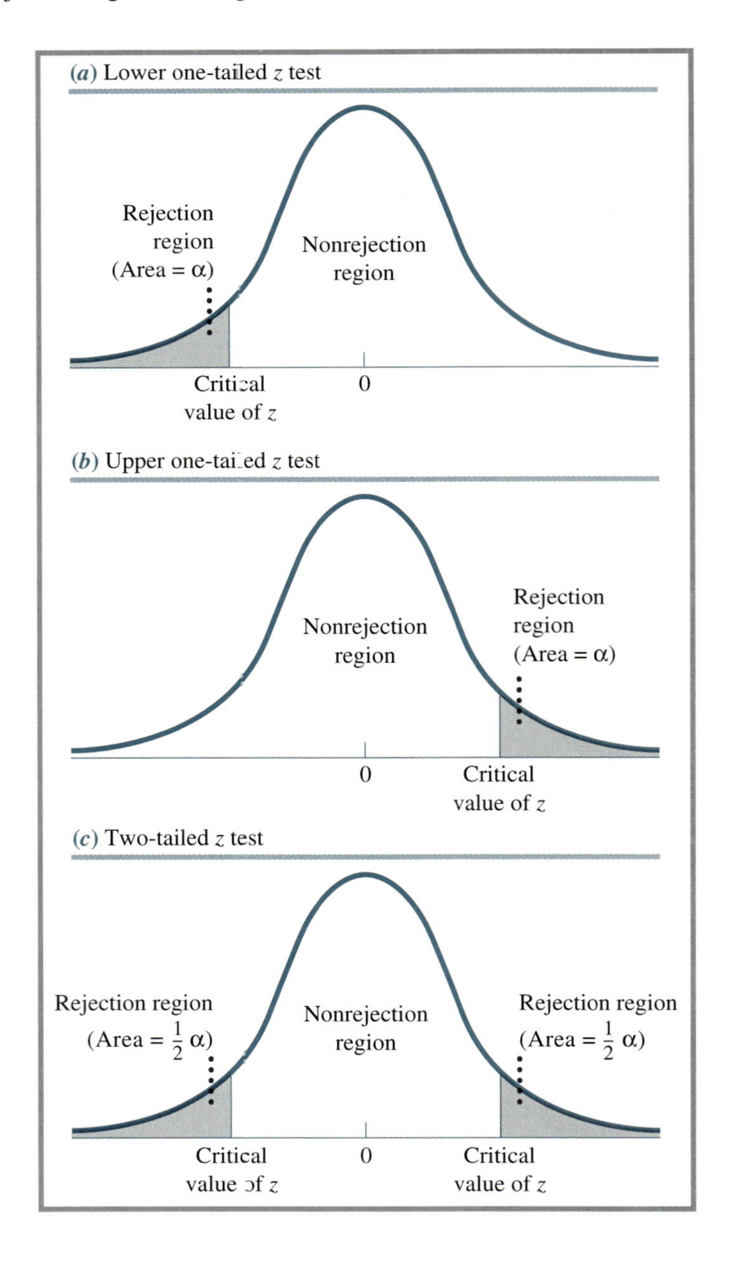

Null and Alternative Hypotheses

Research hypotheses usually state that an experimental treatment *will* have an effect, or that a difference *does* exist between populations, or that two or more variables *are* related. Although researchers test the null hypothesis of *no* effect, *no* difference, or *no* relationship, they usually hope to reject the null hypothesis in favor of the research hypothesis. In statistical tests, the research hypothesis plays the role of the ***alternative hypothesis*** (H_A), the conclusion to be reached if H_0 is rejected. In normal curve tests (and many other tests as well), it is common practice to begin by stating not only the null hypothesis but the alternative hypothesis as well.

In a lower one-tailed test, H_0 can be rejected only if the sample mean is sufficiently smaller than μ_H; and the conclusion reached if H_0 is rejected is that $\mu < \mu_H$. Therefore, the alternative hypothesis in a lower one-tailed test is H_A: $\mu < \mu_H$. In an upper one-tailed test, where H_0 can be rejected only if the sample mean is sufficiently greater than μ_H, the alternative hypothesis is H_A: $\mu > \mu_H$. In a two-tailed test, a sample mean sufficiently different from μ_H in either direction results in rejection of H_0; therefore the alternative hypothesis is $\mu < \mu_H$ or $\mu > \mu_H$—that is, H_A: $\mu \neq \mu_H$.

Note that in a lower one-tailed test, with all of the rejection region in the lower tail and none of the rejection region in the upper tail, no sample mean greater than μ_H, no matter how much greater, can lead to rejection of H_0. Therefore, it is impossible to conclude that $\mu > \mu_H$. Since sample means greater than μ_H inevitably result in nonrejection of H_0 in a lower one-tailed test, the possibility of $\mu > \mu_H$ is combined with $\mu = \mu_H$, and the null hypothesis is stated: $\mu \geq \mu_H$. Thus, in a lower one-tailed test, the null and alternative hypotheses are as follows:

$$H_0 : \mu \geq \mu_H$$
$$H_A : \mu < \mu_H$$

Statement of H_o *and* H_A *in a lower one-tailed test*

For example, the medical researcher's null and alternative hypotheses are

$$H_0 : \mu \geq 5$$
$$H_A : \mu < 5$$

If his test results in rejection of H_0, he concludes that $\mu < 5$; that is, with the new drug, patients require fewer than 5 days of treatment, on the average. But if his test results in nonrejection of H_0, he cannot conclude that the new drug is effective in reducing treatment time.

In an upper one-tailed test, with all of the rejection region in the upper tail and none of the rejection region in the lower tail, no sample mean smaller than μ_H, no matter how much smaller, can lead to rejection of H_0. Therefore, it is impossible to conclude that $\mu < \mu_H$. Since sample means smaller than μ_H always result in nonrejection of H_0 in an upper one-tailed test, the possibility of $\mu < \mu_H$ is combined with $\mu = \mu_H$, and the null hypothesis is stated: $\mu \leq \mu_H$. Thus, in an upper one-tailed test, the null and alternative hypotheses are as follows:

$$H_0 : \mu < \mu_H$$
$$H_A : \mu > \mu_H$$

Statement of H_o *and* H_A *in an upper one-tailed test*

For example, the college professor's null and alternative hypotheses are

$$H_0 : \mu \leq 60$$
$$H_A : \mu > 60$$

If her test results in rejection of H_0, she can conclude that the new instructional method does increase final exam performance. If, on the other hand, the test results in nonrejection of H_0, she does not have sufficient evidence to conclude that final exam performance is improved by the new instructional method.

In two-tailed tests, with the rejection region split between the tails, H_0 is rejected if the sample mean is sufficiently far below or above μ_H. Only sample means sufficiently near to μ_H result in nonrejection of H_0. Therefore the null hypothesis is simply: $\mu = \mu_H$. The null and alternative hypotheses of a two-tailed normal curve test are

$$H_0 : \mu = \mu_H$$
$$H_A : \mu \neq \mu_H$$

Statement of H_0 *and* H_A *in a two-tailed test*

For example, the consumer psychologist's null and alternative hypotheses are

$$H_0 : \mu = 25$$
$$H_A : \mu \neq 25$$

Note three things about H_0 and H_A:

1. H_0 and H_A are mutually exclusive and exhaustive. They cannot both be true at the same time; and one or the other must be true because, between them, they include all possibilities.
2. Since H_0 and H_A are mutually exclusive and exhaustive, rejection of H_0 provides support for H_A; nonrejection of H_0, on the other hand, means that there is not sufficient evidence to support H_A.
3. Since H_A is usually equivalent to the research hypothesis, rejection of Ho provides support for the research hypothesis; nonrejection of H_0 means that evidence is insufficient to support the research hypothesis.

Deciding Whether to Reject the Null Hypothesis in One-Tailed and Two-Tailed Tests

As you can see in Figure 7–5, values of z needed to reject H_0 depend on whether the test is lower one-tailed, upper one-tailed, or two-tailed. In a lower one-tailed test, only negative values of z sufficiently far below zero can result in the decision to reject H_0; in upper one-tailed tests, only positive values of z sufficiently far above zero can result in the decision to reject H_0; in two-tailed tests, values of z that are sufficiently far from zero in either direction result in the decision to reject H_0.

The following procedure is used to decide whether to reject H_0 in a lower one-tailed test:

1. Check to make sure that z is negative.
2. In the normal curve table, find $p(z <$ the calculated value). This is the area beyond z.
3. If $p < \alpha$, reject H_0.

EXAMPLE In a lower one-tailed test with $\alpha = .05$, you have calculated $z = -1.85$. Can you reject H_0?

z is negative, as required. According to the normal curve table, the area beyond $z = -1.85$ is .0332; $p(z < -1.85) = .0332$. Since .0332 is less than .05, you can reject H_0.

The following procedure is used to decide whether to reject H_0 in an upper one-tailed test:

1. Check to make sure that z is positive.
2. In the normal curve table, find $p(z >$ the calculated value). This is the area beyond z.
3. If $p < \alpha$, reject H_0.

EXAMPLE In an upper one-tailed test with $\alpha = .05$, you have calculated $z = 1.38$. Can you reject H_0?

z is positive, as required. According to the normal curve table, the area beyond $z = 1.38$ is .0838; $p(z > 1.38) = .0838$. Since .0838 is greater than .05, you cannot reject H_0.

The evaluation of z is a little more complex in a two-tailed test. The sign of z is irrelevant. You need to find the probability of a $|z|$ this far from zero in *either* the lower tail *or* the upper tail. For example, if you have calculated $z = -2.15$, you need to find $p(z < -2.15$ *or* $z > 2.15)$. Similarly, if you have calculated $z = +1.54$, you need to find $p(z < -1.54$ or $z > 1.54)$. That is, you need to find the area beyond $|z|$ in both tails combined. Use the following procedure:

1. In the normal curve table, find the area beyond the calculated z in one tail.
2. Double the area. The result, the area in both tails combined, is p.
3. If $p < \alpha$, reject H_0.

EXAMPLE In a two-tailed test with $\alpha = .05$, you have calculated $z = -2.15$. Can you reject H_0?

According to the normal curve table, the area beyond z in one tail is .0158. The area beyond $|z|$ in both tails combined is $2(.0158) = .0316$. Therefore, $p = .0316$. Since .0316 is less than .05, you can reject H_0.

EXERCISE 7–5

1. In each of the following, decide whether a one-tailed or two-tailed test is required. Then state H_0 and H_A, and draw a picture showing where the rejection region lies.
   a. Over the past 10 years the kindergarten students at Harrison Elementary School have had a mean score of 58 on a reading readiness test (regard this as a population mean). The kindergarten teacher wonders whether a newly instituted educational program has succeeded in raising children's scores.
   b. I know that the mean score of statistics students nationwide on a standardized statistics test is 75. I am interested in finding out whether statistics students at my university score at a different level.
   c. The mean score of depressed patients on a certain personality inventory is 32. Dr. Smith hypothesizes that his clients come from a population that has a mean lower than 32.
2. In each of the following cases, find the needed probability, and decide whether to reject H_0.
   a. In an upper one-tailed test with $\alpha = .05$, $z = 1.82$.
   b. In a two-tailed test with $\alpha = .05$, $z = 1.82$.
   c. In a two-tailed test with $\alpha = .05$, $z = -1.82$.
   d. In a lower one-tailed test with $\alpha = .01$, $z = -1.82$.

Now we are ready to illustrate the entire z test procedure for one-tailed and two-tailed tests.

An Example of a Lower One-Tailed Test

Roberta Watkins, a child psychologist, is working with children in a remote community in rural Appalachia. She believes that the extremely impoverished environment has lowered the children's IQ scores. She decides to test the hypothesis that the mean IQ of the children in this community is less than the general population mean, 100. Thus, Roberta's research hypothesis is $\mu < 100$. The appropriate test is a lower one-tailed test. We show all the steps in the test of the hypothesis.

1. *State* H_0 *and* H_A.

$$H_o : \mu \geq 100$$
$$H_A : \mu < 100$$

(Note that the alternative hypothesis, H_A, is equivalent to Roberta's research hypothesis.)

2. *Select a level of* α.
Roberta decides to use $\alpha = .05$.
3. *Collect data to test* H_0 *and calculate* z.
Roberta administers an IQ test to a random sample of 25 community children and finds the mean (M) = 94. This is below 100, as she expected. But is it far enough below 100 to reject H_0? In order to find out she must convert M into a value of the test statistic (z) using the formula:

$$z = \frac{M - \mu_H}{\sigma_M}$$

She already knows that $\mu_H = 100$ and $M = 94$. She needs to calculate σ_M. Since the general population of IQ scores has standard deviation (σ) = 15 and her sample size (N) is 25

$$\sigma_M = \frac{\sigma}{\sqrt{N}} = \frac{15}{\sqrt{25}} = \frac{15}{5} = 3.0$$

Therefore,

$$z = \frac{94 - 100}{3.0} = -2.00$$

4. *Find the probability of* z.
Roberta must find $p(z < -2.00)$ in the standard normal curve. She finds that the area below (beyond) $z = -2.00$ is .0228. Therefore, $p(z < 2.00) = .0228$, which is less than $\alpha = .05$. Figure 7–6 shows the results of Roberta's test.
5. *Decide whether to reject* H_0.
Since $p < .05$, Roberta rejects H_0 on the grounds that, if H_0 is true, a result such as hers would occur by chance less than 5 times in 100. Her sample's mean IQ is significantly lower than 100 at the .05 level. As she suspected, the population of children she is working with appears to have a mean IQ less than 100.

Before leaving Roberta Watkins, let us consider two other possible outcomes of her study. First, suppose that the mean IQ of her sample of 25 children were 97 instead of 94. In this case, in Step 3 of the test

$$z = \frac{97 - 100}{3.0} = \frac{-3}{3.0} = -1.00$$

Figure 7–6

Roberta Watkins's lower one-tailed test. Roberta calculated $z = -2.00$. The figure shows that $p(z < -2.00) = .0228$.

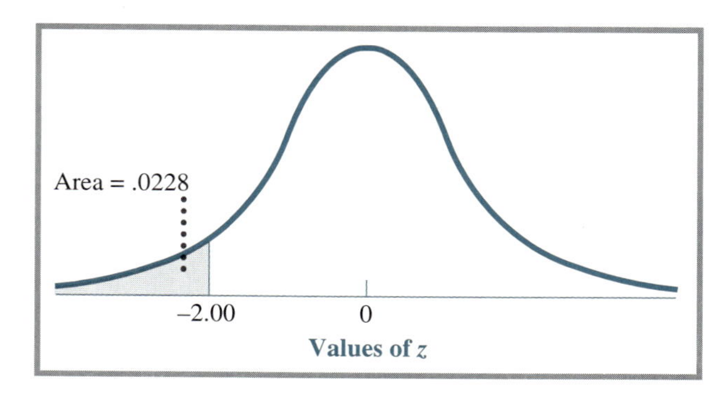

The probability of $z < -1.0$ is .1587, which is larger than .05. Therefore, Roberta cannot reject H_0. Even though M is less than 100, it is not far enough below 100 to reject H_0. A sample mean only 3 points below the hypothesized population mean, in a sample size of 25, would occur by chance more than 5 times in 100 ($p > .05$) if H_0 is true. Under these circumstances, Roberta would conclude that the mean IQ of children in the community could be equal to or greater than 100, as the null hypothesis states; the impoverished environment has not significantly reduced their IQ scores.

Second, suppose that Roberta's sample mean M had been 110, well above the hypothesized μ_H. In this case

$$z = \frac{110 - 100}{3.0} = \frac{10}{3.0} = 3.33$$

A z score this far from zero occurs very seldom in random sampling from a normal distribution. Nevertheless, Roberta cannot reject H_0. Why not? Because she decided ahead of time to reject H_0 only if M was sufficiently far *below* 100. In a lower one-tailed test, any positive value of z, no matter how large, falls in the nonrejection region and results in nonrejection of H_0.

An Example of an Upper One-Tailed Test

Arthur Fingerhut has devised a program for teaching intellectual skills that he believes will increase children's IQ scores. He plans to select a representative sample of children and provide them with his training program. He will use the normal curve test to test his hypothesis that the mean IQ of children who have undergone his program is greater than 100 ($\mu > 100$). The appropriate test, an upper one-tailed test, is as follows:

1. *State* H_0 *and* H_A.

$$H_0 : \mu \leq 100$$
$$H_A : \mu > 100$$

2. *Select a level of* α.
   Arthur decides to use $\alpha = .01$.
3. *Collect data to test* H_0 *and calculate* z.
   Ten children undergo Arthur's training program. Their IQ scores after training are 113, 124, 116, 104, 97, 92, 119, 100, 105, 135. Arthur calculates the mean IQ of the sample:

$$M = \frac{\Sigma X}{N} = \frac{113 + 124 + \cdots + 135}{10} = 110.5$$

The sample mean is above 100, but is it far enough above 100 to reject the null hypothesis? To find out, Arthur must calculate z.

First, he calculates σ_M. With $\sigma = 15$ and $N = 10$

$$\sigma_M = \frac{15}{\sqrt{10}} = 4.75$$

(Why is his value of σ_M larger than Roberta's?)

Now he is ready to calculate z:

$$z = \frac{110.5 - 100}{4.75} = 2.21$$

4. *Find the probability of* z.
   Since this is an upper one-tailed test, Arthur must find $p(z > 2.21)$ in the standard normal curve. From the normal curve table, he finds that $p(z > 2.21)$ is .0136, which is greater than $\alpha = .01$. Figure 7–7 shows the results of his test.
5. *Decide whether to reject* H_0.
   Since $p > .01$, Arthur cannot reject H_0 at the .01 level. His sample mean is not significantly greater than 100 at the .01 level. Therefore, he must conclude: "I do not have sufficient evidence to state that my training program is effective in increasing children's IQs."

See if you can answer the following questions yourself: (1) If Arthur had used $\alpha = .05$ instead of $\alpha = .01$, would he have reached the same conclusion? (2) If his 10 children had had a mean IQ of 85 after training, what conclusion would he have drawn? The answers can be found after the example of a two-tailed test.

An Example of a Two-Tailed Test

Marcia Mangelbrot, a school superintendent, wonders whether the children in her school district have a mean IQ different from the population mean of U.S. children, 100. Since she is equally interested in any difference from 100, in either a positive or negative direction, she uses a two-tailed test.

1. *State* H_0 *and* H_A.

$$H_O: \mu = 100$$
$$H_A: \mu \neq 100$$

2. *Select a level of* α.
   Marcia decides to use $\alpha = .01$.
3. *Collect data and calculate* z.
   From the school district records, Marcia selects a random sample of 500 children. She calculates their mean IQ (M) to be 97.5. Next, she calculates σ_M and z.

$$\sigma_M = \frac{15}{\sqrt{500}} = 0.67$$

$$z = \frac{97.5 - 100}{0.67} = -3.73$$

4. *Find the probability of* z.
   Marcia must find the probability of a z this far from zero, in either direction. That is, she must find $p(z < -3.73 \text{ or } z > 3.73)$. In the normal curve table, she finds that

Figure 7–7

Arthur Fingerhut's upper one-tailed test. Arthur calculated $z = 2.21$. The figure shows that $p(z > 2.21) = .0136$.

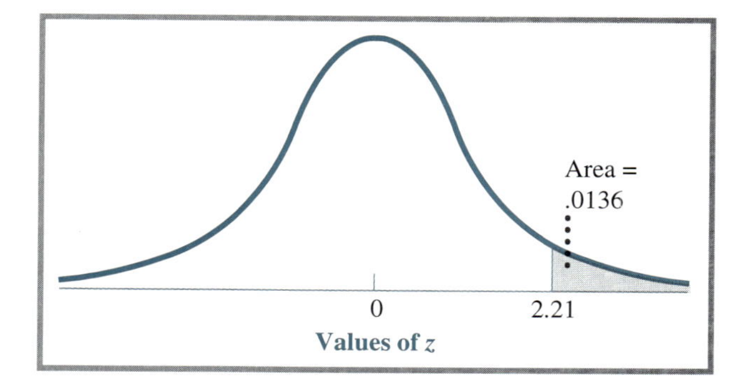

Figure 7–8

Marcia Mangelbrot's two-tailed test. Marcia calculated $z = -3.73$. The figure shows that $p(z < -3.73 \text{ or } z > 3.73) = .0001 + .0001 = .0002$.

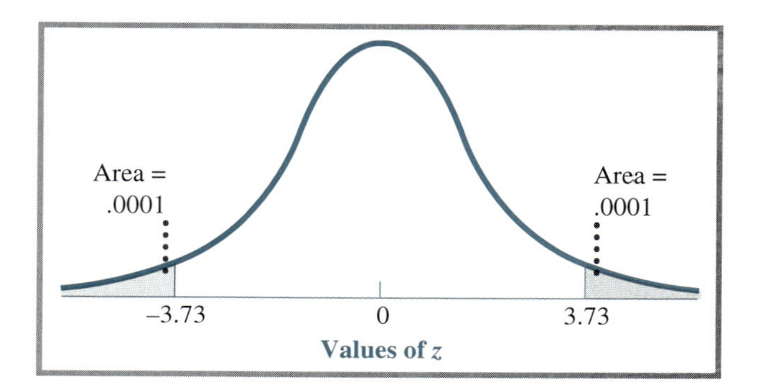

$p(z < -3.73) = .0001$ and $p(z > 3.73) = .0001$. Therefore, $p(z < -3.73 \text{ or } z > 3.73) = 2(.0001) = .0002$, which is less than .01. Figure 7–8 shows the result of Marcia's test.

5. *Decide whether to reject* H_0.
   Since $p < .01$, Marcia rejects H_0. The mean IQ of the sample is significantly different from 100 at the .01 level. She concludes that the children in her school district have a mean IQ different from 100. (And since her sample mean was 97.5, the difference seems to be in the negative direction.)

Here is another question to answer: (3) Both Arthur and Marcia used $\alpha = .01$. The mean IQ of Arthur's sample, 110.5, was much farther from 100 than the mean IQ of Marcia's sample, 97.5. Yet her sample mean was significantly different from 100 at the .01 level, whereas his was not. Why?

Answers to Questions 1, 2, and 3

1. If Arthur had used $\alpha = .05$, he would have rejected H_0, because, although the probability of his sample outcome, .0136, is greater than .01, it is less than .05. His sample mean is significantly greater than 100 at the .05 level but not at the .01 level.
2. If Arthur's 10 children had had a mean IQ of 85, he would not have rejected H_0. Remember: In an upper one-tailed test, the entire rejection region is in the upper tail. No value of M less than 100, no matter how much less, can result in rejection of H_0.
3. Arthur used a sample of size 10. The means of samples this small vary a great deal from one sample to another. Therefore σ_M is large (in his case, 4.75).† Marcia had a sample of size 500. Means of samples of size 500 vary little from sample to sample,

†Remember that σ_M, the standard error of the mean, is a measure of the variability of sample means. The more sample means vary around the population mean, the greater is the value of σ_M.

and σ_M is small (in her case, 0.67). Since $z = (M - \mu_H)/\sigma_M$, the smaller the value of σ_M, the larger the value of z. Even though Arthur's $M - \mu_H$ was larger than Marcia's, this was more than offset by her much smaller value of σ_M. Consequently, Marcia got a much larger value of $|z|$ than Arthur.

EXERCISE 7–6

1. A certain normal population has $\mu = 10$ and $\sigma = 5$.
   a. A researcher believes that his population differs from this one in an unspecified way. He selects a random sample from his population. The scores in the sample are 6, 9, 1, and 6. Test the researcher's hypothesis. Use $\alpha = .05$ and show all the steps.
   b. Suppose that a random sample of size 25 had the same mean as the sample in part (a). Test the hypothesis with $\alpha = .05$.
   c. How do you account for the difference in your results and conclusions in parts (a) and (b)?
2. On an inventory measuring self-esteem, the scores of boys in general are normally distributed with a mean of 20 and a standard deviation of 8. A personality theorist believes that boys from large families have a higher mean. The inventory is administered to a group of 16 boys from large families. Their mean score is 24.6. Test the appropriate hypothesis. Use $\alpha = .01$.
3. The mean number of treatment days with a certain drug is 8.5 days with standard deviation = 2.4. It is hoped that a new drug will reduce treatment time. In a sample of size 50, the mean treatment time under the new drug is 7.6 days. Test the appropriate hypothesis, with $\alpha = .05$.

Postscript on One-Tailed Versus Two-Tailed Tests

In most statistical tests, rejection of the null hypothesis (H_0) provides support for the research hypothesis. Therefore, researchers usually hope to reject H_0. In one-tailed tests, H_0 can be rejected only if the sample mean (M) differs from the hypothesized population mean (μ_H) in a particular direction; a difference in the opposite direction, no matter how large, results in nonrejection of H_0. Why, then, do researchers use one-tailed tests? The reason is that a smaller value of $|z|$ is needed to reject H_0 in a one-tailed test than in a two-tailed test with the same alpha, thus making it "easier" to reject H_0 in a one-tailed test. For example, with $\alpha = .05$, $z = 1.80$ results in rejection of H_0 in an upper one-tailed test because $p(z > 1.80) = .0359$. But if the test is two-tailed, H_0 is not rejected because $p(z < -1.80 \text{ or } z > 1.80) = 2(.0359) = .0718$.

If one-tailed tests make it easier to reject H_0, why don't researchers simply conduct lower one-tailed tests whenever $M < \mu_H$ and upper one-tailed tests whenever $M > \mu_H$? The answer is: The decision to use a two-tailed test, an upper one-tailed test, or a lower one-tailed test must be made on the basis of the research hypothesis *before* sample data are collected and M is calculated. If the research hypothesis is directional, a one-tailed test (in the direction specified by the hypothesis) is called for. But a nondirectional research hypothesis always calls for a two-tailed test.

Some Final Comments About Normal Curve Tests

The normal curve test is not widely used because it requires that we know the standard deviation of the hypothesized population, a condition that is rarely met in practice. In Chapter 8 we will introduce a test of hypotheses about population means when population standard deviations are unknown and must be estimated from sample data. This test, called the *t test*, is very similar to the z test. In particular, the statement of hypotheses and rationale for one-tailed and two-tailed tests are exactly the same.

It is well to remember that the conclusions drawn from a normal curve test, as from any test, are probabilistic. When we reject a null hypothesis H_0, we do so because a sample result like ours is *unlikely* to occur if H_0 is true. But "unlikely" does not mean impossible. Sometimes we reject a true H_0, a Type I error of inference. Similarly, when we do not reject H_0, it is because our sample result is not too unlikely if H_0 is true; but "not too unlikely" does not necessarily mean very likely. Sometimes we fail to reject a false H_0, a Type II error of inference.

In Chapter 5, we discussed some of the factors that influence the probability of making Type I and Type II errors of inference. We will come back to these in Chapter 8 and discuss them in more detail, particularly as they apply to tests of hypotheses about population means.

SUMMARY

The normal curve is a family of theoretical distributions of great importance in statistics. Each possible combination of a mean and a standard deviation defines one curve in the family. The normal distribution with mean = 0 and standard deviation = 1 is called the ***standard normal curve***. Any normal distribution can be transformed into the standard normal distribution by converting score values into standard (z) scores.

The table of the standard normal curve (Table D–2 in Appendix D) shows proportions of area between each z score and the mean and beyond each z score. The table can be used to find the proportion of area below, above, or between z scores or to find z scores cutting off certain proportions of area. Proportions of area, in turn, provide information about proportions, percents, probabilities, and expected frequencies of observations.

Given a normal population with mean μ and standard deviation σ, questions about proportions, percents, probabilities, and expected frequencies of particular score values (X) are answered by, first, converting score values to z scores using the formula

$$z = \frac{X - \mu}{\sigma}$$

Then the relevant areas (proportions, percents, etc.) are found in the normal curve table. If scores that cut off given proportions, percents, and the like are desired, the table is used to find the z score(s) cutting off the given area. Then z scores are converted to score values by the formula

$$X = \mu + z(\sigma)$$

Normal curve procedures can be used to find approximate ***binomial probabilities***, providing that the number of repetitions (N) is sufficiently large. In a binomial distribution with elementary probability = p_E, the mean is Np_E, and the standard deviation is $\sqrt{Np_E(1 - p_E)}$. Since the normal distribution is continuous whereas the binomial distribution is discrete, a ***correction for continuity*** is applied to outcome values, X, before calculating z, unless N is at least 20.

If sample size is reasonably large, the sampling distribution of the mean is approximately normal, with mean (μ_M) = μ and standard deviation (standard error of the mean, σ_M) = $\frac{\sigma}{\sqrt{N}}$. Probabilities of particular sample means (M) can be found by converting the sample means into z scores by the formula

$$z = \frac{M - \mu}{\sigma_M}$$

and then referring to the normal curve table. To find sample mean values that cut off given proportions (percents, probabilities), the table is used to find z; then z is converted to a sample mean by the formula

$$M = \mu + z(\sigma_M)$$

The normal curve test, or z test, is used to test hypotheses about population means when the population standard deviation is known. The steps in hypothesis testing and the methods for drawing conclusions are the same as in any statistical test. After the null hypothesis (H_0) and alternative hypothesis (H_A) have been stated and a level of alpha selected, data are collected, and the sample mean (M) is converted into a value of the test statistic, z, using the formula

$$z = \frac{M - \mu_H}{\sigma_M}$$

Then the probability of the calculated z is found in the normal curve table. If $p < \alpha$, H_0 is rejected, and the difference between M and μ_H is significant at level α; if $p > \alpha$, H_0 is not rejected, and the difference is not significant at level α.

Normal curve tests can be two-tailed or one-tailed. If the researcher is interested in detecting any kind of difference from the hypothesized population mean, μ_H, a ***two-tailed test*** is called for. The rejection region is divided equally between the two tails of the standard normal curve, and H_0 is rejected if the probability of z this different from zero in either direction is less than alpha.

If the researcher is interested only in detecting differences from μ_H in the positive direction, an ***upper one-tailed test*** is used. The entire rejection region is located in the upper tail, and H_0 is rejected if $p(z >$ calculated value) is less than alpha. If the researcher is interested only in detecting differences from μ_H in the negative direction, a ***lower one-tailed test*** is used. The entire rejection region is located in the lower tail, and H_0 is rejected if $p(z <$ calculated value) is less than alpha.

One-tailed tests have the advantage that a smaller value of $|z|$ is needed to reject H_0. However, a one-tailed test is appropriate only for testing directional research hypotheses. The decision to use a one-tailed or a two-tailed test depends on the nature of the research question or hypothesis, not on the sample outcome.

As in all tests, inferences from normal curve tests are always probabilistic. Sometimes a test results in rejection of a true null hypothesis (a Type I error); and sometimes a test results in nonrejection of a false null hypothesis (a Type II error). Factors affecting Type I and Type II errors will be discussed more fully in Chapter 8.

QUESTIONS

1. A normally distributed population has mean = 80, SD = 12.

   a. What percentage of the population has scores between 70 and 80?
   b. In a group of size 250 from the population, how many individuals are expected to score above 95?
   c. What is the 30th percentile?
   d. What are the score values between which the middle 80% of the population falls?
   e. What is the probability of a score value less than 75?
   f. If a sample of size 25 is selected randomly from the population, what is the probability of a sample mean less than 75?
   g. Only 12% of the population scores below what value?
   h. Only 12% of samples of size 25 have means below what value?

2. An instructor plans to grade on the curve. She will give an A to the top 10% of the students, a B to the next 20%, a C to the middle 40%, a D to the next 20%, and an F to the lowest 10%. If her test is approximately normally distributed with mean = 35 and SD = 5, what are the cutoff points between successive letter grades?

3. In a normal distribution with mean = 65 and SD = 10, calculate the semi-interquartile range.

4. Figure 4–9a (p. 114) shows the binomial distribution with elementary probability = .5 and number of repetitions = 5.

   a. Find the exact probability of getting either one or two heads in tossing five fair coins.

b. Use the normal curve to find the approximate probability of one or two heads in tossing five fair coins. Compare your answer with the actual probability.
c. If the number of coins were 12, would you expect the actual probability and the normal approximation to be more or less similar than with 5 coins? Why?
d. Use the normal curve to find the approximate probability of 8 or more heads in tossing 12 fair coins.

5. My white rat Samuel has no right-turning or left-turning preference. Use the normal curve to find the probability that he will turn right on fewer than 5 out of 20 trials.

6. In Chapter 4 we sampled from a population consisting of 100 members. (The population distribution is shown in Figure 4–4, p. 97). The mean of the population is 8 and the standard deviation is 1.66. If the distribution were normal but had the same mean and *SD*, how many of 100 individuals would be expected to have each score? (*Note:* In order to answer this question, you must treat a score of 3 as "less than 3.5," 4 as "3.5–4.5," etc.)

7. In the general adult population, the mean score on an inventory measuring depression is 70 with $SD = 14$. Dr. Hartman-Jones expects the mean of psychiatric patients to be lower than that of the general population. She finds that the mean of a sample of 50 psychiatric patients is 65.5. Test the appropriate hypothesis. Use $\alpha = .05$.

8. According to the almanac, the town of Smithville experiences an average of 18.5 rainy days each July, with $SD = 2.3$. The city fathers believe their town actually gets less rain in July than the figure given in the almanac. They review the records of the years 1990 to 1994 and find the following data:

Year	**Number of July Days with Rain**
1990	16
1991	19
1992	17
1993	10
1994	16

Conduct the appropriate test. Use $\alpha = .05$.

9. A school district is testing a new mathematics curriculum on a sample of 100 fifth-grade students. The district wants to find out whether the new curriculum makes a difference in scores on the districtwide end-of-year math test. Over the last 10 years the mean test score has been 68.4, with $SD = 13.5$. The sample of 100 students taught by the new curriculum achieves an average score of 67.6. What should the school district conclude? Use $\alpha = .05$.

10. An economist wonders whether the mean family income in a certain region differs from the national average for comparable communities. The national average is \$31,560, with $SD = \$3500$. A sample of 100 families from the region of interest has a mean income of only \$30,930. The economist decides to conduct a lower one-tailed test because the sample mean is smaller than the national average.

a. Should he conduct a lower one-tailed test? If not, why not?
b. Test the hypothesis using a lower one-tailed test with $\alpha = .05$. Then conduct a two-tailed test with $\alpha = .05$. Explain the difference in the results and conclusions.

Using the *t* Distribution to Make Inferences About a Population Mean

INTRODUCTION

In Chapter 7 we discussed the use of the normal curve for testing hypotheses about a population mean, μ. In the normal curve test, it is assumed that the population standard deviation, σ, is known. This is often unrealistic. Just as we must estimate μ from sample data, we usually must also estimate σ. However, when the sample estimator of σ is substituted for the actual value, the standard normal curve cannot be used to find probabilities of various sample means. A distribution called the t distribution is required instead. This chapter begins with a discussion of the t distribution and its use for testing hypotheses and for computing confidence intervals for a population mean. After a brief discussion of confidence intervals for a population percentage, we take up additional issues related to hypothesis testing in general: errors of inference, power, and parametric versus nonparametric statistical tests.

Objectives

When you have completed this chapter, you will be able to

- Define the estimated standard error of the mean, s_M, and compare it with the actual standard error of the mean, σ_M.
- Compute s_M from sample data.
- Explain how a t distribution is generated.
- Compare the shape of the z and t distributions, and describe how the shape of the t distribution varies with degrees of freedom (df).
- Use the t test to test H_0: $\mu = \mu_H$, using one-tailed or two-tailed tests as needed.
- Calculate and interpret confidence intervals for a population mean, and describe the relation between confidence intervals and tests of hypotheses.
- Calculate and interpret confidence intervals for a population percentage.
- Define Type I errors and Type II errors.
- List and discuss the factors that affect the probability of Type I and Type II errors.
- Define the power of a test and describe the factors that affect power.
- Define parametric and nonparametric tests, and describe the relative advantages and disadvantages of each.

Introduction to the t Distribution

In Chapter 7 you learned to use the standard normal curve test (z test) to test hypotheses about population means. In the test, you calculated a value of $z = \frac{M = \mu_H}{\sigma_M}$, where σ_M is the standard error of the mean. Then you referred to the table of the standard normal curve to determine whether the probability of a z score like yours was less than alpha (reject H_0) or greater than alpha (do not reject H_0).

Since σ_M, the denominator of the z ratio, is equal to the population standard deviation, σ, divided by the square root of the sample size, the normal curve test rests on the assumption that σ is known. For this reason, the normal curve test is most useful for testing hypotheses that samples have been drawn from known populations, with known means and standard deviations. Often, however, we want to test hypotheses about means of populations whose standard deviations are unknown. In cases like this, we must calculate an estimate of σ from sample data and use the estimate to calculate an estimate of σ_M.

Calculating the Estimated Standard Error of the Mean

The actual standard error of the mean (σ_M) is calculated by the formula

$$\sigma_M = \frac{\sigma}{\sqrt{N}}$$

The sample estimator of σ_M is called the ***estimated standard error of the mean*** (symbolized by s_M). It is calculated by substituting the sample estimator for σ in the formula above:

$$s_M = \frac{\text{estimator of } \sigma}{\sqrt{N}}$$

In Chapter 6 we used the symbol s to represent the sample estimator of σ. Therefore we can write

$$s_M = \frac{s}{\sqrt{N}}$$ *Formula for the estimated standard error of the mean*

For example, if a sample of size 20 has $s = 6.3$, $s_M = \frac{6.3}{\sqrt{20}} = 1.41$.

Let us review the calculation of s, the sample estimator of σ. The formula for calculating s is

$$s = \sqrt{\frac{SS}{N-1}}$$

The term $N - 1$ in the denominator is the number of degrees of freedom (df) for estimating the population standard deviation from sample data. As you will see, it is also the number of df for the t test. The numerator term, SS, stands for *sum of squares*, short for "sum of the squared deviations from the mean." The sum of squares can be calculated by either of two methods:

Deviation method: $SS = \Sigma(X - M)^2$

Raw score method: $SS = \Sigma X^2 - \frac{(\Sigma X)^2}{N}$

When calculating by hand, the raw score method is usually easier.

The estimated standard error of the mean, s_M, can be calculated from sample data in three steps, as follows:

1. Calculate the sum of squares, SS, by either the deviation or raw score method.
2. Calculate the estimated standard deviation, $s = \sqrt{\frac{SS}{N-1}}$.
3. Calculate $s_M = \frac{s}{\sqrt{N}}$.

Here is an illustration for a sample of size 8 ($N = 8$) with scores of 8, 6, 7, 4, 10, 5, 6, 3. We will use the raw score method to calculate SS.

Step 1: $\Sigma X = 8 + 6 + 7 + \cdots + 3 = 49$

$$\Sigma X^2 = 8^2 + 6^2 + \cdots + 3^2 = 64 + 36 + \cdots + 9 = 335$$

$$SS = \Sigma X^2 - \frac{(\Sigma X)^2}{N} = 335 - \frac{(49)^2}{8} = 34.875$$

Step 2: $s = \sqrt{\frac{SS}{N-1}} = \sqrt{\frac{34.875}{7}} = \sqrt{4.982} = 2.232$

Step 3: $s_M = \frac{s}{\sqrt{N}} = \frac{2.232}{\sqrt{8}} = \frac{2.232}{2.828} = 0.789$

All three steps are incorporated in the following one-step formula for s_M:

$$s_M = \sqrt{\frac{\Sigma X^2 - \frac{(\Sigma X)^2}{N}}{N(N-1)}}$$

One-step formula for calculating the estimated standard error of the mean

For this example

$$s_M = \sqrt{\frac{335 - \frac{(49)^2}{8}}{8(7)}} = \sqrt{\frac{34.875}{56}} = \sqrt{0.62277} = 0.789, \text{ as before}$$

Two brief notes:

1. In Chapter 6 we distinguished between the descriptive standard deviation of a data set and the sample estimator of a population standard deviation. In this chapter we are concerned with estimating population standard deviations rather than describing sample variability. Therefore, whenever we refer to a "sample standard deviation" or "sample SD," we will mean the sample estimator $s = \sqrt{SS/(N-1)}$. Similarly, when we refer to a "standard error of the mean," we will mean the estimated standard error, s_M.
2. Do not confuse s and s_M. Both are sample estimators. But they are estimators of different parameters. The s (without the subscript M) is an estimator of σ, the population standard deviation; s_M is an estimator of σ_M, the actual standard error of the mean.

EXERCISE 8–1

1. Without referring to the text, explain the difference between s and s_M. (Name each one and tell what each one is an estimator of.)
2. Calculate the estimated standard error of the mean of the following sample: 2, 3, 1, 1, 4, 2, 0, 2, 5, 3.
3. The standard deviation (s) of a sample of size 25 is 12.4. Calculate the standard error of the mean, s_M.

Comparison of the t Distribution and the Standard Normal Curve

In Chapter 7 we showed that if a population is normal or N is reasonably large, the distribution of $(M - \mu)/\sigma_M$, calculated over all possible samples, forms a standard normal curve. However, if s_M is substituted for σ_M, the ratio $M - \mu/s_M$, calculated over all possible samples, is not normally distributed. The shape is that of another probability distribution called the ***t distribution***. In order to examine the t distribution and see how it differs from the standard normal curve, we consider three hypothetical sampling experiments.

Sampling Experiment 1

Suppose that we select all possible samples of size 5 from a normal population and compute the mean M of each sample. Since the population is normal, the sampling distribution of the mean is normal with mean $(\mu_M) = \mu$ and SD (the standard error of the mean $\sigma_M) = \sigma/\sqrt{N}$. Next, we convert each sample mean into a z score:

$$z = \frac{M - \mu}{\sigma_M}$$

The distribution of the z scores over all samples is a standard normal distribution with mean = 0 and SD = 1. Therefore, probabilities of various sample means can be found by referring to the table of the standard normal curve. This distribution of z scores is the basis for the procedures we used in Chapter 7 to test hypotheses about population means.

Sampling Experiment 2

Again, we select all possible samples of size 5 from the same normal population and compute M in each sample. Again, the sampling distribution of the mean is normal with mean = μ and SD $(\sigma_M) = \sigma/\sqrt{N}$. But this time, instead of computing $(M - \mu)/\sigma_M$ in each sample, we do the following:

1. In each sample, we compute s_M, the sample estimator of σ_M. Based as it is on sample data, s_M varies from one sample to another, just like any other statistic.
2. Next, we calculate the following ratio in each sample:

$$\frac{M - \mu}{s_M}$$

When we examine the distribution of this ratio over all possible samples, we find that the shape is not a standard normal curve. Instead, the shape is a probability distribution called the t distribution. So we can write:

$$t = \frac{M - \mu}{s_M}$$

How does the distribution of t compare with the distribution of z? In some ways, the two distributions are similar. Both are symmetrical and both have mean = 0. But the t distribution is markedly lower in the middle and higher in the tails than the standard normal distribution, and its standard deviation is greater than 1. That is, less of the area is near the center (near $t = 0$) and more of the area is in the tails as shown in Figure 8–1.

Because of the difference in shape of the z and t distributions, probabilities of various values of t are not the same as those of the same values of z. For example, the probability of a z score above 1.96 is .025; but the probability of a t above 1.96 is greater than .025 because the t distribution has a greater proportion of its area in the tail above

Figure 8–1

A comparison of the distribution of $t = \frac{M - \mu}{s_M}$ for $N = 5$ (solid line) with the standard normal curve (broken line).

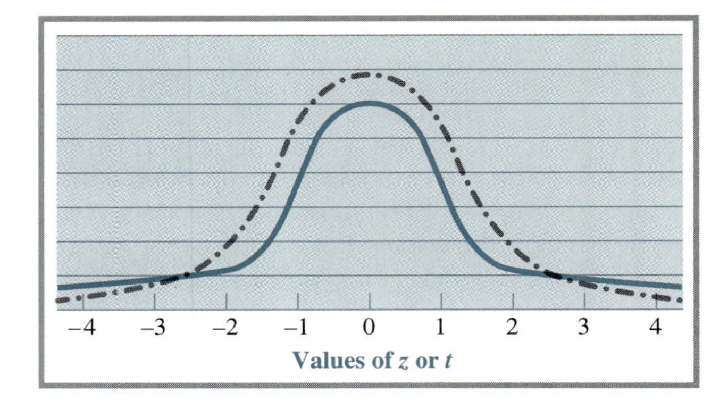

1.96. The probabilities found in the normal curve table do not apply to the t distribution, and were we to use the table to find probabilities of various sample means, such as in testing hypotheses about μ, our conclusions might well be incorrect.

Sampling Experiment 3

Again, we select all possible samples from the same normal population and compute M in each sample, but this time the size of each sample is 25 rather than 5. Again, the sampling distribution of the mean is normal with mean = μ and SD (σ_M) = $\sigma/\sqrt{N}$ (though, of course, σ_M is smaller than when N was 5).

We continue as in Sampling Experiment 2. First, s_M is computed in each sample; then each sample mean is converted into a value of t. Surprisingly, when we look at the distribution of t over all samples we find that it is not identical with the t distribution based on samples of size 5 in Sampling Experiment 2. It is higher in the center and lower in the tails. In fact, it looks a great deal more like the standard normal distribution, as you can see in Figure 8–2. Though probabilities of values of t in this t distribution are not identical with those of the same values of z, they are much more similar to standard normal probabilities than are those in the t distribution for $N = 5$.

A comparison of the t distributions generated in Sampling Experiments 2 and 3 reveals an important difference between the t distribution and the standard normal curve. Whereas there is only one standard normal distribution, there are many t distributions with different shapes. The shape of the t distribution depends on N (or, more accurately, on degrees of freedom). When df is very small, the distribution of t is markedly non-normal—quite low in the center and very high in the tails. As df increases, the distribution of t becomes more and more like a normal distribution. Thus, the t distribution bears some resemblance to both the binomial and the chi square distributions, with which you are already familiar. Like the binomial, the t distribution approaches the normal as N increases; and, like the chi square distribution, the t distribution is a family of distributions varying in degrees of freedom.

What are the implications of the three sampling experiments for statistical inferences about population means? The implications are that, when s_M is computed in a sample instead of σ_M, the t distribution rather than the z distribution must be used to determine probabilities. For example, when testing hypotheses using the test statistic $t = \frac{M - \mu}{s_M}$, instead of $z = \frac{M - \mu_H}{\sigma_M}$, we must evaluate the calculated t against critical values of t, not of z. The use of probabilities based on t rather than z is particularly crucial when N is small, because then probabilities of particular values of t are quite different from the probabilities of the same values of z.

Figure 8–2

A comparison of the distribution of $t = \frac{M - \mu}{s_M}$ for $N = 25$ (solid line) with the standard normal curve (broken line).

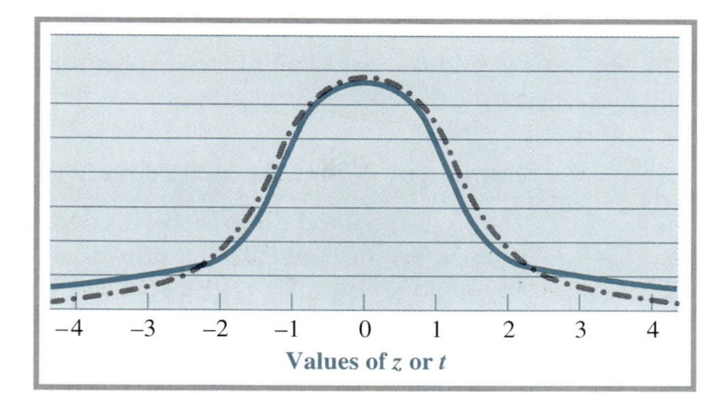

EXERCISE 8–2

1. I have selected a random sample of size 8 from a normal population and computed $M = 73.4$ and $\frac{M - \mu}{s_M} = 1.20$. According to the normal curve table, the area above a z score of 1.20 is .1151.
   a. Why can't I say that the probability of $M > 73.4$ is .1151?
   b. Is the area above $t = 1.20$ greater or less than .1151?
2. If my sample size were 50 instead of 8, would I be more or less in error than for $N = 8$ in stating that $p(M > 73.4) = .1151$? Why?

A Historical Note

The full name of the t distribution is *Student's t distribution.* The man who discovered the t distribution in the early 20th century and worked out its details was not named Student but William Gosset. Gosset worked for a brewery that did not permit its employees to publish under their own names, probably for fear of disclosing industrial secrets. Not wishing to jeopardize his job but recognizing the importance of his findings, Gosset published under the name "Student." Apparently, Gosset was highly successful not only as a statistician but also as a brewer, for he continued to publish under the pseudonym Student, while rising rapidly in his company's ranks until his death in 1937.

Addendum

Statistics textbooks sometimes differentiate between "large-sample" and "small-sample" statistics. They advocate using normal curve procedures for making inferences about μ when $N \geq 30$ and using t when $N < 30$. This reasoning is based on the fact that when $N \geq 30$, the t distribution is so close to the normal that the probabilities of various values of t are almost the same as those of the corresponding z scores. For example, when N is large, the area above $t = 1.96$ is only a little more than .025. Therefore, no significant error is made in interpreting the ratio $\frac{M - \mu}{s_M}$ as though it were a z score rather than a value of t.

This is certainly true. No great harm results if the normal curve table is used for finding probabilities of sampling outcomes and making inferences about μ when $N \geq 30$. However, to say that $\frac{M - \mu}{s_M} = z$ when $N \geq 30$ is just plain wrong! $\frac{M - \mu}{s_M}$ is always distributed as t, not z. As N increases, the distribution of $\frac{M - \mu}{s_M}$ becomes more and more like the distribution of $z = \frac{M - \mu}{\sigma_M}$. But nothing magical happens when $N = 30$

that suddenly transforms the ratio from a t into a z. Strictly speaking, whenever the sample estimator s_M is substituted for the actual standard error of the mean σ_M, the result is a t distribution, no matter how large the sample size.

Using the Table of the *t* Distribution

The t table, Table D–3 in Appendix D, gives critical values of t that cut off specified areas in one or both tails of t distributions with degrees of freedom ranging from 1 to ∞ (infinity).† The table is very similar to the one you used with the chi square test. Each row provides information about the t distribution with a particular number of degrees of freedom. At the top of each column are two areas (probabilities). The upper area is the proportion of area cut off in *one* tail of the t distribution by the critical values in that column. The lower area is the proportion of area cut off in *both* tails by the critical values in that column.

The t distribution is symmetrical around a mean of zero and includes both positive and negative values of t. Although negative values are not given in the table, each critical value can be regarded as positive, negative, or both ($\pm$) as needed. As an example, find the entry 2.500 in the row for 23 degrees of freedom. Now look at the two column headings. The upper one is .01. This means that in the t distribution with 23 degrees of freedom, $t = 2.500$ cuts of an area of .01 in the upper tail, and $t = -2.500$ cuts of an area of .01 in the lower tail. [In other words, $p(t > 2.500) = .01$, and $p(t < -2.500) = .01$.] The lower column heading is .02. This means that in the t distribution with 23 degrees of freedom, the two values of $t = \pm\ 2.500$ cut off a total area of .02 in the two tails combined. [That is, $p(t < -2.500 \text{ or } t > 2.500) = .02$.]

Each time you conduct a t test you will calculate a value of t and use the table to find its probability. To find the probability, use the following strategy:

1. If your test is one-tailed, make sure that t has the correct sign. In an upper one-tailed test, t must be positive. In a lower one-tailed test, t must be negative. If t has the "wrong" sign, H_0 cannot be rejected, regardless of the value of t. (In a two-tailed test, t can be positive or negative.)
2. In the row for the degrees of freedom of your test, find the two critical values that bracket your calculated t between them. Then, at the top of those two columns, find the probability of each critical value. Be sure to look at the upper column headings if your test is one-tailed and the lower column headings if your test is two-tailed. The probability (p) associated with your calculated t falls between the two column headings.

As in any hypothesis test, you will reject H_0 if $p < \alpha$. Here are two examples.

Example 1: In a lower one-tailed test with $\alpha = .01$ and 20 degrees of freedom, you have calculated $t = -2.96$. Should you reject H_0?

Step 1. t is negative, as it must be since the test is lower one-tailed. You can proceed.

Step 2. In the row for 20 degrees of freedom, the two critical values bracketing -2.96 are $(-)$ 2.845 and $(-)$ 3.850. The two *upper* column headings—probabilities for a one-tailed test—are .005 and .0005. The probability of your t is between .005 and .0005 ($p < .005$). Since p is less than alpha, .01, you can reject H_0.

†Since the t distribution approaches the normal distribution as df increases, the t distribution at $df = \infty$ is the standard normal curve. Critical values of t for $df = \infty$ are critical values of z in the standard normal curve.

EXAMPLE 2: In a two-tailed test with 9 degrees of freedom and $\alpha = .05$, you have calculated $t = 2.08$. Can H_0 be rejected?

Since the test is two-tailed, Step 1 is not needed.

Step 2. In the row for 9 degrees of freedom, the two critical values bracketing 2.08 are 1.833 and 2.262. The two *lower* column headings—the probabilities for a two-tailed test—are .10 and .05. Therefore the probability of $t = 2.08$ is between .10 and .05. Since p is greater than .05, you cannot reject H_0.

EXERCISE 8–3

1. You are conducting an upper one-tailed test with $\alpha = .05$ and 7 *df*. If the calculated $t = 2.02$, do you reject H_0?
2. You are conducting a lower one-tailed t test with $\alpha = .05$, $df = 12$. If the calculated $t = -2.02$, do you reject H_0?
3. If $df = 14$, what is the minimum value of $|t|$ needed to reject H_0 with $\alpha = .001$ (two-tailed test)?

Conducting the t Test

The rationale and procedure for t tests of hypotheses about population means are basically the same as for the normal curve test. First, the null hypothesis (H_0) and alternative hypothesis (H_A) are stated. As in the normal curve test, the formats of the null and alternative hypotheses for two-tailed and one-tailed t tests are

Two-tailed test:	H_0: $\mu = \mu_H$
	H_A: $\mu \neq \mu_H$
Upper one-tailed test:	H_0: $\mu \leq \mu_H$
	H_A: $\mu > \mu_H$
Lower one-tailed test:	H_0: $\mu \geq \mu_H$
	H_A: $\mu < \mu_H$

Next, a level of significance α is selected. Then the test statistic is calculated and evaluated by the methods discussed in the previous section. If $p < \alpha$, H_0 is rejected; the difference between M and μ_H is significant at level α. If $p > \alpha$, H_0 is not rejected; the difference between M and μ_H is not significant at level α.

The t test differs from the normal curve test in the following ways:

1. The estimated standard error of the mean, s_M, is computed instead of σ_M.
2. The test statistic is $t = \dfrac{M - \mu}{s_M}$ instead of $z = \dfrac{M - \mu_H}{\sigma_M}$.
3. The t table is used instead of the normal curve table to find the probability of the obtained t. The number of degrees of freedom for the test is $N - 1$.

Here are two examples.

EXAMPLE 1: The national mean on a fifth-grade reading test is 100. An Elmville teacher wonders whether the mean of fifth graders at his school differs from the national average. He gives the test to a random sample of 10 fifth graders. Their scores are: 96, 110, 114, 101, 98, 124, 117, 95, 105, 110. Test the appropriate hypothesis. Use $\alpha = .05$.

1. State the null hypothesis and the alternative hypothesis.

   Since the teacher is interested in detecting any kind of difference from the national average, a two-tailed test is called for.

   $$H_o: \mu = 100$$
   $$H_A: \mu \neq 100$$

2. Select a level of alpha: $\alpha = .05$.
3. Collect data to test the hypothesis, and calculate M, s_M, and t.

   $$N = 10,\ \Sigma X = 1070,\ \Sigma X^2 = 115{,}332$$

   (You can confirm these if you like.)

   $$M = \frac{\Sigma X}{N} = \frac{1070}{10} = 107.0$$

   $$s_M = \sqrt{\frac{\Sigma X^2 - \frac{(\Sigma X)^2}{N}}{N(N-1)}} = \sqrt{\frac{115{,}332 - \frac{(1070)^2}{10}}{10(9)}} = 3.059$$

   $$t = \frac{M - \mu_H}{s_M} = \frac{107.0 - 100}{3.059} = \frac{7.0}{3.059} = 2.29$$

4. Find the probability of t.

   $df = N - 1 = 10 - 1 = 9$. In the row for 9 df, 2.29 falls between the critical values 2.262 and 2.821, which have two-tailed probabilities of .05 and .02. Therefore $p < .05$.

5. Decide whether to reject H_O.

   Since $p < .05$, the teacher rejects H_O. The sample mean, 107.0, is significantly different from 100 at the .05 level. The teacher concludes that the mean reading test score of Elmville fifth graders is not equal to 100.

EXAMPLE 2: The mean score of the general adult population on a neuroticism scale is 25. A personality theorist believes that the mean score of college students is greater than that of the general adult population. She administers the scale to 15 college students. The sample mean is 27.8, with $SD = 5.80$. Is the sample mean significantly greater than 25 at the .01 level?

Since the theorist is interested in detecting a difference from 25 only in the positive direction, an upper one-tailed test is required.

1. State H_O and H_A.

   $$H_o: \mu \leq 25$$
   $$H_A: \mu > 25$$

2. Select a level of alpha: $\alpha = .01$.

3. Calculate M, s_M, and t.

$$M = 27.8$$

$$s_M = \frac{s}{\sqrt{N}} = \frac{5.80}{\sqrt{15}} = \frac{5.80}{3.87} = 1.50$$

$$t = \frac{M - \mu_H}{s_M} = \frac{27.8 - 25}{1.50} = 1.87$$

4. Find the probability of t.

$df = 15 - 1 = 14$. In the row for 14 df, the calculated t falls between 1.761, with $p = .05$, and 2.145, with $p = .025$. Therefore p is between .05 and .025.

5. Decide whether to reject H_0.

Since $p > .01$, H_0 is not rejected. The sample mean is not significantly greater than 25 at the .01 level. The personality theorist cannot conclude that the population of college students has a mean greater than 25 on the neuroticism test.

EXERCISE 8–4

1. In each of the following, decide whether the hypothesis is directional or nondirectional. State the null hypothesis H_0 and alternative hypothesis H_A, and tell in which tail or tails the rejection region lies.
   a. In population X the mean on a certain task is 50. I have a sample of people from a population that I believe has a mean different from that of population X.
   b. The mean score of fourth graders nationwide on a math test is 54. I have chosen a random sample of fourth graders from the school district of Patootieburg and given them the test. I hypothesize that performance in this school district is above the national level.
   c. With traditional therapy, a certain phobic disorder clears up in 25 sessions, on the average. A behavior modification therapist decides to use a new type of therapy with a sample of phobic patients, claiming that the phobia will disappear faster (in fewer sessions) with the new therapy.
2. Suppose that the study described in problem 1(c) has been done. Twelve patients were successfully treated, with a mean of 15.3 sessions per patient and a sample standard deviation of 6.2. Test the appropriate hypothesis. Use $\alpha = .05$.
3. You have selected a random sample from a normal population. The sample scores are 7, 9, 6, 11, 8. Conduct a test to find out whether the sample mean is significantly different from 10 at the .01 level.

Estimating the Population Mean: Confidence Intervals

Thus far we have used hypothesis-testing procedures for making inferences about a population mean, μ. Sometimes, however, we want to infer the probable value of μ without testing a particular hypothesis. Suppose, for example, that you are interested in the mean of some population. You select a random sample and compute its mean, 18.4. You have no particular hypothesis to test, so a t test is not appropriate, but you want to make a reasonable estimate of μ. What should that be? Is "$\mu = 18.4$" a reasonable estimate? After all, a random sample usually has a mean (M) somewhere in the neighborhood of μ. But sample means are rarely exactly equal to the population mean. Therefore stating that μ is exactly equal to 18.4 is not reasonable.

Let us try another approach. I have said that M is usually "somewhere in the neighborhood" of μ. Is there a way to make the notion of "somewhere in the neighborhood" more precise? Can we establish a range or interval of values around M, and state with some specified degree of confidence that μ is probably in that interval? Yes we can. The procedure is called ***interval estimation***, and the interval is called a ***confidence interval for the population mean***. The limits of the interval are called ***confidence limits***.

Confidence intervals can be calculated at various ***levels of confidence***. The level of confidence is expressed as a percentage, such as 95% or 99%. The higher the percentage, the greater the level of confidence. We will develop the rationale for the 95% confidence interval first, then generalize to other levels of confidence.

Rationale and Computation of Confidence Intervals

The 95% Confidence Interval

Suppose that we are sampling from a population with mean μ. If we select all possible samples of a certain size N and convert each sample mean to a value of t using the formula $\frac{M-\mu}{s_M}$, only 5% of the sample means will have t values beyond the two critical values of t for $\alpha = .05$ in a two-tailed test. Therefore, 95% of the sample means will have t values between the critical values. Let us call these two critical values $-t_{.05/2}$ and $t_{.05/2}$ to indicate that they are the critical values of t for $\alpha = .05$, two-tailed. This information is depicted graphically in Figure 8–3.

We have established that in 95% of samples

$$-t_{.05/2} < t < t_{.05/2}$$

Since $t = \frac{M-\mu}{s_M}$, we can write:

$$\text{In 95\% of samples, } -t_{.05/2} < \frac{M-\mu}{s_M} < t_{.05/2}$$

Multiplying each term by s_M:

$$\text{In 95\% of samples, } -t_{.05/2}(s_M) < M-\mu < t_{.05/2}(s_M)$$

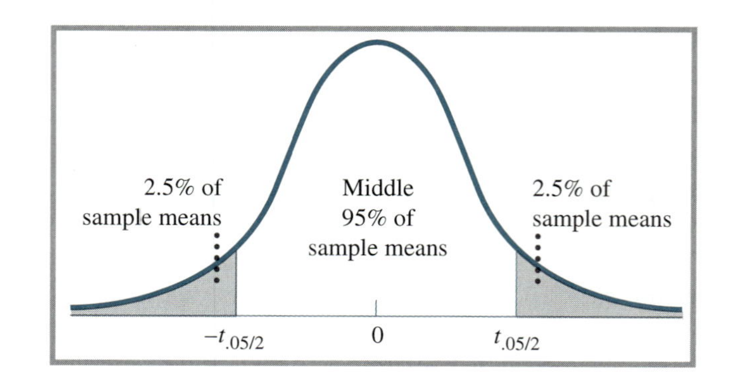

Figure 8–3

A t distribution. The middle 95% of sample means have values of t between $-t_{.05/2}$ and $t_{.05/2}$.

Now we subtract M from all terms:

$$\text{In 95\% of samples,}\quad -M - t_{.05/2}(s_M) < -\mu < -M + t_{.05/2}(s_M)$$

We want to change the center term, $-\mu$, to $+\mu$. To do so, we multiply through by -1. However, this changes the direction of the inequality signs[†] and we obtain the following statement:

$$\underbrace{M + t_{.05/2}(s_M)}_{\substack{\text{Upper 95\% confidence}\\ \text{limit for } \mu}} \quad > \quad \mu \quad > \quad \underbrace{M - t_{.05/2}(s_M)}_{\substack{\text{Lower 95\% confidence}\\ \text{limit for } \mu}}$$

The value $M - t_{.05/2}$ (s_M) is called the *lower 95% confidence limit* for μ, and the value $M + t_{.05/2}$ (s_M) is called the *upper 95% confidence limit.* The interval between these two values is the *95% confidence interval for the population mean.*

Thus, to calculate 95% confidence limits for the population mean from a sample of size N, do the following:

1. Calculate the sample mean M and the estimated standard error of the mean s_M.
2. In the t table, find $t_{.05/2}$ (the critical value of t for $\alpha = .05$, two-tailed test) for $N - 1$ *df.*
3. Calculate the limits: Lower confidence limit = $M - t_{.05/2}$ (s_M)
   Upper confidence limit = $M + t_{.05/2}$ (s_M)

You can now state that you are 95% confident that the population mean μ lies between the lower and upper limits.

EXAMPLE 1: A random sample of size 25 is selected from a normal population. The sample mean is 18.4, and $SD = 7.0$. Calculate the 95% confidence interval for the population mean.

$M = 18.4$, $s = 7.0$, $N = 25$

$$s_M = \frac{s}{\sqrt{N}} = \frac{7.0}{\sqrt{25}} = \frac{7.0}{5} = 1.4$$

$df = N - 1 = 24$. According to the t table, $t_{.05/2}$ for 24 $df = 2.064$. (Confirm this yourself.)

Lower confidence limit = 18.4 − (2.064)(1.4) = 18.4 − 2.89 = 15.51

Upper confidence limit = 18.4 + (2.064)(1.4) = 18.4 + 2.89 = 21.29

We are 95% confident that the population mean μ lies between 15.51 and 21.29.

Generalization to Other Levels of Confidence

To calculate the 95% level of confidence, we used the critical value of t for $\alpha = .05$, two-tailed test. You will probably not be surprised to learn that, for the 99% level of confidence, we use $\alpha = .01$ (two-tailed); for the 90% level of confidence, $\alpha = .10$ (2-tailed);

[†]To understand why multiplying all terms by -1 changes the direction of the inequality signs, consider the following true statement: $-5 < -3 < -1$. If you multiply all terms by -1 but leave the inequality signs alone, you get the false statement: $5 < 3 < 1$. But if you change the direction of the inequality signs, the statement becomes a true one: $5 > 3 > 1$.

and so on. If we let LOC stand for level of confidence, the value of α associated with any LOC can be found by the formula

$$\alpha = 1 - \frac{\text{LOC}}{100}$$

For example, for the 92% level of confidence

$$\alpha = 1 - \frac{92}{100} = 1 - .92 = .08$$

In general, to obtain confidence limits for any level of confidence, first determine the value of α associated with that level of confidence. Then use the t table to find $t_{\alpha/2}$, the critical value of t for that level of α in a two-tailed test. The confidence limits are then calculated by the formula

Confidence limits for a given $\alpha = M \pm t_{\alpha/2}\,(s_m)$

Example 2: Calculate the 99% confidence interval for the data in Example 1.

$M = 18.4$, $s_M = 1.4$, and $N = 25$, as before

For the 99% level of confidence, $\alpha = .01$.

According to the t table, $t_{.01/2}$ for 24 $df = 2.797$. (Confirm this yourself.)

Lower confidence limit = $18.4 - (2.797)(1.4) = 18.4 - 3.92 = 14.48$

Upper confidence limit = $18.4 + (2.797)(1.4) = 18.4 + 3.92 = 22.32$

We are 99% confident that μ lies between 14.48 and 22.32.

Notice that the 99% confidence interval is wider than the 95% confidence interval, which was 15.51 to 21.29.

Interpretation of Confidence Intervals

What does it mean to say that we are "95% confident" that μ is between two numbers? It means that if we use the procedures just described to compute the 95% confidence interval in all possible samples from the population, the interval will include μ in 95% of the samples and will exclude μ in only 5% of the samples. We are betting that the interval we have calculated in our sample is one of the 95% that include μ and not one of the 5% that do not, and the odds in our favor are 95 to 5.

Suppose that you draw many, many random samples from a population with $\mu = 20$. In each sample, you calculate the 95% confidence interval. Since M and s_M vary from one sample to another, the confidence limits also vary over samples. Nevertheless, in 95% of the samples, the value 20 will be included within the confidence interval; only 5% of the confidence intervals you construct will not include the value 20. If you calculate the 99% confidence limits in each sample rather than the 95% confidence limits, the value 20 will be found within the interval in 99% of the samples and outside the interval in only 1%.

There is one thing you must be very clear about: The population mean, μ, is a constant. Although M and s_M vary from one sample to another, μ is fixed and unchanging. Avoid making statements that imply that the interval you have computed is fixed and μ flits about from one sample to another, such as the following statement: "95% of the

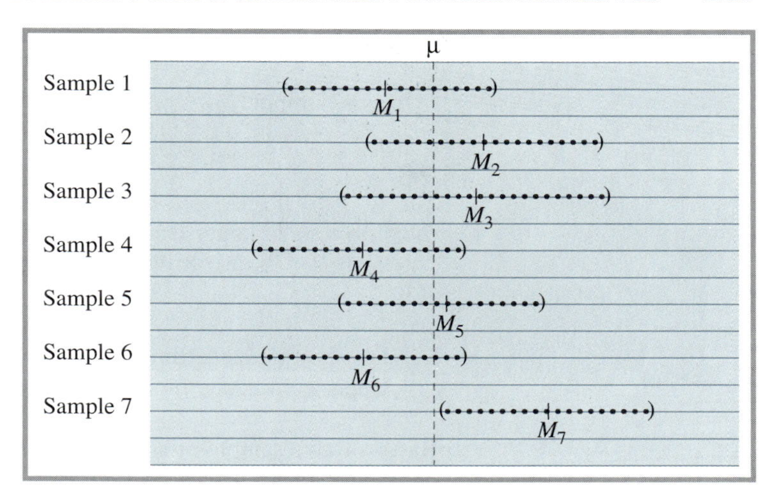

Figure 8–4

Diagram illustrating the correct interpretation of a confidence interval. For each of seven samples, the line between the parentheses is the confidence interval. In Samples 1 to 6, μ is within the confidence interval; in Sample 7, it is not. At the 95% level of confidence, only 5% of samples are like Sample 7. At the 99% level of confidence, only 1% of samples are like Sample 7.

time μ lies between 15.51 and 21.29, and 5% of the time it doesn't." This makes it sound as though μ is sometimes at one location and sometimes another, which is impossible. Either μ is in the interval or it is not, but it is always at one unvarying point. *It is the confidence interval that varies from one sample to another, not* μ. The diagram in Figure 8–4 illustrates the correct interpretation; the parentheses around each sample mean, *M*, indicate the lower and upper confidence limits.

In the figure, Samples 1 to 6 all have a mean (*M*) sufficiently close to the (fixed) value of μ so that μ is "captured" within the confidence interval around *M*. In Sample 7, however, *M* is so far from μ that the confidence interval around *M* does not include the value of μ. When the 95% interval is computed, outcomes like that for Sample 7 occur in only 5% of all samples; when the 99% interval is computed, such outcomes occur in only 1% of all possible samples.

EXERCISE 8–5

1. A sample from a normal population has scores of 13, 12, 15, 9, 15. Calculate the 90% confidence interval for μ.
2. A student has calculated a 95% confidence interval for μ of 17.3 to 21.5. He interprets the results as follows: "In 95% of samples, μ falls between 17.3 and 21.5." Explain what is wrong with this interpretation, and give a correct one.

FACTORS AFFECTING THE WIDTH OF CONFIDENCE INTERVALS

Two factors affect the width of confidence intervals for the mean of a population: the level of confidence and the sample size. First, for any given sample, the 99% confidence interval is always wider than the 95% interval, as we observed in the two examples. This makes sense: In order to be more confident that you have "captured" μ in the interval, you must include a wider range of possible values of μ. In fact, to be 100% confident the interval would have to extend from $-\infty$ to $+\infty$!

Now let us examine the second factor, sample size. In Examples 1 and 2, we computed the 95% and 99% confidence intervals for a sample of size 25 with $M = 18.4$ and $s = 7.0$. Let us retain the same values of *M* and *s*, but change the sample size. In Examples 3 and 4, we calculate 95% and 99% confidence intervals for a smaller sample ($N = 9$) and a larger sample ($N = 100$), respectively.

Example 3: $N = 9$, $M = 18.4$, $s = 7.0$.

First, we compute s_M:

$$s_M = \frac{s}{\sqrt{N}} = \frac{7.0}{\sqrt{9}} = \frac{7.0}{3} = 2.33$$

(Note that s_M is larger than it was for $N = 25$.)

With $N = 9$, $df = 8$. For the 95% level of confidence, $t_{.05/2}$ for 8 df = 2.306.

Lower confidence limit = 18.4 − (2.306)(2.33) = 13.03
Upper confidence limit = 18.4 + (2.306)(2.33) = 23.77

Based on this sample of size 9, we are 95% confident that μ lies between 13.03 and 23.77.

For the 99% level of confidence, $t_{.01/2}$ for 8 df = 3.355.

Lower confidence limit = 18.4 − (3.355)(2.33) = 10.58
Upper confidence limit = 18.4 + (3.355)(2.33) = 26.22

We are 99% confident that μ lies between 10.58 and 26.22.

Example 4: $N = 100$, $M = 18.4$, $s = 7.0$.

First, we compute s_M:

$$s_M = \frac{s}{\sqrt{N}} = \frac{7.0}{\sqrt{100}} = 0.7$$

(Note that with larger N, s_M is smaller.)

With $N = 100$, $df = 99$. For the 95% level of confidence, $t_{.05/2}$ for 99 df is approximately 1.99.†

Lower confidence limit = 18.4 − (1.99)(0.7) = 17.01
Upper confidence limit = 18.4 + (1.99)(0.7) = 19.79

We are 95% confident that μ lies between 17.01 and 19.79.

For the 99% level of confidence, $t_{.01/2}$ for 99 df is approximately 2.64.†

Lower confidence limit = 18.4 − (2.64)(0.7) = 16.55
Upper confidence limit = 18.4 + (2.64)(0.7) = 20.25

We are 99% confident that μ lies between 16.55 and 20.25.

The effect of sample size (N) on the width of the confidence interval can be seen in the following table.

†Though 99 df is not listed in the table, $t_{.05/2}$ is about halfway between that for 60 df (2.000) and that for 120 df (1.980). Similarly, $t_{.01/2}$ is about halfway between 2.660 and 2.617.

	N = 9 (Example 3)	N = 25 (Examples 1 and 2)	N = 100 (Example 4)
95% interval	13.03–23.77	15.51–21.29	17.01–19.79
99% interval	10.58–26.22	14.48–22.32	16.55–20.25

Note that for a given level of confidence, as N increases, the width of the confidence interval decreases. This makes sense: The larger the sample size, the less the variation of sample means from one sample to another. In other words, with large samples, M is more often very close to μ. Therefore, even though we can never specify μ exactly from a single sample, we can pinpoint it more and more precisely as N increases.

Relation Between Interval Estimation and Hypothesis Testing

Although interval estimation and hypothesis testing are two different approaches to inference, there is a very close relation between them. Specifically, a confidence interval may be viewed as a simultaneous test of all possible hypotheses about a population mean. When we state that μ probably lies between two limits with a certain level of confidence, we are, in a sense, not rejecting any hypothesis for a value of μ between the two limits, and rejecting all hypotheses for values of μ outside the limits. For example, with $M = 18.4$, $s = 7.0$, and $N = 25$, we calculated 95% confidence limits 15.51 and 21.29. We are thus saying, with 95% confidence, that μ *could* be any number between 15.51 and 21.29; at the .05 level, we do not reject any hypothesis that μ_H is a number between 15.51 and 21.29 (such as H_0: $\mu = 16$ or H_0: $\mu = 20$). At the same time, we are saying, with 95% confidence, that μ is *not likely* to be any number below 15.51 or above 21.29; at the .05 level, we reject all hypotheses that $\mu_H < 15.51$ or $\mu_H > 21.29$ (such as H_0: $\mu = 12$ or H_0: $\mu = 23$).

To illustrate, let us test H_0: $\mu = 16$ (a value within the 95% confidence interval) and H_0: $\mu = 15$ (a value outside the interval), using $\alpha = .05$ and a two-tailed test. According to our discussion, the first test should result in nonrejection of H_0 and the second in rejection of H_0.

Example 1: $H_o: \mu = 16$

$H_A: \mu \neq 16$

$\alpha = .05$

$M = 18.4,\ s = 7.0,\ N = 25$

$$s_M = \frac{s}{\sqrt{N}} = \frac{7.0}{\sqrt{25}} = 1.4$$

$$t = \frac{M - \mu_H}{s_M} = \frac{18.4 - 16}{1.4} = \frac{2.4}{1.4} = 1.714$$

In the row of the t table with 24 degrees of freedom, the two critical values that bracket $t = 1.714$ are 1.711, with $p = .10$, and 2.064, with $p = .05$. Since $p > .05$, H_0 is not rejected. The sample mean does not differ significantly from the hypothesized population mean at the .05 level.

Example 2: $H_o: \mu = 15$

$H_A: \mu \neq 15$

$\alpha = .05$

$M = 18.4$ and $s_M = 1.4$, as before

$$t = \frac{M - \mu_H}{s_M} = \frac{18.4 - 15}{1.4} = \frac{3.4}{1.4} = 2.43$$

In the row of the t table with 24 degrees of freedom, the two critical values on either side of 2.43 are 2.064, with $p = .05$, and 2.492, with $p = .02$. Since $p < .05$, H_0 is rejected. The sample mean differs significantly from the hypothesized population mean at the .05 level.

In summary, if you have calculated a confidence interval with a certain α and then test a hypothesis about μ, using a two-tailed test with the same α, you will reject H_0 if the hypothesized population mean, μ_H, lies outside the confidence interval, but not reject H_0 if μ_H lies within the interval. By the same token, if a two-tailed test with a given α results in nonrejection of H_0, μ_H will lie within the confidence interval for the same α; but if the test results in rejection of H_0, μ_H will fall outside the interval. For example, if you have rejected H_0: $\mu = 20$ at the .05 level (two-tailed test), the value 20 will not be in the 95% confidence interval around the sample mean.

EXERCISE 8–6

1. a. A sample of size 25 from a normal population has a mean of 78.4 and $SD = 9.83$. Compute the 99% confidence interval for the mean.
   b. Answer before calculating: Will the 95% confidence interval be narrower or wider? Then calculate the 95% confidence interval.
   c. Answer before calculating: If the sample size were 16 instead of 25, would the 99% confidence interval be narrower or wider than for $N = 25$? Then calculate the 99% confidence interval for $N = 16$.
   d. If you test H_0: $\mu = \mu_H$ with $\alpha = .01$, two-tailed test, using the data in part (a) which of the following hypotheses will you reject?
      (1) H_0: $\mu = 70$
      (2) H_0: $\mu = 75$
      (3) H_0: $\mu = 80$
      (4) H_0: $\mu = 85$
2. The 95% confidence interval for a mean is −13.5 to −9.3. Is the sample mean significantly different from −10 at the .05 level? Is it significantly different from −15 at the .05 level?
3. A two-tailed test of H_0: $\mu = 25$ has resulted in nonrejection of H_0 at the .01 level. Is the value 25 within the 99% confidence interval for μ?
4. You have conducted a two-tailed test of H_0: $\mu = 60$ with $\alpha = .05$ and concluded that the sample mean is significantly different from 60. Is the value 60 within the 95% confidence interval for μ?

Confidence Intervals for Population Percentages

We have seen that confidence intervals for the mean are calculated by the formula $M \pm t_{\alpha/2}\, s_M$. Confidence intervals can be calculated for other parameters besides the population mean. Whatever the parameter, confidence intervals always take the form

$$\begin{bmatrix}\text{Sample statistic (the} \\ \text{estimator of the} \\ \text{parameter)}\end{bmatrix} \pm \begin{bmatrix}\text{Critical value from the} \\ \text{relevant probability} \\ \text{distribution}\end{bmatrix}\begin{bmatrix}\text{Standard} \\ \text{error of} \\ \text{the statistic}\end{bmatrix}$$

The probability distribution varies depending on the statistic, as does the formula for calculating the standard error of the statistic. Nevertheless, the purpose, rationale, and interpretation are the same as for the mean.

Surveys and opinion polls often report confidence intervals for population percentages, such as the percent of voters favoring a certain candidate or the percent of consumers preferring a particular product. After calculating the percent in a sample from the population, the polling organization typically presents the results in a statement like, "The percentage of consumers favoring Brand X is 61 ± 4%." The "61%" is the sample percentage, and the "4%" is a margin of error obtained by multiplying a critical value (at some level of confidence) by the standard error of the percentage. Since confidence intervals for percentages are so widely used, we discuss their calculation here.

We will use P to stand for a sample percentage and s_p to stand for the standard error of the percentage. The standard error of the percentage is calculated by the following formula:

$$s_p = \sqrt{\frac{P(100-P)}{N}} \qquad \textit{Standard error of the percentage}$$

The probability distribution that describes the sampling variation of P is the standard normal curve (the z distribution), provided that the following condition is met: Both NP and $N(100 - P)$ must be at least equal to 1000. If either NP or $N(100 - P)$ is less than 1000, the procedure described in this section should not be used.

Provided that the condition is met, a confidence interval for a population percentage is calculated as follows:

$$\textit{Confidence interval} = P \pm z_{\alpha/2} s_p$$

The critical value of z depends only on the level of confidence (LOC). If, for example, LOC = 95, $\alpha = .05$, and $z_{.05/2} = 1.96$, because z scores of ± 1.96 cut off a total area of .05 in the two tails of the standard normal curve. Critical values of z for other levels of confidence can be found either in the normal curve table or in the row of the t table with $df = \infty$.

As an illustration, we will use data on the results of a mayoral election poll in the town of Tooterville. In the poll, 113 of 200 voters (56.5%) stated their intention to vote for Candidate Smith. In Chapter 5 we used the chi square test to test the null hypothesis that the population percentage is 50%. Our test resulted in nonrejection of the null hypothesis. Now we will calculate the 95% confidence interval for the population percentage.

First, we check to make sure that NP and $N(100 - P)$ are both at least 1000:

$$NP = 200(56.5) = 11{,}300$$

$$N(100 - P) = 200(100 - 56.5) = 200(43.5) = 8700$$

Both products are well over 1000. We can proceed.

Because we are using LOC = 95%, the critical value of z, $z_{.05/2}$, is 1.96. We need to calculate s_p:

$$s_p = \sqrt{\frac{P(100-P)}{N}} = \sqrt{\frac{(56.5)(43.5)}{200}} = 3.5055$$

The confidence interval is 56.5 ± (1.96)(3.5055)

Lower confidence limit = 56.5 − (1.96)(3.5055) = 56.5 − 6.87 = 49.63
Upper confidence limit = 56.5 + (1.96)(3.5055) = 56.5 + 6.87 = 63.37

We are 95% confident that the percentage of all Tooterville voters who favor Candidate Smith is between 49.63% and 63.37%. Notice that the interval includes the value 50%. Based on the confidence interval, we cannot reject the hypothesis that the population percentage is 50%. This agrees with our conclusion in the chi square test.

The interpretation of confidence intervals for percentages is the same as for all confidence intervals. Because sample percentages vary over samples, we cannot be certain that our confidence interval includes the true population percentage. We say that we are "95% confident" because intervals constructed by this method include the population percentage in 95% of samples.

EXERCISE 8–7

1. In a random sample of 150 business executives from large companies, 62 agree with the statement "Downsizing is essential for companies that want to remain competitive in the global market." Calculate the 99% confidence interval for the population percentage.
2. The margin of error in the 95% confidence interval for the Tooterville poll was 6.87. Since the margin of error = ($z_{\alpha/2}\ s_p$), and since s_p decreases as sample size increases, large samples result in smaller margins of error. Suppose that the Tooterville pollsters wish to reduce the margin of error to 5% at the 95% level of confidence. Assuming that the sample percentage remains 56.5%, how large a sample is needed?

HINT: Margin of error $= z_{\alpha/2} s_p = z_{\alpha/2}\sqrt{\dfrac{P(100-P)}{N}}$. If we solve for N, we get

$$N = \left(\frac{z_{\alpha/2}}{\text{Margin of error}}\right)^2 P(100-P).$$

Errors of Inference

Type I and Type II Errors

Inferences from samples to populations are always probabalistic. We can never be certain that our inferences are correct. We may be 95% confident in an interval estimate; we may choose $\alpha = .01$ in testing a hypothesis; but we can never be 100% confident or choose $\alpha = 0$. Whatever decision we make or conclusion we reach, there is always a possibility that we are wrong. Reaching a wrong conclusion is called an *error of inference.*

We introduced errors of inference in Chapter 5, where we noted two types of errors of inference: *Type I errors* and *Type II errors.* A Type I error occurs whenever we reject a null hypothesis that is actually true.

Type I error = rejection of a true null hypothesis

For example, suppose that the mean of the population from which you are sampling is 50. However, your test of H_0: $\mu = 50$ results in rejection of H_0, and you conclude that μ is not equal to 50. You have made a Type I error.

A Type II error occurs if we fail to reject a null hypothesis that is actually false.

Type II error = nonrejection of a false null hypothesis

For example, suppose that the true mean (true μ) of a population is 50. You test H_0: $\mu = 55$ and do not reject H_0. You have made a Type II error.

To summarize: Some state of affairs exists in the population: Either H_0 is true or H_0 is false. We conduct a test and, based on the result, either reject or do not reject H_0. If H_0 is true and we do not reject it, we have made a correct inference. If, however, H_0 is true but we reject it, we have made a Type I error. If H_0 is false and we reject H_0, we have made a correct inference. But if H_0 is false and we do not reject H_0, we have made a Type II error. All the possibilities are shown in the following table.

		Decision Based on Test	
		Reject H_0	Do Not Reject H_0
Actual State of Affairs in the Population	H_0 is true	Type I error	Correct inference
	H_0 is false	Correct inference	Type II error

In a given situation we can never know whether or not an error has occurred because we don't know the state of affairs in the population. (If we did, we wouldn't need a statistical test.) If our test has resulted in rejection of H_0, we may have done so correctly, or we may have made a Type I error. Which of these outcomes has actually occurred we do not know. If our test has resulted in nonrejection of H_0, we may have done so correctly or we may have made a Type II error. Again, we do not know which of these outcomes has occurred. Fortunately, however, we can, to some degree, assess and control the *probability* of each type of error.

The probability of a Type I error is *alpha* (α), the level of significance selected for the test.

$$\alpha = p\text{ (Type I error)} = p\text{ (rejecting a true } H_o\text{)}$$

Suppose, for example, that you choose $\alpha = .01$. You are saying, "I will reject H_0 if my sample outcome has a probability of .01 or less *when* H_0 *is true*." Thus, by choosing a level of α, you are predetermining the probability of a Type I error.

The probability of a Type II error is called *beta* (β).

$$\beta = p\text{ (Type II error)} = p\text{ (not rejecting a false } H_o\text{)}$$

Though we do not ordinarily preselect β when conducting a statistical test, there are a number of factors that affect β, as we will discuss.[†]

A graphic representation of Type I and Type II errors and their probabilities is shown in Figure 8–5. The figure illustrates an upper one-tailed test; however, the representation would be similar for lower one-tailed and two-tailed tests.

Figure 8–5 shows two sampling distributions of the mean. Distribution (a), centered around μ_H, is the distribution of sample means that is expected if H_0 is true (i.e., if μ really is equal to μ_H). To test the null hypothesis, we determine where our obtained sample mean (M) falls in Distribution (a) by calculating $z = \frac{M - \mu_H}{\sigma_M}$ or $t = \frac{M - \mu_H}{s_M}$. Since

[†]It is possible to preselect β under certain conditions. How this is done is beyond the scope of this book.

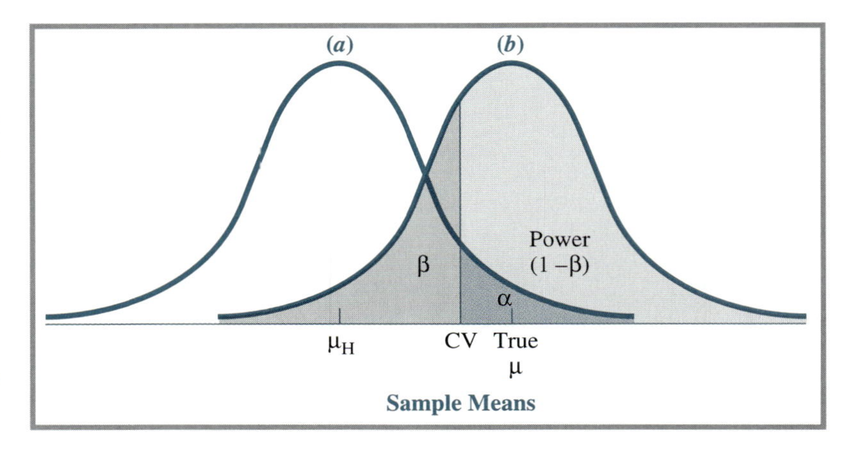

Figure 8–5

The probability of a Type I error (α), the probability of a Type II error (β), and the power of a test (1 − β) in an upper one-tailed test. Distribution (a) is the distribution of sample means around the hypothesized population mean, μ_H. Distribution (b) is the distribution of sample means around a different value, labeled "true μ," a possible value of μ if the null hypothesis is false.

the figure portrays an upper one-tailed test, the rejection region, with area α, is in the upper tail of Distribution (a). The cut-off point between sample means falling in the nonrejection and rejection regions is labeled "CV" in the figure. We reject H_0 if our sample mean is greater than CV.

Suppose that H_0 is true. If so, sample means are actually distributed as in Distribution (a). If we select a sample with a mean less than CV, we correctly do not reject H_0. If, however, we happen to select one of those relatively few samples with a mean greater than CV, we reject H_0, even though H_0 is true. We have made a Type I error. The probability of a Type I error is area α in Distribution (a).

Now suppose that H_0 is false and the population mean is not μ_H but another value, which we call *true* μ. If so, sample means are distributed not around μ_H but around true μ, as in Distribution (b) of Figure 8–5. If, in our test of H_0, we select a sample with a mean greater than CV, we reject H_0—a correct inference, since H_0 is false. However, suppose that we select a sample with a mean less than CV. In this case, we do not reject H_0, even though H_0 is false, and we make a Type II error. The probability of obtaining a sample mean less than CV, if the population mean is equal to true μ, is area β in distribution (b). Area β, then, represents the probability of a Type II error.

Factors Affecting the Probability of Type I and Type II Errors

Although Type I and Type II errors cannot be eliminated, they can, to a considerable degree, be controlled or, at least, assessed. Let us examine the factors that affect the probability of Type I and Type II errors.

Factors Affecting the Probability of a Type I Error

As we have already noted, the probability of a Type I error is equal to alpha, the level of significance chosen for the test. If you choose $\alpha = .05$ for a test, the probability of a Type I error is .05; if you choose $\alpha = .001$, the probability of a Type I error is .001, and so on. *No other factors influence the probability of a Type I error.*

Since the only factor affecting the probability of a Type I error is α, you may wonder why we don't make α very, very small. Why not, for example, select $\alpha = .0001$? Then there is only one chance in 10,000 of making a Type I error. The problem is that, as the probability of a Type I error decreases, the probability of a Type II error increases. By choosing a very small α, you reduce the probability of rejecting the null hypothesis if it is true. But, at the same time, you increase the probability of not rejecting the null hypothesis if it is false. Thus, there is a trade-off between Type I and Type II errors, if other factors are equal.

However, other factors are not always equal. Although the probability of a Type I error is completely determined by the level of α used in the test, several factors affect the probability of a Type II error.

Factors Affecting the Probability of a Type II Error

Five factors affect the probability of a Type II error: (1) the difference between the true value of μ and the hypothesized value (μ_H); (2) the variability of scores in the population; (3) the level of α used in the test; (4) sample size (N); and (5) the selection of a one-tailed versus a two-tailed test.

1. *As the difference between true μ and μ_H increases, the probability of a Type II error decreases.* For example, if you are testing H_0: $\mu = 50$, but H_0 is false, you are less likely to make a Type II error if true $\mu = 60$ than if true $\mu = 55$.
2. *As the variability of scores in the population decreases, the probability of a Type II error decreases.* For example, a Type II error is less likely if the population from which you are sampling has variance (σ^2) = 100 than if $\sigma^2 = 200$.
3. *As α increases, the probability of a Type II error decreases*, as we have already seen. For example, if other factors are equal, a test with $\alpha = .05$ has a smaller probability of a Type II error than a test with $\alpha = .01$.
4. *As sample size (N) increases, the probability of a Type II error decreases.* If, in fact, H_0 is false, you can reduce the probability of failing to reject H_0 (a Type II error) by increasing the sample size.
5. *The probability of a Type II error is smaller with a one-tailed test than a two-tailed test using the same α (assuming that a one-tailed test is appropriate for the given research question).*

We will examine reasons why these factors affect the probability of a Type II error later in the chapter, in connection with our discussion of the power of a test.

Choosing a Level of Alpha

In all the tests we have conducted so far, I have set a level of α, usually .05 or .01, without justifying it. Now we are ready to consider how one chooses a level of α. Basically, one decides what kind of error, Type I or Type II, would be more serious in the research to be conducted. If a Type I error would be more serious, it is better to make α quite small, such as .01 rather than .05, in order to reduce the probability of a Type I error. If, on the other hand, a Type II error would be more serious, a larger level of α, .05 or even .10, can be used; in addition, it may be wise to increase N or to take still other steps to reduce the probability of a Type II error.

Deciding whether a Type I or a Type II error would be more serious in a given case is not based on statistical manipulations or mathematical reasoning. It is based on knowledge of the field of research and consideration of the theoretical and/or practical consequences of drawing incorrect conclusions from the test.

For example, suppose that you are conducting an experiment on the effectiveness of a new method of training auto mechanics. The null hypothesis states that the new method is not more effective than the old. If you reject H_0 and conclude that the new method is better, the company for which you work will switch over to the new method. If you do not reject H_0 and conclude that the new method is not better than the old, the old method will be retained.

Two kinds of errors are possible. You may mistakenly reject H_0 and conclude that the new method is more effective when it actually is not (Type I error), or you may mistakenly conclude that the new method is not more effective, though it actually is (Type II error). Which is more serious? Let us examine the consequences of both errors.

If you mistakenly conclude that the new method is better, though it really isn't, the company will go to a great deal of unnecessary expense—retraining instructors, buying new equipment and instructional materials and supplies, and so on. This is a very serious error from the company's standpoint, an error you want to avoid. On the other hand, if you mistakenly conclude that the new method is not better, but it really is, the company will continue to use the old method. If the old method is reasonably effective, failing to replace it with the new method may not be a very serious error, especially if the difference between the two methods is not great.

As you can see, in this situation a Type I error is more serious for the company than a Type II error. You would do well to make the probability of a Type I error very small by choosing a small value of α, for example, $\alpha = .01$.

On the other hand, suppose that you are a medical researcher testing a new treatment for cystic fibrosis. The null hypothesis states that the new treatment, drug X, is not more effective than the old, drug Y. If you mistakenly conclude that drug X is more effective though it really is not (Type I error), drug X will be used to treat cystic fibrosis patients in the future. How serious is this error? Assuming that the new drug, X, has no more dangerous side effects than the old drug, Y, and is not less effective, switching to drug X does no harm. On the other hand, suppose that you mistakenly conclude that drug X is not more effective than drug Y, even though it really is (Type II error). As a result, patients will continue to be treated with a less effective drug, with potentially serious effects including, possibly, death. Clearly, in this case, a Type II error is more serious than a Type I error, and the level of α chosen should be relatively large, .05 or even .10. Though this increases the probability of a Type I error, it reduces the probability of a more serious Type II error.

In summary, before testing any statistical hypothesis, always consider the consequences of Type I and Type II errors and decide which type of error would be more serious. If a Type I error is more serious, select a small value of α. If a Type II error is more serious, select a relatively large level of α. Finally, if both types of errors are serious (as they often are), set an appropriately low level of α and take other steps, such as increasing sample size, to decrease the probability of a Type II error as well.

EXERCISE 8–8

1. a. The mean of a population is 25. You test H_0: $\mu = 20$ and do not reject H_0. Have you made an error of inference? If so, what kind?
   b. The mean of a population is 25. In a test of H_0: $\mu = 25$, you select a sample with a mean of 20 and reject H_0. Have you made an error of inference? If so, what kind?
2. The probability of a Type I error is α/β. (Choose one.)
3. a. Increasing N while holding α constant will increase/decrease/not affect the probability of a Type II error.
   b. Increasing N while holding α constant will increase/decrease/not affect the probability of a Type I error.

4. a. Let us say that with $\alpha = .05$, two-tailed test, $\beta = .40$. If I change α to .01, β will be (1) equal to .40, (2) less than .40, (3) greater than .40
   b. If I use a one-tailed test instead of a two-tailed test, with $\alpha = .05$ (assuming that a one-tailed test is appropriate), β will be (1) equal to .40, (2) less than .40, (3) greater than .40
5. Suppose that you are testing H_0: $\mu = 20$. Are you more likely to make a Type II error if true μ is 15 or if true μ is 22?
6. Suppose that you are testing H_0: $\mu = 20$. Are you more likely to make a Type II error if the population standard deviation is 3 or if the population standard deviation is 7?
7. Think of an example of research where a Type I error would be more serious than a Type II error. Think of an example of research where a Type II error would be more serious than a Type I error.

The Power of a Test

Recall from Chapter 4 that the probability that event A will not occur, p (not A), is equal to $1 - p(A)$. The probability of not rejecting a false null hypothesis is β. Therefore, the probability of correctly rejecting a false null hypothesis is $1 - \beta$. $1 - \beta$ is called the ***power*** of a test.

$$\text{Power of a test} = \text{Probability of rejecting a false null hypothesis} = 1 - \beta$$

For example, if $\beta = .20$, power $= 1 - .20 = .80$.

In plain English, the power of a test is the probability of detecting a difference, such as that between true μ and μ_H, if a difference really exists. It is clear that since researchers generally hypothesize that a difference exists, they desire tests with high power.

The power of a test and its relation to α and β are illustrated in Figure 8–5. As we have already observed, if H_0 is false and the population mean is not μ_H but true μ, sample means are distributed as in Distribution (b) of the figure. If an obtained sample has a mean (M) less than CV, H_0 is not rejected (a Type II error); the probability of a Type II error is area β of Distribution (b). However, if the sample mean is greater than CV, H_0 is correctly rejected. The area to the right of CV in Distribution (b), $1 - \beta$, is the power of the test.

Factors Affecting the Power of a Test

Since the power of a test is equal to $1 - \beta$, the same factors that affect β also affect power but in the opposite direction. Anything that decreases the probability of a Type II error, β, increases power, and vice versa.

1. *As the difference between true μ and μ_H increases, power increases.* In other words, a test is more likely to detect a difference between true μ and μ_H if the difference is large than if it is small.
2. *As population variability decreases, power increases.* A test is more likely to detect a difference between true μ and μ_H if the population standard deviation is small than if it is large.

3. *As α increases, power increases.* A test is more likely to detect a difference between true μ and μ_H with $\alpha = .05$ than with $\alpha = .01$.
4. *As N increases, power increases.* The larger the sample size, the more likely that the test will detect a difference between true μ and μ_H.
5. *A one-tailed test (if appropriate) is more powerful than a two-tailed test.* A one-tailed test is more likely to detect a difference between true μ and μ_H (provided that the difference is in the hypothesized direction) than a two-tailed test.

Let us consider why these factors affect power as they do. The explanation is based on the following chain of reasoning:

Step 1: Power refers to the probability of rejecting H_0 when it is false. Therefore, anything that increases the probability of rejecting H_0 increases power.

Step 2: H_0 is rejected if the calculated z or t falls in a tail of the z or t distribution, beyond the critical value of z or t for the chosen alpha. That is, H_0 is rejected if
z test: $|z| > |\text{critical value of } z|$
t test: $|t| > |\text{critical value of } t|$

Step 3: Therefore anything that increases the calculated $|z|$ or $|t|$ or decreases the | critical value | increases the power of the test.

We must ask ourselves, then, what are the factors that increase $|z|$ or $|t|$? And what are the factors that decrease the | critical value | ?

Factors That Increase Power by Increasing |z| or |t|

z and t are both ratios:

$$z = \frac{M - \mu_H}{\sigma_M} \text{ and } t = \frac{M - \mu_H}{s_M}$$

The value of a ratio increases if the numerator increases or if the denominator decreases. Therefore, anything that increases the value of $M - \mu_H$ or decreases the value of σ_M or s_M increases the power of the test.

First, consider the numerator, $M - \mu_H$. If H_0 is false, the average value of M over all possible samples is not μ_H but true μ. Therefore, the average value of $M - \mu_H$ over all samples is equal to (true $\mu - \mu_H$). The larger the value of | true $\mu - \mu_H$ |, the greater is the value of $|M - \mu_H|$ in most samples and, therefore, the larger is the value of $|z|$ or $|t|$. This is why the power of the test increases as | true $\mu - \mu_H$ | increases.

The denominator of z is $\sigma_M = \sigma/\sqrt{N}$, and the denominator of t is $s_M = s/\sqrt{N}$, where σ is the population standard deviation and s is the sample estimator of σ. The value of the denominator decreases and $|z|$ or $|t|$ increases, if either σ decreases or N increases. This is why the power of the test increases as σ decreases or N increases.

Factors That Increase Power by Decreasing the |Critical Value| of z or t

Several factors affect the size of the | critical value | of z or t. One factor is α, the level of significance of the test. As α increases, the | critical value | decreases. For example, in an upper one-tailed z test with $\alpha = .01$, the critical value of z is 2.33; but if $\alpha = .05$, the critical value of z is only 1.645. This is why the power of the test increases as α increases.

Another factor that influences the | critical value | of z or t is the choice of a one-tailed versus a two-tailed test. The critical value is always closer to zero in a one-tailed test than in a two-tailed test with the same level of α. For example, with $\alpha = .05$, the critical values of z in a two-tailed test are $\pm$ 1.96, but the critical value of z in a one-tailed test with $\alpha = .05$ is only -1.645 or $+1.645$. This is why one-tailed tests, if appropriate, are more powerful than two-tailed tests.

In the t test, but not the z test, one additional factor—degrees of freedom—influences the critical value. The | critical value | of t decreases as df increases. Since df increases as sample size (N) increases, the | critical value | of t decreases as sample size increases. Thus, in the t test, increasing sample size increases power not only by increasing the calculated $|t|$ but also by decreasing the | critical value | .

Controlling the Power of a Test

The first two factors affecting the power of a test—the difference between true μ and μ_H and the population variability—are difficult or impossible for researchers to control.† The other factors are more easily controllable. Thus, one way to increase the power of a test is to increase α, another way is to increase sample size, N, and a third way is to be sure to use a one-tailed test rather than a two-tailed test whenever a directional research hypothesis is to be tested.

All these factors must be considered in the design and analysis of research studies. For example, if the true population mean is thought to be very different from the hypothesized population mean, especially if the population variability is small, a statistical test may be able to detect the difference (reject H_0) even with small α and small N. However, if the difference between true μ and μ_H is believed to be small and/or population variability is large, and if α must be kept small in order to guard against a Type I error, a much larger sample size, N, may be needed to detect the difference.

EXERCISE 8–9

1. Explain the concept of *power* and its relationship to the probability of a Type II error in your own words.
2. In each of the following cases, tell which of the two tests, Test A or Test B, is more powerful. Assume that all other factors are equal in each case.

Test A	Test B
a. Population variance = 100	Population variance = 200
b. $\alpha = .05$	$\alpha = .01$
c. $\mu_H = 30$, true $\mu = 28$	$\mu_H = 30$, true $\mu = 35$
d. $N = 50$	$N = 30$
e. Research hypothesis: $\mu < 10$	Research hypothesis: $\mu \neq 10$

3. If α is left unchanged but N changes from 20 to 50, the probability of a Type I error increases/decreases/remains the same. (Be careful, this is a trick question.)
4. You are testing H_0: $\mu = 30$. Are you more likely to make a Type II error if true μ is 25 or if true μ is 40?
5. The difference between true μ and μ_H is 5 points. What can you do to make it more likely that you will detect this difference?

†There are sometimes steps that researchers can take to design their experiments in such a way as to increase $|\text{true } \mu - \mu_H|$ or to decrease population variability. These often involve sophisticated research methods and are beyond the scope of this book.

Postscripts

Postscript 1: Some Common Errors in Interpretation

Sometimes students do the following. Having calculated a 95% confidence interval for μ, based on a small sample, a student says, "I am not as confident that μ is between the limits I have calculated as I would be if N were larger." Or, having tested a hypothesis with a small sample, obtained a value of t significant at the .05 level, the student says, "However, I can't really be confident that the difference is significant; if N had been larger with the same level of significance, I could reject H_0 more confidently." This reasoning is incorrect. The level of confidence or the level of significance of a test result determines the confidence one can have in estimating μ or in rejecting H_0; N has nothing to do with it.

Does that mean that sample size is unimportant? Of course not. In computing confidence intervals, the width of the interval for a given level of confidence decreases as N increases. For example, if, with $N = 10$, the 95% confidence interval around a sample mean is 47.5 to 54.5, then with $N = 50$ and the same sample mean, the interval may be reduced to 50.0 from 52.0. In other words, as N increases we can pinpoint the population mean more and more precisely. However, if we have selected the same level of confidence, for example, 95%, the odds that the confidence interval we have calculated includes the population mean are the same, 95 to 5, whether N is 10 or 50 or any other number.

Similarly, if a test results in rejection of H_0 at the .05 level, the probability of a Type I error is the same, .05, whether N is small or large. We can be equally confident in rejecting H_0 in either case. Sample size *does* make a difference in the sense that, in order to get a value of $|z|$ or $|t|$ sufficiently large to reject H_0, the difference between the sample mean and the hypothesized population mean must be larger when N is small than when N is large.

If, on the other hand, a test results in nonrejection rather than rejection of H_0 at a given level of significance, then we can be more confident that our decision was correct if N is large than if N is small. This is because the probability of incorrectly failing to reject a false hypothesis (a Type II error) does indeed decrease as N increases. The statement "The difference was not significant, but might be significant if the sample size were increased" is a valid statement.

Postscript 2: Parametric and Nonparametric Statistical Tests

I have emphasized the importance of certain basic assumptions underlying the use of the z or t distribution to compute confidence limits or test hypotheses about population means. The assumptions are that the population is normal or, if it is not, the sample size is reasonably large. Let us review the reasons for these assumptions:

1. The sampling distribution of the mean is normal only if the population is normal or if N is reasonably large.
2. The distribution of $\frac{M-\mu}{\sigma_M}$ is normal only if the sampling distribution of the mean is normal. Similarly, the distribution of $\frac{M-\mu}{s_M}$ is a t distribution only if the sampling distribution of the mean is normal.
3. The normal curve table accurately describes probabilities of various sample outcomes only when the distribution of $\frac{M-\mu}{\sigma_M}$ is normal. Similarly, the t table accurately describes probabilities of various sample outcomes only when $\frac{M-\mu}{s_M}$ is distributed as t.

Statistical tests of hypotheses about population parameters that rest on normal curve assumptions are called ***parametric tests***. Thus, the z test and t test are parametric tests. Tests that do not require normal curve assumptions are called ***nonparametric tests***.† You have already learned to use one nonparametric statistical test, the chi square test. Recall that the chi square test is used with categorical data to compare distributions of obtained frequencies with distributions of expected (hypothesized) frequencies. The chi square test makes no assumptions regarding the shape of distributions of expected frequencies.

Sometimes two or more different tests may be used to answer the same question or solve the same problem. For example, certain hypotheses dealing with proportions may be tested with either the chi square test (nonparametric) or the normal approximation to the binomial (parametric). Under what circumstances should we choose one rather than another?

It might be argued that, given a choice, nonparametric tests should be applied since they require no assumptions concerning population shape. In real life, we rarely know the shape of a population. It is possible that many important population distributions are not normal. We can minimize errors of inference due to the use of inappropriate probability distributions by using nonparametric tests.

However, parametric tests have one major advantage over nonparametric tests: They are usually (not always) more powerful. That is, if the null hypothesis being tested is false, the probability that it will be correctly rejected is greater if a parametric test rather than a nonparametric test is applied. Because, as already mentioned, rejection of a false null hypothesis is usually considered very important, researchers usually prefer to use parametric tests.

Fortunately, studies conducted by statisticians have shown that, even if the assumption of population normality is not met, decisions made on the basis of the normal curve test or t test are likely to be reasonably accurate, unless the departure from normality is very marked and N is small. Parametric tests are said to be ***robust***. That means that they usually work pretty well, even when their underlying assumptions are violated to some degree.

What are the implications for your choice of a parametric versus a non-parametric test? If you are sampling from a clearly non-normal population, especially if your sample size is small, it may be better to avoid using a parametric test such as a t test. However, if N is reasonably large and the population is not markedly non-normal, a parametric test (z or t) can be applied with little risk of error and is likely to be more powerful.

Needless to say (I hope), the assumption of random sampling (or, at least, some method of sampling that is likely to yield representative samples) underlies all tests, parametric and nonparametric alike. Biased sampling invalidates the conclusions of any statistical test.

Using a Computer to Conduct *t* Tests and Calculate Confidence Intervals

Most statistical packages include programs for conducting t tests for population means. If a data set is large, using a computer considerably reduces computational labor. In addition, the computer output includes not only the value of t but also its probability, thus saving you the trouble of referring to a t table.

One word of caution: Though some statistical packages conduct upper or lower one-tailed tests, as requested, others only do two-tailed tests. However, this is not an insur-

†The term *distribution-free tests* is also used. Nonparametric/distribution-free tests are discussed in Chapter 14.

mountable difficulty. Whether a test is two-tailed or one-tailed, the calculated t is the same. Only the probability is different. If the program you are using gives the probability of the computed t for a two-tailed test but you are conducting a one-tailed test, simply divide the computer-supplied probability by 2. For example, if the computer output is $p = .046$ (two-tailed probability), the one-tailed probability is .023.

Many computer t test programs automatically calculate a confidence interval (usually the 95% interval) as well. If not, a separate program is available for doing so. If you want an interval other than the 95% interval, you can easily specify this.

Always remember that, although the computer can save you time and effort, you alone can judge the appropriateness of the test or the confidence interval for your data. Never blindly submit a set of data for computer analysis. Always examine it carefully first. If, for example, the sample size is small and you find that the distribution is very skewed or has outliers, it may be wise to avoid using a t test or confidence interval and to look for other methods of analysis instead.

SUMMARY

If the standard deviation σ of the population from which one is sampling is not known, the ***estimated standard error of the mean***, s_M, calculated from sample data, is substituted for the actual standard error of the mean, σ_M, and the ratio $\frac{M-\mu}{s_M}$ is substituted for the ratio $z=\frac{M-\mu}{\sigma_M}$. If sample means are normally distributed, $(M - \mu)/s_M$ is distributed as t with $N - 1$ degrees of freedom (df). Therefore, when σ is unknown, tests of hypotheses about population means use the t distribution rather than the standard normal distribution.

The ***t distribution***, like the standard normal distribution, is symmetrical with a mean of zero. However, it is lower near the center and higher in the tails, so that larger values of $|t|$ are needed to cut off given areas in the tails. The shape of the t distribution depends on degrees of freedom. As df increases, the t distribution approaches the standard normal curve.

Like z tests, t tests can be two-tailed or upper one-tailed or lower one-tailed. The rationale and procedure for the test are much the same as for the z test, except that the probability of the calculated t is found in the t table rather than the table of the standard normal distribution.

Sometimes we want to estimate the value of a population mean without testing a particular hypothesis. To do so, we calculate a ***confidence interval*** for the population mean. A confidence interval is a range of values around the sample mean that is believed, at a certain level of confidence, to include the population mean. Confidence intervals may be calculated at any desired level of confidence, such as 95% ($\alpha = .05$) or 99% ($\alpha = .01$). The level of confidence is the percentage of samples in which the calculated interval includes the population mean. The formula for a confidence interval is $M \pm t_{\alpha/2}(s_M)$, where $t_{\alpha/2}$ is the critical value of t for the given level of alpha, two-tailed test. As the level of confidence increases, the width of the confidence interval increases. As sample size (N) increases, the width of the confidence interval decreases.

Calculating a confidence interval at level alpha is similar to testing all possible hypotheses about the population mean at the same alpha. Any hypothesis that μ is a value within the interval is not rejected, while all hypotheses that μ is a value outside the interval are rejected.

Confidence intervals for other parameters take the same form as for the mean. For example, the formula for calculating a confidence interval for a population percentage is $P \pm z_{\alpha/2}\, s_P$, where P stands for the population percentage and s_P for the standard error of the percentage = $\sqrt{P(100 - P)/N}$. The interpretation of confidence intervals for percentages is similar to that for means.

If the conclusion reached by a statistical test does not correspond with the actual state of affairs in the population, an ***error of inference*** has occurred. Rejecting a true null hypothesis is a ***Type I error***; not rejecting a false null hypothesis is a ***Type II error***. The probability of a Type I error is α, and the probability of a Type II error is β.

The only factor that affects the probability of a Type I error is α, the level of significance chosen for the test. The probability of a Type II error decreases (1) as the difference between true μ and μ_H increases, (2) as population variability (σ) decreases, (3) as α increases, (4) as N increases, and (5) if a one-tailed test is used rather than a two-tailed test. Since the probability of a Type I error (α) increases as the probability of a Type II error (β) decreases, and vice versa, selection of a level of α for a test should be based on a consideration of the relative seriousness of the two types of errors.

The probability of rejecting a false H_0, $1 - \beta$, is called the ***power*** of the test. All factors that decrease the probability of a Type II error increase the power of a test.

Parametric tests, such as the z test and the t test, rest on the assumption that the population is normal. Tests that do not require the assumption of population normality are called ***nonparametric*** tests. Nonparametric tests are usually less powerful than parametric tests. Fortunately, the z test and t test are relatively ***robust***; inferences from the tests are about as likely to be correct whether the normality assumption is true or not, unless the population shape is markedly nonnormal and N is small.

Most statistical packages include programs for conducting t tests for population means and/or calculating confidence intervals. As always, human judgment is necessary to decide whether the methods are appropriate for the data as well as to interpret the results.

QUESTIONS

1. Use the data on the heights of 28 statistics students in Table 2–13 (p. 49) to answer the following questions.

   a. Conduct a test to determine whether the mean height of the 28 students differs significantly from 67 inches at the .05 level.
   b. Is it possible that you have made a Type I error?
   c. Is it possible that you have made a Type II error?
   d. Before calculating, predict whether the 95% confidence interval around the sample mean will include the value 67.
   e. Calculate the 95% confidence interval for the mean height of the population.

2. Use the quiz scores of the 28 students to answer the following questions:

   a. Calculate the 99% confidence interval for the mean quiz score of the population.
   b. Use the confidence interval to determine whether the sample mean is significantly different from 16 at the .01 level.
   c. Use the confidence interval to determine whether the sample mean is significantly different from 10 at the .01 level.
   d. Without conducting the test, is the sample mean significantly different from 10 at the .05 level?

3. In an experiment on problem solving, 15 subjects required the following amounts of time, in minutes, to reach a solution: 12, 13, 15, 8, 9, 12, 11, 5, 10, 16, 13, 10, 8, 10, 16. Conduct a test to determine whether the mean time to solution is significantly less than 12 minutes at the .05 level.

4. A sample of size 16 has mean = 24.6 and standard deviation = 7.25.

   a. Is the sample mean significantly greater than 23 at the .05 level?
   b. Is it possible that you have made a Type I error?
   c. Is it possible that you have made a Type II error?

5. A sample of size 100 has the same mean and SD as in question 4.

   a. Is the sample mean significantly greater than 23 at the .05 level?

b. How do you account for the difference between your conclusions in questions 4(a) and 5(a)?

6. A sample of size 25 has mean = 13.5 and SD = 4.2. Is the sample mean significantly different from 15? Use α = .01.

7. A supermarket manager wondered how many items were purchased, on the average, on one visit to his supermarket. He counted the number of items purchased by 10 customers with the following results: 18, 7, 15, 25, 16, 18, 10, 21, 20, 20.

   a. Calculate the 95% confidence interval for the mean number of purchases by the population of all customers.
   b. If the sample mean had been the same, but number of items purchased had ranged from 12 to 20 instead of 7 to 25, would the 95% confidence interval have been wider or narrower?

8. A sample of size 25 has a mean equal to 95 and standard deviation of 20.

   a. Calculate the 95% confidence interval for the population mean.
   b. Calculate the 99% confidence interval for the mean.

9. The mean of a population is 45.

   a. You test H_0: μ = 45 and do not reject H_0. Have you made an error of inference? If so, what kind?
   b. You test H_0: μ = 47 and do not reject H_0. Have you made an error of inference? If so, what kind?
   c. You test H_0: μ = 45 and reject H_0. Have you made an error of inference? If so, what kind?
   d. You test H_0: μ = 47 and reject H_0. Have you made an error of inference? If so, what kind?

10. You are planning to conduct a test of H_0: μ = 30 with N = 20 and α = .05.

   a. Which of the following changes will decrease the probability of rejecting H_0 if it is true?
      (1) Increase N
      (2) Decrease α to .01
      (3) Increase α to .10
   b. Which of the following changes will decrease the probability of not rejecting H_0 if it is false?
      (1) Increase N
      (2) Decrease α to .01
      (3) Increase α to .10

11. In a survey of 135 randomly selected students at Allegro College, 42 said that they planned to go to professional or graduate school.

   a. Calculate the 90% confidence interval for the percentage of all Allegro students who plan to go to professional or graduate school.
   b. At the .10 level, can you reject the hypothesis than one-fourth of all Allegro students plan to go to professional or graduate school?
   c. How large a sample would be needed to reduce the margin of error, at the 90% level of confidence, to ± 5%? (Assume that the sample percentage remains the same.)

12. Which of the following tests is most powerful? (Assume that all other factors are equal.)

   Test A: N = 20, α = .01
   Test B: N = 20, α = .05
   Test C: N = 50, α = .01
   Test D: N = 50, α = .05

13. a. You are conducting a two-tailed normal curve test of H_0: μ = 90 with α = .05. If the population standard deviation (σ) is 15 and the sample mean is 92.5, how many subjects do you need in order to reject H_0? (*Hint:* To reject H_0 at the .05 level, you need a z of at least 1.96. Calculate the value of σ_M needed to get z = 1.96 with M = 92.5 and μ_H = 90. Then calculate the N needed to get that value of σ_M if σ = 15.)
   b. Suppose that you test H_0: μ = 90 with an upper one-tailed test instead of a two-tailed test. Again, α = .05, σ = 15, and M = 92.5. How many subjects do you need in order to reject H_0?

Correlation: Measuring Relationships Between Variables

INTRODUCTION

In Chapters 6 through 8 we dealt primarily with a single variable measured in one group of individuals, objects, or events. Often researchers measure not one but two (or more) variables in a group in order to study the relationship between the variables. Thus, a teacher measures reading ability and self-esteem in his class to find out how the children's self-esteem is related to their reading ability. An industrial psychologist, interested in finding out how absenteeism and productivity are related, measures number of days absent from work and sales in a group of salespersons. Both the teacher and the psychologist need a measure of the relation between two variables. In this chapter, we deal with the calculation, interpretation, and use of measures of relationship.

Objectives

When you have completed this chapter, you will be able to

- Define positive and negative correlations between two variables. Given information about how two variables co-vary, decide whether the correlation is positive, negative, or neither.
- Construct scatter diagrams and, from them, estimate the direction and strength of the correlation.
- Define the covariance of two variables, CoVar_{xy}.
- Define the Pearson correlation coefficient, r, as a standardized covariance.
- Describe the range of values of r; given a value of r, describe the direction and strength of the correlation.
- Calculate r using the covariance formula and the computational formula.
- Describe factors affecting the size of r. Given information about these factors, predict the effect on r.
- Explain why a correlation between X and Y does not necessarily imply that X causes Y.
- Give a components-of-variance explanation of the squared correlation coefficient, r^2.
- Define and describe the sampling distribution of r and its relationship to the population correlation coefficient, ρ, and sample size, N.

- Conduct one-tailed and two-tailed tests of the null hypothesis: $\rho = 0$.
- Calculate the Spearman rank-order correlation coefficient, the point biserial correlation coefficient, and the phi coefficient, and test them for significance. Explain the types of data for which these coefficients are appropriate, as well as their relationship to the Pearson r.

Basic Concepts

The Meaning of Correlation

Wondering whether his students' self-esteem is related to their academic achievement, Mr. Snyder, a sixth-grade teacher, gives tests of self-esteem and reading ability to his class. Dr. Bosworth, an industrial psychologist at Ajax Sales, decides to examine the relationship between absenteeism and productivity of Ajax salespersons. Toward this end, she collects data on number of days absent from work and amount of sales of the company's employees over a 2-month period. Mr. Ranger, a college admissions officer, wonders whether there is a relationship between applicants' Verbal Scholastic Assessment Test (SAT) scores and the number of extracurricular activities in which they engage; he examines the records of recent applicants to find out. All three of these people are interested in the relationship, or correlation, between two variables.

What does it mean to say that two variables are correlated? It means that the variables *co-vary* in some systematic fashion. That is, as one variable increases, the other tends to increase or to decrease. We will see what this means by examining Mr. Snyder's, Dr. Bosworth's, and Mr. Ranger's data.

In situations such as these, researchers usually collect data on large numbers of subjects. In order to keep the discussion and examples simple, we will use data from only eight subjects in each case. The concepts are easily generalized to larger groups.

Table 9–1 shows the self-esteem and reading scores of Mr. Snyder's eight students, as well as the mean, variance,† and standard deviation of each variable. To make it easier to detect co-variation, the students are listed in order in terms of their self-esteem scores, from lowest to highest. Self-esteem scores vary from 4 to 13, and reading ability varies from 10 to 17. Do self-esteem and reading ability co-vary, and, if so, how?

Table 9–1 A Positive Correlation

Child	Self-Esteem	Reading Ability
1	4	13
2	6	10
3	7	16
4	8	13
5	10	17
6	11	12
7	13	14
8	13	17
Mean	9.0	14.0
Variance	9.5	5.5
SD	3.08	2.345

†In this chapter we use the formula SS/N to calculate the variance.

Look down the column of reading ability scores in Table 9–1. Notice how, as self-esteem increases, reading scores tend to increase, too. Of the four children who are below the mean on self-esteem, three are below the mean on reading ability as well. Of the four children who are above the mean on self-esteem, two are above the mean and one is at the mean of reading ability. Thus, low self-esteem scores tend to be associated with low reading ability and high self-esteem scores with high reading ability. When two variables co-vary in the same direction, as in this example, there is a *positive correlation* between them. Thus, self-esteem and reading ability are positively correlated in these eight children.

We can see the positive correlation more vividly in a graph of the data. A graph that displays the relationship between two quantitative variables is called a ***scatter diagram*** or ***scatter plot***. A scatter diagram of the data on self-esteem and reading ability is shown in Figure 9–1.

The steps in constructing a scatter diagram are as follows:

1. One of the variables is assigned to the abscissa, the other to the ordinate. We will call the variable assigned to the abscissa X and the variable assigned to the ordinate Y. When the purpose is simply to describe the relationship, it does not matter which variable is designated X and which one Y. In Figure 9–1, self-esteem is variable X and reading ability is Y. Note that both axes are labeled with the variable name.
2. Values of the variables are marked on the axes. Since self-esteem varied from 4 to 13, values of 4, 5, . . . , 13 are placed at equal distances on the abscissa. Reading ability varied from 10 to 17; values of 10, 11, . . . , 17 are marked at equal intervals on the ordinate. In a scatter diagram, it is not necessary to start from zero on either axis.
3. Next, the data are plotted. Each pair of scores is represented by a single point, above the X value on the abscissa, at a height equal to the Y value on the ordinate. For example, to plot a point for Child 1, with self-esteem $(X) = 4$ and reading ability $(Y) = 13$, find the value 4 on the abscissa, go up to the value 13 on the ordinate, and place a point at height 13 above the value 4, like this:

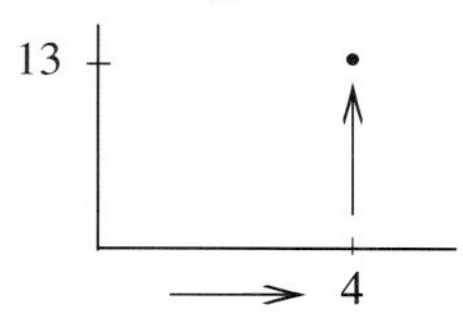

The remaining pairs of scores are plotted in the same manner.

Since each pair of scores is represented by one point, the number of points is equal to the number of pairs of scores or the number of individuals, in this case, 8.

Now look at the trend of the points in Figure 9–1. The points tend to cluster around a diagonal line from the lower left of the scatter diagram to the upper right. Whenever the correlation between two variables is positive, as here, the points in the scatter diagram flow from lower left to upper right.

Next, let us examine Dr. Bosworth's data on absenteeism and sales of eight employees, shown in Table 9–2. The employees are listed by their absenteeism scores, from lowest to highest. Days absent vary from 0 to 10 and sales from \$3500 to \$7500. Let us see whether and how they co-vary.

Note that, of the four employees whose absenteeism is below the mean, three have sales above the mean; of the four employees whose absenteeism is above the mean, three have sales below the mean. Thus, low values on each variable are associated with high values on the other; as absenteeism increases, sales tend to decrease. When two variables co-vary inversely, as in this example, the correlation between them is a *negative correlation*. Absenteeism and sales are negatively correlated in these eight employees.

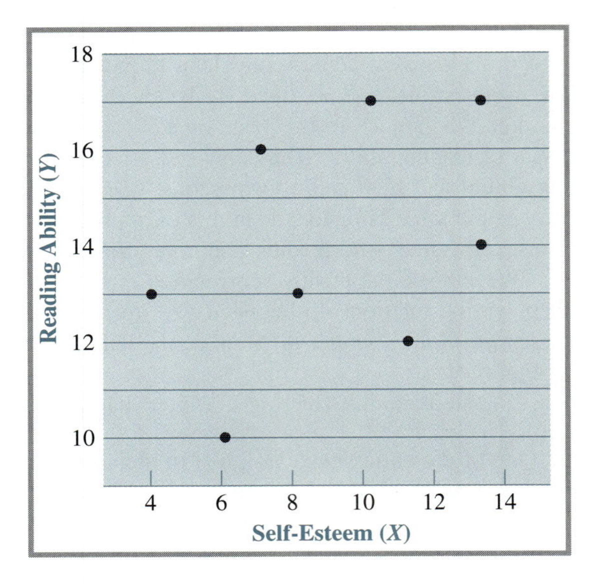

Figure 9–1

Scatter diagram of the relation between self-esteem and reading ability.

Table 9–2 A Negative Correlation

Employee	Days Absent	Sales (in hundreds of dollars)
1	0	75
2	1	65
3	3	65
4	4	50
5	6	60
6	7	40
7	9	50
8	10	35
Mean	5.0	55.0
Variance	11.5	162.5
SD	3.39	12.75

The scatter diagram of the relationship between absenteeism and sales is shown in Figure 9–2. The variable days absent has been designated X and sales Y. Each point in the scatter diagram represents the paired data for one employee. As you can see, the points tend to cluster about a diagonal line from upper left to lower right. When the correlation between two variables is negative, as here, the points in the scatter diagram flow from upper left to lower right.

Also note that in both Figures 9–1 and 9–2, the diagonal line that fits the trend of the points is a straight line. When a straight line fits the data points in a scatter diagram reasonably well, the relationship between the variables is said to be *linear*. In this chapter we are concerned primarily with the measurement and interpretation of linear relationships.

Finally, consider Mr. Ranger's data on Verbal SAT scores and number of extracurricular activities of eight college applicants (Table 9–3). The applicants are listed from

Figure 9–2

Scatter diagram of the relation between absenteeism and sales.

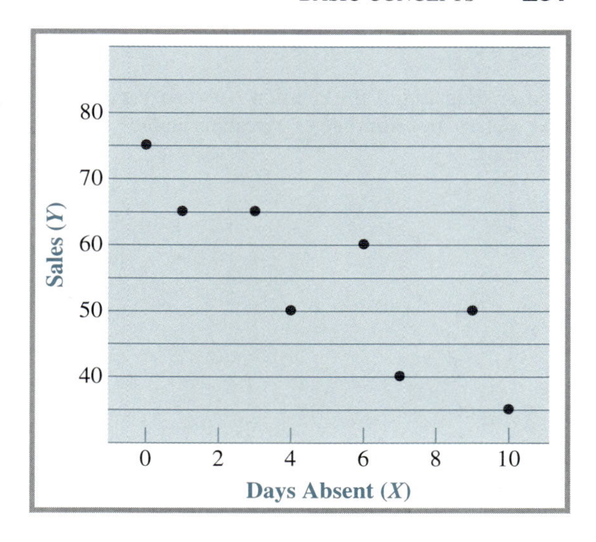

Table 9–3 **A Lack of Relationship**

Applicant	Verbal SAT Score	Number of Extracurricular Activities
1	350	5
2	380	7
3	440	0
4	500	3
5	510	1
6	550	8
7	620	6
8	650	2
Mean	500	4.0
Variance	10,000	7.5
SD	100	2.74

lowest to highest on the SAT. SAT scores vary from 350 to 650, and extracurricular activities vary from 0 to 8. Do they co-vary?

Examination of the data reveals that, of the four applicants with SAT scores at or below the mean, two have extracurricular activities above and two below the mean. Similarly, of the four applicants with SAT scores above the mean, two are above and two below the extracurricular activities mean. Extracurricular activities neither increase nor decrease as SAT scores increase. Although both SAT scores and number of activities vary, they do not *co*-vary. Thus, SAT scores and number of extracurricular activities are uncorrelated in Mr. Ranger's group of applicants.

Figure 9–3 shows the scatter diagram for the relation between SAT scores (X) and extracurricular activities (Y). As you can see, the points do not lie along a diagonal from lower left to upper right or from upper left to lower right; they are, quite literally, scattered in no discernible pattern. When there is no correlation between two variables, as here, the points in the scatter diagram do not flow in any particular direction.

Figure 9–3

Scatter diagram of the relation between Verbal SAT scores and number of extracurricular activities.

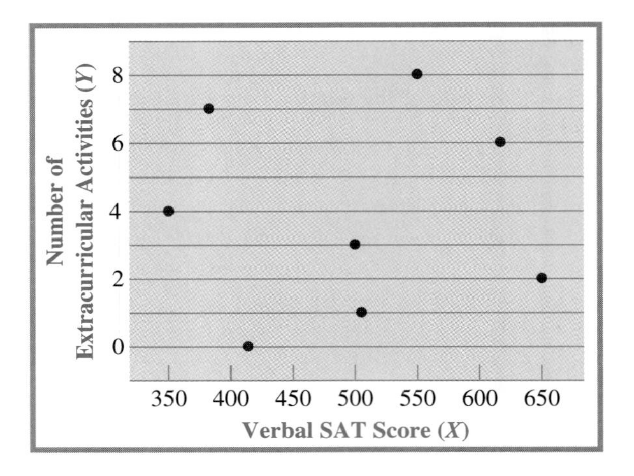

From these examples, we know the following:

1. Two variables are positively correlated if, as one variable increases, the other tends to increase as well—that is, if high values on one variable tend to be associated with high values on the other, and low values on the first tend to be associated with low values on the second. Here are some additional examples of variables that are positively correlated:

   Time spent studying and performance in school
   Religiosity and attendance at church services
   Number of martinis consumed and tendency to act silly

2. Two variables are negatively correlated if, as one variable increases, the other tends to decrease—that is, if high values on one variable tend to be associated with low values on the other, and low values on the first tend to be associated with high values on the second. Some additional examples of negative correlation include

   Number of hours spent practicing and number of mistakes in playing the piano
   Number of martinis consumed and ability to walk a straight line
   Belief in free enterprise and tendency to vote for the Socialist Party candidate

3. Two variables are uncorrelated if they do not co-vary systematically. As one variable increases, the other neither increases nor decreases in a regular fashion. Low values as well as high values on one variable are equally often associated with high, moderate, and low values of the other. Here are some additional examples of variables that are uncorrelated:

   Time spent studying and religiosity
   Liking for broccoli and grades in a statistics course
   Interest in classical music and cooking skill

Do not confuse negative correlation with lack of correlation. A negative correlation involves systematic covariation of the two variables, although in an inverse direction. A lack of correlation means that there is *no* systematic covariation, direct or inverse, between the variables.

EXERCISE

9–1

1. In each of the following, would you expect a positive correlation, a negative correlation, or no correlation?
   a. Likelihood of playing college basketball and height
   b. Skill at playing basketball and neatness of handwriting
   c. Number of disruptive behaviors and reading ability in grade-school children
   d. Self-confidence and tendency to blush
   e. Typing speed and interest in baseball
   f. Typing speed and finger dexterity
   g. Willingness to work and grade in a statistics course
2. Ms. Jones, an algebra teacher, found a positive correlation between time required to complete an exam and scores on the exam. Did students who turned their exams in early tend to score higher or lower than their classmates who took longer to finish?
3. Your introductory psychology textbook states that, as children grow older, their fear of monsters declines. Do age and fear of monsters co-vary positively, negatively, or not at all?
4. From your own experience or reading, name some pairs of variables that are positively correlated, some that are negatively correlated, and some that are uncorrelated.

The Strength of Correlations

Correlations between two variables differ not only in direction (positive versus negative) but also in strength. The strength of a correlation is independent of its direction; both positive and negative correlations can be strong or weak. Graphically, the strength of a correlation is revealed by how the points in the scatter diagram cluster around a diagonal line from lower left to upper right (positive correlation) or upper left to lower right (negative correlation). The more closely the points "hug" the diagonal, the stronger is the correlation between the variables. Figure 9–4 presents scatter diagrams for strong, moderate, and weak positive and negative correlations.

By comparing the scatter diagrams of Mr. Snyder's data (Figure 9–1) and Dr. Bosworth's data (Figure 9–2) with those in Figure 9–4, we see that the correlation between self-esteem and reading is a moderate positive correlation, whereas the correlation between absenteeism and sales is a strong negative one.

The strong positive and negative correlations portrayed in Figure 9–4, parts (a) and (d), are not the strongest that are possible. The strongest possible positive correlation is illustrated in Table 9–4 and Figure 9–5 for the variables "height in inches" and "height in centimeters," measured in five individuals. Since one inch = 2.54 centimeters, height in centimeters increases by exactly 2.54 units for each unit increase of height in inches. In the scatter diagram, all the points lie on a straight line from lower left to upper right. The relationship between height in inches and height in centimeters is called a *perfect positive correlation.*

The strongest possible negative correlation is depicted in Table 9–5 and Figure 9–6 for the variables "distance traveled" on a 10-mile hike and "distance remaining." As you can see, for every increase in distance traveled, there is an equal decrease in distance remaining. The points in the scatter diagram all lie on a straight line from upper left to lower right. The correlation is a *perfect negative correlation.*

In Tables 9–4 and 9–5, z scores as well as raw scores are listed for both variables. Note that in Table 9–4 (perfect positive correlation) each z score of height in inches

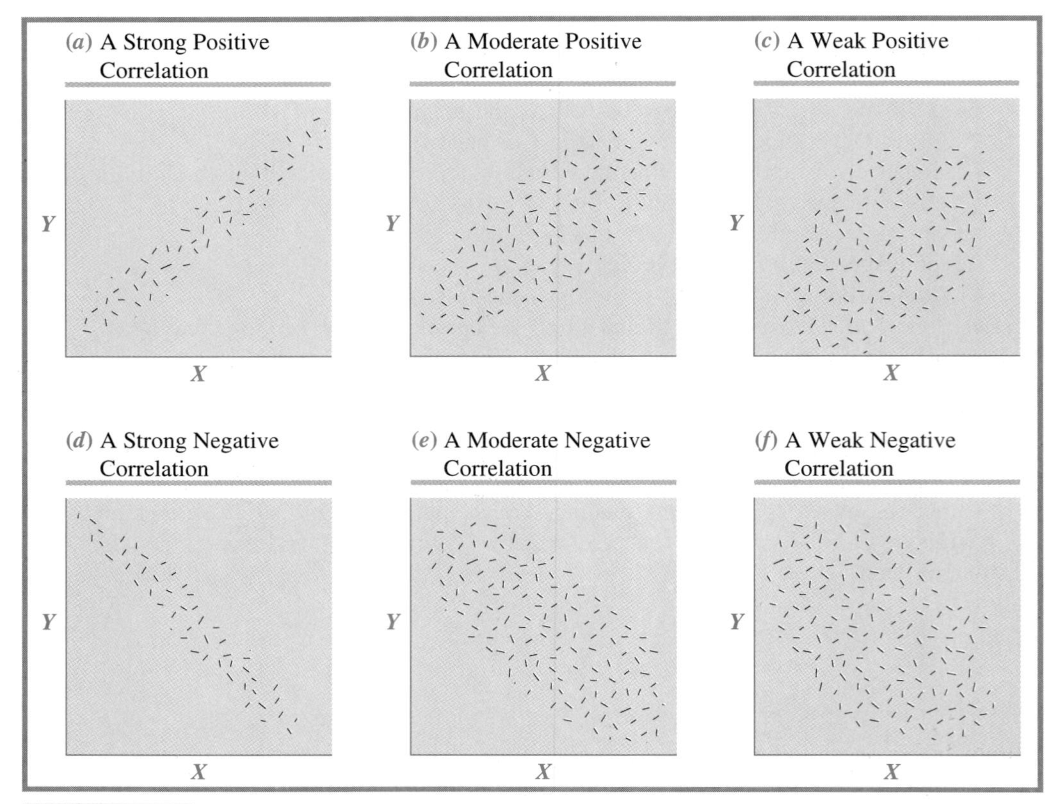

Figure 9–4 **Scatter diagrams of strong, moderate, and weak positive and negative correlations.**

Table 9–4 **A Perfect Positive Correlation Between Height in Inches and Height in Centimeters**

Individual	Height in Inches (X)	Height in Centimeters[a] (Y)	z_X	z_Y
1	64	162.56	−0.59	−0.59
2	71	180.34	1.47	1.47
3	68	172.72	0.59	0.59
4	61	154.94	−1.47	−1.47
5	66	167.64	0	0
Mean	66	167.64		
SD	3.406	8.651		

[a] One inch = 2.54 centimeters.

is identical with the paired z score of height in centimeters. This is always the case in a perfect positive correlation. In the perfect negative correlation (see Table 9–5), the two z scores in each pair have the same absolute value but opposite signs. This is always the case with perfect negative correlations. We will return to these facts later in the chapter.

Figure 9–5

Scatter diagram of a perfect positive correlation.

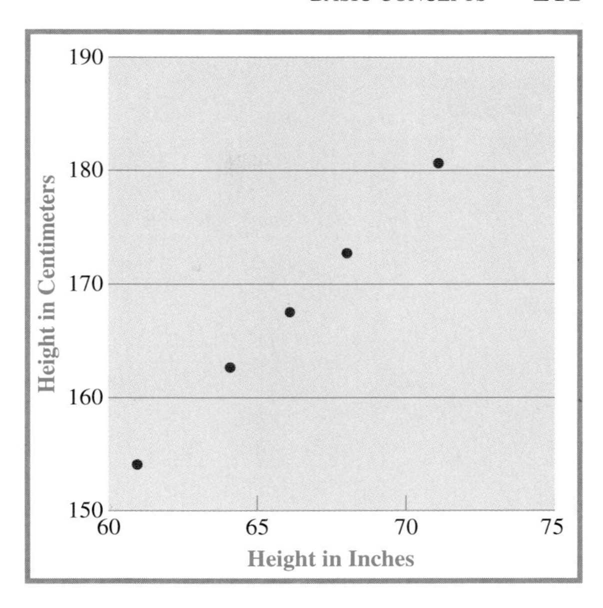

Table 9–5 **A Perfect Negative Correlation Between Distance Traveled and Distance Remaining on a 10-Mile Hike**

Paired Distances	Distance Traveled (X)	Distance Remaining (Y)	z_X	z_Y
1	1	9	–1.41	1.41
2	3	7	–0.71	0.71
3	5	5	0	0
4	7	3	0.71	–0.71
5	9	1	1.41	–1.41
Mean	5	5		
SD	2.83	2.83		

Figure 9–6

Scatter diagram of a perfect negative correlation.

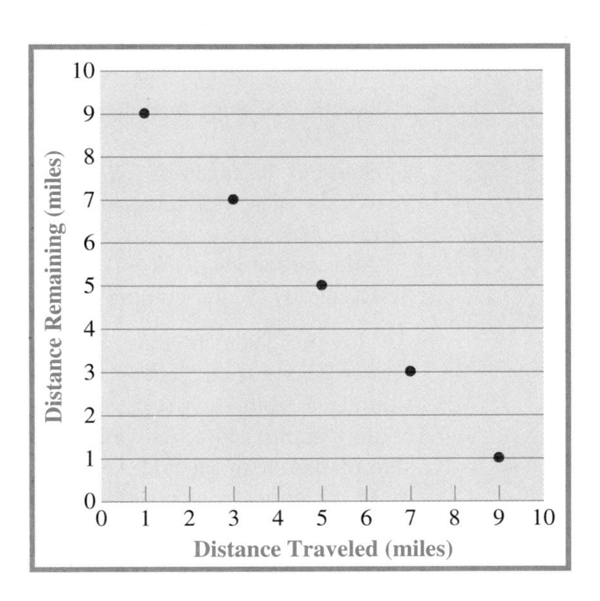

EXERCISE 9–2

1. Draw scatter diagrams of the following three data sets.

A			B			C		
Individual	*Fame*	*Fortune*	*Individual*	*Apples Eaten per Month*	*Number of Doctor Visits*	*Individual*	*Beauty*	*Character*
1	8	14	1	10	4	1	10	16
2	12	16	2	7	8	2	5	18
3	8	17	3	13	2	3	7	13
4	11	19	4	10	5	4	14	18
5	5	11	5	5	8	5	8	17
6	11	15	6	8	6	6	5	15
7	6	15	7	13	4	7	12	14
8	10	17	8	7	4	8	12	11
9	5	13	9	5	7	9	10	13
10	8	11	10	10	2	10	11	18

2. Describe each correlation in terms of its direction and strength.
3. Which of the three correlations is strongest? Which is weakest?

Measuring Covariation

Now we are ready to quantify the concepts of positive and negative and strong and weak correlations. That is, we are ready to discuss the measurement of correlations. We begin by considering the ***covariance*** of two variables.

The Covariance

The covariance of two variables, CoVar_{xy}, is defined as follows:

$$\text{CoVar}_{xy} = \frac{\Sigma(X - M_x)(Y - M_y)}{N}$$

Calculation of the covariance entails the following steps:

1. Subtract the mean of X (M_x) from each X and the mean of $Y(M_y)$ from each Y.
2. Within each individual, multiply $(X - M_x)$ by $(Y - M_y)$ to get the product $(X - M_x)(Y - M_y)$.
3. Add the values of $(X - M_x)(Y - M_y)$. The sum is $\Sigma(X - M_x)(Y - M_y)$.
4. Divide by N, the number of individuals or pairs of scores.

Let us calculate the covariance of Mr. Snyder's self-esteem and reading ability scores. The first three steps are shown in Table 9–6. First, the mean of self-esteem, 9, is subtracted from each self-esteem score, and the mean of reading ability, 14, from each reading ability score. As we have already observed, most of the children fall on the same side of the mean on both variables. Therefore, in most cases, their deviations from the mean are either both positive or both negative. Thus, the product of the two deviations, $(X - M_x)(Y - M_y)$, is usually positive, and the sum of the products, $\Sigma(X - M_x)(Y - M_y)$, is a positive number, 25. Finally, $\Sigma(X - M_x)(Y - M_y)$ is divided by 8 to get the covariance of self-esteem and reading ability:

Table 9–6 **Calculation of $\Sigma(X - M_x)(Y - M_y)$ for Self-Esteem–Reading Ability**

	Self-Esteem	Reading Ability			
Child	X	Y	$X - M_x$	$Y - M_y$	$(X - M_x)(Y - M_y)$
1	4	13	−5	−1	5
2	6	10	−3	−4	12
3	7	16	−2	2	−4
4	8	13	−1	−1	1
5	10	17	1	3	3
6	11	12	2	−2	−4
7	13	14	4	0	0
8	13	17	4	3	12
					25
					$\Sigma(X - M_x)(Y - M_y)$

$$\text{CoVar}_{xy} = \frac{25}{8} = 3.125$$

Table 9–7 shows the calculation of $\Sigma(X - M_x)(Y - M_y)$ for Dr. Bosworth's data. First, the mean of absenteeism, 5, is subtracted from each value of days absent, and the mean of sales, 55, is subtracted from each value of sales. Since most employees are on opposite sides of the mean on the two variables, the product $(X - M_x)(Y - M_y)$ is negative in most cases, and the sum of the products is negative, −300. Finally, $\Sigma(X - M_x)\ (Y - M_y)$ is divided by 8 to get the covariance of absenteeism and sales:

$$\text{CoVar}_{xy} = \frac{-300}{8} = -37.5$$

The calculation of $\Sigma(X - M_x)(Y - M_y)$ for Mr. Ranger's SAT–extracurricular activities data is shown in Table 9–8. As we have already observed, pairs of scores on the two variables are about equally likely to be on the same side or on opposite sides of the mean. As a result, the product $(X - M_x)(Y - M_y)$ is positive about half the time and negative half the time. $\Sigma(X - M_x)(Y - M_y)$ is −160, and the covariance is

$$\text{CoVar}_{xy} = \frac{-160}{8} = -20$$

Table 9–7 **Calculation of $\Sigma(X - M_x)(Y - M_y)$ for Days Absent–Sales**

	Days Absent	Sales			
Employee	X	Y	$X - M_x$	$Y - M_y$	$(X - M_x)(Y - M_y)$
1	0	75	−5	20	−100
2	1	65	−4	10	−40
3	3	65	−2	10	−20
4	4	50	−1	−5	5
5	6	60	1	5	5
6	7	40	2	−15	−30
7	9	50	4	−5	−20
8	10	35	5	−20	−100
					−300
					$\Sigma(X - M_x)(Y - M_y)$

Table 9–8 Calculation of $\Sigma(X - M_x)(Y - M_y)$ for SAT–Extracurricular Activities

	SAT Score	Extracurricular Activities			
Applicant	X	Y	$X - M_x$	$Y - M_y$	$(X - M_x)(Y - M_y)$
1	350	5	−150	1	−150
2	380	7	−120	3	−360
3	440	0	−60	−4	240
4	500	3	0	−1	0
5	510	1	10	−3	−30
6	550	8	50	4	200
7	620	6	120	2	240
8	650	2	150	−2	−300
					−160 $\Sigma(X - M_x)(Y - M_y)$

The Pearson Correlation Coefficient

A problem with the covariance as a measure of relationship is that it is affected by the units of measurement of both variables. We know, for example, that the correlation between self-esteem and reading ability is moderately strong and that SAT scores and extracurricular activities are uncorrelated. Yet CoVar_{xy} of self-esteem and reading ability, 3.125, is smaller than $|\text{CoVar}_{xy}|$ of SAT scores and extracurricular activities, 20. This is because SAT scores have a much larger unit of measurement than self-esteem or reading ability.

The effect of the unit of measurement can be eliminated by "standardizing" the covariance. This is done by dividing CoVar_{xy} by the standard deviations of both variables, S_x and S_y. The result is a unit-free index of correlation called the ***Pearson correlation coefficient***, symbolized $\boldsymbol{r}$. Thus

$$r = \frac{\text{CoVar}_{xy}}{S_x S_y}$$

The Pearson correlation coefficient as a standardized covariance

Shortly, we will calculate r for Mr. Snyder's, Dr. Bosworth's, and Mr. Ranger's data. First, let us discuss some facts about r.

The Pearson correlation coefficient r is an index of the *linear relation* between two quantitative variables. There are several other kinds of correlation coefficients, some of which are discussed later in the chapter. For now, when we refer to a "correlation coefficient," we will mean Pearson's r.

The correlation coefficient can take values ranging from −1 to +1. The *sign* of r provides information about the *direction* of the relationship. If X and Y are positively correlated, r is positive; if X and Y are negatively correlated, r is negative. A correlation of zero or near zero means that the variables are unrelated.

The *magnitude* of $|r|$ provides information about the *strength* of the relationship. The nearer the value of r to ± 1, the stronger is the relationship. In a perfect positive correlation, $r = 1$; for example, the correlation between height in inches and height in centimeters in Table 9–4 and Figure 9–5 is +1. In a perfect negative correlation, $r = -1$; the correlation between distance traveled and distance remaining in Table 9–5 and Figure 9–6 is −1. Here is a schematic diagram of the range of values of r and their meaning:

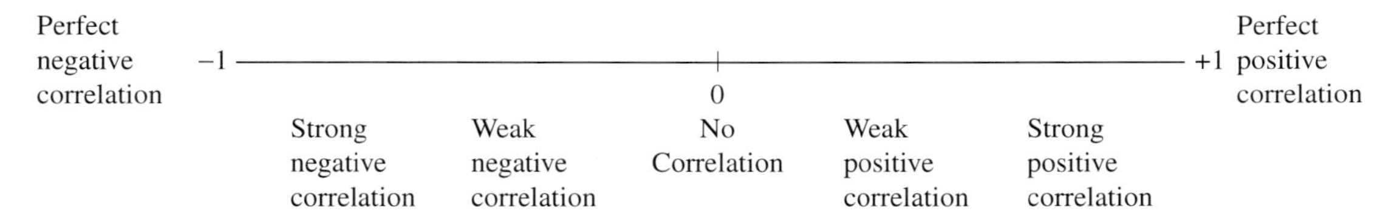

Do not confuse the direction of a relationship with its strength. Direction and strength are independent properties of a relationship. A positive correlation can be strong, moderate, or weak. A negative correlation can be strong, moderate, or weak. A correlation of +.70 is a stronger positive correlation than one of +.50. A correlation of −.80 is a stronger negative correlation than one of −.55. A correlation of −.80 is a stronger correlation than one of +.70. Correlations of +.45 and −.45, although opposite in direction, are equally strong.

Now let us calculate the correlation coefficient between self-esteem and reading ability. We have calculated $\text{CoVar}_{xy} = 3.125$. Table 9–1 lists the two standard deviations: $S_x = 3.08$ and $S_y = 2.345$. Therefore,

$$r = \frac{3.125}{(3.08)(2.345)} = .43$$

We already knew that the correlation is a moderate positive one. Now we know that its exact value is .43.

Next, we calculate the correlation coefficient between absenteeism and sales. Since we know that the correlation is a strong negative one, we expect r to be close to −1. We have calculated $\text{CoVar}_{xy} = -37.5$. According to Table 9–2, $S_x = 3.39$ and $S_y = 12.75$. Therefore,

$$r = \frac{-37.5}{(3.39)(12.75)} = -.87$$

As we expected, the correlation is close to −1.

In the case of SAT scores and extracurricular activities, we expect a correlation coefficient close to zero. We have calculated $\text{CoVar}_{xy} = -20$. From Table 9–3, $S_x = 100$ and $S_y = 2.74$. Therefore,

$$r = \frac{-20}{(100)(2.74)} = -.07$$

A correlation of −.07 is so close to zero as to be negligible.

I pointed out earlier that the correlation coefficient is a unit-free index of the relation between two variables. This makes it possible to compare correlation coefficients from different studies or between different pairs of variables. For example, despite the differences in variables and their methods of measurement, we can state that the correlation between absenteeism and sales is stronger than that between self-esteem and reading ability (at least in the samples studied).

Because the correlation coefficient is unit-free, it is unaffected if a constant is added to or subtracted from each score on one or both variables, or if each score is multiplied or divided by a constant. Consider the Verbal SAT scores of Mr. Ranger's study; they are all multiples of 10. Mr. Ranger could make his calculations easier by dividing each SAT score by 10 before calculating r. The correlation between the transformed SAT scores and extracurricular activities would still be equal to −.07. (You may wish to confirm this by dividing each SAT score by 10 and calculating r yourself.)

The Correlation Coefficient as the Ratio of Covariance to Average Variance

We have defined the correlation coefficient as a standardized measure of covariance:

$$r = \frac{\text{CoVar}_{xy}}{S_x S_y}$$

Since the standard deviation of a variable is the square root of the variance, this formula can be rewritten as

$$r = \frac{\text{CoVar}_{xy}}{\sqrt{(\text{Var}_x)(\text{Var}_y)}}$$

The denominator of the formula is the square root of the product of the variances of X and Y. The square root of the product of two numbers is a kind of average called the *geometric mean* of the numbers. This gives us another interpretation of r:r is the *ratio of the covariance of X and Y to the geometric mean of the two variances.*

The formula shows that if $r = \pm 1$, the covariance of X and Y is equal to the average of the two variances. The weaker the relationship between X and Y, the smaller is their covariance relative to the average of the variances.

The Correlation Coefficient in Terms of z Scores

By a slight algebraic manipulation of the covariance formula for the correlation coefficient, we can obtain a formula expressing r as a function of z scores on the X and Y variables.

$$r = \frac{\text{CoVar}_{xy}}{S_x S_y} = \frac{\frac{\Sigma(X - M_x)(Y - M_y)}{N}}{S_x S_y} = \frac{\Sigma\left(\frac{X - M_x}{S_x}\right)\left(\frac{Y - M_y}{S_y}\right)}{N}$$

Remember that a z score is equal to $\frac{Score - Mean}{SD}$. Therefore, $\frac{X - M_x}{S_x} = z_x$ and $\frac{Y - M_y}{S_y} = z_y$, and r can be expressed as

$$r = \frac{\Sigma z_x z_y}{N} \qquad \textit{A z score formula for r}$$

The z score formula can be used to show why the maximum possible value of r is $|1|$. We noted earlier that when two variables have a perfect positive correlation, $z_x = z_y$ in each pair (see, for example, Table 9–4). Therefore, $z_x z_y$ in each pair is equal to $z_x z_x$, or $z_x{}^2$. Then $\Sigma z_x z_y = \Sigma z_x^2$, and $r = \Sigma z_x^2/N$. But $\Sigma z_x^2/N$ is the variance of the z score distribution, and the variance of a distribution of z scores is always equal to 1. Therefore, the value of r for a perfect positive correlation is 1.

When two variables have a perfect negative correlation, $z_x = -z_y$ in each pair (see, for example, Table 9–5). Therefore, $z_x z_y$ in each pair is equal to $z_x(-z_x) = -z_x z_x = -z_x^2$. Then $\Sigma z_x z_y = -\Sigma z_x{}^2$, and $r = -\left(\frac{\Sigma z_x^2}{N}\right)$. Again, since $\frac{\Sigma z_x^2}{N}$ is equal to 1, the value of r for a perfect negative correlation is -1.

Try calculating $\Sigma z_x z_y$ for the data in Tables 9–4 and 9–5. You will find that $\Sigma z_x z_y$ in Table 9–4 is 5 and $r = \frac{5}{5} = 1$. In Table 9–5, $\Sigma z_x z_y = -5$ and $r = \frac{-5}{5} = -1$.

EXERCISE 9–3

1. Sam, Joe, and Clare calculated the Pearson r for data they had collected as part of a class project. Comment on the following conversation among the three students:

   Sam: My value of r is $-.60$. My variables are unrelated.
   Joe: I got $r = +.40$. My correlation is stronger than Sam's.
   Clare: My correlation coefficient is the strongest of all. I got $r = 1.20$.

2. The standard deviation of Variable A is 8, and the standard deviation of Variable B is 10.
   a. Calculate the correlation between A and B if their covariance is
      (1) −50
      (2) 65
      (3) 5
   b. What is the covariance between Variables A and B if their correlation is +1?
3. Here are aptitude test scores and supervisor ratings of 10 employees.

Employee	Aptitude Test Score	Supervisor Rating
1	10	4
2	8	4
3	9	2
4	13	3
5	12	5
6	7	2
7	11	3
8	11	2
9	10	1
10	8	1

   a. Draw a scatter diagram of the data and, without calculating r, describe the direction and strength of the relation between aptitude test scores and supervisor ratings.
   b. Predict the effect on the correlation if each employee's aptitude test score is increased by 5 points (with supervisor ratings unchanged).
   c. Add 5 to each employee's aptitude test score, and draw a scatter diagram. Compare the two graphs. Did adding 5 points to each employee's aptitude test score change the correlation between test scores and supervisor ratings?

Alternative Methods for Calculating the Correlation Coefficient

The covariance and z score formulas for r are rarely used for calculation because they require considerable computational labor, especially if N is large. Correlation coefficients are easily obtained using either statistical calculators or computer software. One simply enters the data and either presses an appropriate key (calculator) or types in a command (computer). For example, suppose that Mr. Snyder has typed his data into a Minitab data file, with identification numbers in Column 1 (c1), self-esteem scores in Column 2 (c2), and reading scores in Column 3 (c3). Now he simply types the "correlate" command at the Minitab prompt, like this:

```
MTB > Correlate c2 c3
```

The following output appears immediately on his computer screen:

```
Correlation of c2 and c3 = 0.432
```

When r must be calculated by hand, there is a computational formula that operates directly upon raw scores X and Y and has the added advantage of not involving negative quantities (except, in the case of negative correlations, near the end of the calculations). The formula is

$$r = \frac{N(\sum XY) - (\sum X)(\sum Y)}{\sqrt{\left[N(\sum X^2) - (\sum X)^2\right]\left[N(\sum Y^2) - (\sum Y)^2\right]}} \qquad \textit{Computational formula for } r$$

Horrendous as this formula may seem, it is very easy to use. The terms in the denominator are old friends.

$N(\Sigma X^2)$ and $N(\Sigma Y^2)$ mean:	(1) Square each score
	(2) Add the squared scores
	(3) Multiply the sum by N, the number of pairs
$(\Sigma X)^2$ and $(\Sigma Y)^2$ mean:	(1) Add the scores, then
	(2) Square the sum

The terms $N(\Sigma XY)$ and $(\Sigma X)(\Sigma Y)$ in the numerator of the formula are new. Let us consider these terms one at a time.

$N(\Sigma XY)$ means:	(1) Multiply each X by the paired Y to get XY
	(2) Add over all pairs to get ΣXY
	(3) Multiply the sum by N, the number of pairs
$(\Sigma X)(\Sigma Y)$ means:	(1) Add the X's to get ΣX
	(2) Add the Y's to get ΣY
	(3) Multiply the two sums

To illustrate use of the computational formula, we calculate r for the absenteeism–sales data. Using the covariance formula, we calculated $r = -.87$. Table 9–9 lists values of XY, X^2, and Y^2 for each employee and shows the calculation of all the needed sums.

Let us first calculate the numerator and denominator of r separately:

$$\text{Numerator: } N(\sum XY) - (\sum X)(\sum Y) = 8(1900) - (40)(440) = 15{,}200 - 17{,}600 = -2400$$

$$\text{Denominator: } \sqrt{[N(\sum X^2) - (\sum X)^2][N(\sum Y^2) - (\sum Y)^2} = \sqrt{[8(292) - (40)^2][8(25{,}500) - (440)^2]}$$

$$= \sqrt{(2336 - 1600)(204{,}000 - 193{,}600)} = \sqrt{(736)(10{,}400)} = \sqrt{7{,}654{,}400} = 2766.66$$

$$\text{Then: } r = \frac{-2400}{2766.66} = -.87, \text{ as before}$$

Errors sometimes occur in calculating r by hand because of the large number of terms in the computational formula. Here are several suggestions that may help you to avoid errors.

1. Before calculating r, take a few minutes to draw a scatter diagram. It will tell you what kind of correlation coefficient to expect.
2. Remember that N always refers to the number of pairs of scores, not the total number of scores.

Table 9–9 Worktable for Calculating r for Absenteeism and Sales Using the Computational Formula

Employee	Days Absent (X)	Sales (Y)	XY	X^2	Y^2
1	0	75	0	0	5,625
2	1	65	65	1	4,225
3	3	65	195	9	4,225
4	4	50	200	16	2,500
5	6	60	360	36	3,600
6	7	40	280	49	1,600
7	9	50	450	81	2,500
8	10	35	350	100	1,225
Sum	40	440	1900	292	25,500
	ΣX	ΣY	ΣXY	ΣX^2	ΣY^2

3. The sign of the numerator determines the sign of r. If $N(\Sigma XY) - (\Sigma X)(\Sigma Y)$ is positive, r is positive; if $N(\Sigma XY) - (\Sigma X)(\Sigma Y)$ is negative, r is negative. Be careful not to "lose" the minus sign in the latter case.
4. Both terms in the denominator are always positive. If either of them turns out to be negative, you have made a mistake in your calculations.
5. Don't forget to take the square root of the product of $[N(\Sigma X^2) - (\Sigma X)^2]$ and $[N(\Sigma Y^2) - (\Sigma Y)^2]$ in the denomination before dividing into the numerator.
6. The value of r is always between -1.00 and $+1.00$. If you get a number less than -1.00 or greater than $+1.00$, check your calculations; you have made an error.
7. Check your value of r against the scatter diagram you have drawn. If the computed value does not make sense, check your calculations again. For example, if the points in the scatter diagram are widely scattered about a diagonal from upper left to lower right (weak negative correlation) and your computations result in $r = +.67$, there is something wrong.

EXERCISE 9–4

1. Use the computational formula to calculate the correlation between self-esteem and reading ability.
2. Here are data on the number of symptoms and scores on a depression inventory of 12 depressed patients. Draw a scatter diagram and estimate the probable value of r from the graph. Then calculate r, using the computational formula. Check to make sure that your result is consistent with the scatter diagram.

Patient	Number of Symptoms	Score on a Depression Inventory
1	10	14
2	8	16
3	8	11
4	5	10
5	12	15
6	6	15
7	7	13
8	10	11
9	9	12
10	11	13
11	10	16
12	6	13

Interpreting Correlation Coefficients

What Is a "High" Correlation?

People often ask, "How large must a correlation coefficient be in order to be regarded as high (strong)? Is $r = \pm.70$ strong or is it moderate? Is $r = \pm.35$ moderate or weak?" There is no simple answer. The answer depends on the types of variables being correlated and on the purpose for which r has been computed.

Consider the field of psychological testing as an example. When new tests are developed, or when existing tests are considered for new uses, testers generally calculate two types of correlation coefficients. First, they give the test to a single group of individuals on two occasions and calculate r between the two sets of scores; this r is called a *reliability coefficient*.[†] Second, they calculate the correlation between the test and some other variable, a ***criterion variable***, that the test has been designed to predict, such as performance on a job; this r is called a *validity coefficient*. Reliability coefficients of well-constructed tests tend to be .85 or higher. Therefore, a reliability coefficient of .70 would almost certainly be considered unacceptably low. But a validity coefficient of .70 would be considered high because correlations between tests and real-life criteria are usually lower than this.

In research, too, the same correlation coefficient may be considered high or low, depending on the nature and purpose of the research study. In a large-scale, well-controlled study designed to test a theory that a strong relationship exists between two variables, a correlation of $\pm.35$ would probably be considered too weak to support the theory; on the other hand, researchers doing preliminary studies in a new field where relationships are largely unknown and where many uncontrolled factors may be operating might be delighted with a correlation of $\pm.35$.

Still another factor that influences the interpretation of the strength of a correlation coefficient is sample size, as you will see later in the chapter. Like other statistics, correlation coefficients vary over different samples from the same population. Sample-to-sample variation decreases as sample size increases. Therefore, given a particular value of r, such as .30, we can be more confident that the two variables are truly related (in the population) if $N = 100$ than if $N = 20$.

Factors Affecting the Correlation Coefficient

Several factors affect the size and interpretation of a correlation coefficient. We will discuss three of them: the linearity of the relationship between the variables, the pattern of scatter of points around the diagonal in a scatter diagram, and restriction of range.

Linearity of Relationship

In all the cases of positive and negative correlation considered earlier in this chapter, we noted that the points in the scatter diagrams tended to lie along a *straight line* extending either from lower left to upper right (positive relationship) or from upper left to lower right (negative relationship). When the points in a scatter diagram cluster about a straight line, the relationship is said to be *linear*. The Pearson r is an index of the linear relationship between two variables. If two variables have a nonlinear rather than linear relationship, the Pearson r underestimates the strength of the relationship between the variables.

[†]There are additional procedures for measuring reliability. We are not concerned with them here.

As an example of a nonlinear relationship between two variables, consider the relation between level of anxiety and success in solving difficult problems. People with very low levels of anxiety are not sufficiently motivated to put forth the effort needed to solve the problems and generally do not do well. People with very high anxiety cannot concentrate and generally do not do well. Success is most probable with a moderate level of anxiety—high enough to be motivating but not high enough to interfere with concentration. The relation between anxiety and problem-solving success is shown in Figure 9–7. A relationship like this one is said to be *curvilinear* because a curved line fits the data better than a straight one.

As you can see, the correlation between anxiety and problem-solving success is positive between low and moderate levels of anxiety, but negative between moderate and high levels of anxiety. If you were to calculate r for the data in Figure 9–7, the positive and negative trends would cancel each other out, and r would be about zero. If you simply calculated r without examining the data or drawing a scatter diagram, you might conclude that anxiety and problem-solving success are unrelated. This conclusion would be incorrect. They are, in fact, quite strongly related, but the relationship is not linear. The Pearson r is not an accurate index of the correlation between two variables when the relationship is not linear.

The Pearson r for data having a curvilinear relationship is not necessarily zero. Consider the example in Figure 9–8. Although the relationship between variables X and Y is curvilinear, the points in the scatter diagram do tend to flow from lower left to upper right, as the figure shows. Therefore, r is positive. Nevertheless, the value of r underestimates the strength of the relationship, because r measures the degree to which the points are scattered about the straight line in the figure rather than about the curved line which better fits the trend of the points.

Curvilinear relationships like those in Figures 9–7 and 9–8, although not common, do occasionally happen. You should always examine data carefully and construct a scatter diagram before calculating r, so that your interpretation of r can take curvilinearity, if it occurs, into account. Statistical packages for computers include commands for constructing scatter diagrams. For example, in Minitab, if self-esteem scores are in c2 and reading scores are in c3, the command "plot c3 c2" yields a scatter diagram like that in Figure 9–1.

Figure 9–7

A curvilinear relationship. Pearson $r = 0$.

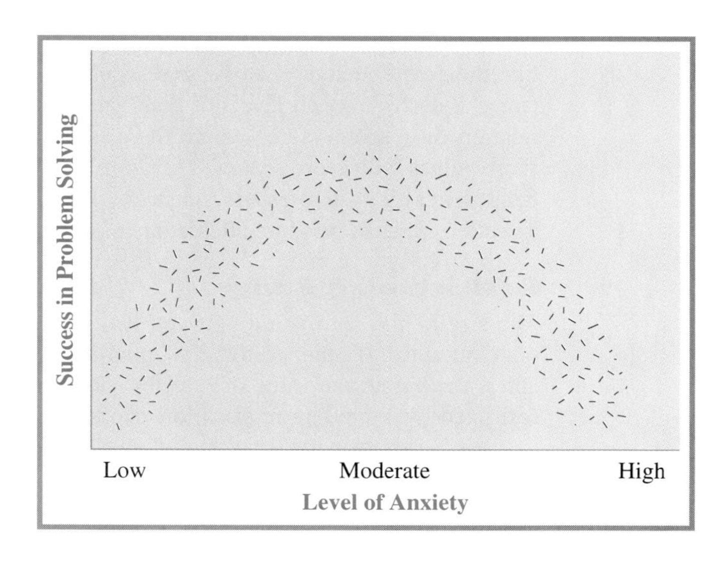

Figure 9–8

A curvilinear relationship. Pearson *r* is positive but underestimates the strength of the relationship.

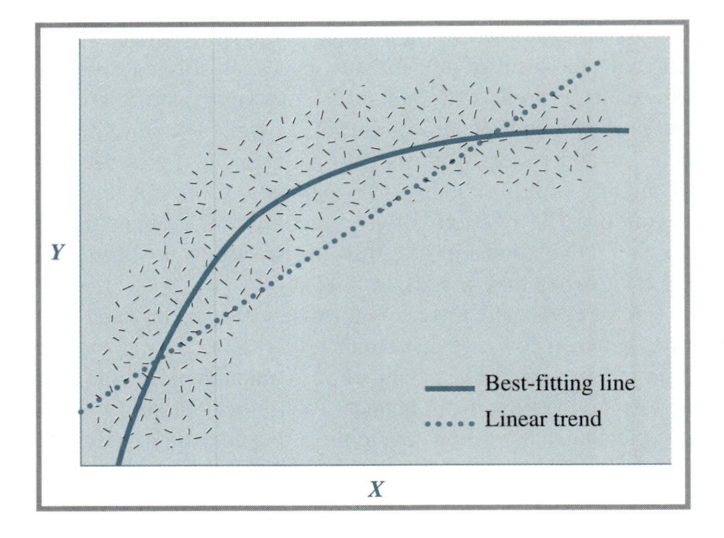

The Pattern of Scatter

In most cases of linear relationship, the points in the scatter diagram tend to fall within an ellipse, as shown in Figure 9–4. Relationships like these are said to display ***homoscedasticity***. This daunting expression is simply a combination of the Latin words *homo*, meaning "same," and *scedasticity*, meaning "scatter." In other words, *homoscedasticity* means that the degree of scatter (variation) of *Y* scores is the same at all values of *X*, and the degree of scatter (variation) of *X* scores is the same at all values of *Y*.

Sometimes, however, the points of a scatter diagram fall within a conelike or pearlike shape, as in Figures 9–9a and 9–9b. These two scatter diagrams display ***heteroscedasticity***. *Hetero* means "different." In other words, the amount of scatter (variation) of *Y* scores differs at different values of *X*, and the amount of scatter (variation) of *X* scores differs at different values of *Y*. In Figure 9–9a, there is little scatter of *Y* scores at low values of *X*, but a large amount at high values of *X*. In Figure 9–9b, scores are more scattered at low values of both variables than at high levels.

When heteroscedasticity exists, the strength of the relationship between the variables differs from one level to another. In Figure 9–9a, for example, the relation between *X* and *Y* is strong at low levels of *X* and *Y* but weak at high levels of *X* and *Y*. In cases like this, the Pearson *r* can be said to measure the *average* strength of the relation between *X* and *Y* over all levels of both variables. But it underestimates the strength of the relation over score values where there is little scatter and overestimates its strength at score values with more scatter. Therefore, *r* is not an accurate index of relationship when heteroscedasticity is present. Again, the importance of examining data and constructing a scatter diagram prior to calculating *r* is clear.

Restriction of Range

The size of the correlation between two variables is affected by the range of scores on both variables. If other factors are equal, the greater the variability of scores on the variables, the higher the value of *r*. If for some reason the range on one or both variables is restricted by excluding high or low scores or both, the value of $|r|$ decreases.

We illustrate with some hypothetical data consisting of scores on tests of analytic reasoning and creativity in 50 high school students. The scores of the students on the two tests are listed in Table 9–10; Figure 9–10 displays a scatter diagram of the data.

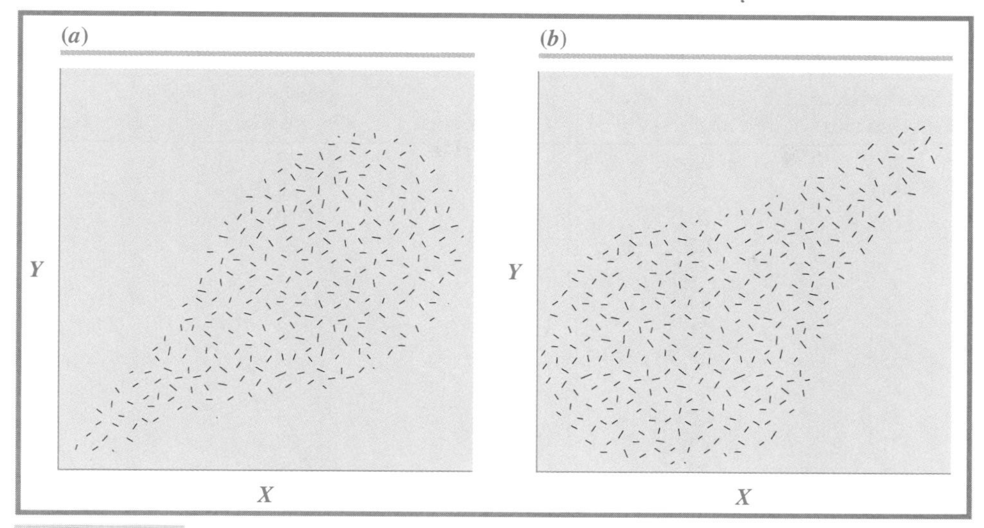

Figure 9–9 **Two scatter diagrams illustrating heteroscedasticity.**

Most points fall within a relatively narrow ellipse extending from the lower left to the upper right of the graph. The correlation is positive and moderately strong. In fact, the value of r is .65.

Now let us examine the correlation between analytic reasoning and creativity just in the subgroup of students scoring 27 or less on the analytic reasoning test. We are restricting the range of analytic reasoning scores to 21 to 27, as compared to the original range of 21 to 34. In the scatter diagram of Figure 9–10, a vertical line divides the points representing students with analytic reasoning scores of 27 or less from those of students with scores above 27. Use a piece of paper or your hand to cover the part of the diagram to the right of the vertical line, and examine the distribution of points to the left of the line. Note that, though the points tend to fall within an ellipse from lower left to upper right, the ellipse is fatter relative to its length than the ellipse enclosing the entire data set. Therefore, r is weaker. In fact, the correlation between analytic reasoning skills and creativity in the group of students scoring 27 or less on the analytical reasoning test is only .49.

You may be thinking, "Well, the correlation is lower in this subgroup of students because of their low scores on the analytic reasoning test." This is not so. *Correlation coefficients are affected by the range of scores but not by their average value.* To demonstrate, let us calculate the correlation between analytic reasoning and creativity for just the subgroup of students with analytic reasoning scores above 27—a group whose average score on the reasoning test is high, Again, we find that r is smaller than .64; it is .51. Again, because of the restricted *range* of scores, the correlation coefficient is reduced below the value of r over the entire range.

In basic and applied research, correlation coefficients are often computed in groups whose range on one or both variables is restricted; consequently, r is relatively weak. For example, the correlation between SAT scores and grade point average (GPA) at some selective colleges is only about .20. Does this mean that there is little relationship between SAT scores and college success? Not necessarily. Since SAT scores were probably one requirement of admission, the range of SAT scores at selective colleges is relatively narrow. The restriction of range of SAT scores lowers the correlation with GPA below what it would be if the SAT score range were wider. If these colleges were to

Table 9–10 Scores of 50 High School Students on Tests of Analytic Reasoning and Creativity

Student	Analytic Reasoning	Creativity	Student	Analytic Reasoning	Creativity
1	30	25	26	32	27
2	28	19	27	28	22
3	24	20	28	30	21
4	29	22	29	22	20
5	23	17	30	28	20
6	26	24	31	24	16
7	32	21	32	33	24
8	23	23	33	24	22
9	26	20	34	31	26
10	34	26	35	24	20
11	31	24	36	27	20
12	31	22	37	23	19
13	26	22	38	29	24
14	21	16	39	25	17
15	32	23	40	28	24
16	32	26	41	30	22
17	24	18	42	26	20
18	31	24	43	22	17
19	30	19	44	27	24
20	27	21	45	29	26
21	22	18	46	27	18
22	27	24	47	25	19
23	30	27	48	34	27
24	25	19	49	25	21
25	26	18	50	29	20

Figure 9–10

Scatter diagram of the relation between analytic reasoning and creativity. A vertical line divides scores of 27 or less from scores above 27 on analytic reasoning.

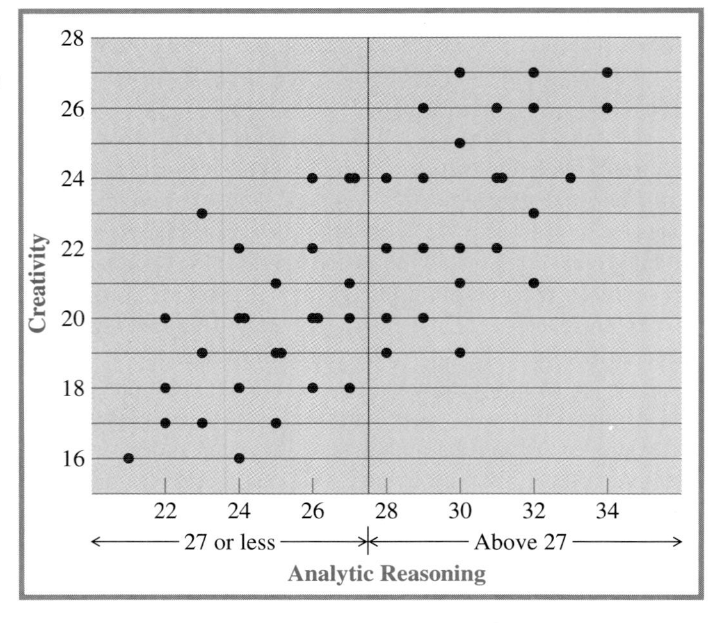

admit students with a wider range of SAT scores, the correlation between SAT scores and GPA would increase.

By the same token, it is impossible to compare correlations between the same two variables computed in two groups, unless both groups are about equally variable. Suppose, for example, that two investigators have studied the relationship between reading and mathematics test scores. One has tested children in a large, consolidated school serving diverse communities and calculated $r = .65$. The other has tested children in a private school, whose students had to pass difficult reading and math exams before being admitted, and has obtained $r = .30$. The first investigator argues that there is a moderately strong positive relationship between reading and math scores; the second maintains that the relationship is weak. Actually, they are both right—for the particular populations with which they are dealing.

The moral is: When interpreting a correlation coefficient, always take the range of values of both variables into account. Remember that the correlation would increase with increased range and decrease with decreased range. When collecting data for the purpose of computing r and describing the relationship between two variables, make certain that the range of scores on both variables is appropriate for the decisions to be made and conclusions to be drawn.

EXERCISE 9–5

1. A human resources manager gives two tests of tool use to a group of industrial workers. The two tests are the same length, and both have a mean of 25. He notices that workers with low scores on Test A receive a score on Test B that is within 2 or 3 points of their Test A score. However, workers with high scores on Test A have scores on Test B ranging from 10 points below to 10 points above their Test A score. The correlation between the Test A and B scores is .65.
   a. Draw a scatter diagram showing the relationship between scores on Tests A and B.
   b. Comment on the interpretation of r in a case like this.
2. You have computed the correlation between manual dexterity and eye-hand coordination in a group of high school freshmen and in a group of master carpenters. Which r do you expect to be higher? Why?
3. Joe Blow found the correlation between height and weight to be .87 in a group of children ranging in age from 6 to 13 years. But John Doe got a correlation between height and weight of only .43 in a group of fourth graders. How do you account for the difference?
4. From your reading or your own experience, give examples of pairs of variables that are likely to have a curvilinear relationship with one another. Draw scatter diagrams showing what you think the relationships would look like.

Correlation and Causal Relationships

When one has obtained a substantial correlation between two variables, it is tempting to conclude that one variable caused the other. This is not necessarily true. A correlation coefficient different from zero tells us that two variables are related. But it does not tell us why they are related.

Consider Mr. Snyder's correlation, .43, between self-esteem and reading ability. Let us suppose that this correlation coefficient was calculated on 80 students rather than 8, so that we can be more confident that there truly is a positive correlation between the two variables. Mr. Snyder may interpret the correlation to mean that self-esteem is a causal factor in reading ability, concluding that children who are good readers owe their reading skill to their high self-esteem, whereas poor readers have failed to acquire these skills because of their low self-esteem. If we use an arrow to indicate causality, Mr. Snyder's conclusion can be diagrammed like this

Self-esteem → Reading ability

Perhaps, however, Mr. Snyder has it backward. The correlation may be due to the fact that reading ability influences self-esteem. Specifically, reading well may cause children to think well of themselves, whereas reading poorly may cause children to think poorly of themselves.

Reading ability → Self-esteem

There is still another possibility. Perhaps there is no direct causal relation between self-esteem and reading ability at all. Instead, there may be a third variable, unmeasured in Mr. Snyder's study, that influences both self-esteem and reading ability. As a result, they vary together, even though neither one of them causes the other. An example of such a third variable might be parents' use of praise. Perhaps frequent praise by parents both produces high self-esteem in children and motivates the children to learn to read well, whereas infrequent praise by parents both produces low self-esteem in children and lowers their motivation to learn to read.

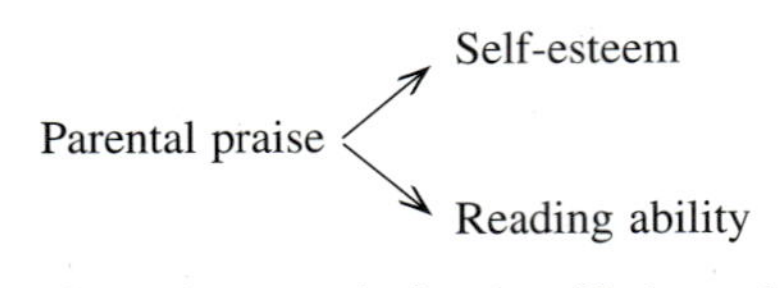

Based on the correlation coefficient alone, we do not know which of the three possibilities is the correct explanation of the relationship.

In summary, if a correlation exists between two variables, X and Y, we know that the variables are related, but we do not know why. It is possible that X causes Y. It is possible that Y causes X. It is possible that another variable or set of variables causes both X and Y.

You may recall that, in the early years of research on the relationship between cigarette smoking and lung cancer, many smokers continued to insist that there was no conclusive evidence that smoking *causes* lung cancer, despite increasing correlational data. And the smokers were correct. In the years since then, of course, other types of evidence of the link between smoking and lung cancer have surfaced, and now medical professionals (and even heavy smokers) are convinced that smoking is a causal factor in lung cancer. Correlation coefficients alone, however, no matter how strong they may be, are not sufficient to establish a causal relationship such as this.

The Correlation Coefficient and Components of Variance

Because a correlation coefficient looks so much like a proportion, people sometimes interpret it as one. Thus, one sometimes hears statements such as "The correlation between X and Y is .50. Therefore, there is a 50% relationship between X and Y." This is just plain wrong! A correlation coefficient is in no sense a proportion.

However, the square of a correlation coefficient, r^2, can be interpreted as a proportion of sorts; specifically, *r^2 is the proportion of each variable's variance that is accounted for by differences on the other variable.*†

†"Accounted for" means accounted for statistically; it does not mean "caused."

To understand what this means, let us review the concept of ***components of variance***, introduced in Chapter 6. Recall that the variance of a set of scores can sometimes be apportioned into components, each representing one factor that accounts for individual differences on that variable. Correlational analysis provides one means of apportioning the variance of two variables, X and Y, into components.

Consider the example of self-esteem and reading ability. The two variables are related in this group of children ($r = .43$). Some of the self-esteem variance is covariance with reading ability, and some of the reading ability variance is covariance with self-esteem. Yet, because the correlation between them is not +1, part of the variance of each variable is independent of the other. What proportion of each variable's variance is covariance with the other variable? The answer is $(.43)^2$, or .18. Eighteen percent of the self-esteem variance in this group of children is accounted for by their differences in reading ability, and 18 percent of their reading ability variance is accounted for by differences in self-esteem. The remaining 82 percent of the variance of each variable is independent of the other variable.

For the variables absenteeism and sales, we calculated $r = -.87$. Since $r^2 = (-.87)^2 = .76$, we can say that 76 percent of the sales variance is accounted for by absenteeism and 24 percent by other factors.

Unless $r = \pm 1$, the square of r is always less than $|r|$. Therefore, correlation coefficients must be strong to account for a substantial proportion of variance. For example, a correlation coefficient of at least $\pm.71$ is required to account for half the variance of each variable. We get a better (and more sobering) picture of the true strength of relationship between two variables by examining the squared correlation coefficient than by examining the correlation coefficient itself.

Because r^2 is equal to the proportion of variance of either variable that is accounted for by the other variable, r^2 is called the ***coefficient of determination***. Similarly, $1 - r^2$, the proportion of variance unaccounted for, is called the ***coefficient of nondetermination***. Thus, if $r = .8$, the coefficient of determination is $(.8)^2 = .64$, and the coefficient of nondetermination is $1 - (.8)^2 = 1 - .64 = .36$.

EXERCISE 9–6

1. Dr. Bosworth, who collected the data on absenteeism and sales, argues that the strong negative correlation proves that missing many days of work causes sales to decrease. What do you think about this argument?
2. A psychologist asked 50 fifth-grade boys what TV programs they watched regularly and counted the number of violent programs watched per week by each boy. The boys' teachers rated their level of aggressiveness on a scale from 1 (totally nonaggressive) to 10 (extremely aggressive). The correlation between number of violent programs watched and rated aggressiveness was .55.
   a. Can the psychologist conclude that watching violent TV programs causes boys to behave aggressively?
   b. Give two other possible explanations for the positive correlation between number of violent programs watched and aggressiveness.
   c. What proportion of the boys' aggressiveness variance is accounted for by the number of violent TV programs watched? What proportion of the boys' aggressiveness variance is not accounted for by the number of violent TV programs watched? What is the coefficient of determination for these data? What is the coefficient of nondetermination?

EXERCISE 9–6

3. Sally calculated the following two correlation coefficients:

$$r_{AB} = .40$$
$$r_{CD} = .80$$

She reports: "The correlation between C and D is twice as strong as that between A and B." Comment on Sally's statement.

4. What correlation coefficient is needed between variables X and Y to conclude that 30% of the X variance is accounted for by Y?

Drawing Conclusions About Population Correlation Coefficients From Sample Data

When we calculate a correlation coefficient between two variables, X and Y, we usually do so not simply to describe the relationship between X and Y in the sample but to obtain information about the relationship between X and Y in the population from which the sample came. We will refer to a population correlation coefficient as ρ (rho, the Greek letter r). Thus, we use the sample r to make inferences about ρ.

Like any other statistic, r varies from one sample to another. If we wish to draw conclusions about the correlation between variables X and Y in the population, we need to know something about how r varies from one sample to another, and how r is related to ρ. That is, we need to consider the *sampling distribution of the correlation coefficient.*

The Sampling Distribution of the Correlation Coefficient

Recall from earlier chapters that a sampling distribution is a distribution of a statistic over all possible samples of a given size from a population. For example, to obtain a sampling distribution of the correlation coefficient for samples of size 25, we would (1) draw (or list) all possible samples of size 25 from the population, and (2) calculate r in each sample. The distribution of the values of r would be the sampling distribution of r for samples of size 25 from this population.

Fortunately, we do not need to draw all possible samples of size N in order to derive the properties of the sampling distribution of r. It can be shown that, if both X and Y are normally distributed in the population, the properties of the sampling distribution of r are determined by just two factors, the value of the population correlation coefficient, ρ, and sample size, N.

First let us consider the effect of the population correlation coefficient, ρ, on the sampling distribution of r. If $\rho = 0$ (i.e., if the two variables are uncorrelated in the population), sample correlation coefficients vary around zero and are equally likely to be positive or negative. The sampling distribution of r is close to normal and has an average value equal to zero, as Figure 9–11 shows.

If the correlation between X and Y in the population is not equal to zero, the sampling distribution of r is skewed. Suppose, for example, that the population correlation, ρ, is equal to .85, a strong positive correlation. Sample values of r will vary around .85. However, sample correlation coefficients can vary from .85 in the positive direction only up to +1; the possible range of variation in the negative direction is much greater. Consequently, the sampling distribution of r is negatively skewed when $\rho = .85$, as Figure 9–12 shows.

Figure 9–11

A sampling distribution of r when $\rho = 0$.

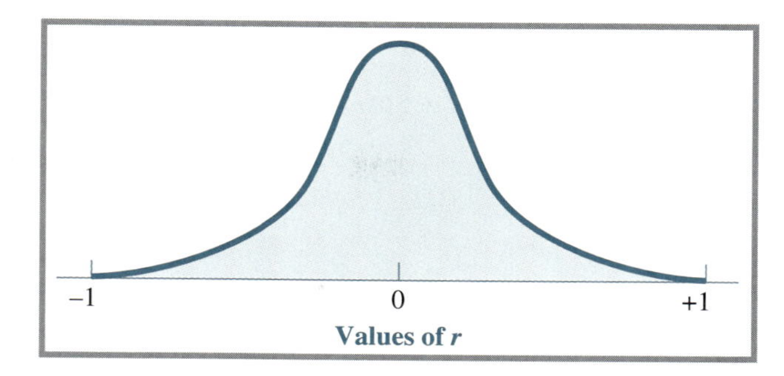

Figure 9–12

A sampling distribution of r when $\rho = .85$.

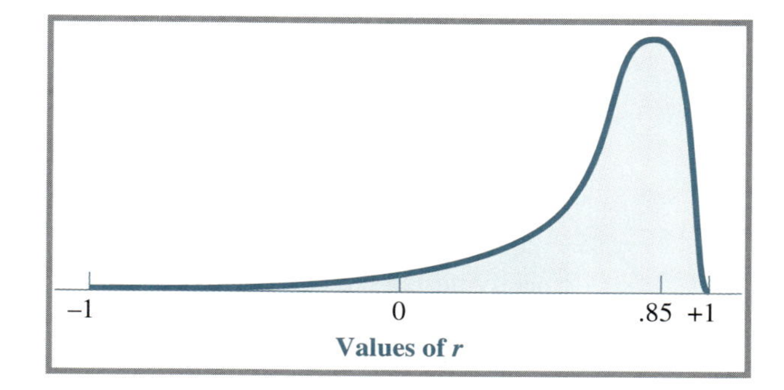

Figure 9–13

A sampling distribution of r when $\rho = -.50$.

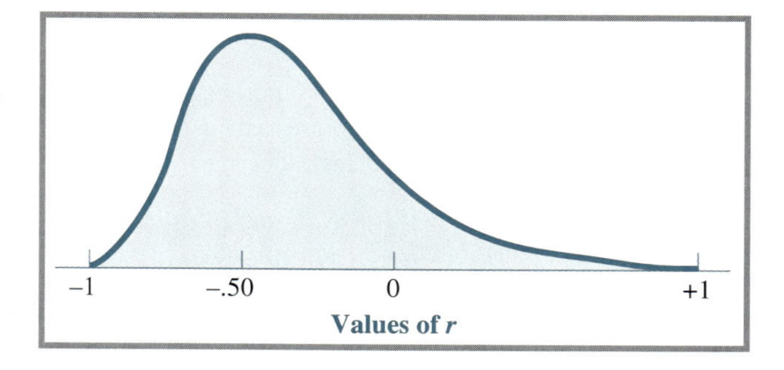

By the same token, if ρ is negative, the sampling distribution of r is positively skewed because the range of possible variation of r is greater in the positive than the negative direction. Figure 9–13 shows a sampling distribution of r if $\rho = -.50$.

The second factor that influences the sampling distribution of r is sample size N. As with other statistics, the larger the sample size, the less r varies from one sample to another. If $\rho = 0$, for example, values of r from samples of size 10 vary more widely around zero than values of r from samples of size 50, as Figure 9–14 shows. For example, even though the variables are uncorrelated in the population, sample correlation coefficients greater than .5 are not unusual in samples of size 10. But sample correlation coefficients greater than .5 are rare in samples of size 50.

Testing Hypotheses About Population Correlation Coefficients

Tests of hypotheses about population correlation coefficients rest on the kinds of sampling distributions of r discussed in the previous section. In theory, it is possible to hypothesize that ρ is any value, from -1 to $+1$. In practice, however, tests about population

Figure 9–14

Sampling distributions of r when $\rho = 0$ for $N = 10$ and $N = 50$.

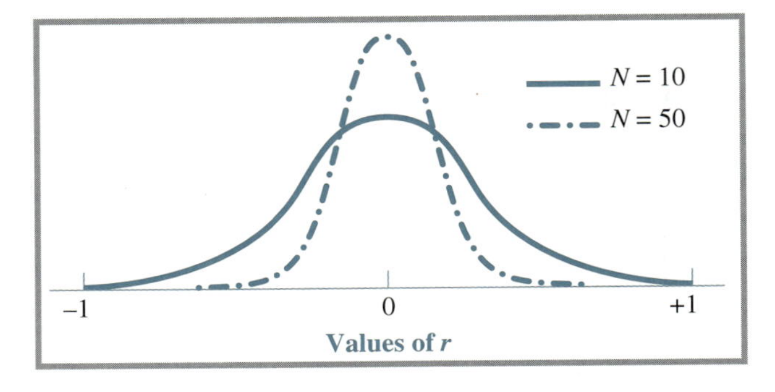

correlation coefficients almost always test the null hypothesis (H_o) that X and Y are *unrelated*, that is, that $\rho = 0$. If the test results in rejection of H_o, the researcher concludes that the two variables are related in the population.

Like tests of hypotheses about population means, tests of H_o: $\rho = 0$ can be one-tailed or two-tailed. If the research hypothesis states that X and Y are related, without specifying the direction of relationship, or if the researcher is interested in detecting any kind of relationship between X and Y, a two-tailed test is called for. The null hypothesis (H_o) and alternative hypothesis (H_A) are

H_o: $\rho = 0$ *Null and alternative hypotheses*
H_A: $\rho \neq 0$ *for a two-tailed test*

If the research hypothesis states that X and Y are positively related, or if the researcher is only interested in detecting a positive correlation, the appropriate test is an upper one-tailed test. The null and alternative hypotheses are

H_o: $\rho \leq 0$ *Null and alternative hypotheses*
H_A: $\rho > 0$ *for an upper one-tailed test*

If the research hypothesis states that X and Y are negatively related, or if the researcher is interested only in detecting a negative correlation, the appropriate test is a lower one-tailed test. The null and alternative hypotheses are

H_o: $\rho \geq 0$ *Null and alternative hypotheses*
H_A: $\rho < 0$ *for a lower one-tailed test*

The logic and steps in testing hypotheses about ρ are the same as in any hypothesis test. After H_o and H_A have been stated (Step 1), the remaining steps are as follows:

2. A level of significance (α) is chosen.
3. A sample (ideally, a random sample) is selected from the population, and the correlation between variables X and Y is calculated. The test statistic is then calculated.
4. The calculated value of the test statistic is compared with one or more critical values to determine whether the probability of a sample r like the one obtained is less than or greater than α. If the probability (p) is less than α, H_o is rejected; the correlation between the variables is significant at level alpha, and we conclude that the variables are related in the population. If the probability is greater than α, H_o is not rejected; the correlation is not significant at level alpha, and we conclude that the variables may be unrelated in the population.

What is the test statistic in tests of hypotheses about the correlation coefficient, and how are critical values found (Steps 3 and 4)? There are two alternative methods of conducting the test. One method involves the calculation of t from sample data and comparison of the calculated t with critical values of t. The other method bypasses the calculation of t and permits the direct comparison of the sample r with critical values of r.

The *t* Test

If X and Y are normally distributed in the population, and if H_o is true, the ratio $r\sqrt{N-2}/\sqrt{1-r^2}$, calculated in all possible samples of size N, is distributed as t with $N-2$ degrees of freedom (df). Therefore, the test statistic in the t test for the significance of r is

$$t = \frac{r\sqrt{N-2}}{\sqrt{1-r^2}}$$

The test statistic in the t test of H_o: $\rho = 0$

After t has been calculated using this formula, it is compared with critical values of t for $N-2$ df. Critical values of t for one-tailed and two-tailed tests are found in Table D–3 in Appendix D.

Here is an example of a t test of a hypothesis about a population correlation coefficient.

According to a certain personality theory, anxiety (A) and extraversion (E) are negatively correlated in the adult population. Assume that both A and E are normally distributed in the population. A psychologist measures A and E in a sample of 25 adults and obtains $r = -.32$. Do the results support the theory? Use $\alpha = .05$.

Since the theory specifies a negative correlation, a lower one-tailed test is called for. We show all the steps in the test.

H_o: $\rho \geq 0$

H_A: $\rho < 0$

$\alpha = .05$

$r = -.32$ and $N = 25$

$$t = \frac{r\sqrt{N-2}}{\sqrt{1-r^2}} = \frac{-.32\sqrt{25-2}}{\sqrt{1-(-.32)^2}} = -1.62$$

Since $N = 25$, $df = 25 - 2 = 23$. Referring to the row for 23 degrees of freedom in the t table, we find that $|-1.62|$ falls between the critical values 1.319 with $p = .10$ and 1.714 with $p = .05$. Since $p > .05$, H_o cannot be rejected; the correlation between A and E is not a significant negative correlation. We do not have enough evidence to conclude that anxiety and extraversion are negatively correlated in the population.

EXERCISE 9–7

If the sample size were 50 instead of 25, with the same value of r as in the example just given, would your conclusion be the same as in the example? Conduct a test to answer this question.

Using a Table of Critical Values of *r*

A simpler way to conduct tests of the significance of r is to use a table of critical values of r. In this version of the test, the relevant probability distribution is the sampling distribution of r. The sample correlation coefficient, r, is the test statistic and can be compared, without further calculation, to critical values of r for $df = N - 2$. As in the t test, it is assumed that X and Y are normally distributed in the population.

Table D–4 in Appendix D lists critical values of r for one-tailed and two-tailed tests with df ranging from 1 to 100. The use of the table is similar to the use of the t table. In the row for $N - 2$ df, one finds the two critical values that bracket the calculated $|r|$ and finds their probabilities at the top of the table (in the upper row of p values if the test is one-tailed or the lower row if the test is two-tailed). The probability of $|r|$ as far from zero as the calculated r falls between those values of p.

Here is an example of a test of the significance of a correlation coefficient using the table of critical values of r.

> A correlation coefficient of .48 has been calculated in a sample of size 18. Is the correlation significantly different from zero at the .05 level?
>
> The question implies an interest in any kind of difference from zero. Therefore, a two-tailed test is required. First, we state the null and alternative hypotheses and choose a level of significance, α.
>
> $$H_o: \rho = 0$$
> $$H_A: \rho \neq 0$$
> $$\alpha = .05$$
>
> Since $N = 18$, $df = 16$. Referring to the row for $df = 16$ of Table D–4, we find that $r = .48$ falls between .468 with $p = .05$ and .542 with $p = .02$. Since $p < .05$, H_o is rejected. The sample correlation coefficient is significantly different from zero at the .05 level. We conclude that the two variables are related in the population.

In the previous section, we conducted a t test to test the hypothesis that anxiety and extraversion are negatively related. Now let us use the table of critical values of r to conduct the same test. Recall that the sample r was $-.32$, and N was 25. Once again, we conduct a lower one-tailed test with $\alpha = .05$.

> $$H_o: \rho \geq 0$$
> $$H_A: \rho < 0$$
> $$\alpha = .05$$
>
> Since $N = 25$, $df = 23$. According to the row with 23 df in the table of critical values of r, $|-.32|$ falls between the critical values .265 with $p = .10$ and .337 with $p = .05$. Since $p > .05$, we do not reject H_o. The sample correlation, although negative, is not strong enough to conclude that anxiety and extraversion are negatively related in the population.

As you can see, we reached the same conclusion using the t test and the table of critical values of r. This is always the case if both methods are applied to the same data, because they are simply computational variations of a single test.

Statistical and Nonstatistical Significance: The Effect of Sample Size

Let us examine the information in the table of critical values of r more closely. By looking down any column of the table, you can see that, as df increases, the critical value of r decreases. For example, with a sample size of 10 ($df = 8$), the critical value of r for $\alpha = .05$ in a two-tailed test is $\pm.632$; only sample correlations stronger than $\pm .632$ are significantly different from zero at the .05 level. With a sample of size 30 ($df = 28$), the critical value is $\pm .361$, and sample correlations stronger than $\pm .361$ are significant at the .05 level. Thus, for example, with $\alpha = .05$, a correlation of $-.5$ is not significant in a sample of size 10 but is significant in a sample of size 30.

With sufficiently large sample sizes, even very weak sample correlation coefficients may be significantly different from zero. Suppose, for example, that you have calculated $r = .20$ in a sample of size 102. Now you conduct a two-tailed test to find out whether r differs significantly from zero. Since r falls between the critical values for $p = .05$ and .02 (with 100 *df*), the correlation is significantly different from zero at the .05 level. However, the relationship may be too weak to be of theoretical or practical importance. For example, only $(.20)^2$ or 4 percent of the variance of Y is accounted for by X.

The moral is: In drawing conclusions about a population correlation coefficient from a sample, one must, of course, determine whether the sample r is significantly different from zero at level α. But the size of r should also be considered. For example, although correlations of .65 and .20 in samples of size 100 are both significant at the .05 level, we would regard the correlation of .65 as, potentially, more important, because it suggests that the relation between the variables may be quite strong.

This discussion is intended to illustrate the distinction between the statistical significance and the importance of a test result. Sometimes, especially when sample sizes are very large, results are obtained that are statistically significant but essentially meaningless. To avoid this, researchers planning correlational studies often ask themselves, before collecting data, "What is the minimum correlation between variables X and Y that would be theoretically or practically important?" Then they select a sample size large enough so that the minimally important r will be significant at the chosen alpha but not so large that much weaker values of r will also be significant.

As an example, suppose that you are planning to conduct an upper one-tailed test with $\alpha = .05$, and you have decided that the minimum correlation worth detecting is .35. You can use the table of critical values of r to help decide what sample size to use. If you go down the column labeled "Level of Significance for One-tailed Test .0500," you will find that the smallest number of *df* for which $r = .35$ is significant is $df = 22$, with critical value = .344. That is, if you obtain $r \geq .35$ in a sample of size 24, you will be able to conclude that your two variables are positively correlated. In samples smaller than size 24, a sample correlation coefficient of .35 is not significant at the .05 level. Therefore, you are well advised to use a sample of at least size 25 or 30 in your study. On the other hand, there is no need to use a sample of size 100 or more, in which even much smaller values of r than .35 will be significantly greater than zero.

EXERCISE 9–8

1. a. Use the table of critical values of r to find out whether Mr. Snyder's correlation coefficient of .43 between self-esteem and reading ability of 8 children is significantly greater than zero at the .05 level.
   b. What is the smallest sample size for which a correlation of .43 is significantly greater than zero at the .05 level?
2. A social researcher wonders whether women's socioeconomic status (SES) and attitudes toward abortion are related. She measures SES and attitudes toward abortion in 30 women and obtains $r = -.42$.
   a. If high attitude scores indicate more favorable attitudes toward abortion, are low-SES women more or less likely to be pro-choice than high-SES women?
   b. Is a two-tailed test or a lower one-tailed test called for?
   c. Conduct the appropriate test. Use $\alpha = .05$.
   d. Is SES significantly related to attitudes toward abortion at the .05 level?

EXERCISE 9–8

3. What is the minimum sample size needed to reject the null hypothesis in each of the following?
   a. Two-tailed test with $\alpha = .05$; r is expected to be about $\pm.30$.
   b. Upper one-tailed test with $\alpha = .01$; r is expected to be about .40.
   c. Lower one-tailed test with $\alpha = .05$; r is expected to be about $-.25$.

Other Types of Correlation Coefficients

Pearson's r is an index of the linear relationship between two quantitative variables, both of which include a range of values. Sometimes we want to know the correlation between two sets of ranks, or between a dichotomous (two-category) variable and a quantitative variable, or between two dichotomous variables. We discuss procedures for calculating these three types of correlation coefficients in this section.

The Spearman Rank-Order Correlation Coefficient

Suppose that five applicants for a job have been interviewed by two persons, Judges X and Y. Each judge has ranked the five applicants from 1 (best) to 5 (worst). Now the judges want to know how well their rankings agree. Do Judges X and Y consistently rank the same applicants as good and poor (a positive correlation)? Or is there little relation between their rankings (a correlation close to zero)? Or—they fervently hope not—are applicants ranked as good by Judge X usually ranked as poor by the Judge Y, and vice versa (a negative correlation)? Table 9–11 shows the rankings of the two judges.

Many years ago, a British statistician named Spearman developed procedures for calculating correlation coefficients for ranked data. Consequently, the correlation coefficient is referred to as the ***Spearman rank-order correlation coefficient, r_s***. The formula for the Spearman r_s is

$$r_s = 1 - \frac{6\Sigma(R_x - R_y)^2}{N(N^2 - 1)}$$

The Spearman rank-order correlation coefficient

N represents the number of individuals being ranked (in the example, $N = 5$). R_x is the rank of each individual on variable X (in this case, Judge X), while R_y is the rank on variable Y (Judge Y in the example). To obtain $\Sigma(R_x - R_y)^2$:

1. Subtract R_y from R_x for each individual.
2. Square the difference to get $(R_x - R_y)^2$.
3. Add the $(R_x - R_y)^2$ values to get $\Sigma(R_x - R_y)^2$.

The calculation of $\Sigma(R_x - R_y)^2$ is shown in Table 9–11; $\Sigma(R_x - R_y)^2 = 6$. Then

$$r_s = 1 - \frac{6(6)}{5(5^2 - 1)} = 1 - \frac{36}{120} = 1 - .30 = .70$$

Apparently, the judges agreed reasonably well in their rankings of the five applicants.

When the number of cases is small, the formula for r_s is easily used to calculate a correlation coefficient between two sets of ranks. If, however, the data set is large, and especially if a computer or statistical calculator is available, it is better to treat the ranks

Table 9–11 **Worktable for Calculating the Spearman Rank-Order Correlation Coefficient (r_s)**

Job Applicant	Rank Assigned by Judge X (R_x)	Rank Assigned by Judge Y (R_y)	$R_x - R_y$	$(R_x - R_y)^2$
Ann	3	2	1	1
Bob	1	3	−2	4
Chris	4	4	0	0
Dave	5	5	0	0
Edith	2	1	1	1
				6 $\Sigma(R_x - R_y)^2$

as scores and to calculate the Pearson r, because the Spearman formula is simply a shortcut computational method for calculating the Pearson r when the data are in the form of ranks. If, for example, you apply the computational formula for Pearson's r to the data of Table 9–11, you will get $r = .70$.

One more item of information before we leave Spearman's r_S: Ties sometimes occur when data are ranked. For example, in a contest, two persons may both come in second. The question then becomes: What ranks should be given to the tied cases, as well as to the remaining members of the group? The following reasoning is used: If these two persons were not tied, they would have ranks of 2 and 3. Therefore, each of them is given the rank 2.5, which is the average of 2 and 3. The next lower person then receives rank = 4, and so on. Here is an example that includes two sets of tied ranks in a group of eight people.

Highest	Joe	Sue	Carol	Ted	Art	Mary	Nan	Tom	Lowest
		tied			tied				
Rank	1	2.5	2.5	4	6	6	6	8	

When there are tied ranks on either variable X or Y, the values of r_s (calculated by Spearman's formula) and the Pearson r may be slightly different. The Pearson r is generally considered a more accurate index of the relationship under these circumstances (Howell, 1992).

The Point Biserial Correlation Coefficient

Sometimes, we want to measure the correlation between a continuous variable and a variable that is dichotomous, that is, takes only two values. The dichotomous variable may be a qualitative categorical variable, such as gender (male, female), or it may be quantitative, such as scores on one item of a test (correct = 1 point, incorrect = 0 points). For example, we may be interested in the correlation between gender and anxiety scale scores, or between scores on one item of a test and scores on the test as a whole. The correlation coefficient between a dichotomous variable and a continuous one is called a ***point biserial correlation coefficient,*** $\boldsymbol{r_{pb}}$.

As an example, consider data from 10 students who have taken a 25-item test. We are interested in the correlation between their scores on Item 1 and the total test. Table 9–12(a) lists the paired item (X) and test (Y) scores of the 10 students. Item scores are 1 and 0; test scores range from 12 to 25.

The first step in calculating r_{pb} is to rearrange the data by listing the test scores of individuals with each item score, as shown in Table 9–12(b). The test scores of the six

Table 9–12 Paired Scores of 10 Students on One Item of a Test and on the Total Test

(a)

Student	Item Score (X)	Test Score (Y)
1	1	21
2	0	23
3	1	25
4	1	18
5	1	20
6	0	14
7	0	15
8	1	15
9	0	20
10	1	24

(b)

	Item Score X = 1	Item Score X = 0
Test Scores	21	23
	25	14
	18	15
	20	20
	15	
	24	
Sum	123	72
Mean	20.5	18.0

students who answered Item 1 correctly are listed under "$X = 1$," while the test scores of the four students who answered Item 1 incorrectly are listed under "$X = 0$." Note that the mean test score of students who answered the item correctly is higher than that of students who answered the item incorrectly. In other words, higher item scores tend to be associated with higher test scores and lower item scores with lower test scores. We know, therefore, that r_{pb} will be positive.

The formula for calculating r_{pb} is

$$r_{pb} = \left(\frac{M_{y\cdot 1} - M_{y\cdot 0}}{SD_y}\right)\sqrt{p(1-p)}$$

where

$M_{y\cdot 1}$ = mean of Y for the subgroup with $X = 1$
$M_{y\cdot 0}$ = mean of Y for the subgroup with $X = 0$
SD_y = standard deviation of all the Y scores
p = proportion of the entire group with $X = 1$

For the item-test score data, $M_{y\cdot 1} = 20.5$, $M_{y\cdot 0} = 18.0$, SD_y (the standard deviation of the test scores) = 3.72, and p (the proportion of the group answering the item correctly) = 6/10 = .6. Therefore,

$$r_{pb} = \left(\frac{20.5 - 18.0}{3.72}\right)\sqrt{(.6)(.4)} = .33$$

The correlation between scores on Item 1 and on the total test is .33.

Item scores, though they take only two values, are quantitative. The point biserial correlation coefficient can also be calculated if the dichotomous variable is qualitative (e.g., gender). One simply codes one value of the variable as 1 and the other as 0 and applies the method just demonstrated. The sign of r_{pb} in this case depends on which value was coded as 1 and which one as 0. For example, suppose that, in a study on the relation between gender and scores on a test, you code male as 1 and female as 0 and calculate $r_{pb} = .6$. If you were to code male as 0 and female as 1 instead, r_{pb} would be $-.6$. Since either coding of gender is equally correct, the sign of r_{pb} is meaningless, and r_{pb} gives information only about the strength of the relationship between the variables when the dichotomous variable is qualitative.

Despite the special formula for calculation, r_{pb}, like r_s, is a Pearson correlation coefficient in disguise. If you apply the formula for Pearson r to the data in Table 9–12(a), you will get $r = .33$. The advantage of the special formula is ease of calculation by hand; statistical calculators and computers use the Pearson r method for calculating r_{pb}.

The Phi Coefficient

Sometimes we want to measure the correlation between two dichotomous categorical variables, such as between gender (male, female) and smoking (smoker, non-smoker), or between scores on two items of a test. The relevant correlation coefficient is the ***phi coefficient,* ϕ**. To describe and illustrate the calculation, we again use data from the 10 students who took a 25-item test. This time we are interested in the correlation between their performance on Items 1 and 2. The students' scores on the two items are displayed in Table 9–13(a).

The first step is to rearrange the data in a 2 × 2 contingency table, as shown in Table 9–13(b). Each cell of the table gives the number of students with one particular combination of scores on the two items. For example, the number of students with a score of 1 on Item 1 and 0 on Item 2 is two. As you can see, most students with high scores (1) on Item 1 had high scores on Item 2 as well, and most students with low scores (0) on Item 1 also scored low on Item 2. Therefore, we know that the correlation is positive.

To calculate the phi coefficient, we label the entries in the contingency table like this:

a	*b*
c	*d*

For example, in Table 9–13(b), $a = 4$, $b = 2$, $c = 1$, and $d = 3$. The formula for calculating the phi coefficient is

$$\phi = \frac{ad - bc}{\sqrt{(a+b)(c+d)(a+c)(b+d)}}$$

In the example,

$$\phi = \frac{(4)(3)-(2)(1)}{\sqrt{(4+2)(1+3)(4+1)(2+3)}} = \frac{12-2}{\sqrt{(6)(4)(5)(5)}} = \frac{10}{\sqrt{600}} = \frac{10}{24.5} = .41$$

Scores on the two items have a moderate positive correlation.

The phi coefficient can also be calculated for data sets in which one or both dichotomous variables are qualitative rather than quantitative. In such cases, the sign of ϕ depends on how data are arranged in the contingency table. Consider, for example, a study of the relationship between children's gender and their choice of toys, which we examined in Chapter 5. Two arrangements of the results are shown in Table 9–14. The phi coefficient calculated from Table 9–14(a) is

$$\phi = \frac{(25)(20)-(9)(17)}{\sqrt{(25+17)(9+20)(25+9)(17+20)}} = \frac{+347}{1237.8} = .28$$

The phi coefficient calculated from Table 9–14(b) is

$$\phi = \frac{(17)(9)-(20)(25)}{\sqrt{(17+25)(20+9)(17+20)(25+9)}} = \frac{-347}{1237.8} = -.28$$

Table 9–13 **Paired Scores of 10 Students on Two Items of a Test**

(a)

Student	Item 1	Item 2
1	1	0
2	0	1
3	1	1
4	1	1
5	1	1
6	0	0
7	0	0
8	1	0
9	0	0
10	1	1

(b)

Number of Students with Each Combination of Scores on Two Items

		Item 2 Score 1	Item 2 Score 0
Item 1 Score	1	4	2
	0	1	3

Since either arrangement of the data is equally valid, the sign of ϕ is meaningless, and the coefficient provides information only about the strength of the relationship between the variables.

The phi coefficient, like Spearman's r_s and the point biserial coefficient, is, computationally, a Pearson r. If, for example, you were to apply the formula for the Pearson r to the data in Table 9–13(a), for which we calculated $\phi = .41$, you would get $r = .41$. The advantage of the special formula for ϕ is ease of computation when manual calculation is necessary. Statistical calculators and computers use the same method for calculating all the correlation coefficients discussed in this chapter.

Tests of Significance

Sometimes, we want to test a correlation coefficient other than a Pearson r for significance. For Spearman rank-order correlation coefficients and point biserial correlation coefficients, the test is the same as that for the Pearson r, providing that N is at least 10. One can either calculate t by the formula

$$t = \frac{r_s\sqrt{N-2}}{\sqrt{1-r_s^2}} \quad \text{or} \quad t = \frac{r_{pb}\sqrt{N-2}}{\sqrt{1-r_{pb}^2}}$$

or consult a table of critical values of r (Table D–4 in Appendix D). In either case, the number of degrees of freedom is $N - 2$, where N is the number of pairs of ranks or scores.

For example, we calculated $r_{pb} = .33$ between one item of a test and the total test score in a group of 10 students. Therefore,

$$t = \frac{.33\sqrt{8}}{\sqrt{1-(.33)^2}} = \frac{.933}{.944} = 0.99$$

In an upper one-tailed test with 8 degrees of freedom, $t = .99$ falls between the critical values .889 with $p = .20$ and 1.108 with $p = .15$. Since $p > .05$, the correlation coefficient is not significantly greater than zero at the .05 level. Using the table of critical values of r instead, we find that the minimum value of r that is significant at the .05 level in an upper one-tailed test, with 8 degrees of freedom, is .549; again, we conclude that the obtained correlation, .33, is not significant.

Table 9–14 **Two Arrangements of Data From a Study of the Relationship Between Children's Gender and Their Choice of Toys**

(a)

Gender		Nature of Toys Selected: Warlike	Nonwarlike
	Girls	25	9
	Boys	17	20

(b)

Gender		Nature of Toys Selected: Warlike	Nonwarlike
	Boys	17	20
	Girls	25	9

The appropriate test for the significance of a phi coefficient is the chi square (χ^2) test of independence, which was discussed in Chapter 5. χ^2 can be calculated using the methods of Chapter 5 or by the following formula:

$$\chi^2 = N\phi^2$$ *Relationship between χ^2 and the phi coefficient*

The number of degrees of freedom for the test is 1. Critical values of χ^2 can be found in Table D–1 in Appendix D.

Again, consider the choice-of-toys example. We calculated $\phi = .28$. The number of children in the study (N) was 71. Therefore,

$$\chi^2 = 71(.28^2) = 71(.0784) = 5.57^{\dagger}$$

Referring to the chi square table in the row with $df = 1$, we find that the probability of a χ^2 this large is between .02 and .01. Since $p < .05$, the relationship between gender and choice of toys is significant at the .05 level. However, the relationship does not appear to be very strong; with $\phi = .28$, only $(.28)^2 = .08$ of the choice-of-toy variance is accounted for by gender.

EXERCISE 9–9

1. Use the formula for the Pearson r to calculate the correlation coefficients for the data in Tables 9–11, 9–12, and 9–13. You should get the same values as those obtained by using the special formulas for r_s, r_{pb}, and ϕ.
2. A high school algebra teacher finds that, on the final exam, the point biserial correlation between scores on a particular item (correct = 1, incorrect = 0) and scores on the total exam is −.50. Assuming that the test as a whole was a good measure of knowledge of algebra, do you think that this item is a good item? Why or why not?
3. The table below shows the relation between responses to two survey questions. Calculate ϕ. Then conduct a χ^2 test to determine whether ϕ is significantly different from zero at the .05 level.

Question 6		Question 10: Yes	No
	Yes	23	7
	No	12	8

†In Chapter 5, we calculated $\chi^2 = 5.61$ for the same data; the difference between the two values of χ^2 is due to rounding.

SUMMARY

If two variables, X and Y, co-vary systematically, they are said to be *correlated*. If Y increases as X increases, the correlation is positive. If Y decreases as X increases, the correlation is negative. Correlations between two variables can differ in strength as well as in direction.

The direction and strength of the relationship between X and Y can be shown in a graph called a ***scatter diagram*** in which each pair of scores is represented by a point. If the points flow along a diagonal from the lower left to the upper right, the correlation is positive. In a negative correlation, the points flow from the upper left to the lower right of the diagram. The more closely the points cluster about the diagonal, the stronger the relationship.

The ***covariance*** of two variables, X and Y, is defined as

$$\text{CoVar}_{xy} = \frac{\Sigma(X - M_x)(Y - M_y)}{N}$$

However, the covariance is affected by the units of measurement on both variables. A unit-free index of correlation, called the ***Pearson correlation coefficient, r***, is obtained by dividing the covariance by both standard deviations.

$$r = \frac{\text{CoVar}_{xy}}{S_x S_y}$$

The Pearson r can take values from −1 (a perfect negative correlation) through zero (no relation) to +1 (a perfect positive correlation). The sign of r indicates the direction of the relation, and the absolute value of r indicates its strength.

The correlation coefficient can be interpreted as the ratio of the covariance of X and Y to the geometric mean of their variances. Pearson's r can also be expressed in terms of z scores:

$$r = \frac{\Sigma z_x z_y}{N}$$

Correlation coefficients can be easily obtained using a statistical calculator or computer software. For calculation by hand, the following computational formula can be used:

$$r = \frac{N(\Sigma XY) - (\Sigma X)(\Sigma Y)}{\sqrt{[N(\Sigma X^2) - (\Sigma X)^2][N(\Sigma Y^2) - (\Sigma Y)^2]}}$$

Whether a given value of $|r|$ is considered strong or weak depends on the types of variables and the purpose for which r is computed. Since the Pearson r measures the linear relationship between X and Y, r underestimates the strength of a curvilinear relationship. The presence of ***heteroscedasticity*** (unequal scatter) also complicates the interpretation of r. If the range of one or both variables is restricted, $|r|$ is reduced.

A positive or negative correlation coefficient between two variables tells us that the variables are related but not why. Three possible reasons are: X causes Y; Y causes X; both X and Y are influenced by a third variable, Z, and have no direct causal relationship.

The square of the correlation coefficient, r^2, is called the ***coefficient of determination***. It is the proportion of the variance of each variable that is accounted for by the other. $1 - r^2$ is the proportion of each variable's variance that is not accounted for by the other variable and is called the ***coefficient of nondetermination***.

Inferences about population correlation coefficients from sample data are based on the ***sampling distribution of r***. If the population correlation coefficient ρ is equal to zero, and if both X and Y are normally distributed in the population, sample correlation coefficients are approximately normally distributed around the value zero. As N increases, the variability of the sampling distribution of r decreases.

One-tailed and two-tailed tests of the null hypothesis (H_o) that $\rho = 0$ can be conducted by converting the sample r into a value of t and comparing t with critical values of t for $N - 2$ degrees of freedom, or by referring to a table of critical values of r. If $p < \alpha$, H_o is rejected and the correlation between X and Y is significant at level alpha.

The ***Spearman rank-order correlation coefficient, r_s*** can be calculated between two sets of ranks. A correlation coefficient between a dichotomous and a continuous variable is the ***point***

biserial correlation coefficient, r_{pb}. A correlation coefficient between two dichotomous variables is called the ***phi (ϕ) coefficient***. Special formulas exist for their calculation; however, all three coefficients can also be calculated by the standard formulas for the Pearson r.

The test for the significance of a Pearson r can be applied to Spearman's r_s and to point biserial r_{pb} as well, provided that N is at least 10. The appropriate test of significance for phi coefficients is the chi square test of independence with 1 degree of freedom.

QUESTIONS

1. Estimate the direction and strength of the relationship between the two variables in each of the following pairs.

a. Study habits and grades in college courses.
b. Carelessness and scores on an arithmetic test.
c. Carelessness and number of errors on an arithmetic test.
d. Batting averages and salaries of professional baseball players.
e. Inches of annual rainfall and number of umbrellas sold in 50 American cities.
f. Inches of annual rainfall and number of sunny days per year in 50 American cities.

2. In each of the following, tell whether the relationship is positive, negative, or close to zero.

a. Days spent apart and fondness if "absence makes the heart grow fonder."
b. Days spent apart and fondness if "out of sight, out of mind."
c. Degree of attraction and similarity of attitudes if "opposites attract."
d. Wealth and happiness if "money can't buy happiness."

3. To find out which of two tests is more strongly related to success on the job, a human resources manager gives both tests to a group of employees and calculates the correlation between each test and job performance. Test A has a correlation of .42 with job performance; the correlation between Test B and job performance is −.63.

a. Do employees with high scores on Test A tend to receive higher or lower performance scores than employees with low test scores?
b. On Test B, do employees with high test scores tend to receive higher or lower performance scores than employees with low test scores?
c. Which of the two tests is more strongly related to job performance?

4. Here are two sets of paired data, A and B. The same X scores and Y scores are used in both sets, so that M_x, M_y, SD_x, and SD_y are the same in both sets. However, the X and Y scores are paired differently in Sets A and B.

Set A	X	11	13	11	7	1	17
	Y	1	28	11	14	13	17
Set B	X	11	13	11	7	1	17
	Y	17	13	11	14	28	1

a. Draw a scatter diagram of each data set and estimate the value of r.
b. Calculate the mean of X, the mean of Y, the standard deviation of X, and the standard deviation of Y. (Reminder: These are the same in both sets.) Then calculate the covariance and the correlation of X and Y and r in each set.

5. If the standard deviation of extraversion is equal to 5, the standard deviation of sociability is 7, and the covariance of extraversion and sociability is 20, calculate the correlation between extraversion and sociability.

6. In a group of college students, the standard deviation of scores on a measure of math anxiety was 12.5, and the standard deviation of scores on an algebra test was 7.4. The correlation between math anxiety and algebra scores was −.24.

a. In words, describe the covariation between math anxiety scores and algebra scores.
b. Calculate the covariance of the math anxiety and algebra scores.

7. In Table 2–13 (p. 49), heights and quiz scores of 28 statistics students are listed. Calculate the correlation between height and quiz scores.

8. In a group of 100 men and women, the correlation between height and weight is .80. If you calculate the correlation between height and weight of the women only, do you expect r to be less than .80, greater than .80, or about equal to .80? Do you expect the correlation in the men to be less than .80, greater than .80, or about equal to .80?

9. Here are paired data on two variables:

X	5	10	2	1	3	13
Y	2	1	6	12	4	1

a. Calculate r.
b. Construct a scatter diagram. What does the graph suggest concerning your interpretation of r?

10. A developmental psychologist measures eye-hand coordination in children ranging in age from 6 months to 5 years. She finds that the average level of eye-hand coordination increases in a linear fashion with age. However, the variability of eye-hand coordination decreases with age; for example, 5-year-olds are more alike in their eye-hand coordination than 3-year-olds, who in turn are more alike than 1-year-olds.

a. Draw a scatter diagram that is in accord with the information given.
b. Over all age levels, the correlation between eye-hand coordination and age is .48. Comment on the interpretation of this correlation coefficient, given the nature of the data.

11. The relative importance of heredity and environment in determining intelligence is a controversial question in psychology. One approach to answering the question is to measure the intelligence of adopted children, their biological parents, and their adoptive parents and then compare the correlation between IQs of the children and their biological parents with the correlation between the IQs of the children and their adoptive parents. Here are hypothetical data from such a study.

Family	Child's IQ	Biological Parent's IQ	Adoptive Parent's IQ
1	92	80	121
2	93	90	102
3	120	102	94
4	107	83	120
5	104	90	97
6	100	103	112
7	113	92	122
8	110	110	107
9	115	105	98
10	124	112	125

a. Calculate r between the children's IQs and the IQs of their biological parents. Then calculate r between the children's IQs and the IQs of their adoptive parents.
b. Calculate the proportion of the children's IQ variance that is accounted for by the IQs of their biological parents and the proportion accounted for by the IQs of their adoptive parents.
c. What do your calculations in parts a and b suggest concerning the contributions of the biological parents and the adoptive parents to the children's intelligence?
d. Calculate the mean IQ of the children, of the biological parents, and of the adoptive parents, and compare the three means. Do the means suggest a contribution to the children's IQs by the adoptive parents that was not apparent from the correlation coefficients?

12. A psychologist hypothesized that age and openness to new experiences are negatively related. He calculated the correlation between these two variables in a group of 30 subjects and got $r = -.35$. Test the appropriate hypothesis at the .05 level. Do the results support the psychologist's hypothesis?

13. Which of the following correlation coefficients are significantly different from zero at the .05 level (two-tailed test)?
   a. $r = -.75$, $N = 5$
   b. $r = -.75$, $N = 10$
   c. $r = -.75$, $N = 15$
   d. $r = .43$, $N = 22$
   e. $r = .43$, $N = 20$
   f. $r = .43$, $N = 37$

14. Which of the correlation coefficients in question 13 are significantly different from zero at the .01 level (two-tailed test)?

15. As you can see in Table D–4 in Appendix D, a correlation coefficient of .40 in a sample of size 20 is significant at the .05 level in an upper one-tailed test but not in a two-tailed test. Explain.

16. In a two-tailed test with $\alpha = .01$, what is the minimum sample size needed to reject the null hypothesis if $r = .53$?

17. You plan to conduct an upper one-tailed test with $\alpha = .05$. Collecting data is an expensive proposition, and you do not want to use more subjects than necessary. However, you want to use enough subjects to reject H_o if variable X accounts for at least 20% of the variance of variable Y. What is the minimum sample size you should use?

18. Twelve graduate students in psychology were ranked by two members of the faculty. Calculate the correlation between the two sets of ranks. Then conduct a test to determine whether the correlation coefficient is significantly greater than zero at the .05 level.

	Ranks	
Student	**Faculty Member 1**	**Faculty Member 2**
1	5	3
2	9	9
3	3	1
4	8	7
5	7	11
6	4	6
7	12	12
8	2	5
9	11	10
10	6	2
11	10	8
12	1	4

19. Use the data in Table 2–13 (p. 49) to calculate the correlation between gender and height. Conduct a test for the significance of the correlation coefficient.

20. Here are data on gender (M, F) and smoking behavior (smoker = S, nonsmoker = N) for 20 army trainees.

Trainee	Gender	Smoking Behavior
1	M	S
2	M	S
3	F	N
4	M	N
5	F	N
6	F	S
7	F	S
8	M	S
9	F	N
10	M	N
11	M	S
12	F	S
13	M	S
14	F	N
15	F	N
16	M	N
17	M	S
18	F	N
19	M	S
20	F	S

   a. Calculate the phi coefficient between gender and smoking behavior.
   b. Does the sign of ϕ (positive or negative) have any meaning in this situation?
   c. Test ϕ for significance at the .05 level.

10 Analyzing Correlational Data: Linear Regression

INTRODUCTION

A practical problem that faces admissions officers in colleges and universities, as well as personnel officers in industry, is the prediction of future performance. What grade point average can we expect each applicant to next year's freshman class to achieve? Which job applicants should be selected to fill a limited number of vacancies? Researchers, too, are often concerned with prediction. Can we use the level of stress that psychiatric patients report to predict the severity of their overt symptoms? Can voters' level of confidence in government be used to predict their support for certain legislative initiatives?

It may have occurred to you, from our discussion of relationships and their measurement in Chapter 9, that knowledge of the correlation between two variables allows one to predict one variable from the other. If the correlation between Scholastic Assessment Test (SAT) scores and grade point average (GPA) in last year's freshman class was .60, we can predict that, of this year's applicants, those with higher test scores are more likely to do well if admitted. If, in the past, patients reporting higher levels of stress generally had more symptoms, we can predict that current patients with high levels of stress will probably have more symptoms than those with low stress. In this chapter, we deal with techniques for making predictions under circumstances like these. We will discuss the rationale, calculation, and use of prediction equations called *regression equations*, as well as the measurement and use of errors in prediction.

Calculations in regression are laborious if done manually, and use of statistical calculators or computers is almost mandatory, especially for large data sets. As mentioned before, however, step-by-step calculation of a new procedure (at least once!) results in better understanding. To keep the computations manageable, relatively small data sets are used. Finally, we discuss and demonstrate the use of a computer program to conduct a regression analysis.

Objectives

When you have completed this chapter, you will be able to

- Develop and use regression equations for predicting values of a criterion variable from values of a predictor variable.

- Explain the relation between the regression equation and the regression line. Explain in your own words why the regression line is called the *least squares line* or *line of best fit*.
- Describe how the slope of the regression lines varies with the correlation coefficient, r, and explain the term *regression toward the mean*.
- Compute and interpret the residual variance and the standard error of estimate, and explain their relation to r.
- Show how regression can be used to divide criterion score variation into explained and unexplained components. Describe the relationship among the explained variance, the residual variance, the total variance, and r.
- Describe the relationship between regression analyses in samples and in populations.
- Calculate and interpret prediction intervals for criterion scores.
- Define multiple regression. Given a multiple regression equation, use it to predict criterion scores.
- Explain how correlations of predictors with the criterion and with one another influence the predictors' weights in multiple regression equations; given information about these correlations, identify the relative contribution of each predictor to the prediction of the criterion.
- Define the multiple correlation coefficient, R, and describe its relation to the set of predictors.

The Regression Equation and Its Properties

General Form of the Regression Equation

The Acme Insurance Company uses a short 5-item interview to hire salespersons. On the basis of follow-ups of previously interviewed and hired salespersons, the human resources department has developed the following equation for predicting sales from interview responses:

$$\hat{Y} = 6.59 + 4.40X$$

where

$\hat{Y}$ = predicted number of policies sold
X = number of correct interview responses

Here is how the equation is used: An applicant for a sales position is interviewed. His or her number of correct responses is entered into the equation, and the applicant's predicted number of policies sold is calculated. Only applicants with predicted sales of 15 or more are hired. Here are the results for two applicants:

Applicant 1, Helen, answered 3 of the 5 interview questions correctly. We enter $X = 3$ into the equation

$$\hat{Y} = 6.59 + 4.40(3) = 19.79$$

We predict that Helen will sell 19.79 policies; Helen is hired.

Applicant 2, Joe, answered one question correctly. We enter $X = 1$ into the equation

$$\hat{Y} = 6.59 + 4.40(1) = 10.99$$

Joe's predicted sales, 10.99 policies, falls short of the required 15. Joe is not hired.

A prediction equation like Acme's is called a ***regression equation***. More precisely, it is called the *regression equation of Y on X*, and it is used for predicting Y from X. It is also possible to develop a regression equation for predicting X from Y. For example, we could write an equation to predict interview scores from sales. This would be the regression equation of X on Y. However, we are usually interested in predicting in just one direction.

In regression, the variable being predicted is called the ***criterion variable***, and the one used for making predictions is called the ***predictor variable***. The criterion variable at Acme is sales, and the predictor variable is interview scores. Y is used to represent the criterion variable and X to represent the predictor variable.

Let us examine the components of Acme's regression equation. The only term on the left is $\hat{Y}$, the predicted criterion score (sales). The "hat" (^) is used to differentiate the predicted score from the actual score, which we don't know. Not everyone will achieve exactly the criterion score predicted from the equation. For instance, Helen may sell 18 policies or 20 or 16 or 24 rather than 19.8. However, on the basis of the known relationship between interview scores and sales, *the most likely* sales for employees with an interview score of 3 is 19.8. What "most likely" means in this context will be explained more fully later.

The first term on the right of the equal sign is 6.59. The particular number varies from one equation to another. It is symbolized a. Thus, in Acme's equation, $a = 6.59$. The second term on the right of the equal sign is $4.40X$. In every regression equation, this term consists of some number times the predictor score, X (the number varies from one equation to another). The number that multiplies X is called the ***regression coefficient*** and is symbolized b. In the Acme equation, $b = 4.40$.

Every regression equation takes the same general form as the one used by Acme.

$$\hat{Y} = a + bX \qquad \textit{General form of regression equation}$$

The calculation of a and b will be dealt with later. For now, let us look at some more examples of regression equations.

EXAMPLE 1: What is the regression equation if $a = -19.3$ and $b = .75$?

Answer: $\hat{Y} = -19.3 + .75X$

EXAMPLE 2: What is the regression equation if $a = .024$ and $b = .0005$?

Answer: $\hat{Y} = .024 + .0005X$

EXAMPLE 3: $b = -1.32$, $a = 8.17$. What is the regression equation?

Answer: $\hat{Y} = 8.17 + (-1.32)X$, or $\hat{Y} = 8.17 - 1.32X$

EXAMPLE 4: In a certain regression equation, $a = -17$ and $b = .63$. Al's predictor score is 10. Predict his score on the criterion.

First, we write the regression equation:

$$\hat{Y} = -17 + .63X$$

Second, we substitute Al's predictor score, 10, for X:

$$\hat{Y} = -17 + .63(10) = -10.7$$

Answer: Al's predicted criterion score is -10.7.

EXERCISE

1. If I want to predict burnout from years of teaching experience, which variable is the predictor and which is the criterion?
2. The regression equation for predicting severity of emotional disturbance from scores on a diagnostic test is $\hat{Y} = -4.2 + 12X$.

a. Which of the two variables is X and which is Y?
b. What are the values of a and b?
c. Predict the severity of emotional disturbance for someone with diagnostic test score = 10.

3. The regression equation for predicting assertiveness from anxiety is $\hat{Y} = 16.3 - 1.5X$.
   a. Which variable is X and which is Y?
   b. What are the values of a and b?
   c. Predict assertiveness for someone whose level of anxiety is 5.
4. Write the regression equation if $b = -.07$ and $a = 14.92$.
5. Write the regression equation if $b = 10.3$ and $a = -71.2$.
6. Use the Acme Insurance Company's regression equation to predict sales for the following applicants:
   a. Tom answered two questions correctly.
   b. Sally answered none of the questions correctly.
   c. Ted answered all five questions correctly.
   d. Mary answered four questions correctly.
   e. Using the requirement of at least 15 predicted sales, which of these applicants will Acme hire?

A Graphic Approach to Regression: The Regression Line

We have not yet answered one very important question: Where do regression equations come from? Specifically, how are a and b computed? We arrive at the answer by, first, developing the concept of regression as a running mean, then as a line in a scatter diagram, and finally as an equation of the line.

Let us return to the Acme Insurance Company. Assume that Acme's regression equation is based on data that Acme collected on interview scores and sales of 26 salespersons. (In real life, samples much larger than 26 are usually used to develop regression equations. We are using $N = 26$ to keep things simple. The reasoning and computations would be the same for larger samples.) The data for the 26 salespersons are given in Table 10–1. Table 10–1 also includes summary statistics (means, sums of squares, variances, etc.) to which we will refer later in the chapter.

A scatter diagram showing the relationship between interview scores and sales is displayed in Figure 10–1. Notice that the predictor variable (interview scores) is assigned to the abscissa and the criterion variable (sales) to the ordinate. In plotting data for regression, the predictor variable is always on the abscissa, the criterion on the ordinate. The scatter diagram reveals that the correlation between interview scores and sales is moderately strong and positive. In fact, $r = .76$, as Table 10–1 shows.

Because sales tend to increase as interview scores increase, we would predict higher sales for people who answer many questions correctly than for people who answer few correctly. But how can we make this notion more precise? What specific sales should we predict for each interview score?

Our first approximation to the prediction problem is as follows. For each interview score we predict that value of sales which is the mean of sales for all persons having that interview score. Thus, for anyone with interview score = 3, we predict sales equal to the mean sales of individuals with interview score = 3. From the data table (and from the scatter diagram), we see that five persons had an interview score of 3. Their sales scores were 34, 25, 21, 16, and 27. The mean sales of these five people is

$$\frac{34+25+21+16+27}{5} = \frac{123}{5} = 24.6$$

We reason that because people are more likely to score near the mean than farther away, a prediction of sales = 24.6 for all people scoring 3 on the interview is reasonable.

The process is repeated for each interview score. The 26 pairs of scores are rearranged in Table 10–2. Each column of the table lists the sales scores (Y) for individuals with a particular interview score (X). The set of Y scores for a given value of X is called an *array*. The mean sales of each array is shown at the bottom of the array. Next, we graph the means of sales as a function of interview scores on the scatter diagram, and we connect the means with straight lines. The result is the broken line in Figure 10–2.

Table 10–1 **Acme Insurance Company Data**

Salesperson	Number of Correct Interview Responses	Number of Policies Sold	Salesperson	Number of Correct Interview Responses	Number of Policies Sold
1	3	34	14	0	14
2	1	7	15	2	16
3	4	19	16	1	3
4	2	3	17	5	21
5	1	19	18	4	16
6	0	5	19	5	35
7	5	30	20	3	21
8	4	32	21	3	16
9	1	10	22	4	24
10	2	8	23	0	9
11	3	25	24	5	27
12	0	2	25	3	27
13	1	17	26	2	13

Summary Statistics

X = Interview scores

$\sum X = 64$

$\sum X^2 = 230$

$M_x = \frac{64}{26} = 2.46$

$SS_x = 230 - \frac{(64)^2}{26} = 72.46$

$S_x^2 = \frac{SS_x}{N} = \frac{72.46}{26} = 2.787$

$S_x = \sqrt{2.787} = 1.669$

$s_x^2 = \frac{SS_x}{N-1} = \frac{72.46}{25} = 2.898$

$s_x = \sqrt{2.898} = 1.70$

Y = Number of policies sold (sales)

$\sum Y = 453$

$\sum Y^2 = 10{,}331$

$M_y = \frac{453}{26} = 17.42$

$SS_y = 10{,}331 - \frac{(453)^2}{26} = 2438.346$

$S_y^2 = \frac{SS_y}{N} = \frac{2438.346}{26} = 93.783$

$S_y = \sqrt{93.783} = 9.684$

$s_y^2 = \frac{SS_y}{N-1} - \frac{2438.346}{25} = 97.534$

$s_y = \sqrt{97.534} = 9.876$

$\sum XY = 1434$

$$r = \frac{26(1434) - (64)(453)}{\sqrt{[26(230) - (64)^2][26(10{,}331) - (453)^2]}} = .76$$

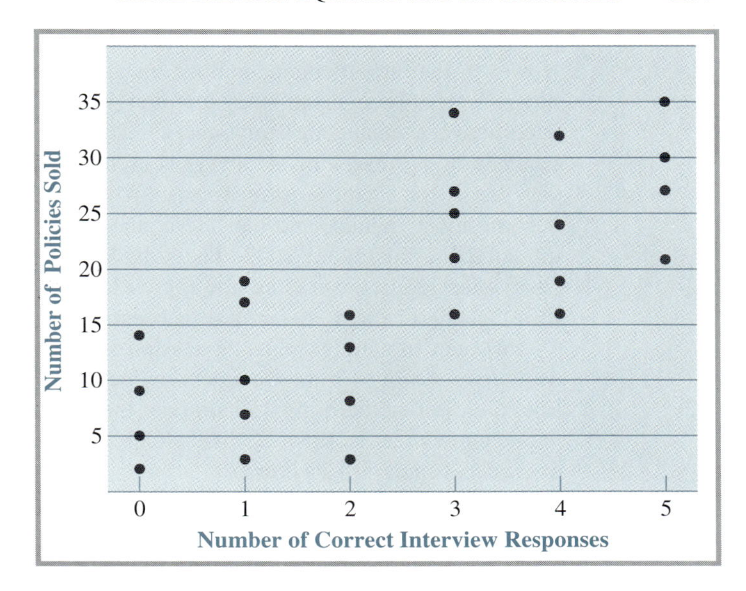

Figure 10–1 Scatter diagram of Acme interview–sales data.

Table 10–2 Sales of Acme Employees as a Function of Interview Scores (X)

	$X = 0$	$X = 1$	$X = 2$	$X = 3$	$X = 4$	$X = 5$
	5	7	3	34	19	30
	2	19	8	25	32	21
	14	10	16	21	16	35
	9	17	13	16	24	27
		3		27		
Mean of sales	7.50	11.20	10.00	24.60	22.75	28.25

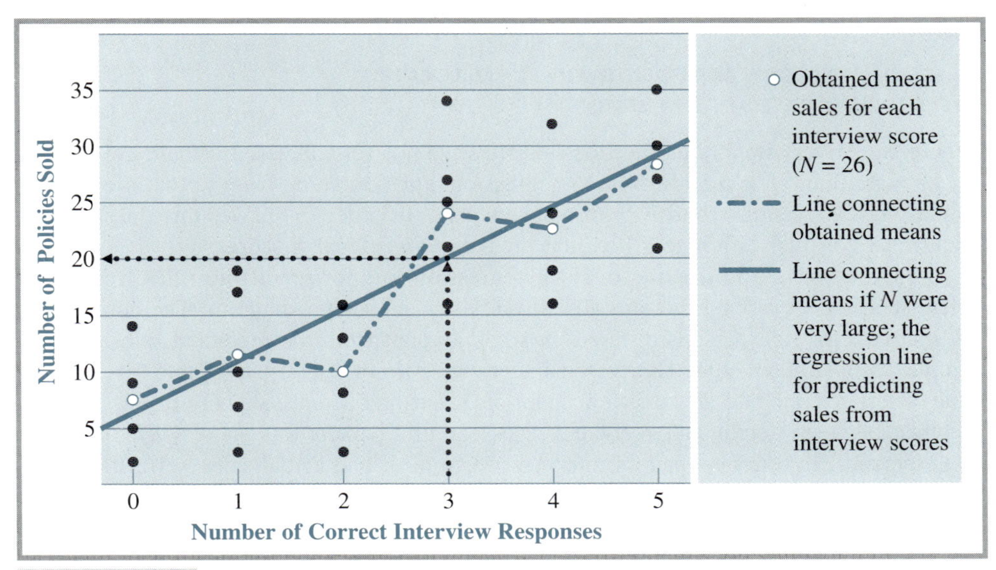

Figure 10–2 Obtained mean sales (broken line) and hypothetical mean sales, assuming a very large N (the regression line [solid line]). Use of the line for predicting sales for interview score = 3 is shown.

As you can see, the mean of sales tends to increase as interview scores increase. However, the rate of change is not constant from one array to the next. For example, the difference between mean sales for interview scores of 1 and 2 is −1.20; but the difference between mean sales for interview scores of 2 and 3 is 14.60. As a result, the line connecting the means of the arrays is crooked. It is reasonable to suppose that if N were very large, the increase in mean sales from one interview score value to another would become more regular, and the line connecting the means would become straighter, like the solid line in Figure 10–2. The solid line is the ***regression line*** for predicting sales from interview scores. Thus, the regression line is a kind of running or floating mean. It is the line of mean Y values at successive values of X.

We can use the graphed regression line to predict sales from interview scores. To do so, we simply locate the interview score in question on the abscissa, go up to the line, then left to the ordinate, and read the value of predicted sales. For example, for an applicant with a score of 3, like Helen, the predicted sales is about 20 (19.79 to be exact), as Figure 10–2 shows.

Although the graphed regression line can be used to find the predicted criterion score ($\hat{Y}$) for each predictor score (X), it is generally easier to use algebraic methods. From elementary algebra, we know that any straight line drawn on a pair of X, Y axes has an equation that takes the form $Y = a + bX$, where b is the slope of the line and a is its Y-intercept (the point at which the line crosses the Y-axis). This formula should look familiar because it is the general form of the regression equation. In other words, the regression equation is the equation of the regression line.

In light of this information, let us reexamine Acme's regression equation: $\hat{Y} = 6.59 + 4.40X$. We now know that 4.40, the regression coefficient b, is the slope of the regression line in Figure 10–2. The slope of a line refers to the amount of change in Y for each unit increase in X. In other words, for each increase of one point in interview score, predicted sales increases by 4.40 policies. Thus, an applicant with interview score = 3 is predicted to sell 4.40 more policies than an applicant with a score of 2, and 4.40 fewer policies than an applicant with a score of 4. The Y-intercept, a, is 6.59. This means than an applicant with an interview score of zero ($X = 0$) is predicted to sell 6.59 policies.

Calculating the Slope and the Y–Intercept of the Regression Equation

There is one very important question that we have not yet addressed: Where did the values $b = 4.40$ and $a = 6.59$ in Acme's regression equation come from? What are the formulas for calculating a and b from a set of data? In order to answer this question, we first must consider differences between predicted and actual Y scores.

As you can see in Figure 10–2, the regression line for predicting sales from interview scores does not go through all the points in the scatter diagram. The actual sales of many Acme salespersons differed by some amount from the predicted sales. For example, the predicted sales ($\hat{Y}$) for an interview score of 3 is 19.79. Table 10–1 shows that Salesperson 1 had an interview score of 3, but had actual sales (Y) of 34. The difference between actual and predicted sales for this person was $34 - 19.79 = 14.21$. Salesperson 21, who also had an interview score of 3, had actual sales = 16; the difference between actual and predicted sales was $16 - 19.79 = -3.79$.

The difference between actual and predicted Y values for a given individual is called an ***error of estimate*** or a ***residual***.

$$Y - \hat{Y} = \text{error of estimate or residual}$$

Since no straight line can go through all the points of a scatter diagram (unless $r = \pm 1$), errors of estimate are inevitable. However, the regression line can be positioned in such a way as to minimize errors of estimate over the data as a whole. Specifically, the regression line is fitted to the data in such a way that the sum of the squared errors of estimate, $\Sigma(Y-\hat{Y})^2$, is a minimum. For this reason, the regression line is called the *least squares line*, or, sometimes, the *line of best fit*.

By using a little calculus, it is possible to derive formulas for the slope and the Y-intercept of the line that minimizes $\Sigma(Y-\hat{Y})^2$ for a given set of paired X,Y scores. The formulas are

$$b = r\left(\frac{S_y}{S_x}\right)$$ *Formula for the regression coefficient or slope of the regression line*

$$a = M_y - b\,M_x$$ *Formula for the* Y*-intercept of the regression line*

where

r = the Pearson correlation coefficient between X and Y
M_x = mean of the predictor scores
M_y = mean of the criterion scores
S_x = standard deviation of the predictor scores
S_y = standard deviation of the criterion scores.

Note that b must be calculated first, because b is used in the calculation of a.

Now we can show why the Acme regression equation is $\hat{Y} = 6.59 + 4.40X$. First, we go to Table 10–1 to find the means and standard deviations of the interview scores (X) and policies sold (Y). We find that $M_x = 2.46$, $S_x = 1.669$, $M_y = 17.42$, and $S_y = 9.684$. The correlation (r) between interview scores and sales is .76. Therefore

$$b = r\left(\frac{S_y}{S_x}\right) = .76\left(\frac{9.684}{1.669}\right) = 4.40$$

$$a = M_y - b\,M_x = 17.42 - (4.40)(2.46) = 6.59$$

The regression equation is $\hat{Y} = a + bX$ or $\hat{Y} = 6.59 + 4.40X$, as we have already seen.

Interview scores and sales at the Acme Company are positively correlated. Let us construct a regression equation for two variables that have a negative correlation.

> Anxiety and aggressiveness have been measured in a group of adolescents. The mean anxiety score is 20 with a standard deviation of 4. The mean aggressiveness score is 50 with a standard deviation of 7. The correlation between anxiety and aggressiveness is –.80. What is the regression equation for predicting aggressiveness from anxiety?
>
> Because we want to predict aggressiveness from anxiety, anxiety is the predictor variable, X, and aggressiveness is the criterion variable, Y.
>
> $M_x = 20,\ S_x = 4,$
>
> $M_y = 50,\ S_y = 7,$
>
> and $r = -.80$.
>
> $$b = r\left(\frac{S_y}{S_x}\right) = -.80\left(\frac{7}{4}\right) = -1.4$$
>
> $$a = M_y - bM_x = 50 - (-1.4)(20) = 50 - (-28) = 50 + 28 = 78$$

The regression equation for predicting aggressiveness from anxiety is

$$\hat{Y} = 78 - 1.4X$$

The regression line for predicting aggressiveness from anxiety is graphed in Figure 10–3. Note that when r is negative, the slope of the regression line, b, is negative. Consequently, as predictor scores increase, predicted criterion scores decrease. In the case of anxiety and aggressiveness, for each increase of 1 point in anxiety, predicted aggressiveness decreases by 1.4 points.

Raw Score Formulas for *b* and *a*

If calculations must be performed manually, it may be convenient to calculate b and a directly from raw scores. Raw score formulas for b and a are the following:

$$b = \frac{N(\sum XY) - (\sum X)(\sum Y)}{N(\sum X^2) - (\sum X)^2}$$

Raw score formulas for b *and* a

$$a = \frac{\sum Y - b(\sum X)}{N}$$

For example, at Acme, $N = 26$, $\Sigma XY = 1434$, $\Sigma X = 64$, $\Sigma Y = 453$, and $\Sigma X^2 = 230$. Therefore,

$$b = \frac{26(1434) - (64)(453)}{26(230) - (64)^2} = 4.40 \quad \text{as before}$$

$$a = \frac{453 - (4.40)(64)}{26} = 6.59 \quad \text{as before}$$

Another example of the use of raw score formulas to calculate b and a and construct a regression equation is shown in Table 10–3. The table presents data on reported level of stress and number of overt symptoms in 10 psychiatric patients. The researcher

Figure 10–3

The regression line for predicting aggressiveness (Y) from anxiety (X) (regression equation: $\hat{Y} = 78 - 1.4X$).

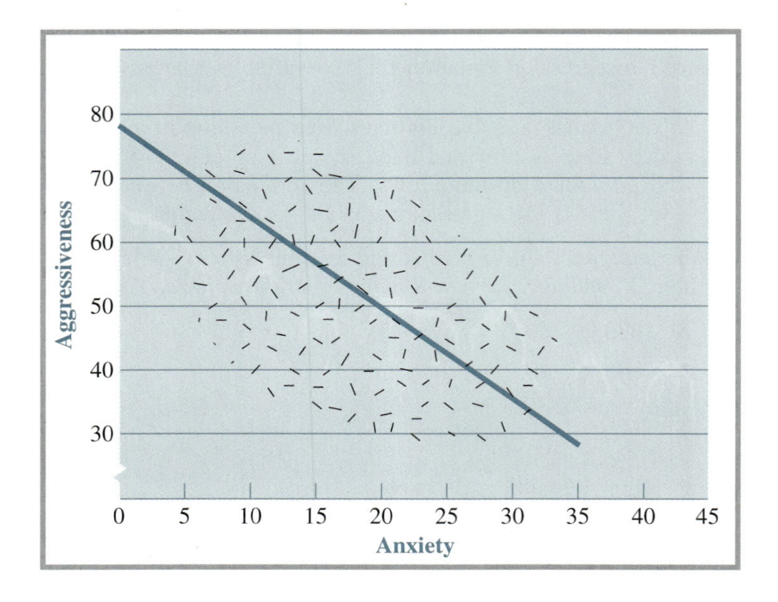

Table 10–3 Use of Raw Score Formulas to Construct a Regression Equation for Predicting Number of Symptoms From Level of Stress

Patient	Level of Stress (X)	Number of Symptoms (Y)	XY	X^2	Y^2
1	7	4	28	49	16
2	10	3	30	100	9
3	9	5	45	81	25
4	10	0	0	100	0
5	7	2	14	49	4
6	12	2	24	144	4
7	8	1	8	64	1
8	11	5	55	121	25
9	7	0	0	49	0
10	5	1	5	25	1
	86	23	209	782	85
	ΣX	ΣY	ΣXY	ΣX^2	ΣY^2

$$b = \frac{10(209) - (86)(23)}{10(782) - (86)^2} = \frac{112}{424} = .264$$

$$a = \frac{23 - \left[\frac{112}{424}\right](86)}{10} = .028$$

Regression equation: $\hat{Y} = .028 + .264X$

wishes to predict number of symptoms (Y) from level of stress (X). As you can see, $b = .264$ and $a = .028$, and the regression equation is $\hat{Y} = .028 + .264X$. Thus, a patient with level of stress = 10 is predicted to have .028 + .264(10) = 2.668 symptoms.

EXERCISE 10–2

1. The mean score on a computer programming aptitude test is 31.5, with $SD = 6.8$. On the final exam at the end of a programming course, the mean is 71.3 with $SD = 12.5$. The correlation between the two tests is .55.
   a. Construct the regression equation for predicting final exam scores from aptitude test scores.
   b. Predict the final exam score of a student whose aptitude test score is 60.
2. Here are the scores of five children on a reading and a math test.

Child	Reading Score	Math Score
1	7	10
2	4	4
3	3	5
4	6	12
5	8	9

   a. Use the raw score formulas for b and a to generate the regression equation for predicting math scores from reading scores.
   b. If Johnny takes the reading test and scores 5, what score do you predict for him on the math test?

Some Facts About Regression

In this section we discuss and demonstrate the following facts about regression:

1. The equation for predicting X from Y is a different equation than that for predicting Y from X. One cannot predict in both directions from a single regression equation (unless $r = \pm 1$).
2. When the value of the predictor variable is the mean of X, the predicted criterion score is the mean of Y.
3. If $r = 0$, the predicted Y for all values of X is the mean of Y.
4. Unless $r = \pm 1$, the predicted Y for any given X is closer to the mean of Y (in standard deviation units) than X is to the mean of X. This is called ***regression toward the mean***. The weaker the correlation between X and Y, the greater the amount of regression toward the mean.

The Regression Equation for Predicting X From Y

Suppose that instead of predicting number of symptoms from level of stress for the data in Table 10–3, the researcher decides to predict level of stress from number of symptoms. In other words, he wants to predict variable X from variable Y rather than predicting variable Y from variable X. Should he simply solve the regression equation $\hat{Y} = .028 + .264X$ for X? The answer is no. An entirely new regression equation must be calculated from the data. The regression equation will have the form $\hat{X} = a + bY$.

To calculate a and b for predicting X from Y, we simply reverse the X's and Y's in the formulas. The values of b and a for predicting level of stress (X) from number of symptoms (Y) are

$$b = \frac{N(\sum XY) - (\sum Y)(\sum X)}{N(\sum Y^2) - (\sum Y)^2} = \frac{10(209) - (23)(86)}{10(85) - (23)^2} = \frac{112}{321} = .35$$

$$a = \frac{\sum X - b(\sum Y)}{N} = \frac{86 - (.35)(23)}{10} = 7.80$$

The regression equation for predicting level of stress from number of symptoms is $\hat{X} = 7.80 + .35Y$. Notice that this is not an algebraic transformation of the equation $\hat{Y} = .028 + .264X$ but an entirely different equation.

Why is the equation for predicting X from Y different from that for predicting Y from X? The equations are different because they are equations of two different lines. The regression line for predicting Y from X is the line for which $\Sigma(Y-\hat{Y})^2$ is a minimum; that is, it is the line that minimizes errors in predicting Y. The regression line for predicting X from Y, on the other hand, is the line for which $\Sigma(X-\hat{X})^2$ is a minimum; it is the line that minimizes errors in predicting X. Unless $r = \pm 1$, the line that minimizes errors in predicting Y is not the same line that minimizes errors in predicting X.

Figure 10–4 shows the two regression lines for the data in Table 10–3 superimposed on the scatter diagram. The regression lines cross at the point $X = 8.6$, $Y = 2.3$. These are the respective means of X and Y. Whenever both regression lines are constructed for a set of data, the lines always cross at the point $X = M_x$, $Y = M_y$.

Note the angle separating the two regression lines in Figure 10–4. The size of the angle between the two regression lines depends on the strength of the correlation between the variables, $|r|$. As $|r|$, increases, the angle between the lines decreases. For the data in the figure, r is only .30, and the angle separating the lines is quite large. If r were greater than .30, the angle between the two regression lines would be smaller. If r were equal to ± 1, all the points in the scatter diagram would lie on a straight line, and

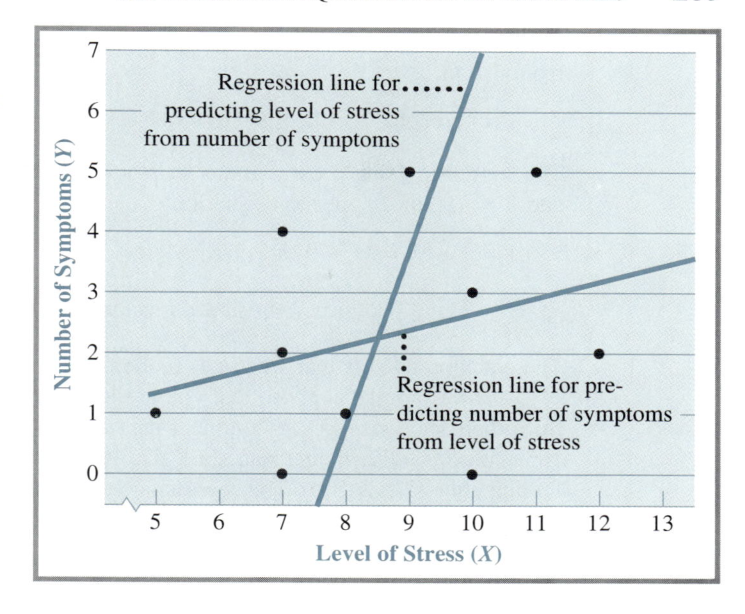

Figure 10–4

Regression lines for predicting Y from X and X from Y for the data in Table 10–3.

that line would be both the regression line for predicting Y from X and also the regression line for predicting X from Y (and there would be no errors of estimate). Only when $r = \pm 1$ can the same regression equation be used for predicting in both directions.

We have constructed both regression equations for the level of stress–number of symptoms data in order to show that an equation developed to predict in one direction cannot be used to predict in the other. As we pointed out earlier, however, in research as well as in practical applications of regression, interest almost always lies in predicting in just one direction. In the rest of this chapter we concern ourselves only with the prediction of variable Y from variable X.

The Predicted *Y* for Individuals at the Mean of *X*

We noted that the regression line for predicting number of symptoms (Y) from level of stress (X) goes through the point (M_x, M_y). This is true not only for the stress–symptoms data but for all regression lines: No matter what the means, the standard deviations, and the correlation of the two variables, the regression line always goes through the point (M_x, M_y). Therefore, for individuals who fall at the mean of X, the predicted Y is always the mean of Y.

At the Acme Insurance Company, for example, the mean interview score (M_x) is 2.46, and the regression equation for predicting sales from interview scores is $\hat{Y} = 6.59 + 4.40X$. Suppose an applicant had an interview score of 2.46. The applicant's predicted sales would be 6.59 + 4.40 (2.46) = 17.42, the mean of sales.

The Regression Line and the Regression Equation When *r* = 0

Suppose that the correlation between two variables, X and Y, is zero. In other words, the two variables are uncorrelated. Where is the regression line located, and what does the regression equation look like? To answer these questions, let us calculate the slope, b, and Y-intercept, a, of the regression line when $r = 0$.

$$\text{Slope:} \quad b = r\left(\frac{S_y}{S_x}\right) = 0\left(\frac{S_y}{S_x}\right) = 0$$

When $r = 0$, the slope of the regression line is zero. A line with a slope of zero is a horizontal line.

$$Y\text{-intercept:} \quad a = M_y - b(M_x) = M_y - 0 \cdot (M_x) = M_y$$

The regression line crosses the Y-axis at a height equal to the mean of Y. Since $b = 0$ and $a = M_y$, the regression equation is

$$\hat{Y} = a + bX = M_y - 0 \cdot X$$

$$\hat{Y} = M_y \qquad \textit{Regression equation when } r = 0$$

The equation tells us that when $r = 0$, the predicted Y for all values of X is the mean of Y.

As an example, suppose that you are interested in predicting college students' IQs (Y) from their shoe size (X). You collect data on 100 students and find that their mean IQ is 105 with $SD = 12$, their mean shoe size is 7½ with $SD = 2$, and the correlation between IQ and shoe size is zero. The regression equation for predicting IQ from shoe size is

$$\hat{Y} = 105$$

Predicted IQ = mean IQ

Figure 10–5 shows the regression line for predicting IQ from shoe size. It is a horizontal line at a height equal to 105. As you can see from both the equation and the line, the predicted IQ of all students is 105, regardless of their shoe size.

You may think it odd to predict the same Y score for everyone, regardless of their score on X. But it makes sense when the correlation between X and Y is zero. After all, if X and Y are unrelated, knowing X does not improve our prediction of Y. Therefore, predicting Y from X is just like predicting Y without any other information. And with no information, the best single predicted Y for everyone is the mean of Y, because most observations in a distribution fall near the mean (at least if the distribution of Y is reasonably symmetrical).

Regression Toward the Mean

Regression toward the mean stands for the fact that unless $r = \pm 1$, the predicted Y score for any X is closer to the mean of Y, in standard deviation units, than the X score is to the mean of X. For example, a person who scores two standard deviations above or

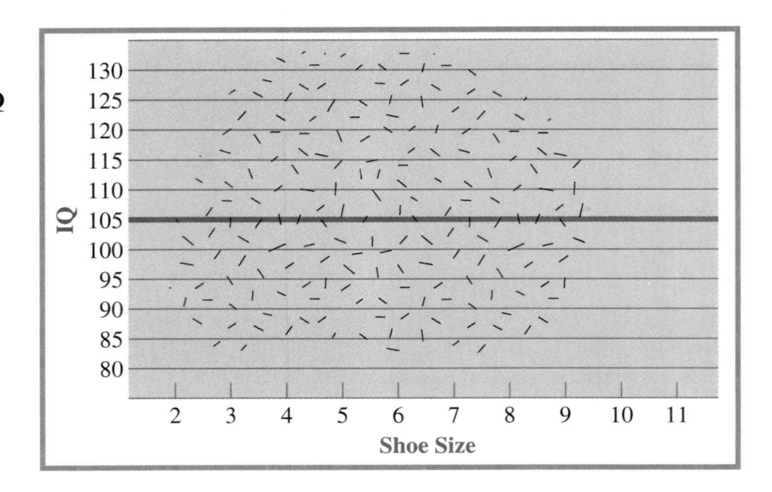

Figure 10–5

The regression line for predicting IQ from shoe size.

below the mean of X is predicted to score *less* than two standard deviations above or below the mean of Y.

In order to examine regression toward the mean systematically, let us express the regression equation in terms of z scores rather than raw scores. That is, we will write the equation so that we can predict each individual's z score on the Y variable from his or her z score on X. Remember that the mean of any distribution of z scores is zero and the standard deviation is 1. Thus, when X and Y are expressed as z scores, M_x and M_y are both equal to zero, and S_x and S_y are both equal to one. Therefore

$$b = r\left(\frac{S_y}{S_x}\right) = r\left(\frac{1}{1}\right) = r$$

$$a = M_y - b(M_x) = 0 - r\,(0) = 0$$

and the regression equation is

$$\hat{z}_y = rz_x \qquad \textit{Regression equation expressed in terms of z scores}$$

In other words, for any z score on the X variable, the predicted z score on the Y variable is the correlation coefficient times the z score on X.

Recall that z scores express the standard deviation distance of a score from the mean. For example, $z = -1.5$ means a score that is 1.5 *SD* below the mean; $z = 0.75$ means a score that is .75 *SD* above the mean. Thus, the z score form of the regression equation tells us that the *SD* distance of a predicted criterion score from the mean criterion score is r times the *SD* distance of the predictor score from the mean predictor score; and since $|r|$ is always less than or equal to 1, the predicted criterion score's *SD* distance from the mean is always less than or equal to the predictor score's *SD* distance. Moreover, the closer the value of $|r|$ to zero, the smaller is the difference between the predicted Y and the mean of Y, relative to the difference between X and the mean of X.

As an example, suppose that $r = .5$. The z score regression equation is $\hat{z}_y = .5\,z_x$. Therefore, each predicted Y score is half as far from the mean of Y (in *SD* units) as the X score is from the mean of X. A person who scores one *SD* above the mean of X is predicted to score only a half *SD* above the mean on Y; a person who is two *SD*'s below the mean of X is predicted to score only one *SD* below the mean of Y, and so on.

Suppose that $r = -.5$ instead of $+.5$. In this case, the regression equation is $\hat{z}_y = -.5z_x$. Again, each person is predicted to score half as far from the mean on Y as his or her distance from the mean of X, but on the opposite side of the mean. For example, a person who scores two *SD*'s *above* the mean on X is predicted to score only one *SD* *below* the mean on Y.

We have already seen that if $r = 0$, everyone is predicted to score at the mean of Y, no matter what his or her score on X. When $r = 0$, regression toward the mean is complete.

Let us consider a real-life example of regression toward the mean: the prediction of boys' adult heights from the heights of their fathers.[†] The mean height of adult males is about 70 inches. The correlation between the heights of fathers and their sons is positive and strong but not perfect. A hypothetical scatter diagram is shown in Figure 10–6. The regression line for predicting sons' heights from the heights of their fathers is included in the figure.

[†]The example is not based on actual data, so particular numbers may not be accurate. However, the actual numbers would be close to those given here.

Figure 10–6

Illustration of regression toward the mean. Predictions of sons' heights for fathers who are 63 and 82 inches tall are shown.

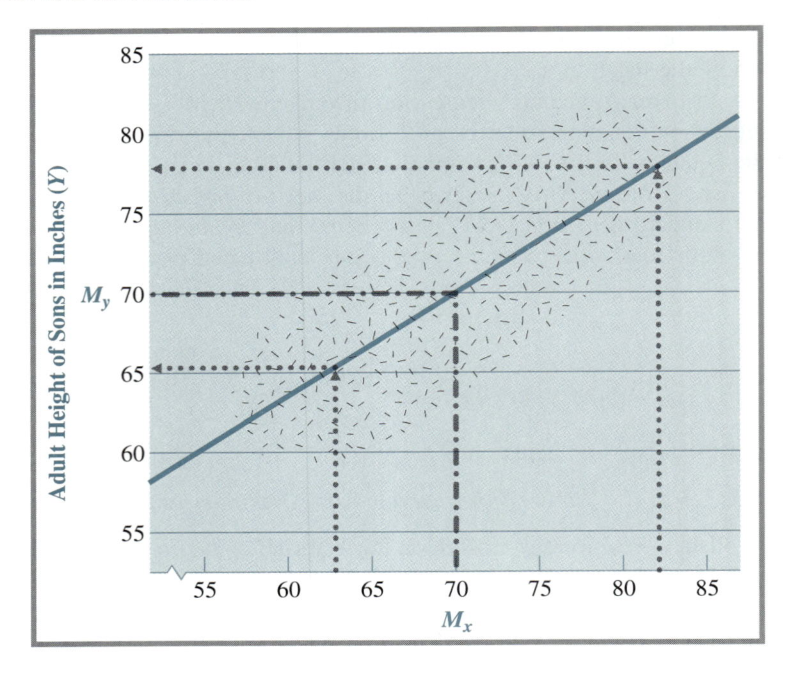

As you can see, taller than average fathers tend to have taller than average sons, and shorter than average fathers tend to have shorter than average sons. However, the predicted adult height of sons of taller than average fathers is less than their fathers' heights, and the predicted adult height of sons of shorter than average fathers is greater than their fathers' heights. For example, fathers who are 82 inches tall (12 inches above the mean) are predicted to have sons whose height is 78 inches (only 8 inches above the mean), whereas fathers who are 63 inches tall (7 inches below the mean) are predicted to have sons whose height is 66 inches (only 4 inches below the mean). Thus, predicted heights of sons are closer to the mean height of adult males than are the heights of their fathers. (This does not, of course, imply that the heights of all sons are closer to the mean than their fathers' heights. Remember that the regression line is a kind of running average. Regression toward the mean tells us that, *on the average*, heights of sons are closer to the mean than their fathers' heights; however, sons' actual heights vary around their predicted heights, and some sons are as far or farther from the mean than their fathers. For example, fathers who are 82 inches tall have sons with an average height of 78 inches, but, as you can see in Figure 10–6, the sons' heights vary from about 71 to 85 inches.)

We can see regression toward the mean operating in the Acme Insurance Company data, too. Consider Max, a salesperson with an interview score of 1. Because the mean interview score at Acme was 2.46 and the standard deviation was 1.669, Max's interview z score is $\frac{1-2.46}{1.669} = -0.87$; Max's interview score is about nine-tenths of a standard deviation below the mean interview score. Max's predicted sales is $6.59 + 4.40(1) = 10.99$. Because the mean of sales at Acme is 17.42 with a standard deviation of 9.684, Max's predicted sales z score is $\frac{10.99-17.42}{9.684} = -0.66$. Although Max's interview score was ninth-tenths *SD* below the mean, his predicted sales is only about two-thirds *SD* below the mean.

Regression toward the mean sometimes creates problems for the interpretation of research results. In some educational research studies, for example, children who score very low on a pretest are provided with special instruction, then posttested to find out whether their performance has improved. The correlation between the pretest and posttest scores, though usually strong, is less than perfect. Therefore, children who score far below the mean on the pretest would be expected to score somewhat nearer to the mean (in other words, higher) on the posttest than on the pretest, even if no special educational program intervened. It is often difficult to tell how much of the "improvement" in scores of such children is really due to the educational program and how much is just regression toward the mean.

EXERCISE 10–3

1. Ms. Jones, a fifth-grade teacher, read in a teacher's guide that the equation for predicting achievement in social studies from achievement in reading is $\hat{Y} = -5 + 1.6X$. She wanted an equation to predict reading achievement from social studies achievement. She solved the equation for X and obtained $X = 3.125 + .625\hat{Y}$. Therefore, she said, the equation for predicting reading achievement from social studies achievement is $\hat{X} = 3.125 + .625Y$. Is this correct?
2. The mean of test X is 50, the mean of test Y is 40, and the correlation between X and Y is $-.6$.
   a. Josh scored 68 on X. Is Josh's predicted Y less than 40, equal to 40, or greater than 40?
   b. Buffy scored 50 on X. What is Buffy's predicted score on Y?
   c. If tests X and Y were uncorrelated, what score would you predict for Josh and Buffy on test Y?
3. The mean IQ of Americans is 100. The correlation between the IQ of parents and their children is about .50.
   a. If a parent has IQ = 125, would you predict the child's IQ to be less than 125, equal to 125, or greater than 125?
   b. If a child has IQ = 80, would you predict the parent's IQ to be less than 80, equal to 80, or greater than 80?

Measuring Variability Around the Regression Line

I pointed out earlier that the difference between an actual Y score and the predicted score, $Y - \hat{Y}$, is called an *error of estimate* or a *residual*. The regression equation minimizes errors of estimate, but it does not eliminate them (unless $r = \pm 1$). The weaker the correlation between X and Y, the larger the average size of errors of estimate. In other words, as $|r|$ decreases, the variation of actual scores around predicted scores increases.

To see the relationship between r and the variation of actual Y scores around predicted Y scores, look at Figure 10–7. The figure shows scatter diagrams, with regression lines included, for values of r ranging from ± 1 to zero. The total variation of the Y scores, over the entire data set, is the same in each scatter diagram and is indicated by the length of the double arrow drawn along the ordinate. The variation of actual Y scores around predicted Y scores is represented by the lengths of the double arrows above selected values of X.

When $r = \pm 1$, as in Figures 10–7a and 10–7b, all the points in the scatter diagram lie on the regression line; and there is no variation of Y scores around predicted scores.

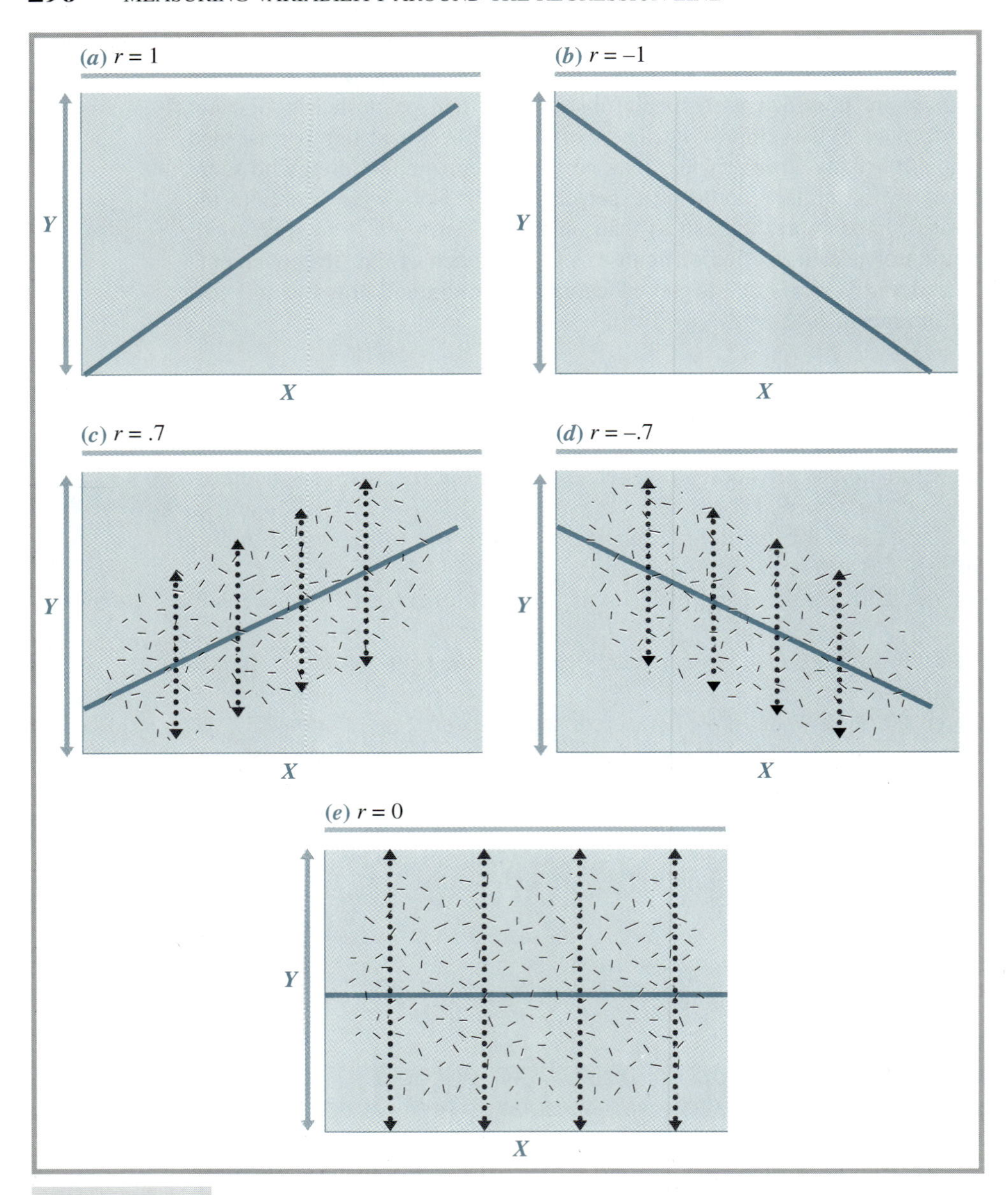

Figure 10–7 Variation of *Y* scores around the regression line.

In other words, all predictions are exactly correct, and all errors of estimate are equal to zero. When r is moderately strong (diagrams c and d), not all predictions are correct, and Y scores at particular values of X (i.e., within particular arrays) vary around predicted Y scores. However, the variation within arrays is less than the total variation of Y scores. When $r = 0$ (diagram e), the variation of Y scores around predicted scores within each array is as great as the total Y variation.

A regression analysis is not complete without information about the amount of variation of actual criterion scores around predicted scores. Let's consider how this variation is measured.

The Residual Variance and the Standard Error of Estimate

We measure errors in prediction by calculating the variance and standard deviation of actual Y scores around predicted Y scores. The variance of actual Y scores around predicted Y scores is called the ***residual variance*** S^2_{res}, and the standard deviation of Y scores around predicted scores, is called the ***standard error of estimate,*** S_{est}. The residual variance is equal to the sum of the squared errors of estimate divided by N.

$$S^2_{res} = \frac{\Sigma(Y - \hat{Y})^2}{N}$$

Deviation formula for the residual variance; variance of actual Y *scores around predicted* Y *scores*

The standard error of estimate is the square root of the residual variance.

$$S_{est} = \sqrt{S^2_{res}} = \sqrt{\frac{\Sigma(Y - \hat{Y})^2}{N}}$$

Deviation formula for the standard error of estimate; standard deviation of actual Y *scores around predicted* Y *scores*

Table 10–4 illustrates the use of the formulas for calculating the residual variance and the standard error of estimate for the Acme Insurance Company data. Interview scores (X) and obtained sales scores (Y) are listed in columns (a) and (b) of the table. The Acme regression equation, $\hat{Y} = 6.59 + 4.40X$, was used to calculate the predicted sales ($\hat{Y}$) of each salesperson, column (c). Columns (d) and (e) list errors of estimate or residuals ($Y - \hat{Y}$), and squared residuals ($Y - \hat{Y})^2$, respectively. The sum of the squared residuals, $\Sigma(Y - \hat{Y})^2$, is 1034.6786.

Then

$$S^2_{res} = \frac{\Sigma(Y - \hat{Y})^2}{N} = \frac{1034.6786}{26} = 39.795$$

$$S_{est} = \sqrt{39.795} = 6.31$$

Let us compare the residual criterion variance with the total sales variance at Acme. According to Table 10–1, the total variance of sales, S^2_Y, is equal to 93.783. Acme's residual variance, 39.795, is smaller than the total variance. That is, Y scores vary less around the predicted Y scores than around the mean of all the Y scores, as is to be expected since the correlation between interview scores and sales (.76) is moderately strong.

S^2_{res} and S_{est} measure the variability of Y scores around predicted scores over the entire data set. If all the arrays of the data set have equal variability (i.e., if the data display homoscedasticity), the residual variance can be interpreted as the variance of criterion scores within each array, and the standard error of estimate as the standard deviation of criterion scores within each array. If, for example, Acme had collected data on several hundred salespersons rather than just 26, and if the amount of variability in sales were the same for all interview scores, we could make statements such as "The standard deviation of sales for persons with interview score = 2 (or any other specific interview score) is 6.31."

Since the standard error of estimate is a kind of standard deviation, it can be interpreted and used in much the same way. We know, for example, that in a normal distribution 68% of observations fall within one standard deviation of the mean (between mean ± 1 SD), and about 95% of observations fall within two standard deviations of the mean (between mean ± 2 SD). Therefore, if criterion scores are normally distributed, 68% of obtained criterion scores fall within one standard error of estimate of predicted

Table 10–4 Worktable for Calculating the Residual Variance, S^2_{res}, by the Formula $S^2_{\text{res}} = \frac{\Sigma(Y - \hat{Y})^2}{N}$

Salesperson	(a) Interview Score (X)	(b) Sales Score (Y)	(c) Predicted Sales Score ($\hat{Y}$)	(d) Error of Estimate (Residual) $Y-\hat{Y}$	(e) $(Y-\hat{Y})^2$
1	3	34	19.79	14.21	201.9241
2	1	7	10.99	–3.99	15.9201
3	4	19	24.19	–5.19	26.9361
4	2	3	15.39	–12.39	153.5121
5	1	19	10.99	8.01	64.1601
6	0	5	6.59	–1.59	2.5281
7	5	30	28.59	1.41	1.9881
8	4	32	24.19	7.81	60.9961
9	1	10	10.99	–0.99	0.9801
10	2	8	15.39	–7.39	54.6121
11	3	25	19.79	5.21	27.1441
12	0	2	6.59	–4.59	21.0681
13	1	17	10.99	6.01	36.1201
14	0	14	6.59	7.41	54.9081
15	2	16	15.39	0.61	0.3721
16	1	3	10.99	–7.99	63.8401
17	5	21	28.59	–7.59	57.6081
18	4	16	24.19	–8.19	67.0761
19	5	35	28.59	6.41	41.0881
20	3	21	19.79	1.21	1.4641
21	3	16	19.79	–3.79	14.3641
22	4	24	24.19	–0.19	0.0361
23	0	9	6.59	2.41	5.8081
24	5	27	28.59	–1.59	2.5281
25	3	27	19.79	7.21	51.9841
26	2	13	15.39	–2.39	5.7121
					1034.6786 $\Sigma(Y-\hat{Y})^2$

criterion scores, and about 95% of criterion scores fall within two standard errors of estimate of predicted criterion scores.

In 68% of individuals, Y is between $\hat{Y} \pm S_{\text{est}}$.
In about 95% of individuals, Y is between $\hat{Y} \pm 2\ S_{\text{est}}$.

To continue with the Acme example, assume, again, that N is very large and that the distribution of sales at Acme is approximately normal. The regression equation for predicting sales from interview scores is, again, $\hat{Y} = 6.59 + 4.40X$, and the predicted sales for salespersons with interview score $(X) = 3$ is $6.59 + 4.40(3) = 19.79$. Since $S_{\text{est}} = 6.31$, we can now state

68% of persons with interview score = 3 sell between 19.79 ± 6.31, or between 13.48 and 26.10 policies.
About 95% of persons with interview score = 3 sell between 19.79 ± 2(6.31), or between 7.17 and 32.41 policies.

EXERCISE 10–4

1. Here are anxiety and extraversion scores for seven high school students.

Student	Anxiety	Extraversion
1	4	4
2	2	6
3	4	1
4	3	2
5	5	3
6	1	6
7	2	7

The regression equation for predicting extraversion from anxiety is $\hat{Y} = 7 - X$.

a. Calculate the predicted extraversion score for each student. Then calculate the residual variance and the standard error of estimate.

b. Answer without calculating: Is the residual variance equal to, less than, or greater than the total variance of the extraversion scores? After you have answered this question, calculate the variance of the extraversion scores to check your answer.

2. The regression equation for a certain set of data is $\hat{Y} = 10 + .75X$. The standard error of estimate is 3. Scores on Y are normally distributed, and the relation between X and Y displays homoscedasticity.

a. 68% of individuals with $X = 12$ have Y scores between what two values?

b. About 95% of individuals with $X = 8$ have Y scores between what two values?

A Return to Components of Variance

We have seen that, unless $r = \pm 1$, prediction is not perfect. Even when we use the relationship between X and Y to minimize errors in prediction of Y, there remains residual (unexplained) variation of actual criterion scores around the predicted values. We have measured this unexplained variation of Y scores around predicted Y scores by calculating the residual variance, S^2_{res}, or its square root, the standard error of estimate, S_{est}.

But part of the total variation of Y scores is covariation with X; it is variation that we expect because of the correlation between X and Y. For example, if r is positive, we expect the Y scores of individuals who fall above the mean on X to vary from the mean of Y in the positive direction, and we expect the Y scores of individuals who fall below the mean on X to vary from M_y in the negative direction. At Acme, for example, the mean of sales (M_y) is 17.42. However, the predicted scales ($\hat{Y}$) for a person with interview score = 3 is not 17.42 but 19.79. Because this person's interview score is above the mean, we expect him or her to sell 2.37 more policies than the average salesperson. Thus, for any given individual, $\hat{Y} - M_y$ is a measure of *expected* variation, the amount of variation in Y that is explained by X. To obtain a measure of the explained Y variation over the entire data set, we calculate the variance of predicted Y scores ($\hat{Y}$) around the mean of Y (M_y). We symbolize this variance $S^2_{\hat{Y}}$.

$$S^2_{\hat{y}} = \frac{\Sigma(\hat{Y} - M_y)^2}{N}$$

Explained variance; variance of predicted Y *scores around the mean of* Y

Table 10–5 shows the calculation of $S^2_{\hat{Y}}$ for the Acme data. Predicted sales are listed in column (a). Column (b) shows the deviation of each salesperson's predicted sales from the mean of sales, 17.42 Each deviation is squared in column (c). As you can see, $\Sigma(\hat{Y} - M_y)^2 = 1402.8554$. Then $S^2_{\hat{Y}} = 1402.8554/26 = 53.956$. The explained Y variance

Table 10–5 Worktable for Calculating the Variance of Predicted Scores ($S^2_{\hat{y}}$) for the Acme Insurance Company Data

Salesperson	(a) Predicted Sales Score $\hat{Y}$	(b) $\hat{Y} - M_Y$	(c) $(\hat{Y} - M_Y)^2$
1	19.79	2.37	5.6169
2	10.99	–6.43	41.3449
3	24.19	6.77	45.8239
4	15.39	–2.03	4.1209
5	10.99	–6.43	41.3449
6	6.59	–10.83	117.2889
7	28.59	11.17	124.7689
8	24.19	6.77	45.8329
9	10.99	–6.43	41.3449
10	15.39	–2.03	4.1209
11	19.79	2.37	5.6169
12	6.59	–10.83	117.2889
13	10.99	–6.43	41.3449
14	6.59	–10.83	117.2889
15	15.39	–2.03	4.1209
16	10.99	–6.43	41.3449
17	28.59	11.17	124.7689
18	24.19	6.77	45.8329
19	28.59	11.17	124.7689
20	19.79	2.37	5.6169
21	19.79	2.37	5.6169
22	24.19	6.77	45.8329
23	6.59	–10.83	117.2889
24	28.59	11.17	124.7689
25	19.79	2.37	5.6169
26	15.39	–2.03	4.1209
			1402.8554 $\Sigma(\hat{Y}-M_y)^2$

Note: $M_y = 17.42$.

is 53.956; we expect this much variance of the Y scores around the mean of Y based on the correlation between X and Y.

Recall that the residual (unexplained) sales variance, S^2_{res}, at Acme was 39.795. If we add the explained and residual variances, we get 53.956 + 39.795 = 93.751. Except for a small difference due to rounding, this is equal to the total variance of sales (93.78). Thus,

S^2_y	=	$S^2_{\hat{y}}$	+	S^2_{est}
Total variance	=	Explained variance	+	Residual (unexplained) variance

This relationship holds true not just at Acme but for all data sets. Thus, regression analysis lets us apportion the total criterion variance into two components: a component that is explained by the predictor X and a component that is not explained by X.

Because $S^2_y = S^2_{\hat{y}} + S^2_{\text{res}}$, the ratio $S^2_{\hat{y}}/S^2_y$ is the proportion of the total Y variance that is explained by variable X. At Acme, $S^2_{\hat{y}}/S^2_y = 53.956/93.783 = .5753$; about 58% of Acme's sales variance is explained by differences in interview scores.

If we take the square root of .5753, we get ± .76. Recall that the correlation between interview scores and sales at Acme was .76. In other words, the correlation coefficient, r, is the square root of $S_{\hat{y}}^2/S_y^2$.

$$r^2 = \frac{S_{\hat{y}}^2}{S_y^2}$$

Similarly, it can also be shown that

$$r^2 = \frac{S_{\hat{x}}^2}{S_x^2}$$

This is true not just for Acme but for all data sets. Thus, as you learned in Chapter 9, the square of the correlation coefficient between two variables is the proportion of the variance of each variable that is explained by the other. You may recall that, for this reason, r^2 is called the *coefficient of determination.*

In sum, we have shown that the total Y variance, S_y^2, can be apportioned into two components, one component that is predicted (explained) by X and one component that is not. The explained component is the variance of predicted Y scores around the mean of Y, $S_{\hat{y}}^2$. The unexplained component is the variance of actual Y scores around predicted Y scores, S_{res}^2. The proportion of the total Y variance that is explained, $S_{\hat{y}}^2/S_y^2$, is equal to r^2, the squared correlation between X and Y.

Alternative Formulas for the Residual Variance and the Standard Error of Estimate

Unless N is very small, the deviation formulas for calculating the residual variance, S_{res}^2, and the standard error of estimate, S_{est}, are computationally laborious because they require the calculation of $Y - \hat{Y}$ for each individual. We can use the relationship between $S_{\hat{y}}^2$, S_y^2, and r to develop alternative formulas for S_{res}^2 and S_{est}. We begin with

$$r^2 = \frac{S_{\hat{y}}^2}{S_y^2}$$

In the previous section, we showed that $S_y^2 = S_{\hat{y}}^2 + S_{\text{res}}^2$. Therefore, $S_{\hat{y}}^2 = S_y^2 - S_{\text{res}}^2$. Substituting in the equation above, we now have

$$r^2 = \frac{S_y^2 - S_{\text{res}}^2}{S_y^2}$$

If we solve this equation for S_{res}^2, we get

$$S_{\text{res}}^2 = S_y^2(1 - r^2)$$ *Alternative formula for the residual variance*

Taking the square root of both sides

$$S_{\text{est}} = S_y\sqrt{1 - r^2}$$ *Alternative formula for the standard error of estimate*

Let us apply these formulas to Acme's data. At Acme, the variance of sales, S_y^2, is equal to 93.783, and r is equal to .76. Therefore,

$$S_{\text{res}}^2 = 93.783\,[1 - (.76)^2] = 39.614$$

$$S_{\text{est}} = 9.684\,\sqrt{1 - (.76)^2} = 6.29$$

Except for a small difference due to rounding, the results are the same as those we obtained using the deviation formulas.

The formula for S^2_{res} can be used to show why the residual variance decreases as r increases, and why its maximum value is equal to the variance of Y, as we observed earlier.

$$\text{If } r = \pm 1: \quad S^2_{\text{res}} = S^2_y[1 - (\pm 1)^2] = S^2_y(1-1) = S^2_y(0) = 0$$

When $r = \pm 1$, all the points lie on the regression line, and there is no residual variance, as we saw in Figures 10–7a and 10–7b.

$$\text{If } r = 0: \quad S^2_{\text{res}} = S^2_y(1 - 0^2) = S^2_y(1) = S^2_y$$

When $r = 0$, the variance of Y scores around predicted Y scores is equal to the total Y variance (see Figure 10–7e). This is the maximum possible value of S^2_{res}. For values of r between $\pm$ 1 and zero, S^2_{res} is a number between zero and S^2_y; and the residual variance, though not zero, is less than the total Y variance, as you can see in Figures 10–7c and 10–7d.

EXERCISE 10–5

1. The correlation between scores on a test of clerical aptitude and secretarial skills at a certain company is .30. The standard deviation of secretarial skills is 7.5.
   a. Calculate the standard error of estimate in predicting secretarial skills from the clerical aptitude test.
   b. Assume that the relation between aptitude test scores and secretarial skills displays homoscedasticity. What is the standard deviation of secretarial skills for secretaries with aptitude test score = 25?
   c. Calculate the explained variance of secretarial skills. (Hint: Remember that $S^2_y = S^2_{\hat{y}} + S^2_{\text{res}}$.)
   d. Suppose that the company finds a better clerical aptitude test, one whose correlation with secretarial skills is .60. First, predict the effect on the standard error of estimate, relative to your answer in part a. Then calculate the standard error of estimate for the new test.
2. At West Podunk University, the mean SAT score of freshmen is 425, with *SD* = 75; the freshmen's mean GPA is 2.20, with *SD* = 0.50. The correlation between SAT and GPA is .60.
   a. Generate the regression equation for predicting GPA from SAT scores.
   b. Calculate the residual variance and the standard error of estimate.
   c. Assume that GPA is normally distributed in the students, and that the relation between SAT and GPA displays homoscedasticity. Find the GPA values between which the middle 95% of freshmen with SAT = 350 fall.

Going Beyond Sample Regression: Making Predictions in the Population

The Relation Between Sample and Population Regression Analyses

Up to now our emphasis has been on developing regression equations and measuring variation around predicted criterion scores in samples of data. Usually, we want to generalize beyond samples to the populations from which the samples came. The Acme

Company, for example, is really not interested in predicting sales of the 26 current salespersons; after all, their sales figures are known. Acme wants to predict sales of applicants who have not yet been hired and whose sales, therefore, are presently not known. Can Acme use the results of the sample regression analysis for this purpose? In order to answer this question, we need information about the relation between the results of a regression analysis in a sample and the results of the regression analysis in the population. We will consider, first, the relation between the sample regression equation and the regression equation in the population. Then we will examine the relation of the sample's residual variance and standard error of estimate to those in the population.

The Relation Between Sample and Population Regression Equations

The sample regression equation for predicting Y from X includes two sample statistics, the slope or regression coefficient b and the Y-intercept a. Like any statistic, b and a vary over samples. Therefore, the regression equation differs from one sample to another. Fortunately, however, though b and a vary over samples, they are unbiased estimators of the population regression coefficient and Y-intercept. That is, the average value of b over all samples is equal to the slope of the population regression equation; and the average value of a over all samples is equal to the Y-intercept of the population regression equation. Therefore, the sample regression equation, $\hat{Y} = a + bX$, is the best estimator of the population regression equation and can be used without change or adjustment to predict criterion scores in the population.

However, it is well to keep in mind that the values of b and a calculated in a sample minimize errors of estimate *in that sample*. The values of b and a calculated in another sample may be quite different, especially if N is small. In real-life applications of regression, the regression equation developed in one sample is usually tried out in a second, comparable sample before being endorsed for general use. The process of trying out a regression equation on a second sample to see whether its usefulness extends beyond the original sample is called *cross-validation*.

The Relation Between Sample and Population Residual Variances

Unfortunately, even though a and b are unbiased estimators of the corresponding population values, the sample residual variance, $S_{\text{res}}^2 = \dfrac{\Sigma(Y - \hat{Y})^2}{N}$, is a biased estimator of the residual variance in the population. Specifically, it tends to underestimate the population residual variance. An unbiased estimator of the population's residual variance, which we will symbolize s_{res}^2, is obtained by dividing $\Sigma(Y-\hat{Y})^2$ by $N - 2$ rather than by N:

$$s_{\text{res}}^2 = \frac{\Sigma(Y - \hat{Y})^2}{N - 2}$$ *Unbiased estimator of the residual variance in the population*

At Acme, for example, where $N = 26$ and $\Sigma(Y-\hat{Y})^2 = 1034.6786$, $s_{\text{res}}^2 = 1034.6786/24 = 43.112$; we estimate that the residual variance of sales in the population of all Acme salespersons is 43.112.

The square root of the estimated residual variance, s_{est}, is the best available estimate of the population standard error of estimate:

$$s_{\text{est}} = \sqrt{\frac{\Sigma(Y - \hat{Y})^2}{N - 2}}$$ *Estimated population standard error of estimate*

At Acme, for example, the estimated standard error of estimate of sales in the population is $\sqrt{43.112} = 6.57$.

Why does the formula for the estimated population residual variance use $N - 2$ in the denominator instead of N? $N - 2$ is the number of degrees of freedom for estimating the residual population variance from sample data. Recall that the degrees of freedom of an estimator is equal to the sample size, N, minus the number of previously calculated sample statistics on which the estimator depends. The estimated residual variance is dependent on values of $\hat{Y}$. $\hat{Y}$, in turn, is dependent on two sample statistics, b and a. Consequently, the degrees of freedom for estimating the residual variance of the population from sample data is $N - 2$.

When the sample size is small, the estimated population residual variance, s^2_{res}, may be quite different from the sample residual variance, S^2_{res}. For example, at Acme, the estimated population residual variance, $s^2_{\text{res}} = 43.112$, is considerably larger than the sample residual variance, $S^2_{\text{res}} = 39.795$. If N is small, as it is at Acme, the substitution of $N - 2$ for N is vital when inferences about the population are to be made from sample data.

If, however, N is large, as it often is in regression analyses, the difference between S^2_{res} and s^2_{res} is small. For example, if $\Sigma(Y - \hat{Y})^2 = 5000$ and $N = 200$, $S^2_{\text{res}} = 5000/200 = 25$ and $s^2_{\text{res}} = 5000/198 = 25.25$. The difference between the sample standard error of estimate, S_{est}, and the estimated population standard error of estimate, s_{est}, is even smaller; $S_{\text{est}} = \sqrt{25} = 5.00$ and $s_{\text{est}} = \sqrt{25.25} = 5.02$. With large samples, therefore, it makes little difference whether formulas incorporating N or $N - 2$ are used to calculate the residual variance and the standard error of estimate.

The estimated residual variance and standard error of estimate can also be calculated by the following formulas, which are computationally less laborious than the deviation formulas:

$$s^2_{\text{res}} = S^2_y(1 - r^2)\left(\frac{N}{N-2}\right)$$

$$s_{\text{est}} = S_y\sqrt{(1 - r^2)\left(\frac{N}{N-2}\right)}$$

Computational formulas for s^2_{res} *and* s_{est}, *assuming* $S^2_y = SS_y / N$

In these formulas, S^2_y is the sample Y variance, SS_y/N. If the estimated population variance, $s^2_y = SS_y/(N - 1)$, is used instead, the formulas for s^2_{res} and s_{est} are

$$s^2_{\text{res}} = s^2_y(1 - r^2)\left(\frac{N-1}{N-2}\right)$$

$$s_{\text{est}} = s_y\sqrt{(1 - r^2)\left(\frac{N-1}{N-2}\right)}$$

Computational formulas for s^2_{res} *and* s_{est}, *assuming* $s^2_y = SS_y /(N - 1)$

At the Acme Company, for example, the sample sales variance, S^2_y, is equal to 93.783, and the sample standard deviation, S_y is equal to 9.684. The correlation between interview scores and sales is .76. Therefore,

$$s^2_{\text{res}} = S^2_y(1 - r^2)\left(\frac{N}{N-2}\right) = 93.783[1 - (.76)^2]\left(\frac{26}{24}\right) = 43.149^{\dagger}$$

$$s_{\text{est}} = S_y\sqrt{(1 - r^2)\left(\frac{N}{N-2}\right)} = 9.684\sqrt{[1 - (.76)^2]\left(\frac{26}{24}\right)} = 6.57$$

†The small difference from the answer obtained by using the formula $s^2_{\text{res}} = \dfrac{\Sigma(Y - \hat{Y})}{N - 2}$ is due to rounding.

Since the difference between the biased estimator S^2_{est} and the unbiased estimator s^2_{est} is small when N is large, the term $[N/(N - 2)$ or $(N - 1)/(N - 2)]$ can be dropped from the calculation when working with large samples.

Calculating Prediction Intervals for Criterion Scores

If criterion scores in the population are assumed to be normally distributed, and if the relation between X and Y displays homoscedasticity, it is possible to calculate ***prediction intervals for criterion scores***. These are intervals constructed around predicted criterion scores within which we are confident, at some level, that actual criterion scores lie. Any level of confidence can be chosen; however, the 95% and the 99% levels of confidence are the most commonly used. Each level of confidence has an associated level of alpha (α). For the 95% level of confidence, $\alpha = .05$; for the 99% level of confidence, $\alpha = .01$; and so on.

The formula for calculating the criterion score prediction interval (PI) for a given predictor score, X_i, is

$$\text{PI} = \hat{Y} \pm t_{\alpha/2} s_{\text{est}} \sqrt{1 + \frac{1}{N} + \frac{(X_i - M_x)^2}{SS_x}}$$

Formula for calculating a prediction interval for a criterion score

Let us consider the terms to the right of the equal sign, one at a time:

$\hat{Y}$ is the predicted criterion score for $X = X_i$, obtained by plugging the value $X = X_i$ into the regression equation.

$t_{\alpha/2}$ is the critical value of t in a two-tailed test with the chosen α and $N - 2$ degrees of freedom. Values of $t_{\alpha/2}$ can be found in the t table, Table D–3 in Appendix D.

s_{est} is the estimated population standard error of estimate, calculated by the procedures discussed in the previous section.

The term $\sqrt{1 + \frac{1}{N} + \frac{(X_i - M_x)^2}{SS_x}}$ is a kind of correction factor. It is needed because sample-to-sample variation of criterion scores around predicted criterion scores increases as $X_i - M_x$ increases. Consequently, in order to be equally confident that we have "captured" the actual criterion score within an interval around $\hat{Y}$, we have to make the interval wider if the predictor score X_i is far from the mean than if it is near the mean. The term $\sqrt{1 + \frac{1}{N} + \frac{(X_i - M_x)^2}{SS_x}}$ makes the needed adjustment.

As N increases, the correction factor approaches 1. Therefore, if N is very large, this term can be dropped, and the prediction interval can be calculated by the simpler formula:

$$\text{PI} = \hat{Y} \pm t_{\alpha/2} s_{\text{est}}$$

Prediction interval for a criterion score if N *is very large*

Suppose that the Acme Company wants to calculate the 95 percent prediction interval for criterion scores (sales) of applicants for sales positions. Since Acme's regression equation was developed in a small sample ($N = 26$), the correction factor must be included in the calculation.

With $N = 26$, $df = 24$. From the t table, we find that $t_{\alpha/2}$, the critical value of t for a two-tailed test with 24 df and $\alpha = .05$, is 2.064. In a previous section, we calculated

s_{est} = 6.57. From Table 10–1, we find M_x = 2.46 and SS_x = 72.46. Therefore the formula for Acme's 95% prediction interval is

$$PI = \hat{Y} \pm (2.064)(6.57)\sqrt{1 + \frac{1}{26} + \frac{(X_i - 2.46)^2}{72.46}}$$

$$PI = \hat{Y} \pm 13.56\sqrt{1.0385 + \frac{(X_i - 2.46)^2}{72.46}}$$

The values of $\hat{Y}$ and X_i, of course, vary from one applicant to another. X_i is the applicant's interview score, and $\hat{Y}$ is the predicted sales obtained by plugging the value X_i into Acme's regression equation, $\hat{Y} = 6.59 + 4.40X$.

Let us calculate the 95% prediction interval for two Acme applicants, Carl, whose interview score 5 is far from the mean, and Cindy, whose interview score 3 is near the mean. First, we use Acme's regression equation, $\hat{Y} = 6.59 + 4.40X$, to predict sales ($\hat{Y}$) for Carl and Cindy:

Carl: $\hat{Y} = 6.59 + 4.40(5) = 28.59$

Cindy: $\hat{Y} = 6.59 + 4.40(3) = 19.79$

The 95% prediction interval for Carl is

$$PI = 28.59 \pm 13.56\sqrt{1.0385 + \frac{(5 - 2.46)^2}{72.46}} = 28.59 \pm 14.399 = 14.191 \text{ to } 42.989$$

We are 95% confident that Carl will sell between 14.19 and 42.99 policies.

The prediction interval for Cindy is

$$PI = 19.79 \pm 13.56\sqrt{1.0385 + \frac{(3 - 2.46)^2}{72.46}} = 19.79 \pm 13.845 = 5.945 \text{ to } 33.635$$

We are 95% confident that Cindy will sell between 5.945 and 33.635 policies. Note that because Carl's interview score is farther from the mean than Cindy's, the prediction interval for his criterion score is wider.

What do we mean when we say that we are "95% confident" that an individual criterion score falls within the interval we have calculated? The interpretation is similar to that for confidence intervals for a population mean. Prediction intervals, like confidence intervals, vary from one sample to another. However, the criterion score for any given individual with $X = X_i$ will be included within the prediction interval for $X = X_i$ in 95% of samples.

EXERCISE 10–6

1. First predict whether the 99% prediction interval for a criterion score will be wider or narrower than the 95% interval. Then calculate the 99% prediction interval for sales of Carl and Cindy.
2. A mental health center collected data on 25 clients and developed the following regression equation for predicting number of treatment sessions (Y) from scores on an intake inventory (X): $\hat{Y} = 2.29 + .15X$. The following statistics were also calculated: $M_x = 35.4$, $SS_x = 1350$, $s_{est} = 2.57$.

a. Mr. Smith and Ms. Jones are new clients. Mr. Smith scored 15 on the intake interview; Ms. Jones scored 40. Which client will have a wider 95% prediction interval for number of treatment sessions?
b. Calculate the 95% prediction interval for Mr. Smith's and Ms. Jones's number of treatment sessions.

Another Example of Regression

We have followed the fortunes of the Acme Insurance Company throughout this chapter. We began by developing the regression equation for predicting sales from interview scores, went on to calculate and interpret the standard error of estimate, discussed the relation between the sample regression equation and regression for future applicants, and finished with the calculation and interpretation of prediction intervals for sales of future employees. As a means of summarizing the various components of regression analyses, we work through all the steps in another example.

Let us suppose that a school district gave an aptitude test to 100 eighth-grade students and then measured their grade point average in ninth grade. The mean aptitude test score of the 100 students was 38.0, with a standard deviation of 7.42. In ninth grade, the students achieved a mean GPA of 2.30, with a standard deviation of .55. The correlation between aptitude test scores and GPA was .65. The school district wants to use these data to develop a regression equation for predicting ninth-grade GPA from aptitude test scores, as well as to calculate 95 percent prediction intervals, for future students. We will assume that the relation between aptitude test scores and GPA is linear, that GPA is normally distributed in the population of ninth-grade students, and that the relation between the two variables displays homoscedasticity (i.e., the amount of variation of GPA's is assumed to be the same at all score levels on the aptitude test).

Calculation

Let X = aptitude test scores
Let Y = GPA

$M_x = 38.0$ $\quad M_y = 2.30$
$s_x = 7.42$ $\quad s_y = .55$
$r = .65$

Explanation

The aptitude test is the predictor, X.
The criterion variable, Y, is GPA.

The information given verbally in the paragraph above is translated into symbols.

Note: It is not clear whether the two standard deviations were calculated by the formula $\sqrt{SS/N}$ or $\sqrt{SS/(N-1)}$. We will assume the latter. However, since N = 100, the difference between the two values is small.

Calculation of Regression Equation

$$b = r\left(\frac{S_y}{S_x}\right) = .65\left(\frac{.55}{7.42}\right) = .048$$

$a = M_y - bM_k = 2.30 - (0.48)(38.0) = .476$
Regression equation: $\hat{Y} = .476 + .048X$

The regression equation for predicting GPA from aptitude test scores is generated by first calculating the regression coefficient or slope, b, and the Y-intercept, a. Then b and a are plugged into the general form of the regression equation: $\hat{Y} = a + bX$, where $\hat{Y}$ = predicted GPA and X = aptitude test score.

Calculation of Standard Error of Estimate

$$s_{\text{est}} = s_y\sqrt{(1-r^2)\left(\frac{N-1}{N-2}\right)}$$

$$= .55\sqrt{[1-(.65)^2]\left(\frac{99}{98}\right)}$$

$$= .42$$

Because the district wants to predict GPA for future students, the estimated population standard error of estimate, s_{est}, is calculated.

Interpretation of $s_{\text{ext}} = .42$: Because s_{est} is the *SD* of each array of *Y* scores, the *SD* of GPA's for students with any given aptitude test score is estimated to be .42.

Calculation of 95% Prediction Interval

$$t_{\alpha/2} = 1.98$$

$$SS_x = (N-1)\ s_x^2 = 99(7.42)^2$$

$$= 5450.58$$

Besides s_{est} which has already been calculated. The values of $t_{\alpha/2}$ and SS_x are needed.

$$\text{PI} = \hat{Y} \pm t_{\alpha/2} s_{\text{est}}\sqrt{1+\frac{1}{N}+\frac{(X_i-M_x)^2}{SS_x}}$$

$$= \hat{Y} \pm (1.98)(.42)\sqrt{1+\frac{1}{100}+\frac{(X_i-38.0)^2}{5450.58}}$$

For the 95% level of confidence, $\alpha = .05$. Since $N = 100$, $df = 98$. The critical value of t for a two-tailed test with 98 *df* and $\alpha = .05$ is 1.98.

$$\text{Since } s_x^2 = \frac{SS_x}{N-1},\ SS_x = (N-1)s_x^2$$

$$= \hat{Y} \pm .8316\sqrt{1.01+\frac{(X_i-38.0)^2}{5450.58}}$$

This is the formula for calculating the 95% prediction interval for GPA for any given aptitude test score, X_i.

Example of Calculation of the Prediction Interval for a Student with an Aptitude Test Score of 22

$$\hat{Y} = .476 + .048(22) = 1.532$$

First, we use the regression equation to calculate the student's predicted GPA. For $X_i = 22$, $\hat{Y} = 1.532$.

$$\text{PI} = 1.532 \pm .8316\sqrt{1.01+\frac{(22-38.0)^2}{5450.58}}$$

$$= 1.532 \pm .855 = .677 \text{ to } 2.387$$

$X_i = 22$ and $\hat{Y} = 1.532$ are entered into the formula for the prediction interval.

We are 95% confident that a student with an aptitude test score of 22 will attain a GPA between .677 and 2.38.

A Comparison of Predictions With and Without the Use of Regression

The calculations involved in regression analyses are complex and laborious (though considerably less so with a statistical calculator or computer). Why do researchers and practitioners regard all this labor as worthwhile? To answer this question, let us compare predictions of a criterion variable made with and without the use of regression. Again, we use Acme as an example.

First, let us consider what predictions of sales would look like if Acme had not developed a regression equation. Under these circumstances, if you were to ask Ms. Woods, Acme's manager of human resources, to predict the number of policies sold by Carl, who had an interview score of 5, Ms. Woods would have to answer: "Well, the mean number of policies sold by our salespeople is 17.42. Therefore, I predict that he will sell about 17.42 policies." If you were to ask Ms. Woods for some kind of error estimate, such as "17.42 ± how many?" she might answer, "Since the estimated standard deviation of policies sold by Acme salespeople is 9.88, the chances are about 95 in 100 that Carl will sell somewhere between 17.42 ± 2(9.88) policies, that is, between −2.34 and 37.18 policies."

By using Acme's regression equation, on the other hand, we predict that the most likely sales for Carl is 6.59 + 4.40(5) = 28.59. This prediction is better than 17.42 because it takes into account the fact that Carl had an interview score well above the mean and, given the positive correlation between interview scores and sales, will probably sell more than the mean number of policies.

Again, suppose you insist that Ms. Woods attach an error estimate to her prediction. To do this, she reports Carl's 95% prediction interval (which we calculated earlier) and responds, "The chances are about 95 in 100 that Carl will sell somewhere between 14.19 and 42.99 policies."

Compare the two predictions:

Without use of regression:	−2.34 to 37.18 policies (a range of 39.52)
With use of regression:	14.19 to 42.99 policies (a range of 28.8)

Not only is the prediction interval centered around a more accurate probable value of sales when regression is taken into account, but the width of the interval is narrower as well. In other words, the amount of Carl's probable sales is pinpointed more closely.

A skeptic might argue, "Yes, but look how wide Carl's prediction interval is, even though the correlation between interview scores and sales, .76, is quite high. So using regression doesn't result in highly accurate predictions of individual scores. There is still a large margin of error."

It is true that predictions for individuals, based on regression equations, are not highly accurate (unless $|r|$ is very large). However, use of regression techniques can considerably enhance prediction over groups. Consider the Acme example again. Assume that before Acme began using interview scores for selection of applicants, only 60% of the salespeople it hired sold an "adequate" number of policies. Assume also that the company hires 50% of its applicants. If, in the future, Acme uses its regression equation and hires only the 50% of applicants with the highest predicted sales, the percentage of hired salespeople who sell an "adequate" number of policies will increase from 60% to 86%.[†] Thus, use of regression analyses can enhance institutional decision-making to a considerable degree.

A Few Cautionary Notes: When Not to Use Regression Analyses

Given the interdependence between correlation and regression, regression analyses of the type discussed in this chapter have the same limitations as do the correlational methods

[†]The combination of 60% success rate and 50% hiring rate was selected arbitrarily. The results would be similar for other combinations. The success rate of 86% for a correlation of .76 was obtained from a table generated by Taylor and Russell (1939).

discussed in Chapter 9. First, fitting a regression line to paired values on two variables is only appropriate if the relation between the variables is linear. In addition, if heteroscedasticity (unequal scatter) is present, predictions from regression are more accurate at some values of the predictor variable and less accurate at others. Outliers can also have a substantial effect on the location of the regression line, especially if N is small. You should always construct a scatter diagram before undertaking any kind of correlational analysis, including regression. If curvilinearity, heteroscadasticity, or outliers are present, proceed with caution, if at all.

Also keep in mind that inference about population regression based on sample data is based on the assumption that the distribution of the criterion variable is normal with equal variability at all values of the predictor. Inferential procedures such as calculation of prediction intervals should be undertaken only if the assumptions appear reasonable for the data.

A Nontechnical Introduction to Multiple Regression

When businesses or educational institutions make predictions of future performance, they rarely do so on the basis of a single predictor variable. Besides SAT scores (both Verbal and Mathematical), colleges use high school GPA, high school rank, and other measures as well. Businesses use interviews, tests, recommendations, and the like. In research, too, interest usually lies in predicting a criterion variable from a set of two or more predictor variables. For example, a psychologist studying factors predictive of aggressive behavior in children is likely to measure many predictor variables, such as TV preferences, activity level, cognitive abilities, social skills, and aspects of the home environment. Equations that are developed to predict values of a criterion variable from a combination of two or more predictors are called ***multiple regression equations***.

Multiple regression equations take the general form:

$$\hat{Y} = a + b_1X_1 + b_2X_2 + \cdots + b_kX_k$$

General form of the multiple regression equation

where

$X_1, X_2, \cdots, X_k$ are scores on k predictors
$b_1, b_2 \cdots$ are regression coefficients for predictors X_1, X_2, etc.
a is the Y-intercept of the regression line

Here is an example of a multiple regression equation that includes three predictors ($k = 3$):

$$\hat{Y} = 32.35 + 0.43X_1 + 1.16X_2 - 0.22X_3$$

A person scoring 24 on X_1, 7 on X_2, and 95 on X_3 would have a predicted criterion score of $32.35 + 0.43(24) + 1.16(7) - 0.22(95) = 29.89$.

As you can see, multiple regression equations are very similar to those for single predictors, and they are easy to use. However, the calculations of b_1, b_2, etc., are considerably more complicated. In fact, no one in his right mind would do a multiple regression analysis by hand. Fortunately, many packaged computer programs for multiple regression are available. Therefore, the purpose of this section is not to teach you how to do a multiple regression analysis but to explain the underlying rationale and the proper interpretation of results.

FACTORS THAT INFLUENCE THE MULTIPLE REGRESSION EQUATION

Predictions of a criterion variable based on two or more predictors are more accurate than those based on just one, provided that:

1. Each predictor is correlated with the criterion.†
2. Each predictor measures something different from the other predictors; that is, each predictor adds some additional information to the prediction, beyond that provided by the other predictors.

To illustrate, suppose that you are trying to predict students' performance in a word processing course. You have identified two potential predictors, a perceptual speed test (X_1) and a test of language usage (X_2). You calculate the correlation of each test with course performance (Y) as well as the correlation between the tests. The correlation between the perceptual speed test and course performance is r_{x_1y}; the correlation between the language usage test and course performance is r_{x_2y}; and the correlation between the two tests is $r_{x_1x_2}$. Let us consider two possible outcomes.

Case 1

The correlations among the variables are as follows: $r_{x_1y} = .55$, $r_{x_2y} = .40$, and $r_{x_1x_2} = 10$. Thus, each test is correlated with the criterion, but the tests are essentially uncorrelated with one another. This means that the two tests are predicting different aspects of criterion performance. The situation is depicted in Figure 10–8.

Because of the small overlap between the predictors, each contributes something to criterion prediction that the other does not. In this case, prediction of course performance based on both tests is likely to be considerably more accurate than prediction based on either test alone.

Case 2

The correlations among the variables are as follows: $r_{x_1y} = .55$, $r_{x_2y} = .40$, and $r_{x_1x_2} = .80$. The correlation of each test with the criterion is the same as in Case 1; but this time the two tests have a high correlation with one another. Accordingly, they are, to a great extent, predicting the same aspects of criterion performance, as depicted in Figure 10–9.

Because of the large overlap between the two predictors, the amount of criterion variance that is predicted by the combination of the two tests is not much greater than that predicted by either one alone. Therefore, prediction of the criterion based on a combination of both tests is not much improved over prediction based on just one of them.

Because a predictor's usefulness depends on its relations with other predictors as well as with the criterion, the regression coefficient, b, of each predictor variable in a multiple regression equation is dependent not only on that predictor's correlation with the criterion but also on its correlations with the other predictors. In other words, a predictor's regression coefficient reflects what that predictor *adds* to the prediction of the criterion, beyond the contributions made by other predictors.

†This is somewhat oversimplified. Under some circumstances, a predictor that is not correlated with the criterion may contribute to prediction. However, this situation is beyond the scope of our discussion.

Figure 10–8

Prediction of criterion (Y) performance by two predictors (X_1 and X_2) with a low intercorrelation.

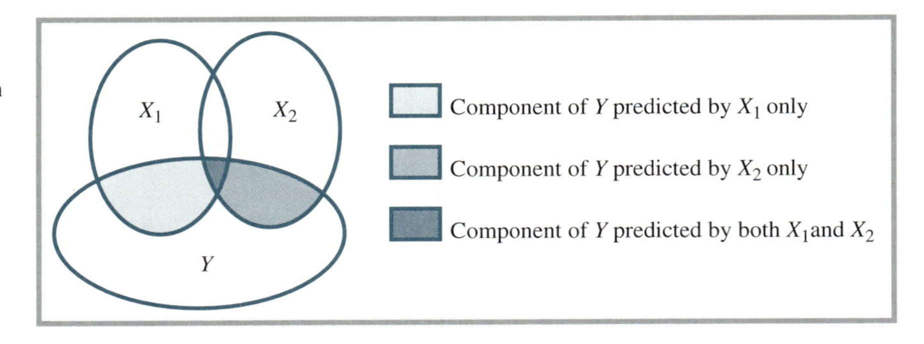

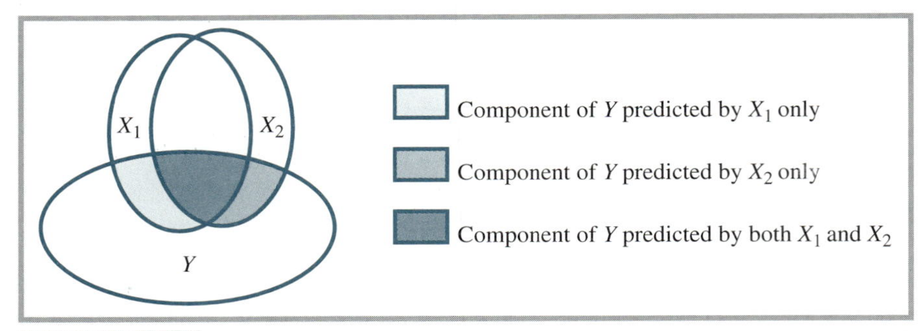

Figure 10–9 **Prediction of criterion (Y) performance by two predictors (X_1 and X_2) with a high intercorrelation.**

The Multiple Correlation Coefficient

It is possible to calculate a correlation coefficient that measures the linear relationship between the criterion variable, Y, and the set of predictors, X_1, X_2, $\cdots$. The correlation coefficient is called a ***multiple correlation coefficient***, symbolized $\boldsymbol{R}$. R is the correlation between the criterion and the weighted combination of the predictors. Thus it is not the correlation between Y and $(X_1 + X_2 + \cdots + X_k)$ but the correlation between Y and $(b_1X_1 + b_2X_2 + \cdots + b_kX_k)$.

Unlike simple correlation coefficients (r), which may be positive or negative, the multiple correlation coefficient, R, is always positive. (If any or all predictors are negatively related to Y, they are negatively weighted, so that their correlation with Y becomes positive.) The multiple R is always at least as large as the largest of the simple correlations of individual predictors with the criterion; in Case 1 and Case 2, for example, the correlation of the first test with the criterion is .55. Therefore, the multiple R of the two tests with the criterion is at least .55. How much larger R is than the largest predictor-criterion correlation depends on how much the remaining predictors add to the prediction of Y. In Case 1, where the second test adds substantially to prediction, R may be equal to .70 or more. In Case 2, on the other hand, where the second test adds little to prediction, R may be only .56 or .57.

Just as the squared correlation between two variables, r^2, is the proportion of criterion variance explained by the predictor, R^2 can be interpreted as the proportion of the criterion variance explained by the weighted combination of all the predictors. If, for example, the multiple correlation R between course performance and the two tests in Case 1 is .70, then the proportion of the variance of course performance which is explained by the weighted combination of the two tests is $(.70)^2 = .49$.

Measuring Errors of Prediction in Multiple Regression

In multiple regression, as in simple regression, the residual variance and the standard error of estimate measure variability of actual criterion scores around predicted scores. The formulas for calculation are the same as for simple regression, except that R is substituted for r. For example, the sample standard error of estimate (S_{est}) is equal to $\sqrt{\frac{\Sigma(Y-\hat{Y})^2}{N}}$, or $S_y\sqrt{1-R^2}$, and the estimated population standard error of estimate (s_{est}) is equal to $\sqrt{\frac{\Sigma(Y-\hat{Y})^2}{N-2}}$, or $s_y\sqrt{1-R^2\left(\frac{N-1}{N-2}\right)}$.

As R increases, the standard error of estimate decreases, just as in simple regression. In multiple regression, as in simple regression, prediction intervals for criterion scores can be calculated around predicted criterion scores. The formulas, procedures, and interpretations are the same as in simple regression.

EXERCISE 10–7

1. A company has developed the following equation for predicting productivity (Y) from a verbal test (X_1), a job knowledge test (X_2), and an interview evaluation (X_3):

   $$\hat{Y} = -200 + 2.7X_1 + 4.5X_2 + 0.6X_3$$

   a. Predict the most likely productivity for
      (1) Joe, with verbal score = 50, job knowledge score = 25, and interview score = 8.
      (2) Max, with verbal score = 30, job knowledge score = 40, and interview score = 5.
   b. How do you account for the fact that Max has a higher predicted productivity, despite the fact that both his verbal and interview scores were lower than Joe's?
2. Assume that reading ability, mathematics ability, and mechanical comprehension are about equally correlated with success in pilot training. If reading and mathematics ability are highly intercorrelated but mechanical comprehension has little relationship with either one, which of the following combinations will yield better predictions of success in pilot training?
   a. Reading ability plus mathematics ability
   b. Reading ability plus mechanical comprehension
3. In a group of 500 persons, the multiple correlation between a criterion and a set of four predictors is .64. The standard deviation of criterion scores is 7.
   a. What is the proportion of criterion variance that is accounted for by the weighted combination of the four predictors?
   b. Calculate the estimated population standard error of estimate.
   c. Assuming homoscedasticity and a normal distribution of criterion scores, calculate the 95% prediction interval for a person with a predicted criterion score of 21. (*Note:* Because N is large, you can omit the correction factor.)

Using a Computer to Conduct a Regression Analysis

Both simple and multiple regression are available on a large number of statistical software packages. As usual, the most tedious part of the procedure is typing the data into a data file. Once this has been done, the computer can calculate all the components of a regression analysis discussed in this chapter, and more.

We illustrate with the Acme Company data. The program used in the example is Minitab; however, the results would be the same with other programs. First, the identification numbers, interview scores, and sales of Acme's 26 employees were typed into columns 1 through 3 (c1, c2, c3), respectively, of a Minitab data file. Then the following sequence of one command with two subcommands (SUBC) was issued:

```
MTB  > regress c3    1    c2;
SUBC > predict 5;
SUBC > predict 3.
```

The command "regress c3 1 c2" instructed Minitab to regress the variable in c3 (sales) on *one* (1) predictor variable in c2 (interview scores). The two subcommands asked Minitab to give predicted sales and prediction intervals for interview scores of 5 (Carl's score) and 3 (Cindy's score). The output is displayed in Table 10–6.

The Minitab regression analysis output (and that from most other computer programs) provides a great deal of information, including some which we have not dealt with in this chapter. Let us work our way through the output to identify those parts which are relevant for us.

The first component in the output is the regression equation: (predicted) sales = 6.59 + 4.40 interview. The small table just below the equation provides additional information about the Y intercept a ("Constant") and the slope b ("interview"). Under "Coef," the values a and b are given to three or four decimal places. The remaining columns give results of tests for the significance of the Y-intercept and the slope. For example, the very small p for "interview" ($p = .000$ to three decimal places, or $p < .001$) tells us that the slope is significantly different from zero. Although we have not discussed this test, it is equivalent to the test of significance of the correlation coefficient (.76) between interview scores and sales, which you learned in Chapter 9. A test of H_o: $\rho = 0$ on the Acme data would result in $p < .001$.

Table 10–6 Minitab Output for the Regression Analysis of the Acme Insurance Company Data

```
The regression equation is sales = 6.59 + 4.40 intview

Predictor        Coef        Stdev      t-ratio        p
Constant        6.589        2.294         2.87    0.008
intview        4.4013       0.7713         5.71    0.000

s = 6.566      R-sq = 57.6%      R-sq(adj) = 55.8%

Analysis of Variance

SOURCE         DF        SS        MS         F          p
Regression      1    1403.7    1403.7     32.56      0.000
Error          24    1034.7      43.1
Total          25    2438.3

Unusual Observations
Obs.   intview    sales     Fit   Stdev.Fit   Residual   St.Resid
  1       3.00    34.00   19.79        1.35      14.21      2.21R

R denotes an obs. with a large st. resid.

  Fit    Stdev.Fit       95% C.I.            95% P.I.
28.60         2.34    (23.76, 33.43)     (14.20, 42.99)
19.79         1.35    (17.00, 22.59)      (5.95, 33.63)
```

The value "$s = 6.566$" is the estimated population standard error of estimate, which we have symbolized s_{est}. We calculated the same value earlier in the chapter. "R–sq" is r^2, the squared correlation coefficient. The Minitab output gives $r^2 = 57.6\%$, or .576. The square root of .576 is .76, the correlation between interview scores and sales.

The "Analysis of Variance" in the Minitab output is yet another test of the significance of the slope. Notice that p is, once again, $<.001$. (Computer outputs are often repetitive, presenting the same information in several different ways.) The layout of the table and the interpretation of its contents will become clear to you after you have read about analysis of variance in Chapter 12.

Next, Minitab tells us that there is one "unusual observation" in the data set, that is, an individual whose paired interview–sales scores deviate from the pattern displayed by most of the group. This is the individual with interview score = 3, sales = 34. This person is unusual because the "standardized residual," 2.21, is relatively large. In the scatter diagram of Figure 10–2, you can see that this individual not only falls rather far from the regression line but also relatively far from other individuals with the same interview score. Unusual observations alert the analyst to outliers that may affect the results of the regression analysis. In this particular case, we judge that the individual does not depart so markedly from the general trend as to affect the analysis.

The last two lines of the output give us the predictions for individuals with interview scores of 5 and 3. The values in the "Fit" column are the predicted sales of these two individuals, 28.60 and 19.79, respectively. The last column, "95% P.I.," gives the 95% prediction interval for each individual. The intervals are, of course, the same as those we calculated in an earlier section for Carl and Cindy, respectively.

SUMMARY

If two variables are correlated, it is possible to develop an equation called a ***regression equation*** for predicting the ***criterion variable***, Y, from the ***predictor variable***, X. The general form of the regression equation is $\hat{Y} = a + bX$. $\hat{Y}$ is the predicted criterion score; a is the Y-intercept, and b (the ***regression coefficient***) is the slope of the ***regression line***. The regression line is the line for which $\Sigma(Y-\hat{Y})^2$ is a minimum. Both b and a are calculated from sample data. The formulas for calculating b and a are

$$b = r\left(\frac{S_y}{S_x}\right) \quad \text{or} \quad b = \frac{N\sum XY - (\sum X)(\sum Y)}{N(\sum X^2) - (\sum X)^2}$$

$$a = M_y - bM_x \quad \text{or} \quad a = \frac{\sum Y - b\sum X}{N}$$

Some facts about regression include

1. Unless $r = \pm 1$, the regression equation for predicting X from Y is a different equation than the equation for predicting Y from X.
2. In all regression equations, the predicted Y for persons falling at the mean on X is the mean of Y.
3. If r is positive, the regression line has a positive slope; as X increases, $\hat{Y}$ increases. If r is negative, the slope of the regression line is negative; as X increases, $\hat{Y}$ decreases. If $r = 0$, the slope is zero, and the predicted Y for all values of X is the mean of Y.
4. Unless $r = \pm 1$, the predicted criterion score is always closer to the mean of Y, in *SD* units, than the predictor score is to the mean of X. This phenomenon is called ***regression toward the mean***.

The ***residual variance***, S^2_{res}, and its square root, the ***standard error of estimate***, S_{est}, measure variation of actual criterion scores around predicted criterion scores. S^2_{res} is calculated by the formula

$$S^2_{res} = \frac{\Sigma(Y-\hat{Y})^2}{N} \quad \text{or} \quad S^2_{res} = S_y\sqrt{1-r^2}$$

If $r = \pm 1$, actual scores and predicted scores are equal, and S^2_{res} and S_{est} are equal to zero. As $|r|$ decreases, S^2_{res} and S_{est} increase. When $r = 0$, the residual variance is equal to the total Y variance, S^2_Y, and the standard error of estimate is equal to the standard deviation of Y. If the variability of Y is equal at all values of X, S_{est} can be interpreted as the standard deviation of arrays of Y scores at fixed values of X.

The residual variance, S^2_{res}, is the unexplained component of the total Y variance. The variance of predicted scores around the mean of Y, $S^2_{\hat{Y}}$, is the explained component of the total variance. The total Y variance, S^2_Y, is equal to $S^2_{\hat{Y}} + S^2_{\text{res}}$. The proportion of total variance that is explained variance, $S^2_{\hat{Y}}/S^2_Y$, is equal to r^2.

The sample regression equation is an unbiased estimator of the population regression equation. However, the sample residual variance, S^2_{res}, is a biased estimator of the population residual variance. An unbiased estimator, s^2_{res}, is obtained by substituting $N - 2$, the number of degrees of freedom, for N in the calculation of the residual variance.

Prediction intervals for criterion scores can be constructed around predicted scores at any chosen level of confidence by utilizing the t distribution and the estimated standard error of estimate, s_{est}. Because the amount of sample-to-sample variation of predicted Y scores increases as the distance of X from the mean of X increases, the calculation of prediction intervals includes a correction factor that takes $X_i - M_X$ into account.

An equation for predicting a criterion from two or more predictors ($X_1, X_2, \cdots, X_k$) is called a ***multiple regression equation*** and takes the form $\hat{Y} = a + b_1X_1 + b_2X_2 + \cdots + b_kX_k$. The regression coefficients (b_1, b_2, etc.) are dependent on each predictor's unique contribution toward the prediction of Y. The correlation coefficient between Y and the weighted combination of predictors is called the ***multiple correlation coefficient***, R. R^2 is the proportion of Y variance accounted for by the combined weighted predictors. The calculation and interpretation of standard errors of estimate and prediction intervals for criterion scores are similar in multiple and simple regression.

Regression analyses are usually conducted using statistical software for computers. Computer outputs typically include not only the regression equation and the standard error of estimate but also tests for the significance of the slope and Y-intercept as well as other information. An example of a Minitab regression analysis was discussed.

QUESTIONS

1. A regression equation is $\hat{Y} = 13 + 1.6X$.
   a. At what value of Y does the regression line cross the Y-axis?
   b. As X increases, does $\hat{Y}$ increase or decrease?
   c. How much does $\hat{Y}$ change for each unit increase in X?
   d. Is the correlation between X and Y positive or negative?

2. An inventory measuring depression was administered to 500 clients in several mental health centers. Each client also completed a checklist of stressful life events. The mean score on the depression inventory was 28 with a standard deviation of 6. The mean number of stressful life events checked was 7, with a standard deviation of 2.5. The correlation between depression scores and number of stressful life events was .40.
   a. Generate the regression equation for predicting scores on the depression inventory from number of stressful life events.
   b. Calculate the standard error of estimate.
   c. Construct 95% prediction intervals for depression scores of clients reporting 3 and 8 stressful life events.

3. Here are pretest and posttest quiz scores of 10 students. Construct the regression equation for predicting posttest scores from pretest scores.

Pretest:	6	9	2	7	4	7	10	3	5	8
Posttest:	10	10	8	7	4	20	18	9	16	15

4. Here are means and standard deviations of measures of impulsivity and math skills in 300 first-grade children.

	Impulsivity	**Math Skills**
Mean	30	25
SD	5	8

a. Generate the regression equation for predicting math skills from impulsivity if the correlation between impulsivity and math skills is −.60.
b. Generate the regression equation if the correlation between impulsivity and math skills is −.20.
c. Johnny has an impulsivity score of 40. Before calculating, decide whether his predicted math skills score is higher if $r = -.60$ or if $r = -.20$.
d. Use the regression equations you calculated in parts a and b to predict Johnny's math skills if $r = -.60$ and if $r = -.20$. How do you account for the difference?
e. Before calculating, state whether the standard error of estimate is larger with $r = -.60$ or $-.20$. Then calculate the standard error of estimate in each case.

5. A regression equation is $\hat{Y} = 28.5$. What is the correlation between X and Y?

6. What is the predicted criterion z score for a predictor z score of 1.5 if

a. $r = -.85$?
b. $r = .30$?
c. $r = 0$?

7. A predictor has a mean of 20. The criterion mean is 50. The predictor scores of three persons are: Tim, 10, Alex, 20, Leslie, 25.

a. If $r = -.3$, which person has the highest predicted criterion score?
b. What is Alex's predicted criterion score?
c. If the predictor's standard deviation = 5, how many standard deviations from the criterion mean does each person's predicted criterion score fall?
d. If $r = 0$, what is Tim's predicted criterion score?

8. A researcher reports a residual variance equal to the total criterion variance. What is the correlation between the predictor and the criterion?

9. The total variance of criterion scores is 80 and the residual variance is 25.

a. What is the correlation between X and Y?
b. What proportion of criterion score variance is accounted for by the predictor?

10. Why is it not appropriate to calculate prediction intervals for Y by the usual methods if Y is not normally distributed?

11. Why is it not appropriate to calculate prediction intervals for Y by the usual methods if the relation between X and Y displays heteroscedasticity?

12. The correlations of three predictors with a criterion are .63, .42, and .37, respectively.

a. What is the minimum possible value of the multiple correlation coefficient R?
b. Will R be higher if the three predictors are strongly or weakly intercorrelated?

13. The correlation between X_1 and a criterion is .36. The multiple R of X_1 and X_2 with the criterion is .45.

a. What proportion of criterion variance is predicted by the combination of X_1 and X_2?
b. What proportion of criterion variance is predicted by X_2 but not by X_1?
c. What proportion of criterion variance is predicted by neither variable?
d. If the standard deviation of the criterion is 12, what is the standard error of estimate when both X_1 and X_2 are used for prediction?

11 Making Inferences About Differences Between Population Means

INTRODUCTION

Of all statistical tests, one of the most widely used is the t test for the significance of a difference between means. The purpose of the t test is to determine whether the difference between two sample means is large enough for us to conclude that the populations from which the samples came have different means. The t test and related topics are the subjects of this chapter.

To begin, we consider the research context in which the t test is commonly applied. In particular, we discuss differences between independent and correlated sample designs. Then we take up the sampling distribution that underlies the t test—the sampling distribution of the difference between means. We consider the relation of the sampling distribution to parameters of the two populations, to sample size, and to the research design.

The sampling distribution of the difference between means is the basis for t tests for independent and correlated samples, which are discussed and illustrated. Additional topics covered in this chapter include the calculation and interpretation of confidence intervals for the difference between means, the power of the t test, and measurement of the strength of the relation between the independent and dependent variables.

Objectives

When you have completed this chapter, you will be able to

- Describe research designs to which the t test is applied.
- Describe methods for selecting independent and correlated samples from two or more populations; given relevant information, decide whether two samples are independent or correlated.
- Define the sampling distribution of the difference between means, and describe its shape, mean, and standard deviation (the standard error of the difference between means).
- Describe differences between the sampling distribution of the difference between means for independent and correlated samples.
- Explain why the t distribution is the probability distribution underlying tests for comparing two means.

- Describe the assumptions, null and alternative hypotheses, and steps of the *t* test.
- Calculate the estimated standard error of the difference between means of independent samples and of correlated samples.
- Plan and carry out one-tailed and two-tailed *t* tests for independent and for correlated samples.
- Compute and interpret confidence intervals for the difference between means.
- Discuss factors influencing the power of a *t* test. Compare the power of the *t* tests for independent and correlated samples.
- Calculate a measure of the strength of the relation between the independent and dependent variables.
- Discuss implications of violating the assumptions underlying the *t* test.

The Research Context

Many research investigations in psychology and related fields are directed toward discovering or exploring differences between the means of two groups, such as males and females, or under two treatment conditions, such as two teaching methods. The following four examples are typical:

- Millie Gaynor, a graduate student in psychology, is planning to compare the mean level of comprehension of text under two cue conditions, pictorial and verbal.
- Millie's fellow graduate student, Tom Willis, is designing a study to compare the mean number of faint tones detected in the presence and absence of white noise.
- Jim Hillman, a developmental psychologist, wants to test a theory that preadolescent boys exhibit less helping behavior, on the average, than preadolescent girls.
- Ann Lowy, a school administrator, plans to compare the mean level of reading comprehension of children who have and have not attended preschool.

We will follow these four examples throughout the chapter.

Millie, Tom, Jim, and Ann are all examining the difference between two means in order to throw light on the relationship between two variables, an independent variable and a dependent variable. Recall that the ***independent variable*** in a study is the one whose effect is of interest; and the ***dependent variable*** is the variable affected by the independent variable. In Millie's study of text comprehension, the independent variable is the type of cues provided to subjects, and the dependent variable is their level of comprehension. In Tom's study, the independent variable is the amount of white noise present, and the dependent variable is the number of faint tones that are detected. In Jim's research, the independent variable is gender, and the dependent variable is the amount of helping behavior. Ann's independent variable is the amount of preschool education, and her dependent variable is first-grade reading achievement.

A difference between means on the dependent variable provides evidence that the independent and dependent variables are related. For example, if Ann finds that the mean reading achievement of preschool attenders is significantly higher than that of nonattenders, she will conclude that preschool attendance is related to reading achievement in first grade.

Despite the similarities among the four studies, there are several important differences. For one thing, two of the studies—Millie's and Tom's—are experiments; the other two are not. In an experiment, the researcher *manipulates* the independent variable; that is, he or she produces differences on the independent variable that did not exist prior to the study. Millie will manipulate types of cues by having some subjects read text containing pictorial cues, while other subjects read the same text containing verbal cues. Tom will manipulate the amount of white noise by requiring subjects to listen for faint tones with a white noise present or absent. In experiments, the independent variable is often called the ***treatment variable***, and each level of the independent variable is called a *treatment*.

In Jim's and Ann's studies, on the other hand, the levels of the independent variable refer not to treatments but to preexisting differences between groups. Jim will compare two groups of subjects who already differ in gender before his study begins; and Ann will compare first graders who already differ in amount of preschool education (some vs. none). As you will see, this difference has implications for the interpretation of research results.

The four studies also differ with respect to the ***design*** that the researchers plan to use for collecting data. In her experiment, Millie will have two groups of subjects, a pictorial-cue group and a verbal-cue group. But Tom plans to use a single group of subjects, each subject to be tested under both white-noise-on and white-noise-off conditions. Jim and Ann will both select samples from two populations, preadolescent boys and girls in Jim's case and preschool attenders and nonattenders in Ann's case. But Ann plans to match her preschool attenders and nonattenders on intelligence, whereas Jim does not plan to use matching.

The reason that we need to consider designs for collecting data is that some designs result in independent samples of data, whereas others result in dependent or correlated samples of data. The difference is important, because the calculations and, to some extent, the rationale of tests comparing two or more samples of data differ, depending on whether the samples are independent or correlated. Let us consider designs for producing independent and correlated samples.

Independent Sample Designs

Two or more samples are independent if assignment of subjects to one sample does not influence, and is not influenced by, assignment of subjects to the other(s). In experimental research, the most common method for obtaining independent samples is the ***randomized groups design***, in which subjects are randomly subdivided into two (or more) treatment groups. Because random subdivision ensures that assignments of particular subjects to one group do not influence assignments of other subjects to the same group or to any other, the samples that result are independent. The randomized groups design is illustrated in Figure 11–1.

Millie is using a randomized groups design in her experiment on text comprehension. She proceeds as follows: She obtains the names of 16 college students who are available for her study. Using a random number table, she assigns a number to each student. Students with even numbers are assigned to the pictorial-cue group, and students with odd numbers are assigned to the verbal-cue group. Because random numbers are, by definition, unrelated, Millie's two groups (samples) are independent.

In nonexperimental research, random assignment of subjects to groups is impossible. Jim obviously cannot start with a genderless group of preadolescents and randomly (or otherwise) assign some to the gender "female" and others to the gender "male." Instead, samples must be selected from preexisting populations.

Samples selected from two or more preexisting populations are independent if the selection of members from one population does not affect the selection of members from

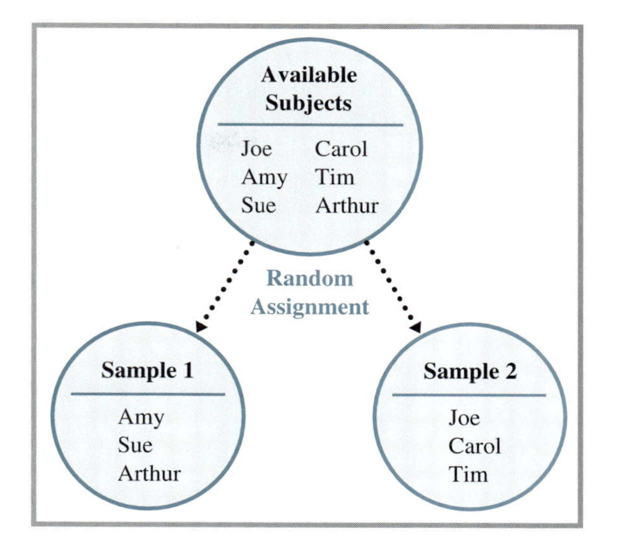

Figure 11–1

Independent samples in experimental research: randomized groups.

the other(s). The ideal way to obtain independent samples from two or more populations is to select members randomly from each one. In many cases, including Jim's, random selection from the populations concerned is not possible. Jim uses the following procedure to obtain independent samples of boys and girls: First, he arranges with the principal of a local school to use sixth-grade students whose parents have given permission for their participation. Fifteen boys and 18 girls meet these requirements, and they become the subjects in Jim's study. Because the selection of particular boys did not influence the selection of girls, and vice versa, Jim's two samples are independent. Selection of independent samples from preexisting populations is illustrated in Figure 11–2.

In both experimental and nonexperimental research using independent samples, the independent variable is said to vary *between subjects* or *between groups*. Thus, in Millie's study, type of cues varies between groups; and, in Jim's study, gender varies between groups.

Note: Do not confuse independent *variables* with independent *samples*. Millie's, Jim's, Tom's, and Ann's studies all include an independent variable, but not all of them use independent samples, as you will see.

Correlated Sample Designs

Two (or more) samples are dependent or correlated if assignment of subjects to one sample is contingent upon (influenced by) assignment to the other(s). There are two types of designs for selecting correlated samples: repeated measures designs and matching designs.

Repeated Measures

As the name implies, ***repeated measures designs*** are those in which a single group of subjects is measured under two (or more) conditions. Occasionally, this design is necessitated by the nature of the research question. For example, studies comparing achievement test scores before and after instruction, or comparing attitudes before and after an attitude-change program, must necessarily compare "before" and "after" scores obtained in the same subjects. However, even in experiments where independent samples could be used, researchers sometimes choose to compare the responses of a single

Figure 11–2

Selection of independent samples from two preexisting populations.

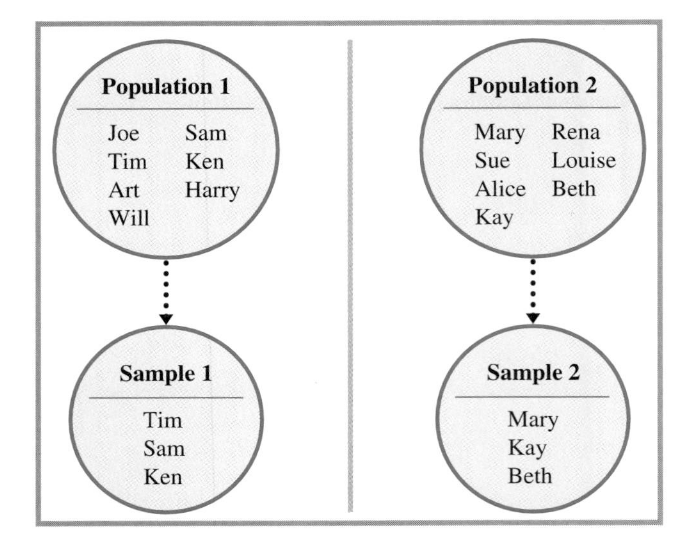

group of subjects to two or more treatments, rather than using a different group for each treatment. The repeated measures design is diagrammed in Figure 11–3.

As an example of a repeated measures design, consider Tom's experiment on the detection of faint tones with white noise present and absent. Instead of assigning different subjects to white-noise-present and white-noise-absent conditions, Tom decides to test each of 10 subjects under both conditions. Like Millie and Jim, Tom will have two samples of data: one collected with white noise present and the other collected with white noise absent. But both samples will be based on the same subjects. Obviously, assignment of a particular subject to one sample predetermines assignment of the same subject to the second sample. Therefore, Tom's two samples are not independent but correlated.

When a single group of subjects is measured under both (or all) treatment levels of an independent variable, the variable is said to vary *within subjects*. Thus, amount of white noise varies within subjects in Tom's experiment.

Repeated measures designs for comparing two or more experimental treatments have one important limitation. They should be used only if little or no carryover effect is expected from one treatment to another. A ***carryover effect*** occurs if a treatment administered early in a series has long-lasting effects that influence responses to later treatments. If, for example, exposure to white noise in Tom's experiment leaves a lingering ringing in the ears that is still present during a subsequent test with white noise absent, the ringing may change the subjects' responses to the white-noise-absent condition. Tom would be well advised to ensure an adequate time interval between the two treatment conditions to allow any such lingering effects to dissipate. As an additional control, he should test half his subjects with white noise present first and absent second, and reverse the order for the other subjects.

In Tom's experiment, carryover effects, if any, are small, brief, and easy to control. In many kinds of research, carryover effects are large and enduring. Suppose, for example, that you are comparing the effectiveness of two treatments for phobia. You cannot treat a phobic client first with Treatment A and then with Treatment B (or vice versa). Assuming that an effective treatment has lasting effects, as it should, any response during the second treatment period would not be a response to that treatment

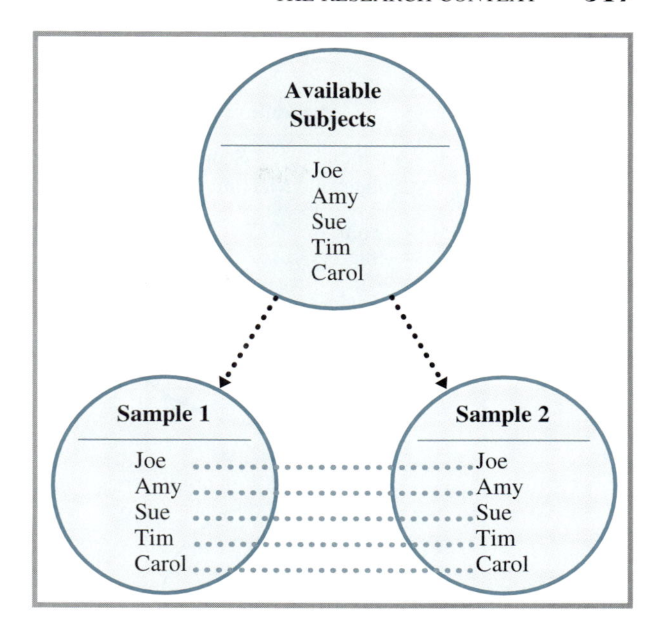

Figure 11–3

Repeated measures.

alone but to that treatment plus the carryover effects of the previous treatment. Obviously, you must use different subjects in your two treatment groups.

Many kinds of research in the social and biological sciences study variables that are expected to have long-lasting effects. These include most research on learning and therapeutic effects, among others. Repeated measures designs cannot be used in cases like these.

Matching

Another strategy for producing correlated samples entails matching each member of one sample with one and only one member of the other. The purpose of matching is to control for a variable (or sometimes two or more variables) that, although not itself of interest to the researcher, could affect performance on the dependent variable. The variable being controlled is referred to as the ***matching variable***. In studies that use matching, the matching variable is first measured in all potential subjects. Then scores on the matching variable are used to pair each member of one sample with a member of the other sample. Matching results in correlated samples because the assignment of a member to one sample is contingent upon the assignment of the matched member to the other.

As an example, consider Ann's study on the effect of preschool attendance on reading achievement in first grade. Ann plans to compare reading test scores of first graders who did and did not attend preschool. However, Ann knows that reading achievement is related to intelligence. In order to ensure that differences in reading achievement between her two groups are not due to differences in intelligence, she decides to match the two groups on intelligence. To this end, Ann obtains intelligence test scores of first graders who are eligible for her study from the school district records. Next, she pairs each preschool attender with a nonattender of approximately equal intelligence, insofar as this is possible; nonmatchable children are excluded from her study. When she is finished, she has two samples (25 preschool attenders and 25 nonattenders) who are matched on intelligence. Ann's design is illustrated in Figure 11–4.

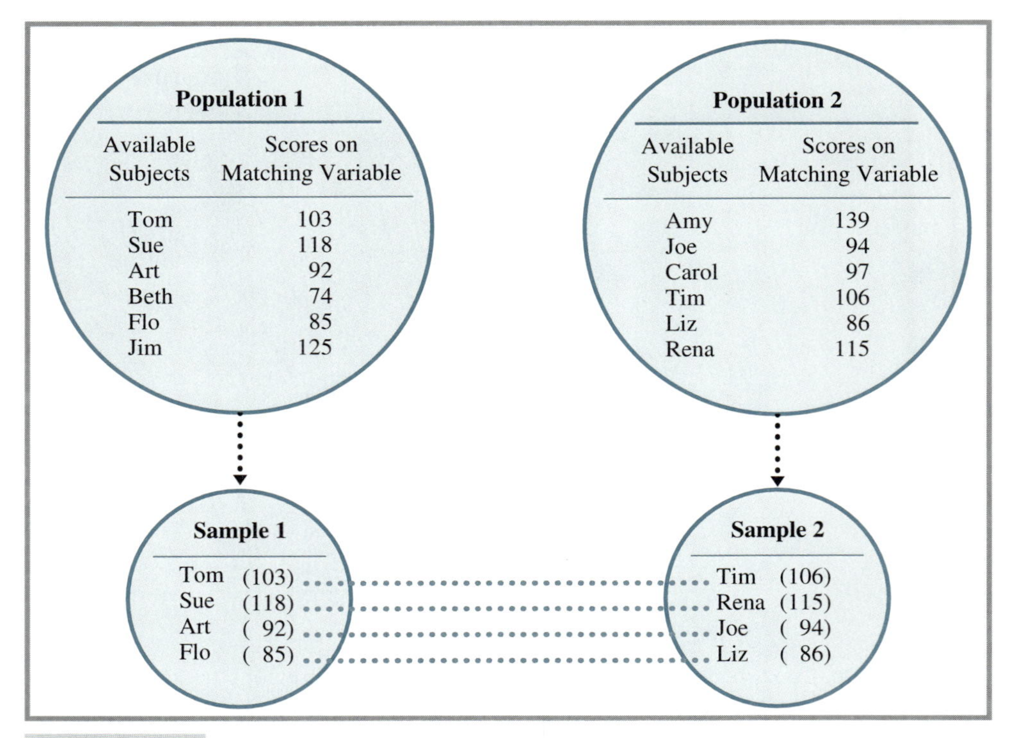

Figure 11–4 Matching subjects from two populations.

Matching designs, once popular in behavioral research, are less common today for several reasons. First, it is often difficult to find good matching variables, ones that are known to be related to the dependent variable. Second, the loss of unmatchable subjects presents problems in both sampling and in generalization of the results to the population. Third, the ready availability of computer packages for data analysis makes it possible to use alternative analysis procedures that, though computationally more complex, control for matching variables without the disadvantages of matching.

Distinguishing Between Independent and Correlated Data Sets

We have discussed the difference between independent and correlated samples in considerable detail because procedures for analyzing data differ, depending on whether an independent- or correlated-samples design was used. Obviously, given a pair of samples to be compared, you must first decide whether they are independent or correlated, so that you can apply the appropriate procedure. A flowchart that can help you to make this decision is presented in Figure 11–5.

Another clue that may help you decide whether two samples are independent or correlated is the sizes of the samples. In both repeated measures and matching designs, each score in the first sample is paired with one and only one score in the second. Therefore, both sample sizes are necessarily equal. If samples are independent, on the other hand, sample sizes may be the same or they may be different. Therefore, if the samples to be compared are not the same size, you know that they must be independent.

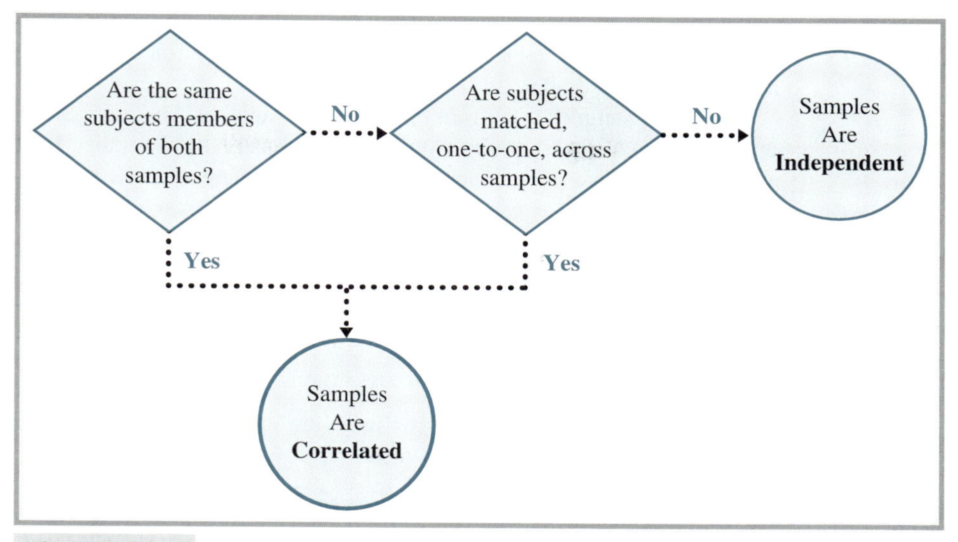

Figure 11–5 **Flowchart for deciding whether samples are independent or correlated.**

EXERCISE 11–1

In each of the following cases, decide (1) whether the study is an experiment, and (2) whether the samples are independent or correlated.

1. A cognitive psychologist plans to compare two methods of teaching archery. Since upper-arm strength is a factor in archery, she first measures the upper-arm strength of all students enrolled in an archery course and forms matched pairs of students based on the results. One student in each pair is assigned to teaching method A and the other to teaching method B.
2. To study the effect of safety instruction on number of industrial accidents, an industrial psychologist randomly assigns each of a group of assembly-line workers to a training or a no-training condition.
3. Is math anxiety related to scores on statistics exams? A statistics instructor selects 50 students with high math anxiety and 50 with low math anxiety and compares their mean statistics exam scores.
4. To determine whether reaction time differs to auditory and visual stimuli, an experimenter presents a visual stimulus on half the trials and an auditory stimulus on the other half to each of a group of 30 subjects. He then compares the mean reaction time to auditory versus visual stimuli.
5. Concept learning is to be compared under two different information loads. Subjects are assigned to the two conditions as follows: As each subject enters the laboratory, the investigator tosses a coin. The subject is run under a low information load if the coin comes up heads and under a high information load if the coin comes up tails.

The *t* Test for the Difference Between Means

Although researchers like Millie, Tom, Jim, and Ann will calculate the difference between two sample means, $M_1 - M_2$, they are really interested in the difference between the means of the two populations from which the samples came, $\mu_1 - \mu_2$. Millie's research question, for example, concerns the effects of pictorial versus verbal cues not just in the 16 subjects in her experiment but in the population of adult readers from which the subjects were drawn.

Means of samples vary from one sample to another. Therefore, the difference between sample means, $M_1 - M_2$, varies from one pair of samples to another. In order to understand how researchers can draw conclusions about the difference between two population means from the difference between sample means in just one pair of samples, we need to know how values of $M_1 - M_2$ vary from one pair of samples to another and how these values are related to the actual difference between population means, $\mu_1 - \mu_2$. In other words, we need to examine the sampling distribution of the difference between means.

The Sampling Distribution of the Difference Between Means

Consider the following hypothetical sampling experiment: You have access to two populations. The first population has mean μ_1 and variance σ_1^2, and the second population has mean μ_2 and variance σ_2^2. You select a sample of size n_1 from the first population and a sample of size n_2 from the second population, calculate the mean of each sample, and calculate the difference between the sample means, $M_1 - M_2$. Then you take a second pair of samples from the two populations, a sample of size n_1 from the first and a sample of size n_2 from the second, and you calculate $M_1 - M_2$ of the second pair of samples. You continue the sampling process until you have calculated $M_1 - M_2$ in all possible pairs of samples of size n_1 and n_2 from the two populations. The distribution of values of $M_1 - M_2$ over all possible pairs of samples is called the ***sampling distribution of the difference between means***.

There are three things we need to know about the sampling distribution of the difference between means: its shape, its mean, and its standard deviation. As you will see, these properties depend on the means and variances of the two populations, on the sample sizes, and on whether the samples within each pair of samples are independent or correlated.

The Shape of the Sampling Distribution of the Difference Between Means

The shape of the sampling distribution of the difference between means depends on the shapes of the populations from which the samples were drawn and on sample size. *If both populations are normal, the sampling distribution is normal, regardless of the sample size.* For example, if you draw all possible pairs of samples of size 5 and 10, respectively, or size 2 and 6, or size 13 and 13, or any other combination of sample sizes, from two normal populations, and calculate $M_1 - M_2$ in each pair, the distribution of values of $M_1 - M_2$ over all possible pairs of samples will be normal.

If the populations are not normal, the sampling distribution of the difference between means is not precisely normal, but it approaches a normal distribution as sample size increases. Suppose, for example, that you draw all possible pairs of samples of size $n_1 = 5$ and $n_2 = 5$ from two highly skewed populations and calculate $M_1 - M_2$ in each pair. The distribution of values of $M_1 - M_2$ will be noticeably different from a normal distribution. But if n_1 and n_2 are 25 instead, the distribution of values of $M_1 - M_2$ will be so nearly normal as to be indistinguishable from a normal distribution.

The fact that the sampling distribution of the difference between means is close to normal unless sample sizes are small is very important. It allows us to use procedures based on normal curves for making inferences about differences between population means, regardless of the shape of the populations, provided that our samples are rea-

Figure 11–6

An approximately normal sampling distribution of the difference between means.

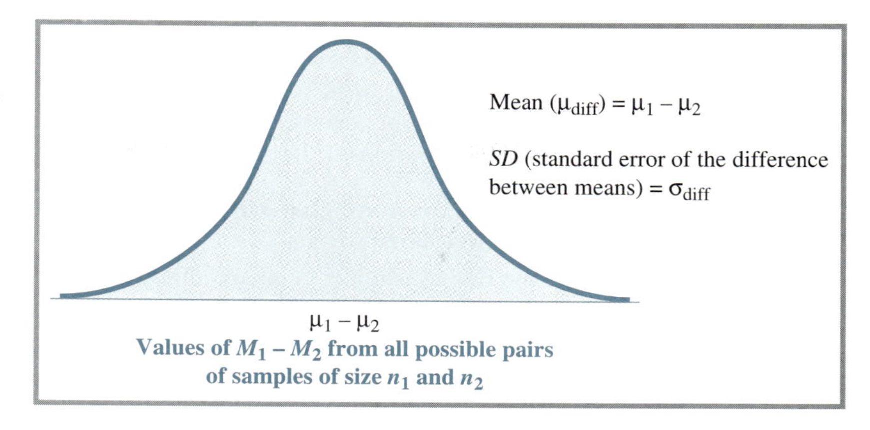

sonably large. A normally shaped sampling distribution of the difference between means is displayed in Figure 11–6.

The Mean of the Sampling Distribution of the Difference Between Means

We will use the symbol μ_{diff} to stand for the mean of the sampling distribution of the difference between means—the average value of $M_1 - M_2$ over all pairs of samples. The value of μ_{diff} is dependent only on the means of the two populations from which the samples were drawn. Specifically, μ_{diff} *is equal to the difference between the two population means,* μ_1 *and* μ_2.

$$\mu_{\text{diff}} = \mu_1 - \mu_2$$ *Mean of the sampling distribution of the difference between means*

In other words, even though $M_1 - M_2$ varies from one pair of samples to another, the average difference between the sample means is equal to the actual difference between the population means. For example, if the mean of the first population is 10 and the mean of the second population is 8, the average value of $M_1 - M_2$ over all pairs of samples is $10 - 8 = 2$.

Of course, the fact that $\mu_{diff} = \mu_1 - \mu_2$ does not mean that $M_1 - M_2 = \mu_1 - \mu_2$ in every pair of samples. Because M_1 varies over samples from the first population and M_2 varies over samples from the second population, $M_1 - M_2$ varies from one pair of samples to another. For example, if $\mu_1 = 10$ and $\mu_2 = 8$, $M_1 - M_2$ in one pair of samples may be equal to 3.5; in another pair of samples, $M_1 - M_2$ may be equal to 1.3; and so on. Despite this variation, however, the average of the values of $M_1 - M_2$ over all pairs of samples will be exactly equal to 2, the difference between μ_1 and μ_2. Figure 11–6 shows how values of $M_1 - M_2$ vary around $\mu_1 - \mu_2$.

The Standard Deviation of the Sampling Distribution of the Difference Between Means

We now know that the mean value of $M_1 - M_2$ over all possible pairs of samples is $\mu_1 - \mu_2$. But how much do values of $M_1 - M_2$ vary around $\mu_1 - \mu_2$? We use the standard deviation of the sampling distribution of the difference between means to measure the amount of variability of values of $M_1 - M_2$.

The standard deviation of the sampling distribution of the difference between means is called the ***standard error of the difference between means***, symbolized σ_{diff}. (Keep in mind that the term *error* in statistics means chance sampling variability. The standard deviation of the distribution of values of $M_1 - M_2$ is called a standard "error" because it measures how much differences between sample means vary by chance.)

One factor that determines the value of the standard error of the difference between means is the sampling design, that is, whether the samples are independent or correlated. We consider each case in turn.

The Standard Error of the Difference Between Means of Independent Samples

The formula for the standard error of the difference between means of independent samples is

$$\sigma_{\text{diff}} = \sqrt{\sigma^2_{M_1} + \sigma^2_{M_2}}$$

Standard error of the difference between means of independent samples

where σ_{M_1} is the standard error of the mean for samples of size n_1 from the first population, and σ_{M_2} is the standard error of the mean for samples of size n_2 from the second population.

Recall that the squared standard error of the mean, σ^2_M, for samples from a single population is equal to the population variance, σ^2, divided by the sample size, n. Thus, we can rewrite the formula for the standard error of the difference between means of independent samples like this:

$$\sigma_{\text{diff}} = \sqrt{\frac{\sigma^2_1}{n_1} + \frac{\sigma^2_2}{n_2}}$$

For example, if the variances of the two populations are 100 and 150, respectively, and the size of samples from each population is 25

$$\sigma_{\text{diff}} = \sqrt{\frac{100}{25} + \frac{150}{25}} = 3.16$$

The formula shows that σ_{diff} increases as σ^2_1 and σ^2_2 increases. In other words, differences between sample means vary more from one pair of samples to another if the population variances are large than if they are small. On the other hand, σ_{diff} decreases as n_1 and n_2 increase. Thus, the larger the samples, the less differences between sample means vary over pairs of samples from the two populations.

The Standard Error of the Difference Between Means of Correlated Samples

The formula for the standard error of the difference between means of correlated samples is

$$\sigma_{\text{diff}} = \sqrt{\sigma^2_{M_1} + \sigma^2_{M_2} - 2\rho_{M_1M_2}\sigma_{M_1}\sigma_{M_2}}$$

Standard error of the difference between means of correlated samples

The first part of the formula, $\sigma^2_{M1} + \sigma^2_{M2}$, is the same as for independent samples. Therefore, σ_{diff} for correlated samples, as for independent samples, increases as the population variances increase and decreases as sample sizes increase. However, the formula for correlated samples includes an additional term: $2\rho_{M_1M_2}\sigma_{M_1}\sigma_{M_2}$, where $\rho_{M_1M_2}$ stands for the correlation between the means of samples from Population 1 and the means of samples from Population 2.

Let us see what effect the inclusion of this term has on the standard error of the difference between means.

1. If the correlation between the sample means is positive ($\rho_{M_1M_2} > 0$), σ_{diff} is smaller for correlated samples than for independent samples. For example, if $\sigma_{M_1} = 3$, $\sigma_{M_1} = 4$, and $\rho_{M_1M_2} = .5$.

   Independent samples: $\sigma_{\text{diff}} = \sqrt{3^2 + 4^2} = 5$

   Correlated samples: $\sigma_{\text{diff}} = \sqrt{3^2 + 4^2 - 2(.5)(3)(4)} = 3.61$

2. If the correlation between the sample means is negative ($\rho_{M_1M_2} < 0$), σ_{diff} is larger for correlated than for independent samples. For example, if $\sigma_{M_1} = 3$, $\sigma_{M_2} = 4$, and $\rho_{M_1M_2} = -.5$:

   Independent samples: $\sigma_{\text{diff}} = \sqrt{3^2 + 4^2} = 5$

   Correlated samples: $\sigma_{\text{diff}} = \sqrt{3^2 + 4^2 - 2(-.5)(3)(4)} = 6.08$

3. If the sample means are uncorrelated ($\rho_{M_1M_2} = 0$), the value of σ_{diff} is the same for correlated and for independent samples. For example, if $\sigma_{M_1} = 3$, $\sigma_{M_2} = 4$, and $\rho_{M_1M_2} = 0$:

   Independent samples: $\sigma_{\text{diff}} = \sqrt{3^2 + 4^2} = 5$

   Correlated samples: $\sigma_{\text{diff}} = \sqrt{3^2 + 4^2 - 2(0)(3)(4)} = 5$

Hypothetically, then, the standard error of the difference between means of correlated samples can be smaller than, equal to, or greater than the standard error of the difference between means of independent samples. In practice, however, the standard error of the difference between means is usually smaller for correlated samples. Here is why: If a sample from the first population has a mean (M_1) greater than μ_1, the correlated sample from the second population, consisting of the same subjects or matched subjects, probably has a mean (M_2) greater than μ_2. Similarly, if a sample from the first population has M_1 less than μ_1, the correlated sample from the second population probably has M_2 less than μ_2.

We know that when high values on one variable are associated with high values on another, and low values on the first with low values on the second, the correlation between the two variables is positive. Therefore, in repeated measures and matching designs, M_1 and M_2 are usually positively correlated over all possible pairs of samples ($\rho_{M_1M_2} > 0$). For this reason, the standard error of the difference between means of correlated samples is usually smaller than that of independent samples if other factors are equal. We will return to this fact later in our discussion of the power of the t test.

The similarities and differences between the sampling distribution of the difference between means of independent and correlated samples are summarized in Figure 11–7. The figure assumes that the pairs of samples came from the same two populations, and that sample sizes are the same in both cases. It also assumes that, in the case of correlated samples, $\rho_{M_1M_2}$ is positive. As you can see, the two sampling distributions have the same shape and the same mean ($\mu_1 - \mu_2$). But the variability of values of $M_1 - M_2$ around $\mu_1 - \mu_2$ is less in the sampling distribution for correlated samples than in the sampling distribution for independent samples.

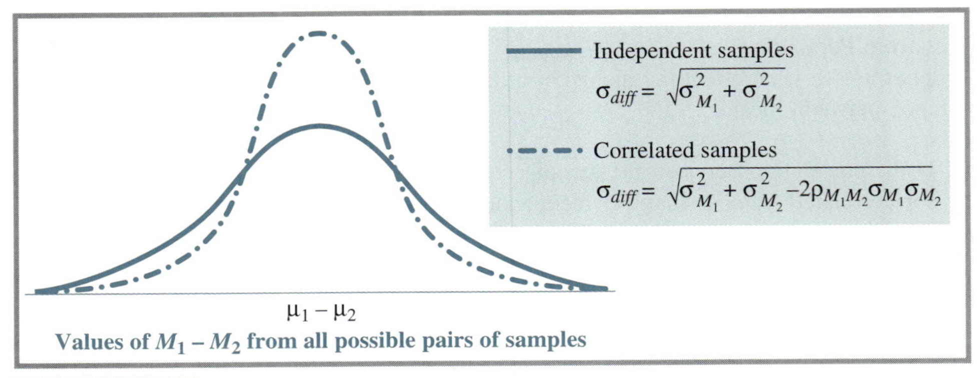

Figure 11–7 **Sampling distributions of the difference between means of independent and correlated samples drawn from the same two populations. The figure assumes the same values of n_1 and n_2 in both cases and $\rho_{M_1M_2} > 0$.**

EXERCISE 11–2

1. You draw all possible pairs of independent samples from two normal populations. The size of samples from each population is 16. The first population has a mean equal to 50 and standard deviation = 10, and the second has a mean equal to 30 and standard deviation = 8. In each pair of samples, you calculate the difference between means. Now you examine the distribution of these differences.
   a. What is the distribution called?
   b. What is its shape? Why?
   c. Compute the mean of the distribution.
   d. What is the standard deviation of the distribution called?
   e. Compute the standard deviation of the distribution.
   f. If the size of samples from both populations were 100, would the standard deviation of the distribution increase or decrease?
   g. Calculate the standard deviation for sample sizes of 100.
2. Again you draw all possible pairs of samples from the same two populations, but with one difference: Each time you select a sample from the first population, you select a matched sample from the second population. Again, the size of samples from each population is 16.
   a. Which of your answers to question 1 will change? Why?
   b. What additional information do you need in order to compute the answers that have changed?

Important: Answer question 2 before you look at question 3.

3. The correlation between the sample means is .60. Now compute the answers that have changed from question 1 to question 2.
4. From two rectangular populations you first draw all possible pairs of independent samples of size 5 and 8, respectively, and calculate $M_1 - M_2$ in each pair. Then you draw all possible pairs of independent samples of size 40 and 35, respectively, and calculate $M_1 - M_2$ in each pair. Now you examine your two distributions of values of $M_1 - M_2$.
   a. Which of the distributions of $M_1 - M_2$ is more nearly normal?
   b. Which distribution has a smaller standard deviation?

Rationale of the *t* Test

As you know, values of a variable can be converted into z scores by the formula

$$z = \frac{\text{Variable value} - \text{mean}}{SD}$$

The variable in the sampling distribution of the difference between means is $M_1 - M_2$; the mean of the sampling distribution is $\mu_1 - \mu_2$; and the standard deviation is σ_{diff}. Therefore, values of $M_1 - M_2$ are converted into z scores by the formula $z = \frac{(M_1 - M_2) - (\mu_1 - \mu_2)}{\sigma_{\text{diff}}}$. Moreover, if the sampling distribution of the difference between means is approximately normal, the distribution of z scores over all pairs of samples forms a standard normal distribution. Therefore, if we could convert differences between sample means into z scores, we could use the table of the standard normal curve to answer questions like "What is the probability of a difference between sample means as large as this one?"

In order to calculate z, however, we need to know σ_{diff}. And, to calculate σ_{diff}, we need to know both population variances, σ_1^2 and σ_2^2, as well as, in the case of correlated samples, the correlation between the sample means, $\rho_{M_1M_2}$. Because these parameters are generally unknown, we cannot calculate the actual value of σ_{diff}. Instead, we proceed by calculating an estimator of σ_{diff} from sample data. The sample estimator of σ_{diff} is called the ***estimated standard error of the difference between means*** and is symbolized s_{diff}. The calculation of s_{diff} will be dealt with later. For now, let us examine the effect of substituting s_{diff} for σ_{diff}. We do so by considering a hypothetical sampling experiment.

Suppose that we select (or list) all possible pairs of samples of size n_1 and n_2 from two populations. (We assume that the populations are normal or n_1 and n_2 are reasonably large.) Within each pair of samples, we calculate the difference between the sample means, $M_1 - M_2$, and the estimated standard error of the difference between means, s_{diff}. Then we calculate the following ratio in each pair of samples: $\frac{(M_1 - M_2) - (\mu_1 - \mu_2)}{s_{\text{diff}}}$. When we examine the distribution of values of this ratio over all pairs, we find that it is not a standard normal distribution. Instead, the ratio is distributed as t. So we can write

$$t = \frac{(M_1 - M_2) - (\mu_1 - \mu_2)}{s_{\text{diff}}}$$

Because of this, tests of hypotheses about differences between populations means use the t distribution.

The steps in the t test for the difference between means are the same as in any statistical test. They are as follows:

1. State the hypothesis to be tested.
2. Select a level of alpha (α).
3. Collect data and calculate M_1, M_2 and s_{diff}.
   Calculate the test statistic, t, by the formula

   $$t = \frac{(M_1 - M_2) - (\mu_1 - \mu_2)_{\text{H}}}{s_{\text{diff}}}$$

   where $(\mu_1 - \mu_2)_{\text{H}}$ stands for the hypothesized value of $\mu_1 - \mu_2$.

4. Compare the calculated t with critical values of t to determine whether its probability (p) is less than α or greater than α. If $p < \alpha$, reject the hypothesis; if $p > \alpha$, do not reject the hypothesis.

In principle, it is possible to test the hypothesis that $\mu_1 - \mu_2$ is any value. In practice, however, the question of interest is almost always: Is there a difference (other than zero) between μ_1 and μ_2? Therefore, the hypothesis that is tested is the null hypothesis (H_0) of no difference, the hypothesis that $\mu_1 - \mu_2 = 0$. Because the hypothesized difference between population means, $(\mu_1 - \mu_2)_H$, is zero in tests of the null hypothesis, the test statistic becomes

$$t = \frac{(M_1 - M_2) - 0}{s_{\text{diff}}}$$

$$t = \frac{M_1 - M_2}{s_{\text{diff}}}$$ *The test statistic in the* t *test*

If the test results in rejection of the null hypothesis of no difference between population means, the researcher concludes that a real difference exists.

One-Tailed and Two-Tailed Tests

The t test for the difference between means, like the t test for a single population mean, may be one-tailed or two-tailed, depending on whether the researcher's hypothesis is directional or nondirectional. If the research hypothesis is nondirectional, or if the analyst is interested in detecting a difference in either direction, a two-tailed test is called for. For example, consider Millie's experiment on the effect of type of cues on text comprehension. Because Millie wants to know whether either type of cue, pictorial or verbal, is more effective than the other, she should conduct a two-tailed test.

The null and alternative hypotheses for a two-tailed test are

$$H_o: \mu_1 - \mu_2 = 0$$
$$H_A: \mu_1 - \mu_2 \neq 0$$

The rejection region is split between the two tails of the t distribution, with an area equal to $\frac{1}{2}\alpha$ in each tail. There are two critical values of t separating the rejection and nonrejection regions, one negative and one positive. The null hypothesis is rejected if the calculated t is less than (below) the negative critical value or greater than the positive critical value. In either case, the difference between the means is significant at the α level. If, on the other hand, the calculated t falls between the two critical values, the null hypothesis is not rejected, and the difference between the means is not significant at the α level. A diagrammatic representation of a two-tailed t test is shown in Figure 11–8a.

If the research hypothesis states that μ_1 is greater than μ_2 (that is, $\mu_1 - \mu_2 > 0$), or if the analyst is only interested in detecting a difference in this direction, an upper one-tailed test is required. For example, in her comparison of first graders who have and have not attended preschool, Ann expects that preschool attenders (Group 1) will have higher reading achievement, on the average, than nonattenders (Group 2). Because Ann's research hypothesis states that $\mu_1 > \mu_2$, she should use an upper one-tailed test.

The null and alternative hypotheses for an upper one-tailed test are

$$H_o: \mu_1 - \mu_2 \leq 0$$
$$H_A: \mu_1 - \mu_2 > 0$$

In an upper one-tailed test, the entire rejection region, with area = α, is located in the upper tail of the t distribution, and the critical value of t that separates the rejection and nonre-

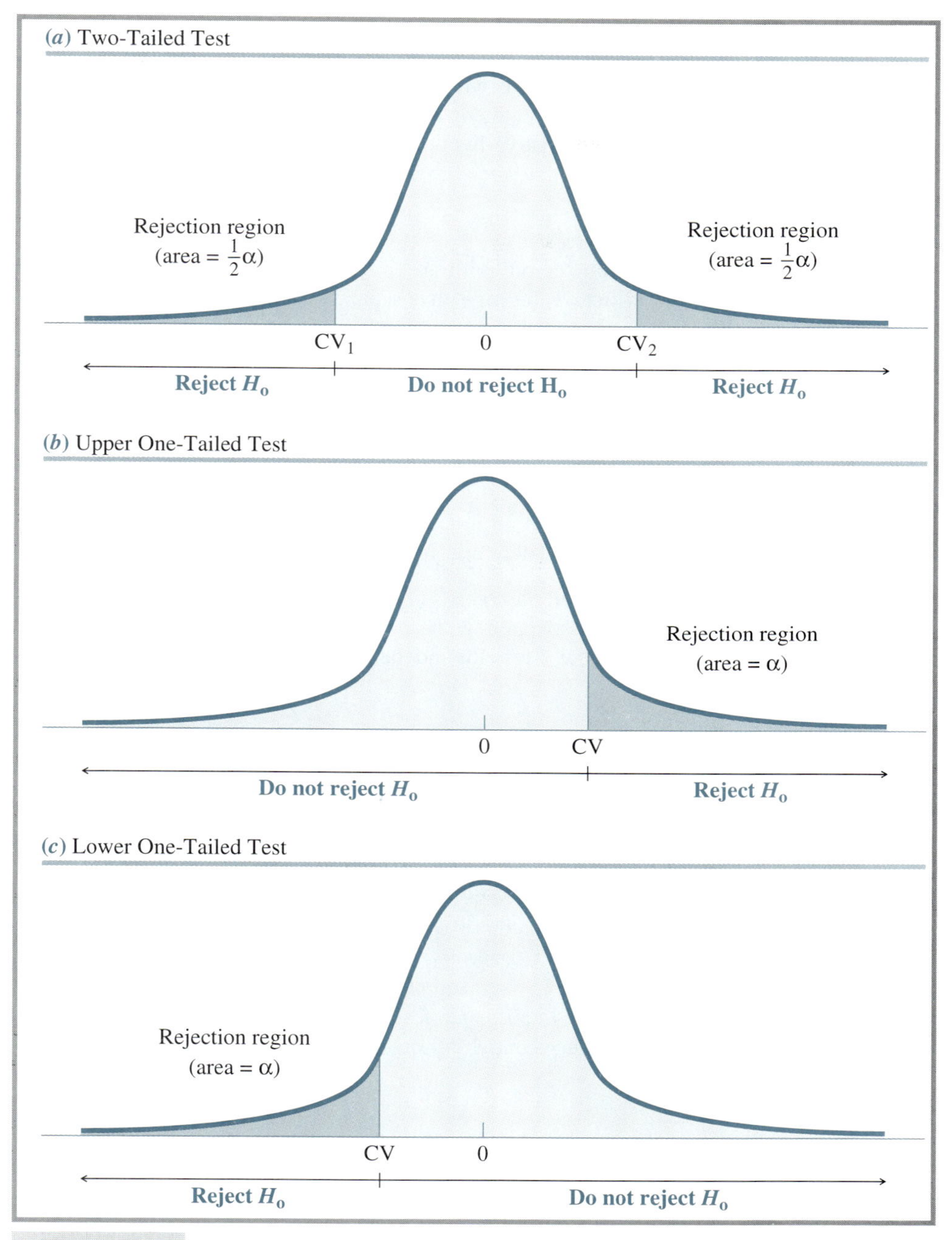

Figure 11–8 **Schematic representations of two-tailed and one-tailed t tests (CV = critical value of t for chosen α).**

jection regions is positive. If the calculated t is greater than the critical value, the null hypothesis is rejected, and the difference between the means is significant at the α level. If, on the other hand, the calculated t is less than the critical value, the null hypothesis is not rejected, and the difference between the means is not significant at the α level. Figure 11–8b shows an upper one-tailed test.

If the research hypothesis states that μ_1 is less than μ_2 (that is, $\mu_1 - \mu_2 < 0$), or if the analyst is only interested in detecting a difference in this direction, a lower one-tailed test is required. For example, in his experiment on the detection of tones with and without the presence of white noise, Tom has hypothesized that presentation of white noise (Treatment 1) will interfere with (decrease) detection of faint tones relative to the absence of white noise (Treatment 2). Therefore, he should use a lower one-tailed test.

The null and alternative hypotheses for a lower one-tailed test are

$$H_o: \mu_1 - \mu_2 \geq 0$$
$$H_A: \mu_1 - \mu_2 < 0$$

The entire rejection region, with area = α, is located in the lower tail of the t distribution, and the critical value of t that separates the rejection and nonrejection regions is negative. If the calculated t is less than the critical value, the null hypothesis is rejected, and the difference between the means is significant at the α level. If, on the other hand, the calculated t is greater than the critical value, the null hypothesis is not rejected, and the difference between the means is not significant at the α level. Figure 11–8c shows a lower one-tailed test.

The types of hypotheses that are tested and the sequence of steps in the test are the same in t tests for both independent and correlated samples. In both cases the test statistic is $t = \frac{M_1 - M_2}{s_{\text{diff}}}$. There are, however, two differences between the t tests for independent and correlated samples: (1) the calculation of s_{diff}, and (2) the number of degrees of freedom associated with the test. In the sections to follow, we discuss and demonstrate, first, the t test for independent samples, and then the t test for correlated samples.

EXERCISE 11–3

In each of the following, decide whether a two-tailed test, an upper one-tailed test, or a lower one-tailed test is required. Then state H_0 and H_A.

1. Research hypothesis: Patients who receive presurgical information about the amount of pain to be expected (Condition 1) will report a lower level of pain, on the average, than patients not receiving such information (Condition 2).
2. Research question: Is there a difference between the mean number of hours of studying per week of fraternity members and independents?
3. Research question: Do subjects who are offered a monetary incentive for achieving high scores on an intelligence test score higher, on the average, than subjects who are not offered an incentive? (Let monetary incentive be Condition 1.)

The t Test for the Difference Between Means of Independent Samples

Assumptions Underlying the Test

Two assumptions underlie the t test for independent samples:

1. Both populations are normal (or, if not, the samples are relatively large).
2. The variances of both populations are the same ($\sigma_1^2 = \sigma_2^2$). This assumption is referred to as the ***homogeneity-of-variance*** assumption.

The first assumption is necessary to ensure that the ratio $\frac{M_1 - M_2}{s_{\text{diff}}}$ is distributed as t. The homogeneity-of-variance assumption affects the calculation of s_{diff} and the number of degrees of freedom in the test, as you will see.

Calculating the Estimated Standard Error of the Difference Between Means

Recall that the standard error of the difference between means of independent samples, σ_{diff}, is equal to $\sqrt{\frac{\sigma_1^2}{n_1}+\frac{\sigma_2^2}{n_2}}$. Because the test assumes that $\sigma_1^2 = \sigma_2^2$, we can use the symbol σ^2 without the subscript to stand for both σ_1^2 and σ_2^2, and we write

$$\sigma_{\text{diff}} = \sqrt{\frac{\sigma^2}{n_1}+\frac{\sigma^2}{n_2}} = \sqrt{\sigma^2\left(\frac{1}{n_1}+\frac{1}{n_2}\right)}$$

The estimated standard error of the difference between means, s_{diff}, is obtained by substituting a sample estimator of σ^2 in the formula, like this:

$$s_{\text{diff}} = \sqrt{(\text{Estimator of } \sigma^2)\left(\frac{1}{n_1}+\frac{1}{n_2}\right)}$$

The estimator of σ^2 is calculated by combining data over the two samples. In statistics, combining data over two or more samples is referred to as *pooling*. Therefore, the estimator of σ^2 is called the ***pooled variance estimate***, symbolized s_p^2. So we can write

$$s_{\text{diff}} = \sqrt{s_p^2\left(\frac{1}{n_1}+\frac{1}{n_2}\right)}$$

The pooled variance estimate is a weighted average of the two sample variances. It is equal to the pooled sums of squares divided by the pooled degrees of freedom.

$$s_p^2 = \frac{\text{pooled } SS}{\text{pooled } df} = \frac{SS_1 + SS_2}{df_1 + df_2}$$

$df_1 = n_1 - 1$ and $df_2 = n_2 - 1$. Therefore,

$$s_p^2 = \frac{SS_1 + SS_2}{(n_1 - 1) + (n_2 - 1)}$$

$$s_p^2 = \frac{SS_1 + SS_2}{n_1 + n_2 - 2}$$ *Formula for the pooled variance estimate*

To calculate s_p^2, you must first calculate the sum of squares, SS, of each sample. If you are working with raw data, you can use either of the following methods for calculating each sum of squares:

Deviation method: $SS = \Sigma(X - M)^2$

Raw score method: $SS = \Sigma X^2 - \frac{(\Sigma X)^2}{n}$

If the two sample standard deviations (s_1 and s_2) have already been calculated, use the following formulas:

$$SS_1 = (n_1 - 1)s_1^2$$
$$SS_2 = (n_2 - 1)s_2^2$$

Now we return to the estimated standard error of the difference between means, s_{diff}. s_{diff} can be calculated in two steps, by calculating s_p^2 by one of the methods described earlier and substituting in the formula for s_{diff}:

$$s_{\text{diff}} = \sqrt{s_p^2\left(\frac{1}{n_1} + \frac{1}{n_2}\right)}$$

Or s_{diff} can be calculated in one step using the following formula:

$$s_{\text{diff}} = \sqrt{\frac{SS_1 + SS_2}{n_1 + n_2 - 2}\left(\frac{1}{n_1} + \frac{1}{n_2}\right)}$$

Formula for calculating the estimated standard error of the difference between means of independent samples

Here are two examples:

EXAMPLE 1 Summary data from two independent samples:

	Sample 1	Sample 2
n	9	8
ΣX	90	64
ΣX^2	1350	812

First, we calculate the sum of squares of each sample:

$$SS_1 = \Sigma X_1^2 - \frac{(\Sigma X_1)^2}{n_1} = 1350 - \frac{(90)^2}{9} = 450$$

$$SS_2 = \Sigma X_2^2 - \frac{(\Sigma X_2)^2}{n_2} = 812 - \frac{(64)^2}{8} = 300$$

Then

$$s_{\text{diff}} = \sqrt{\frac{450 + 300}{9 + 8 - 2}\left(\frac{1}{9} + \frac{1}{8}\right)} = 3.436$$

EXAMPLE 2 The standard deviation of a sample of size 20 is 6, and the standard deviation of a second independent sample of size 15 is 8.

$$SS_1 = (n_1 - 1)s_1^2 = (20 - 1)(6^2) = 19(36) = 684$$

$$SS_2 = (n_2 - 1)s_2^2 = (15 - 1)(8^2) = 14(64) = 896$$

$$s_{\text{diff}} = \sqrt{\frac{684 + 896}{20 + 15 - 2}\left(\frac{1}{20} + \frac{1}{15}\right)} = 2.364$$

Degrees of Freedom

The degrees of freedom in the t test for independent samples is the pooled degrees of freedom, $n_1 + n_2 - 2$. In Example 1, $df = 9 + 8 - 2 = 15$; in Example 2, $df = 20 + 15 - 2 = 33$.

Conducting the *t* Test for Independent Samples

Now we are ready to demonstrate the t test for independent samples by analyzing the data from Millie's study which compared the effects of pictorial and verbal cues on comprehension. Recall that Millie randomly subdivided a group of 16 subjects into two experimental groups. Table 11–1 lists the comprehension scores and shows the calculation of ΣX and ΣX^2 for each group.

Because Millie is interested in detecting any kind of difference between visual and verbal cue treatments, a two-tailed test is required. The null and alternative hypotheses are

Table 11–1 Comprehension Scores of Subjects in Millie's Experiment

Pictorial Cues Group		Verbal Cues Group	
X	X^2	X	X^2
17	289	17	289
20	400	15	225
16	256	15	225
18	324	13	169
17	289	16	256
18	324	11	121
13	169	16	256
19	361	15	225
138	2412	118	1766
ΣX	ΣX^2	ΣX	ΣX^2

$$H_o: \mu_1 - \mu_2 = 0$$
$$H_A: \mu_1 - \mu_2 \neq 0$$

Millie decides to use $\alpha = .05$.

First, Millie calculates the mean comprehension score, M, of each group. In the pictorial cue group, $M = 138/8 = 17.25$. In the verbal cue group, $M = 118/8 = 14.75$. Next, Millie calculates the estimated standard error of the difference between means, s_{diff}. To calculate SS of each sample, she uses the formula $\Sigma X^2 - \frac{(\Sigma X)^2}{n}$.

$$SS_1 = 2412 - \frac{(138)^2}{8} = 31.5$$

$$SS_2 = 1766 - \frac{(118)^2}{8} = 25.5$$

$$s_{\text{diff}} = \sqrt{\frac{SS_1 + SS_2}{n_1 + n_2 - 2}\left(\frac{1}{n_1} + \frac{1}{n_2}\right)} = \sqrt{\frac{31.5 + 25.5}{8 + 8 - 2}\left(\frac{1}{8} + \frac{1}{8}\right)} = 1.009$$

Now she is ready to calculate t:

$$t = \frac{M_1 - M_2}{s_{\text{diff}}} = \frac{17.25 - 14.75}{1.009} = 2.48$$

Finally, Millie evaluates t by comparing it with the critical values in the t table (Table D–3 in Appendix D). The number of degrees of freedom for Millie's test is $n_1 + n_2 - 2 = 8 + 8 - 2 = 14$. Referring to the row with 14 df, Millie finds that $t = 2.48$ falls between the critical values 2.145 with two-tailed $p = .05$ and 2.624 with two-tailed $p = .02$. Therefore, p is between .05 and .02. Because $p < .05$, Millie rejects the null hypothesis. The difference between the pictorial and verbal cue means is significant at the .05 level. Based on the fact that the pictorial cue mean is higher, Millie concludes that pictorial cues are more effective for enhancing comprehension of text than verbal cues.

Commentary: Millie has, of course, not proven that pictorial cues are more effective than verbal cues. What she has shown is that, if the two types of cues are equally effective (as H_0 states), the probability of a difference between sample means as great as hers is less than .05. In other words, a difference this large is unlikely to occur by chance and is more likely due to the experimental treatment.

As another example of the *t* test for independent samples, let us analyze the data from Jim's study on differences in helping behavior of preadolescent boys and girls. According to the theory Jim is testing, boys display less helping behavior than girls. Jim's research hypothesis, therefore, is that the mean number of helping behaviors of boys (Population 1) is less than that of girls (Population 2). A lower one-tailed test is required, with null and alternative hypotheses:

$$H_o: \mu_1 - \mu_2 \geq 0$$
$$H_A: \mu_1 - \mu_2 < 0$$

Because Jim is particularly concerned about reducing the probability of a Type I error, he decides to use $\alpha = .01$.

Summary data from Jim's study are displayed in Table 11–2. As you can see, the difference between the means is in the expected direction; Jim's sample of boys exhibited fewer helping behaviors, on the average, than his sample of girls. But is the difference large enough to be significant at the .01 level?

First, Jim calculates the sum of squares (*SS*) of each sample, using the formula $SS = (n - 1)s^2$:

$$SS_1 = (14)(3.41)^2 = 162.79$$
$$SS_2 = (17)(2.95)^2 = 147.94$$

Substituting these values in the formula for s_{diff}, he calculates

$$s_{\text{diff}} = \sqrt{\frac{162.79 + 147.94}{15 + 18 - 2}\left(\frac{1}{15} + \frac{1}{18}\right)} = 1.11$$

Now he is ready to calculate *t*:

$$t = \frac{M_1 - M_2}{s_{\text{diff}}} = \frac{7.4 - 8.5}{1.11} = -0.99$$

The degrees of freedom for Jim's test is 15 + 18 − 2 = 31. Since $df = 31$ is not listed in the *t* table, he uses the critical values for $df = 30$ and finds that his $|t|$ is between the critical values 0.854 with $p = .20$ and 1.055 with $p = .15$. Because $p > .15$, Jim cannot reject the null hypothesis. The mean number of helping behaviors of boys is not significantly smaller than that of girls at the .01 level (or even at the .15 level). Based on his test results, Jim cannot conclude that preadolescent boys display less helping behavior than preadolescent girls.

EXERCISE

1. Do first-born infants cry more, on the average, than later-born infants? A child psychologist asks 17 mothers of young infants how many hours, on the average, their children cry per day. He also asks whether the infant is a first child or a later-born child. Here are the results. Conduct the appropriate test. Use $\alpha = .05$.

   First-borns: 5, 7, 10, 6, 12, 8, 10
   Later-borns: 6, 3, 8, 10, 7, 4, 10, 5, 4, 7

2. A company provided different types of training to two independent samples of clerical workers. Their performance on a posttraining test follows. Determine whether the means differ significantly at the .05 level.

	Training Method	
	A	B
n	12	17
Mean	20.5	24.6
SD	8.51	7.6

Table 11–2 Helping Behaviors of Preadolescent Boys and Girls—Summary Statistics

	Boys	Girls
n	15	18
M	7.4	8.5
SD	3.41	2.95

An Alternative Nonpooled t Test for Independent Samples

The pooled variance t test assumes equality of the two population variances. The test yields reasonably valid results even when the homogeneity-of-variance assumption is violated to some degree. Therefore, researchers usually use the pooled procedure even if their sample data suggest the possibility of unequal variances. The one condition under which the pooled variance procedure should not be used is when (1) the two sample standard deviations are very different, and (2) sample sizes are small and unequal.

Here is an example: Suzie, a student researcher, is testing the hypothesis that the handwriting of left-handed students is more slanted, on the average, than the handwriting of right-handed students. As subjects, she is using 28 students in her statistics class. Since many more people are right-handed than left-handed, her two sample sizes are very unequal: n (left-handed) = 5 and n (right-handed) = 23. Suzie has calculated the following summary statistics:

	Group 1 (left-handed)	Group 2 (right-handed)
Mean slant	30.6	22.1
SD of slant	10.4	2.7

The discrepancy between the sample standard deviations, 10.4 versus 2.7, suggests that the population variances may be very different. The combination of small and unequal sample sizes and widely different standard deviations precludes use of the pooled variance t test.

An alternative, nonpooled t test procedure can be used in circumstances like these. The procedure bypasses the calculation of a pooled variance estimate and substitutes the two sample variances, s_1^2 and s_2^2, directly into the formula for the estimated standard error of the difference between means, like this:

$$s_{\text{diff}} = \sqrt{\frac{s_1^2}{n_1} + \frac{s_2^2}{n_2}}$$

Formula for s_{diff} *based on two nonpooled sample variances*

Then t is equal to $\frac{M_1 - M_2}{s_{\text{diff}}}$, as usual.

The degrees of freedom in the nonpooled t test is a number between (smaller n) − 1 and $n_1 + n_2 - 2$, the exact value depending on the standard deviations and sizes of both samples. A rather dreadful formula exists for calculating *df*, but it is not given here. If the t test is performed by statistical software, the computer calculates and uses the exact value. For tests conducted by hand, use of df = (smaller n) − 1 is recommended.

Let us apply the nonpooled t test procedure to Suzie's data. Suzie has hypothesized that the mean slant of handwriting is greater in left-handed students ($\mu_1 > \mu_2$). She needs to conduct an upper one-tailed test. The null and alternative hypotheses are

$$H_o: \mu_1 - \mu_2 \leq 0$$
$$H_A: \mu_1 - \mu_2 > 0$$

Suzie decides to use $\alpha = .05$. She calculates s_{diff} and t:

$$s_{\text{diff}} = \sqrt{\frac{s_1^2}{n_1} + \frac{s_2^2}{n_2}} = \sqrt{\frac{(10.4)^2}{5} + \frac{(2.7)^2}{23}} = 4.685$$

$$t = \frac{M_1 - M_2}{s_{\text{diff}}} = \frac{30.6 - 22.1}{4.685} = 1.814$$

Because the smaller of the two group sizes is 5, Suzie evaluates t with 5 − 1 = 4 degrees of freedom. Referring to the t table, she finds that 1.814 falls between the critical values 1.533, with $p = .10$, and 2.132, with $p = .05$. Since $p > .05$, the difference is not significant at the .05 level. Suzie cannot confidently conclude that the mean slant of left-handed students' writing is greater.

As we have seen, one obstacle to the use of the nonpooled procedure is the difficulty in calculating the exact degrees of freedom. This calculation is no problem for a computer. With many statistical software packages, the user can choose either the pooled or the nonpooled procedure. If the nonpooled procedure is selected, the program calculates and uses a more accurate value of *df* than (smaller n) − 1.

Some computer programs (e.g., SPSS) automatically conduct both tests. The output of an SPSS t test analysis is shown in Table 11–3. The test involves a comparison of male and female means on the PITTSSS, a scale measuring a personality trait called sen-

Table 11–3 SPSS Output for an Independent-Samples t Test

```
t-tests for independent samples of SEX

Variable     Number of Cases    Mean            SD        SE of Mean
--------------------------------------------------------------------
PITTSSS
SEX M                   96      118.5313      22.335          2.280
SEX F                  137      108.7664      18.490          1.580
--------------------------------------------------------------------
Mean Difference = 9.7648

Levene's Test for Equality of Variances: F = 2.151 P = .144

t-test for Equality of Means
                                                SE of          95%
Variances   t-value     df      2-Tail Sig    Diff      CI for Diff
--------------------------------------------------------------------
Equal         3.64     231         .000       2.683   (4.477, 15.053)
Unequal       3.52     179.27      .001       2.773   (4.291, 15.239)
--------------------------------------------------------------------
```

sation-seeking. The first part of the output is a table of descriptive statistics (the size, mean, *SD*, and standard error of the mean of each group). The male mean is higher by 9.7648 points. Beneath this table is the statement "Levene's Test for Equality of Variances: F = 2.151, P = 1.44." Levene's is one of several tests that can be used to check the homogeneity-of-variance assumption. The relatively large *P* (probability) means that the sample variances are not significantly different. However, even if the sample variances had differed significantly, use of the pooled procedure would be justified in this case because of the large sample sizes.

The table at the bottom of the printout presents the results of the pooled ("Equal Variances") and nonpooled ("Unequal Variances") t tests. The degrees of freedom for the pooled test is $n_1 + n_2 - 2 = 96 + 137 - 2 = 231$. The df for the nonpooled test is 179.27, which is less than $n_1 + n_2 - 2$ but greater than (smaller n) $-$ 1. The pooled procedure results in a slightly larger value of t and a slightly smaller level of significance, p. However, because the sample standard deviations are similar and the sample sizes are large, the results of the two tests are very similar.

In general, use of the pooled variance t test procedure is recommended, unless the following conditions prevail: (1) small and unequal sample sizes, and (2) markedly different sample standard deviations. In the rest of this chapter references to the "independent-samples t test" can be taken to mean the pooled variance procedure.

EXERCISE 11–5

1. Conduct Suzie's test using the pooled variance t test and compare the result with that of the nonpooled t test. (Keep in mind that, in this case, the nonpooled procedure is more appropriate.)
2. Decide which t test procedure is more appropriate in each of the following situations:

		Sample 1	Sample 2
a.	n	54	55
	s	3	8
b.	n	6	16
	s	4	5
c.	n	18	6
	s	10	3

The t Test for the Difference Between Means of Correlated Samples

The procedures discussed in the preceding sections apply only to data from independent samples. This section deals with procedures for calculating the estimated standard error of the difference between means and conducting the t test for data resulting from repeated measures or matching designs, designs that produce correlated samples.

Calculating the Estimated Standard Error of the Difference Between Means

Earlier, the following formula was given for the standard error of the difference between means of correlated samples:

$$\sigma_{\text{diff}} = \sqrt{\sigma^2_{M_1} + \sigma^2_{M_2} - 2\rho_{M_1M_2}\sigma_{M_1}\sigma_{M_2}}$$

Because σ_{M_1} is equal to $\sigma_1/\sqrt{n_1}$, and σ_{M_2} is equal to $\sigma_2/\sqrt{n_2}$, we can rewrite the formula as

$$\sigma_{\text{diff}} = \sqrt{\frac{\sigma_1^2}{n_1} + \frac{\sigma_2^2}{n_2} - \frac{2\rho_{M_1M_2}\sigma_1\sigma_2}{\sqrt{n_1}\sqrt{n_2}}}$$

In correlated samples, n_1 and n_2 are always equal. Therefore, we can use n without the subscript to represent the size of each sample and write

$$\sigma_{\text{diff}} = \sqrt{\frac{\sigma_1^2}{n} + \frac{\sigma_2^2}{n} - \frac{2\rho_{M_1M_2}\sigma_1\sigma_2}{n}} = \sqrt{\frac{1}{n}\left(\sigma_1^2 + \sigma_2^2 - 2\rho_{M_1M_2}\sigma_1\sigma_2\right)}$$

One way to calculate the estimated standard error of the difference between means is to substitute a sample estimator for each population parameter in the formula above. The sample estimators of σ_1^2 and σ_2^2 are the two sample variance estimates, s_1^2 and s_2^2, respectively, and the sample estimator of $\rho_{M_1M_2}$ is r_{12}, the correlation between the Sample 1 and Sample 2 scores. Therefore,

$$s_{\text{diff}} = \sqrt{\frac{1}{n}(s_1^2 + s_2^2 - 2r_{12}s_1s_2)}$$

Formula for calculating the estimated standard error of the difference between means of correlated samples

EXAMPLE In a study of attitude change with 10 subjects, the standard deviations of pretest and posttest scores are 6.5 and 8.3, respectively. The correlation between the pretest and posttest is .35. The estimated standard error of the difference between means is

$$s_{\text{diff}} = \sqrt{\frac{1}{10}[(6.5)^2 + (8.3)^2 - 2(.35)(6.5)(8.3)]} = 2.709$$

The formula just illustrated is convenient to use if the sample standard deviations and the correlation coefficient have been calculated in advance. However, if calculations must be performed on raw scores, there is a computationally simpler method.

We have already noted that when two samples are correlated, each score in the first sample (X_1) is paired with one and only one score in the second sample (X_2). Now calculate the difference between X_1 and X_2 in each pair and call the difference D: $D = X_1 - X_2$. Because there is one difference score for each pair of scores, the result is a sample of n difference scores, as in Table 11–4.

It can be shown that for any set of paired X_1 and X_2 scores, $s_1^2 + s_2^2 - 2\,r_{12}s_1s_2$ is equal to s_D^2, the variance of the difference scores. Therefore,

Table 11–4 **Calculation of Difference Scores for Paired X_1 and X_2 Scores**

Subject or Matched Pair	Sample 1 Scores (X_1)	Sample 2 Scores (X_2)	Difference Scores ($D = X_1 - X_2$)
A	3	5	–2
B	8	7	1
C	4	8	–4
D	2	6	–4
E	8	12	–4

$$s_{\text{diff}} = \sqrt{\frac{1}{n}\left(s_1^2 + s_2^2 - 2r_{12}s_1s_2\right)} = \sqrt{\frac{1}{n}\left(s_D^2\right)}$$

The variance of the difference scores, s_D^2, is equal to the sum of squares of the difference scores SS_D, divided by $n - 1$, where n is the number of difference scores.

$$s_D^2 = \frac{SS_D}{n-1}$$

SS_D can be calculated by either of the following formulas:

Deviation method: $SS_D = \Sigma(D - M_D)^2$

Raw score method: $SS_D = \Sigma D^2 - \frac{(\Sigma D)^2}{n}$

The raw score method is usually easier to use. Substituting into the formula for s_{diff}, we get

$$s_{\text{diff}} = \sqrt{\frac{1}{n}(s_D^2)} = \sqrt{\frac{1}{n}\left(\frac{SS_D}{n-1}\right)} = \sqrt{\frac{1}{n}\left[\frac{\Sigma D^2 - \frac{(\Sigma D)^2}{n}}{n-1}\right]}$$

$$s_{\text{diff}} = \sqrt{\frac{\Sigma D^2 - \frac{(\Sigma D)^2}{n}}{n(n-1)}}$$

Formula for calculating the estimated standard error of the difference between means of correlated samples using difference scores

To illustrate use of the difference score method for calculating s_{diff}, we use the data of Table 11–4. The calculation of ΣD and ΣD^2 is shown in Table 11–5. $\Sigma D = -13$, $\Sigma D^2 = 53$, and $n = 5$. Therefore

$$s_{\text{diff}} = \sqrt{\frac{53 - \frac{(-13)^2}{5}}{5(4)}} = 0.98$$

If you were to calculate s_1^2, s_2^2, and r_{12} for the data in Tables 11–4 and 11–5 and then substitute those values into the formula $s_{\text{diff}} = \sqrt{\frac{1}{n}\left(s_1^2 + s_2^2 - 2r_{12}s_1s_2\right)}$, you would get $s_{\text{diff}} = 0.98$, just as we did using the difference score formula. I think you will agree that, for small data sets, calculating s_{diff} from difference scores is considerably easier.

The two methods of calculating the estimated standard error of the difference between means for correlated samples are summarized here:

1. If the two sample standard deviations, s_1 and s_2, and the correlation between the two sets of scores, r_{12}, are readily available:

$$s_{\text{diff}} = \sqrt{\frac{1}{n}\left(s_1^2 + s_2^2 - 2r_{12}s_1s_2\right)}$$

2. When calculating from raw scores, calculate $D = X_1 - X_2$ within each pair of scores. Then

$$s_{\text{diff}} = \sqrt{\frac{\sum D^2 - \frac{(\sum D)^2}{n}}{n(n-1)}}$$

In both cases, n is the number of pairs of scores (the number of difference scores).

Degrees of Freedom

The number of degrees of freedom in the t test for correlated samples is $n - 1$. For example, for the data in Tables 11–4 and 11–5, $df = 5 - 1 = 4$.

Conducting the *t* Test for Correlated Samples

To illustrate the t test for correlated samples, let us analyze the data from Tom's repeated measures experiment on detection of faint tones with white noise present and absent. Recall that Tom hypothesized that the presence of white noise (Treatment 1) would decrease detection below that in the absence of white noise (Treatment 2). Tom needs to use a lower one-tailed test. His null and alternative hypotheses are

$$H_o: \mu_1 - \mu_2 \geq 0$$

$$H_A: \mu_1 - \mu_2 < 0$$

Tom decides to use $\alpha = .05$.

Table 11–5 Calculation of ΣD and ΣD^2 for the Data in Table 11–4

Pair	X_1	X_2	D	D^2
A	3	5	−2	4
B	8	7	1	1
C	4	8	−4	16
D	2	6	−4	16
E	8	12	−4	16
Sum			−13	53
			ΣD	ΣD^2

Table 11–6 Number of Tones Detected by 10 Subjects Under White-Noise-Present and White-Noise-Absent Conditions

Subject	White Noise Present (X_1)	White Noise Absent (X_2)	D	D^2
1	6	9	−3	9
2	4	6	−2	4
3	6	5	1	1
4	7	10	−3	9
5	4	10	−6	36
6	5	8	−3	9
7	5	7	−2	4
8	7	5	2	4
9	6	6	0	0
10	4	8	−4	16
Sum	54	74	−20	92
Mean	5.4	7.4		

Table 11–6 lists the number of tones detected by each of Tom's 10 subjects under both treatments and shows some of the calculations required in the test. The mean number of tones detected with the white noise on was 5.4, and the mean with the white noise off was 7.4. The direction of the difference between the means is in accord with Tom's hypothesis. However, a test is needed to determine whether the difference is significant at the .05 level.

Tom uses the D score method to calculate s_{diff}. Table 11–6 shows $\Sigma D = -20$ and $\Sigma D^2 = 92$. Therefore

$$s_{\text{diff}} = \sqrt{\frac{\sum D^2 - \frac{(\sum D)^2}{n}}{n(n-1)}} = \sqrt{\frac{92 - \frac{(-20)^2}{10}}{10(9)}} = .760$$

Now Tom is ready to calculate t:

$$t = \frac{M_1 - M_2}{s_{\text{diff}}} = \frac{5.4 - 7.4}{.760} = -2.63$$

With 10 difference scores, the number of degrees of freedom for Tom's test is $10 - 1 = 9$. Consulting the t table, Tom finds that his $|t|$, 2.63, falls between 2.262 with $p = .025$ and 2.821 with $p = .01$. Because $p < .05$, Tom rejects the null hypothesis. As he expected, his subjects detected significantly fewer tones with white noise present than with white noise absent. The results support Tom's research hypothesis that white noise interferes with detection of faint tones.

As a second example of a t test for correlated samples, let us analyze the data from Ann's study of the relation between preschool attendance and reading ability in first grade. Ann hypothesized that the average first-grade reading achievement of preschool attenders is greater than that of nonattenders who are matched on intelligence. If she calls the attenders Group 1, her research hypothesis is $\mu_1 > \mu_2$. Ann needs to use an upper one-tailed test. Her null and alternative hypotheses are

$$H_o: \mu_1 - \mu_2 \leq 0$$
$$H_A: \mu_1 - \mu_2 > 0$$

Ann decides to use $\alpha = .05$.

Ann first calculates the mean and standard deviation of each group and the correlation between the two sets of scores. The mean reading score of the 25 preschool attenders is 65.9 with $SD = 10.5$; the mean of the 25 nonattenders is 61.5, with $SD = 11.8$; the correlation between the matched reading scores of attenders and nonattenders is .40.

Now Ann is ready to calculate the estimated standard error of the difference between means (s_{diff}) and t.

$$s_{\text{diff}} = \sqrt{\frac{1}{n}\left(s_1^2 + s_2^2 - 2r_{12}s_1s_2\right)} = \sqrt{\frac{1}{25}\left[(10.5)^2 + (11.8)^2 - 2(.4)(10.5)(11.8)\right]} = 2.45$$

$$t = \frac{M_1 - M_2}{s_{\text{diff}}} = \frac{65.9 - 61.5}{2.45} = \frac{4.4}{2.45} = 1.796$$

The number of degrees of freedom in Ann's test is $25 - 1 = 24$. According to the t table, Ann's t, 1.796, falls between the critical values 1.711 with $p = .05$ and 2.064 with $p = .025$. Since $p < .05$, the mean reading achievement of the preschool attenders is significantly higher than that of the nonattenders at the .05 level. Ann rejects the null hypothesis and concludes that children who attended preschool read better at the end of first grade than equally intelligent children who did not attend preschool.

EXERCISE

1. A teacher administered a 15-item pretest before starting a new unit of work and a 15-item posttest after the unit was completed. Here are the pretest and posttest scores of the eight students in the class. Conduct a test to determine whether test scores were significantly higher after instruction than before.

Student	Pretest	Posttest
1	4	7
2	8	7
3	6	12
4	8	14
5	5	8
6	10	12
7	5	6
8	7	13

2. A class of 20 first-grade children was taught reading by the phonics method. A second class of 20 children, matched with the first on reading readiness scores, was taught reading by the whole-word method. The table summarizes the performance of the two classes on a reading test at the end of the year. Was either method significantly more effective than the other?

	Phonics	Whole-Word
Mean	38.8	35.2
SD	7.45	7.75

Correlation between paired reading scores of the two groups = .45

3. a. Treat the two groups of children in Ann's study as independent rather than matched samples. Assuming the same means and standard deviations, is the preschool attenders' mean significantly higher than that of the nonattenders at the .05 level? Conduct the appropriate test to answer this question.
   b. How do you account for the difference between the outcome of the analyses for matched and independent samples?

Postscript: Drawing Conclusions From t Tests in Experimental and Nonexperimental Research

In the preceding sections we have applied the t test to data from both experimental and nonexperimental studies. For example, we used the t test for independent samples to analyze data from Millie's experiment on text comprehension and Jim's nonexperimental comparison of helping behavior of boys and girls. We used the same procedures for calculating t and for deciding whether to reject the null hypothesis in both cases. Does this mean that differences between experimental and nonexperimental research are unimportant? No, it does not. Although the procedures for calculating and evaluating t are the same, conclusions concerning causes of differences on the dependent variable are possible only in experimental research.

To see why this is so, let us reexamine the results of Millie's experimental study and Ann's nonexperimental study. The difference between means was significant in both studies. However, conclusions concerning reasons for the difference between means differ. Millie can argue as follows: "Since I divided my 16 subjects randomly into two groups, the comprehension of the groups probably did not differ initially. But after I had manipulated the independent variable (type of cues), their comprehension did differ. Because all other factors except type of cues were the same for both groups, I conclude that different types of cues *caused* differences in comprehension."

Ann, on the other hand, cannot conclude that preschool attendance caused increased reading achievement in first grade, even though her preschoolers were better readers, on the average, than her nonattenders. The reason is that the preschool attenders and nonattenders in her study may differ on other variables (besides preschool attendance) that influence reading achievement. Because she matched her two samples on intelligence, she knows that the difference between their mean reading achievement was not due to intelligence. But there are countless other factors that may be responsible, at least in part, for the difference. Here are a few possibilities:

- Perhaps preschool attenders read better in first grade not because they attended preschool but because their parents read to them more frequently at home.
- Perhaps preschool attenders read better in first grade not because they attended preschool but because their parents play more word games with them.
- Perhaps preschool attenders read better in first grade because parents are more likely to send children to preschool if the children display an interest in learning to read, and it is their interest in reading rather than their preschool attendance that improves their achievement.

In our discussion about interpreting the results of correlational research in Chapter 9, we pointed out that a correlation between two measured variables, X and Y, cannot be taken as evidence that X causes Y. Nonexperimental comparisons of two means are like correlational research in that both the independent and dependent variables are measured variables. Just as a significant correlation coefficient between X and Y tells us that X and Y are related but not why they are related, a significant difference between means of preexisting groups tells us that the independent and dependent variables are related but not why they are related. Only experimental research, in which an independent variable is not just measured but manipulated, allows us to conclude with confidence that the relation between the variables is a causal one.

Estimating the Difference Between Two Population Means–Confidence Intervals

Sometimes we compare two sample means not to test a hypothesis but to estimate the size of the difference between the population means. In such cases, calculation of a confidence interval (*CI*) for the difference between population means is appropriate.

In Chapter 8 we calculated confidence intervals for the mean, μ, of a single population using the formula $M \pm t_{\alpha/2}(s_M)$. Confidence intervals for a difference between means, $\mu_1 - \mu_2$, take a similar form:

$$CI = (M_1 - M_2) \pm t_{a/2} s_{\text{diff}}$$ *Confidence interval for the difference between means*

In the calculation of the confidence interval, s_{diff}, the estimated standard error of the difference between means, is calculated by the procedures described earlier in this chapter. The value of $t_{\alpha/2}$ is found in the t table and depends on two factors: the chosen level of confidence and the number of degrees of freedom (df). Each level of confidence has an associated value of α. For the 95 percent level of confidence, $\alpha = .05$; for the 99 percent level of confidence, $\alpha = .01$; and so on. $t_{\alpha/2}$ is the critical value of t for the relevant α (two-tailed test) and the appropriate number of degrees of freedom. As in the t test, df for independent samples is $n_1 + n_2 - 2$, and df for correlated samples is $n - 1$.

To illustrate the procedure, we calculate confidence intervals for the difference between means in Jim's study of helping behavior of boys and girls (independent samples) and in Tom's perception study (correlated samples).

Calculating the 99% Confidence Interval in Jim's Study of Helping Behavior

Since Jim used $\alpha = .01$ in his statistical test, we will calculate the 99% confidence interval for the difference between means.

$$M_1 \text{ (boys)} = 7.4 \text{ and } M_2 \text{ (girls)} = 8.5$$

Jim calculated $s_{\text{diff}} = 1.11$.

With $n_1 = 15$ and $n_2 = 18$, $df = 15 + 18 - 2 = 31$. $t_{\alpha/2}$, the critical value of t for $\alpha = .01$ (two-tailed test) and 31 degrees of freedom, is 2.75.

The 99% confidence interval is $(7.4 - 8.5) \pm (2.75)(1.11) = -1.1 \pm 3.05$
$= -4.15$ to 1.95

Jim is 99% confident that the true difference between the mean helping behaviors of all preadolescent boys and girls is between 4.15 points favoring girls and 1.95 points favoring boys.

Confidence intervals for $\mu_1 - \mu_2$, like any other statistic, vary from one pair of samples to another. If, for example, Jim were to select another sample of boys and another sample of girls from the same two populations, the two sample means, as well as the estimated standard error of the difference between means, would differ somewhat from those of his original samples. Therefore, the 99% confidence interval would differ, too. However, Jim knows that in 99% of sample pairs the calculated confidence interval would include the true difference between the population means, and only 1% of sample pairs would "miss" the true difference. Therefore, the odds are 99 to 1 that $\mu_1 - \mu_2$ is somewhere within the interval Jim has calculated, –4.15 to 1.95.

Notice that the 99% confidence interval in Jim's study includes the value zero. Because of this, he cannot reject the null hypothesis that $\mu_1 - \mu_2 = 0$ at the .01 level. Our t test on the data resulted in the same conclusion. In general, if a confidence interval at a certain level of confidence (α) includes the value zero, the difference between the means is not significant at level α. If, on the other hand, a confidence interval at level α does not include the value zero, the difference between the means is significant at level α.

Calculating the 95% Confidence Interval for Tom's Perception Study

The 10 subjects in Tom's perception study detected a mean of 5.4 tones under white-noise conditions and a mean of 7.4 tones under no-white-noise conditions. Tom calculated $s_{\text{diff}} = .760$.

For the 95% level of confidence, $\alpha = .05$. In a repeated measures study with 10 subjects, $df = 10 - 1 = 9$. $t_{\alpha/2}$ for $\alpha = .05$ (two-tailed test) and 9 df is 2.262.

The 95% confidence interval is $(5.4 - 7.4) \pm (2.262)(.760) = -2.0 \pm 1.72$
$= -3.72$ to -0.28.

Tom is 95% confident that, in the population concerned, between 3.72 and 0.28 fewer tones are detected, on the average, with white noise present than absent.

Tom's 95% confidence interval does not include the value zero. Therefore, he rejects the null hypothesis that $\mu_1 - \mu_2 = 0$ and concludes that the difference between the two treatments is significant at the .05 level, the same conclusion that he reached in his t test.

EXERCISE 11–7

1. a. Calculate the 95% confidence interval for Millie's comprehension study. Compare the interval with the result of the t test. Do you reach the same conclusion in both cases?
   b. Calculate the 99% confidence interval for Millie's data. Based on the interval, is the difference between the means significant at the .01 level?
2. a. A 99% confidence interval for $\mu_1 - \mu_2$ extends from 2.75 to 9.31. Is the difference between the means significant at the .01 level?
   b. The difference between two means is not significant at the .05 level in a two-tailed test. Which of the following could be the 95% confidence limits for $\mu_1 - \mu_2$?
      (1) −8.15 and −0.63
      (2) 1.15 and 4.35
      (3) −0.75 and 7.54

Power, Robustness, and the Strength of Relationships

The Power of the *t* Test

Although researchers test the null hypothesis—the hypothesis that no difference exists between two population means—they usually hope to reject this hypothesis. Most research, after all, is undertaken with the expectation that two populations do differ or that two experimental treatments have different effects. Therefore, researchers try to maximize the probability of detecting a difference if it does, indeed, exist.

In Chapter 8 we defined the probability that a test will detect a true difference (i.e., that it will result in rejection of a false null hypothesis) as the ***power*** of the test, and we discussed factors that affect the power of tests of H_0: $\mu = \mu_H$. Much the same factors affect the power of the t test for the difference between two means. We list and discuss these factors below.

To understand how each factor influences the power of the test, remember that (1) power increases as the probability of rejecting the null hypothesis increases, and (2) the null hypothesis is rejected if $|t|$ is greater than the | critical value |. Therefore, anything that increases $|t|$ or decreases the | critical value | increases the power of the test. Specifically,

1. The larger the difference between the population means, $|\mu_1 - \mu_2|$, the greater the power of the test. In other words, the t test is more likely to detect a large difference between means than a small one.

   The reason that power increases as $|\mu_1 - \mu_2|$ increases is that, as $|\mu_1 - \mu_2|$ increases, the average size of differences between sample means, $|M_1 - M_2|$, increases. Because

$t = \frac{M_1 - M_2}{s_{\text{diff}}}$, increasing $|M_1 - M_2|$ increases the value of $|t|$, thereby increasing the power of the test.

2. The smaller the variance, σ^2, of the two populations, the greater the power of the test.

   The reason that power increases as σ^2 decreases is that, as σ^2 decreases, the standard error of the difference between means decreases. The smaller the value of s_{diff}, the greater the value of $|t|$. Therefore, as σ^2 decreases, the power of the test increases.

3. As n_1 and n_2 increase, the power of the test increases. Thus, if you wish to increase the likelihood of detecting a difference between population means, a good way is to increase sample size.

   The power of the t test increases as n_1 and n_2 increase for two reasons. First, as n_1 and n_2 increase, the value of s_{diff} decreases and, as s_{diff} decreases, $|t|$ increases. Second, as n_1 and n_2 increase, the degrees of freedom (df) for the test increase, and, as df increases, the | critical value | of t decreases. Thus, increasing sample size increases power both by increasing $|t|$ and by decreasing the | critical value |.

4. As α increases, the power of the test increases. For example, you are more likely to detect a difference between population means if you use $\alpha = .05$ than if you use $\alpha = .01$.

   The reason that power increases as α increases is that, as α increases, the | critical value | of t for $p = \alpha$ decreases.

5. A one-tailed test, if appropriate, is more powerful than a two-tailed test if other factors are equal.

   The reason that a one-tailed test is more powerful is that, for any given combination of α and df, the | critical value | of t is smaller in a one-tailed test than in a two-tailed test.

The Relative Power of the *t* Tests for Independent and Correlated Samples

To compare the power of the t tests for independent and correlated samples, we need to consider two factors: the degrees of freedom (df) for the test and the size of the standard error of the difference between means (σ_{diff}). We consider degrees of freedom first.

In the pooled variance t test for independent samples, $df = n_1 + n_2 - 2$. In the test for correlated samples, df is only half as large: $n - 1$. For example, with 10 subjects in each of two independent samples, $df = 10 + 10 - 2 = 18$; but with 10 subjects in each of two matched groups, $df = 10 - 1 = 9$. Because the | critical value | of t decreases as df increases, the fact that the t test for independent samples has a greater number of degrees of freedom increases its power relative to the test for correlated samples, assuming an equal number of observations or scores.

However, a comparison of σ_{diff} in independent and correlated samples paints a different picture. The formulas for σ_{diff} are

$$\text{Independent samples:} \quad \sigma_{\text{diff}} = \sqrt{\sigma_{M_1}^2 + \sigma_{M_2}^2}$$

$$\text{Correlated samples:} \quad \sigma_{\text{diff}} = \sqrt{\sigma_{M_1}^2 + \sigma_{M_2}^2 - 2\rho_{M_1M_2}\sigma_{M_1}\sigma_{M_2}}$$

As we observed earlier, if σ_1, σ_2, n_1, and n_2 are the same in both cases, and if the correlation between sample means, $\rho_{M_1M_2}$, is positive, σ_{diff} is smaller in correlated than in independent samples, and, as a result, $|t|$ is larger. This factor increases the power of the t test for correlated samples relative to the power of the t test for independent samples, provided that $\rho_{M_1M_2}$ is positive.

Under what circumstances can we expect $\rho_{M_1M_2}$ to be positive and to increase the power of the test for correlated samples? In repeated measures designs, $\rho_{M_1M_2}$ is almost always positive and often very large. Therefore, a repeated measures design usually results in a more powerful test than an independent samples design (assuming an equal number of observations in both cases). However, as we have seen, carryover effects prevent the use of repeated measures designs in many kinds of research.

In matching designs, carryover effects are not a problem. However, matching results in positive values of $\rho_{M_1M_2}$ (and greater power) only if the matching variable is positively related to the dependent variable. In Ann's study of first-grade reading achievement, the matching variable, intelligence, was known to have a substantial positive correlation with reading achievement. Therefore, Ann could be confident that matching on intelligence would result in a more powerful test. But in many studies, matching variables related to the dependent variable are hard to find. And matching on an unrelated variable is not only a waste of time but results in a test with fewer degrees of freedom and hence less power. Because of the difficulties in identifying good matching variables, and due to other factors discussed earlier, matching is rarely the design of choice.

Statistical Versus Practical or Theoretical Significance

As we have observed, the power of the t test increases as sample size increases. In fact, with large enough samples, even a very small difference between sample means may result in a relatively large, statistically significant value of t. This brings us to the issue of statistical versus practical significance. Not all differences that are statistically significant are of theoretical or practical importance.

Suppose, for example, that a new method of training factory workers increases productivity from an average of 1586 gizmos produced per month to 1587 gizmos produced per month, and that the company regards this as a meaninglessly small change. If a very large number of workers is used in an experimental comparison of the two methods, the difference between the sample means may turn out to be significant at the .05 or even the .01 level. If so, the difference, although statistically significant, is not significant in theoretical or practical terms.

There is no point to detecting differences that, although statistically significant, are not worth detecting. Statistical procedures are, after all, tools for advancing theory and practice, not ends in themselves. Therefore, when research is designed to answer questions about differences between population means, researchers should ask themselves not "How many subjects do I need to detect any difference, however small?" but "How many subjects do I need to detect a difference of ______ or more?" (For example, a difference of 5 or more points between means.)

Both algebraic methods and tables, called *power tables*, are available to help researchers identify the minimum sample size needed to detect a difference of a given size. The procedures involved are beyond the scope of this text. Relevant discussions can be found in more advanced statistics textbooks as well as books on experimental design. There is, however, a relatively simple way to assess the importance of an obtained difference. It involves calculating a measure of the strength of the relationship between the independent and dependent variables.

Measuring the Strength of the Relation Between the Independent and Dependent Variables

When we want to measure the strength of a relationship between two variables, we calculate a correlation coefficient. Then we square the correlation coefficient to find out what proportion of the variance of each variable is accounted for by the other variable. If we can calculate the squared correlation coefficient between the independent and dependent variables in a research study, we will know what proportion of the variance of the dependent variable is accounted for by the independent variable.

In studies like those discussed in this chapter, the independent variable is dichotomous; that is, it has only two values (e.g., visual cues and verbal cues, boys and girls). The dependent variable is quantitative with a range of values. The correlation coefficient for measuring the relation between a dichotomous and a quantitative variable is called the ***point biserial correlation coefficient,*** $\boldsymbol{r_{pb}}$. The point biserial correlation coefficient, then, is the type of correlation coefficient that is appropriate for measuring the relation between the independent and dependent variables.

A formula for calculating r_{pb} was presented and demonstrated in Chapter 9. Fortunately, however, if you have already conducted a t test, you do not need to calculate r_{pb}^2 from the data, because there is a precise relation between t and r_{pb}^2 in independent samples. Specifically,

$$r_{pb}^2 = \frac{t^2}{t^2 + df}$$

where $df = n_1 + n_2 - 2$.

In Millie's study, for example, we calculated $t = 2.48$; n_1 and n_2 were both equal to 8, so that $df = 8 + 8 - 2 = 14$. Therefore,

$$r_{pb}^2 = \frac{(2.48)^2}{(2.48)^2 + 14} = .31$$

Type of cues accounts for 31% of the comprehension test variance; the remaining 69% is accounted for by other factors.

Note that if t remains the same but n_1 and n_2 increase, $r^2{}_{pb}$ decreases. Suppose, for example, that Millie had used 50 subjects in each of her two groups instead of 8, and suppose that she got $t = 2.48$, as before.† Then $df = 50 + 50 - 2 = 98$ and

$$r_{pb}^2 = \frac{(2.48)^2}{(2.48)^2 + 98} = .06$$

With 50 subjects per group and $t = 2.48$, the independent variable accounts for only about 6% of the variance on the dependent variable, a much less impressive percentage than 31.

The effect of sample size on r_{pb}^2 provides additional support for a point that was made earlier: If you collect enough data, you may get a statistically significant value of t, even though the actual difference between means is meaninglessly small. By calculating r_{pb}^2 after you have obtained a significant difference between means in a t test, you get a better indication of the "importance" of the difference.

EXERCISE 11–8

1. In each of the following studies, decide which test, A or B, has greater power. Assume that other factors are equal in each study.

	Test A	Test B
a.	$\mu_1 = 15, \mu_2 = 25$	$\mu_1 = 15, \mu_2 = 20$
b.	$n_1 = n_2 = 12$	$n_1 = n_2 = 24$
c.	$\alpha = .01$	$\alpha = .005$
d.	H_0: $\mu_1 - \mu_2 \geq 0$	H_0: $\mu_1 - \mu_2 = 0$
	H_A: $\mu_1 - \mu_2 < 0$	H_A: $\mu_1 - \mu_2 \neq 0$

†With 50 subjects in each group, a considerably smaller difference between sample means would result in $t = 2.48$.

e. A single group of 10 subjects is run under both experimental conditions. (Assume that $\rho_{M_1M_2}$ is large and positive and that there is no carryover effect.) — Each of 20 subjects is randomly assigned to one of two experimental conditions.

f. $\sigma_1^2 = \sigma_2^2 = 70$ — $\sigma_1^2 = \sigma_2^2 = 250$

2. Did the use of repeated measures rather than independent samples reduce s_{diff} in Tom's perception study? (To answer this question, treat the white-noise-present and white-noise-absent data as independent samples and calculate s_{diff}.)
3. In an experiment, subjects are randomly assigned to two experimental conditions. The t test comparing the means on the dependent variable yields $t = -7.83$.
   a. Calculate and interpret r_{pb}^2 if $n_1 = n_2 = 10$.
   b. Calculate and interpret r_{pb}^2 if $n_1 = n_2 = 100$.
   c. Is the effect of the independent variable more impressive with the smaller or the larger sample size?

The Assumptions Underlying the *t* Test and the Robustness of the Test

Use of the t distribution for conducting tests and calculating confidence intervals for differences between means rests on the following assumptions:

1. The population distributions are normal or, failing that, n_1 and n_2 are reasonably large.
2. For the pooled variance t test, the population variances are equal.

We have already seen that violation of the second assumption is not a problem unless sample sizes are small and unequal. What are the effects of violating the first assumption? The answer is, once again, not very serious, unless the populations are highly skewed and sample sizes are very small. For example, suppose you want to compare the means of two samples of size 15 each, and stemplots of the two samples show a slight but definite skew (positive or negative). Despite this, you can safely conduct a t test using the procedures of this chapter.

When a test is relatively "immune" to violations of the underlying assumptions, the test is said to be ***robust***. Thus, the t test for the difference between means is a robus*t* test.

Use of the t distribution should be avoided, however, if there is evidence of extreme skew and/or outliers, especially when sample sizes are small. As you learned in Chapter 3, under these conditions the mean is not a good measure of central tendency. Therefore, tests comparing means are not appropriate, and it may be better to use tests that make no assumptions about population shape, tests called *nonparametric tests*. Some common nonparametric tests are discussed in Chapter 14.

SUMMARY

The t test for the difference between means is used to decide whether the difference between two sample means, $M_1 - M_2$, is sufficiently large to reject the null hypothesis that the means of the two populations, μ_1 and μ_2, are equal. The two samples whose means are to be compared may be ***independent*** or ***correlated***. Samples are independent if the membership of one sample is unrelated to the membership of the other. Samples are correlated if the same subjects are members of both samples (repeated measures) or if subjects in the two samples are matched. Degrees of freedom and computational procedures differ for independent and correlated samples.

The sampling distribution that underlies the t test is the ***sampling distribution of the difference between means***—the distribution of values of $M_1 - M_2$ over all possible pairs of samples of size n_1 and n_2 from two populations. The shape of the sampling distribution is approximately normal, provided that both populations are normal or n_1 and n_2 are reasonably large. The mean of the sampling distribution, μ_{diff}, is equal to the difference between the population means, $\mu_1 - \mu_2$. The standard deviation of the sampling distribution of the difference between means, σ_{diff}, is called the ***standard error of the difference between means***. The formulas for σ_{diff} are

Independent samples:

$$\sigma_{\text{diff}} = \sqrt{\sigma_{M_1}^2 + \sigma_{M_2}^2}$$

Correlated samples:

$$\sigma_{\text{diff}} = \sqrt{\sigma_{M_1}^2 + \sigma_{M_2}^2 - 2\rho_{M_1M_2}\sigma_{M_1}\sigma_{M_2}}$$

where σ_{M_1} and σ_{M_2} are the standard errors of the mean of samples from the two populations, respectively, and $\rho_{M_1M_2}$ is the correlation between sample means. If the sample estimator of σ_{diff}, symbolized s_{diff}, is substituted for σ_{diff}, the ratio $\frac{(M_1 - M_2) - (\mu_1 - \mu_2)}{s_{\text{diff}}}$ is distributed as t.

The t test for the difference between means tests the null hypothesis (H_0) that the two population means are equal ($\mu_1 - \mu_2 = 0$). The t test may be two-tailed, upper one-tailed, or lower one-tailed, depending on the research question or hypothesis. The steps and the reasoning in the test are the same as in any statistical test. After the null and alternative hypotheses have been stated and a level of α has been selected, t is calculated by the formula

$$t = \frac{M_1 - M_2}{s_{\text{diff}}}$$

The calculated t is compared with critical values of t for the relevant degrees of freedom, and if $p < \alpha$, H_0 is rejected. If H_0 is rejected, the difference between the means is significant at level α.

The t test for independent samples assumes that the population variances are equal. A ***pooled variance estimate***, s_p^2, is calculated by combining data from both samples, and s_p^2 is used to calculate s_{diff}. The t test based on the pooled procedure has $n_1 + n_2 - 2$ degrees of freedom. If the assumption of equal variances is unreasonable and sample sizes are small, an alternative procedure can be used. The alternative method has degrees of freedom between (smaller n) $- 1$ and $n_1 + n_2 - 2$; use of $df =$ (smaller n) $- 1$ is recommended if the exact degrees of freedom are unknown.

In the t test for correlated samples, t can be calculated either from the sample means and standard deviations and the correlation between the two sets of scores, or t can be calculated based on difference scores. The test for correlated samples has degrees of freedom equal to the number of pairs of scores minus one.

If we wish to estimate the size of the difference between two population means, $\mu_1 - \mu_2$, we construct a confidence interval for the difference between means. The formula for the confidence interval (CI) is

$$CI = (M_1 - M_2) \pm t_{\alpha/2} s_{\text{diff}}$$

$t_{\alpha/2}$ is the critical value of t for a two-tailed test with the chosen level of confidence (α) and the appropriate degrees of freedom.

The power of the t test refers to the probability that it will detect a difference between μ_1 and μ_2, assuming a difference exists. The power of the t test increases as $\mu_1 - \mu_2$ increases, as σ^2 decreases, as α increases, and as n_1 and n_2 increase. One-tailed tests, if appropriate, are more powerful than two-tailed tests. Tests for correlated samples tend to be more powerful than tests for independent samples, provided that $\rho_{M_1M_2}$ is positive.

If sample sizes are very large, the t test may be powerful enough to detect even very small and possibly unimportant differences between means. To measure the actual strength of the relationship between the independent and dependent variables in studies using independent samples, the squared point biserial correlation can be calculated by the formula

$$r_{pb}^2 = \frac{t^2}{t^2 + df}$$

where $df = n_1 + n_2 - 2$. r_{pb}^2 is the proportion of the dependent variable's variance that is accounted for by the independent variable.

The assumptions underlying the t test are: (1) normal populations or reasonably large sample size, and (2) (for the pooled t test) $\sigma_1^2 = \sigma_2^2$ (the homogeneity-of-variance assumption). The t test is relatively ***robust*** to violations of the second assumption unless n_1 and n_2 are small and very different. Violation of the first assumption also presents no problem unless nonnormality is marked or outliers are present, and sample sizes are small.

QUESTIONS

1. You select all possible pairs of independent samples of size 20 each from two populations. The mean of the first population is 75 with $SD = 10$; the mean of the second population is 82 with $SD = 8$. In each pair of samples you calculate the difference between the sample means, $M_1 - M_2$.

a. What is the distribution of values of $M_1 - M_2$ called?
b. What is the mean of the distribution?
c. What is the standard deviation of the distribution called? Calculate its value.
d. If the samples are correlated with $\rho_{M_1M_2} = .5$, which of your answers to parts a–c will change? Calculate the changed answer(s).

2. From two highly skewed populations, you select all possible pairs of samples of size 5 each and calculate $M_1 - M_2$ in each pair. Call the distribution of values of $M_1 - M_2$ "Distribution A." Then you select all possible pairs of samples of size 10 each from the same populations and calculate $M_1 - M_2$ in each pair. Call this distribution of values of $M_1 - M_2$ "Distribution B." Which of the two distributions, A or B, is more nearly normal?

3. If two populations have the same mean, what is the average value of $M_1 - M_2$ over all possible pairs of samples?

4. You conduct two experiments to compare the mean reaction time to two complex displays. In the first experiment, you test a single group of 25 subjects with both displays. In the second experiment, you randomly assign 25 subjects to Display A and 25 to Display B.

a. In which experiment are you using independent samples, and in which are you using correlated samples?
b. If other factors are equal, and assuming no carryover effects, which experiment will probably result in a larger value of t?

5. A social psychologist hypothesizes that creation of cognitive dissonance will increase group cohesiveness in discussion groups. Sixteen students are randomly divided into two discussion groups of 8 students each. Dissonance is produced in one but not the other. Afterward, all students fill out a questionnaire designed to measure degree of group cohesiveness. Here are the results:

Group Cohesiveness Scores

Dissonance Group	Nondissonance Group
3	3
8	6
6	4
7	3
5	5
6	2
7	6
6	4

Did dissonance significantly increase group cohesiveness? Use $\alpha = .05$.

6. A clinical psychologist measured intensity of headaches of 12 migraine sufferers both before and after relaxation training. The mean intensity of headaches before training, on a scale of 1 to 10, was 6.5, with $SD = 2.61$. After training the mean was 4.2 with $SD = 2.22$. The correlation between headache intensity before and after training was .72. Conduct a test to determine whether relaxation training significantly reduced the intensity of headaches of migraine sufferers. Use $\alpha = .05$.

7. Conduct a t test to determine whether strenuous exercise significantly reduces strength of grip.

Strength of Grip (Pounds)

Student	Before Exercise	After Exercise
1	20	15
2	17	14
3	12	12
4	15	12
5	9	10
6	15	8
7	13	9

8. The manufacturers of two hormone creams both claim that their product is the most effective for reducing facial wrinkles due to aging. To determine whether the creams differ in effectiveness, an independent consumer agency randomly divides a group of 22 elderly women in half and has each group use a different cream for 3 months. At the end of this time, the number of facial wrinkles of each woman is measured. The results are summarized here.

Number of Wrinkles

Cream A	Cream B
$n = 11$	$n = 11$
$\Sigma X = 1067$	$\Sigma X = 616$
$\Sigma X^2 = 111{,}971$	$\Sigma X^2 = 42{,}244$

a. Conduct a test to determine whether the creams differ in effectiveness. Use $\alpha = .05$.

b. If yes, which cream appears to be more effective at reducing facial wrinkles?

c. Without calculating: Will the 95% confidence interval for the difference between means include the value zero?

d. Calculate the 95% confidence interval.

e. Calculate the 99% confidence interval.

f. Use the confidence interval to decide whether the creams' effectiveness differs significantly at the .01 level.

g. What proportion of the facial wrinkle variance is accounted for by the type of cream used?

9. In a group of beginning first graders, a psychologist first formed six matched pairs of children based on their reading readiness scores. One member of each pair was randomly assigned to a phonics reading class, the other to a whole-word class. After 3 weeks of reading instruction, all the children were tested to see how many of a list of 20 new words they could read. The scores were as follows:

	Number of Words Read	
Pair of Children	Phonics Method	Whole-Word Method
1	12	8
2	8	7
3	15	13
4	10	8
5	7	9
6	12	12

a. Was one method more effective than the other? Use $\alpha = .05$.

b. Answer before calculating: Will the 95% confidence interval for the difference between means include the value zero?

c. Calculate the 95% confidence interval.

10. Here are data from two matched samples:

	Matched Pair							
	1	2	3	4	5	6	7	8
Sample 1	5	4	6	3	7	5	7	4
Sample 2	4	7	7	3	5	9	8	1

a. Calculate t. Use the difference-score method.

b. Assume that Samples 1 and 2 are independent instead of matched, and calculate t. Explain the difference between the values of t in parts a and b.

c. Here are data from another two matched samples:

	Matched Pair							
	1	2	3	4	5	6	7	8
Sample 1	5	6	3	7	4	6	4	6
Sample 2	4	1	4	4	6	7	2	4

Calculate t using the difference score method.

d. Treat the two samples in part c as independent samples and calculate t. Compare the values of t in parts c and d.

e. How do you account for the difference between your comparisons of t in part a versus part b and in part c versus part d?

11. In each of the following cases (a–e), tell which test, A or B, is more powerful, if other factors are equal.

Test A	Test B
a. $\alpha = .05$	$\alpha = .01$
b. $\mu_1 = 30$, $\mu_2 = 32$	$\mu_1 = 30$, $\mu_2 = 26$
c. $\sigma_1^2 = \sigma_2^2 = 70$	$\sigma_1^2 = \sigma_2^2 = 50$
d. two-tailed test	one-tailed test
e. $n_1 = n_2 = 60$	$n_1 = n_2 = 40$

12. Dr. Smith was interested in the relationship between anxiety and problem solving. She randomly assigned 30 subjects to each of two groups. One group was given anxiety-arousing instructions for a problem-solving task, while the other group received non-anxiety-arousing instructions. She used a t test to compare the mean task performance of the two groups. Another investigator, Dr. Jones, was also interested in the relationship between anxiety and problem solving. She gave a test measuring anxiety to 60 subjects and conducted a t test to compare the mean problem-solving performance of the 30 most anxious subjects and the 30 least anxious subjects. Both Dr. Smith and Dr. Jones found a significant difference favoring the low-anxiety group.

a. Can Dr. Smith and Dr. Jones both conclude that anxiety and problem-solving performance are related?
b. Can Dr. Smith and Dr. Jones both conclude that anxiety interferes with problem solving? If not, why not?

13. At the company where you work, you decide to compare the work attitudes of workers who have been employed by the company for less than 5 years with the attitudes of workers who have been employed for over 5 years. You plan to interview 20 workers from each population. Assume that you can either randomly select subjects from each population or you can match the groups on age.

a. If age is strongly related to work attitudes, which of your two methods of sampling is more likely to result in a significant value of t?
b. If age is unrelated to work attitudes, which method of sampling is more likely to result in a significant value of t?

14. Two randomized groups experiments both yielded $t = 2.84$. Experiment 1 used 5 subjects in each group; Experiment 2 used 15 subjects in each group.

a. In which experiment was the difference between the sample means larger?
b. Without calculating: In which experiment did the independent variable account for a greater proportion of the variance of the dependent variable?
c. Calculate the proportion of variance accounted for by the independent variable in each experiment.

15. The point biserial correlation between an independent and a dependent variable is .6. What is the value of t in the t test if

a. $n_1 = n_2 = 10$.
b. $n_1 = n_2 = 30$.

12 Simple or One-Way Analysis of Variance

INTRODUCTION

Research questions in the social and behavioral sciences are often concerned with differences among three or more means, such as differences in achievement under three instructional conditions. When the number of means to be compared is greater than two, t tests no longer suffice. The procedure that is used in these cases is the *analysis of variance*, or *ANOVA*.

The analysis of variance is a family of tests for analyzing data from studies that range from simple designs with a single independent variable to highly complex designs with multiple variables. In this chapter we deal with the most basic type of ANOVA, called *simple* or *one-way ANOVA*. It is called "one-way" because it analyzes data from studies with a single independent variable. We discuss the rationale, the computational procedures, and the display and interpretation of the results in this chapter.

As you will see, despite its name, the analysis of variance is a test for comparing means. It compares the obtained variation of a set of sample means with the amount of variation to be expected by chance alone. If the obtained variance of the means is greater than would be expected by chance, we conclude that the means of the populations from which the samples were drawn are not all the same.

Sometimes, either before or after we have conducted an ANOVA comparing the means of three or more groups, we also want to compare particular pairs of means. If so, we conduct either preplanned tests or tests selected after we have examined the results of the ANOVA. The last part of the chapter discusses tests that can be used to make comparisons of these sorts.

Objectives

When you have completed this chapter, you will be able to

- Explain why multiple t tests are inadvisable for comparing three or more means.
- State the null and alternative hypotheses of the one-way ANOVA.
- Explain how the ANOVA analyzes differences between means by comparing two variance estimates, the mean square between groups (*MS* between) and the mean square within groups (*MS* within).

- Show how the total sum of squares of a data set (*SS* total) can be partitioned into two components: the sum of squares within groups (*SS* within), and the sum of squares between groups (*SS* between).
- Use deviation formulas or computational (raw score) formulas to calculate *SS* total, *SS* between, and *SS* within.
- Calculate the degrees of freedom (*df*) between and the *df* within, the mean square (*MS*) between and the *MS* within, and the test statistic, *F*.
- Explain how the *F* test comparing two variance estimates provides information about differences between means, and be able to interpret *F* and draw appropriate conclusions.
- Prepare and interpret ANOVA summary tables.
- Compare the *t* test and the ANOVA for analyzing the difference between the means of two independent samples.
- Describe factors affecting the power of ANOVA. Given information about these factors in two or more tests, decide which test is more powerful.
- Estimate the proportion of the total data variance accounted for by the independent variable.
- Define multiple comparisons. Differentiate between orthogonal (independent) and nonorthogonal comparisons, and be able to decide whether a given pair of comparisons is orthogonal or not.
- Define a family of comparisons. Explain how the family of interest affects the choice of a test.
- Explain the difference between planned comparisons and post hoc tests.
- Conduct tests for planned comparisons.
- Describe the family of comparisons for each of the following post hoc tests: the Scheffé test and the Tukey test. Conduct Scheffé and Tukey tests and interpret the results.

Rationale of the Analysis of Variance

The Case Against Multiple *t* Tests

Suppose that you have collected data from three independent samples, such as three experimental groups or samples from three populations. You want to compare the three means. How should you proceed? Perhaps you perform several *t* tests. You could conduct a *t* test comparing the Group 1 and Group 2 means, a second *t* test comparing the Group 1 and Group 3 means, and a third *t* test comparing the Group 2 and Group 3 means. You might also want to compare the mean of Group 1 with that of Groups 2 and 3 combined, the mean of Group 2 with that of Groups 1 and 3 combined, and the mean of Group 3 with that of Groups 1 and 2 combined. As you can see, six *t* tests are possible when three means are to be compared.

The number of possible *t* tests multiplies rapidly when more than three means are to be compared. For example, with four groups, the number of possible *t* tests is 25. This suggests that one problem with using *t* tests to analyze multiple-group data is the sheer number of tests required. A far more serious problem, however, is something called the ***Type I error rate***—the probability of a Type I error over the entire set of tests.

As you know, in any one test, the probability of a Type I error (the probability of rejecting a true null hypothesis, H_0) is equal to alpha (α). For example, if H_0 is true and you conduct a test with $\alpha = .05$, the probability that you will incorrectly reject H_0 is .05.

However, if you conduct two or more tests, each with a = .05, and if all the null hypotheses are true, the probability that you will incorrectly reject H_0 in at least one test is greater than .05. In other words, with multiple tests, the Type I error rate is greater than the alpha level selected for each test.

In order to understand why this is so, recall the addition law of probability (see Chapter 4). Given two or more mutually exclusive events, the probability that at least one of them will occur is the sum of the separate probabilities. For example, if $p(\text{A}) = .05$, $p(\text{B}) = .05$, and $p(\text{C}) = .05$, then $p(\text{A or B or C}) = .05 + .05 + .05 = .15$. If the events in question are not mutually exclusive, the probability that at least one will occur is somewhat less than the sum of the probabilities but still considerably greater than the individual probabilities. For example, given three *t* tests of true null hypotheses, each with $\alpha = .05$, the probability that H_0 will be rejected in at least one of them is approximately equal to .14.

Let us go back to our three-group example. As you will recall, comparing all possible pairs of means requires six *t* tests. Suppose that you conduct all six tests and that one of the differences is significant at the .05 level. How confident can you be in rejecting H_0? Since the probability of making at least one Type I error with six tests is close to .3, there are about 3 chances in 10 that you are rejecting a true null hypothesis.

In the analysis of variance (ANOVA), the Type I error rate is controlled by conducting a single test for the overall difference among the entire set of means, at a level of alpha chosen by the researcher, usually .05 or .01. If H_0 is rejected, the researcher concludes that somewhere within the set of means a true difference exists; and the probability that this conclusion is false (i.e., that a Type I error has occurred) is equal to the chosen alpha. Then, the researcher can conduct subsequent tests, called ***post hoc tests***, to determine which particular means are significantly different from which others.

Before we examine the ANOVA itself, let me mention that multiple *t* tests are not always a bad thing. If a researcher has decided, in advance of data collection, to test for a small number of specific differences between means, these can be tested before, or in lieu of, an ANOVA. For example, an experimenter who is comparing two experimental groups and one control group may know in advance that he is interested in two particular differences: the difference between the two experimental groups and the difference between the combined experimental groups and the control group. Thus, only two rather than six comparisons are to be conducted, and these two have been selected in advance. Comparisons like these are called ***planned comparisons***. With preplanned tests, the Type I error rate is of less concern than with post hoc tests, as you will see.

The Rationale of ANOVA

Simple or one-way ANOVA is used to analyze data from studies with a single independent variable varying between two or more independent samples. We will use *k* to represent the number of means to be compared. The null hypothesis of the ANOVA states that all *k* population means are equal.

$$H_o\colon \mu_1 = \mu_2 = \cdots = \mu_k$$

The alternative hypothesis in the ANOVA—the conclusion reached if H_0 is rejected—is that not all the population means are equal.

$$H_A\colon \text{ Not all } \mu \text{ are equal.}$$

Note that the alternative hypothesis does *not* state that all the population means are different; it states that *at least one* population mean differs from the others. For example, if $k = 5$ and μ_1, μ_2, μ_3, and μ_4 are all equal to 10, but $\mu_5 = 12$, H_0 is false and H_A is true.

If the ANOVA is a test for the difference between means, why isn't it called analysis of "means" rather than analysis of "variance"? It is called analysis of variance because the comparison of means rests on partitioning the total variance of the data set into two components: (1) a component based on differences between the sample means, and (2) a component based on variation of scores within each sample around the sample mean. The statistical test of the ANOVA compares the two variance components.

Let me expand on this a little. Suppose that we are testing a true null hypothesis, that is, $\mu_1 = \mu_2 = \cdots = \mu_k$. We select a sample from each population. Because statistics vary from one sample to another, the k sample means ($M_1, M_2, \cdots, M_k$) are expected to differ somewhat from one another, just by chance. For example, if $k = 3$ and $\mu_1 = \mu_2 = \mu_3 = 8$, we do not expect all three samples to have M exactly equal to 8. But because sample means are usually reasonably close to the population mean, we expect the three sample means to be close to 8.

If, on the other hand, the null hypothesis is false and not all the population means are equal, we expect considerably more variation among the k sample means. For example, if $\mu_1 = 6$, $\mu_2 = 8$, and $\mu_3 = 10$, we expect M_1 to be close to 6, M_2 to be close to 8, and M_3 to be close to 10. The three sample means are expected to be more different from one another if the three population means are different than if they are the same.

In the one-way ANOVA, we calculate two variance estimates, which are referred to as *mean squares*. (Recall from Chapter 3 that "mean square" is another name for "variance.") One of the mean squares is called the ***mean square between groups (*MS *between)***, because it is based on the variation between the sample means. The other is called the ***mean square within groups (*MS *within)***, because it is based on variation of scores within samples around their own means.

Later in the chapter we will show that if all the population means are equal (i.e., if H_0 is true), the *MS* between and the *MS* within are expected to be approximately equal. But if not all the population means are equal (H_0 is false), the greater than chance variation among the sample means increases the value of the *MS* between, without affecting the *MS* within. Consequently, the *MS* between is expected to be greater than the *MS* within when H_0 is false.

The statistical test of the ANOVA compares the *MS* between with the *MS* within. If *MS* between is not too different from *MS* within, the variation between the sample means is judged to be due to chance, and H_0 is not rejected. If, on the other hand, the *MS* between is much greater than the *MS* within—so much greater that the probability of a difference this great is less than alpha—H_0 is rejected, and H_A, the hypothesis that not all the population means are equal, is regarded as confirmed.

Like any statistical test, the ANOVA has certain underlying assumptions. They are the following:

1. All the populations are normal.
2. All the population variances are equal. That is, $\sigma_1^2 = \sigma_2^2 = \cdots = \sigma_k^2$ (the homogeneity of variance assumption).

Note that the assumptions underlying the ANOVA are the same as those underlying the t test for the difference between means. The t test and the ANOVA are closely related, as you will see.

EXERCISE 12–1

1. In your own words, explain what is meant by the statement "As the number of statistical tests conducted on a set of data increases, the Type I error rate increases."
2. Which of the following is/are correct statements of the alternative hypothesis (H_A) in ANOVA?
   a. At least one population mean differs from the others.
   b. The population variances are different.
   c. All the populations have different means.
   d. $\mu_1 \neq \mu_2 \neq \cdots \neq \mu_k$
   e. Not all the population means are the same.
   f. Not all the population variances are equal.
3. Put >, <, or = in each space: If H_0 is true, *MS* between is expected to be _____ *MS* within; if H_0 is false, *MS* between is expected to be _____ *MS* within.

Rationale and Steps of the ANOVA

Partitioning the Total Variation of a Data Set

The total variation of a data set can be partitioned into two components, a component based on variation of scores within samples and a component based on variation between sample means. We will demonstrate how this is done using two data sets. The first (Set A) consists of three samples drawn from normal populations with equal means and equal variances. Thus, in Set A, we know in advance that H_0 is true. The three samples of Set B are drawn from normal populations with different means (but equal variances). Thus, in Set B, we know that H_0 is false and H_A is true.

Data Sets A and B are displayed in Table 12–1. Each set includes three independent samples of size 8. Using n to stand for the size of a sample, $n_1 = n_2 = n_3 = 8$. The total size of each data set, which we symbolize N_T, is 24. The three samples of Set A were randomly selected† from populations with mean (μ) = 10 and variance (σ^2) = 9. Thus, in Set A, $\mu_1 = 10$, $\mu_2 = 10$, and $\mu_3 = 10$. The three samples of Set B were randomly selected† from populations with means of 8, 10, and 12, respectively. Thus, $\mu_1 = 8$, $\mu_2 = 10$, and $\mu_3 = 12$. As in Set A, the variance of all three populations (σ^2) is 9.

The means of the samples in both data sets are displayed at the bottom of the table. In Set A, $M_1 = 9$, $M_2 = 11$, and $M_3 = 10$. Since $\mu_1 = \mu_2 = \mu_3$, we know that the variation among the sample means is simply due to chance. The mean of all 24 scores in Set A, which we call the ***grand mean*** (***GM***), is equal to 10. In Set B, $M_1 = 7$, $M_2 = 11$, and $M_3 = 12$. The variation of the sample means is greater in Set B than in Set A, as we expect, since the three population means are different. The grand mean of Set B, like that of Set A, is equal to 10.

We will show that in both Sets A and B, the sum of squared deviations of all 24 scores from the grand mean consists of two components: (1) the sum of squared deviations of scores within samples from the sample means, and (2) the sum of squared deviations of sample means from the grand mean. The sum of squared deviations of all scores from the grand mean is called the ***total sum of squares (SS total)***. The sum of squared deviations of scores within samples from the sample means is called the ***sum of squares within groups (SS within)***; and the sum of squared deviations of sample means

†Sampling was not completely random; some adjustments were made to simplify calculations. However, the departure from random sampling was minimal.

Table 12–1 Data Sets A and B

(a) Data Set A

$\mu_1 = \mu_2 = \mu_3 = 10$

$\sigma_1^2 = \sigma_2^2 = \sigma_3^2 = 9$

	X_1	X_2	X_3	
	8	12	13	
	10	9	11	
	5	10	9	
	8	14	12	
	14	11	3	
	7	6	9	
	9	12	8	
	11	14	15	
Sum	72	88	80	
M	9.0	11.0	10.0	*GM* = 10.0

(b) Data Set B

$\mu_1 = 8, \mu_2 = 10, \mu_3 = 12$

$\sigma_1^2 = \sigma_2^2 = \sigma_3^2 = 9$

	X_1	X_2	X_3	
	11	15	13	
	3	14	14	
	6	10	6	
	10	12	10	
	7	12	12	
	8	6	12	
	9	9	16	
	2	10	13	
Sum	56	88	96	
M	7.0	11.0	12.0	*GM* = 10.0

from the grand mean is called the ***sum of squares between groups (SS between)***. Thus, we will demonstrate that

$$SS \text{ total} = SS \text{ within} + SS \text{ between}$$

The calculations for Data Set A are shown in Table 12–2, and those for Data Set B in Table 12–3.

The Total Sum of Squares

The total sum of squares is equal to the sum of the squared deviations of all N_T scores from the grand mean, *GM*. Thus

$$SS \text{ total} = \sum^{N_T} (X - GM)^2$$ *Deviation formula for the total sum of squares*

where $\sum^{N_T}$ means "add over all N_T scores."

The calculation of the total sum of squares of Sets A and B is shown in Columns (1) and (2) of Tables 12–2 and 12–3. In each set, we subtract the grand mean, 10, from each score, square each deviation, then add the squared deviations. As you can see, in Set A, *SS* total = 212, and in Set B, *SS* total = 304.

The Sum of Squares Within

The sum of squares within is the sum of squared deviations of scores from their sample means across all N_T scores. That is, to calculate *SS* within, we must calculate $\Sigma(X - M)^2$ within each sample, then add over the *k* samples.

$$SS \text{ within} = \Sigma(X_1 - M_1)^2 + \Sigma(X_2 - M_2)^2 + \cdots + \Sigma(X_k - M_k)^2$$

Because $\Sigma(X - M)^2$ in any one sample is the sample sum of squares, we can write

$$SS \text{ within} = SS_1 + SS_2 + \cdots + SS_k$$

Thus, *SS* within is a pooled sum of squares across the *k* samples.

The calculation of the sum of squares of each sample in Sets A and B is shown in Columns (3) and (4) of Tables 12–2 and 12–3. In Set A, $SS_1 = 52$, $SS_2 = 50$, and $SS_3 = 94$. To obtain the sum of squares within, we add the sample sums of squares.

Table 12–2 Calculation of Sums of Squares in Data Set A (GM = 10)

		(1)	(2)	(3)	(4)	(5)	(6)
	X	$X - GM$	$(X - GM)^2$	$X - M$	$(X - M)^2$	$M - GM$	$(M - GM)^2$
Sample 1	8	−2	4	−1	1	−1	1
	10	0	0	1	1	−1	1
	5	−5	25	−4	16	−1	1
$n_1 = 8$	8	−2	4	−1	1	−1	1
$M_1 = 9$	14	4	16	5	25	−1	1
	7	−3	9	−2	4	−1	1
	9	−1	1	0	0	−1	1
	10	1	1	2	4	−1	1
Sample sum					$52 = SS_1$		$8 = 8(1) = n_1(M_1 - GM)^2$
Sample 2	12	2	4	1	1	1	1
	9	−1	1	−2	4	1	1
	10	0	0	−1	1	1	1
$n_2 = 8$	14	4	16	3	9	1	1
$M_2 = 11$	11	1	1	0	0	1	1
	6	−4	16	−5	25	1	1
	12	2	4	1	1	1	1
	14	4	16	3	9	1	1
Sample sum					$50 = SS_2$		$8 = 8(1) = n_2(M_2 - GM)^2$
Sample 3	13	3	9	3	9	0	0
	11	1	1	1	1	0	0
	9	−1	1	−1	1	0	0
$n_3 = 8$	12	2	4	2	4	0	0
$M_3 = 10$	3	−7	49	−7	49	0	0
	9	−1	1	−1	1	0	0
	8	−2	4	−2	4	0	0
	15	5	25	5	25	0	0
Sample sum					$94 = SS_3$		$0 = 8(0) = n_3(M_3 - GM)^2$
Total Sum			212 SS total		196 SS within		16 SS between

$$\text{Set A: } SS \text{ within} = 52 + 50 + 94 = 196$$

In Set B, SS_1 = 72, SS_2 = 58, and SS_3 = 62.

$$\text{Set B: } SS \text{ within} = 72 + 58 + 62 = 192$$

The Sum of Squares Between

The sum of squares between is the sum of squared deviations of the sample means from the grand mean across all N_T scores. To obtain SS between, we calculate $\Sigma(M - GM)^2$ in each sample, then add over the samples. The calculations are shown in Columns (5) and (6) of Tables 12–2 and 12–3.

As you can see in the tables, $(M - GM)^2$ within each sample has the same value over all n members of the sample. To get $\Sigma(M - GM)^2$ within a sample, we must add $(M - GM)^2$ n times. For example, in Set A, each member of Sample 1 has $(M - GM)^2$

Table 12–3 Calculation of Sums of Squares in Data Set B ($GM = 10$)

	X	(1) $X - GM$	(2) $(X - GM)^2$	(3) $X - M$	(4) $(X - M)^2$	(5) $M - GM$	(6) $(M - GM)^2$
Sample 1	11	1	1	4	16	−3	9
	3	−7	49	−4	16	−3	9
	6	−4	16	−1	1	−3	9
$n_1 = 8$	10	0	0	3	9	−3	9
$M_1 = 7$	7	−3	9	0	0	−3	9
	8	−2	4	1	1	−3	9
	9	−1	1	2	4	−3	9
	2	−8	64	−5	25	−3	9
Sample sum					$72 = SS_1$		$72 = 8(9) = n_1(M_1 - GM)^2$
Sample 2	15	5	25	4	16	1	1
	14	4	16	−3	9	1	1
	10	0	0	−1	1	1	1
$n_2 = 8$	12	2	4	1	1	1	1
$M_2 = 11$	12	2	4	1	1	1	1
	6	−4	16	−5	25	1	1
	9	−1	1	−2	4	1	1
	10	0	0	−1	1	1	1
Sample sum					$58 = SS_2$		$8 = 8(1) = n_2(M_2 - GM)^2$
Sample 3	13	3	9	1	1	2	4
	14	4	16	2	4	2	4
	6	−4	16	−6	36	2	4
$n_3 = 8$	10	0	0	−2	4	2	4
$M_3 = 12$	12	2	4	0	0	2	4
	12	2	4	0	0	2	4
	16	6	36	4	16	2	4
	13	3	9	1	1	2	4
Sample sum					$62 = SS_3$		$32 = 8(4) = n_3(M_3 - GM)^2$
Total Sum			304 *SS* total		192 *SS* within		112 *SS* between

$= 1$. Therefore, $\Sigma(M - GM)^2$ of Sample $1 = 1 + 1 + 1 + 1 + 1 + 1 + 1 + 1 = 8(1) = n_1(M_1 - GM)^2$. Similarly, $\Sigma(M - GM)^2$ of Sample $2 = n_2(M_2 - GM)^2$; and so on.

The sum of squares between is the sum of $n(M - GM)^2$ over all samples:

$$SS \text{ between} = n_1(M_1 - GM)^2 + n_2(M_2 - GM)^2 + \cdots + n_k(M_k - GM)^2$$

Deviation formula for the sum of squares between groups

In Set A, $n_1(M_1 - GM)^2 = 8$, $n_2(M_2 - GM)^2 = 8$, and $n_3(M_3 - GM)^2 = 0$. Therefore,

$$\text{Set A: } SS \text{ between} = 8 + 8 + 0 = 16$$

In Set B, $n_1(M_1 - GM)^2 = 72$, $n_2(M_2 - GM)^2 = 8$, and $n_3(M_3 - GM)^2 = 32$. Therefore

$$\text{Set B: } SS \text{ between} = 72 + 8 + 32 = 112$$

Table 12–4 lists the *SS* total, *SS* within, and *SS* between of both data sets. As you can see, in Set A, *SS* within + *SS* between = 196 + 16 = 212 = *SS* total. Similarly, in

Table 12–4 The Sums of Squares of Sets A and B

	Set A	Set B
SS total	212	304
SS within	196	192
SS between	16	112

Set B, *SS* within + *SS* between = 192 + 112 = 304 = *SS* total. Moreover, this would be the case with any other data set as well. Thus, as stated earlier, the total variation of a set of data from k independent samples is the sum of two components: one based on variation of scores within samples and the other based on variation of the sample means.

Now let us compare the sums of squares of Sets A and B. The *SS* total of Set B is larger than that of Set A. But this is entirely due to Set B's larger *SS* between; the *SS* within is about the same size in both sets. Thus, the fact that the samples in Set B came from populations with different means did not affect *SS* within; *SS* within reflects only variation within each sample, independent of any difference between the samples. But the greater variation of the sample means in Set B resulted in a substantially larger value of *SS* between than in Set A.

Remember that the three samples in Set A were selected from populations with the same mean. Therefore, we know that the variation of the sample means, reflected in *SS* between = 16, is entirely due to chance. On the other hand, the three samples of Set B were selected from populations with different means. Therefore, we know that the variation of the sample means, reflected in *SS* between = 112, is not entirely due to chance; some of the variation of the Set B means is due to actual differences between the population means.

With real data, of course, we do not know in advance whether the population means are equal (H_0 true) or not (H_0 false). Therefore, we need a statistical test to determine whether the variation among a set of sample means is or is not within the limits to be expected by chance. We consider next the calculation of the mean square within and the mean square between and the test for comparing their values.

EXERCISE 12–2 Here are data from four independent samples.

Sample 1	Sample 2	Sample 3	Sample 4
2	5	9	3
6	4	8	5
5	7	12	1
1	2	6	4
6	7	5	2

Calculate *SS* total, *SS* within, and *SS* between. Show that *SS* total = *SS* within + *SS* between.

Mean Squares and Degrees of Freedom

As stated earlier, the test of significance in the ANOVA compares a variance estimate based on variation between the sample means (the mean square between) with a variance estimate based on variation within samples (the mean square within). If H_0 is true, *MS* between and *MS* within are expected to be approximately equal. But if H_0 is false,

MS between is expected to be larger than *MS* within. As you may already suspect, *MS* between is related to *SS* between, and *MS* within is related to *SS* within. We start by considering *MS* within.

The Mean Square Within

Recall that when we conducted t tests for two independent samples, we calculated an estimate of the population variance called the *pooled variance estimate*, s_p^2, by dividing the pooled sum of squares over the two samples by the pooled degrees of freedom (pooled *df*).

$$s_p^2 \text{ of } t \text{ test} = \frac{\text{pooled } SS}{\text{pooled } df} = \frac{SS_1 + SS_2}{(n_1 - 1) + (n_2 - 1)} = \frac{SS_1 + SS_2}{n_1 + n_2 - 2}$$

In the ANOVA, the mean square within, like s_p^2 of the t test, is a pooled variance estimate over the k samples in the data set.

$$MS \text{ within} = \frac{\text{pooled } SS}{\text{pooled } df} = \frac{SS_1 + SS_2 + \cdots + SS_k}{(n_1 - 1) + (n_2 - 1) + \cdots + (n_k - 1)} = \frac{SS \text{ within}}{(n_1 + n_2 + \cdots + n_k) - k} = \frac{SS \text{ within}}{\text{N}_\text{T} - k}$$

The pooled degrees of freedom, $N_T - k$, is called the ***degrees of freedom within*** (***df within***). Thus, the mean square within is equal to the sum of squares within divided by the degrees of freedom within.

$$MS \text{ within} = \frac{SS \text{ within}}{df \text{ within}}$$ *Calculation of mean square within groups*

Let us calculate *df* within and *MS* within for Sets A and B. In both sets, $N_T = 24$ and $k = 3$, so that *df* within = 24 − 3 = 21. In Set A, *SS* within = 196. Therefore,

$$\text{Set A: } MS \text{ within} = \frac{196}{21} = 9.33$$

In Set B, *SS* within = 192. Therefore

$$\text{Set B: } MS \text{ within} = \frac{192}{21} = 9.14$$

Like s_p^2 in the t test, *MS* within is an estimate of the population variance, σ^2. Thus, we estimate that the Set A samples came from populations with variance $(\sigma^2) = 9.33$; and we estimate that the Set B samples came from populations with $\sigma^2 = 9.14$.

Ordinarily, of course, we do not know the true population variance. In the case of Sets A and B, however, we know that $\sigma^2 = 9$. In both data sets, the mean square within is very close to 9; in fact, the departure from the true value of σ^2, 9, is just due to chance. If you were to draw all possible sets of three samples from the three populations of Data Sets A and B, and if you calculated *MS* within in each set, you would find that the average value of *MS* within, over all sets of samples, would be exactly equal to the population variance, 9.

The Mean Square Between

I said earlier that, if H_0 is true, the mean square between is expected to be approximately equal to the mean square within. This is because, when H_0 is true, *MS* between is also an estimate of σ^2, but one based on variation between sample means rather than variation of scores within samples. To explain this, we must go back to the central limit theorem (see Chapter 6). The theorem tells us that if we select all possible samples of size n from a population with mean μ and variance σ^2 and calculate the mean (M) of each sample, the variance of the sample means, σ_M^2, is equal to the population variance divided by the sample size.

$$\sigma_M^2 = \frac{\sigma^2}{n}$$

Multiplying both sides by n,

$$n\,\sigma_M^2 = \sigma^2$$

Thus, the population variance is equal to n times the variance of the sample means.

Now suppose that we select from the population not all possible samples of size n but a limited number of k samples. For example, if we select just five samples, $k = 5$. If we calculate M of each sample, as well as the grand mean GM of the data set, we can calculate an *estimate* of σ_M^2 as follows:

$$\text{Estimate of } \sigma^2 = \frac{\sum^{k}(M - GM)^2}{k - 1}$$

where $k - 1$ is the number of degrees of freedom for estimating σ_M^2 from k samples. Then, because $n\,\sigma_M^2 = \sigma^2$, we can obtain an estimate of the population variance σ^2 by multiplying the estimate of σ_M^2 by n:

$$\text{Estimate of } \sigma^2 = n\left[\frac{\sum^{k}(M - GM)^2}{k - 1}\right] = \frac{\sum^{k}n(M - GM)^2}{k - 1}$$

Now we return to the case of k samples not from one population but from k populations. If all the populations have the same mean (if H_0 is true), and the same variance, σ^2 (as the ANOVA assumes), drawing one sample from each of k populations is equivalent to drawing k samples from one population. Therefore, if H_0 is true, the previous formula yields an estimate of σ^2 based on variation among the means of k samples. And this estimate is called the mean square between (*MS* between).

$$MS \text{ between} = \frac{\sum^{k}n(M - GM)^2}{k - 1}$$

Note that the numerator of this formula is the sum of squares between. The denominator, $k - 1$, is called the ***degrees of freedom between* (df *between*)**. Thus, the mean square between is equal to the sum of squares between divided by the degrees of freedom between.

$$MS \text{ between} = \frac{SS \text{ between}}{df \text{ between}}$$ *Calculation of the mean square between groups*

Let us calculate the mean square between of Data Set A. In Set A, $k = 3$. Therefore, df between $= 3 - 1 = 2$. Because SS between of Set A = 16

$$\text{Set A: } MS \text{ between} = \frac{16}{2} = 8$$

In Set A, H_0 is true. Because *MS* between is an estimator of σ^2 when H_0 is true, the *MS* between, 8, is an estimate of σ^2 based on variation between the sample means. We know that, in fact, $\sigma^2 = 9$. The difference between *MS* between of Set A and the true population variance, 9, occurred by chance. If we were to take not just one set but all possible sets of three samples from the three populations of Set A and calculate *MS* between in each set, the average value of *MS* between, over all sets, would be exactly equal to the population variance, 9.

Next, let us calculate *df* between and *MS* between for Set B. In Set B, $k = 3$; therefore, *df* between = 3 − 1 = 2. Because *SS* between of Set B = 112,

$$\text{Set B: } MS \text{ between} = \frac{112}{2} = 56$$

As you can see, the *MS* between of Set B is considerably larger than the population variance, 9. This is because, when H_0 is not true, as in Set B, the sample means vary more than expected by chance, so that the mean square between is not an estimate of σ^2 but of a larger value. If you were to take all possible sets of three samples from the three populations of Set B and calculate *MS* between in each set, the average value of *MS* between over all sets would have a value substantially larger than the variance of each population, 9.

Partitioning the Degrees of Freedom

Before we consider the test for comparing the mean squares between and within, let us reexamine the degrees of freedom of the ANOVA. We have defined the degrees of freedom between and within as follows:

$$df \text{ within} = N_T - k$$

$$df \text{ between} = k - 1$$

It is also possible to define the ***total degrees of freedom (df total)*** of a data set:

$$df \text{ total} = N_T - 1$$

For example, in Sets A and B, *df* total = 24 − 1 = 23.

Note that

$$\begin{array}{ccccc} N_T - k & + & k - 1 & = & N_T - 1 \\ df \text{ within} & + & df \text{ between} & = & df \text{ total} \end{array}$$

Thus, the total degrees of freedom of a data set is partitioned into within-groups and between-groups components in just the same way as the total sum of squares.

EXERCISE 12–3

1. For the data in Exercise 12–2
   a. Calculate *df* within and *df* between. Show that *df* within + *df* between = *df* total.
   b. Calculate *MS* within and *MS* between.
2. Calculate the degrees of freedom between, within, and total for each of the following examples:
   a. Five independent samples with $n_1 = 12$, $n_2 = 10$, $n_3 = 12$, $n_4 = 8$, and $n_5 = 8$.
   b. An experiment in which 35 subjects are randomly subdivided into seven groups.

Comparing *MS* Between and *MS* Within: The F Test

We have seen that when H_0 is true, *MS* between and *MS* within are expected to be approximately equal. However, when H_0 is false, *MS* between is expected to be greater than *MS* within. The statistical test of the ANOVA compares the *MS* between and the *MS* within of the data set. H_0 is rejected if *MS* between is significantly greater than *MS* within. The test is called the **F *test*** because it is based on a probability distribution called the **F *distribution***.

The *F* Distribution

Assume that we are sampling from two populations with the same variance. That is, $\sigma_1^2 = \sigma_2^2$. We take all possible pairs of independent samples from the two populations. In each pair, we use the data from Sample 1 to calculate the estimated variance of Population 1 and the data from Sample 2 to calculate the estimated variance of Population 2. Then we calculate the ratio of the two variance estimates in each pair of samples, like this:

$$\frac{\text{Variance estimate 1}}{\text{Variance estimate 2}}$$

Like any statistic, Variance estimate 1 and Variance estimate 2 vary from one pair of samples to another, so that the ratio varies as well. However, when we examine the distribution of (Variance estimate 1)/(Variance estimate 2) over all possible pairs of samples, we find that the distribution takes the shape of a probability distribution called the *F* distribution.

$$F = \frac{\text{Variance estimate 1}}{\text{Variance estimate 2}} \qquad \text{if } \sigma_1^2 = \sigma_2^2$$

The *F* distribution has the following properties:

1. Like the *t* distribution, the *F* distribution is a family of distributions whose shape depends on both the degrees of freedom of Variance estimate 1 and the degrees of freedom of Variance estimate 2. The combination of degrees of freedom for any given *F* distribution is usually indicated by the notation (numerator *df*, denominator *df*).
2. Since the two populations are assumed to have the same variance, the average value of Variance estimate 1 is equal to the average value of Variance estimate 2. Therefore, the average value of *F* over all possible pairs of samples—the mean of the F distribution—is equal to 1, for all combinations of degrees of freedom.
3. Since variances are always positive, the minimum possible value of *F* is zero; but *F* has no upper limit. Therefore, all *F* distributions are positively skewed.

Two *F* distributions are shown in Figure 12–1.

The *F* Test of the ANOVA

Recall that the mean square between and the mean square within are both variance estimates. If the null hypothesis of the ANOVA is true, both mean squares are estimates of the same population variance. Moreover, they are independent variance estimates, be-

Figure 12–1

Two *F* distributions.

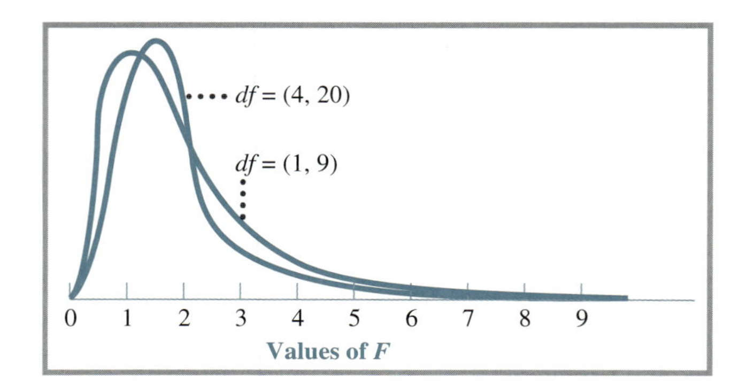

cause *MS* between depends only on between-sample variation, whereas *MS* within depends only on within-sample variation. Therefore, if H_0 is true, the ratio of the mean square between to the mean square within is distributed as F with df = (df between, df within).

To test the null hypothesis in an ANOVA, we calculate the test statistic

$$F = \frac{MS \text{ between}}{MS \text{ within}} \quad \text{with } (k-1,\ N_T - k)\ df \qquad \textit{Test statistic in the ANOVA}$$

If the calculated F is smaller than one or "not too much larger," we do not reject the null hypothesis on the grounds that the *MS* between and *MS* within could well be estimates of the same population variance. In other words, the variation among the sample means could be due to chance alone. If, on the other hand, *MS* between is "substantially" greater than *MS* within and F is "sufficiently" larger than one, we reject H_0 on the grounds that the variation among the sample means is greater than would be expected by chance.

How large must the calculated F value be in order to reject H_0? That depends on the level of alpha we have selected for the test (as well as on df between and df within). For example, if we have selected $\alpha = .05$, we reject H_0 if we get a value of F so much greater than one that it would occur less than 5 times in 100 when H_0 is true.

Because we only reject H_0 if F is sufficiently *greater* than one, the entire rejection region in the F test of the ANOVA is in the upper tail of the F distribution, as Figure 12–2 shows. The critical value of F for a given alpha is that value of F that cuts off an area equal to α in the upper tail.

Critical values of F can be found in the F table (Table D–5 in Appendix D). The columns of the F table are labeled "Degrees of freedom for numerator mean square"; in the one-way ANOVA, the numerator mean square is the mean square between. The rows of the table are labeled "Degrees of freedom for denominator mean square" (the mean square within). Each cell of the table contains two numbers. The upper number, in regular type, is the critical value of F for $\alpha = .05$ for the given combination of numerator and denominator df. The lower number, in boldface, is the critical value of F for $\alpha = .01$ for the same combination of df.[†]

Suppose, for example, that you are conducting an ANOVA of data from five independent samples ($k = 5$) and a total of 30 subjects ($N_T = 30$), so that df between $= k - 1 = 5 - 1 = 4$, and df within $N_T - k = 30 - 5 = 25$. You are using $\alpha = .05$. You have calculated $F = 3.42$. Is your result significant at the .05 level? Can you reject H_0? In Table D–5, you find the column marked 4 and the row marked 25. In the cell at the intersection of the row and the column, you find the entries $\substack{2.76 \\ \mathbf{4.18}}$. Your F falls between the critical values of F at the .05 and .01 levels. Therefore, $p < .05$. Your test result is significant at the .05 level. You can reject H_0.

In general, if the F calculated in an ANOVA is less than the upper entry for the relevant (df between, df within), this means that $p > .05$, and the result is not significant at the .05 level. If the calculated value falls between the two entries (as in the previous example), $.01 < p < .05$; the result is significant at the .05 but not the .01 level. If the calculated F is greater than the lower (boldface) entry, $p < .01$; the result is significant at the .01 level.

Now we are ready to complete the analysis of Data Sets A and B. Let us use $\alpha = .05$. Recall that in Set A we calculated *MS* between = 8 and *MS* within = 9.33. In Set

[†]More extensive F tables with critical values for additional levels of alpha can be found in books of statistical tables and textbooks with more extensive discussions of ANOVA.

Figure 12–2

The rejection and nonrejection regions of the F distribution in the ANOVA.

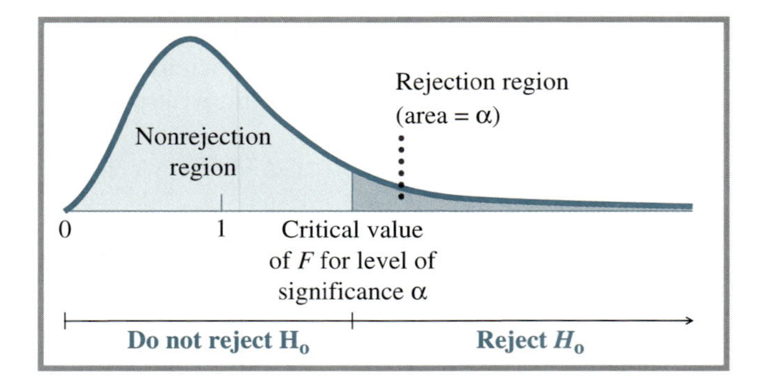

B, MS between = 56 and MS within = 9.14. The two values of F = (MS between)/(MS within) are

$$\text{Set A:}\quad F = \frac{8}{9.33} = 0.86$$

$$\text{Set B:}\quad F = \frac{56}{9.14} = 6.13$$

In both sets, df between = 2 and df within = 21. According to the F table, the critical values of F for $\alpha = .05$ and $\alpha = .01$ and (2,21) df are 3.47 and 5.78, respectively. Because the F value of Set A is less than 3.47 ($p > .05$), we do not reject H_0, and we conclude that all three populations could well have the same mean. This is, of course, an entirely appropriate decision, because we know that all three populations do have the same mean. The F value of Set B is greater than 5.78 ($p < .01$); therefore, we reject H_0 and conclude that not all the population means are equal—a correct conclusion, because we know that, in fact, the population means are different.

Summary of Reasoning and Procedure in the One–Way ANOVA

We have come a long way with Data Sets A and B. Because the reasoning and computations extended over many pages, we summarize the main points here.

1. The ANOVA tests H_0: $\mu_1 = \mu_2 = \cdots = \mu_k$, against H_A: Not all μ are equal. (In Sets A and B, we knew in advance whether H_0 was true or not. Ordinarily, we do not.)
2. The assumptions underlying the one-way ANOVA are (a) normal populations and (b) equal population variances ($\sigma_1^2 = \sigma_2^2 = \cdots = \sigma_k^2$—the homogeneity of variance assumption.
3. The one-way ANOVA partitions the total variation of scores around the grand mean of a data set (the total sum of squares) into two components: (a) variation of scores within samples around the sample means (the sum of squares within) and (b) the variation of sample means around the grand mean (the sum of squares between): SS total = SS within + SS between.

   To calculate the sums of squares we have used the following formulas:

   $$SS\text{ total} = \sum^{N_T} (X - GM)^2$$

   $$SS\text{ within} = SS_1 + SS_2 + \cdots + SS_k,\ \text{where } SS \text{ of each sample} = \Sigma(X - M)^2$$

   $$SS\text{ between} = n_1(M_1 - GM)^2 + n_2(M_2 - GM)^2 + \cdots + n_k(M_k - GM)^2$$

4. Two variance estimates, the mean square within groups (MS within) and the mean square between groups (MS between), are obtained by dividing SS within and SS between by their respective degrees of freedom. Thus,

$$MS \text{ within} = \frac{SS \text{ within}}{df \text{ within}}$$

$$MS \text{ between} = \frac{SS \text{ between}}{df \text{ between}}$$

$$df \text{ within} = N_T - k, \text{ and } df \text{ between} = k - 1$$

$$df \text{ within} + df \text{ between} = N_T - 1 = df \text{ total}$$

5. MS within is a pooled variance estimate; it is an estimate of σ^2 based on variation of scores within samples.

   If H_0 is true, MS between is also an estimate of σ^2, but one based on variation between the sample means. Thus, if H_0 is true, MS between and MS within are both estimates of the same parameter and are expected to be approximately equal.

   If, however, H_0 is false, the greater than chance variability of the sample means increases MS between relative to MS within. Consequently, MS between is expected to be greater than MS within if H_0 is false.
6. To find out whether the variation among the sample means is greater than would be expected by chance, we calculate F:

$$F = \frac{MS \text{ between}}{MS \text{ within}}$$

   The calculated F is compared with critical values of F for $df = (k - 1, N_T - k)$. If $p < \alpha$, H_0 is rejected; we conclude that the population means are not all equal. If $p > \alpha$, we do not reject H_0; we conclude that all the samples may have been drawn from populations with the same mean.

EXERCISE 12–4

1. Summarize the reasoning behind the ANOVA in your own words.
2. Complete the analysis of variance of the data in Exercise 12–2. Use $\alpha = .05$ to determine whether the variation among the sample means is significantly greater than would be expected by chance.

Displaying the Results of an ANOVA: Summary Tables

The results of an ANOVA are commonly displayed in what is called a ***summary table***. Table 12–5 presents a summary table for Data Set A. Here are some things you should notice about the summary table:

1. Each row represents one source of variation. In one-way ANOVA, the sources are between and within. Between is always listed first.

   Within is often referred to as "error" because it measures chance variation of scores within samples, unaffected by differences between sample means.
2. The "SS" column can be listed before or after the "df" column. However, the order of the remaining columns is fixed.
3. Sums of squares and degrees of freedom are entered for Total as well as for Between and Within. However, only Between and Within mean squares are included since the Total mean square is irrelevant to the analysis.

Table 12–5 Summary Table for Data Set A

Source of Variation	*SS*	*df*	*MS*	*F*	*p*
Between	16	2	8	0.86	n.s.
Within (error)	196	21	9.33		
Total	212	23			

Table 12–6 Incomplete Summary Table for Data Set B

(1) Source of Variation	(2) *SS*	(3) *df*	(4) *MS*	(5) *F*
Between	112	2		
Within (error)	192	21		
Total	304	23		

4. The value of F is always entered in the "Between" row.
5. The column labeled p (probability) indicates the level of significance attained. In Set A, the result was not significant at the .05 level; "n.s." stands for "not significant."
   The "p" column is optional. The level of significance can be indicated by footnotes instead; the footnote method will be illustrated with Data Set B.

When you conduct an ANOVA manually, it is a good idea to prepare the rows and columns of the summary table before beginning calculations, because many calculations can be done right in the table. Degrees of freedom are calculated and entered first. Remember: df between $= k - 1$; df within $= N_T - k$; and df total $= N_T - 1$. In Data Set B, for example, df between $= 3 - 1 = 2$, df within $= 24 - 3 = 21$, and df total $= 24 - 1 = 23$. Next, calculate and enter SS total and SS between. In Set B, we calculated SS total $=$ 304, SS within $= 192$, and SS between $= 112$. Table 12–6 shows how the summary table for Set B looks after we have entered the degrees of freedom and sums of squares. Try to complete the calculations and fill in the remaining slots in the table. Then read on.

Here is how you should have proceeded:

1. Each of the two mean squares, Between and Within, is equal to its SS (Column 2) divided by its df (Column 3). MS between = 112 divided by 2 = 56. MS within = 192 divided by 21 = 9.14. Place "56" and "9.14" in the appropriate slots in Column 4.
2. F is equal to MS between divided by MS within = 56/9.14 = 6.13. Place "6.13" under F in the row marked "Between."
3. In the F table, find the critical values of F for $df = (2,21)$ and compare the calculated F with the critical values. Because F is greater than the critical value for $\alpha = .01$, the difference between the means is significant at the .01 level ($p < .01$). Indicate this either by entering the value ".01" in an additional column labeled "p," or use the footnote method illustrated in Table 12–7.

The completed summary table for Set B is shown in Table 12–7.

Table 12–7 Completed Summary Table for Data Set B

Source of Variation	*SS*	*df*	*MS*	*F*
Between	112	2	56	6.13*
Within (error)	192	21	9.14	
Total	304	23		

*$p < .01$.

EXERCISE 12–5

Complete the following summary table.

Source of Variation	*SS*	*df*	*MS*	*F*	*p*
Between	698.2	3			
Within					
Total	1491.8	19			

How many groups were compared in this study?
If there was an equal number of subjects per group, what was the size of each group?

Computational Procedures

Conducting One-Way ANOVA by Computer

As we progress from the relatively simple procedures in early chapters to more complex procedures and analyses, the calculations involve more steps, and, if done manually, absorb increasing amounts of time. With the ready availability of statistical software packages for data analysis, manual calculations of procedures as complex as ANOVA are becoming a thing of the past. Computer programs not only do all the necessary calculations very fast and without error (assuming that the data have been entered correctly), they also construct and display summary tables. In addition, many programs provide additional information such as confidence intervals for population means.

Table 12–8 displays the output from an ANOVA of Data Set B by the computer package Minitab. In the summary table, the term "Factor" is between, and "Error" is within. The degrees of freedom, sums of squares, mean squares, and *F* are, of course, the same as those we calculated. Using the *F* table, we were able to determine that *p* was less than .01, but not how much less. The computer analysis gives the exact value of *p*, .008. In addition to the summary table, Minitab calculates the mean and standard deviation of each sample and the 95% confidence interval for the mean of each of the three populations.

Computational Formulas for Manual Calculation

Occasionally it may be necessary to conduct a one-way ANOVA by hand. If the data set is not too large, manual calculation is relatively easy. However, the deviation methods we used earlier to calculate the sums of squares can be computationally laborious even for small data sets, especially when the means are not whole numbers. In this section

Table 12–8 Minitab ANOVA of Data Set B

```
ANALYSIS OF VARIANCE
SOURCE      DF        SS        MS        F        P
FACTOR       2    112.00     56.00     6.12    0.008
ERROR       21    192.00      9.14
TOTAL       23    304.00
                                     INDIVIDUAL 95 PCT CI'S FOR MEAN
                                     BASED ON POOLED STDEV
LEVEL        N      MEAN     STDEV   ------+----------+---------+---------+-
X1           8     7.000     3.207     (------*------)
X2           8    11.000     2.878              (------*------)
X3           8    12.000     2.976                  (------*------)
                                     ------+----------+---------+---------+-
POOLED STDEV =     3.024                  6.0        9.0      12.0      15.0
```

we present and illustrate computational formulas based on raw scores rather than on deviation scores.

The raw score formulas for calculating the total sum of squares, the sum of squares between, and the sum of squares within utilize sums of scores and sums of squared scores within each sample as well as over the entire data set. We will use subscripts to denote the group whose sum (or squared sum) is under consideration. For example, ΣX_3 means the sum of scores in Group (Sample) 3; ΣX_2^2 means the sum of squared scores in Group 2; ΣX_T means the total sum, that is, the sum of all the scores over all groups.

We will present the raw score formulas and illustrate their use with Data Set B. The raw scores are shown in Table 12–1b. The values of ΣX and ΣX^2 are as follows:

	Group 1	Group 2	Group 3	Total
ΣX	56	88	96	240
ΣX^2	464	1026	1214	2704

The Total Sum of Squares

The raw score formula for the total sum of squares is

$$SS \text{ total} = \Sigma X_T^2 - \frac{(\Sigma X_T)^2}{N_T}$$

Raw score formula for the total sum of squares

In Set B, $\Sigma X_T^2 = 2704$, $\Sigma X_T = 240$, and $N_T = 24$. Therefore,

$$SS \text{ total} = 2704 - \frac{(240)^2}{24} = 304$$

the same value as that obtained using the deviation formula.

The Sum of Squares Between

The raw score formula for SS between is

$$SS \text{ between} = \frac{(\Sigma X_1)^2}{n_1} + \frac{(\Sigma X_2)^2}{n_2} + \cdots + \frac{(\Sigma X_k)^2}{n_k} - \frac{(\Sigma X_T)^2}{N_T}$$

Raw score formula for the sum of squares between

In Data Set B, $\Sigma X_1 = 56$, $\Sigma X_2 = 88$, $\Sigma X_3 = 96$, and each sample size is 8. Therefore,

$$SS \text{ between} = \frac{(56)^2}{8} + \frac{(88)^2}{8} + \frac{(96)^2}{8} - \frac{(240)^2}{24} = 112, \text{ as before}$$

The Sum of Squares Within

As we have already observed, SS within = $SS_1 + SS_2 + \cdots + SS_k$. The raw score formula for SS of a sample of size n is $\sum X^2 - \frac{(\sum X)^2}{n}$. In Data Set B

$$SS_1 = 464 - \frac{(56)^2}{8} = 72$$

$$SS_2 = 1026 - \frac{(88)^2}{8} = 58$$

$$SS_3 = 1214 - \frac{(96)^2}{8} = 62$$

$$SS \text{ within} = 72 + 58 + 62 = 192, \text{ as before}$$

Recall that SS total = SS within + SS between. This fact can be used to save a little computational labor. SS total and SS between are somewhat easier to calculate than SS within, especially if there are many groups. Therefore, after calculating SS total and SS between from the data, you can obtain SS within by subtraction:

$$SS \text{ within} = SS \text{ total} - SS \text{ between}$$

In Set B, for example, after you have calculated SS total = 304 and SS between = 112, you can calculate SS within = 304 − 112 = 192. If you do this, however, be sure that your calculations of SS total and SS between are correct; if either one is incorrect, your value of SS within will also be incorrect.

EXERCISE 12–6

1. Use the raw score formulas to calculate the three sums of squares of Set A.
2. Use the raw score formulas to calculate the three sums of squares of the data in Exercise 12–2. Complete the analysis of variance and present the results in a summary table. Use alpha = .05 to evaluate F. Do you reject H_0? Is the variation among the sample means significant at the .05 level?

Another Example of One-Way ANOVA

As another illustration of an ANOVA, including all calculations and preparation of a summary table, we use hypothetical data from a study comparing the relative effectiveness of four types of therapy for treating anxiety. In the study, 35 clients who came to a mental health clinic for treatment of anxiety disorders were randomly assigned to four types of therapy: psychoanalytic, humanistic, behavioral, and rational-emotive. Each client kept a record of anxiety attacks experienced over the year of treatment. The number of anxiety attacks reported by each client is shown in Table 12–9. We will conduct a test to determine whether the four therapies differ in effectiveness for treatment of anxiety.

The null hypothesis is that the therapies are equally effective; that is, that the mean number of anxiety attacks is the same under all four treatments.

$$H_o: \mu_1 = \mu_2 = \mu_3 = \mu_4$$

Table 12–9 Number of Anxiety Attacks Reported by Clients Under Four Therapy Conditions

	Type of Therapy				
	Psychoanalytic (X_1)	Humanistic (X_2)	Behavioral (X_3)	Rational–Emotive (X_4)	All Groups Combined
	2	4	1	25	
	16	8	12	9	
	28	6	0	3	
	2	9	1	0	
	21	2	2	13	
	2	1	7	27	
	2	10	7	5	
	19	13	13	1	
	39		0	2	
n	9	8	9	9	35
ΣX	131	53	43	85	312
ΣX^2	3379	471	417	1643	5910
M	14.6	6.6	4.8	9.4	

The alternative hypothesis is that the therapies are not all equally effective; that is, that the mean number of anxiety attacks is not the same under all treatments.

H_A: Not all μ are equal.

The means of the four therapy groups, shown in the last row of Table 12–9, are 14.6, 6.6, 4.8, and 9.4. They certainly look different. However, these are just sample means. The question is: Is the difference between the sample means larger than would be expected by chance if H_0 is true? We need to conduct an ANOVA to answer this question. We will use $\alpha = .05$.

The sample size (n), ΣX, and ΣX^2 for each group as well as for the total data set are displayed at the bottom of Table 12–9. You may wish to calculate them yourself, for extra practice.

First, we determine the number of degrees of freedom between, within, and total. Since $N_T = 35$ and $k = 4$:

$$\begin{aligned} df\text{ between} &= k-1 = 4-1 = 3 \\ df\text{ within} &= N_T - k = 35-4 = 31 \\ df\text{ total} &= N_T - 1 = 35-1 = 34 \end{aligned}$$

Notice that, as always, df between + df within = df total.

Next, we calculate the total sum of squares and the sum of squares between.

$$SS\text{ total} = \Sigma X_T^2 - \frac{(\Sigma X_T)^2}{N_T} = 5910 - \frac{(312)^2}{35} = 3128.74$$

$$SS\text{ between} = \frac{(\Sigma X_1)^2}{n_1} + \frac{(\Sigma X_2)^2}{n_2} + \frac{(\Sigma X_3)^2}{n_3} + \frac{(\Sigma X_4)^2}{n_4} - \frac{(\Sigma X_T)^2}{N_T}$$

$$= \frac{(131)^2}{9} + \frac{(53)^2}{8} + \frac{(43)^2}{9} + \frac{(85)^2}{9} - \frac{(312)^2}{35} = 484.87$$

At this point, you may wish to prepare a summary table, enter the numbers we have calculated, and complete the calculations in the table. Then compare your results with those we describe here.

The remaining calculations are

$$SS \text{ within} = SS \text{ total} - SS \text{ between} = 3128.74 - 484.87 = 2643.87$$

$$MS \text{ between} = \frac{SS \text{ between}}{df \text{ between}} = \frac{484.87}{3} = 161.623$$

$$MS \text{ within} = \frac{SS \text{ within}}{df \text{ within}} = \frac{2643.87}{31} = 85.286$$

$$F = \frac{MS \text{ between}}{MS \text{ within}} = \frac{161.623}{85.286} = 1.895$$

The completed summary table is shown in Table 12–10.

To evaluate F, we go to the F table to find the critical values of F for (3, 31) df. The critical value for $\alpha = .05$ is 2.91. Because the calculated F is less than 2.91, we cannot reject H_0. The variation among the treatment means is not significant at the .05 level. We cannot conclude that any of the four treatments is more effective than the others for treating anxiety.

In the summary table, you can indicate the decision made either by adding a column labeled "p" and entering the value "> .05" or "n.s.," or by putting an asterisk by the F like this: "1.895*" and placing a footnote under the table reading "$p > .05$" or "not significant."

You may wonder how differences as large as those among the four sample means—14.6 for psychoanalytic therapy, 6.6 for humanistic therapy, 4.8 for behavioral therapy, and 9.4 for rational-emotive therapy—can fail to be significant. To understand this, look at the spread of scores within each therapy group. The large amount of variation within groups resulted in a large MS within. With this much variation within groups, the differences among the sample means would have to be considerably greater than they were to attain significance.

EXERCISE 12–7

A college dean believes that students with different majors may respond differently to academic frustration. She constructs an "academic frustration inventory" and administers it to random samples of majors in four departments: English, psychology, mathematics, and business. Here are the academic frustration scores of students in the four samples. Perform an ANOVA and prepare a summary table. Then give a verbal interpretation of the results.

English Majors	Psychology Majors	Mathematics Majors	Business Majors
12	16	14	4
9	15	9	4
9	12	13	6
10	11	9	12
15	7	15	3
10	13	18	5
11	7	3	
	15		

Table 12–10 ANOVA Summary Table of the Anxiety Treatment Study

Source of Variation	SS	df	MS	F
Between Treatments	484.87	3	161.623	1.895
Within Treatments	2643.87	31	85.286	
Total	3128.74	34		

The Relationship Between ANOVA and the *t* Test

In Chapter 11 we used *t* tests to compare the means of two samples. The one-way ANOVA is used to compare the means of two or more independent samples. When two means are to be compared, either the *t* test or the ANOVA can be applied. Are the two tests related? Is there any advantage to using one or the other?

We have already pointed out one relationship between the *t* test and the ANOVA: The mean square within of the ANOVA is a pooled variance estimate, just like s_p^2 of the *t* test. In fact, when there are only two samples, s_p^2 and *MS* within are exactly the same.

$$\text{When } k = 2,\ s_p^2 \text{ of the } t \text{ test} = \frac{SS_1 + SS_2}{n_1 + n_2 - 2} = \frac{SS_1 + SS_2}{N_T - 2} = \frac{SS_1 + SS_2}{N_T - k} = \frac{SS \text{ within}}{df \text{ within}} = MS \text{ within}$$

This is not the only similarity between the *t* test for independent samples and the one-way ANOVA. In order to examine the relationship between the tests, let us use ANOVA to analyze the data from Millie Cartwright's study of comprehension, which we analyzed using a *t* test in Chapter 11. As in the *t* test, we will use $\alpha = .05$. Millie's data are reproduced in Table 12–11. ΣX and ΣX^2 for each group, as well as for the total data set, are shown at the bottom of the table.

In her two-tailed *t* test with $\alpha = .05$, Millie calculated $t = 2.478$. Her *t* was greater than the critical value for $\alpha = .05$ and 14 *df* (2.145), so that *p* was less than .05. Therefore, she rejected H_0 and concluded that type of cues does affect comprehension of text.

Now we use an ANOVA to analyze Millie's data. First, we calculate the degrees of freedom between and within. Since $k = 2$, *df* between $= 2 - 1 = 1$; *df* within $= N_T - k = 16 - 2 = 14$; *df* total $= 16 - 1 = 15$. Note that the *df* within of the ANOVA, 14, is equal to the number of *df* in the *t* test.

Next, we calculate the sums of squares:

$$SS \text{ total} = \Sigma X_T^2 - \frac{(\Sigma X_T)^2}{N_T} = 4178 - \frac{(256)^2}{16} = 82$$

$$SS \text{ between} = \frac{(\Sigma X_1)^2}{n_1} + \frac{(\Sigma X_2)^2}{n_2} - \frac{(\Sigma X_T)^2}{N_T} = \frac{(138)^2}{8} + \frac{(118)^2}{8} - \frac{(256)^2}{16} = 2$$

$$SS \text{ within} = SS \text{ total} - SS \text{ between} = 82 - 25 = 57$$

The two mean squares are

$$MS \text{ between} = \frac{SS \text{ between}}{df \text{ between}} = \frac{25}{1} = 25$$

$$MS \text{ within} = \frac{SS \text{ within}}{df \text{ within}} = \frac{57}{14} = 4.07, \text{ the value of } s_p^2 \text{ in Millie's } t \text{ test}$$

Table 12–11 Millie's Recall Study Data

	Pictorial Cues (Group 1)	Verbal Cues (Group 2)	
	17	17	
	20	15	
	16	15	
	18	13	
	17	16	
	18	11	
	13	16	
	19	15	
n	8	8	$N_T = 16$
ΣX	138	118	$\Sigma X_T = 256$
ΣX^2	2412	1766	$\Sigma X_T^2 = 4178$
M	17.25	14.75	

Table 12–12 Summary Table of Millie's ANOVA

Source of Variation	*SS*	*df*	*MS*	*F*
Between types of cues	25	1	25	6.14*
Within groups	57	14	4.07	
Total	82	15		

*$p < .05$.

Finally

$$F = \frac{MS\text{ between}}{MS\text{ within}} = \frac{25}{4.07} = 6.14$$

On the surface, there appears to be no relation between the calculated F, 6.14, and the value of t calculated in the t test, 2.478. But there is a relation: The square root of 6.14 is 2.478 and, of course, $(2.478)^2 = 6.14$. When a t test and an ANOVA are applied to the same data, $t^2 = F$.

The last step in the ANOVA of Millie's data is the evaluation of F. Referring to the F table, we find that F falls between the critical values for $\alpha = .05$ and $\alpha = .01$ in the F distribution with (1, 14) degrees of freedom. Because $p < .05$, we reject H_0. The difference between the means is significant at the .05 level. Our conclusion is the same as in the t test: Type of prior information does affect recall. The ANOVA is summarized in Table 12–12.

Before we leave Millie's study, let us compare the critical values for $\alpha = .05$ in the two tests. The critical value of F was 4.60; the critical value of t for the same level of α, two-tailed test, was 2.145, which is the square root of 4.60. When a two-tailed t test and an ANOVA are applied to the same data, (critical value of $t)^2$ = critical value of F.

In summary, we have observed the following relationships between a t test and a one-way ANOVA applied to the same data for two independent samples:

1. MS within of the ANOVA = s_p^2 of the t test.
2. df within in ANOVA = df in the t test.
3. F of the ANOVA = t^2 of the t test.
4. If the t test is two-tailed, the critical value of F is the square of the critical value of t for any given alpha.

Because calculated F = (calculated $t)^2$ and critical value of F = (critical value of $t)^2$, the conclusion drawn from an ANOVA for a given set of data is always the same as that drawn from a two-tailed t test for the same data, provided that the same level of α is used. If ANOVA results in rejection of H_0, so does the t test; if the difference between means is not significant according to a t test, it will not be significant according to ANOVA either.

Because of the relationships between the t test and ANOVA, the choice of one or the other for testing a nondirectional research question or hypothesis is largely a matter of preference. The t test does, however, have two advantages:

1. As we have seen, the critical value of F in the ANOVA is the square of the critical value of t in a two-tailed t test with the same level of α. The ANOVA is, in a sense, a two-tailed test of the difference between two means because any greater than chance difference between the means, in either direction, increases the value of MS between relative to MS within. For example, if other factors are equal, $M_1 - M_2 = +10$ and $M_1 - M_2 = -10$ will result in the same value of MS between. Therefore, if the research question or hypothesis is directional, a one-tailed t test is more appropriate than an ANOVA.
2. The t test is more obviously a test for comparing two *means*. The ratio $t = \frac{M_1 - M_2}{s_{\text{diff}}}$ explicitly displays the two sample means in the numerator. One can conduct an ANOVA, however, without calculating M_1 or M_2. Therefore it is easy to lose sight of the fact that means are being compared.

Here is an example that illustrates the hazards of forgetting that the ANOVA is a test for comparing means. As I write this, I have in front of me a copy of an article by authors who will, mercifully, be nameless. The authors used a packaged ANOVA computer program to compare the mean level of test performance under two experimental conditions. The printout of the results provided the following information: $F = .01$, $p = .938$ (in other words, p considerably greater than .05). Obviously, the F is not significant. In fact, to get an F value this small, the two sample means would have to be almost identical. The authors, unfortunately, reversed the F and p values. They misread "$F = .01$" as "$p = .01$" and concluded that the difference between the means was significant at the .01 level. Then they went on for several pages discussing the importance of this nonexistent "difference." Evidently, the authors never thought to look at the two means whose difference was the subject of the analysis.

EXERCISE 12–8

Here are data from an experiment utilizing two independent groups:

Group 1	Group 2
6	3
10	3
9	7
9	5
7	6

1. Use a two-tailed t test with $\alpha = .05$ to analyze the difference between the means.
2. Analyze the data using one-way ANOVA. Use $\alpha = .05$. Compare the results with the results of the t test.

Power, Robustness, and Strength of Relations

The Power of the ANOVA

The power of a test is the probability of rejecting the null hypothesis (H_0) when it is false. In the ANOVA, H_0 is rejected if F is greater than the critical value. Therefore, any factor that increases F or decreases the critical value increases the power of the test. The factors that affect the power of an ANOVA are the following:

1. The amount of variation among the population means, $\mu_1, \mu_2, \cdots, \mu_k$

   The greater the differences among the population means, the greater is the likely value of *MS* between. And, as *MS* between increases, F increases. Therefore *the power of the ANOVA increases as the variation among the population means increases.* (In plain English, the ANOVA is more likely to detect a difference between means if the difference is large than if it is small.) For example, if other factors are equal, you are more likely to correctly reject H_0 if $\mu_1 = 7$, $\mu_2 = 9$, and $\mu_3 = 12$ than if $\mu_1 = 8$, $\mu_2 = 9$, and $\mu_3 = 11$.
2. The population variance, σ^2

   In the ANOVA, the *MS* within is an estimator of σ^2. As σ^2 decreases, *MS* within decreases. Because $F =$ *MS* between/*MS* within, decreasing *MS* within increases the value of F. Therefore, *the power of the ANOVA increases as the population variance decreases.* For example, you are more likely to reject H_0 if $\sigma_1^2 = \sigma_2^2 = \cdots = \sigma_k^2 = 20$ than if $\sigma_1^2 = \sigma_2^2 = \cdots = \sigma_k^2 = 30$.
3. The sample size

   Sample size affects both the calculated F and the critical value of F. Because *MS* between $= \dfrac{\sum^k n(M - GM)^2}{k-1}$, *MS* between increases as $n_1, n_2, \cdots, n_k$ increase; and as *MS* between increases, F increases. In addition, because df within $= N_T - k$, df within increases as N_T increases; and as df within increases, the critical value of F decreases.

   For both reasons, *the power of the ANOVA increases as sample size increases.* As is true with other tests, the probability of rejecting a false H_0 is greater if you use large rather than small samples.
4. The level of significance (α)

   As α increases, the critical value of F decreases. Therefore, *the power of the ANOVA increases as a increases.* For example, the probability of rejecting H_0 is greater if $\alpha = .05$ than if $\alpha = .01$.

EXERCISE 12–9

In each of the following cases, decide whether ANOVA A or ANOVA B is more powerful. Assume that other factors are equal in each case.

ANOVA A	ANOVA B
(a) $n_1 = n_2 = n_3 = 20$	$n_1 = n_2 = n_3 = 10$
(b) $\alpha = .025$	$\alpha = .001$
(c) $\sigma^2 = 100$	$\sigma^2 = 50$
(d) $\mu_1 = 24, \mu_2 = 27, \mu_3 = 28$	$\mu_1 = 6, \mu_2 = 12, \mu_3 = 16$

Measuring the Strength of the Relationship Between the Independent and Dependent Variables

The most common method researchers use to increase the power of their tests is to increase N. Researchers often consult power tables to help them decide what sample size they need in order to detect a difference of some preselected size. This is an important part of research planning; after all, a nonsignificant difference represents not just a disappointment for the researcher but, if H_0 is false, an erroneous conclusion.

Keep in mind, however, that use of very large samples may increase the power of a test to the point that it will detect very small differences between means—differences that may not be worth detecting. The distinction between statistical significance and theoretical or practical importance must always remain at the forefront of researchers' minds.

One way to evaluate the importance of a statistically significant difference is to calculate a measure of the strength of the relationship between the independent and dependent variables. This involves calculating the proportion of the total data variance that is accounted for by the independent variable. We can do this simply by dividing *SS* between by *SS* total. For example, in Set B, we calculated *SS* between = 112 and *SS* total = 304. Therefore, the independent variable accounts for 112/304 = .37 (37%) of the total variance of Data Set B.

Although the ratio (*SS* between)/(*SS* total) is the proportion of the total *sample* variance accounted for by the independent variable, it tends to overestimate the proportion of variance accounted for by the independent variable in the population. Because we are generally interested in drawing conclusions about the populations from which samples were drawn, we need a more accurate, unbiased estimate of the proportion of the total population variance accounted for by the independent variable.

The proportion of population variance accounted for by the independent variable is symbolized ω^2 (the Greek letter omega). An unbiased estimate of ω^2, which we will symbolize $\hat{\omega}^2$, can be obtained by using the following formula:

$$\hat{\omega}^2 = \frac{SS \text{ between} - (k-1)\, MS \text{ within}}{SS \text{ total} + MS \text{ within}}$$

Formula for calculating the estimated proportion of the population variance that is accounted for by the independent variable.

In Set B, for example, $k = 3$ and *MS* within = 9.14. Therefore,

$$\hat{\omega}^2 = \frac{112-(2)(9.14)}{304+9.14} = .30$$

We estimate that in the population from which Data Set B was drawn, the independent variable accounts for about 30% of the total variance.

Let us see how $\hat{\omega}^2$ is affected by sample size. Suppose that you have conducted two ANOVAs, each comparing the means of four independent samples. In the first study, there were 20 subjects; the number of subjects in the second was 100. The results of both ANOVAs are shown in Table 12–13.

The value of F is identical in both ANOVAs and is significant in the .05 but not the .01 level. Judging from the level of significance attained, both results are equally significant. But let us compare the two values of $\hat{\omega}^2$:

$$N_T = 20: \qquad \hat{\omega}^2 = \frac{630-3(60)}{1590+60} = .27$$

$$N_T = 100: \qquad \hat{\omega}^2 = \frac{630 - 3(60)}{6390 + 60} = .07$$

In the study with 20 subjects, the independent variable is estimated to account for 27% of the total variance. In the study with 100 subjects, the estimated percent of variance accounted for is only 7. Clearly, the effect of the independent variable was greater in the first study. In fact, despite the statistical significance of the study with 100 subjects, we might question whether it is significant in any theoretical or practical sense.

The Robustness of the ANOVA

As we have seen, the ANOVA rests on the same assumptions as the t test, specifically,

1. Normal population distributions
2. Equal population variances

The ANOVA, like the t test, is robust with respect to violation of these assumptions. Unless the populations are very skewed and sample sizes are small, or the population variances are very different and the groups differ greatly in size, the conclusions reached using the ANOVA are approximately as likely to be correct when one or more assumptions are violated as when they are not.

EXERCISE 12–10

1. In Exercise 12–4, you conducted an ANOVA comparing the responses to academic frustration of college students with four different majors. The overall F of the ANOVA was significant at the .05 level. Some of the summary statistics are SS total = 479.25, SS between = 165.345, MS within = 13.079. Estimate the proportion of academic frustration variance that is accounted for by major.
2. John Doe and Harry Coe both got F = 8.74 in their ANOVAs. John used 15 subjects and Harry used 75 subjects. Which result is more impressive?
3. Marcy Roe's F was barely significant at the .05 level, whereas Amanda Low's was significant at the .01 level. In both cases, $\hat{\omega}^2$ was .45. Which researcher had a larger sample size?

Table 12–13 Two ANOVA Summary Tables

(a) $N_T = 20$

Source of Variation	*SS*	*df*	*MS*	F
Between	630	3	210	3.50*
Within	690	16	60	
Total	1590	19		

*$p < .05$.

(b) $N_T = 100$

Source of Variation	*SS*	*df*	*MS*	F
Between	630	3	210	3.50*
Within	5760	96	60	
Total	6390	99		

*$p < .05$.

Multiple Comparisons

Introduction

The null hypothesis of the one-way ANOVA is $\mu_1 = \mu_2 = \cdots = \mu_k$. We can refer to this as the *complete* or *overall* null hypothesis. Often, however, we are interested in differences between particular means or combinations of means. That is, we want to test particular ***subhypotheses***. For example, if we are interested in the difference between the means of treatments 3 and 4 in an experiment, our subhypothesis is H_0: $\mu_3 - \mu_4 = 0$. A test of a particular subhypothesis is called a ***comparison***.

Sometimes, comparisons are planned in advance of conducting an ANOVA. Such comparisons are called ***planned comparisons***. At other times, we decide which comparisons to test after examining the results of the ANOVA. Comparisons of this sort are called ***post hoc comparisons*** or ***post hoc tests*** (*post hoc* is Latin for "after the fact").

How a comparison is tested depends, first, on whether the comparison is planned or post hoc. In planned comparisons, which are usually small in number, the Type I error rate (though greater than the alpha of each test) is usually reasonably small, and little or no adjustment is needed to reduce it. In post hoc tests, however, the number of potential comparisons is considerably larger. Not only that, but post hoc tests may capitalize on chance differences; selecting a post hoc comparison because the difference between the sample means is large is a little like betting on a horse race after it is over. Therefore, post hoc tests must control the Type I error rate more stringently than tests for planned comparisons.

Post hoc tests also need to take into account the number of potential comparisons. For example, as we have already observed, in a study with four groups, the total number of possible comparisons is 25. Suppose, however, that the researcher is interested only in ***pairwise*** differences between means: M_1 vs. M_2, M_1 vs. M_3, M_1 vs. M_4, M_2 vs. M_3, M_2 vs. M_4, and M_3 vs. M_4. As you can see, there are only six possible pairwise comparisons, and if we use a given alpha for each comparison, the Type I error rate is smaller for pairwise comparisons than for all possible comparisons.

A set of potential comparisons is called a ***family of comparisons***. A researcher contemplating all possible comparisons for a set of data is concerned with a different family of comparisons than a researcher interested only in pairwise comparisons. Each post hoc test is appropriate for a particular family of comparisons and controls the Type I error rate for that family.

Subhypotheses and Weights Assigned to Sample Means

In order to test a comparison between means or combinations of means, it is necessary to assign weights to all k sample means. How weights are assigned depends on the subhypothesis to be tested. To illustrate, we consider all possible comparisons of three means ($k = 3$). With $k = 3$, the six possible comparisons and their corresponding subhypotheses are as follows:

1. Comparison of M_1 vs. M_2.

$$H_o\text{: } \mu_1 - \mu_2 = 0$$

2. Comparison of M_1 vs. M_3.

$$H_o\text{: } \mu_1 - \mu_3 = 0$$

3. Comparison of M_2 vs. M_3.

$$H_o\text{: } \mu_2 - \mu_3 = 0$$

4. Comparison of combination of M_1 and M_2 vs. M_3.

$$H_o\text{: Mean of } \mu_1 \text{ and } \mu_2 = \mu_3\text{, or } \frac{\mu_1 + \mu_2}{2} - \mu_3 = 0$$

5. Comparison of combination of M_1 and M_3 vs. M_2.

$$H_o\text{: Mean of } \mu_1 \text{ and } \mu_3 = \mu_2\text{, or } \frac{\mu_1 + \mu_3}{2} - \mu_2 = 0$$

6. Comparison of combination of M_2 and M_3 vs. M_1.

$$H_o\text{: Mean of } \mu_2 \text{ and } \mu_3 = \mu_1\text{, or } \frac{\mu_2 + \mu_3}{2} - \mu_1 = 0$$

In a comparison, weights (w) are assigned to the means in such a way that they (1) reflect the subhypothesis being tested and (2) sum to zero. Any mean not involved in a given comparison is given a weight of zero. The remaining means are weighted in accordance with the statement of the corresponding null subhypothesis.

For example, in the comparison of M_2 and M_3, M_1 is not included; therefore, its weight is zero. Because the subhypothesis being tested is $\mu_2 - \mu_3 = 0$, M_2 is assigned a weight of 1, and M_3 is assigned a weight of –1. Thus, the weights assigned to the three means in the comparison of M_2 and M_3 are

	M_1	M_2	M_3
Weight (w)	0	1	–1

Note that $w_1 + w_2 + w_3 = 0$.

Now consider the comparison of the combination M_1 and M_2 vs. M_3. All three means are included in this comparison; therefore, none of them has a weight of zero. To determine their weights, we look at the corresponding subhypothesis: $\frac{\mu_1 + \mu_2}{2} - \mu_3 = 0$. This can be rewritten as $\frac{1}{2}\mu_1 + \frac{1}{2}\mu_2 - \mu_3 = 0$. Therefore, the weights assigned to the three sample means are

	M_1	M_2	M_3
Weight (w)	½	½	–1

Once again $\Sigma w = 0$. The weights for all six comparisons in the three-sample case are shown in Table 12–14.

Because the only restriction on the values of w for a given comparison is that $\Sigma w = 0$, we can get rid of fractional values, as in subhypotheses 4, 5, and 6 of Table 12–14, by multiplying all weights by a constant. Thus, for these three subhypotheses, we multiply the weights given in Table 12–14 by 2, with the results shown in Table 12–15. These new weights can be used instead of the original fractional ones; the results of the comparisons will be the same.

The process of assigning weights in cases involving more than three means is basically the same. For example, suppose that an experiment has included five groups, the first three being experimental groups and the last two being control groups. The researcher is interested in comparing the combination of the three experimental groups

Table 12–14 Assignment of Weights for Tests of All Possible Subhypotheses for Three Groups

Subhypothesis	Weight (w) M_1	M_2	M_3	Sum
1. H_0: $\mu_1 - \mu_2 = 0$	1	−1	0	0
2. H_0: $\mu_1 - \mu_3 = 0$	1	0	−1	0
3. H_0: $\mu_2 - \mu_3 = 0$	0	1	−1	0
4. H_0: $\frac{\mu_1 + \mu_2}{2} - \mu_3 = 0$	½	½	−1	0
5. H_0: $\frac{\mu_1 + \mu_3}{2} - \mu_2 = 0$	½	−1	½	0
6. H_0: $\frac{\mu_2 + \mu_3}{2} - \mu_1 = 0$	−1	½	½	0

Table 12–15 Another Possible Set of Weights for Subhypotheses 4, 5, and 6

Subhypothesis	Weight (w) M_1	M_2	M_3	Sum
4. H_0: $\frac{\mu_1 + \mu_2}{2} - \mu_3 = 0$	1	1	−2	0
5. H_0: $\frac{\mu_1 + \mu_3}{2} - \mu_2 = 0$	1	−2	1	0
6. H_0: $\frac{\mu_2 + \mu_3}{2} - \mu_1 = 0$	−2	1	1	0

with the combination of the two control groups. The relevant subhypothesis is $H_o: \frac{\mu_1 + \mu_2 + \mu_3}{3} - \frac{\mu_4 + \mu_5}{2} = 0$. One possible set of weights is

	M_1	M_2	M_3	M_4	M_5	Sum
Weights	⅓	⅓	⅓	−½	−½	0

Or, if each weight above is multiplied by 6:

	M_1	M_2	M_3	M_4	M_5	Sum
Weights	2	2	2	−3	−3	0

Orthogonal and Nonorthogonal Comparisons

Any two comparisons on a set of sample means may be independent or nonindependent. In mathematics, the word "orthogonal" means independent. Therefore, if two comparisons are independent of one another, they are called ***orthogonal comparisons***.

Sometimes, it is easy to tell whether or not two comparisons are orthogonal. For example, given a set of four means, the comparisons $M_1 - M_2$ and $M_3 - M_4$ are clearly orthogonal, because the result of the first comparison has no bearing on the result of the second, and vice versa. But is the comparison $M_1 - M_3$ orthogonal to the comparison $M_2 - M_3$? Is the comparison $\frac{M_2 + M_3}{2} - M_4$ orthogonal to the comparison $M_2 - M_3$?

Table 12–16 **Testing the Comparisons $M_1 - M_3$ and $M_2 - M_3$ for Orthogonality**

	Weight (w)				
Comparison	M_1	M_2	M_3	M_4	**Sum**
$M_1 - M_3$	1	0	−1	0	0
$M_3 - M_3$	0	1	−1	0	0
Cross products of weight in Comparison 1 × weight in Comparison 2	0	0	1	0	1 Sum of cross products

Table 12–17 **Testing the Comparisons $\frac{M_2 + M_3}{2} - M_4$ and $M_2 - M_3$ for Orthogonality**

	Weight (w)				
Comparison	M_1	M_2	M_3	M_4	**Sum**
$\frac{M_2 + M_3}{2} - M_4$	0	½	½	−1	0
$M_2 - M_3$	0	1	−1	0	0
Cross products of weights	0	½	−½	0	0

There is a simple test to determine whether two comparisons are orthogonal. One simply multiplies each mean's weight in Comparison 1 by the weight of the same mean in Comparison 2. Then one adds the cross products. If the sum of the cross products is equal to zero, the comparisons are orthogonal; otherwise, they are nonorthogonal.

Let us apply this test to see whether the comparisons $M_1 - M_3$ and $M_2 - M_3$ are orthogonal. The calculations are shown in Table 12–16. Because the sum of the cross products is not equal to zero, these two comparisons are not orthogonal.

In Table 12–17, we test to see whether the comparisons $\frac{M_2 + M_3}{2} - M_4$ and $M_2 - M_3$ are orthogonal or not. The sum of the cross products is equal to zero. Therefore, these two comparisons are orthogonal.

It can be shown that the maximum number of possible orthogonal comparisons for a set of k means is $k - 1$. For example, if $k = 4$, a set of orthogonal comparisons can include, at most, three comparisons. Any additional comparison is necessarily nonorthogonal to at least one of the three. This does not mean, however, that there is only one possible set of $k - 1$ orthogonal comparisons. As an example, Table 12–18 lists two sets of orthogonal comparisons for four means. (You may wish to test each pair of comparisons within each set to assure yourself that they are truly orthogonal.)

When the number of comparisons to be made is small, orthogonal comparisons are generally preferred over nonorthogonal ones, because they provide nonredundant (nonoverlapping) information. If two comparisons are nonorthogonal, the total amount of information obtained is less than from two orthogonal comparisons. For this reason, it is generally recommended that planned comparisons be orthogonal. In post hoc tests, on the other hand, the number of potentially interesting comparisons may be well over $k - 1$; if so, the comparisons obviously cannot all be orthogonal to one another.

If a set of orthogonal comparisons is planned, the set to be tested must be chosen with care, in accordance with the research hypotheses and the design of the study. As an

Table 12–18 Two Sets of Orthogonal Comparisons Among Four Means

(a)	Set 1	(b)	Set 2
Comparison 1	$\frac{M_1 + M_2 + M_3}{3} - M_4$	Comparison 1	$\frac{M_1 + M_2}{2} - \frac{M_3 + M_4}{2}$
Comparison 2	$\frac{M_1 + M_2}{2} - M_3$	Comparison 2	$M_1 - M_2$
Comparison 3	$M_1 - M_2$	Comparison 3	$M_3 - M_4$

example, suppose that you have conducted an experiment with four groups. Groups 1, 2, and 3 are experimental groups and Group 4 is a control group. One comparison you would probably want to make is the combined experimental groups versus the control group. Therefore, Set 1 in Table 12–18 could be an appropriate set of comparisons, but Set 2 would not be appropriate.

EXERCISE

12–11

1. Assign weights to the means in each of the following comparisons on a set of five means.
   a. $M_3 - M_4$
   b. Combination of M_1 and M_4 versus M_5
   c. M_2 vs. combination of M_1, M_3, M_4, and M_5
2. Test each pair of comparisons in problem 1 for orthogonality.
3. You are comparing five methods of learning prose material. The first three methods are "meaningful" and the last two are "rote." Of the three meaningful methods, the first two are verbal and the third is nonverbal.
   a. Design a set of comparisons to answer the following questions:
      (1) Is there a difference between meaningful and rote learning?
      (2) Is there a difference between meaningful verbal and meaningful nonverbal learning?
      (3) Is there a difference between the two methods of rote learning?
      (4) Is there a difference between the two methods of meaningful verbal learning?
   b. Test the four comparisons for orthogonality.

Tests for Planned Comparisons

In planned comparisons, a value of F is calculated for each comparison, and the calculated F is compared with critical values for the relevant degrees of freedom. We need to consider the calculation of F, the number of degrees of freedom, and the selection of alpha.

Calculating F for a Comparison

The F ratio for a particular comparison includes a *comparison term*, symbolized C. C is calculated as follows:

$$C = w_1M_1 + w_2M_2 + \cdots + w_kM_k$$ *Formula for calculating a comparison term*

Suppose, for example, that we are planning comparisons on three means: $M_1 = 50$, $M_2 = 49$, and $M_3 = 45$. If we want to compare M_1 and M_2, the weights are

	M_1	M_2	M_3
w	1	−1	0

Therefore, $C = (1)(M_1) + (-1)(M_2) + 0(M_3) = 1(50) + (-1)49 + (0)(45) = 50 - 49 = 1$.

For the comparison of $M_1 + M_2$ versus M_3, a possible set of weights is

	M_1	M_2	M_3
w	1	1	−2

For this comparison

$$C = 1(50) + 1(49) + (-2)(45) = 50 + 49 - 90 = 9$$

The value of F for a given comparison is calculated as follows:

$$F = \frac{C^2}{MS\text{ within}\left(\frac{w_1^2}{n_1} + \frac{w_2^2}{n_2} + \cdots \frac{w_k^2}{n_k}\right)}$$

Formula for calculating F *for a comparison*

where MS within is that from the overall ANOVA of the data set.

To illustrate the calculation of F for comparisons, let us use Data Set B, on which we conducted an ANOVA earlier in the chapter. Set B includes three samples of size 8, with means of 7, 11, and 12, respectively. Let us assume that the following two comparisons were planned in advance: mean 1 versus mean 2 and means 1 and 2 combined versus mean 3. The weights assigned to the means in the two comparisons are as follows:

	M_1	M_2	M_3
Comparison 1	1	−1	0
Comparison 2	1	1	−2

As you can see, the two comparisons are orthogonal.

In the ANOVA, we calculated MS within = 9.14. (See Table 12–7 for a summary of the ANOVA.) Now we calculate C and F for each comparison.

Comparison 1

$$C = (1)(7) + (-1)(11) + 0(12) = 7 - 11 = -4$$

$$F = \frac{(-4)^2}{9.14\left[\frac{1^2}{8} + \frac{(-1)^2}{8} + \frac{(0)^2}{8}\right]} = \frac{16}{9.14(.25)} = \frac{16}{2.285} = 7.00$$

Comparison 2

$$C = (1)(7) + (1)(11) + (-2)(12) = 7 + 11 - 24 = -6$$

$$F = \frac{(-6)^2}{9.14\left[\frac{1^2}{8} + \frac{(1)^2}{8} + \frac{(-2)^2}{8}\right]} = \frac{36}{9.14(.75)} = \frac{36}{6.885} = 5.25$$

Evaluating F

In planned comparisons, the number of comparisons is small, and the Type I error rate is not a great concern. Therefore, it is permissible to use the same level of alpha for each comparison as that used in the overall ANOVA. Assume that we decided to use $\alpha = .05$ in advance of conducting the ANOVA of Set B. We will evaluate each comparison F using $\alpha = .05$.

To find critical values in the F table, we need to know the numerator *df* and denominator *df*. Because a comparison always compares just two means (or two combinations of means), the numerator *df* is always $2 - 1 = 1$. The denominator *df* is *df* within. In the case of Set B, *df* within = 21 (see Table 12–7). Therefore, we need to evaluate F for each comparison against critical values for (1, 21) degrees of freedom.

According to the F table, the critical values of F for $\alpha = .05$ and .01 and $df = (1, 21)$ are 4.32 and 8.02, respectively. Both comparison F values fall between 4.32 and 8.02, so that $p < .05$. Therefore, we can reject the null subhypothesis in both cases. We conclude that the Group 1 mean is significantly different from the Group 2 mean at the .05 level, and the mean of Groups 1 and 2 combined is significantly different from the mean of Group 3 at the .05 level.

A brief addendum: Even if the number of planned tests is small, the Type I error rate is greater than the level of alpha in each test. For example, with two planned tests, each using $\alpha = .05$, the probability of a Type I error in one or both tests is close to .10. Therefore, some statisticians suggest using a level of alpha in each test that is equal to the desired error rate divided by the number of tests. For example, with two tests and desired overall alpha = .05, we would use $\alpha = .05/2 = .025$ in each test.

Let us see what happens if we follow this guideline in our analysis of the two Set B comparisons. The critical value of F for $\alpha = .025$ and (1, 21) *df* is 5.85.† Because only our first comparison had F greater than 5.85, we reject H_0 only in that comparison, and we conclude that, although the means of Groups 1 and 2 are significantly different, the mean of Groups 1 and 2 combined does not differ significantly from the mean of Group 3.

EXERCISE 12–12

You have conducted an experiment with two experimental groups and two control groups. There were five subjects in each group. Before conducting the ANOVA, you planned the following comparisons: (a) mean of experimental groups versus mean of control groups, (b) comparison of the two experimental groups, and (c) comparison of the two control groups. The mean square within of the ANOVA was 30. The four group means were as follows:

Experimental Groups		Control Groups	
1	2	1	2
24	18	12	15

1. Show that the three comparisons are orthogonal.
2. Calculate F for each comparison, and evaluate each F using $\alpha = .05$. In which comparisons, if any, is the difference between the means significant at the .05 level?
3. If you want to hold the Type I error rate over all three comparisons to .05, what level of α should you use in each test?

Post Hoc Tests

In some post hoc tests, a value of F is calculated for each comparison by the same method as for planned comparisons. What is different is the evaluation of F. Because

†This critical value was found in a more comprehensive table of the F distribution than the table in Appendix D.

the number of potential comparisons in post hoc tests may be very large, using $\alpha = .05$ (or even $\alpha = .01$) in each test is not appropriate, because the Type I error rate would be unacceptably large. On the other hand, using α = (Desired Type I error rate)/(Number of tests) may yield so small a value of alpha for each test that it is almost impossible to reject H_0, even when H_0 is false. In other words, it may result in tests of comparisons that have almost no power to detect true differences between means.

Because of this, special tests have been designed for post hoc comparisons. They are generally intended to be used only if the overall F of the ANOVA is significant; that is, if we know from the ANOVA that, somewhere within the set of means there is or are significant differences, and now we want to locate them.

There are many post hoc tests. Each test is appropriate for a particular family of comparisons and controls the Type I error rate for that family. We will consider two widely used post hoc tests, the Scheffé test for the family of all possible comparisons and the Tukey "honestly significant difference" test for the family of pairwise comparisons.

The Scheffé Test

Suppose that your ANOVA of a data set has resulted in a significant F. You know, therefore, that the entire set of k means varies more than would be expected by chance alone. But, where exactly do differences lie? Because you had not planned any particular comparisons in advance, you examine the sample means and test those differences that look as though they may be "large enough" to be significant. You are interested in testing not only for pairwise differences but for differences involving combinations of means as well. In fact, the family of comparisons of potential interest to you is all possible comparisons. The appropriate test is the ***Scheffé test***.

In conducting a Scheffé test, you select a level of alpha, usually the one that was used in the ANOVA. The alpha you choose is the Type I error rate over the family of all possible comparisons. For example, if you choose $\alpha = .05$ in a Scheffé test, you know that the probability of at least one Type I error over the family of all possible comparisons is .05.

If the overall F of an ANOVA is significant at level alpha, the Scheffé test always yields a significant difference for at least one of the possible comparisons with the same alpha. If, on the other hand, the overall F was not significant, there is no point to applying the Scheffé test, because none of the comparisons will be significant.

In conducting a Scheffé test, a value of F is calculated for each comparison by the same method as for planned comparisons. But instead of comparing F with the critical value for the chosen alpha and $df = (1, df \text{ within})$, a special critical value is calculated as follows:

1. In the F table, find the critical value of F for the level of alpha used in the ANOVA and $df = (k - 1, df \text{ within})$.
2. Multiply the tabled value by $k - 1$. The result is the critical value used for each comparison in the Scheffé test.

To illustrate the procedure, let us again use Data Set B. The overall F of the ANOVA was significant at the .01 level. We will use $\alpha = .01$ in the Scheffé test as well. The three group means in Set B are 7, 11, and 12, respectively. You examine the means and say to yourself, "It looks like the difference between Group 1 and the combination of Groups 2 and 3 may be significant, so I will test that comparison. I'll also test Group 1 versus Group 2 alone."

Note, incidentally, that these two comparisons are not orthogonal. There is no requirement for comparisons to be orthogonal in post hoc tests. The weights (w) for the two comparisons of interest are as follows:

	M_1	M_2	M_3
Comparison 1	2	−1	−1
Comparison 2	1	−1	0

First, we calculate F for each comparison.

Comparison 1

$$C = 2(7) + (-1)(11) + (-1)(12) = 14 - 11 - 12 = -9$$

$$F = \frac{(-9)^2}{9.14\left[\frac{(2)^2}{8} + \frac{(-1)^2}{8} + \frac{(-1)^2}{8}\right]} = \frac{81}{9.14(.75)} = \frac{81}{6.855} = 11.82$$

Comparison 2

$$C = 1(7) + (-1)(11) + 0(12) = 7 - 11 = -4$$

$$F = \frac{(-4)^2}{9.14\left[\frac{(1)^2}{8} + \frac{(-1)^2}{8} + \frac{(0)^2}{8}\right]} = \frac{16}{9.14(.25)} = \frac{16}{2.285} = 7.00$$

Next, we calculate the critical value against which the comparison F's will be evaluated. The critical value of F for $\alpha = .01$ and $df = (2, 21)$ is 5.78. To obtain the critical value for the Scheffé test, we multiply 5.78 by $k - 1 = 2$; the critical value is $2(5.78) = 11.56$. Only the first comparison had F greater than 11.56. Therefore, the difference between M_1 and the combination of M_2 and M_3 is significant at the .01 level, but the difference between M_1 and M_2 is not.

EXERCISE 12–13

1. The Scheffé test guarantees that if the overall F of the ANOVA is significant at level α, at least one of the possible comparisons will be significant at the same level. We have found one significant difference in Set B. There may be others. Select any of the remaining four possible comparisons and use the Scheffé test to test the relevant null subhypothesis.
2. An ANOVA of an experiment with 25 subjects in five independent groups yields $F = 5.15$ ($p < .01$). MS within = 17. The sizes and means of the five groups are as follows:

Group	1	2	3	4	5
n	6	4	5	5	5
M	16	12	18	25	23

a. Calculate the critical value of F for Scheffé test comparisons. Use $\alpha = .01$.

b. Before calculating F for any of the possible comparisons, complete the following statement: For this data set, the calculated F value for at least one comparison will be greater than ________ .

c. Calculate and evaluate F for the following comparisons:

(1) Group 2 vs. Group 4

(2) Group 2 vs. combination of Groups 1 and 3

THE TUKEY "HONESTLY SIGNIFICANT DIFFERENCE" (*HSD*) TEST

If a researcher is interested only in pairwise comparisons (i.e., comparisons that involve only two means), an appropriate test is the ***Tukey "honestly significant difference" (HSD)*** test. The family of tests to which the Tukey test applies is the family of all pairwise comparisons. As we observed earlier, this family is smaller than the set of all possible comparisons. In the Tukey test, the level of alpha chosen by the researcher is the Type I error rate over the family of pairwise comparisons. For example, in a Tukey test with $\alpha = .05$, the probability of at least one Type I error over all pairwise comparisons is .05.

One limitation of the Tukey test is that it, unlike the Scheffé test, requires equal group sizes. However, since most studies include the same number of subjects in each group, this is usually not a problem.

Tukey's test is based on a test statistic called the *studentized range statistic*, symbolized q, and defined as follows:

$$\text{For any set of } k \text{ means, } q = \frac{\text{largest } M - \text{smallest } M}{\sqrt{\dfrac{MS \text{ within}}{n}}}$$

where n is the size of each group. As in the Scheffé test, *MS* within is the mean square within of the ANOVA.

Tukey showed that if the complete null hypothesis of the ANOVA is true, values of q over all possible sets of k samples form a sampling distribution whose shape is dependent only on k and on *df* within. The Tukey test, then, involves calculating $q = \dfrac{M_1 - M_2}{\sqrt{\dfrac{MS \text{ within}}{n}}}$ for each pair of means (where M_1 is the larger of the two means). The calculated q is evaluated against critical values of q for the given k and *df* within.

Critical values of q for $\alpha = .05$ and $\alpha = .01$ are listed, for various combinations of k and *df* within, in Table D–6 of Appendix D. If the calculated q for a comparison exceeds the critical value for the chosen α, the difference is significant at level α and the null hypothesis for the comparison is rejected. For example, for $\alpha = .05$, $k = 4$, and *df* within = 24, the critical value of q, in the table, is 3.90. A difference between two group means is significant at the .05 level if the calculated q is greater than 3.90.

One way to conduct the Tukey test is to calculate q for each comparison of interest. An easier way is based on the following reasoning:

$$\text{In order to be significant, } \frac{M_1 - M_2}{\sqrt{\dfrac{MS \text{ within}}{n}}} \text{ must be greater than } q_{\text{crit}}.$$

Multiplying both terms of the statement above by $\sqrt{\dfrac{MS \text{ within}}{n}}$, we get the statement:

$$\text{In order to be significant, } M_1 - M_2 \text{ must be greater than } q_{\text{crit}}\sqrt{\frac{MS \text{ within}}{n}}.$$

The value of $q_{\text{crit}}\sqrt{\dfrac{MS \text{ within}}{n}}$, the minimum significant difference between two means, is called the *honestly significant difference (HSD)*.

The usual procedure in the Tukey test involves, first, calculating *HSD*. Then each pairwise difference is compared with *HSD*. If the difference between the means is greater than *HSD*, the difference is significant; otherwise, the difference between the means is not significant.

As an example of the Tukey test, consider a 5-group experiment with 10 subjects per group. The overall F of the ANOVA was significant at the .05 level. The *MS* within was 9.7. Now we want to compare each pair of group means.

The means of the five groups, arranged in order by size, are shown in the rows and the columns of Table 12–19. The entries in the table are the differences between the means of each pair. First, we find q_{crit}. Because the overall ANOVA was significant at the .05 level, we use $\alpha = .05$; $k = 5$ and df within = 50 – 5 = 45. According to Table E, q_{crit} for $\alpha = .05$, $k = 5$, and $df = 45$ is 4.04.

Next we calculate *HSD*:

$$HSD = q_{crit}\sqrt{\frac{MS\text{ within}}{n}} = 4.04\sqrt{\frac{9.7}{10}} = 3.98$$

All pairwise differences greater than 3.98 are significant at the .05 level. As you can see in Table 12–18, the following differences between means are significant:

M_1 vs. M_3
M_1 vs. M_4
M_1 vs. M_5
M_2 vs. M_4
M_2 vs. M_5

The remaining pairwise differences are not significant.

Because the family of comparisons in the Tukey test is smaller than that in the Scheffé test, the Tukey test is more powerful. In other words, if true pairwise differences exist among the population means, you are more likely to detect them with the Tukey than with the Scheffé test.

To illustrate this, let us apply the Scheffé test to all the pairwise comparisons in Table 12–18 and compare the results with those of the Tukey test. First, we assign weights to each comparison and calculate F by the formula

$$F = \frac{C^2}{MS\text{ within}\left[\frac{w_1^2}{n_1} + \frac{w_2^2}{n_2} + \cdots + \frac{w_5^2}{n_5}\right]}.$$

In a pairwise comparison, C is always equal to the difference between the two means, $M_1 - M_2$. The weights of the two means are always 1 and –1, respectively, and all the other means have a weight of zero. Therefore, in all pairwise comparisons

$$F = \frac{(M_1 - M_2)^2}{MS\text{ within}\left(\frac{1}{n_1} + \frac{1}{n_2}\right)}$$

In our five-group example, *MS* within = 9.7 and $n_1 = n_2 = \cdots = n_5 = 10$. Therefore, for each comparison

$$F = \frac{(M_1 - M_2)^2}{9.7\left(\frac{1}{10} + \frac{1}{10}\right)} = \frac{(M_1 - M_2)^2}{1.94}$$

Table 12–19 Pairwise Differences Between the Means of Five Groups

		Mean				
		5.4	5.7	9.5	10.5	12.0
	5.4	0	0.3	4.1	5.1	6.6
	5.7		0	3.8	4.8	6.3
Mean	9.5			0	1.0	2.5
	10.5				0	1.5
	12.5					0

Table 12–20 *F* Ratios for All Pairwise Comparisons of Five Means

		Mean				
		1	2	3	4	5
	1		.05	8.66	15.03	22.45
	2			7.44	11.88	20.46
Mean	3				0.52	3.22
	4					1.16
	5					

For example, in the comparison of the Group 1 and Group 2 means

$$F = \frac{(5.4 - 5.7)^2}{1.94} = \frac{.09}{1.94} = .05$$

The remaining F ratios are calculated in the same way. The resulting F ratios are shown in Table 12–20.

According to the F table, the critical value of F for $\alpha = .05$ and $df = (4{,}45)$ is approximately equal to 2.61. Therefore, the critical value for the Scheffé test is 4(2.61) = 10.44. The following comparisons have $F > 10.44$ and are significant at the .05 level:

M_1 vs. M_4
M_1 vs. M_5
M_2 vs. M_4
M_2 vs. M_5

The difference between M_1 and M_3 was significant in the Tukey test but is not significant in the Scheffé test. If μ_1 and μ_3 are not equal, the Tukey test was powerful enough to detect the difference, but the Scheffé test was not.

In general, one should always use the most powerful post hoc test for the family of comparisons that are of interest. If the family includes only pairwise comparisons, the Tukey test is a better choice than the Scheffé test. If, however, the family of interest includes nonpairwise as well as pairwise comparisons, the Scheffé should be used to test all comparisons, including pairwise ones.

EXERCISE 12–14 In an ANOVA of four groups, the overall F was significant at the .01 level. Each group had 10 subjects, and the four group means were 17.2, 19.4, 15.8, and 19.0. The MS within was 3.06. Use the Tukey test, with $\alpha = .01$, to test all pairwise comparisons.

SUMMARY

The ***one-way analysis of variance (ANOVA)*** is a test for analyzing differences among the means of two or more independent samples. The ANOVA tests the null hypothesis (H_0): $\mu_1 = \mu_2 = \cdots = \mu_k$ against the alternative hypothesis: Not all μ are equal. If the variation among the sample means is greater than would be expected by chance, H_0 is rejected.

The assumptions underlying the ANOVA are normal populations and equal population variances. However, the ANOVA is relatively robust to violations of the assumptions, unless sample sizes are small and unequal and the population variances are quite different.

Given a set of N_T scores in k independent samples, the total sum of squared deviations of all N_T scores from the grand mean (GM), called the total sum of squares (SS total), can be partitioned into two components: (1) the sum of squared deviations of scores within samples from the sample means (SS within), and (2) the sum of squared deviations of the sample means from the grand mean (SS between). That is

$$SS \text{ total} = SS \text{ within} + SS \text{ between}$$

The sums of squares can be calculated by either deviation formulas or raw score formulas. The deviation formulas are

$$SS \text{ total} = \sum^{N_T} (X - GM)^2$$

$$SS \text{ within} = SS_1 + SS_2 + \cdots + SS_k, \text{ where}$$

$$SS \text{ of each sample} = \Sigma(X - M)^2$$

$$SS \text{ between} = n_1(M_1 - GM_1)^2 + n_2(M_2 - GM)^2 + \cdots + n_k(M_k - GM)^2$$

The raw score formulas are

$$SS \text{ total} = \Sigma X_T^2 - \frac{(\Sigma X_T)^2}{N_T}$$

$$SS \text{ within} = SS_1 + SS_2 + \cdots + SS_k, \text{ where}$$

$$SS \text{ of each sample} = \Sigma X^2 - \frac{(\Sigma X)^2}{n}$$

$$SS \text{ between} = \frac{(\Sigma X_1)^2}{n_1} + \frac{(\Sigma X_2)^2}{n_2} + \cdots + \frac{(\Sigma X_k)^2}{n_k} - \frac{(\Sigma X_T)^2}{N_T}$$

If SS total and SS between are calculated first, SS within can be calculated by the formula SS within = SS total – SS between.

Mean squares (MS) are calculated by dividing the sums of squares within and between by their respective degrees of freedom.

$$MS \text{ within} = \frac{SS \text{ within}}{df \text{ within}}$$

$$MS \text{ between} = \frac{SS \text{ between}}{df \text{ between}}$$

df within = $N_T - k$, and df between = $k - 1$. df within + df between = $N_T - 1$ = df total.

If H_0 is true, both MS within and MS between are estimates of the population variance, σ^2, and they are expected to be approximately equal. If, on the other hand, H_0 is false, MS between is expected to be larger than MS within. To test H_0, we calculate the ratio $F = MS$ between/MS within and compare F with critical values of F to find its probability (p). If $p < \alpha$, we reject H_0 and conclude that not all the population means are equal.

The results of an ANOVA are usually displayed in a ***summary table*** that lists the sources of variation, sums of squares, degrees of freedom, mean squares, F, and level of significance attained. When calculating by hand, it is useful to prepare a summary table before conducting the analysis because many calculations can be performed in the table. Outputs of statistical software packages include summary tables, the exact p value of F, and often other information as well.

The difference between the means of two independent samples can be tested either with a t test or an ANOVA. Because $t^2 = F$ and (critical value of $t)^2$ for a two-tailed test = critical value of F for the same alpha, a two-tailed t test and an ANOVA of the same data always lead to the same conclusion.

The power of the ANOVA increases as (1) the variation among the population means increases, (2) the population variance decreases, (3) N_T increases, and (4) alpha increases. If N_T is sufficiently large, even small and possibly unimportant differences between means may result in a significant value of F. A measure of the strength of the relationship between the independent and dependent variables can be calculated by the formula

$$\hat{\omega}^2 = \frac{SS \text{ between} - (k-1)\, MS \text{ within}}{SS \text{ total} + MS \text{ within}}$$

$\hat{\omega}^2$ is an estimate of the proportion of the total population variance accounted for by the independent variable.

An ANOVA tells us whether the variation among all the sample means is greater than would be expected by chance. Researchers often want to know whether or not differences between particular pairs of means or combinations of means are significant. Tests of differences like these are called ***comparisons***. Comparisons may be planned in advance (***planned comparisons***) or may be decided upon after examining the results of the ANOVA (***post hoc tests***). Any two comparisons on a set of means may be ***orthogonal*** (independent) or nonorthogonal. Planned comparisons should generally be orthogonal to one another.

A test of a comparison involves calculating a *comparison term*, C, and a value of F. In planned comparisons, F is evaluated against a critical value of F for $(1, N_T - k)$ degrees of freedom, using either the same α as in the ANOVA or, possibly, a level of alpha equal to the alpha of the ANOVA divided by the number of tests.

When comparisons have not been preplanned, the number of potential tests, the ***family*** of comparisons, may be very large, so that the *Type I error rate*—the probability of at least one Type I error over all tests—is substantially larger than the alpha level for a single test. Each test for post hoc comparisons controls the Type I error rate for a particular family of comparisons.

When the set of potential comparisons is the family of all possible tests, the appropriate post hoc test is the ***Scheffé test***, in which the comparison F ratio is evaluated against an adjusted critical value. If the researcher is interested in pairwise comparisons only, the ***Tukey "honestly significant difference" (HSD)*** test can be used. In the Tukey test, *HSD*—the minimum difference between means that is significant—is calculated, and the difference between each pair of sample means is compared with the *HSD*.

QUESTIONS

1. Your friend Tilly conducted an ANOVA on data from three samples and obtained an F significant at the .01 level. She concluded that all three population means differ from one another.

   a. Explain to Tilly what is wrong with her conclusion.

   b. It is possible that, as Tilly stated, $\mu_1 \neq \mu_2 \neq \mu_3$. List three other possibilities that are compatible with the alternative hypothesis.

2. An experimenter randomly subdivided 32 subjects into four groups. After applying her treatments, she calculated SS total = 270 and SS between = 60. Prepare a summary table and complete the analysis. Evaluate F using $\alpha = .05$.

3. Each of five classes of first-grade children was taught a different strategy for remembering the events in a story. Then the children's ability to recall a story's events was tested. Here are the recall scores of the five classes.

1	2	3	4	5
10	12	7	6	16
1	10	16	10	10
2	15	15	4	15
5	16	12	4	18
7	15	12	3	16
8	7	13	9	17
5	13	10	8	
3	13	9	7	
5	14		4	
7	12			
	9			

a. Calculate the five means. Based on the means and other aspects of the data, do you expect to obtain a significant difference between the means? Why or why not?
b. Conduct an ANOVA, prepare a summary table, and evaluate F at the .05 level. Are the results in accordance with your expectations?
c. Estimate the proportion of recall score variance that is accounted for by type of strategy.

4. Here are means and standard deviations for two studies, A and B. If the sample sizes are the same in both studies, which is more likely to result in a significant value of F? Why?

Case A

Group	1	2	3	4
Mean	18.5	19.3	12.5	16.4
SD	7.2	5.8	6.1	5.4

Case B

Group	1	2	3	4
Mean	18.5	19.3	12.5	16.4
SD	3.5	2.7	4.1	3.7

5. Here are the results of an ANOVA.

Source of Variation	*SS*	*df*	*MS*	F
Between	140	2	70	14.0*
Within	1985	397	5	
Total	2125	399		

$^*p < .01$.

a. Calculate $\hat{\omega}^2$.
b. Explain the apparent discrepancy between the large value of F and the relatively small $\hat{\omega}^2$.

6. An ANOVA of two independent samples, of size 10 each, resulted in $F = 4.33$. *MS* within was equal to 35.

a. If a t test were conducted on the same data, what would be the value of t?
b. What is the estimated standard error of the difference between means (s_{diff})?
c. What is the difference between the two means ($M_1 - M_2$)?

7. A consumer research agency compared the effectiveness of several over-the-counter painkillers for reducing headaches. Thirty subjects who complained of frequent headaches were randomly assigned to six treatments: Aspirin A, Aspirin B, regular-strength Ibuprofen C, regular-strength Ibuprofen D, extra-strength Ibuprofen E, and a placebo. The subjects, who did not know which drug they were taking, used the prescribed medication for headaches over a 1-month period. At the end of the month, each subject rated the amount of relief he/she experienced on a scale of 0 (no effect) to 10 (complete relief). Here are the results:

Aspirin A	Aspirin B	Ibuprofen C
6	6	9
8	3	6
4	8	8
9	6	7

Ibuprofen D	Extra-strength E	Placebo F
9	8	3
7	10	5
7	6	2
9	4	6
5	7	4

The following five comparisons had been planned in advance:

Painkiller vs. placebo
Aspirin vs. Ibuprofen
Aspirin A vs. Aspirin B
Regular-strength Ibuprofen vs. extra-strength Ibuprofen
Ibuprofen C vs. Ibuprofen D

a. Conduct an ANOVA. Use $\alpha = .05$.
b. Assign weights for each comparison, and test each pair of comparisons for orthogonality.
c. Conduct tests for the comparisons using $\alpha = .05$ for each test.
d. Test the comparisons using $\alpha = .01$ for each one. (*Note:* .01 is equal to .05 divided by the number of tests.)

8. Fifteen fourth-grade children who were having difficulty in math were randomly assigned to three groups: (1) tutoring by a teacher, (2) tutoring by a peer (another student), or (3) no tutoring (control). Here are their scores on a math test 3 months later:

Teacher Tutoring	Peer Tutoring	No Tutoring
20	22	11
15	11	8
10	17	3
13	9	9
18	14	5

a. Conduct an ANOVA. Use $\alpha = .05$.
b. You wonder which pairwise differences between means are significant. Conduct a Tukey test of all pairwise comparisons.
c. Suppose that you are also interested in the question: Is there a significant difference between tutoring (of any kind) versus no tutoring? What is the null hypothesis for this comparison?
d. Assuming that none of your comparisons was planned in advance, what is the appropriate test for the set of comparisons that includes not only pairwise comparisons but also the comparison of part c?
e. Use the appropriate test for the following set of three comparisons:
(1) Teacher tutoring vs. no tutoring
(2) Peer tutoring vs. no tutoring
(3) Tutoring vs. no tutoring

9. The attitudes of subjects toward a liberal immigration policy were measured after they had been presented with one of four types of arguments for immigration. The scores of the four groups are listed below (higher scores indicate more favorable attitudes).

Oral/ Factual	Oral/ Emotional
17	20
12	24
14	14
10	18

Oral/Factual and Emotional	Written/Factual and Emotional
15	8
25	12
20	9
23	10

a. Conduct an ANOVA. Use $\alpha = .01$.
b. The researcher examines the results of the ANOVA and decides on the following three comparisons:
(1) Oral vs. written
(2) Factual vs. factual and emotional
(3) Oral factual vs. oral emotional
Are the three comparisons orthogonal?
c. Conduct the appropriate tests for the comparisons.

10. An ANOVA of five independent samples of size 10 each yielded the following results:

Source of Variation	*SS*	*df*	*MS*	*F*
Treatments	180	4	45	3*
Within	675	45	15	
Total	855	399		

$^*p < .05$.

The five group means were

M_1	M_2	M_3	M_4	M_5
17	18	20	15	15

Conduct a Tukey test for all pairwise comparisons. Use $\alpha = .05$.

13 Analyses of Variance for Factorial Designs and Repeated Measures

INTRODUCTION

The one-way or simple analysis of variance, discussed in Chapter 12, is used to analyze results of studies with a single independent variable that varies between independent groups of subjects. With the advent of high-speed computers for fast, reliable data analysis, researchers no longer feel constrained to limit themselves to such simple designs. Increasingly, researchers are using designs in which two or more independent variables are manipulated simultaneously, or in which variables are manipulated within rather than (or in addition to) between subjects.

The ANOVA is readily extended to the analysis of more complex designs. In each case, the total variance of the data set is simply apportioned into appropriate components, a mean square is calculated for each component, and F tests are used to compare treatment mean squares with "error" mean squares. The factors that differ from one design to another are how the total variance is partitioned, and, in some cases, the component or components that measure chance (error) variation.

In this chapter we deal with the analyses of two common designs. One is the factorial design with two independent variables, both varying between subjects, and referred to as two-way ANOVA. The other is the one-way repeated measures (correlated samples) design, in which a single independent variable varies within subjects.

Objectives

When you have finished this chapter, you will be able to

- Describe how treatments are combined and subjects are assigned to groups in the two-way factorial design.
- Explain what is meant by an interaction between variables.
- State the null and alternative hypotheses and the assumptions underlying the two-way ANOVA.
- Describe the partitioning of sums of squares and degrees of freedom.
- Calculate sums of squares, mean squares, and F ratios. Be able to interpret the results and draw appropriate conclusions.

- Be able to graph the results of two-way factorial design studies and interpret the graphs.
- Describe factors affecting the power of the two-way ANOVA.
- Calculate and interpret the proportion of variance accounted for by each independent variable and by the interaction.
- Conduct planned comparisons and post hoc tests for a factorial design.
- State the null and alternative hypotheses and the underlying assumptions for the repeated-measures ANOVA.
- Describe the partitioning of sums of squares and degrees of freedom.
- Calculate sums of squares, mean squares, and F. Be able to interpret the results and draw appropriate conclusions.
- Describe the relation between the t test and the ANOVA for analyzing the difference between means of two correlated samples.
- Compare the power of one-way ANOVA and repeated measures ANOVA; describe factors that influence the relative power of the two tests.
- Conduct planned comparisons and post hoc tests for a repeated measures study.

Two-Way Analysis of Variance

What are the effects of diet and exercise on weight loss? How does adult intelligence vary with age and level of education? How are the grade point averages of college students affected by scholastic aptitude and study habits? What are the effects of financial incentives and nonfinancial incentives on job satisfaction? All of these questions ask how a single dependent variable is affected by two independent variables or *factors*. The experimental design in which two (or more) independent variables are manipulated simultaneously is called a ***factorial design***, and the method of analysis is called ***two-way analysis of variance***.

Constructing a Factorial Design

What is a factorial design, and how is it constructed? To answer this question, we consider the following hypothetical example. The English teachers at Westville High School have designed an SAT-Verbal preparation course for college-bound seniors. Before they insert it into the high school curriculum, however, they want to conduct a study to find out whether the course is effective. The teachers decide to try the course out on a small sample of seniors who have taken the SAT once and compare their gains on the Verbal test with the gains of students who have not taken the course.

At a planning meeting, Mr. Pershing, the guidance counselor, points out that the effects of the course may differ in low-scoring and high-scoring students. For example, the course may raise scores of low-scoring students but not high-scoring students, or vice versa. When the effects of a variable differ, depending on the level of another variable, the two variables are said to ***interact***. Thus, Mr. Pershing is suggesting that SAT preparation may interact with students' initial SAT scores.

Ms. Marton, one of the teachers, proposes that two studies be conducted, the first study to include only students with initially low SAT scores and the second to include only students whose initial scores were high. In each study, a random half of the students will take the SAT preparation course, and the other half will not. Then a t test will be used to compare the mean gain scores of coached and uncoached students in each study.

Although the two proposed studies would reveal the effects of the preparation course in high-scoring and in low-scoring students, the design has two drawbacks. First, it is generally less efficient to conduct two studies rather than just one to answer a given research question. Second, if the results indicate that the course has different effects in low- and high-scoring students, there is no way to quantify the amount of the interaction or to measure its strength.

Fortunately, Ms. Claymore, another teacher, is currently taking a course on experimental design. "Let us conduct the study," she says, "as a factorial design. Then we can manipulate both variables, initial SAT scores and provision of instruction, in a single study. Moreover, the analysis will tell us not only whether each variable has an effect by itself but also whether they interact." She then explains the procedure to be used. The gist of her explanation follows.

We use the letters A and B to stand for the two independent variables or factors. In this study, we let A = initial SAT score level and B = provision of instruction. Numerical subscripts, 1, 2, . . . , are used to denote the levels of each variable. We let A_1 represent low initial SAT score level and A_2 represent high initial SAT score level; similarly, B_1 represents instruction provided and B_2 represents no instruction provided. There are four possible combinations of a level of A with a level of B:

A_1B_1: Low initial score level, instruction provided
A_1B_2: Low initial score level, no instruction provided
A_2B_1: High initial score level, instruction provided
A_2B_2: High initial score level, no instruction provided

Each combination defines one group of subjects. Thus, there will be four groups in the SAT preparation study.

A factorial design is commonly referred to in terms of the number of levels or treatments on each variable. If we let *a* represent the number of levels of variable A and *b* represent the number of levels of variable B, the design is an $a \times b$ ("*a* by *b*") factorial design. The SAT study, with two levels of each variable, is a 2×2 factorial design. If there were three levels of initial SAT scores instead of two (e.g., high, moderate, and low), the study would be a 3×2 factorial design. If four levels or types of instruction were to be compared instead of two, it would be a 2×4 factorial design.

Note that the total number of groups in a factorial design is equal to *a* times *b*. Thus, in a 2×2 design, there are four groups; a 3×4 design would have 12 groups; and so on.

EXERCISE 13–1

1. A biopsychologist is studying effects of drugs on activity levels in rats. Both type of drug (Prozac, Paxil, and Xanax) and drug dosage level (high, medium, low) will be manipulated in the study. Let A stand for the type of drug and B stand for dosage level.
   a. List all possible combinations of a level of A with a level of B.
   b. If 10 rats are assigned to each combination, how many rats will be used in the study?
2. Your developmental psychology textbook states, "Heredity and environment interact to determine behavior." Explain what this means.

Rationale of the Two-Way ANOVA

Components of Variance

In the two-way ANOVA, as in the one-way ANOVA, the total sum of squares of the data set is first partitioned into a sum of squares between groups (*SS* between), which measures variation between the group means, and a sum of squares within groups (*SS*

Figure 13–1

Partitioning the total sum of squares in the two-way ANOVA.

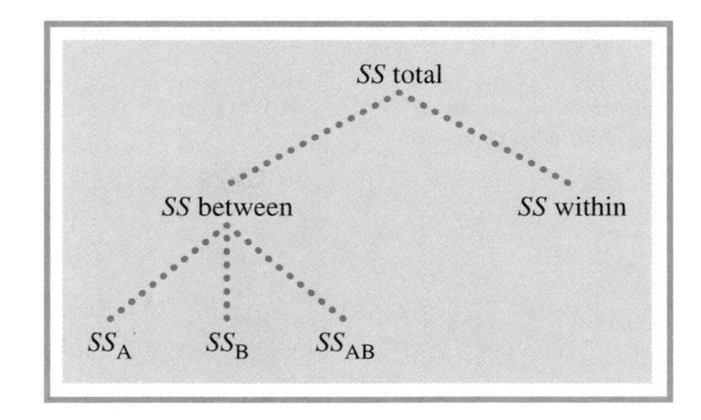

within), which measures variation of scores within groups around the group means. In the factorial design, however, variation between groups is due in part to Factor A, in part to Factor B, and in part to the interaction between Factors A and B. Therefore, the sum of squares between is further partitioned into a sum of squares for Factor A (SS_A), a sum of squares for Factor B (SS_B), and a sum of squares for the interaction between A and B (SS_{AB}), as shown in Figure 13–1.

The total degrees of freedom are partitioned in the same way as the total sum of squares. As in the one-way ANOVA, *df* total = $N_T - 1$; *df* between = number of groups minus one; and *df* within = N_T − number of groups. Because the number of groups = *ab*, *df* between = $ab - 1$ and *df* within = $N_T - ab$. In the SAT study, for example, if five subjects are used in each of the four groups, so that the total number of subjects (N_T) is 20:

$$df\text{ total} = 20 - 1 = 19$$
$$df\text{ between} = 4 - 1 = 3$$
$$df\text{ within} = 20 - 4 = 16$$

The *df* between is further partitioned into df_A, df_B, and df_{AB}. df_A is the number of levels of A minus one, or $a - 1$; df_B is the number of levels of B minus one, or $b - 1$; the interaction *df*, df_{AB}, is the product of df_A and df_B, or $(a - 1)(b - 1)$. In the 2×2 SAT study:

$$df_A = 2 - 1 = 1$$
$$df_B = 2 - 1 = 1$$
$$df_{AB} = (1)(1) = 1$$

Note that $df_A + df_B + df_{AB} = (a - 1) + (b - 1) + (a - 1)(b - 1) = ab - 1$, the degrees of freedom between.

$$df\text{ between} = df_A + df_B + df_{AB}$$

The partitioning of the total degrees of freedom is summarized in Figure 13–2.

Null and Alternative Hypotheses

The two-way ANOVA tests three null hypotheses simultaneously. The first null hypothesis states that the population means under all levels of Factor A are equal. The alternative to the first null hypothesis is that not all the Factor A means are equal.

1. H_o: $\mu_{A_1} = \mu_{A_2} = \cdots = \mu_{A_a}$
   H_A: Not all μ_A are equal

Figure 13–2

Partitioning the total degrees of freedom in the two-way ANOVA.

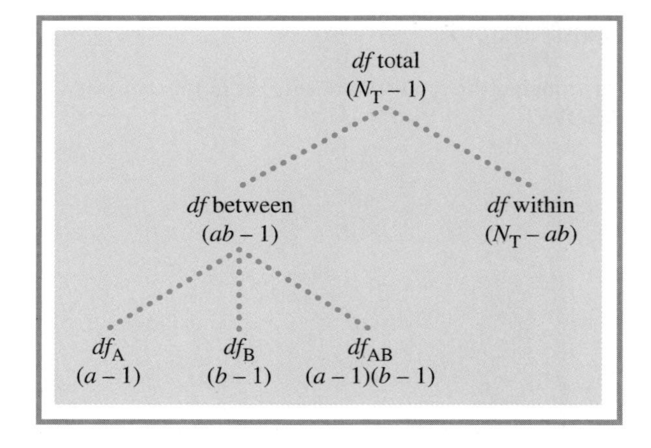

The second null hypothesis states that the population means under all levels of Factor B are equal; the alternative hypothesis states that not all the Factor B means are equal.

2. H_o: $\mu_{B_1} = \mu_{B_2} = \cdots = \mu_{B_b}$
   H_A: Not all μ_B are equal

As you can see, each of the first two null hypotheses states that one factor (A or B) has no effect. The effect (if any) of one particular variable in a factorial design is called a ***main effect***. Therefore, the first two null hypotheses can be stated as follows:

1. H_o: No main effect of Factor A in the population
2. H_o: No main effect of Factor B in the population

The third null hypothesis states that there is no interaction between A and B. The alternative hypothesis states that an interaction exists.

3. H_o: No A × B interaction effect in the population
   H_A: A and B have an interactive effect in the population

Testing the Null Hypotheses

In the analysis of the factorial design, a mean square (*MS*) is calculated for each effect. As in the one-way ANOVA, each mean square is obtained by dividing a sum of squares by the associated degrees of freedom. Thus,

$$MS_A = \frac{SS_A}{df_A}$$

$$MS_B = \frac{SS_B}{df_B}$$

$$MS_{AB} = \frac{SS_{AB}}{df_{AB}}$$

The mean square within is calculated, as in the one-way ANOVA.

$$MS \text{ within} = \frac{SS \text{ within}}{df \text{ within}}$$

If the null hypothesis for Variable A is true (i.e., if there is no A effect in the population), the average value of MS_A, over all possible sets of samples, is equal to MS within.

If H_o for A is true, $MS_A = MS$ within, on the average.

If, however, the null hypothesis is false and there is a true A effect, the average value of MS_A is greater than MS within.

If H_o for A is false, $MS_A > MS$ within, on the average.

The null hypothesis for A is tested by calculating F:

$$F_A = \frac{MS_A}{MS \text{ within}}$$

The probability (p) of F_A is found by comparing F_A with critical values for the chosen alpha and ($a - 1$, $N_T - ab$) degrees of freedom. If $p < \alpha$, the main effect of A is significant at level alpha. H_o is rejected, and the analyst concludes that Variable A has an effect in the population. If, on the other hand, $p > \alpha$, the main effect of A is not significant at level α, and H_o is not rejected.

The reasoning is the same for the main effect of B and for the interaction. In both cases, if the null hypothesis is true, the average value of the effect mean square, over all possible sets of samples, is equal to MS within, but if the null hypothesis is false, the effect mean square is greater than MS within, on the average.

For Variable B:

If H_o for B is true, $MS_B = MS$ within, on the average.
If H_o for B is false, $MS_B > MS$ within, on the average.

For the interaction between A and B:

If H_o for the interaction is true, $MS_{AB} = MS$ within, on the average.
If H_o for the interaction is false, $MS_{AB} > MS$ within, on the average.

Therefore the test of significance for each effect involves calculating F.

$$\text{Variable B:} \quad F_B = \frac{MS_B}{MS \text{ within}}$$

$$\text{Interaction:} \quad F_{AB} = \frac{MS_{AB}}{MS \text{ within}}$$

The probability (p) of F_B is found by comparing F_B with critical values for ($b - 1$, $N_T - ab$) degrees of freedom; p for F_{AB} is found by comparing F_{AB} with critical values for $[(a - 1)(b - 1), N_T - ab]$ degrees of freedom. In both cases, if $p < \alpha$, the effect is significant at level alpha, and the null hypothesis is rejected. If, on the other hand, $p > \alpha$, the effect is not significant at level alpha, and the null hypothesis is not rejected.

EXERCISE 13–2

Consider again the study of the effects of three drugs (Prozac, Paxil, and Xanax) and three dosage levels (high, medium, low) on the activity levels of rats. Let A be type of drug and B be dosage level. Ten rats are assigned to each combination.

1. State the three null and alternative hypotheses.
2. Calculate the degrees of freedom for each main effect, the degrees of freedom for the interaction, and the degrees of freedom within.

Conducting the Analysis

Analyses of data from factorial designs are rarely conducted by hand. However, a 2 × 2 design with a small number of subjects per group, such as the SAT study, is small enough for hand calculation. In accordance with my conviction that procedures are better understood if done step by step at least once, the calculations are described and carried out in this section.

Display of Data and Related Symbols

The data from an $a \times b$ factorial design can be displayed in a two-way table, as in Table 13–1. Each row of the table displays data for one level of Factor A, and each column displays data for one level of Factor B. Each cell includes the data from one group.

Certain sums needed for the analysis are shown in Table 13–1. These include the sum of the scores in each cell (group), as well as the sum across each row and down each column. We will represent the sum of the scores in a cell as ΣAB, and use subscripts to stand for the level on Factors A and B. For example, the sum of scores in group A_3B_5 is ΣA_3B_5.

Table 13–1 Schematic Data Matrix for an $a \times b$ Factorial Design

	Levels of Factor B					
	B_1	B_2	. . . .	B_b	Sum of Scores	Number of Scores
A_1	Scores of the n subjects in Group A_1B_1. Sum = ΣA_1B_1	Scores of the n subjects in Group A_1B_2. Sum = ΣA_1B_2		Scores of the n subjects in Group A_1B_b. Sum = ΣA_1B_b	ΣA_1	bn
A_2	Scores of the n subjects in Group A_2B_1. Sum = ΣA_2B_1	Scores of the n subjects in Group A_2B_2. Sum = ΣA_2B_2		Scores of the n subjects in Group A_2B_b. Sum = ΣA_2B_b	ΣA_2	bn
Levels of Factor A						
A_a	Scores of the n subjects in Group A_aB_1. Sum = ΣA_aB_1	Scores of the n subjects in Group A_aB_2. Sum = ΣA_aB_2		Scores of the n subjects in Group A_aB_b. Sum = ΣA_aB_b	ΣA_a	bn
Sum of Scores	ΣB_1	ΣB_2		ΣB_b	ΣX_T	
Number of Scores	an	an		an	N_T =	abn

The sum of scores within any given row, ΣA, is the sum for that level of A, over all levels of B. For example, ΣA_4 is the sum of the A_4 scores over all levels of Factor B ($\Sigma A_4 = \Sigma A_4B_1 + \Sigma A_4B_2 + \cdots + \Sigma A_4B_b$). Similarly, the sum of scores in any given column, ΣB, is the sum for that level of B, over all levels of A; for example, $\Sigma B_3 = \Sigma A_1B_3 + \Sigma A_2B_3 + \cdots + \Sigma A_aB_3$). As in one-way ANOVA, ΣX_T represents the sum of all the scores in the data set, and ΣX_T^2 represents the sum of all the squared scores.

We will use n to represent the number of scores within each cell (i.e., within each group). As in the one-way ANOVA, N_T stands for the number of scores in the entire data set. Because there are ab groups, each of size n, $N_T = nab$. The number of scores in any given row (i.e., for each level of A) is equal to n times the number of columns, or bn. The number of scores in each column (each level of B) is equal to n times the number of rows, or an.

For example, a 3×4 factorial design includes 3 levels of Factor A ($a = 3$) and 4 levels of Factor B ($b = 4$), for a total of 12 combinations or groups. If each group contains eight subjects ($n = 8$), the total number of subjects (N_T) is $8 \times 3 \times 4 = 96$. Condition A_1 includes the eight subjects in group A_1B_1, the eight subjects in group A_1B_2, the eight subjects in group A_1B_3, and the eight subjects in group A_1B_4, so that the number of subjects in Condition A_1 is $\underset{\uparrow \atop b}{4} \times \underset{\uparrow \atop n}{8} = 32$. This is also the number of subjects in Conditions A_2 and A_3. The subjects in Condition B_1 include the eight subjects in group A_1B_1, the eight in group A_2B_1, and the eight in group A_3B_1, so that the number of subjects in Condition B_1 (as well as in Conditions B_2, B_3, and B_4) is $\underset{\uparrow \atop a}{3} \times \underset{\uparrow \atop n}{8} = 24$.

Now let us return to the SAT gain score study. The English teachers decide to include five subjects in each group ($n = 5$), or 20 subjects in all. To form the groups, they identify 10 students whose initial SAT-Verbal scores were low and randomly subdivide them into instruction and no-instruction conditions. Similarly, 10 high-scoring students are identified and randomly subdivided into instruction and no-instruction conditions.

Over the next 10 weeks, the 10 students assigned to the instruction-provided condition (B_1) participate in the SAT preparation course, and the other 10 students do not. Then all 20 students take the SAT again, and their gain scores (new SAT-Verbal score minus original SAT-Verbal score) are calculated. The gain scores are displayed in Table 13–2. The sums needed for the analysis, as well as the numbers of scores related to various sums, are also shown.

Calculating Sums of Squares

The sums of squares needed for the analysis of a two-way factorial design are SS_A, SS_B, SS_{AB}, and SS within. We begin, as in the one-way ANOVA, by calculating SS total, SS between, and SS within. We will use raw score formulas rather than deviation formulas for the calculations.

The total sum of squares is calculated as in the one-way ANOVA:

$$SS\text{ total} = \sum X_T^2 - \frac{(\sum X_T)^2}{N_T}$$ *Formula for calculating the total sum of squares*

For the gain score data

$$SS\text{ total} = 8700 - \frac{(210)^2}{20} = 6495$$

Table 13–2 SAT-Verbal Gain Scores of 20 High School Students

	Instruction Provided (B_1)	No Instruction Provided (B_2)	Sums
Initially Low-Scoring Students (A_1)	30 25 40 35 50	10 −10 15 0 −10	
	$180 = \Sigma A_1 B_1$ $n = 5$	$5 = \Sigma A_1 B_2$ $n = 5$	$185 = \Sigma A_1$ $10 = bn$
Initially High-Scoring Students (A_2)	10 5 −5 0 15	−10 15 0 −20 15	
	$25 = \Sigma A_2 B_1$ $n = 5$	$0 = \Sigma A_2 B_2$ $n = 5$	$25 = \Sigma A_2$ $10 = bn$
Sums	$205 = \Sigma B_1$ $10 = an$	$5 = \Sigma B_2$ $10 = an$	$210 = \Sigma X_T$ $20 = N_T$

$\Sigma X_T^2 = (30)^2 + (25)^2 + \cdots + (10)^2 + (-10)^2 + \cdots + (10)^2 + \cdots + (10)^2 + (5)^2 + \cdots + (-10)^2 + (15)^2 + \cdots + (15)^2 = 8700$

Next we calculate the sum of squares between groups:

$$SS \text{ between} = \frac{(\Sigma A_1 B_1)^2}{n} + \frac{(\Sigma A_1 B_2)^2}{n} + \cdots + \frac{(\Sigma A_a B_b)^2}{n} - \frac{(\Sigma X_T)^2}{N_T}$$

Formula for calculating the sum of squares between

Notice that *SS* between is a measure of variation among all the groups without regard to their membership on Factors A and B. For the gain score data

$$SS \text{ between} = \frac{(180)^2}{5} + \frac{(5)^2}{5} + \frac{(25)^2}{5} + \frac{(0)^2}{5} - \frac{(210)^2}{20} = 4405$$

As in the one-way ANOVA, *SS* within is the pooled sum of squares of the *ab* groups, and it can be obtained by calculating the sum of squares of each group, then adding. However, because *SS* between + *SS* within = *SS* total, *SS* within is more easily obtained by subtracting *SS* between from *SS* total:

$$SS \text{ within} = SS \text{ total} - SS \text{ between}$$

Formula for calculating the sum of squares within

For the gain scores

$$SS \text{ within} = 6495 - 4405 = 2090$$

Next, we partition the *SS* between. As Figure 13–1 shows, *SS* between = $SS_A + SS_B + SS_{AB}$. The formula for calculating SS_A is

$$SS_A = \frac{(\Sigma A_1)^2}{bn} + \frac{(\Sigma A_2)^2}{bn} + \cdots + \frac{(\Sigma A_a)^2}{bn} - \frac{(\Sigma X_T)^2}{N_T}$$

Formula for calculating SS_A

Note that each term (except the last) is equal to the squared row sum divided by the number of scores in the row. For the SAT gain scores

$$SS_A = \frac{(185)^2}{10} + \frac{(25)^2}{10} - \frac{(210)^2}{20} = 1280$$

The formula for SS_B is

$$SS_B = \frac{(\Sigma B_1)^2}{an} + \frac{(\Sigma B_2)^2}{an} + \cdots + \frac{(\Sigma B_b)^2}{an} - \frac{(\Sigma X_T)^2}{N_T}$$ *Formula for calculating* SS_B

Each term (except the last) is equal to the squared column sum divided by the number of scores in the column. For the gain score data

$$SS_B = \frac{(205)^2}{10} + \frac{(5)^2}{10} - \frac{(210)^2}{20} = 2000$$

Since $SS_A + SS_B + SS_{AB} = SS$ between, SS_{AB} is obtained by subtraction:

$$SS_{AB} = SS \text{ between} - SS_A - SS_B$$ *Formula for calculating* SS_{AB}

For the gain score data

$$SS_{AB} = 4405 - 1280 - 2000 = 1125$$

The formulas for calculating all needed sums of squares are listed, in the order of calculation, in Table 13–3.

The remaining calculations are straightforward. First, we calculate the mean square for each effect and the mean square within.

$$MS_A = \frac{SS_A}{df_A}, \text{ where } df_A = a - 1$$

$$MS_B = \frac{SS_B}{df_B}, \text{ where } df_B = b - 1$$

$$MS_{AB} = \frac{SS_{AB}}{df_{AB}}, \text{ where } df_{AB} = (a-1)(b-1)$$

$$MS \text{ within} = \frac{SS \text{ within}}{df \text{ within}}, \text{ where } df \text{ within} = N_T - ab$$

We have already calculated the degrees of freedom for the gain score study: $df_A = 1$, $df_B = 1$, $df_{AB} = 1$, and df within = 16.

Now we are ready to calculate the mean squares:

$$MS_A = \frac{SS_A}{df_A} = \frac{1280}{1} = 1280$$

$$MS_B = \frac{SS_B}{df_B} = \frac{2000}{1} = 2000$$

$$MS_{AB} = \frac{SS_{AB}}{df_{AB}} = \frac{1125}{1} = 1125$$

$$MS \text{ within} = \frac{SS \text{ within}}{df \text{ within}} = \frac{2090}{16} = 130.625$$

Table 13–3 **Calculation of Sums of Squares in an $a \times b$ Factorial Design**

1.	*SS* total	$= \Sigma X_T^2 - \frac{(\Sigma X_T)^2}{N_T}$
2.	*SS* between	$= \frac{(\Sigma A_1B_1)^2}{n} + \frac{(\Sigma A_1B_2)^2}{n} + \cdots + \frac{(\Sigma A_aB_b)^2}{n} - \frac{(\Sigma X_T)^2}{N_T}$
3.	*SS* within	$= SS \text{ total} - SS \text{ between}$
4.	SS_A	$= \frac{(\Sigma A_1)^2}{bn} + \frac{(\Sigma A_2)^2}{bn} + \cdots + \frac{(\Sigma A_a)^2}{bn} - \frac{(\Sigma X_T)^2}{N_T}$
5.	SS_B	$= \frac{(\Sigma B_1)^2}{an} + \frac{(\Sigma B_2)^2}{an} + \cdots + \frac{(\Sigma B_b)^2}{an} - \frac{(\Sigma X_T)^2}{N_T}$
6.	SS_{AB}	$= SS \text{ between} - SS_A - SS_B$

Next, an F ratio is calculated for each effect by dividing its mean square by *MS* within:

$$F_A = \frac{MS_A}{MS \text{ within}} = \frac{1280}{130.625} = 9.80$$

$$F_B = \frac{MS_B}{MS \text{ within}} = \frac{2000}{130.625} = 15.31$$

$$F_{AB} = \frac{MS_{AB}}{MS \text{ within}} = \frac{1125}{130.625} = 8.61$$

Finally, each F is evaluated by comparing it with the critical values of F for df = (df effect, df within). Because each effect has $df = 1$, the critical values for all three effects are those for $df = (1, 16)$. According to the F table (Table D–5 in Appendix D), the critical value of F for (1, 16) df and $\alpha = .05$ is 4.49, and the critical value for $\alpha = .01$ is 8.53. Because all three F ratios exceed the critical value for $\alpha = .01$, we know that for each one, $p < .01$. Therefore, the main effect of A, initial score level, is significant at the .01 level; the main effect of B, provision of instruction, is significant at the .01 level; and the interaction between initial score level and provision of instruction is significant at the .01 level. A computer analysis of the SAT gain data yielded the following more precise probabilities for the three effects: initial score level, $p = .006$; provision of instruction, $p = .001$; interaction, $p = .010$.

Presentation of Results: The ANOVA Summary Table

The results of an analysis of a factorial design are commonly displayed in a summary table. As in one-way ANOVA, the columns of the table are "Sources of Variation," "Sums of Squares," "Degrees of Freedom," "Mean Squares," "F," and "p." The sources of variation include Variable A, Variable B, Variable A × Variable B (the interaction), and Within. "Within" is often referred to as "error," because it measures the amount of variation to be expected by chance. A summary table of the SAT study ANOVA is presented in Table 13–4.

Table 13–4 Summary Table of the SAT Gain Score Analysis

Source of Variation	*SS*	*df*	*MS*	*F*	*p*
Initial score level (A)	1280	1	1280	9.80	.006
Provision of instruction (B)	2000	1	2000	15.31	.001
A × B	1125	1	1125	8.61	.010
Within groups (error)	2090	16	130.625		
Total	6495	19			

EXERCISE 13–3

The table below gives *sums* of scores in a 2 × 2 factorial design with three subjects per group. $\Sigma X_T^2 = 490$. Conduct an ANOVA and prepare a summary table of the results.

	B_1	B_2
A_1	$\Sigma A_1B_1 = 7$	$\Sigma A_1B_2 = 20$
A_2	$\Sigma A_2B_1 = 18$	$\Sigma A_2B_2 = 21$

Interpreting the Results of a Two-Way ANOVA

The two-way ANOVA tells us which effects, if any, are significant. But it does not tell us about the direction of the effects. In order to describe the nature of the effects, we must calculate and compare means.

Consider the SAT gain score study. From the ANOVA, we know that mean gain scores differed significantly as a function of initial score level. But did low-scoring or high-scoring students gain more? To answer this question, we calculate the mean gain score of the 10 students with initially low scores and the mean gain score of the 10 students with initially high scores by dividing each sum of gain scores by 10.

$$\text{Initially low scores:} \quad \text{Mean gain} = \frac{185}{10} = 18.5$$

$$\text{Initially high scores:} \quad \text{Mean gain} = \frac{25}{10} = 2.5$$

We now know that students with initially low scores gained significantly more on the SAT-Verbal test than students with initially high scores.

The instruction effect was also significant. To find out the direction of the difference, we calculate the mean gain score of the 10 students who received instruction and the mean gain score of the 10 who did not.

$$\text{Instruction provided:} \quad \text{Mean gain} = \frac{205}{10} = 20.5$$

$$\text{No instruction provided:} \quad \text{Mean gain} = \frac{5}{10} = 0.5$$

Students who received instruction gained significantly more on the SAT-Verbal test than students who did not receive instruction.

However, our interpretation of the effect of instruction must be tempered by the interaction between initial score level and provision of instruction. The significant inter-

Table 13–5 Mean Gain Scores of the Four Groups

	Instruction Provided	No Instruction Provided
Initially Low Scores	36	1
Initially High Scores	5	0

action tells us that the effect of instruction differed, depending on students' initial score level. In order to understand this effect, we calculate the mean gain score of each of the four groups by dividing the group's sum of gain scores (ΣAB) by 5. The means are shown in Table 13–5.

Now we compare gains with and without instruction within each score level. The low-scoring students gained an average of 36 points with instruction, but only one without instruction. Thus, they gained 35 more points, on the average, with instruction than without. The high-scoring students gained 5 points, on the average, with instruction, and none without instruction. Thus, they gained only 5 more points with instruction than without. We conclude that instruction was more effective in raising SAT scores of initially low-scoring than initially high-scoring students. The greater effectiveness of the course for low-scoring students suggests that, in the future, if instructional resources at the high school are limited, the teachers may wish to give preference in enrollment to students whose initial SAT-Verbal scores are low.

Graphing the Results of a Factorial Design

The results of a factorial design study can be graphed to show the effect of each variable and the interaction. The SAT study results are graphed in Figure 13–3. The steps in constructing a graph like the one in Figure 13–3 are as follows:

1. Mean values on the dependent variable are displayed on the ordinate. Thus, in the SAT study graph, the ordinate is labeled "Mean Gain." Because mean gain scores of the four groups ranged from 0 to 36, values from 0 to 40 are marked on the ordinate.
2. The abscissa displays the levels of one of the two independent variables. In Figure 13–3, the variable "Initial Score Level" was selected. The levels "Low" and "High" are marked on the abscissa.
3. The mean score of each group is plotted above that group's level on the abscissa variable. Thus, for the SAT study, the mean SAT gain of low-scoring students with instruction provided (mean = 36) and of low-scoring students with no instruction provided (mean = 1) are both plotted above "Low" on the abscissa. Similarly, the mean SAT gain of high-scoring students with instruction provided (mean = 5) and of high-scoring students with no instruction provided (mean = 0) are both plotted above "High" on the abscissa.
4. Finally, the points representing the means of groups with the same level of the second independent variable are connected with straight lines, and each line is labeled by the level of the variable. In Figure 13–3, the means of the two groups who received instruction are connected with a straight line labeled "Instruction provided," and the means of the two no-instruction groups are connected with a line that is labeled "No instruction provided."

The three effects can be easily "read" from the graph. The effect of initial score level is shown by the fact that both lines have a negative slope, thus revealing that the

Figure 13–3

Graph of results of the SAT gain score study.

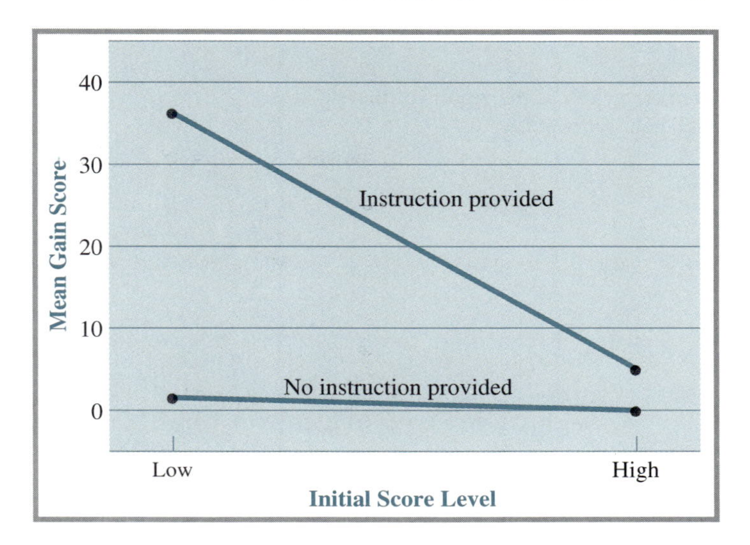

mean gain was less for high-scoring than for low-scoring students. The positive effect of provision of instruction is shown by the fact that the "Instruction provided" line is higher than the "No instruction provided" line. The interaction effect is revealed by the fact that the two lines have very different slopes. If the interaction were not significant, the lines would be more nearly parallel.

In drawing the graph in Figure 13–3, we chose to display the variable "Initial Score Level" on the abscissa. We could have selected "Provision of Instruction" instead. In that case, each initial score level would be represented by a line within the graph. The resulting graph is shown in Figure 13–4.

Despite its different appearance, Figure 13–4 provides the same information as Figure 13–3. The upward (positive) slope of both lines shows the positive effect of provision of instruction. The greater height of the "Initially low scores" line shows the greater average gain of low-scoring students. The difference in slope of the two lines shows the interaction between the independent variables.

Some Other Possible Outcomes of the SAT Study

In our hypothetical SAT gain score study, both main effects and the interaction effect were significant. Let us consider and graph several other possible outcomes.

First, suppose that (1) low-scoring students gained significantly more than high-scoring students, whether they received instruction or not; (2) instruction resulted in significantly greater gains than no instruction in both low-scoring and high-scoring students; but (3) the difference between gains with and without instruction was the same in low-scoring and high-scoring students. In other words, the main effects of initial score level and instruction are significant, but the interaction is not. The resulting mean gains might be those shown in Figure 13–5. Note that both lines in the graph have a negative slope (main effect of initial score level), and the "Instruction provided" line is higher than the "No instruction provided" line (main effect of instruction); but the two lines are parallel (no interaction effect).

Finally, let us consider one other possible outcome. Suppose that low-scoring students gained 20 points, on the average, with instruction, but none without instruction, whereas high-scoring students gained 20 points, on the average, without instruction but none with instruction (a very unlikely result!). The means are displayed and graphed in Figure 13–6.

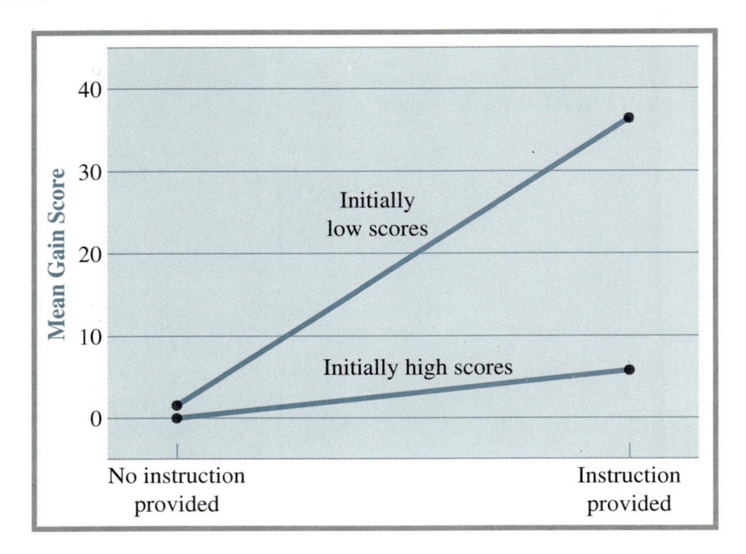

Figure 13–4

Another graph of the results of the SAT gain score study.

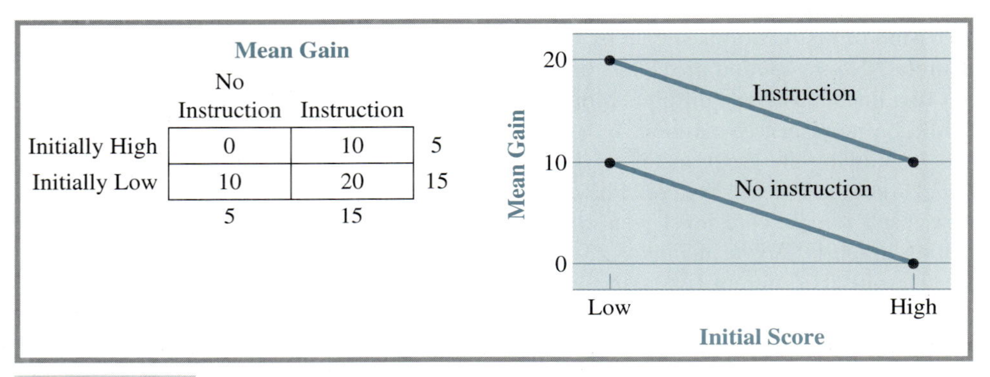

Mean Gain

	No Instruction	Instruction	
Initially High	0	10	5
Initially Low	10	20	15
	5	15	

Figure 13–5 Another possible outcome of the SAT study.

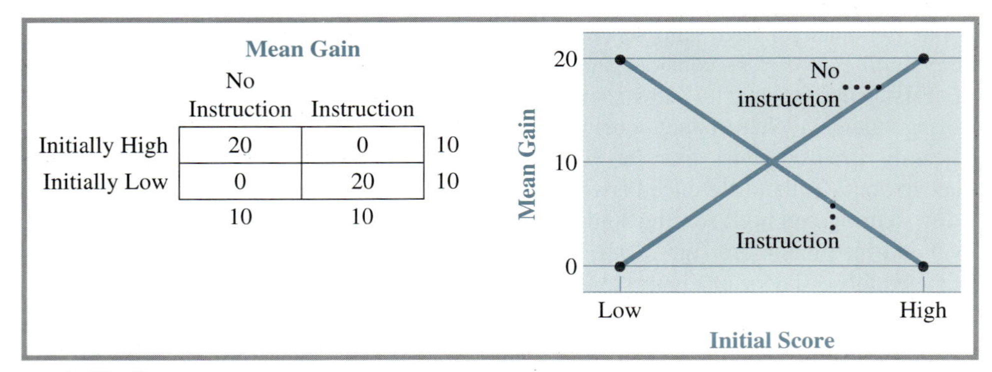

Mean Gain

	No Instruction	Instruction	
Initially High	20	0	10
Initially Low	0	20	10
	10	10	

Figure 13–6 Still another possible—but unlikely—outcome of the SAT study.

The ANOVA of these curious results would show no main effect for initial score level or for provision of instruction. As you can see, the mean gain of initially low and initially high students, over both instructional conditions combined, is the same (mean

= 10), and the mean gain with and without instruction, over initially low and high students combined, is the same (mean = 10). But the interaction effect would almost certainly be significant. If results like these were obtained, the English teachers would do well not only to provide instruction for low-scoring students but to actively withhold it from students whose initial scores were high.

EXERCISE 13–4

For the SAT study, make up a table of means that might result under each of the following conditions. In each case, draw a graph of the results.

1. Significant main effect of provision of instruction (higher mean gain with than without instruction). No main effect of initial score level and no interaction.
2. Significant main effect of provision of instruction and a significant interaction, but no effect of initial score level.

Analysis of a 2 × 3 Factorial Design

As another illustration of two-way ANOVA, we analyze data from a factorial design with two levels of one independent variable and three levels of the other—a 2 × 3 design. The example involves a study of visual tracking of blips on a radar screen. The two independent variables are complexity of display (A_1 = simple, A_2 = complex) and speed of movement of the blips (B_1 = slow, B_2 = moderate, B_3 = fast). Note that $a = 2$ and $b = 3$. The investigator has randomly assigned four subjects to each combination of complexity and speed. Table 13–6 shows the number of errors made by each subject. We wish to know whether the number of errors is affected by complexity, by speed, and by the interaction between complexity and speed.

Table 13–6 Number of Errors as a Function of Complexity and Speed in a 2 × 3 Factorial Design

		Speed			
		Slow (B_1)	Moderate (B_2)	Fast (B_3)	Sum
Complexity	Simple (A_1)	1 6 1 2 $10 = \Sigma A_1B_1$ $n = 4$	11 2 6 4 $23 = \Sigma A_1B_2$ $n = 4$	10 7 7 13 $37 = \Sigma A_1B_3$ $n = 4$	$70 = \Sigma A_1$ $12 = bn$
	Complex (A_2)	4 11 12 10 $37 = \Sigma A_2B_1$ $n = 4$	9 9 9 8 $35 = \Sigma A_2B_2$ $n = 4$	13 10 7 15 $45 = \Sigma A_2B_3$ $n = 4$	$117 = \Sigma A_2$ $12 = bn$
	Sum	$47 = \Sigma B_1$ $8 = an$	$58 = \Sigma B_2$ $8 = an$	$82 = \Sigma B_3$ $8 = an$	$187 = \Sigma X_T$ $24 = N_T$

$\Sigma X_T^2 = 1^2 + 6^2 + \cdots + 7^2 + 15^2 = 1817$

The calculations are shown below.

1. Calculations of sums of squares (formulas for calculation can be found in Table 13–3):

$$SS \text{ total} = 1817 - \frac{(187)^2}{24} = 359.96$$

$$SS \text{ between} = \frac{(10)^2}{4} + \frac{(23)^2}{4} + \frac{(37)^2}{4} + \frac{(37)^2}{4} + \frac{(35)^2}{4} + \frac{(45)^2}{4} + \frac{(187)^2}{24} = 197.21$$

$$SS \text{ within} = 359.96 - 197.21 = 162.75$$

$$SS_A = \frac{(70)^2}{12} + \frac{(117)^2}{12} - \frac{(187)^2}{24} = 92.04$$

$$SS_B = \frac{(47)^2}{8} + \frac{(58)^2}{8} + \frac{(82)^2}{8} + \frac{(187)^2}{24} = 80.08$$

$$SS_{AB} = 197.21 - 92.04 - 80.08 = 25.09$$

2. Calculation of degrees of freedom:

$$df_A = 2 - 1 = 1$$

$$df_B = 3 - 1 = 2$$

$$df_{AB} = (1)(2) = 2$$

$$df \text{ within} = 24 - 6 = 18$$

3. Calculation of mean squares:

$$MS_A = \frac{92.04}{1} = 92.04$$

$$MS_B = \frac{80.08}{2} = 40.04$$

$$MS_{AB} = \frac{25.09}{2} = 12.545$$

$$MS \text{ within} = \frac{162.75}{18} = 9.04$$

4. Calculation of F ratios:

$$F_A = \frac{92.04}{9.04} = 10.18$$

$$F_B = \frac{40.04}{9.04} = 4.43$$

$$F_{AB} = \frac{12.545}{9.04} = 1.39$$

To evaluate F_A, we compare it with critical values of F for $df = (1, 18)$; the critical values are 4.41 ($\alpha = .05$) and 8.28 ($\alpha = .01$). Because F_A is greater than 8.28, the effect of Variable A (complexity) is significant at the .01 level. Both F_B and F_{AB} are evaluated by comparing them with critical values of F for $df = (2, 18)$; the critical values are 3.55 ($\alpha = .05$) and 6.01 ($\alpha = .01$). F_B falls between the critical value for $\alpha = .05$ and $\alpha = .01$. Therefore, the effect of Variable B (speed) is significant at the .05 level. Because F_{AB} is

Table 13–7 **Summary Table of the Analysis of the Visual Tracking Experiment**

Source of Variation	SS	df	MS	F
Complexity	92.04	1	92.04	10.18**
Speed	80.08	2	40.04	4.43*
Complexity × speed	25.09	2	12.545	1.39
Within	162.75	18	9.04	
Total	359.96	23		

*$p < .05$.
**$p < .01$.

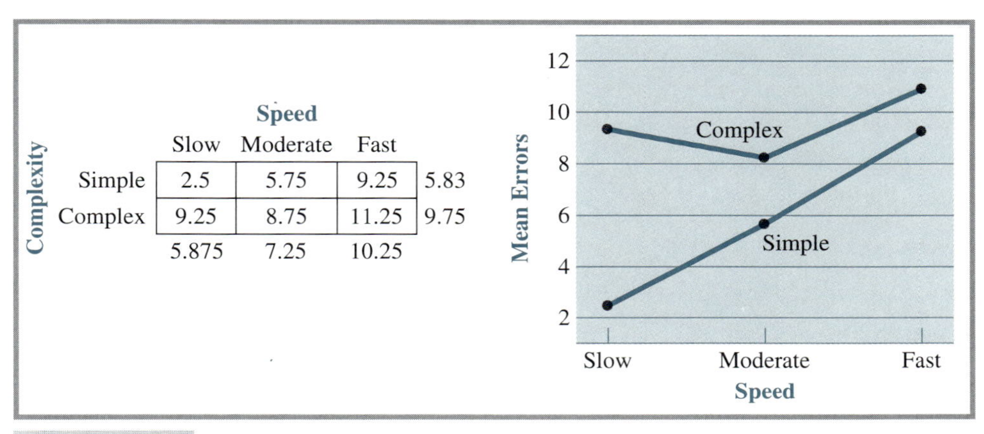

Figure 13–7 **Table of means and graph of the visual tracking experiment results.**

smaller than 3.55, the interaction is not significant at the .05 level. The results are summarized in Table 13–7.

Figure 13–7 presents a table and a graph of the means. As you can see, the mean number of errors was significantly greater for the complex than the simple display, and the number of errors increased with speed. Although the slopes of the two lines appear to be somewhat different, the difference, reflected by the interaction between complexity and speed, was not significant. That is, the rate of increase of errors with speed did not differ significantly at the two levels of complexity.

EXERCISE 13–5

1. Graph the results of the visual tracking study with complexity rather than speed on the abscissa. Compare the graph with Figure 13–7. Which graph do you like better? Why?
2. A developmental psychologist conducted an experiment to determine the effects of type of reward (money, candy, praise) and frequency of reward (every response, every other response, every third response) on children's sharing behavior. Five children were assigned to each combination of experimental treatments. The group means follow.

EXERCISE

13–5

		Frequency of Reward (B)		
		Every Response	*Every Other Response*	*Every Third Response*
Type of Reward (A)	*Money*	4.4	4.0	3.4
	Candy	4.0	3.6	3.0
	Praise	4.0	4.4	5.0

The psychologist calculated the following sums of squares:

SS total = 31.98

SS between = 14.18

$SS_A = 6.58$

$SS_B = 0.84$

a. Complete the analysis and prepare a summary table of the results.
b. Graph the results and describe the effects in your own words.

Assumptions, Power, and Strength of Relationships

Assumptions

The assumptions underlying the two-way ANOVA for the factorial design are the same as those for the one-way ANOVA—normal populations and equal population variances ($\sigma^2_{A1B1} = \sigma^2_{A1B1} = \cdots = \sigma^2_{AaBb}$). Like the one-way ANOVA, the two-way ANOVA is a robust test; the accuracy of conclusions reached by the test is not much changed if the underlying assumptions are violated, unless n is very small and/or the population variances are very different.

The computational procedures presented in this chapter require an equal sample size for all AB combinations. If sample sizes are unequal, the calculations become more complex. Most computer packages for two-way ANOVA can handle both equal and unequal sample sizes.

Power

The power of the two-way ANOVA—the probability that the test will reject a false null hypothesis—is affected by the same factors that affect the power of one-way ANOVA:

1. The greater the difference between population means, the greater the power of the test. For example, an ANOVA has greater power to detect the main effect of A if μ_{A1}, μ_{A2}, $\cdots$, μ_{Aa} are very different from one another than if they are similar.
2. The smaller the variance of the populations, the greater the power of the test. For example, an ANOVA is more powerful if the variance (σ^2) of each population is 50 than if σ^2 is 100.
3. The larger the value of alpha, the greater the power of the test. For example, a test with $\alpha = .05$ is more powerful than one with $\alpha = .01$, if other factors are equal.
4. The power of the test increases as sample size increases. For example, if a true effect exists, the probability of detecting it is greater with five subjects per group than four and greater still with ten subjects per group.

Measuring the Strength of Relationships

The strength of the relation between each independent variable and the dependent variable is measured in the two-way ANOVA, as in the one-way ANOVA, by calculating $\hat{\omega}^2$, the estimated proportion of variance accounted for. In the two-way ANOVA, three values of $\hat{\omega}^2$ can be calculated—one for Variable A, one for Variable B, and one for the interaction. The formulas for $\hat{\omega}^2$ are the following:

$$\hat{\omega}^2_{\text{A}} = \frac{SS_{\text{A}} - (a-1)(MS \text{ within})}{MS \text{ within} + SS \text{ total}}$$

$$\hat{\omega}^2_{\text{B}} = \frac{SS_{\text{B}} - (b-1)(MS \text{ within})}{MS \text{ within} + SS \text{ total}}$$

$$\hat{\omega}^2_{\text{AB}} = \frac{SS_{\text{AB}} - (a-1)(b-1)(MS \text{ within})}{MS \text{ within} + SS \text{ total}}$$

Let us calculate $\hat{\omega}^2$ for all three effects in the 2×2 SAT preparation study. (The needed sums of squares and mean squares can be found in Table 13–4.)

Initial score level (A): $$\hat{\omega}^2 = \frac{1280 - (1)(130.625)}{130.625 + 6495} = .17$$

We estimate that initial score level accounts for 17% of the SAT gain score variance.

Provision of instruction (B): $$\hat{\omega}^2 = \frac{2000 - (1)(130.625)}{130.625 + 6495} = .28$$

We estimate that provision of instruction accounts for 28% of the SAT gain score variance.

Interaction (A × B): $$\hat{\omega}^2 = \frac{1125 - (1)(130.625)}{130.625 + 6495} = .15$$

We estimate that the interaction between initial score level and provision of instruction accounts for 15% of the SAT gain score variance.

Note that the three sources of variation combined account for an estimated .17 + .28 + .15 = .60 (60%) of the SAT gain score variance. The remaining 40% is due to random sampling "error," in other words, to factors that were not measured, manipulated, or controlled in the study.

EXERCISE 13–6

In the 2×3 visual tracking experiment, the variables' complexity and speed had significant effects, but the interaction did not. Calculate $\hat{\omega}^2$ for complexity and speed. The needed sums of squares and mean squares can be found in Table 13–7.

Multiple Comparisons in Two-Way ANOVA

The visual tracking study resulted in a significant effect for speed. This tells us that the mean number of errors made by subjects under slow, moderate, and fast displays vary more than would be expected by chance alone. But just where do the differences lie? For example, is the difference between slow and moderate speeds significant? Did the fast condition result in significantly more errors, on the average, than the slow and moderate conditions combined?

Multiple comparisons can be carried out to answer questions like these. If the comparisons have been planned in advance and are small in number, each one can be evaluated using the same level of alpha as that used in the ANOVA or, more conservatively, the test alpha divided by the number of comparisons. If the comparisons were not preplanned, then post hoc tests, such as the Scheffé test or Tukey's "honestly significant difference" (*HSD*) test, can be used to control the Type I error rate.

Tests for comparisons after a two-way ANOVA use the same rationale and methods as after a one-way ANOVA. The only difference is that, when formulas include values of $n_1, n_2, \ldots,$ these refer to the number of subjects in each of the groups being compared, not to the number in each combination of A and B. For example, in the visual tracking study, although the number of subjects in each complexity-speed combination (n) was 4, the number of subjects at each of the three speed levels was 8. In comparisons among speed levels, the value 8 is substituted for $n_1, n_2, \ldots,$ rather than the number 4.

We will conduct several types of comparisons of the means of the three speed conditions.

Illustration of Planned Comparisons

Let us suppose that the following two orthogonal comparisons were planned in advance of data collection: slow versus moderate speed and slow plus moderate versus fast. The weights assigned to the means in the two comparisons are shown here:

	Weights (w)		
Comparison	M_1 **(slow)**	M_2 **(moderate)**	M_3 **(fast)**
Slow vs. moderate	1	–1	0
Average of slow and moderate vs. fast	1	1	–2

To test each comparison, we calculate F by the formula

$$F = \frac{C^2}{MS \text{ within}\left(\frac{w_1^2}{n_1} + \frac{w_2^2}{n_2} + \frac{w_3^2}{n_3}\right)}$$

Then we evaluate F by comparing it with critical values of F for $df = (1, df \text{ within})$. Because the speed effect was significant at the .05 level, we will use $\alpha = .05$. The number of degrees of freedom within is 18 (see Table 13–7).

In each comparison, C is equal to $w_1M_1 + w_2M_2 + w_3M_3$; n_1, n_2, and n_3 refer to the number of subjects at each level of speed (8); and *MS* within is that calculated in the ANOVA (9.04). The three means are M_1 (slow) = 5.875, M_2 (moderate) = 7.25, and M_3 (fast) = 10.25.

1. Comparison of slow versus moderate

$$C = (1)(5.875) + (-1)(7.25) + 0(10.25) = -1.375$$

$$F = \frac{(-1.375)^2}{9.04\left(\frac{(1)^2}{8} + \frac{(-1)^2}{8} + \frac{0^2}{8}\right)} = \frac{1.89}{2.26} = 0.84$$

Because F is less than one, it is not significant. We conclude that the mean number of errors did not differ significantly under slow and moderate speed conditions.

2. Comparison of slow plus moderate versus fast

$$C = (1)(5.875) + (1)(7.25) + (-2)(10.25) = -7.375$$

$$F = \frac{(-7.375)^2}{9.04\left(\frac{(1)^2}{8} + \frac{(1)^2}{8} + \frac{(-2)^2}{8}\right)} = \frac{54.39}{6.78} = 8.02$$

With $df = (1, 18)$, the critical values of F for $\alpha = .05$ and $.01$ are 4.41 and 8.28, respectively. Because the comparison F falls between the two critical values, we conclude that the difference between the mean number of errors at slow and moderate speeds differs significantly at the .05 level from the mean number of errors made at the fast speed; significantly more errors were made at the fast speed.

Illustration of Scheffé Test

Let us suppose that no particular comparisons were planned in advance. However, after examining the table of means, the researcher decides on the following comparisons: slow versus fast and slow plus moderate versus fast. Because the comparisons were not preplanned (nor are they orthogonal), a post hoc test is appropriate. And, because one of the comparisons is not a pairwise one, the Scheffé test is a good choice.

In the Scheffé test, F ratios are calculated for each comparison by the same formula as in planned comparisons. But the F is evaluated against a special critical value obtained by multiplying the critical value for the level of significance reached in the ANOVA (in this case, .05) by $k - 1$, where k is the number of means being compared. In the ANOVA, the critical value of F for $\alpha = .05$ and $(2, 18)$ df was 3.55. Because there are three speed conditions, the critical value for $\alpha = .05$ in the Scheffé test is $(3 - 1)(3.55) = 7.1$.

In the comparison of slow versus fast, $w_1 = 1$, $w_2 = 0$, and $w_3 = -1$. Because M_1 (slow) = 5.875, M_2 (moderate) = 7.25, and M_3 (fast) = 10.25:

$$C = (1)(5.875) + 0(7.25) + (-1)(10.25) = -4.375$$

$$F = \frac{(-4.375)^2}{9.04\left(\frac{(1)^2}{8} + \frac{(0)^2}{8} + \frac{(-1)^2}{8}\right)} = \frac{19.14}{2.26} = 8.45$$

The calculated F is greater than 7.1. We conclude that significantly more errors were made at the fast speed than at the slow speed.

The comparison of slow plus moderate versus fast is the same as the second of the comparisons illustrated in the previous section (planned comparisons). Therefore, the value of F is the same: 8.02. Because F is greater than the critical value, 7.1, we conclude that the mean number of errors at the fast speed was significantly greater than the mean number of errors at the slow and moderate speeds combined.

Illustration of Tukey *HSD* Test

Suppose that having examined the table of means, the investigator decides to compare each pair of speed condition means. The appropriate test is the Tukey *HSD* test.

In the Tukey test, Table D–6 in Appendix D is used to find the critical value of q (q_{crit}) for the combination of the chosen α, k (the number of means to be compared), and df within of the ANOVA. Then the *HSD* ("honestly significant difference") is calculated by the formula

$$HSD = 3.61\sqrt{\frac{9.04}{8}} = 3.84$$

where n = number of subjects in each of the comparison groups. *HSD* is the minimum significant pairwise difference between means.

For the comparisons of the three speed conditions, $\alpha = .05$, $k = 3$, and df within = 18. According to Table D–6, $q_{crit} = 3.61$. *MS* within = 9.04 and $n = 8$. Therefore,

$$HSD = 3.61\sqrt{\frac{9.04}{8}} = 3.84$$

Differences of 3.84 or more between means are significant at the .05 level. The difference between moderate and slow speeds is 1.375; the difference between fast and slow speeds is 4.375; the difference between fast and moderate speeds is 3.00. Only the difference between fast and slow speeds exceeds *HSD* and is significant at the .05 level.

Main Effects and Interactions as Orthogonal Comparisons

We have discussed multiple comparisons as an adjunct to a two-way ANOVA. However, the effects of A, B, and A × B in the ANOVA may themselves be interpreted as a set of orthogonal comparisons. This is easiest to demonstrate in the 2 × 2 design; we restrict our discussion to this design.

A 2 × 2 factorial design includes four treatment combinations: A_1B_1, A_1B_2, A_2B_1, and A_2B_2. For the moment, let us treat these as four distinct treatments, as in a one-way ANOVA. Now we identify the following set of three orthogonal comparisons:

	Weights			
Comparison	A_1B_1	A_1B_2	A_2B_1	A_2B_2
1. (A_1B_1 and A_1B_2) vs. (A_2B_1 and A_2B_2)	1	1	−1	−1
2. (A_1B_1 and A_2B_1) vs. (A_1B_2 and A_2B_2)	1	−1	1	−1
3. (A_1B_1 and A_2B_2) vs. (A_1B_2 and A_2B_1)	1	−1	−1	1

Let us examine each comparison more closely. The first compares the two A_1 means against the two A_2 means; in other words, it is a comparison of the mean of A_1 and the mean of A_2. Thus, it is a test for the main effect of A. Similarly, the second comparison compares the two B_1 means against the two B_2 means; the second comparison tests for the main effect of B.

To help us interpret the third comparison, we express it in algebraic terms:

$$\begin{aligned}\text{Comparison 3} &= A_1B_1 - A_1B_2 - A_2B_1 + A_2B_2 \\ &= (A_1B_1 - A_1B_2) - (A_2B_1 - A_2B_2) \\ &= A_1(B_1 - B_2) - A_2(B_1 - B_2)\end{aligned}$$

In other words, the third comparison compares the difference between B_1 and B_2 under Condition A_1 with the difference between B_1 and B_2 under Condition A_2. Thus, this comparison corresponds to the interaction between A and B.

We have demonstrated that the tests of the two main effects and the interaction in a 2 × 2 analysis of variance are a set of orthogonal comparisons among the four means. The argument can be extended to analyses with a or b or both greater than two; in all cases, the three tests of the two-way ANOVA are, in fact, orthogonal subanalyses of the overall analysis of variance.

EXERCISE 13–7

In Exercise 13–5, you completed an ANOVA for a 3 × 3 factorial design. Use the results to answer the following questions.

1. The effect of type of reward was significant at the .01 level, and the interaction was significant at the .05 level. Calculate the estimated proportion of sharing behavior variance accounted for by the type of reward and by the interaction.
2. Suppose that you had planned the following two comparisons between types of rewards in advance: (a) tangible reward (money or candy) versus intangible reward (praise); (b) money versus candy. Calculate F for each comparison and evaluate it for significance at the .01 level.
3. Use the Scheffé test (with $\alpha = .01$) to test the same comparisons.
4. Test all pairwise differences among the three types of rewards using the Tukey *HSD* test.

Analysis of Variance for Repeated Measures

Suppose that you want to compare the effects of three noise levels (low, moderate, high) on the productivity of factory workers. If you were to assign different, independent samples of workers to the three noise levels, the appropriate method of analysis would be a simple one-way analysis of variance. But, instead, you decide to test a single group of workers under all three noise levels (a repeated measures design). The three samples of data that result are not independent but correlated. The appropriate analysis is an analysis of variance for repeated measures.

Recall from Chapter 11 that correlated samples can be formed by matching as well as by repeated measures. The ANOVA to be discussed in this section is applicable to both repeated measures on a single group of subjects and to matched sets. Although the discussion and the examples that follow deal with repeated measures, the procedures would be the same for matched sets.

As in one-way ANOVA, we use k to represent the number of treatments. In the study of noise levels, for example, $k = 3$. We use n to represent the number of subjects. As there are k scores for each of n subjects, the total number of scores (N_T) is equal to kn. For example, if 10 workers participate in the noise-level study, $N_T = (3)(10) = 30$.

Data from a repeated measures study are commonly displayed in a two-way table, as shown in Table 13–8. Each row of the table includes scores for one of the n subjects, and each column includes scores under one of the k treatments. Each cell contains the score of one subject under one treatment. Symbols for column and row sums, needed for the calculation of sums of squares, are also shown in the table.

Hypotheses and Assumptions

The ANOVA for the repeated measures design tests the null hypothesis that the population means under all k treatments are equal against the alternative hypothesis that not all the means are equal.

H_o: $\mu_1 = \mu_2 = \cdots \mu_k$
H_o: Not all μ are equal

Table 13–8 Display of Data from a Repeated Measures Design

		Treatments (T)					
		1	2	3	....	k	Sums
	1						ΣS_1
	2						ΣS_2
	3						ΣS_3
	.						.
	.						.
Subjects (S)	.						.
	.						.
	.						.
	.						.
	.						.
	n						ΣS_n
	Sums	ΣT_1	ΣT_2	ΣT_3		ΣT_k	

H_o and H_A can also be stated in terms of treatment effects:

H_o: There is no real (population) treatment effect
H_A: There is a real treatment effect

Three assumptions underlie the test of the null hypothesis. The first two are the same as in one-way ANOVA and ANOVA for the factorial design.

1. The k treatment populations are normal.
2. The variances of the k treatment populations are equal.

However, the analysis of variance for repeated measures has one additional assumption:

3. The covariances of all pairs of treatments are equal.

In effect, this means that the correlation between scores under any two treatments is assumed to be the same as the correlation between scores under any other two treatments. For example, if $k = 3$ and if the correlation between Treatments 1 and 2 in the population (ρ_{12}) is .6, it is assumed that ρ_{13} and ρ_{23} are also equal to .6.

As is the case with the other ANOVAs we have considered, the repeated measures ANOVA is relatively robust in the face of violations of the first two assumptions. Violation of the third assumption can have more serious effects; specifically, the probability of a Type I error may be substantially increased. If intertreatment covariances are known or believed to be unequal in a given case, an adjustment can be made in the analysis. Discussion of the procedure for doing so is beyond the scope of this book and can be found in more advanced texts (e.g., Hays 1988). Most computer programs for repeated measures ANOVA include tests for equality of covariances and incorporate the needed adjustment in the analysis.

Partitioning Sums of Squares and Degrees of Freedom

In the data set from a repeated measures study, the variation among the N_T scores is, in part, due to differences *between* subjects and, in part, to differences *within* subjects. Thus, to begin with, the total sum of squares (*SS* total) can be partitioned into a *sum of squares between subjects (SS subjects)* and a *sum of squares within subjects (SS within subjects)*. The within-subject variation, however, can be further subdivided. Part of the variation of scores within subjects is due to systematic effects of the independent variable (the treatments), and part is due to individual differences in the ways subjects respond to treatments (the interaction between subjects and treatments). Thus, the sum of squares within subjects is further partitioned into a *treatment sum of squares (SS treatments)* and a *subjects × treatments sum of squares (SS subjects × treatments)*. The two-stage partitioning of the total sum of squares is illustrated in Figure 13–8.

The subjects × treatments sum of squares is commonly referred to as the *residual sum of squares (SS residual)*. Thus, the complete partitioning of the total sum of squares is as follows:

$$SS\text{ total} = SS\text{ subjects} + SS\text{ treatments} + SS\text{ residual}$$

The partitioning of the total degrees of freedom parallels the partitioning of the total sum of squares: *df total = df subjects + df treatments + df residual*, where

$$\begin{aligned} df\text{ total} &= N_T - 1 = kn - 1 \\ df\text{ subjects} &= n - 1 \\ df\text{ treatments} &= k - 1 \\ df\text{ residual} &= (df\text{ subjects})(df\text{ treatments}) = (n-1)(k-1) \end{aligned}$$

For example, if 10 subjects are used in the noise-level study

$$\begin{aligned} df\text{ total} &= (3)(10) - 1 = 30 - 1 = 29 \\ df\text{ subjects} &= 10 - 1 = 9 \\ df\text{ treatments} &= 3 - 1 = 2 \\ df\text{ residual} &= (9)(2) = 18 \end{aligned}$$

Note that *df* subjects + *df* treatments + *df* residual = 9 + 2 + 18 = 29 = *df* total.

Testing the Null Hypothesis

Two mean squares are calculated in the repeated measures analysis of variance, the **treatment mean square (*MS* treatments)** and the **residual mean square (*MS* resid-**

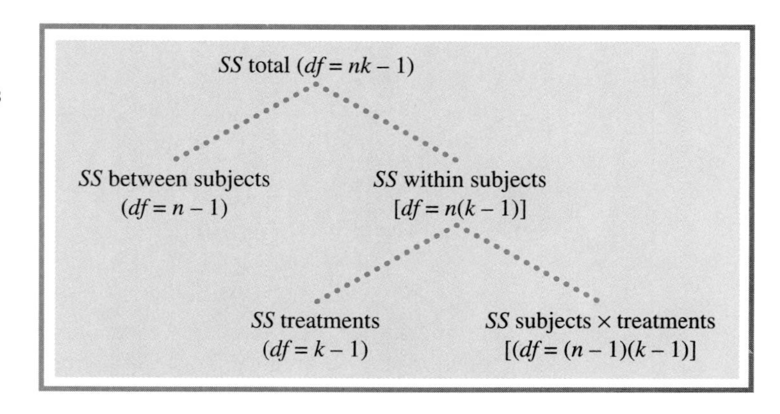

Figure 13–8

Partitioning of the sums of squares and degrees of freedom in the repeated measures ANOVA.

ual). Each mean square is obtained by dividing a sum of squares by the associated degrees of freedom.

$$MS \text{ treatments} = \frac{SS \text{ treatments}}{df \text{ treatments}}$$

$$MS \text{ residual} = \frac{SS \text{ residual}}{df \text{ residual}}$$

It can be shown that if the null hypothesis is true (i.e., if all the population treatment means are equal) and if the underlying assumptions of the analysis are met, the average value of the treatment mean square over all possible samples of size n is equal to the average value of the residual mean square.

If H_o is true, MS treatments = MS residual, on the average.

If, however, the null hypothesis is false and the population treatment means are not all equal, the average value of the treatment mean square over all possible samples of size n is greater than the average value of the residual mean square.

If H_o is not true, MS treatments > MS residual, on the average.

To find out whether MS treatments is significantly greater than MS residual, we calculate F:

$$F = \frac{MS \text{ treatments}}{MS \text{ residual}}$$

The probability of F is found by comparing it with critical values of F for $[k - 1, (n - 1)(k - 1)]$ degrees of freedom. If $p < \alpha$, the null hypothesis is rejected, and the conclusion is drawn that at least one treatment has an effect different from the others. If, on the other hand, $p > \alpha$, the null hypothesis is not rejected. The evidence is regarded as insufficient to conclude that any true differences exist among the treatment means.

EXERCISE 13–8

1. Each of 20 rats is tested under five motivational conditions. Calculate df total, df treatments, df subjects, and df residual.
2. Data from factorial designs and repeated measures are both commonly displayed in two-way tables. In the factorial design, the rows are levels of Factor A and the columns are levels of Factor B. In the repeated measures design, the rows are subjects and the columns are treatments. The partitioning of the total sum of squares in the factorial design is:

$$SS \text{ total} = \underset{\substack{\uparrow \\ SS \text{ rows}}}{SS_A} + \underset{\substack{\uparrow \\ SS \text{ columns}}}{SS_B} + \underset{\substack{\uparrow \\ SS \text{ rows} \times \text{columns}}}{SS_{A \times B}} + SS \text{ within}$$

In the repeated measures design, the partitioning is

$$SS \text{ total} = \underset{\substack{\uparrow \\ SS \text{ rows}}}{SS \text{ subjects}} + \underset{\substack{\uparrow \\ SS \text{ columns}}}{SS \text{ treatments}} + \underset{\substack{\uparrow \\ SS \text{ rows} \times \text{columns}}}{SS \text{ residual}}$$

Why doesn't the repeated measures partitioning include a component comparable to the factorial design's SS within?

Computational Procedures

Data from a hypothetical study on the effect of noise level on worker productivity are shown in Table 13–9. We will use this example to describe and illustrate procedures for calculating the needed sums of squares. Then we will conduct the test to determine whether the variation of the three treatment means is greater than would be expected by chance.

As in other analyses of variance, the total sum of squares is calculated first. The formula is the same as in the one-way and factorial design ANOVAs.

$$SS\text{ total} = \Sigma X_T^2 - \frac{(\Sigma X_T)^2}{N_T} \qquad \textit{Formula for calculating SS total}$$

In the noise-level study, N_T is equal to 30. The sum of the 30 scores, ΣX_T, is equal to 275, and the sum of the squared scores, ΣX_T^2, is equal to 3213. Therefore,

$$SS\text{ total} = 3213 - \frac{(275)^2}{30} = 692.17$$

To calculate *SS* subjects, we use the sum of the *k* scores for each subject, ΣS. The formula is

$$SS\text{ subjects} = \frac{(\Sigma S_1)^2}{k} + \frac{(\Sigma S_2)^2}{k} + \cdots + \frac{(\Sigma S_n)^2}{k} - \frac{(\Sigma X_T)^2}{N_T} \qquad \textit{Formula for calculating SS subjects}$$

In the noise-level study

$$SS\text{ subjects} = \frac{(29)^2}{3} + \frac{(18)^2}{3} + \frac{(42)^2}{3} + \cdots + \frac{(42)^2}{3} - \frac{(275)^2}{30} = 405.50$$

To calculate *SS* treatments, we use the sum of the *n* scores for each treatment, ΣT. The formula is

$$SS\text{ treatments} = \frac{(\Sigma T_1)^2}{n} + \frac{(\Sigma T_2)^2}{n} + \cdots + \frac{(\Sigma T_k)^2}{n} - \frac{(\Sigma X_T)^2}{N_T} \qquad \textit{Formula for calculating SS treatments}$$

In the noise-level study

Table 13–9 Productivity of 10 Workers Under Three Noise Levels

	Noise Level				
Workers	**Low**	**Moderate**	**High**	**Sum**	**Mean**
1	12	8	9	29	9.7
2	9	6	3	18	6.0
3	13	18	11	42	14.0
4	6	8	8	22	7.3
5	19	12	3	34	11.3
6	7	5	1	13	4.3
7	4	5	5	14	4.7
8	6	7	7	20	6.7
9	14	18	9	41	13.7
10	19	10	13	42	14.0
Sum	109	97	69	$275 = \Sigma X_T$	
Mean	10.9	9.7	6.9		

$$SS \text{ treatments} = \frac{(109)^2}{10} + \frac{(97)^2}{10} + \frac{(69)^2}{10} - \frac{(275)^2}{30} = 84.27$$

Because *SS* subjects + *SS* treatments + *SS* residual = *SS* total, *SS* residual can be obtained by subtraction:

$$SS \text{ residual} = SS \text{ total} - SS \text{ subjects} - SS \text{ treatments}$$

Formula for calculating SS *residual*

In the noise-level study

$$SS \text{ residual} = 692.17 - 405.5 - 84.27 = 202.40$$

Now we can complete the analysis of the data from the noise-level study. We calculate *MS* treatments and *MS* residual. (Remember: *df* treatments = $k - 1 = 2$, and *df* residual = $(n - 1)(k - 1) = (9)(2) = 18$.)

$$MS \text{ treatments} = \frac{SS \text{ treatments}}{df \text{ treatments}} = \frac{84.27}{2} = 42.135$$

$$MS \text{ residual} = \frac{SS \text{ residual}}{df \text{ residual}} = \frac{202.4}{18} = 11.244$$

Then we calculate *F*:

$$F = \frac{MS \text{ treatments}}{MS \text{ residual}} = \frac{42.135}{11.244} = 3.747$$

To find out whether *F* is significantly greater than one, we compare it with the critical values of *F* for (2, 18) degrees of freedom. According to Table D–5 in Appendix D, the critical value for $\alpha = .05$ is 3.55, and the critical value for $\alpha = .01$ is 6.01. Because the calculated *F* falls between the two critical values, the result is significant at the .05 level. We conclude that noise level does affect the productivity of factory workers. To describe the nature of the effect, we examine the three means in Table 13–9 and see that productivity decreases as noise level increases.

A summary table of the ANOVA of the noise-level data is presented in Table 13–10. Note that all three sources of variation are listed, as well as their respective degrees of freedom. However, mean squares are calculated only for treatments and residual. The value of *F* is located in the treatment row.

Table 13–10 ANOVA Summary Table for the Noise-Level Study

Source of Variation	*SS*	*df*	*MS*	*F*
Workers (subjects)	405.50	9		
Noise levels (treatments)	84.27	2	42.135	3.747*
Residual (error)	202.40	18	11.244	
Total	692.17	29		

*$p < .05$.

EXERCISE 13–9

1. Here are scores of five children on four verbal puzzles. Conduct an ANOVA to determine whether performance varied significantly over the four puzzles, and present the results in a summary table.

	Puzzle			
Child	**1**	**2**	**3**	**4**
1	3	8	2	4
2	5	7	3	6
3	2	6	0	2
4	4	9	1	5
5	5	5	2	3

2. Each of 25 subjects was tested under six treatments. Complete the summary table of the analysis below and evaluate F.

Source of Variation	*SS*	*df*	*MS*	*F*
Subjects	7,200			
Treatments	4,800			
Residual				
Total	38,160			

Relation Between the Repeated Measures ANOVA and the *t* Test for Correlated Samples

In repeated measures studies comparing just two treatments, the data are commonly analyzed using the t test for correlated samples. However, the repeated measures ANOVA could be applied instead. We will apply both a t test and an ANOVA to a set of data in order to examine the relationship between the two analyses.

Table 13–11 presents data for seven subjects tested under two treatments, T_1 and T_2. Difference scores (D) and sums are listed for each subject. We will conduct, first, a two-tailed t test and then an ANOVA and compare the results. We will use $\alpha = .05$.

Table 13–11 Data for Seven Subjects Tested Under Two Treatments

	Treatment				
Subject	**1**	**2**	***D***	**Sum**	
1	4	3	1	7	
2	3	4	−1	7	
3	8	12	−4	20	$N_T = 14$
4	6	8	−2	14	$\Sigma X_T^2 = 576$
5	2	3	−1	5	
6	4	8	−4	12	
7	5	10	−5	15	
Sum	32	48	−16	$80 = \Sigma X_T$	
Mean	4.57	6.86			

t Test

For a review of computational procedures, see Chapter 11.

$$M_1 = 4.57 \text{ and } M_2 = 6.86$$

$$\sum D = -16 \text{ and } \sum D^2 = 64$$

$$s_{\text{diff}} = \sqrt{\frac{\sum D^2 - \frac{(\sum D)^2}{n}}{n(n-1)}} = \sqrt{\frac{64 - \frac{(-16)^2}{7}}{7(6)}} = .808$$

$$t = \frac{M_1 - M_2}{s_{\text{diff}}} = \frac{4.57 - 6.86}{.808} = -2.83$$

With 6 degrees of freedom, the calculated $|t|$, 2.83, falls between the critical values of t for $\alpha = .05$ and $\alpha = .02$ in a two-tailed test. Since $p < .05$, H_o is rejected. The two treatment means differ significantly at the .05 level.

ANOVA

$$SS \text{ total} = 576 - \frac{(80)^2}{14} = 118.857$$

$$SS \text{ subjects} = \frac{(7)^2}{2} + \frac{(7)^2}{2} + \frac{(20)^2}{2} + \cdots + \frac{(15)^2}{2} - \frac{(80)^2}{14} = 86.857$$

$$SS \text{ treatments} = \frac{(32)^2}{7} + \frac{(48)^2}{7} - \frac{(80)^2}{14} = 18.2857$$

$$SS \text{ residual} = 118.857 - 86.857 - 18.2857 = 13.7143$$

$$df \text{ treatments} = 2 - 1 = 1$$

$$df \text{ residual} = (7-1)(2-1) = (6)(1) = 6 \text{ (this is the same as the } df \text{ in the } t \text{ test)}$$

$$MS \text{ treatments} = \frac{18.2857}{1} = 18.2857$$

$$MS \text{ residual} = \frac{13.7143}{6} = 2.2857$$

$$F = \frac{18.2857}{2.2857} = 8.00$$

With (1, 6) degrees of freedom, the critical values of F for $\alpha = .05$ and .01 are 5.99 and 13.74, respectively. Because F falls between the two critical values, $p < .05$. The two treatment means differ significantly at the .05 level, just as they did in the t test.

Note the following relationships between the two tests:

1. The square of the calculated t, $(-2.83)^2$, is equal to 8.00, the calculated F. $t^2 = F$ not just in this example but whenever a t test and an ANOVA are applied to the same repeated measures data.
2. The square of the critical value of t (two-tailed test) for $\alpha = .05$, $(\pm 2.447)^2$, is equal to 5.99, the critical value of F for $\alpha = .05$. And the square of the critical value of t for $\alpha = .01$ (two-tailed test) $(\pm 3.717)^2$ is equal to 13.74, the critical value of F for $\alpha = .01$. Again, this relationship between the critical values of t and F holds true not

only in this example but in all cases in which a two-tailed t test and an ANOVA are applied to the same repeated measures data.

Because t^2 is always equal to F and (critical value of $t)^2$ is always equal to critical value of F for the same alpha (providing that the t test is two-tailed), the conclusions reached by the two tests are always the same.

As was the case with independent samples, the t test has two advantages over the ANOVA for comparing two means. First, t tests can be one-tailed as well as two-tailed. If the research hypothesis is directional, the t test is a more appropriate test. Second, as pointed out in Chapter 12, because the computations in the ANOVA do not explicitly involve the sample means, it is all too easy to carry out the analysis and draw conclusions without actually examining the means themselves. Because the analysis of variance tests differences between means, interpretation of the results requires a description of the treatment means and their differences (if any).

Comparison of One–Way and Repeated Measures ANOVA

Both the one-way ANOVA and the repeated measures ANOVA analyze data from studies with a single independent (treatment) variable. Which analysis is applied depends on whether the variable varies between subjects (independent samples) or within subjects (correlated samples). Let us consider factors that influence the choice between an independent-samples and a correlated-samples design, as well as the implications of the choice for the analysis.

The choice of an independent versus correlated samples design is often necessitated by the nature of the problem being investigated. For example, a clinical psychologist studying the process of anxiety reduction during the course of a treatment program must necessarily measure the anxiety of a single group of subjects on repeated occasions throughout treatment. Similarly, a study of attitude change as a function of some kind of intervention requires repeated measures of the attitude in question before, during, and after the intervention.

In studies of learning, on the other hand, independent samples are often mandatory. For example, suppose an educator wants to compare three methods of teaching an introductory algebra course. She cannot teach the course to a single class first using Method A, then using Method B, then once again using Method C. Between-subject manipulation of the independent variable is obviously necessary in cases like this.

Repeated measures designs should also be avoided if ***carryover effects*** from one treatment to another are likely—that is, if reactions to later treatments are influenced, in part, by earlier treatments. Consider a study comparing the effects of several drugs on activity level in rats. If each drug has potentially long-lasting effects, it would be unwise to administer the drugs on successive days to a single group of rats; their apparent reaction to the drug they receive on any day after the first may actually be a reaction to that drug plus the ones they received on previous days.

Many research problems, however, can be investigated either with independent samples or with repeated measures. The study of the effect of noise level on worker productivity is an example. Instead of measuring the productivity of one group of workers under all three noise levels, we could have used three groups of workers, each group exposed to just one noise level—an independent-samples design. In cases like this, is there any advantage of one design or the other? Are there statistical factors to consider in choosing between them?

Let us suppose that the noise-level study was conducted by comparing the productivity of three independent samples of workers, each one exposed to just one of the three noise levels. Suppose, further, that each sample included 10 workers and that the productivity scores under each noise level were the same as those in the repeated measures study. The data are shown in Table 13–12. We will analyze the data and compare the results with those of the repeated measures analysis.

Because the three samples are independent, the appropriate method of analysis is one-way ANOVA. The calculations are as follows:

$$SS \text{ total} = \sum X_T^2 - \frac{(\sum X_T)^2}{N_T} = 3213 - \frac{(275)^2}{30} = 692.17$$

$$SS \text{ between} = \frac{(\sum X_1)^2}{n_1} + \frac{(\sum X_2)^2}{n_2} + \frac{(\sum X_3)^2}{n_3} - \frac{(\sum X_T)^2}{N_T}$$

$$= \frac{(109)^2}{10} + \frac{(97)^2}{10} + \frac{(69)^2}{10} - \frac{(275)^2}{30} = 84.27$$

$$SS \text{ within} = SS \text{ total} - SS \text{ between} = 692.17 - 84.27 = 607.90$$

$$MS \text{ between} = \frac{SS \text{ between}}{df \text{ between}} = \frac{SS \text{ between}}{k-1} = \frac{84.27}{2} = 42.135$$

$$MS \text{ within} = \frac{SS \text{ within}}{df \text{ within}} = \frac{SS \text{ within}}{N_T - k} = \frac{607.9}{27} = 22.515$$

$$F = \frac{MS \text{ between}}{MS \text{ within}} = \frac{42.135}{22.515} = 1.87$$

According to the F table, the critical value of F for $\alpha = .05$ and (2, 27) degrees of freedom is 3.35. Because F is less than the critical value, $p > .05$. The result is not significant at the .05 level. We cannot conclude that worker productivity is affected by noise level. The analysis is summarized in Table 13–13.

Table 13–12 Productivity of Three Independent Samples of Workers Under Three Noise Levels

	Noise Level (Group)			
	Low	Moderate	High	
	12	8	9	
	9	6	3	
	13	18	11	
	6	8	8	
	19	12	3	
	7	5	1	
	4	5	5	
	6	7	7	
	14	18	9	
	19	10	13	
	109	97	69	$\Sigma X_T = 275$
ΣX^2	1449	1155	609	$\Sigma X_T^2 = 3213$
				$N_T = 30$

Note: The scores are the same as in Table 13–9.

Table 13–13 ANOVA Summary Table for the Noise-Level Study With Independent Samples

Source of Variation	SS	df	MS	F
Noise levels (between treatments)	84.27	2	42.135	1.87
Within groups (error)	607.90	27	22.515	
Total	692.17	29		

How can we account for the fact that the noise-level effect was significant at the .05 level in the analysis for repeated measures but not in the analysis for independent samples, despite the fact that the three sets of scores were the same in both analyses? To answer this question, compare the sums of squares. First, observe that the total sum of squares is the same in both cases. The sum of squares between treatments is also the same. In the one-way analysis, the difference between *SS* total and *SS* between, the *SS* within, serves as the measure of chance (error) variation. But in the repeated measures ANOVA, the difference between *SS* total and *SS* treatments is further broken down into a sum of squares for subjects and a residual (error) sum of squares. The two partitionings of sums of squares are shown here:

One-way ANOVA: $SS \text{ total} = SS \text{ between treatments} + SS \text{ within treatments}$

Repeated measures ANOVA: $SS \text{ total} = SS \text{ between treatments} + SS \text{ subjects} + SS \text{ residual}$

Thus, the use of repeated measures allows us to remove from the within-treatment variation that part which is due to systematic differences between subjects. When between-subject variation is large, as it is in the repeated measures noise-level data, its removal from the sum of squares within substantially reduces the residual (error) sum of squares. As a result, the error mean square of the repeated measures analysis is smaller than that of the one-way ANOVA, and the value of *F* is larger.

The large value of *SS* subjects in the repeated measures noise-level analysis is accounted for by the strong positive covariation between scores at different treatment levels. As you can see in Table 13–9, most workers with low productivity at one noise level had relatively low productivity at the other noise levels as well (e.g., Workers 4, 6, and 7), and most workers with relatively high productivity at one noise level had relatively high levels of productivity at the other noise levels as well (e.g., Workers 3, 9, and 10). As a result, the sums of scores of individual workers varied widely, from 13 for Worker 6 to 42 for Workers 3 and 10. The large amount of variation between workers resulted in a large *SS* subjects.

Thus, we see that the repeated measures analysis removes from the within-treatment variance that component which is due to covariation between treatments. If the covariation is positive and large, its removal substantially reduces the error mean square and increases the value of *F*. As a result, the repeated measures ANOVA is a more powerful test than the one-way ANOVA.

If, however, there is little or no systematic covariation of scores under different treatments, the repeated measures ANOVA may actually be less powerful than the one-way ANOVA, even if the error mean square is the same in both cases. This is because, given the same values of N_T and k, the degrees of freedom associated with the error mean square is smaller in the repeated measures ANOVA. In the one-way ANOVA, the error *df*, *df* within, is equal to $N_T - k$; in the repeated measures ANOVA, the error *df*, *df* residual, is only equal to $(n - 1)(k - 1) = N_T - k - (n - 1)$. Because the critical value of *F* for a given level of alpha increases as the error *df* decreases, the value of *F* needed for significance at a given level of α is greater for the repeated measures ANOVA.

For example, in the repeated measures ANOVA of the noise-level data, the critical value of F for $\alpha = .05$ and (2, 18) degrees of freedom was 3.55. In the independent samples ANOVA of the same data, the critical value of F for $\alpha = .05$ and (2, 27) degrees of freedom was 3.35. If we had calculated $F = 3.45$ in both cases, the difference between treatment means would be significant in the one-way (independent samples) ANOVA but not in the ANOVA for repeated measures. Thus, if there is little or no covariation among treatments, the repeated measures ANOVA is less powerful than the one-way ANOVA.

If you should find yourself in a position of having to choose between an independent-samples design and a repeated measures design for a study, and assuming that you want the most powerful possible test, you need to consider the factors discussed earlier. If you have reason to expect positive covariance of scores under different treatments, and if carryover effects from one treatment to another are unlikely, a repeated measures design is probably a better choice. If, however, little or no treatment covariance is expected, it may be wise to conduct the study using independent samples instead.

EXERCISE 13–10

1. Table 13–14 presents the same set of scores from a repeated measures design as Table 13–9, but the scores within each column have been randomly reordered. Conduct a repeated measures ANOVA and compare the results with those for the original data. Explain the difference.
2. The t test for correlated samples, like the ANOVA for the repeated measures design, removes from the "error" variation that part which is due to covariation between the two treatments. Refer back to the discussion of the sampling distribution of the difference between means, which underlies the t tests for independent and correlated samples (see Chapter 11). Show how covariation between treatments is removed from the measure of error variation in the t test for correlated samples.

Table 13–14 A Rearrangement of the Data in Table 13–9

	Noise Level		
Worker	Low	Moderate	High
1	19	5	13
2	19	18	7
3	6	8	9
4	13	5	1
5	9	6	9
6	6	10	8
7	14	12	3
8	12	18	11
9	4	7	5
10	7	8	3

Multiple Comparisons in the Repeated Measures Design

Both planned comparisons and post hoc tests are possible with data from a repeated measures study. The procedures are basically the same as for independent samples; the main difference is that when a formula calls for *MS* within, *MS* residual is used instead. We illustrate with the noise level data in Table 13–9. The ANOVA is summarized in Table 13–10.

Example of a Planned Comparison

Assume that you decided, in advance of data collection, to compare the mean level of productivity under low and high noise levels. The comparison, therefore, is a planned one. First, you assign weights to the three means:

	w_1	w_2	w_3
Comparison of M_1 versus M_3	1	0	−1

Next, you calculate the comparison term, C, and F. Since $M_1 = 10.9$, $M_2 = 9.7$, and $M_3 = 6.9$

$$C = w_1M_1 + w_2M_2 + w_3M_3 = 1(10.9) + 0(9.7) + (-1)6.9 = 4.0$$

$$F = \frac{C^2}{MS \text{ residual}\left(\frac{w_1^2}{n} + \frac{w_2^2}{n} + \frac{w_3^2}{n}\right)} = \frac{(4.0)^2}{11.244\left[\frac{(1)^2}{10} + \frac{(0)^2}{10} + \frac{(-1)^2}{10}\right]} = 7.11$$

To evaluate F, you compare it with critical values for df = (1, df residual) = (1, 18). F falls between the critical values for $\alpha = .01$ and $\alpha = .05$. Because $p < .05$, the difference between the mean productivity at low and high noise levels is significant at the .05 level.

Example of Scheffé Test

Suppose that you did not plan any comparisons in advance. You examine the three treatment means and decide on several comparisons, including, but not limited to, pairwise comparisons. Since the family of comparisons includes all possible comparisons, the Scheffé test is an appropriate choice. Let us use the Scheffé test to compare the mean level of productivity under high and low noise levels.

The weights assigned to the three means and the calculation of C and F are the same as in the planned comparison. However, the calculated F is evaluated against a critical value obtained by multiplying the critical value of F for $\alpha = .05$ and (2, 18) degrees of freedom by $k - 1$ or 2. Because the tabled critical value is 3.55, the critical value for the Scheffé test is 2(3.55) = 7.10. The calculated F, 7.11, just exceeds the critical value; the difference between the mean levels of productivity under high and low noise levels is significant at the .05 level.

Example of Tukey *HSD* Test

This time, assume that you are interested only in pairwise post hoc comparisons. The Tukey HSD test is an appropriate choice. First, you calculate $HSD = q_{\text{crit}}\sqrt{\frac{MS \text{ residual}}{n}}$; q_{crit} is found by entering the table of critical values of q (Table D-6 of Appendix D) with $\alpha = .05$, $k = 3$, and df residual = 18. According to the table, $q_{\text{crit}} = 3.61$. Therefore,

$$HSD = 3.61\sqrt{\frac{11.244}{10}} = 3.83$$

All pairwise differences between means that are greater than or equal to 3.83 are significant. Because the difference between the means under high and low noise levels is 4.0, it is significant at the .05 level. But the difference between low and moderate noise levels, 1.2, and the difference between moderate and high noise levels, 2.8, are not significant.

EXERCISE 13–11 For the repeated measures noise-level data, compare the mean productivity under low and moderate noise levels combined, with the mean under a high noise level using (a) a planned comparison and (b) a post hoc test.

Other Analyses of Variance—An Introduction

We have dealt with methods for analyzing data from studies with one or two between-subjects variables or one within-subjects variable. Although these are common designs in the biological and social sciences, they are by no means the only possibilities. Many studies include three or more between-subjects variables or two or more within-subjects variables. Even more common are designs that combine between-subjects variables and within-subjects variables. For example, a researcher studying gender differences in the solution of verbal and mathematical puzzles might give a group of boys and a group of girls both verbal and mathematical puzzles to solve. In this study, gender varies between subjects and type of puzzle varies within subjects.

Despite the great variety of possible designs, the rationale of the ANOVA remains, in principle, the same. In each case, the total data variation is partitioned into components, and one or more values of *F* are calculated by dividing effect mean squares by error mean squares. As an example, Figure 13–9 shows the partitioning of the total sum of squares in the analysis of a design with one between-subjects variable (Variable A) and one within-subjects variable (Variable B). This analysis includes two error mean squares, one used to evaluate the effect of the between-subjects variable and the other to evaluate the effect of the within-subjects variable and the interaction effect.

As the number of variables between and/or within subjects increases, so, too, do the required computations. Today, almost no one conducts any except the simplest analyses by hand. The ready availability of computer programs for microcomputers, as well as mainframes, has enabled researchers to select designs on the basis of their research needs, without regard to computational labor. But the increasing use of ever more complex designs places a special burden on the human user. For, as designs and their analyses grow more complex, the possibility of selecting an inappropriate analysis or of overlooking underlying conditions or of misinterpreting results increases. Now more than ever it is vital that researchers be well grounded in both the theory underlying various statistical procedures and their limitations. Computers have not removed the human element from data analysis. Far from it: The ready availability of computer programs for conducting complex analyses has increased the level of sophistication and expertise needed by the human data analyst.

Figure 13–9

Partitioning the sums of squares in the ANOVA for one between-subjects variable (A) and one within-subjects variable (B).

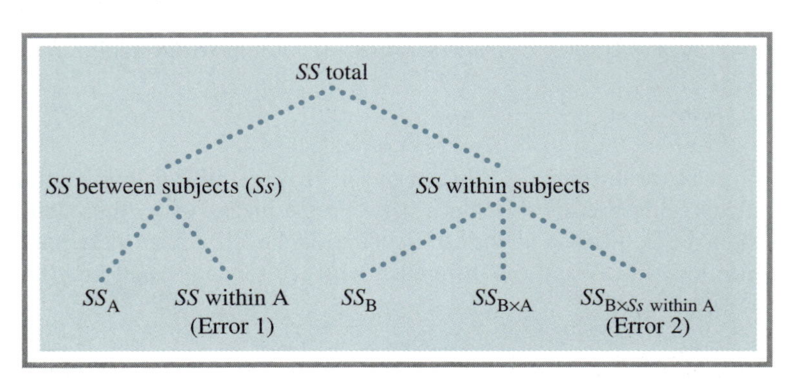

SUMMARY

Designs in which two (or more) between-subjects independent variables are manipulated simultaneously are called ***factorial designs***. Factorial designs are constructed by assigning an equal number of subjects to each combination of treatments on the independent variables.

The analysis of variance applied to data from a two-variable factorial design is called ***two-way analysis of variance***. The two-way ANOVA evaluates not only the ***main effect*** of each variable but also the ***interaction*** between the variables. Two variables are said to interact if the effect of one variable differs, depending on the level of the other.

In the two-way ANOVA, the total sum of squares is partitioned into a sum of squares for the first independent variable (Factor A), a sum of squares for the second independent variable (Factor B), a sum of squares for the interaction between A and B, and a sum of squares within groups.

$$SS \text{ total} = SS_{\text{A}} + SS_{\text{B}} + SS_{\text{AB}} + SS \text{ within}$$

Procedures for calculating sums of squares are summarized in Table 13–3.

The total degrees of freedom are partitioned in the same way:

$$df \text{ total} = df_{\text{A}} + df_{\text{B}} + df_{\text{AB}} + df \text{ within}$$

If we let a stand for the number of levels of Factor A and b for the number of levels of Factor B, the degrees of freedom are as follows:

$$df \text{ total} = N_{\text{T}} - 1$$
$$df_{\text{A}} = a - 1$$
$$df_{\text{B}} = b - 1$$
$$df_{\text{AB}} = (a-1)(b-1)$$
$$df \text{ within} = N_{\text{T}} - ab$$

Mean squares are calculated for A, B, A × B, and within by dividing the sums of squares by their respective degrees of freedom. Then an F ratio is calculated for each effect by dividing the effect mean square by the mean square within. The F ratios are evaluated by comparing them with critical values of F for $df = (df \text{ effect}, df \text{ within})$.

In order to interpret the results of a two-way ANOVA, it is necessary to examine the means. The nature of the effects can be shown in tables or graphs.

The assumptions underlying two-way ANOVA are the same as those underlying one-way ANOVA, as are the factors that affect the power of the test. Like the one-way ANOVA, the two-way ANOVA is robust to violations of the assumptions. The strength of the relation between the dependent variable and each factor can be measured by calculating ω^2, the estimated proportion of variance accounted for by the factor. Procedures for conducting planned comparisons and post hoc tests are essentially the same as those in one-way ANOVA.

If a study includes a single independent variable that varies within rather than between subjects, a repeated measures analysis of variance is called for. We let n stand for the number of subjects and k for the number of treatments. The analysis tests the same null hypothesis as the one-way ANOVA, the hypothesis that all the population means are equal. The assumptions underlying the test are (1) normal populations, (2) equal population variances, and (3) equal covariances between all treatment pairs.

In the repeated measures ANOVA, the total sum of squares is partitioned into a sum of squares between subjects, a treatment sum of squares, and a subjects × treatments (residual) sum of squares.

$$SS \text{ total} = SS \text{ subjects} + SS \text{ treatments} + SS \text{ residual}$$

The total degrees of freedom are partitioned accordingly. The degrees of freedom are df total $= N_{\text{T}} - 1$, df subjects $= n - 1$, df treatments $= k - 1$, df residual $= (n - 1)(k - 1)$.

Treatment and residual mean squares are calculated by dividing each sum of squares by the associated degrees of freedom. The test statistic is

$$F = \frac{MS \text{ treatments}}{MS \text{ residual}}, \quad \text{with } [k-1, \ (n-1)(k-1)] \text{ degrees of freedom}$$

If F exceeds the critical value at a certain level of alpha, the treatment effect is significant at that level.

A repeated measures design comparing two treatments ($k = 2$) can be analyzed by using either a t test for correlated samples or a repeated measures ANOVA. If both tests are conducted on the same data, $t^2 = F$ and, provided that the t test is two-tailed, (critical value of $t)^2$ = critical value of F. Therefore the conclusions drawn from the two tests are the same.

As compared with the one-way ANOVA for independent samples, the repeated measures ANOVA removes from the variation within treatments that part which is due to systematic differences between subjects. If between-subject differences are large (i.e., if there is substantial positive covariance between treatments), the reduction in error variance results in a more powerful test than one-way ANOVA. If, on the other hand, there is little systematic between-subject variation (i.e., intertreatment covariances are near zero), the loss of degrees of freedom may result in a test that is less powerful than one-way ANOVA.

Procedures for carrying out planned comparisons and post hoc tests on repeated measures data are similar to those for other designs. The chapter closed with a brief discussion of the extension of analysis of variance to more complex designs.

QUESTIONS

1. In your own words, tell what is meant by an "interaction" between two variables.

2. A social psychologist examined the effect of group composition (all male, all female, half male and half female) and the number of group meetings (one, three, or five) on attitude change of participants in discussion groups. Ten subjects were assigned to each combination of the two variables. The partially completed summary table is shown below. Complete the table and evaluate all F ratios.

Source of Variation	*SS*	*df*	*MS*	*F*
Group composition (C)	74			
Number of meetings (M)	160			
C × M				
Within	2025			
Total	2419			

3. Here is a summary table of a two-way ANOVA:

Source of Variation	*SS*	*df*	*MS*	*F*
A	840	2	420	4.20*
B	115	1	115	1.15
A × B	180	2	90	0.90
Within	3000	30	100	
Total	4135	35		

*$p < .05$.

Which of the following *could* be a graph of the data?

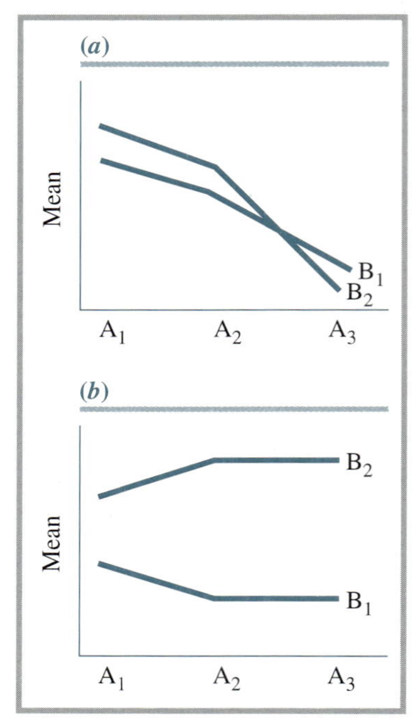

4. The following result was reported in a study of the effect of classroom structure (traditional, open) and children's anxiety (high, low) on achievement in arithmetic. "The average achievement of high- and low-anxious children did not differ significantly. Both groups of children achieved better in the traditional than the open classroom, but classroom structure had a greater effect on the achievement of high-anxious than low-anxious children."

   a. Which effects were significant in this study?

b. Construct a graph that is consistent with the results described.

5. Here are data from a study of the effects of gender and age on dependency behavior of preschool boys and girls.

	Age 3	Age 4	Age 5
Boys	8	8	6
	10	4	4
	7	9	3
	11	10	5
	11	5	2
Girls	5	5	4
	6	9	3
	8	4	9
	5	10	9
	8	9	8

a. Analyze the data and prepare a summary table.
b. Construct a table and graph of the group means. Describe the effects in your own words.
c. Calculate and interpret $\hat{\omega}^2$ for all significant effects.
d. Because the effect of age and the interaction between gender and age are significant at the .05 level, post hoc tests can be conducted to pinpoint the location of the effects more precisely. Conduct appropriate tests to answer the following questions. (Note that the first question is about the age effect and the second is about the gender-age interaction.)
   (1) Does the mean level of dependency at age 5 differ significantly from that at ages 3 and 4 combined?
   (2) Is the difference between boys and girls at age 3 significantly different from the difference between boys and girls at age 4? (*Hint:* If you let boys = A_1 and girls = A_2, age 3 = B_1, age 4 = B_2, age 5 = B_3, then the null subhypothesis for this question is $(\mu_{A_1B_1} - \mu_{A_2B_1}) - (\mu_{A_1B_2} - \mu_{A_2B_2}) = 0$.)

6. The table below gives *sums* of scores in each cell of a 3 × 4 factorial design with four subjects per cell.

	B_1	B_2	B_3	B_4
A_1	53	57	43	49
A_2	35	33	27	40
A_3	19	25	17	16

$\Sigma X_T^2 = 4491$

a. Complete the analysis and prepare a summary table.
b. Calculate all group means and graph the data.
c. Calculate and interpret $\hat{\omega}^2$ for all significant effects.
d. Assume that the following comparisons were planned prior to data collection: (1) mean of B_1 and B_2 versus mean of B_3 and B_4; (2) mean of A_1 versus mean of A_2 and A_3. Conduct the appropriate tests.

7. An educational psychologist divided 80 children into 8 groups of 10 children each. She gave each group either 5 or 10 hours of instruction in the use of one of 4 learning strategies: rote practice, imagery, verbal association, and verbal elaboration. Then she measured the children's performance on a learning task and conducted a two-way ANOVA. The ANOVA summary table and a table of means are shown below.

Source of Variation	*SS*	*df*	*MS*	*F*
Instructional time (T)	210	1	210	4.2*
Strategy (S)	675	3	225	4.5**
T × S	240	3	80	1.6
Within	3600	72	50	
Total	4725	79		

*$p < .05$.
**$p < .01$.

Instructional Time	Strategy: Rote	Imagery	Verbal Association	Verbal Elaboration
5 hours	12	18	22	24
10 hours	18	22	28	26

a. Draw a graph of the means and describe the effects in your own words.

The psychologist had planned the following three comparisons on the strategy variable in advance of data collection: (1) rote versus nonrote; (2) imagery versus verbal; (3) verbal association versus verbal elaboration.

b. Test the three comparisons for orthogonality.
c. Conduct the comparisons, using $\alpha = .01$ for each one.
d. Suppose that no comparisons had been planned in advance. Now the psychologist decides to test all pairwise comparisons between strategies. Conduct an appropriate test.

8. Here are two sets of repeated measures data.

Set A

Subject	Treatment 1	2	3
1	4	5	9
2	7	9	11
3	9	11	11
4	5	7	11
5	3	7	8
6	7	11	13
7	4	8	8
8	7	10	14

Set B

Subject	Treatment 1	2	3
1	7	11	9
2	3	10	8
3	7	7	11
4	4	5	14
5	5	11	8
6	7	9	13
7	9	8	11
8	4	7	11

The same scores are included in both sets, and the treatment sums are the same. Therefore, *SS* total and *SS* treatments are the same in both cases.

a. Which of the two analyses will result in a larger value of F? Why?
b. Conduct both analyses and prepare summary tables of the results.

9. Seven subjects were tested under four experimental conditions. Here are the data:

Subject	Treatment 1	2	3	4
1	8	12	3	15
2	10	12	8	22
3	5	7	4	18
4	12	15	5	20
5	4	10	3	15
6	6	8	2	12
7	7	11	10	17

a. Conduct an ANOVA and prepare a summary table.
b. After examining the means, the researcher decides on the following comparisons: (1) Treatments 1 and 3 versus 2 and 4; (2) Treatments 1, 2, and 3 versus treatment 4. Conduct the appropriate tests.
c. Suppose that the researcher is interested only in pairwise post hoc comparisons. Perform the appropriate test.

10. Marie ran 15 subjects under both an experimental and a control condition. She conducted a two-tailed t test for correlated samples and calculated $t = -1.83$.

a. What is the critical value of t for $\alpha = .05$ in Marie's test?
b. Is the difference between the means significant at the .05 level?
c. If Marie were to conduct a repeated measures ANOVA instead, what would be the value of F?
d. Answer the following questions without consulting the F table:
   (1) Is Marie's F significant at the .05 level?
   (2) What is the critical value of F at the .05 level in Marie's test?

11. Here is a summary table of a repeated measures ANOVA.

Source of Variation	*SS*	*df*	*MS*	F
Subjects	420	14		
Treatments	132	2	66	3.67*
Residual	504	28	18	
Total	1056	44		

$^*p < .05$.

Suppose the researcher had mistakenly analyzed the data using one-way ANOVA (for independent samples). Construct a summary table of his analysis and evaluate F.

14 Some Nonparametric Distribution–Free Tests

INTRODUCTION

Most tests we have considered thus far test hypotheses about parameters such as population means or population correlation coefficients. For this reason, they are called *parametric tests*. The normal curve test and t tests, the test for the significance of a correlation coefficient, and analysis of variance—all are parametric tests.

Occasionally, we test hypotheses not about parameters but about other properties of populations, such as the distribution of membership across categories. To test hypotheses like these we use *nonparametric tests*, and these tests are the subject of this chapter.

Parametric tests usually rest on assumptions about population distributions (and, sometimes, other assumptions as well). The normal curve test, t tests, test for a correlation coefficient, and analysis of variance all assume that the populations from which samples have been drawn are normal. Most nonparametric tests do not make assumptions about the population shape and are, therefore, often referred to as *distribution-free tests*. Hypothetically, there may exist parametric tests that are distribution-free, and nonparametric tests that make assumptions about underlying distributions. The tests to be considered in this chapter, however, are all both nonparametric and distribution-free.

As you will see, nonparametric tests operate upon frequencies in categories or upon ranks rather than upon quantities, amounts, or scores. You have already encountered a widely used nonparametric test for categorical data in Chapter 5—the chi square (χ^2) test. In this chapter you will learn about some nonparametric tests for ranked data.

Objectives

When you have completed this chapter, you will be able to

- Explain how nonparametric tests differ from parametric tests and explain why most nonparametric tests are called distribution-free tests.
- Describe two conditions under which nonparametric tests tend to be applied.
- Describe the types of data for which the following nonparametric tests are appropriate: Mann-Whitney U test, Wilcoxon matched-pairs signed-ranks test, Kruskal-Wallis H test, and Friedman analysis of variance by ranks.

- For each of the nonparametric tests, state the null hypothesis and explain the rationale of the test.
- Conduct each of the four nonparametric tests and interpret the results.
- Discuss the relative power of parametric and nonparametric tests. Describe conditions under which nonparametric tests may be more powerful.

Introduction to Nonparametric Tests

Nonparametric tests operate upon categorical or ranked data rather than upon scores and usually include no assumptions about population shapes. Nonparametric tests tend to be applied under two kinds of conditions:

1. When data are in the form of frequencies in categories or ranks to begin with.

The chi square test, for example, is used to test hypotheses about the distribution of one categorical variable (test of goodness of fit) or about the relation between two categorical variables (test of independence). Similarly, if you have two sets of ranks and you want to know whether the two sets of ranks differ significantly, you need to use a nonparametric test for ranked data.

The other condition under which a nonparametric test may be used is

2. When data, although in the form of scores or measures of amount, are believed to violate the normal curve assumption underlying the relevant parametric test.

The second use is less common than it once was, largely because most parametric tests have been shown to be relatively *robust*. That is, the probability of Type I and Type II errors is usually affected very little, even if one or more assumptions are violated. The conclusions drawn from a t test for the difference between means or from an analysis of variance, for example, are likely to be approximately equally accurate, whether populations are normal or nonnormal, especially if the sample size is reasonably large. Because parametric tests are generally more powerful than nonparametric tests, researchers usually prefer them over nonparametric tests.

In discussing the robustness of t tests and ANOVAs, however, we noted that violations of underlying assumptions may cause serious problems if sample sizes are very small. Suppose, for example, that you have test scores on two groups of five students each, and that you believe that the populations are highly skewed. Under these circumstances, you may be wise to use a nonparametric test rather than a t test to compare the two groups.

There are far too many nonparametric tests to discuss in one chapter. We limit ourselves to four relatively well-known and frequently used tests that operate upon ranks. The tests to be considered include one test for comparing two independent samples, one for comparing two correlated samples, one for comparing k independent samples, and one for comparing k correlated samples. The four tests are the Mann-Whitney U test (two independent samples), the Wilcoxon matched-pairs signed-ranks test (two correlated samples), the Kruskal-Wallis H test (k independent samples), and the Friedman analysis of variance by ranks (k correlated samples). If you are interested in learning about additional nonparametric tests, refer to Siegel (1956) or Walsh (1962–1968).

Four Representative Tests for Ranks

THE MANN-WHITNEY *U* TEST

The *Mann-Whitney U test* is a relatively powerful nonparametric test for comparing two independent samples. Unlike the *t* test for independent samples, which tests hypotheses about a particular parameter, the difference between the two population means, the *U* test tests whether two populations differ in an unspecified way. The null and alternative hypotheses of the *U* test are:

H_o: Populations 1 and 2 are identical

H_A: Populations 1 and 2 are not identical

We start out by discussing and illustrating the Mann-Whitney *U* test with a set of data that are in the form of ranks to begin with. Then we show how the test is applied to data in the form of quantities or scores.

Application of the *U* Test to Data in the Form of Ranks

Suppose that 15 management trainees at the Ajax Manufacturing Company have been ranked by their supervisor on the basis of their job performance. Seven of the trainees are males and eight, females. Examining the rankings, you notice that the male trainees seem to have been given higher ranks, on the average, than the females. Having worked with trainees for these kinds of positions for many years, you know that neither men nor women are, as a group, more competent. Is it possible that the supervisor rankings are biased? You decide to conduct a test to determine whether the male and female ranks differ significantly.

Because the men and women trainees have not been matched or otherwise paired prior to measurement, the samples are independent. An appropriate test for comparing the male and female ranks is the Mann-Whitney *U* test. We will test the null hypothesis that, in the population, the distribution of proficiency evaluations is the same in males and females.

Even if the two population distributions are the same, however, samples from the two populations can be expected to differ somewhat, just by chance. For example, in one pair of independent samples, the males might slightly outrank the females, and in another the females might tend to outrank the males. The Mann-Whitney *U* test is used to find out whether a difference in ranks between two independent samples is within the limits to be expected by chance, or whether the difference is so large that it would occur very infrequently by chance (say, less than 5% of the time). If the test result indicates that the probability of a difference as great as the obtained one between the two samples' ranks has a probability less than alpha when H_o is true, H_o is rejected, and the conclusion is drawn that the two populations are not identical.

Here are the Ajax Company proficiency rankings, from 1 (most proficient) to 15 (least proficient). The male and female ranks are displayed on separate lines to highlight differences between them.

Males ($n = 7$)	1	2		4			7	8	9				13		
Females ($n = 8$)			3		5	6				10	11	12		14	15

The test statistic in the Mann-Whitney *U* test, *U*, is calculated as follows:

1. Add the ranks in one of the two samples. Call this sum of ranks *T*.
   In the males at Ajax, $T = 1 + 2 + 4 + 7 + 8 + 9 + 13 = 44$.

2. Calculate $U = n_1 n_2 + \frac{n_1(n_1+1)}{2} - T$, where Sample 1 is the one in which T was calculated.

$$\text{At Ajax: } U = (7)(8) + \frac{(7)(8)}{2} - 44 = 40$$

(If we had added the ranks of the females instead of the males in Step 1, we would have gotten $T = 76$ and $U = 16$. It does not matter which value of U is used as the test statistic; the decision reached is the same.)

The procedure for evaluating U when samples are small (n_1 and n_2 both less than 20) requires the use of special tables of critical values of U. Unfortunately, a separate table is needed for each level of alpha. Rather than including a large number of tables in this book, we have included only the one for $\alpha = .05$ (two-tailed test) or .025 (one-tailed test). It is Table D–7 in Appendix D. More extensive tables can be found in books such as Siegel (1956).

The row and column headings of the Mann-Whitney U table, N_A and N_B, refer to the sizes of the two samples; it does not matter which sample is called A and which B. Each cell of the table contains two numbers. If the calculated U falls between the two values, H_o is not rejected. But if U is either less than the smaller table value or greater than the larger one, H_o is rejected.

In our Ajax example, $n_1 = 7$ and $n_2 = 8$. The cell at the intersection of the column marked 7 and the row marked 8 (or vice versa) has the entries $^{11}_{45}$. Our calculated value of U, 40, falls between 11 and 45. The difference between the rankings of male and female manager trainees is not significant at the .05 level. We cannot conclude that the Ajax supervisor's rankings are biased.

Before we consider another example of the Mann-Whitney U test, let us explore the rationale of the test statistic, U, further. We will show that the more similar the distributions of ranks of two samples, the more alike are the values of U computed from the two samples; the more different are the sets of ranks, the more different the two values of U.

Suppose that the rankings at Ajax had been as follows:

Males (n = 7)		2		4		6		8		10		12		14	
Females (n = 8)	1		3		5		7		9		11		13		15

The two sets of ranks could not be more similar than this. Now let us calculate U for the males and for the females:

$$\text{Males:} \quad T = 2+4+\cdots+14 = 56$$

$$U = (7)(8) + \frac{(7)(8)}{2} - 56 = 28$$

$$\text{Females:} \quad T = 1+3+\cdots+15 = 64$$

$$U = (8)(7) + \frac{(8)(9)}{2} - 64 = 28$$

The two values of U are identical.

The largest possible difference between the ranks of the males and females at Ajax would be obtained if one of the two samples had all the lowest ranks and the other had all the highest ranks, such as:

Males (n = 7)	1	2	3	4	5	6	7								
Females (n = 8)								8	9	10	11	12	13	14	15

In this case, T and U would be

$$\text{Males:} \quad T = 1 + 2 + \cdots + 7 = 28$$

$$U = (7)(8) + \frac{(7)(8)}{2} - 28 = 56$$

$$\text{Females:} \quad T = 8 + 9 + \cdots + 15 = 92$$

$$U = (7)(8) + \frac{(8)(9)}{2} - 92 = 0$$

The values of U for males and females differ by 56 points. This is the largest possible difference between the two values of U for samples of size 7 and 8.

If the null hypothesis is true, the two values of U are expected to be similar, and the probability of a very large difference is small. In a sense, the table of critical values shows, for each combination of n_1 and n_2, how large the difference between U values must be to have a probability less than .05 when H_o is true. With samples of size 7 and 8, for example, the minimum significant difference between U values is 45 – 11 = 34. The difference between the obtained U values at the Ajax Company was only 40 – 16 = 24. The probability of a difference like this is greater than .05 if H_o is true. Therefore, the difference between the male and female distributions of ranks is not significant at the .05 level.

The Mann-Whitney *U* Test Applied to Quantities or Scores

The Mann-Whitney U test is a popular substitute for the independent-samples t test when sample sizes are small and skew or outliers suggest that the normality assumption underlying the t test is not met. As an example, consider an experiment conducted to find out whether ratings of readability differ as a function of type of font. Two independent samples of adult readers have read a passage printed in one of two fonts and have rated various aspects of readability. Table 14–1 lists the ratings of the two groups. Because of the small group sizes and the likelihood that population ratings are not normally distributed, the experimenter decides to use the Mann-Whitney U test instead of the t test.

In order to apply the Mann-Whitney U test to nonranked data, the data must first be converted into ranks, beginning at either the highest or the lowest score. The experimenter begins by assigning the rank 1 to the lowest rating (6), the rank 2 to the next lowest rating (8), up to the rank 12 to the highest rating. A problem presents itself at rank 4. As you can see, two subjects gave the same, fourth-lowest, rating of 10. Tied scores of this sort are not uncommon when converting data to ranks. The procedure used to deal with ties is shown on the following page.

Table 14–1 **Ratings of Readability of Two Fonts**

	Font A		Font B	
	Rating	Rank	Rating	Rank
	10	4.5	16	9.5
	8	2	14	8
	9	3	10	4.5
	11	6	16	9.5
	6	1	20	12
	13	7	19	11
Sum		23.5		54.5

1. Determine what the ranks of the scores would be if they were not the same. In the example, the two ratings would have ranks of 4 and 5.
2. Give each of the tied scores a rank equal to the average of the ranks they would have, if they were different.

In the example, each rating of 10 is given rank 4.5, the average of 4 and 5. The ranks of all 12 scores are listed in Table 14–1.

The remaining steps are the same as in the Ajax Company example. The experimenter designates the sum of ranks in one of the groups, the Font A group, as T. $T = 23.5$. Next he calculates U.

$$U = n_1 n_2 + \frac{n_1(n_1 + 1)}{2} - T = (6)(6) + \frac{(6)(7)}{2} - 23.5 = 33.5$$

Finally, the table of critical values of U is consulted to determine whether the difference between the two sets of ranks is significant at the .05 level. According to the U table, for N_A and N_B = 6, values of U less than 6 or greater than 30 are significant at the .05 level. Because U is greater than 30, it is significant at the .05 level, and the experimenter concludes that the two fonts differ in readability.

The Mann-Whitney U Test When n_1 or n_2 Is Greater Than 20

Table D–7 in Appendix D includes critical values of U for n_1 and n_2 equal to 20 or less. If either n_1 or n_2 (or both) is greater than 20, and if the two sample sizes are not too different, it is possible to convert U to a value of z and evaluate it using a table of the standard normal curve. We do not discuss the procedure here for two reasons:

1. Ranking procedures are seldom applied to large samples because of the difficulty of doing so. As an example, consider how hard it would be for you to rank order your liking of 50 foods from 1 to 50.
2. If data are not in the form of ranks to begin with and n_1 and n_2 are reasonably large, the t test gives accurate results, even if one or more assumptions are violated.

If you are interested in the Mann-Whitney procedure for large samples, refer to Siegel (1956) or Hays (1988).

EXERCISE

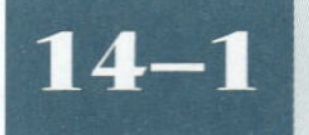

One group of five surgical patients was counseled before surgery about the amount and kind of postsurgical pain to expect. A second group of five patients was not counseled. Here are the patients' ratings, on a scale from 0 (no pain) to 20 (intense pain), of their postsurgical pain:

Counseled patients	2	5	7	0	2
Uncounseled patients	8	16	4	18	12

Use the Mann-Whitney U test to determine whether rated pain differed significantly (at the .05 level) between the two groups.

The Wilcoxon Matched-Pairs Signed-Ranks Test

The *Wilcoxon matched-pairs signed-ranks test*, like the Mann-Whitney U test, tests the null hypothesis that two populations are identical against the alternative hypothesis that they differ in an unspecified way. The Wilcoxon test can be used either to compare two sets of correlated ranks (obtained on the same or matched subjects) or as a substitute for the t test for correlated samples when normality assumptions underlying the t test are not met.

To describe and demonstrate computational procedures, we use data from a hypothetical study on the effect of parents' presence on kindergarten-readiness scores of 5-year-old children. The study involved 10 pairs of children matched on the basis of physical maturity (size, weight, physical skills). One child in each pair took a kindergarten-readiness test with a parent present, and the other was tested without a parent. The readiness scores are listed in Table 14–2. Because the researcher believes that population readiness scores are skewed, she decides to conduct a Wilcoxon test instead of a t test.

In the Wilcoxon test, a signed rank is calculated for each pair of scores by the following procedure:

1. Calculate the difference score, D, within each pair, just as in the t test for correlated samples.
2. Ignore difference scores of zero; they are not included in the analysis.
3. Treat the remaining difference scores as positive numbers, $|D|$. Rank the values of $|D|$ from smallest to largest.

   In Table 14–2, the ranks of the absolute difference scores are listed in the column headed "Rank of $|D|$." The smallest $|D|$, 1, is given rank 1. There are two cases of $|D| = 2$; both are given the average of ranks 2 and 3, or 2.5; similarly, the two cases of $|D| = 3$ both get the average of ranks 4 and 5, or 4.5; and so on. Since there are eight nonzero difference scores, the highest rank assigned is 8.
4. Attach the sign of the difference score, D, to each rank.

   In Table 14–2, Pairs 2, 4, 5, and 8 have negative difference scores. Negative signs are now attached to their absolute ranks. The resulting signed ranks are shown in the last column of the table.

The signed ranks are the basis for the statistical test. The reasoning is as follows: If H_o is true and there is no difference between the population distributions, positive and negative difference scores are expected to occur about equally often and to be about the same size in most pairs of samples. Therefore, the sum of the positive signed ranks is expected to be approximately equal to the sum of the negative signed ranks. The greater

Table 14–2 Kindergarten-Readiness Scores of Two Matched Groups

Pair	Parent Present	Parent Absent	D	$\|D\|$	Rank of $\|D\|$	Signed Rank
1	28	25	3	3	4.5	4.5
2	19	22	−3	3	4.5	−4.5
3	22	22	0			
4	24	25	−1	1	1	−1
5	23	25	−2	2	2.5	−2.5
6	34	26	8	8	8	8
7	20	16	4	4	6	6
8	19	21	−2	2	2.5	−2.5
9	23	18	5	5	7	7
10	24	24	0			

the difference between the two sums, the less likely is such a difference if H_o is true. If the difference between the sum of the positive and the sum of the negative signed ranks is so large that the probability is less than alpha, H_o is rejected, resulting in the conclusion that the two populations are not identical.

The next step of the test, then, is to calculate the sum of the positive ranks and the (absolute) sum of the negative ranks. For the data in Table 14–2:

$$\text{Sum of positive signed ranks} = 4.5 + 8 + 6 + 7 = 25.5$$

$$\text{Sum of negative signed ranks} = 4.5 + 1 + 2.5 + 2.5 = 10.5$$

The test statistic in the Wilcoxon test, W, is the smaller of the two sums. In the kindergarten-readiness example, $W = 10.5$.

Critical values of W for $n = 5$ to 50 are presented in Table D–8 in Appendix D. In the table, n stands for the number of *nonzero* difference scores in the set. In the kindergarten-readiness example, there are eight nonzero values of D, so that $n = 8$. Each entry in the table is the *maximum* value of W which is significant for the given n and level of significance (α); values of W *less than or equal to* the tabled value are significant.

Now we are ready to evaluate $W = 10.5$ in the kindergarten-readiness study. Because the researcher is presumably equally interested in detecting positive or negative differences between the two experimental conditions, a two-tailed (nondirectional) test is called for. For a nondirectional test with $\alpha = .05$ and $n = 8$, the critical value of W is 3. Only values of $W \leq 3$ are significant at the .05 level. The calculated W, 10.5, is not significant. The researcher cannot conclude that kindergarten-readiness scores differ based on whether a parent is present or not.

EXERCISE 14–2

Eight college students rated their interest in skydiving before and after watching a promotional film. Here are the results.

Student	Before Film	After Film
1	0	5
2	3	4
3	1	5
4	3	9
5	2	4
6	5	5
7	7	9
8	4	6

Conduct a Wilcoxon test to determine whether ratings increased significantly after viewing the film. Use $\alpha = .05$.

The Kruskal-Wallis *H* Test

When the number of independent groups (k) is greater than 2, the Mann-Whitney U test cannot be used. An appropriate nonparametric test for ranks, which is, essentially, an extension of the U test, is the Kruskal-Wallis H test. The Kruskal-Wallis test, like the Mann-Whitney and Wilcoxon tests, tests the null hypothesis that the k population distributions are identical against the alternative hypothesis that not all the population distributions are identical. The rationale of the test is that if the null hypothesis is true, ranks are expected to be randomly distributed across groups, and the sum of the ranks in all the groups should be about equal. If differences among the summed ranks are so great that they are unlikely to have occurred by chance, H_o is rejected, and the conclusion is drawn that not all the population distributions are the same.

An example of data to which the Kruskal-Wallis test might be applied is given in Table 14–3. Eighteen subjects were randomly assigned to three groups of six subjects each. The three groups then completed the same logical reasoning tasks under different strategy instructions. The table shows the number of problems successfully completed by each subject. Because of the presence of several extreme outliers, the distribution-free Kruskal-Wallis test was selected in preference to a one-way analysis of variance.

In the Kruskal-Wallis test, scores are converted to ranks, and tied scores are dealt with, in the same way as in the Mann-Whitney U test. Then the sum of ranks, T, is calculated in each group. The ranks of the 18 scores and the sum of ranks of each group are shown in Table 14–3. The test statistic, H, of the Kruskal-Wallis test is calculated as follows:

1. Square T in each group, and divide by the group size n.
2. Add the values of $\frac{T^2}{n}$ over the k groups. The sum is $\sum^{k} \frac{T^2}{n}$.
3. Let N stand for the total number of subjects. ($N = n_1 + n_2 + \cdots + n_k$). To calculate H, use the formula

$$H = \left[\frac{12}{N(N+1)}\right]\left[\sum^{k} \frac{T^2}{n}\right] - 3(N+1)$$

The three values of T in Table 14–3 are very different. But are they sufficiently different to reject H_o? To find out, we calculate H.

1. $\frac{T_1^2}{n_1} = \frac{(30.5)^2}{6} = 155.04$; $\frac{T_2^2}{n_2} = \frac{(62.0)^2}{6} = 640.67$; $\frac{T_3^2}{n_3} = \frac{(78.5)^2}{6} = 1027.04$
2. $\sum^{k} \frac{T^2}{n} = 155.04 + 640.67 + 1027.04 = 1822.75$
3. $H = \left[\frac{12}{18(19)}\right][1822.75] - 3(19) = 6.956$

It can be shown that if the null hypothesis is true and if H is calculated in each possible set of k samples of size $n_1, n_2, \cdots, n_k$, the values of H over all sets of samples are

Table 14–3 **Number of Problems Solved by Subjects in Three Groups**

Group 1	Rank	Group 2	Rank	Group 3	Rank
11	2	12	3	20	12.5
19	11	34	16	16	8.5
14	5.5	27	15	37	17
13	4	23	14	20	12.5
9	1	16	8.5	45	18
15	7	14	5.5	17	10
	$T_1 = 30.5$		$T_2 = 62.0$		$T_3 = 78.5$

approximately distributed as χ^2 with $k - 1$ degrees of freedom (provided that the size of each group is at least 5). Therefore, the value of H is treated as a value of χ^2, and the table of critical values of χ^2 is used to determine whether H is significant at the chosen alpha. The null hypothesis is rejected if H is greater than the critical value of χ^2 for the chosen alpha and $k - 1$ degrees of freedom.

In the example, $k = 3$, so that the number of degrees of freedom is 2. In the chi square table (Table D–1 in Appendix D), $H = 6.956$ falls between the critical values of χ^2 for $\alpha = .05$ and $\alpha = .025$. The result is significant at the .05 level. The experimenter concludes that problem solving is affected by the type of strategy instructions.

If group sizes are less than 5, the chi square distribution is not a good fit to the sampling distribution of H. Under these circumstances, a special table of critical values of H must be consulted to evaluate H. Because sample sizes less than 5 are unusual, the table and its use are not discussed here.

EXERCISE 14–3

In a study of the development of manual dexterity, a developmental psychologist measured the ability of children of four different ages to drop a ball through a hole without touching the sides. Each child's number of successful trials out of 20 attempts is given here:

3-Year-Olds	4-Year-Olds	5-Year-Olds	6-Year-Olds
10	8	16	19
0	1	6	9
4	15	13	18
15	1	3	10
8	16	6	10

Use the Kruskal-Wallis H test to decide whether manual dexterity differs as a function of age over the years tested.

The Friedman Analysis of Variance by Ranks

Despite its name, the *Friedman analysis of variance by ranks* does not analyze variances. It, like the other tests we are considering, operates on ranks and determines whether the obtained distribution of ranks is a likely one or an unlikely one if the null hypothesis is true. The Friedman test is used to compare correlated samples—matched samples or repeated measures—from k populations. The null hypothesis of the test, like that of the Kruskal-Wallis test, is that the k populations are identical; the alternative hypothesis is that not all the populations are identical.

As an example, consider a study of the effect of background color on the recognition of words that are briefly projected on a screen. Subjects are shown 25 randomly chosen five-letter words against each of four color backgrounds. The number of words recognized by each subject against each color is shown in Table 14–4. Because scores within colors are markedly skewed, the researcher decides to use the nonparametric Friedman test instead of the repeated measures analysis of variance.

The first step in conducting the Friedman test is to rank scores *within each row* (i.e., within each subject). The ranks within subjects are shown in Table 14–4.

Table 14–4 Number of Words Recognized Against Four Background Colors

	Background Color				Ranks Within Subjects			
Subject	Blue	Red	Yellow	Green	Blue	Red	Yellow	Green
1	15	14	20	18	2	1	4	3
2	18	19	13	16	3	4	1	2
3	17	15	18	16	3	1	4	2
4	5	6	8	4	2	3	4	1
5	21	12	15	15	4	1	2.5	2.5
6	14	10	12	21	3	1	2	4
7	16	14	18	16	2.5	1	4	2.5
8	6	7	7	5	2	3.5	3.5	1
9	8	10	12	13	1	2	3	4
10	21	18	22	17	3	2	4	1
					25.5	19.5	32.0	23.0
					T_1	T_2	T_3	T_4

The reasoning behind the Friedman test is simple. If the k populations are identical, as H_o states, all possible orders of ranks within rows are equally likely. Therefore, each rank should appear about equally often within each column, and the sum of the ranks in all columns should be about the same. If column sums of ranks differ greatly, this suggests that the populations are not identical, and H_o is rejected.

The remaining steps in the Friedman test are the following:

1. Add the ranks in each column. Call the sum T.
2. Square each sum and add across columns to get $\sum^{k} T^2$.
3. Let n stand for the number of rows (subjects). Calculate the test statistic, χ_r^2, by the formula

$$\chi_r^2 = \frac{12}{nk(k+1)}\left[\sum^{k} T^2\right] - 3n(k+1)$$

As is implied by the symbol used for the test statistic, χ_r^2 is distributed as χ^2 with $k - 1$ degrees of freedom, provided that n is 10 or more. For smaller samples, a special table of critical values, which can be found in books like Siegel (1956), is needed.

Let us complete the test comparing recognition of words projected against four background colors. The sum of ranks in each column, T, is shown in Table 14–4. The remaining steps are

$$\sum^{k} T^2 = (25.5)^2 + (19.5)^2 + (32.0)^2 + (23.0)^2 = 2583.5$$

$$\chi_r^2 = \frac{12}{(10)(4)(5)}(2583.5) - 3(10)(5) = 155.01 - 150 = 5.01$$

To evaluate χ_r^2, we use the chi square table. Referring to the row for $k - 1 = 3$ degrees of freedom, we find that χ_r^2 has a probability between .25 and .15. Since $p > .05$, we cannot reject H_o. Word recognition did not differ significantly as a function of background color.

EXERCISE 14–4

Twelve infants were shown five stimuli varying in complexity. The following table shows how many seconds each infant gazed at each stimulus.

	Level of Complexity				
Infant	**1 (least)**	**2**	**3**	**4**	**5 (most)**
1	1	3	5	8	6
2	5	6	7	11	25
3	2	3	5	11	8
4	4	4	8	10	5
5	1	3	3	5	8
6	2	1	4	1	3
7	4	9	10	10	13
8	3	7	2	5	10
9	7	8	8	7	12
10	2	4	3	4	2
11	7	7	5	12	3
12	10	12	12	15	16

Use the Friedman test to determine whether length of gaze varied as a function of complexity.

The Relative Power of Parametric and Nonparametric Tests

I have stated that parametric tests are usually more powerful than nonparametric tests. In other words, a false null hypothesis is more likely to be (correctly) rejected if a parametric test, rather than a nonparametric test, is used to analyze the data. Because researchers generally hope to reject the null hypothesis, they tend to prefer parametric tests.

A major reason why parametric tests are more powerful than nonparametric tests applied to measures of quantity or amount is that parametric tests utilize more of the information in the data. Suppose, for example, that the three lowest scores of a data set are 50, 54, and 56. A parametric test takes into account the fact that the difference between the scores 50 and 54 is twice as great as that between 54 and 56. But if the scores are converted to ranks of 1, 2, and 3 for purposes of conducting a nonparametric test, the information about the actual differences between the scores is lost.

However, parametric tests are not always more powerful than nonparametric tests. If the assumptions underlying parametric tests are seriously violated, nonparametric tests are sometimes more powerful. This may happen if, for example, the populations are very skewed and sample sizes are small. As an example, consider the data, collected in two independent samples, in Table 14–5. Note that most scores in both groups are 17 or less, but Group 1 includes an extreme outlier, 58. Unusual outliers like this indicate that the population is probably skewed—positively skewed, in this case.

As you can see in Table 14–5, although most of the scores in Group 1 are lower than those in Group 2, the two means are identical. Obviously, there is no point in conducting a t test to compare the means. Let us apply the Mann-Whitney U test, a more appropriate test under the circumstances. The sum of ranks in Group 1, T, is 44. Therefore,

Table 14–5 Scores of Two Independent Samples

	Group 1		Group 2	
	Scores	**Ranks**	**Scores**	**Ranks**
	10	6.5	17	14.5
	9	5	12	9.5
	10	6.5	15	13
	7	3.5	17	14.5
	5	1	11	8
	58	16	14	11.5
	7	3.5	12	9.5
	6	2	14	11.5
Sums	112	44 = T	112	92
Means	14		14	

$$U = (8)(8) + \frac{(8)(9)}{2} - 44 = 56$$

According to the U table, for two samples of size 8 and $\alpha = .05$, H_o should be rejected if U is less than 14 or greater than 50. Because the calculated U is greater than 50, we reject H_o and conclude that the two populations from which the samples were drawn are not identical. Thus, the U test results in rejection of H_o, although the t test does not. At least in this case, the nonparametric U test is more powerful than the parametric t test.

Based on our discussion and on the robustness of parametric tests, we can draw the following conclusions. If samples are large, parametric tests can usually be applied to data, even if the populations are believed to be nonnormal or the population variances are different. These tests are robust with respect to violation of assumptions, and they are usually more powerful than their nonparametric counterparts. If, however, samples are small and evidence suggests that assumptions underlying parametric tests are seriously violated, it may be best to select a nonparametric test instead. Not only can you be more confident that you have met the requirements of the test but, in addition, under some circumstances the nonparametric test may actually be more powerful.

EXERCISE 14–5

1. In Chapter 11, we used a lower one-tailed t test for correlated samples to determine whether subjects detected significantly fewer tones when a white noise was present than when it was absent. Our conclusion was that subjects do indeed detect fewer tones with a white noise present. Here are the data on which the test was based.

Subject	White Noise Present	White Noise Absent
1	6	9
2	4	6
3	6	5
4	7	10
5	4	10
6	5	8
7	5	7
8	7	5
9	6	6
10	4	8

EXERCISE 14–5

Use an appropriate nonparametric test to analyze the data. Compare the results with the results of the t test.

2. In each of the following cases, decide whether you should use a nonparametric or a parametric test, and justify your choice.
   a. You have three correlated samples of size 5 each. As far as you know, the populations are probably approximately normal with equal variances and covariances.
   b. You have two independent samples of size 4 each. It is likely that one or both populations are highly skewed.
   c. You have three independent samples of size 5 each. All but three scores are between 15 and 35, but one sample includes scores of 2 and 4 and another includes a score of 3.

SUMMARY

Tests of hypotheses about properties of populations other than parameters are called *nonparametric tests*. Most such tests make no assumptions about the shape of population distributions and are, therefore, called ***distribution-free tests***. Nonparametric tests operate on categorical or ranked data and are used if either data consist of frequencies in categories or ranks to begin with or assumptions underlying parametric tests are seriously violated.

The most common test for categorical data is the chi square test (see Chapter 5). Common tests for ranked data include the *Mann-Whitney U test*, the *Wilcoxon matched-pairs signed-ranks test*, the *Kruskal-Wallis H test*, and the *Friedman analysis of variance by ranks*. The null hypothesis for all four tests is that the populations from which samples have been drawn are identical.

The Mann-Whitney U test compares ranks in two independent samples. The test involves adding the ranks in one of the samples (sum = T). Then U is calculated by the formula

$$U = n_1 n_2 + \frac{n_1(n_1+1)}{2} - T$$

The calculated U is compared with a table of critical values of U to decide whether to reject the null hypothesis.

The Wilcoxon matched-pairs signed-ranks test is used to compare two correlated samples (repeated measures or matched groups). In the test, a difference (D) score is calculated in each pair. The absolute differences are first ranked from smallest to largest, then given the same signs as the corresponding difference scores. The test statistic, W, is the sum of either the positive or negative signed ranks, whichever is smaller. W is evaluated by referring to a table of critical values of W.

If three or more independent samples are to be compared ($k \geq 3$), an appropriate test is the Kruskal-Wallis H test. After the entire set of scores has been ranked, the sum of ranks, T, is calculated in each group. The test statistic, H, is then calculated by the formula

$$H = \left[\frac{12}{N(N+1)}\right]\left[\sum^{k} \frac{T^2}{n}\right] - 3(N+1)$$

If the size of each group is 5 or more, H is approximately distributed as χ^2 with $k - 1$ degrees of freedom. Therefore, the calculated H is evaluated by referring to the chi square table.

The Friedman analysis of variance by ranks is appropriate for analyzing data from three or more correlated samples. In the test, scores are first ranked within each row (each subject or each matched set). Then the sum of ranks, T, is calculated for each treatment (column). The test statistic, χ_r^2, is calculated by the formula

$$\chi_r^2 = \frac{12}{nk(k+1)}\left[\sum^{k} T^2\right] - 3n(k+1)$$

Provided that the number of rows, n, is 10 or more, χ_r^2 is distributed as χ^2 with $k - 1$ degrees of freedom, and the calculated χ_r^2 is evaluated by referring to the chi square table.

Parametric tests are relatively robust in the face of violations of underlying assumptions and are usually more powerful than nonparametric tests. However, nonparametric tests may be more powerful in some cases where assumptions are seriously violated.

QUESTIONS

1. A Mann-Whitney U test is conducted to compare two groups. The sum of ranks is greater in Group 1 than Group 2, and U is significant at the .05 level. Assuming that the two populations are indeed different (i.e., no Type I error has occurred), list as many differences as you can think of between the two populations that might account for the significant difference between the samples.

In each of the remaining questions, select and conduct an appropriate nonparametric test.

2. Shortly before a campus election, three random groups of students were presented with arguments favoring one of the candidates, John Smith. Six students heard an emotional argument favoring Smith, six heard a rational argument favoring Smith, and six heard an argument that was both emotional and rational. Here are the students' subsequent ratings of Candidate Smith, on a scale of 1 to 20:

Emotional argument	18	16	17	10	19	15
Rational argument	10	7	9	12	15	8
Emotional and rational argument	18	20	19	15	16	17

   Did ratings vary significantly as a function of type of argument?

3. A college admissions official wondered whether satisfaction with college life is related to the type of community from which students come. The official formed three groups of 10 freshmen each, one group from urban high schools, the second from suburban high schools, and the third from rural high schools. The groups were matched on the basis of SAT scores and high school rank. Here are their scores on a scale measuring satisfaction with college life.

Matched Set of Students	Urban	Suburban	Rural
1	10	14	8
2	22	14	15
3	9	18	7
4	10	3	24
5	25	23	9
6	10	15	8
7	19	15	7
8	8	12	24
9	24	25	19
10	9	15	10

   Did satisfaction with college life differ significantly among the three groups?

4. A statistics instructor wondered whether students would do better on his statistics quizzes if they were relieved of time pressures. He formed two groups of students, matched on the basis of previous statistics exam scores. Both groups took the same 20-point quiz, one under timed and the other under untimed conditions. The quiz scores are shown below.

Pair	Timed	Untimed
1	15	17
2	12	11
3	10	15
4	13	10
5	18	18
6	20	18
7	12	16
8	8	12
9	9	8
10	17	15

   Is the difference between scores under timed and untimed conditions significant?

5. A psychology major hypothesized that students become increasingly cynical as they proceed through college. To test her hypothesis, she gave a test measuring cynicism to samples of freshmen, sophomores, juniors, and seniors. Here are the results (higher scores indicate greater cynicism).

Freshmen:	23	5	14	38	11	43	43	49		
Sophomores:	29	21	14	2	7	45	45	36	12	32
Juniors:	38	23	12	8	11	47	47	48	34	
Seniors:	28	37	37	45	15	7	7	38	8	47

Did cynicism differ significantly between classes?

6. Fourteen children took a 25-item pretest before starting a unit of social studies and a 25-item posttest after completing the unit. Was the gain from pretest to posttest significant at the .05 level?

Child	1	2	3	4	5	6	7	8	9	10	11	12	13	14
Pretest	21	18	10	17	15	21	20	15	25	20	21	20	10	5
Posttest	24	25	23	22	25	21	19	12	23	20	18	24	8	21

7. A study was conducted to determine the effect of a model on children's expectations of success at a task. Twenty children watched a model play a ring-toss game. Ten of the children watched a "successful" model who got 15 or more ringers in 25 tosses. The other 10 children watched an "unsuccessful" model who got 10 or fewer ringers in 25 tosses. Then each child was asked, "How many ringers do you think you would get in 25 tosses?" Here are the answers:

Successful model:	21	15	18	10	25	10	25	17	5	25
Unsuccessful model:	20	10	5	0	10	25	15	12	8	18

Did type of model affect the children's answers?

8. Eleven subjects tasted four cereals and rank ordered them from 1 (most liked) to 4 (least liked). Conduct a test to determine whether some of the cereals are preferred over others.

	Cereal			
Subject	**A**	**B**	**C**	**D**
1	3	4	1	2
2	4	2	3	1
3	3	1	4	2
4	4	3	2	1
5	2	4	1	3
6	4	3	1	2
7	3	1	4	2
8	1	3	2	4
9	3	2	1	4
10	4	2	1	3
11	3	4	2	1

9. In a study of the effect of author's reputation on ratings of poems, 14 English majors were randomly subdivided into two groups of seven. Both groups read the same set of poems. One group was told that the poems were by the famous poet Robert Frost; the other group was told that the poems were by an unknown poet, William Rime. Here are the students' ratings of the poems on a scale of 1 to 20:

"Robert Frost"	15	18	20	14	15	20	20
"William Rime"	10	15	18	20	15	10	12

Did ratings differ significantly as a function of author's reputation?

15 Connections

INTRODUCTION

It seems fitting to end an introductory statistics textbook with a chapter about connections—connections among concepts and procedures within statistics, as well as connections between statistics and other aspects of the research process. Much of the material in this chapter is not new and is included to help you synthesize material you have already learned. In addition, however, we provide brief introductions to ideas and procedures that you may encounter in later courses or in your professional work. I hope that these brief glimpses will whet your appetite for further exploration.

We start by considering the question: Given a set of data to analyze, how does one decide what procedure or test is required? The answer, as you will see, depends on a number of factors, including the research question and the type of data, among others. To choose one of many possible analyses for a set of data, you must pay attention to *differences* between research questions and types of data. In the second section we focus on *similarities* instead, examining features shared by many statistical procedures. In the process, we introduce the concept of the *general linear model*, a powerful model that underlies a great number of procedures for analyzing differences and relationships.

Statistical analyses are performed on research data in order to answer research questions. But where do research questions come from? And how are data collected? In other words, what does the research process as a whole entail, and where, exactly, does statistics fit in? The third section of the chapter, a brief survey of the research process, addresses these questions.

The last topic of the chapter is a relative newcomer to the family of statistical procedures called *meta-analysis*. The purpose of meta-analysis is to integrate findings over large numbers of research studies on a given topic. In meta-analysis, the results of individual studies become the data to be analyzed, and the analysis provides information about the average relationship between variables and the interstudy differences that affect the relationship. Meta-analysis is a powerful tool for clarifying research outcomes, especially when results of individual studies are ambiguous or contradictory.

Objectives

When you have completed this chapter, you will be able to

- Given a research question and other relevant information (such as the type of data and number of samples), select an appropriate method of analysis.
- Describe steps in hypothesis testing that are common to all statistical tests.
- Describe the structure of the test statistic in most statistical tests and apply it to particular tests.
- Interpret results of both correlational and experimental research in terms of components of variance.
- Define the general linear model and compare its structure in regression and ANOVA; discuss implications of the common structure for analyses of correlational and experimental data.
- Describe the sequence of activities in conducting a research project and explain the role of statistics in the process.
- Describe the purpose of meta-analysis.
- Define an effect size; given information about the means of two groups or conditions and their average standard deviation, calculate the effect size.

Choosing an Appropriate Procedure or Test

In Chapter 1 three stages of a statistical analysis were identified: (1) selecting an appropriate procedure, (2) applying the procedure (number-crunching), and (3) interpreting the results. As new procedures were introduced in subsequent chapters, their purpose and conditions of use (Stage 1) were described. But, of course, each chapter discussed a relatively small number of concepts and procedures. Therefore, deciding what procedure to apply to problems and questions within the chapter was relatively easy.

In real life, sets of data do not come with labels attached such as "I need a t test for independent samples" or "construct a bar graph." Faced with a set of data to be analyzed, researchers and statisticians must decide which of many possible analyses is (or are) appropriate. Although frequent users of statistics soon become adept at making such decisions, students (and less experienced researchers) often feel overwhelmed by the large number of possibilities. How, they wonder, can one possibly choose?

Given a set of data to be analyzed several factors must be considered in deciding what procedures to apply. The first factor, and the most important, is the *nature of the research question*. In Chapter 1, we grouped research questions into three major classes: (1) questions about the distribution of one variable in one group of individuals, (2) questions about the relation between two or more variables measured in one group of individuals, and (3) questions about differences between two or more groups (or between two or more conditions). The first step in selecting an appropriate analysis is to identify the class of questions to which the given research question belongs.

A second factor in selecting an analysis for a set of data is the *type of data* to be analyzed. For example, categorical data require different analyses than quantitative (amount or score) data, even if the same type of question is being asked. A good strategy, which takes both kinds of factors into account, is the following: (1) identify the class of research questions to which a specific research question belongs; (2) identify the type of data to be analyzed; (3) look for the procedures needed to answer your specific question.

To what class of research questions does this question belong?	→	What kind of data do I have?	→	What specific research question am I trying to answer?	→	What is the appropriate analysis for answering this question for this type of data?

In the sections that follow, we discuss methods used to answer questions within each major class. Then we apply the strategy to several examples.

Questions About the Distribution of One Variable in One Group of Individuals

The goal of some research studies is to describe the status of a single population on a variable. In pursuit of the goal, a sample is selected from the population. Measurements of the variable in the sample constitute the data to be analyzed. Here are some examples:

A survey is undertaken to measure attitudes of American business executives toward free trade.
Preschool children are observed to obtain information about their social development.
A school district administers a national reading test to assess students' reading skills.

In cases like these, researchers usually want to answer questions such as What does the distribution as a whole look like? What is the average? How much variability is there around the average? Based on the sample, what conclusions can be drawn about the population? Table 15–1 summarizes procedures for answering these and other types of questions. As the table shows, the required procedure depends, first, on whether the data are categorical or quantitative and, second, on the specific information being sought. Only procedures discussed in this book are included and chapter references are given.

For both categorical and quantitative data, questions about the appearance of the distribution as a whole are answered by constructing frequency distributions and/or graphs. The particular type of graph depends on the type of data and the features to be displayed. Methods of constructing and interpreting frequency distributions and graphs were discussed in Chapter 2. Questions about averages require the calculation of measures of central tendency. The particular measure required (mean, median, or mode) depends on the type of data, the specific research question, and the shape of the distribution. Questions about variability call for measures of variability (e.g., semi-interquartile range, standard deviation). Each measure of variability is associated with one measure of central tendency. For example, if you have reported the mean of a data set to answer a question about its average value, you are most likely to report the standard deviation (or variance) to answer a question about variability. In Chapter 3 we discussed the calculation of measures of central tendency and variability and factors to be considered in choosing among them.

Another question often asked about distributions of quantitative data has to do with the shape of the distribution: Is the distribution symmetrical or skewed and, if skewed, in what direction? Skew can, of course, be observed in a frequency distribution or graph, or it can be measured either by the relative locations of the first, second, and third quartiles (Q_1, Q_2, and Q_3; see Chapter 2) or by the relative location of the mean and the median (Chapter 3). Sometimes researchers wish to locate particular scores relative to

Table 15–1 Decision Table for Selecting Appropriate Analyses for Questions Concerning the Distribution of One Variable in One Group of Subjects

Type of Data	Specific Question	Analysis or Analyses	Text Chapter
Categorical	What does the distribution as a whole look like?	Frequency distribution Bar graph	2 2
	What is the average?	Modal category	3
	How much variability is there around the average?	Range (number of categories)	3
	What is the distribution in the population?	Chi square test of goodness of fit	5
	What percentage of the population falls in a particular category?	Confidence interval for a population percentage	8
Quantitative (amounts, scores)	What does the distribution as a whole look like?	Frequency distribution Histogram, frequency polygon, and other graphs	2 2
	Is the distribution skewed?	Relative location of Q_1, Q_2, Q_3 Relative location of mean and median	2 3
	Where are specific scores located?	Percentiles and percentile ranks z scores	2 3, 7
	What is the average?	Mean, median, mode	3
	What is the spread of scores around the average?	Variance and standard deviation, interquartile and semi-interquartile range, range	3
	What is the likely population mean (μ)?	Normal curve test of H_o: $\mu = \mu_H$ t test of H_o: $\mu = \mu_H$ Confidence interval for a population mean	7 8 8

others. If so, they can calculate percentiles and percentile ranks (Chapter 2) or z scores (Chapters 3 and 7).

Questions about the population from which a sample was drawn are answered either by conducting a test or constructing a confidence interval. For categorical data, the chi square test of goodness of fit is used to determine how well a hypothesized distribution fits an obtained one (Chapter 5). To estimate the percentage of a population falling in a certain category, a confidence interval for the percentage can be calculated. For quantitative data, questions about the population usually concern the population mean (μ). Hypotheses about μ are tested using either a normal curve test (if the population standard deviation, σ, is known) or a t test (if σ is unknown). To estimate the value of the population mean, a confidence interval for μ can be calculated. The normal curve test was discussed in Chapter 7, and the t test, as well as the construction of confidence intervals, in Chapter 8.

Here are some examples illustrating how selection of an appropriate procedure follows the steps: (1) identify the type of data; (2) seek the answer to the specific question being asked.

EXAMPLE 1: In a telephone survey, 100 randomly selected American business executives are asked, "Do you favor free trade?" The researcher wants to estimate the percentage of all American business executives who favor free trade.

The data consist of answers to a question. Each respondent's answer falls into one of two or more categories (e.g., "Yes" or "No," or "Yes," "No," or "Undecided"). The data are categorical.

The specific question refers to the percentage of the population falling in the category "Yes." To estimate this percentage, the researcher needs to construct a confidence interval for the population percentage around the sample percentage of "Yes" answers.

EXAMPLE 2: Each business executive in the sample is asked an additional 20 questions about free trade. The answer to each question is scored 0 or 1, and the sum of the 20 questions is the total attitude score. The researchers want to know: (1) How are attitude scores distributed over the possible range from 0 to 20?; and (2) What is the average attitude score of all American business executives?

The attitude scores are quantitative. To find out how scores are distributed, the researchers need to construct a frequency distribution and a graph such as a histogram. To complete the description of the obtained distribution, they will calculate measures of central tendency, such as the mean and median, and variability, such as the interquartile range and standard deviation.

To estimate the average attitude score in the population of all American business executives, a confidence interval for the population mean is needed.

EXAMPLE 3: Harry scored 25 on this week's spelling test and 12 on last week's math quiz. What percentage of Harry's classmates scored lower than he did on each occasion?

The data are quantitative. The question concerns the location of specific scores (Harry's) within a distribution (the class). According to Table 15–1, percentiles, percentile ranks, and z scores are calculated to answer this kind of question. Since we want to know the percentage of the class falling below Harry, we need to find Harry's percentile ranks on the exam and on the quiz.

QUESTIONS ABOUT RELATIONS BETWEEN TWO OR MORE VARIABLES MEASURED IN ONE GROUP

Sometimes two (or more) variables are measured in one group of individuals either to answer questions about the relation between the variables or to predict one of the variables from the other(s). Here are some examples:

What is the relationship between college students' math anxiety and performance in science courses?

Are voters' attitudes toward construction of a new library related to their political affiliation?

How well can workers' job performance be predicted from their aptitude test scores and years of experience?

Research designed to answer questions like these is called *correlational research.* Table 15–2 summarizes analyses that are needed to answer specific correlational research questions.

As you can see in Table 15–2, questions about the strength of the relation between two variables are answered by calculating correlation coefficients. In Chapter 9 we discussed the calculation and interpretation of correlation coefficients for data consisting of two sets of scores or amounts (the Pearson r), for two sets of ranks (Spearman's rank-order correlation coefficient, r_s), for two dichotomous categorical variables (the phi coefficient, ϕ), and for one dichotomous and one quantitative variable (the point biserial correlation coefficient, r_{pb}).

The correlation coefficient provides information about the strength of a relationship between two variables. If you want to determine whether the relation is significantly different from zero (i.e., if you want to test the null hypothesis that the variables are unrelated in the population), you need to conduct a test—a chi square test if both variables are dichotomous, a t test for other combinations of variables.

Table 15–2 **Decision Table for Selecting Appropriate Analyses for Questions Concerning the Relation Between Two (or More) Variables Measured in One Group**

Type of Data	Specific Question	Analysis or Analyses	Text Chapter
Two Categorical Variables	Are the two variables related?	Chi square test of independence	5
	What is the strength of the relation (of two dichotomous variables)?	Phi coefficient (ϕ)	9
Two Sets of Ranks	What is the direction/strength of the relation?	Spearman rank-order correlation coefficient (r_s)	9
Two (or More) Sets of Quantitative Data (scores, amounts)	What is the direction/strength of the relation?	Pearson correlation coefficient (r)	9
	Is the relation significantly different from zero?	Test of the significance of r	9
	Prediction of one variable from one or more others.	Regression analysis	10
One Dichotomous Categorical Variable and One Quantitative Variable	What is the strength of the relation?	Point biserial correlation coefficient (r_{pb})	9

Prediction of one variable from one or more others calls for a regression analysis—simple regression if there is just one predictor variable and multiple regression if there are two or more. In either case, the analysis involves the development of a regression equation for predicting the criterion variable from the predictor variable(s) and the calculation of a standard error of estimate to measure the amount of error in the predictions. Prediction intervals for criterion scores can be calculated if needed. Regression was the topic of Chapter 10.

EXAMPLE 1: A psychologist gives a test measuring math anxiety to 75 college students and calculates each student's grade point average (GPA) in science courses. She wants to know how math anxiety and science GPA are related in college students at her university.

Both math anxiety scores and GPA are quantitative variables. To measure the direction and strength of the relation between the variables, the psychologist needs to calculate a Pearson correlation coefficient, r. In addition, since she wants to generalize to the population of college students, she should conduct a test to determine whether r is significantly different from zero.

A few additional comments about this example:

1. Before calculating r, the psychologist should examine the scatter diagram of the relation between math anxiety and science GPA to detect possible nonlinearity, outliers, or heteroscedasticity.
2. The accuracy of her generalization to the population depends on the method by which the sample of 75 students was selected. If the sample was not random, the psychologist needs evidence that a random sample would probably be similar to the one she used.

EXAMPLE 2: To find out whether voters' attitudes toward construction of a new library are related to their political affiliation, the county commissioners poll a random sample of registered voters, asking, "Do you favor construction of a new county library?" and "What is your political affiliation?"

Answers to each question fall into one of two or more categories, such as "Yes" or "No" for library construction and "Democrat" or "Republican" for political affiliation. Both variables are categorical. The test used to find out whether two categorical variables are related is the chi square test of independence.

EXAMPLE 3: A company requires job applicants to take an aptitude test and to fill out a questionnaire about their work history. Once hired, workers are regularly evaluated by their supervisors. The company wonders whether aptitude test scores and years of previous experience can be used to predict job performance ratings of future employees.

There are three variables, all quantitative. The company wants to predict one of the variables, job performance ratings, from the other two, aptitude test scores and previous experience. A regression analysis is needed. Because there are two predictors, multiple regression is called for.

Questions About Differences Between Two or More Groups or Under Two or More Experimental Conditions

Most research questions concern differences between two or more populations or differences between the effects of two or more experimental conditions. Here are some examples:

> Does satisfaction with college life differ between male and female students? If so, how big is the difference?
> Which of three treatments is most effective for reducing anxiety?
> What are the effects of prior knowledge and types of study questions on statistics exam scores?

To answer questions like these, samples are selected from the populations concerned. Inferential procedures are applied to differences between the samples in order to draw conclusions about the differences (if any) between the populations from which the samples came. Therefore, all these questions call for statistical tests for the significance of a difference or for confidence intervals for a difference or both. Relevant procedures occupied our attention over the remainder of this book (Chapters 11 to 14, as well as part of Chapter 5).

As with other types of research questions, the appropriate analysis depends on the type of data and the specific research question. However, two additional factors must also be considered. The first factor is the number of samples. For example, though a t test can be used to compare the means of two samples, a comparison of three or more means requires an analysis of variance (ANOVA). The second factor is whether the samples are independent or correlated. For example, the t test for comparing the means of two correlated samples (repeated measures or matched groups) differs from the t test for comparing the means of two independent samples.

When faced with the need to select an appropriate procedure for comparing two or more samples, a useful strategy is to ask the following questions, in turn:

What kind of data do I have?	→	What specific kind of question am I trying to answer?	→	How many samples are there, and how were they selected?	→	What is the appropriate analysis for answering this question for this type of data, and this number and type of samples?

The strategy is incorporated in the guidelines presented in Table 15–3.

To compare two or more distributions of categorical data or ranks, nonparametric tests are used. If the data are categorical and the samples are independent, the appropriate test is the chi square test of independence (Chapter 5). Comparisons of two or more sets of ranks call for one or another of the nonparametric tests discussed in Chapter 14, the particular test depending on the number of samples and their relationship. (Remember that nonparametric tests can be applied to quantitative data as well, if the scores are first converted to ranks. However, this is not usually done unless assumptions underlying parametric tests are seriously violated and samples are small.)

The majority of tests discussed in this textbook—and by far the most frequently used by researchers and statisticians—are parametric tests for differences between means. If two means are to be compared, the most common method of analysis is the t test for the difference between means. If, however, a researcher wishes to estimate the size of the difference between two population means, he or she needs to construct a con-

Table 15–3 Decision Table for Selecting Appropriate Analyses for Questions About Differences Between Two or More Groups or Under Two or More Conditions

Type of Data	Specific Question	Number of Samples and Method of Selection	Analysis or Analyses	Text Chapter
Categorical	Is there a difference between two or more population distributions?	2 or more independent samples	Chi square test of independence	5
Ranks	Is there a difference between two or more population distributions?	2 independent samples	Mann-Whitney *U* test	14
		2 correlated samples	Wilcoxon test	14
		3 or more independent samples	Kruskal-Wallis *H* test	14
		3 or more correlated samples	Friedman analysis of variance by ranks	14
Quantities, Scores, Amounts	Is there a difference between two or more population means as a function of one independent variable?	2 independent samples	*t* test for independent samples	11
		2 correlated samples	*t* test for correlated samples	11
		3 or more independent samples	One-way ANOVA	12
		3 or more correlated samples	Repeated measures ANOVA	13
	What is the size of the difference between two population means?	2 independent samples	Confidence interval for $\mu_1 - \mu_2$, independent samples	11
		2 correlated samples	Confidence interval for $\mu_1 - \mu_2$, correlated samples	11
	Given two independent variables, *A* and *B*, do population means differ as a function of Variable *A*, Variable *B*, and/or the interaction between *A* and *B*?	Independent samples	Two-way (factorial design) ANOVA	13

fidence interval for the difference between means. Both t tests and confidence intervals for differences between means were discussed in Chapter 11.

Comparisons of three or more means involve some type of analysis of variance (ANOVA), the particular type depending on the number of independent variables (one or two) and the method of sampling. If there is only one independent variable and the samples are independent, a one-way ANOVA is called for (Chapter 12). Data from studies with two independent variables applied to independent samples (factorial designs) require a two-way ANOVA, whereas data from studies with one independent variable but correlated samples (repeated measures or matched groups) call for a repeated measures ANOVA. Both two-way and repeated measures analyses were discussed in Chapter 13. Though not listed in Table 15–3, planned comparisons and/or post hoc tests can be conducted as a supplement to any kind of ANOVA under conditions explained in Chapters 12 and 13.

Example 1: The staff of a student activities office has developed a scale to measure satisfaction with college life. They administer the scale to 100 randomly selected male students and 100 randomly selected female students. They wonder whether satisfaction differs as a function of sex and, if so, how much.

The variable measured in both groups, level of satisfaction, is quantitative. The staff wants to compare the mean satisfaction scores of the males and the females. Since samples were randomly selected from two populations, the samples are independent. The appropriate test is the t test for independent samples.

The staff also wants to estimate the size of the difference between the two population means. They need to construct a confidence interval for the difference between means of two independent samples.

Example 2: Which of three treatments, drug therapy, behavior therapy, or counseling, is most effective for treating anxiety? To answer this question, 18 clients at an anxiety disorder clinic are randomly divided into three treatment groups. After 6 weeks of treatment, an anxiety scale is administered to all the clients.

The question asks about the effect of one independent variable (type of treatment) on a quantitative dependent variable. Since subjects were randomly assigned to groups, the samples are independent. To find out whether there is significant variation among mean anxiety scores of the three groups, a one-way ANOVA should be conducted. If the result is significant, post hoc tests can be used to find out whether any one treatment (and, if so, which one) is most effective.

A few additional comments about this example:

1. The sample sizes are small. If the distributions of anxiety scores within samples are very skewed or if outliers are present, it may be advisable to use a nonparametric test instead of ANOVA. An appropriate nonparametric test is the Kruskal-Wallis H test.
2. Suppose that subjects are assigned to the three treatment groups like this: All 18 clients are given an anxiety pretest. Matched sets of three

clients each are formed based on pretest scores. Within each set, one client is randomly assigned to each treatment.

In this case, the three samples are not independent but correlated. A repeated measures ANOVA is required, followed by appropriate post hoc tests if the result of the ANOVA is significant. The relevant nonparametric test, if assumptions of ANOVA are seriously isolated, would be the Friedman analysis of variance by ranks.

EXAMPLE 3: A pretest of statistical concepts is administered to 40 students at the start of a statistics course. Based on their scores, two subgroups of 20 students each are identified: high prior knowledge (above the median on the pretest) and low prior knowledge (below the median on the pretest). Within each subgroup, 10 randomly selected students are provided with conceptual study questions and 10 with computational study questions. The research question is: What are the effects on statistics exam scores of level of prior knowledge and type of study questions?

The dependent variable, statistics exam scores, is quantitative. The experiment is a factorial design with two independent variables (level of prior knowledge and type of study questions). Because students within each level of prior knowledge were randomly assigned to the study question conditions, the samples are independent. The proper analysis is a two-way ANOVA, which tests not only for the effect of each independent variable but also for the interaction between them.

KEYWORDS

The major factors to be considered in selecting an appropriate analysis have been discussed in the preceding sections. You may have noticed, however, that certain words or phrases often serve as additional clues. The inclusion of one or more such "keywords" in a research question can help you decide on the required analysis. Here are some examples:

1. A college admissions officer wants to *predict* college grade point averages from high school grades and SAT scores.

   The keyword is "predict." To predict one variable from one or more others, researchers construct regression equations. The required analysis is regression (multiple regression in this case because there are two predictors, high school grades and SAT scores).
2. A developmental psychologist wants to *estimate* the average number of words in the vocabulary of 3-year-old children from a sample of 20 3-year-olds.

   The keyword is "estimate." To estimate population parameters from sample statistics, we construct confidence intervals. Here we need a confidence interval for the mean.
3. The research question concerns the relative effectiveness of two methods of teaching reading in first grade. Two groups of children are first *matched* on reading readiness. Then each group is taught by a different method.

   The keyword is "matched." Whenever two or more matched groups are to be compared, a test for correlated samples is required.

Table 15–4 lists keywords. For each keyword, a sample research question is stated and the appropriate analysis is indicated.

Table 15–4 A Sample of Keywords: Clues to Selection of an Appropriate Analysis

Keyword	Example of Use	Analysis or Analyses
Shape	What is the shape of the distribution of anxiety scores?	Frequency distribution, graph, measures of symmetry/skew: comparison of Q_1, Q_2, and Q_3, or comparison of mean and median
Average	What is the average of the experimental group?	Measures of central tendency: mean, median, mode
Individual differences	How great are individual differences in reading ability in third grade?	Measures of variability: *SD*, range, etc.
(Strength of) relation	What is the relation between helping behaviors and popularity in preschool children?	Correlation coefficients: r, etc.
Predict	Can productivity be predicted from job satisfaction?	Regression analysis
Estimate (a parameter)	I want to estimate the mean number of defective widgets per batch from a sample of 10 batches.	Confidence interval
Difference, significant difference	Is the difference between the experimental and control groups significant?	A statistical test
Match, matched, matching	The three groups to be compared were first matched on intelligence.	A test for correlated samples.
Interact, interaction	Is the interaction between number and length of study sessions significant?	Two-way (factorial design) ANOVA

EXERCISE 15–1

1. In each situation, decide what procedure or test is needed to analyze the data.
   a. A certain population has mean = 75. My group of 15 subjects has a mean of 72. Is my group a likely random sample of the population?
   b. You want to estimate the actual difference between the mean scores of all males and females on a mechanical comprehension test, based on a sample of males and females.
   c. A child psychologist wondered whether preschoolers' dependency is affected by their child-care arrangements. The psychologist developed a test to measure dependency. He administered the test to 20 children in a day-care center, 18 children cared for at home by babysitters, and 15 children cared for at home by their parents. Then he compared the dependency scores of the three groups.
   d. Do consumers prefer some methods of packaging a cereal over others? The company that produces the cereal asks a random sample of 150 customers their preference among four packages and counts the number of consumers who prefer each one.
   e. The company wonders whether preferences vary with age. To answer this question, the distribution of preferences of people over and under 25 are compared.
   f. A math class is given a pretest at the start of a unit and a posttest at the end. The teacher wonders whether performance has improved.

g. The teacher also wonders whether there is a relationship between students' pretest and posttest scores.

h. The teacher wants to predict posttest scores from pretest scores.

2. Can you generate other keywords besides those in Table 15–4?

Common Themes

In discussing the selection of appropriate analyses, we focused on *differences* between various tests and procedures. Now we switch our attention to *similarities*. When we look behind procedures that appear, on the surface, to be very different, important unifying concepts become apparent. We will examine three of these unifying concepts. Two of them have been discussed in earlier chapters: the *logic of hypothesis testing* and *components of variance*. The third concept, the *general linear model*, is new.

The Logic of Hypothesis Testing (Again) and a New Look at the Test Statistic

Hypothesis testing is the most common method for making inferences about populations from sample data. Many statistical tests have been discussed in this book, and there are many others that have not. Despite the fact that different tests test different hypotheses, under different conditions, and use different procedures, the logic is the same. In all tests, we

1. State the hypothesis to be tested.
2. Select a *level of significance, alpha* (α). That is, we decide how unlikely our sample outcome must be (if the hypothesis is true) for us to reject the hypothesis. We will reject the hypothesis if the probability (p) of a sample outcome like ours is less than alpha.
3. Collect data and calculate a *test statistic*. That is, we calculate a value in the *probability distribution* that describes the probabilities of various sample outcomes.
4. Find the probability (p) of the calculated test statistic by comparing it with *critical values* of the probability distribution. If $p < \alpha$, reject the hypothesis.

You have seen how this logic is applied in chi square tests, t tests, ANOVA, and others.

As you can see in Step 3, all tests involve the calculation of a test statistic. The particular test statistic and its computation differ from one test to another. For example, different procedures are used to arrive at a value of χ^2 in a chi square test than those used to arrive at a value of F in an ANOVA. But are they really so very different? Let us examine the formulas for calculating test statistics in several tests. We will find a surprisingly common structure.

In most tests, the test statistic is a ratio, that is, a fraction with a numerator and a denominator. For example

Test statistic in the chi square test: $$\chi^2 = \Sigma\left[\frac{(f_o - f_e)^2}{f_e}\right]$$

Test statistic in t tests: $$t = \frac{M - \mu_H}{s_M} \quad \text{or } t = \frac{M_1 - M_2}{s_{\text{diff}}}$$

Test statistic in one-way ANOVA: $$F = \frac{MS \text{ between}}{MS \text{ within}}$$

If we examine the numerators and denominators of these ratios, we notice that

1. In every case, the numerator is a measure of an obtained difference.

Chi square test:	$f_o - f_e$ is the obtained difference between an observed frequency in a category or cell and the expected frequency.
t tests:	$M - \mu_H$ is the obtained difference between the sample mean and the hypothesized population mean.
	$M_1 - M_2$ is the obtained difference between the means of two samples.
One-way ANOVA:	*MS* between, the mean square between, is a measure of the obtained variation (differences) among a set of sample means.

2. In every case, the denominator is a measure of the difference expected by chance.

Chi square test:	f_e (over all cells) is a measure of differences expected among the cell frequencies based on the hypothesis.
t tests:	s_M, the estimated standard error of the mean, is a measure of the chance variation (differences) among means of samples from a population.
	s_{diff}, the estimated standard error of the difference between means, is a measure of chance variation of differences between means ($M_1 - M_2$) from one pair of samples to another.
One-way ANOVA:	*MS* within, the mean square within, is a measure of the variation expected among sample means by chance.

Thus, the test statistic in all these tests is a ratio between an obtained difference and the amount of difference expected by chance.

$$\text{Test statistic} = \frac{\text{Obtained difference}}{\text{Chance difference}}$$

In each test, the hypothesis is rejected if the obtained difference (numerator) is sufficiently greater than that expected by chance (denominator).

In summary, all statistical tests share a common logic and sequence of operations. In many tests, moreover, the structure of the test statistic is the same. Keeping these common features in mind can enhance your understanding not only of tests that were discussed in this book but also of those you may encounter in the future.

EXERCISE 15–2

The t test for the significance of a correlation coefficient, r, tests the null hypothesis that the population correlation, ρ, is zero. The test statistic is $t = \frac{r\sqrt{N-2}}{\sqrt{1-r^2}}$, which can be expressed as $t = \frac{r}{\sqrt{\frac{1-r^2}{N-2}}}$. Assume that this test follows the general rule that test statistic = (obtained difference)/(chance difference).

1. In what sense is the numerator, r, an "obtained difference"?
2. If you were asked to give a name to the denominator term, $\sqrt{\frac{1-r^2}{N-2}}$, what would you call it?

Components of Variance (Again)

In Chapter 6 we introduced the concept of components of variance. We pointed out that many statistical analyses enable researchers to apportion the total variation of their data sets into two or more components, so that they can make statements like "20% of the variance is due to Factor X, 65% to Factor Y, and 15% to other factors." You were informed then that the concept would reappear in many of the following chapters.

Our first application of the concept of components of variance arose in the discussion of correlation and regression (Chapters 9 and 10). In Chapter 9 we stated that the squared correlation coefficient, between two variables, r^2, is the proportion of the variance of one variable that can be attributed to the other variable. For example, if $r = .8$, $(.8)^2 = .64$ (or 64%) of the variance of variable X is accounted for by variable Y, and vice versa. We elaborated on this interpretation of r^2 in Chapter 10.

Any squared correlation coefficient (not just the Pearson r) can be interpreted in terms of components of variance. For example, R^2 (the squared multiple correlation between a criterion and a set of predictors) is equal to the proportion of criterion variance accounted for by the combination of weighted predictors; the square of Spearman's rank-order correlation coefficient (r_s^2) is the proportion of the variance of one set of ranks accounted for by the other; and so on.

In later chapters we observed that components of variance interpretations are not limited to correlational analyses. The results of statistical tests also provide information about components of variance. For example, after conducting a t test for the difference between means of independent samples, we can calculate the squared correlation between the independent variable and the dependent variable by the formula

$$r_{pb}^2 = \frac{t^2}{t^2 + n_1 + n_2 - 2}$$

Or, after conducting a one-way ANOVA, we can calculate $\hat{\omega}^2$ by the formula:

$$\hat{\omega}^2 = \frac{SS\text{ between} - (k-1)\,MS\text{ within}}{SS\text{ total} + MS\text{ within}}$$

Both r_{pb}^2 and ω^2 are squared correlation coefficients, and both tell us what proportion of the dependent variable variance is accounted for by the independent variable.

The concept of components of variance provides a bridge between correlational analyses and statistical tests like t tests and ANOVA. Despite differences in procedure—calculation of correlation coefficients or regression equations in correlational analyses versus calculation and evaluation of test statistics like t or F in statistical tests—they are not as different as they appear on the surface. Both correlational analyses and tests answer questions about the relations between variables. Correlational analyses answer questions about relations between two measured variables, and statistical tests answer questions about relations between independent and dependent variables. And, in both cases, the strength of the relation can be expressed in terms of the proportion of variance of one variable accounted for by the other. Thus, both correlational analyses and statistical tests provide researchers and practitioners with vital information about sources of variation in the populations with which they are concerned.

EXERCISE 15–3

To find out how scores on two variables are related, Dr. Smith gives tests measuring the variables to a group of subjects (a correlational study). Dr. Jones conducts an experiment in which an independent variable is varied among independent groups to determine its effect on a dependent variable. Both Dr. Smith and Dr. Jones want to answer the following two questions: (1) What is the strength of the relation between the variables? (2) Is the relation significant? Describe the procedure each researcher must use to answer each question. (Which of the two questions does Dr. Smith answer first? Which question does Dr. Jones answer first?)

An Introduction to the General Linear Model

We have seen that both statistical tests and correlational analyses provide information about components of variance. Another factor common to statistical tests and correlational analysis is the model underlying the analyses. The word *model* means roughly the same thing as *theory*. For every statistical analysis, there is an underlying model or theory about how an individual's score on a variable, Y, is assumed to be affected by other variables, X. The model takes the form of an algebraic equation that expresses Y as a function of X. The procedures of the statistical analysis and the interpretations of the results can be derived from the model.

The model that underlies both regression analyses for one or more predictor variables and analyses of variance, both simple and complex, as well as a number of other so-called multivariate analyses, is the ***general linear model***. In this section I will try to give you a feeling for what the general linear model is and what it does in a relatively nontechnical way. Extended discussions can be found in more advanced statistical texts.

The Linear Model in Regression Analysis

Let us begin with simple regression. Recall from Chapter 10 that the regression equation for predicting a criterion variable, Y, from a predictor variable, X, is a linear equation with the form

$$\hat{Y} = a + bX$$

$\hat{Y}$ is the predicted Y for a given X; a, the Y intercept of the regression line, is the value that Y is expected to take if $X = 0$, a kind of baseline value; and b is the regression coefficient, a measure of the weight of X in predicting Y.

Unless the correlation between X and Y is perfect, many individuals obtain actual criterion scores that differ to some extent from their predicted criterion scores. The difference between an individual's actual and predicted criterion scores, $Y - \hat{Y}$, is called an *error of estimate, e*. Since $e = Y - \hat{Y}$, $\hat{Y} = Y - e$. Substituting $Y - e$ for $\hat{Y}$ in the regression equation, we get

$$Y - e = a + bX$$

Rearranging terms

$$Y = a + bX + e$$

Thus, we see that an individual's criterion score, Y, is assumed to be a linear combination of a baseline value (a), plus the predictor score weighted by the regression coefficient (bX), plus some amount of error. With the addition of a few assumptions, which we will not go into here, the equation constitutes the linear model underlying simple regression analysis.

The model for simple regression is easily generalized to multiple regression, in which a criterion variable is predicted from two or more predictor variables, $X_1, X_2, \ldots$ If we let k represent the number of predictors, the linear model for multiple regression is

$$Y = a + b_1X_1 + b_2X_2 + \cdots + b_kX_k + e$$

where a is, once again, a baseline value of $\hat{Y}$ when $X_1, X_2, \cdots, X_k = 0$; $b_1, b_2, \cdots, b_k$ are regression coefficients or weights for variables $X_1, X_2, \cdots, X_k$; and e is the error of estimate for this particular individual. Thus, an individual's criterion score, Y, is assumed to be a linear combination of a baseline value (a), plus a sum of weighted predictors ($b_1X_1, b_2X_2, \cdots, b_kX_k$) plus some amount of error.

The Linear Model in Analysis of Variance

Now consider an experiment with one independent variable having k treatment levels. A group of subjects is randomly subdivided into k treatment groups, and, after treatment, the dependent variable is measured in all subjects. The usual method of analyzing data like these is, of course, one-way ANOVA, rather than regression. Nevertheless, if we express a subject's dependent variable score as a function of the experimental treatments and other factors, we wind up with a linear equation (model) that takes exactly the same form as in multiple regression.

We let Y stand for a subject's score on the dependent variable. $X_1, X_2, \cdots, X_k$, represent treatments $1, 2, \cdots, k$, respectively. Every subject has a "score" on each treatment—a score of 1 if the subject was in that treatment group and a score of 0 if he or she was not. For example, in a three-treatment experiment, a subject in Group 2 has the following treatment scores: $X_1 = 0$, $X_2 = 1$, $X_3 = 0$.

Each treatment is assumed to have some effect (which may be positive, negative, or zero). We let t stand for the effect of a treatment.† Thus, t_1 is the effect of Treatment 1, t_2 is the effect of Treatment 2, and so on. As an example, suppose that Treatment 1 raises Y scores by 3 points, Treatment 2 lowers them by 2 points, and Treatment 3 has no effect. Then $t_1 = 3$, $t_2 = -2$, and $t_3 = 0$.

We assume, further, that the effect of a treatment is the same for all members of the treatment group. If we let $\hat{Y}$ stand for a subject's expected dependent variable score, we can now write

$$\hat{Y} = a + t_1X_1 + t_2X_2 + \cdots + t_kX_k$$

where a is a constant that represents the baseline score that is expected without any treatment at all. As an example, in a three-group experiment, the expected dependent-variable score for each subject in Group 2 is $\hat{Y} = a + t_1 (0) + t_2 (1) + t_3 (0)$, or $\hat{Y} = a + t_2$. In plain English, we expect a subject in Group 2 to score at a level equal to the pretreatment baseline plus the effect of Treatment 2.

However, not all subjects in a given treatment group actually obtain the expected dependent variable score, $\hat{Y}$. Individual differences between subjects within groups (e.g., differences in intelligence, motivation, and prior experience) produce within-group variation on the dependent variable. We let e (error) stand for the difference between a subject's actual and expected dependent variable scores.

$$e = Y - \hat{Y}$$

Therefore, as in regression, $\hat{Y} = Y - e$. Substituting in the equation for $\hat{Y}$:

$$Y - e = a + t_1X_1 + t_2X_2 + \cdots + t_kX_k$$

†This is not the same t as in a t test.

Rearranging terms

$$Y = a + t_1 X_1 + t_2 X_2 + \cdots + t_k X_k + e$$

This linear equation is the linear model for one-way ANOVA. With the addition of a few reasonable assumptions, the theory and calculations of one-way ANOVA can be derived from the model.

The General Linear Model

Now let us compare the linear models underlying multiple regression and one-way ANOVA. Note that except for the use of regression coefficients (b_1, b_2, . . .) in the model for multiple regression versus the use of treatment effects (t_1, t_2, . . .) in the model for one-way ANOVA, the equations are identical.

Multiple regression: $Y = a + b_1 X_1 + b_2 X_2 + \cdots + b_k X_k + e$

One-way ANOVA: $Y = a + t_1 X_1 + t_2 X_2 + \cdots + t_k X_k + e$

The linear models underlying more complex analyses of variance, as well as those for other multivariate analyses, are highly similar to the models for multiple regression and one-way ANOVA. The basic format and accompanying assumptions are the general linear model.

The fact that a common model underlies multiple regression and ANOVA has important implications for data analysis. For example, we have used regression procedures for analyzing data from correlational studies while using ANOVA for analyzing data from experimental studies. However, with appropriate coding of the treatment variables, regression procedures can be used to analyze data from experimental studies as well. How this is done is beyond the scope of this book. For further information, see, for example, Judd and McClelland (1989).

Sometimes there are advantages to recoding experimental data and conducting a regression analysis rather than an ANOVA. For example, the two-way ANOVA procedures discussed in Chapter 13 for analyzing data from factorial designs require an equal number of subjects in each group. In multiple regression analyses of factorial design studies, on the other hand, equality of group sizes is not necessary. Another advantage of multiple regression procedures is that they permit the analysis of studies in which some independent variables are dichotomous (like treatments in an experiment) and others are continuous (like predictors in a correlational study). As researchers become more knowledgeable about the general linear model in data analysis, and as computer packages make the model more user-friendly, we can expect to see increased applications of multiple regression procedures to analyze data from both experimental and correlational research in the future.

The Role of Statistics in the Research Process

Statistical analyses are performed on research data. For the statistician, the data are the starting point. For the researcher, however, the statistical analysis comes near the end of the research process. In research, a great deal of thought, planning, and activity take place before data are collected and ready for analysis. What kinds of thinking, planning, and activity are involved? Where does a research study originate? What are the steps in the research process? And just where and how does statistics fit in? These are the kinds of questions to be addressed in this section.

Steps in the Research Process

As we have already seen, there are many methods for conducting research. They include survey methods, correlational methods, experiments, nonexperimental comparisons of preexisting groups, and still others. Despite these differences, however, the sequence of activities in all kinds of research is much the same.

If you have read research reports, such as those in journals, you may have noticed that they follow a standard format. The author first states the purpose of the study, briefly summarizes related theory and research, and then states a research hypothesis or hypotheses. Then a Method section describes the subjects, instrumentation (tests, equipment, etc.), design, and procedure. The final sections present the results and conclusions.

The succession of topics in a research report closely parallels the sequence of activities involved in planning and conducting a research study, as shown in Figure 15–1.

Figure 15–1

Steps in the research process.

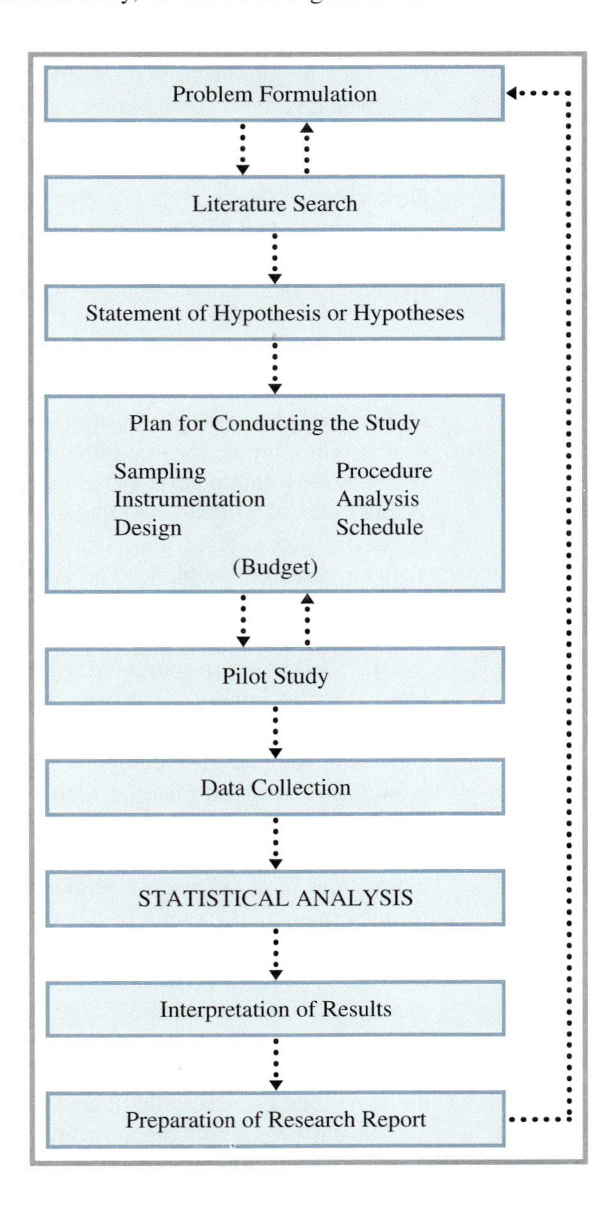

We will discuss each step in turn. To illustrate the process, we will follow the progress of a graduate student named Nestor as he planned and carried out a research project for his master's degree in educational psychology.

Problem Formulation

It goes without saying that before a researcher can begin to investigate a problem, he or she must be aware that a problem worth investigating exists. Where do research problems come from? Research problems originate in many different ways. Some are derived from theories. Others are suggested by the results of previous research. Sometimes they arise empirically, in the form of "I wonder what would happen if . . .?" For people who are knowledgeable about their field, the problem is usually too many research ideas rather than too few.

Nestor's research problem originated in readings and coursework related to his degree program. His area of interest was the effect of parental child-rearing practices on children's emotional development. The research problem that interested him, as he first formulated it, was "What is the relation between parents' behavior and their children's level of anxiety?"

Notice that at this point, Nestor had not specified exactly what was meant by "parents' behavior" or "anxiety," so that his problem was not yet well defined. In this regard, Nestor's case was typical. Research problems, as initially formulated, tend to be fuzzy and overly broad. The variables may be poorly defined or undefined, and the boundaries of the problem are unclear. The next step in the research process is undertaken to clarify the problem, narrow and focus it, and define the variables more precisely.

The Literature Search

Research is not conducted in a vacuum. Research studies almost always have roots in and connections with an existing body of theories, facts, and previously conducted research. The purpose of the literature search is to uncover this body of material.

Theories and reports of research generally exist in print form, in books, monographs, journal articles, or reports of presentations at conferences. Various indexes can be used to locate those particular books, journal articles, and the like that are relevant to a given research problem. The contents of most indexes, such as *Psychological Abstracts*, now exist as computer databases. This makes the search for relevant literature a much quicker and less tedious process than it used to be.

In conducting a computerized literature search, the user enters one or more keywords. For example, if you were searching for literature related to the effect of child abuse on children's performance in school, you might use the keywords "child abuse" and "achievement." The computer then shows on the screen (and prints out, if requested) references to printed material filed under these two keywords.

There is an art to literature searching. Although the computer does most of the clerical work, an intelligent and sometimes creative use of search strategies is necessary to ensure that most, if not all, relevant material has been identified. Other sources of information, such as experts in the field, may also be referred to.

The theories and research results identified by the literature search help the researcher to refine and focus the research problem. For example, by reading articles about related research, the researcher may learn about reliable and valid ways of defining and measuring the major variables of the study, as well as ways of controlling extraneous variables. In Figure 15–1, the return arrow going from "Literature Search" to "Problem Formulation" indicates how the literature search contributes to reformulation of the problem.

Nestor began his computer literature search by entering the keyword "anxiety." He quickly discovered that anxiety was far too broad a concept; hundreds of books, articles, and other materials have been produced about anxiety. After examining various theories of anxiety, he decided that the most fruitful approach for his purpose was that taken by theorists who view anxiety as a trait that influences behavior relatively directly (as opposed to the Freudian concept of anxiety as a defense mechanism that operates unconsciously). A psychologist named Spielberger, Nestor found, had done a great deal of work on the theory and measurement of trait anxiety. Nestor decided to use Spielberger's definition of trait anxiety, as well as his test, the State-Trait Personality Inventory (STAI)[†] in his study.

Next, Nestor began his search of literature related to effects of parents' behavior on child development. He quickly noticed two different approaches to measuring parents' behavior: direct observation on the one hand and assessment of parents' behavior by their children (children's perceptions of parents' behavior), on the other. The second approach seemed more suitable for his purposes. As he reviewed the literature on children's perceptions of parental behavior, Nestor came across a research program conducted by Schaefer and his colleagues.[††] Schaefer identified three dimensions of parental behavior: acceptance versus rejection (warmth), psychological control versus autonomy, and firm control versus lax control. He also constructed an inventory to measure children's perceptions of their fathers' and their mothers' behavior with respect to each dimension, the Children's Reports of Parental Behavior Inventory (CRPBI). Nestor decided to define parents' behavior in terms of Schaefer's three dimensions and to use the CRPBI to measure it.

Having defined his variables, Nestor turned his attention to the population in which he proposed to study the relation between the variables. Both the STAI and the CRPBI were designed for use with older adolescents and young adults, an age range that includes college students. Nestor decided to limit his study to college students. He was now able to reformulate his problem more precisely as "What is the relationship between college students' trait anxiety and their perceptions of their parents' warmth, control, and firmness of control?"

Statement of Hypothesis or Hypotheses

Having formulated a precise problem to be investigated, the researcher is ready to state one or more hypotheses. Hypotheses are statements of the expected results (expected relationships, expected differences), based on theory and previous research.

Note that the hypotheses formulated at this stage are research hypotheses (directional or nondirectional). They are not null hypotheses. Null hypotheses are needed for the statistical analyses to come later, but the conduct of the study is guided by one or more research hypotheses. Based on his literature review, Nestor stated the following four research hypotheses:

1. There is a negative relationship between college students' trait anxiety and their perceptions of acceptance by their fathers and their mothers.
2. There is a positive relationship between college students' trait anxiety and their perceptions of control by their fathers and mothers.
3. There is a positive relationship between college students' trait anxiety and their perceptions of firmness of control by their fathers and mothers.
4. The greater the difference between students' perceptions of their fathers' and mothers' behavior on each dimension, the higher their level of trait anxiety.

[†]Spielberger, C. D. (1970). *The state-trait anxiety inventory manual*. Palo Alto, CA: Consulting Psychologists Press.

[††]Schaefer, E. S. (1965). Children's report of parental behavior: An inventory. *Child Development*, *36*, 413–424.

Plan for Conducting the Study

The researcher cannot go directly from hypothesis statement to data collection. A good bit of planning is needed first. Specifically, the researcher needs to construct plans for sampling, for instrumentation, for the design of the study, for the step-by-step data-collection procedure, and for the analysis that will be applied to the results. Other aspects of planning include preparation of a schedule for the remainder of the study and, sometimes but not always, preparation of a budget.

Sampling. Whatever the research question may be, it is a question about a particular population or populations. How are subjects to be selected from the population? And how many subjects will be required? A sampling plan is formulated to answer these questions.

Instrumentation. Research studies often involve special equipment, instruments (e.g., questionnaires, psychological tests), and other materials, all of which can be referred to as "instrumentation." The researcher must plan and then obtain or construct all the necessary materials. For example, published tests may need to be ordered or stimulus materials prepared.

Design. Particularly in experimental research studies, researchers must decide precisely how to manipulate variables. For example, should the independent variable be varied between or within groups? Should an identified extraneous variable be controlled by matching or by building the variable into the design? Answers to questions like these are referred to as the "design" of the study.

Procedure. After decisions have been made about sampling, instrumentation, and design, the researcher must decide on the exact procedure to be followed in collecting the data. In an experiment, for example, the experimenter must spell out, step by step, what will be done to and by subjects from the time that they enter the laboratory until they leave at the end of the session.

Analysis. Although data analysis cannot be carried out until after data have been collected, it is important at this point to plan the analysis to be applied. Thus, for example, if two independent variables are to be manipulated in a factorial design, the researcher must recognize that the appropriate analysis is a two-way ANOVA. Any preplanned comparisons should also be noted, as well as the types of post hoc tests which may be appropriate, if any. The purpose is to ensure that an appropriate method of analysis exists to test each hypothesis, and also that data to be collected meet the requirements of the planned analysis.

Schedule. Another important aspect of planning is the construction of a schedule for the remainder of the research project, especially if the project is lengthy or includes several phases. By setting up a time line with a projected completion date for each step, the researcher makes it more likely that needed tasks will be completed in a timely manner, and that any deadlines that may exist will be met.

Budget. Conducting research can be very expensive. Although graduate students doing theses or dissertations may be expected to meet all expenses out of their own pockets, funds are often available to help defray some expenses, such as mailing costs. Researchers planning large-scale projects may apply to government agencies or to foundations for large amounts of money. Whether the requested amount is $50 or $250,000, the funding agency will certainly require the submission of a projected budget. Constructing an accurate, detailed list of estimated costs is a time-consuming, difficult job and may require consultation with experts in budget preparation.

Now let us examine Nestor's planning process. We begin with *sampling*. Recall that Nestor had defined his population as "college students." For practical reasons, the population was further restricted to students at the university in which he was enrolled. Although random sampling would be the ideal method of sampling from this population, it was judged to be too difficult and time-consuming to carry out. Nestor decided, instead, to ask several instructors who were teaching undergraduate courses in his department for permission to test their students during class time.† He reasoned that in terms of their level of anxiety and perceptions of parent behavior, these students would be representative of the entire undergraduate student body. Nestor decided that he would need a sample of about 60 subjects. Three classes of about 20 students each would fill his needs.

So far as *instrumentation* was concerned, Nestor had already decided to use the STAI to measure trait anxiety and the CRPBI to measure parent acceptance, control, and firmness. He obtained the needed test booklets, answer sheets, and instructions for administration. No other special materials were needed.

In terms of *design*, Nestor's study was a correlational one and required, simply, that all variables be measured in a single group of subjects. To control for possible order effects, Nestor decided to vary the order of administration of the instruments over subjects, with half the students completing the STAI first and the other half completing the CRPBI first.

The planned *procedure* for data collection followed directly from the design. In each of the three classrooms, Nestor would briefly describe the purpose of his study and request the students' cooperation. Test materials would be distributed; students would be given sufficient time to answer all questions; and test materials would be collected. All the needed data could easily be collected within one meeting of each class. (The ease of data collection is one of the practical advantages of correlational research.)

So far as *data analysis* was concerned, Nestor's study, like most correlational studies, called for the calculation and evaluation of correlation coefficients. Since both the STAI and the CRPBI yield numerical scores, Nestor planned to calculate Pearson correlation coefficients (r). To control for gender (on the chance that the hypothesized relationships might differ between male and female students), Nestor decided that he would calculate separate correlation coefficients for males and females. All correlation coefficients would be tested for significance at the .05 and .01 levels.

As he thought about a *schedule* for the remainder of his project, Nestor realized that data entry into a computer file would require several hours of effort, although the analysis itself would be completed quickly. Nestor knew that the greatest commitment of time would come after data collection and analysis—in the interpretation of the results and the writing of his thesis.

Nestor did not need to prepare a *budget*. His department provided all the needed testing materials, and, as a student, his use of the university computer for data analysis was also free. At this point, Nestor's planning process was complete.

The Pilot Study

Having completed the planning process, the researcher is eager to get on with data collection. But wait a minute. Even though great care has been taken in planning for sampling, instrumentation, and so on, isn't it possible that something has been overlooked? Perhaps the stimulus materials need some changes to make them more useful, or perhaps a different design would produce a better study. If anything isn't working as planned, it is best to find out about it before large-scale data collection begins, while modifications can still be made relatively easily. The purpose of the pilot study is to detect glitches, if any, so that they can be repaired.

†Nestor's study was conducted in the 1970s. His method of obtaining subjects could not be used today. Now, each individual subject must give informed consent for participation.

A pilot study is a small-scale "rehearsal" for the actual data collection. Materials and procedures are tried out on a small number of individuals similar to those who will be used in the study itself. If the tryout suggests a need for changes, such as rewording of questions in a questionnaire or revision of instructions, the changes can usually be made with little difficulty. In Figure 15–1, the return arrow going from "Pilot Study" to "Plan for Conducting the Study" indicates how results of the pilot study feed back into the planning stage.

In his pilot study, Nestor administered the STAI and the CRPBI to a few friends and undergraduate volunteers. He found that subjects were able to complete the inventories without difficulty and that his instructions and time limits were adequate. He also found that the tests were easy to score and that scores showed a satisfactory amount of variability. (*Remember:* If the range of scores on a variable is very small, the possible correlation with other variables is considerably reduced.) Now he was ready to proceed with data collection.

Data Collection

At this point, data are collected according to plan. Except in very small studies, the data are usually entered into a computer file in preparation for analysis. If data are copied from one file to another (e.g., from data sheets to computers) or transformed in any way, it is important to check carefully to make sure that there are no errors in transcription.

Nestor's data collection proceeded smoothly. He tested 59 subjects, including 32 males and 27 females. Next, he scored all the answer sheets, obtaining the following seven scores for each subject: trait anxiety, perceived warmth of father, perceived warmth of mother, perceived control by father, perceived control by mother, perceived firmness of father, perceived firmness of mother. Nestor typed the scores into a computer file and proceeded with his analysis.

Statistical Analysis

The data analysis stage is where the procedures you have learned in this book (Chapters 2 through 14) come in. As you know, analysis may involve constructing graphs or calculating correlation coefficients or conducting statistical tests or still other operations or combinations of several operations. This stage has been adequately discussed in previous chapters; we do not dwell on it further here.

In his analysis, Nestor first examined the distributions of scores on his seven measures and calculated means and standard deviations. The distributions were reasonably symmetrical, with no unusual outliers; therefore he proceeded with his correlational analysis as planned. First he calculated Pearson correlation coefficients (r) between STAI trait anxiety scores and each of the CRPBI scores for male and female subjects. Then he tested all correlation coefficients to determine whether they were significantly different from zero. Because his research hypotheses were directional, Nestor conducted one-tailed tests of the significance of r. Nestor's results are shown in Table 15–5. We discuss Nestor's interpretation of these results in the next section.

Interpretation of Results

The purpose of the statistical analysis is, of course, to test the research hypotheses. Therefore, the first question the researcher asks is, "Did the results of the statistical analysis support the research hypotheses?" If, as is often the case, hypotheses are partially but not entirely supported, it is necessary to consider possible reasons for a lack of fit between hypotheses and results. The researcher may also notice additional research questions that are raised by the results, alternative explanations to be explored, or im-

Table 15–5 Nestor's Pearson Correlation Coefficients Between STAI Trait Anxiety Scores and CRPBI Scores for Male and Female Subjects and for the Total Group

	Warmth		
	Of Father (*F*)	Of Mother (*M*)	Difference Between *F* and *M*
Male subjects	−.19	−.45**	.27
Female subjects	−.11	−.31	.17
Total group	−.15	−.37**	.23*

	Control		
	Of Father (*F*)	Of Mother (*M*)	Difference Between *F* and *M*
Male subjects	.42**	.20	.02
Female subjects	−.11	.14	.17
Total group	−.21	.17	.08

	Firmness of Control		
	By Father (*F*)	By Mother (*M*)	Difference Between *F* and *M*
Male subjects	−.12	.06	.27
Female subjects	.05	.16	−.14
Total group	−.06	.10	.15

*$p < .05$, one-tailed test.
**$p < .01$, one-tailed test.

plications for theory or practice. These, too, belong to the interpretation phase of the study.

How well did Nestor's results support his hypotheses? Nestor's first hypothesis stated that trait anxiety would be negatively correlated with perceptions of parents' warmth. All the correlation coefficients were negative, as expected. However, only the correlations with mothers' warmth, as perceived by males and by the group as a whole, were significant. In female subjects, the correlation with perceived mothers' warmth was not significant; nor were any correlations with perceived fathers' warmth significant.

The second hypothesis stated that trait anxiety would be positively correlated with perceptions of parental control. The correlation between trait anxiety and male subjects' perception of fathers' control was significantly positive; none of the other correlation coefficients was significant.

The third hypothesis stated that trait anxiety would be positively correlated with perceptions of firmness of control. The results showed a mix of positive and negative correlation coefficients, none of which were significant.

The fourth hypothesis stated that trait anxiety would increase with differences between subjects' perceptions of their fathers' and mothers' behavior on each dimension. Only the correlation between trait anxiety and the difference between fathers' and mothers' perceived warmth, over the entire group of subjects, was significant.

Nestor interpreted his results as suggesting that (1) the dimension of parental warmth, especially on the part of mothers, is more salient for the development of anxiety in children than are the two dimensions related to control; and (2) control by the father is more salient for boys than for girls. He drew on several theories of child development to support his conclusions. In addition, he identified several problems that needed further investigation.

Preparation of a Research Report

One more activity is necessary before the study can truly be said to be complete. As mentioned earlier, research is not conducted in a vacuum. Just as the researcher relied on a review of related literature to provide a framework for the present study, the current study must become part of the body of literature available to future researchers to help in their research planning.

The last step in the research process, then, is preparation of a research report. The report may be (and most often is) in the form of an article submitted to a journal for publication. However, it may be part of a program of research described in a book, or it may be a paper presented at a conference, or a thesis or dissertation written as part of the requirement for a postgraduate degree. In Nestor's case, the report took the form of a master's thesis.[†]

As stated earlier, the findings of a research study almost always suggest additional research problems worth investigating. In Figure 15–1, an arrow going from "Preparation of Research Report" back to "Problem Formulation" shows how a new problem suggested by the results of a study can start the cycle of research anew.

The Role of Statistics

In Figure 15–1, statistical analysis is shown as the seventh of nine stages of a research project. It would appear that statistics plays only a minor role. But this is not so. Although analysis itself occurs relatively late in the research process, the (prospective) analysis affects not only subsequent but also earlier events and activities. Let us use Nestor's study to see how this is so.

Not all potentially interesting problems are researchable, at least not with current technology. Early in the process of thinking about his problem and throughout his literature search, Nestor had to ask himself, "Can I find or develop good ways of defining and measuring 'parents' behavior' and 'anxiety,' measures that will yield analyzable data?" Had he not been able to define and measure his variables so as to yield analyzable data, Nestor might have aborted the problem.

The potential statistical analysis affects most aspects of research planning as well. For example, in deciding how many subjects to use, Nestor had to consider the relation between sample size and the likelihood of getting significant results (the power of his tests). In selecting measuring instruments, he had to make sure that they yielded reliable, numerical (analyzable) data. His chosen design and procedure had to ensure that the data collected would meet the requirements (conditions, assumptions) of the planned analysis.

Nestor's pilot study provided information with possible implications for his statistical analysis. For example, conducting parametric tests, as he planned, would be problematical if his score distributions were very skewed. Fortunately, the pilot study data yielded no indication of skew, so that he did not need to alter his plans for analysis. Had

[†]Aristizábal, N. (1980). *The relationship between college students' trait anxiety and their perception of their parents' behavior.* M.A. thesis, University of Pittsburgh.

he found evidence of marked skew, he might have had to consider the use of nonparametric statistics instead.

Nestor's case is quite typical. Statistical analysis, or at least factors related to statistical analysis, permeate almost all aspects of the research planning process. Knowledge of statistics does not, of course, guarantee good research. However, a person without any knowledge of statistics would be hard put to design and conduct valid and reliable research investigations.

EXERCISE 15–4

1. In Nestor's study, a correlation of .27 between male subjects' trait anxiety and the difference between perceptions of fathers' and mothers' warmth was not significant at the .05 level. Yet a correlation of .23 between the same variables over the entire group was significant. How do you account for this?
2. Nestor interpreted the significant correlations between trait anxiety and perceptions of parents' behavior as support for the theory that some kinds of parent behavior *cause* anxiety in their children. Review the discussion of correlation and causality in Chapter 9. Can you give one or more alternative explanations for the relationships Nestor found?
3. Can you think of one or more research questions suggested by Nestor's results?

Analyzing Analyses: Meta-Analysis

Suppose that you are writing a term paper on the relative effectiveness of cognitive-behavioral therapy and drugs for treating phobias. Searching the literature, you locate several dozen studies that compare these two treatments. As you read the research reports, you are puzzled by discrepancies from one study to another. One study finds cognitive-behavioral therapy to be more effective with college students; another finds a difference favoring drug therapy in middle-aged males but not females; still another finds no difference, but apparently used only a small number of subjects. Are there any clear conclusions to be drawn from these inconsistent results?

Until recently, the only recourse for authors of review articles was a rational-logical approach. Using their best judgment and searching for themes and trends among the studies, authors constructed narrative reviews. In their attempts to be objective, they sometimes used a "box-score" method, counting the number of studies favoring Method A, the number favoring Method B, and the number finding no difference, and concluding, if the number favoring A was greater, that A was better than B.

Narrative reviews have been criticized for relying too much on reviewer judgment and for overlooking important details of studies. For example, some studies use many subjects and others use few. Some use shoddy methods, and others are tightly controlled. Some studies get highly significant results, whereas others are barely significant. Shouldn't the quality of research outweigh a simple count of significant results, and shouldn't the number of subjects and levels of significance attained be considered? But how can factors like quality be measured, and how can one take sample sizes and significance levels into account in an objective way?

In the 1970s, Glass and his colleagues (e.g., Glass 1976) introduced a method called ***meta-analysis***. *Meta* used as a prefix for the name of a field or discipline means a new field designed to operate on or deal with the original field. Meta-analysis can be defined as "analysis of analyses." In meta-analysis, individual analyses (studies) become the starting point; they are the raw data to which meta-analytic procedures are applied.

Meta-analysis consists of a set of quantitative operations that can be applied to results of two or more studies to provide information about the overall relations among the variables of concern.

A meta-analyst, planning to synthesize results of many studies, must first deal with the problem that different studies of the same variables often use very different ways to measure the variables. Consider, for example, a meta-analysis on the effect of class size on achievement in arithmetic. Some studies of the relation between these variables used Test A to measure achievement in arithmetic, others used Test B, and others used still other tests. Since different tests have different units of measurement, the scores and the statistics derived from them are not directly comparable from one study to another. For example, a difference of three points between the mean achievement of small and large classes may be a large and important difference on a test with scores ranging from 5 to 15, but is small and trivial on a test with scores ranging from 20 to 200. The first step in a meta-analysis involves expressing the results of all studies in a common metric.

There are several ways to do this. One way, commonly used in studies that compare two sample means (as in the arithmetic achievement example), is to convert the difference between means into a kind of z score called an ***effect size, d***, by the formula

$$d = \frac{M_1 - M_2}{SD}$$ *Calculating an effect size for a difference between means*

where SD is either the standard deviation of the control group (if one of the groups is a control group) or the weighted average of the two group standard deviations. d expresses the difference between the means in standard deviation units.

For example, suppose that in one study of class size and achievement, the mean arithmetic scores of the small and large groups were 66 and 60, respectively, with an average standard deviation of 10. In the second study, the two means were 12 and 9, respectively, with an average standard deviation of 2. The two effect sizes are

$$\text{Study 1:} \quad d = \frac{66 - 60}{10} = 0.6$$

$$\text{Study 2:} \quad d = \frac{12 - 9}{2} = 1.50$$

In the first study, the small group's mean was six-tenths of a standard deviation greater than the large group's mean. In the second study, the small group's mean was one and a half standard deviations greater than the large group's mean. The effect size was larger in Study 2.

In meta-analyses of correlational studies, in which the result of each study is a correlation coefficient (r), conversion to a common metric is not necessary because r is a unit-free index. In other words, r is itself a measure of effect size. However, r is affected by group homogeneity-heterogeneity: As group homogeneity increases (as in a group with a restricted range of scores), r decreases (see Chapter 9). Therefore, if group heterogeneity varies from one study to another, values of r may need to be adjusted to equate the studies on degree of heterogeneity.

After a value of d or r or other measure of effect size has been obtained for each study, various kinds of meta-analytic operations can be applied. If, for example, effect sizes are reasonably similar over most or all studies, it may be enough to calculate the average effect size. In addition, tests can be conducted to determine whether the average effect size is significantly different from zero or to calculate confidence intervals for the true effect size. If effect sizes vary over studies, as is often the case, the studies can be coded (classified) on the basis of various independent variables, such as type of design, length of treatment, gender and age of subjects, or even quality (assuming that quality can be measured in some reasonably objective way). Then multiple regression procedures can be used to determine the degree to which effect size is a function of each independent variable, as well as various combinations of them.

Critics of meta-analysis have pointed out that it, like narrative review, includes subjective elements. This is certainly true. Various stages of conducting a meta-analysis call for judgments on the part of the analyst. For example, in most meta-analyses, very poorly conducted studies are excluded. Two or more meta-analysts might well disagree as to whether or not certain studies are sufficiently poorly conducted to be excluded. In addition, when coding studies in terms of their status on independent variables, there is some subjectivity in deciding, first, what independent variables to include, and, second, the criteria for classification on those variables. Nevertheless, most researchers concur that meta-analysis is a useful and powerful tool, which, though not a replacement for narrative reviews, has been successful in integrating research findings on a large number of topics and will doubtless be used increasingly in the future.

EXERCISE 15–5

1. Here are group means and average standard deviations on three studies comparing Therapy X and Therapy Y.

	Therapy X Mean	Therapy Y Mean	Average *SD*
Study 1	24	30	3
Study 2	95	84	15
Study 3	58	47	3

   Calculate the effect size, *d*, of each study and interpret each result in terms of the standard deviation distance between the means.

2. A study of the relationship between intelligence and creativity in children from a public school serving a diverse neighborhood obtained $r = .36$. The correlation between intelligence and creativity in a group of gifted children was .15. Why can't we regard these two correlation coefficients as comparable statistics?

SUMMARY

This chapter has dealt with concepts and procedures that cut across specific topics in statistics to help you integrate the material you have learned and give you a peek at what lies beyond this book. We started by considering the choice of an appropriate statistical operation or procedure for a given set of data, the first stage in any statistical analysis. Two major factors are the nature of the research question and the type of data that have been collected. Other factors include the number of samples and whether samples are independent or correlated. Strategies that can be used to select an appropriate analysis for various kinds of research questions were presented and discussed (see Tables 15–1 to 15–4).

Next we looked for commonalities among different statistical analyses and found the following:

1. As noted in earlier chapters, the logic of hypothesis testing is the same in all tests.
2. In many tests, the test statistic is the ratio of an obtained difference to the difference expected by chance.

3. Both correlation coefficients and results of tests comparing means can be interpreted in terms of the proportion of one variable's variance accounted for by the other (components of variance).
4. The ***general linear model*** underlies both regression analysis (correlational research) and ANOVA (comparisons of means).

Some implications of these commonalities were explored.

We went beyond statistics to examine the steps in the research process and the role that statistics plays in it. We observed that although statistical analysis is one of the last steps in research, the proposed analysis influences decisions made at earlier steps in the sequence. Figure 15–1 shows the steps of the research process and their interrelationships.

The last topic in the chapter was ***meta-analysis***, the analysis of analyses. In meta-analysis, the results of individual studies about a particular problem become the raw data. An ***effect size*** is calculated for each study, and analyses are conducted to determine the average effect size and to see how effect size is related to aspects of the studies, such as their design, types of subjects, and others.

QUESTIONS

1. Select the appropriate statistical procedure to answer the following questions.
   a. Two methods of teaching algebra were compared. The students in the study were first matched on IQ. The dependent variable was scores on an algebra test at the end of the year.
   b. Three methods of teaching algebra were compared. Students were randomly assigned to groups. The dependent variable was whether students passed (P) or failed (F) the course. The question was: Do the frequencies of P vs. F differ significantly under the three methods?
   c. A developmental psychologist wonders about the extent of individual differences in toddlers' eye-hand coordination.
   d. A study was conducted to determine the effect of length of list and study time on recall of words. Forty subjects were randomly subdivided into four groups. All groups were asked to learn a list of words. Group 1 was given 5 minutes to study a 20-word list, whereas Group 2 had 10 minutes to learn the 20-word list. Group 3 was given 5 minutes to learn a 40-word list, whereas Group 4 had 10 minutes to learn the 40-word list. Then all subjects were tested to find out how many words they could recall.

2. A reading specialist randomly assigned 160 first-grade children to three methods of reading instruction: phonics, whole-word, and mixed. A reading achievement test was administered to all the children in September (pretest) and again in June (posttest). At the end of the year, 120 of the children were promoted to second grade, but 40 were retained in first grade for another year. What analysis must the reading specialist conduct to answer each of the following questions?
   a. Is there a significant difference between the posttest scores of boys and girls?
   b. Are posttest scores (across all groups) significantly higher than pretest scores?
   c. Were posttest scores affected by the method of instruction?
   d. Did the number of children promoted versus retained differ significantly between the methods of instruction?

3. An industrial psychologist at the Acme Company selected a group of younger and a group of older salespersons, matched on years of employment. He obtained measures of leadership, emotional stability, sociability, assertiveness, and productivity in the members of both groups. What must he do to answer the following questions?

a. Is there a difference in productivity between older and younger salespersons?
b. Is productivity significantly related to leadership? To emotional stability? To sociability?
c. Can productivity be predicted from a combination of leadership, emotional stability, and sociability?

4. Consider a statistic called the glumpf. G stands for a sample glumpf, γ stands for a population glumpf. Assume that the sampling distribution of G is normal for all sample sizes, with a mean equal to γ and a standard deviation equal to $1/\sqrt{N+N^2}$. If you want to test a hypothesis about a population glumpf, what is the test statistic?

5. If the correlation between tolerance for new ideas and age in a certain group is −.25, what proportion of the variance of tolerance is accounted for by age?

6. You have conducted an experiment in which each of four groups of rats received a different dosage of a drug to determine the effect of drug dosage on activity level. The results of your ANOVA are summarized in the following table:

Source of Variation	*SS*	*df*	*MS*	*F*
Dosage	849	3	283	9.25
Within groups	1714	56	30.6	
Total	2563	59		

What is the estimated proportion of the population's activity variance accounted for by drug dosage level?

7. Identify a journal that publishes research reports in a field that interests you. Read one article in a recent issue. Make sure that it is a report of one research study. As you read, do the following:

a. Identify the various stages of the research process as reflected in the article.
b. Before reading the Method and Results sections, predict the statistical analysis that seems appropriate, based on the research hypothesis or hypotheses. If the statistical analysis used by the author is different from the one you predicted, can you explain why?
c. Before reading the author's Discussion section, draw your own conclusions from the results of the analysis. Do the author's conclusions agree with yours?

8. In the behavioral and social science departments of most colleges and universities, introductory statistics is taught as a standalone course, often prerequisite to a course in research methods. Some departments, however, teach introductory statistics as part of an integrated research methods–statistics course. Now that you are completing a course in statistics and know something about the role of statistics in the research process, list advantages and disadvantages of both approaches. If you were in a position to choose one approach or the other, which would you choose? Why?

9. Read a meta-analytic review of research on some topic of interest to you. (Your instructor or librarian may be able to help you find one.) You may not be able to understand everything you read, but notice the methodology used in defining and delimiting the topic, selecting studies, and coding the studies for the analysis. What measure of effect size was used in the meta-analysis? Did the results of the meta-analysis clarify the relationship between variables that was the topic under study? Did effect sizes vary systematically as a function of interstudy differences?

A Final Word

In this book, you have been introduced to many new concepts and procedures. Although the exercises within and at the ends of chapters have given you a chance to think about them and practice using them, you probably still feel a little shaky. That is understandable. Just as the ski bunny who has completed a beginner's course in skiing needs to spend a lot of additional time on the slopes to become an adept skier, the student who has completed an introductory statistics course needs to continue to use the concepts that have been learned to become really comfortable with them.

Some of you will be taking more advanced courses in statistics or research methods. If so, you will find that the basic concepts from your introductory course become increasingly meaningful as you learn more advanced techniques. But even if you are not going on in statistics, I hope you will make an effort to be more aware than you were in the past of the role that statistics plays in your life and in society. Notice, for example, how often statistical arguments are used to support one or another side (usually both) on political and social issues. Notice how frequently the popular media—newspapers, TV, magazines—present graphs, tables, and numerical data to describe, illustrate, or explain phenomena or to support particular points of view. Now you can (and should) evaluate these presentations more critically than you did in the past.

Some of you will be involved in research of some sort in the future. If so, there is one more thing I hope you will take with you from this course, besides concepts and skills. That is an appreciation for data and a feeling (I can think of no better word) for what data can tell you. There is usually a great deal more information in a set of data than group means, correlation coefficients, and F ratios. Analyze your data as planned, by all means; you must do this to answer the research questions that were asked. But play with the data, too. Explore them. For example, try different ways of graphing or otherwise visualizing your results. You may discover some interesting things not revealed by the statistical analysis. And these discoveries may suggest additional questions to ask, additional studies to conduct, and additional analyses to perform in the future. In other words, they will make your research more fruitful and, in the process, I hope, more fun.

References

Glass, G. (1976). Primary, secondary, and meta-analysis of research. *Educational Research, 5*, 3–8.

Hays, W. L. (1988). *Statistics*, 4th ed. Forth Worth, TX: Holt, Rinehart and Winston.

Howell, D. C. (1992). *Statistical methods for psychology*, 3rd ed. Boston: PWS-Kent.

Judd, C. M., & McClelland, G. H. (1989). *Data analysis: A model-comparison approach.* San Diego: Harcourt, Brace, Jovanovich.

Siegel, S. (1956). *Nonparametric statistics for the behavioral sciences*. New York: McGraw-Hill.

Taylor, H. C., & Russell, J. T. (1939). The relationship of validity coefficients to the practical effectiveness of tests in selection. *Journal of Applied Psychology, 23*, 565–578.

Walsh, J. E. (1962–1968). *Handbook of nonparametric statistics*. New York: Van Nostrand.

Glossary

Abscissa: the horizontal axis of a graph.

Addition law (of probability): a law stating that if two (or more) events are mutually exclusive, the probability that one or another will occur is the sum of their separate probabilities.

Alpha (α): see *Level of significance*.

Alternative hypothesis (H_A): the alternative to the null hypothesis. The conclusion to be reached if the null hypothesis is rejected; usually equivalent to the research hypothesis concerning the expected effect, difference, or relation.

ANOVA: acronym for analysis of variance.

Analysis of variance: a family of tests for comparing two or more means in studies with one or more independent variables.

Arithmetic mean: *see* Mean.

Average deviation: a measure of variability; the average distance of scores from the mean.

Bar graph: a graph for displaying a distribution of categorical data or discrete quantitative data.

Biased estimator: an estimator of a parameter whose mean value, over all possible samples, is not equal to the parameter.

Biased sample: a sample selected in such a way as to favor certain members of the population over others.

Bimodal distribution: a distribution with the highest frequency of observations at two nonadjacent score values; the graph has two peaks with a valley between.

Binomial distribution: a discrete probability distribution in which an elementary event has two possible outcomes with known probability and the event is repeated one or more times.

Box-and-whisker plot: also called *box plot*; a graph of a frequency distribution in which the middle 50% of observations are contained within a "box," and the outer 50% are represented by "whiskers" extending from the box to the lowest and highest scores.

Carryover effect: in repeated measures designs, an effect that occurs if responses to treatments later in a series are influenced by earlier treatments.

Central limit theorem: the theorem that relates the sampling distribution of the mean to parameters of the population and to sample size; in particular, the theorem states that, as sample size increases, the sampling distribution of the mean approaches a normal distribution.

Central tendency: the point or score value in a distribution around which observations tend to cluster; the center of the distribution.

Chi square (χ^2): a measure of distance of observed frequencies from expected frequencies in a categorical data set; the test statistic in chi square tests.

Chi square distribution: a continuous probability distribution; used in chi square tests.

Chi square tests: tests of hypotheses about distributions of categorical data.

Chi square test of goodness of fit: a procedure for testing hypotheses about the distribution of one categorical variable in one population.

Chi square test of independence: a procedure for testing hypotheses about the relation between two categorical variables.

Class interval size: in a grouped frequency distribution, the range of values within each interval.

Coefficient of determination: the squared correlation coefficient, r^2; the proportion of Y variance that is accounted for (explained by) differences on X.

Coefficient of nondetermination: $1 - r^2$; the proportion of Y variance that is not accounted for (explained by) differences on X.

Command: an instruction to a computer to perform some operation.

Comparative study: research designed to compare two or more populations.

Comparison: in a data set containing three or more conditions or treatments, a test for the difference between any two means or combinations of means.

Components of variance: the result of partitioning the total variance of a data set into two or more parts, each reflecting one source of variation.

Confidence interval: an interval calculated around a sample estimator within which we are confident, at some level, that the parameter lies. The lower and upper limits of the interval are called *confidence limits.*

Contingency table: a table displaying the relation between two categorical variables.

Continuous probability distribution: see *Probability distribution.*

Continuous variable: a variable that can take any value between the lowest and highest possible values.

Correction for continuity: an adjustment applied to discrete data when using the normal approximation to the binomial.

Correlated samples: two or more samples selected in such a way that assignments to one sample depend on assignments to the other(s). Correlated samples are formed by matching or by repeated measures of a single group of individuals.

Correlation coefficient (Pearson correlation coefficient) (r): a standardized index of the linear correlation between two quantitative variables.

Correlational research: research designed to describe the relation between two or more measured variables.

Covariance: a measure of the amount of covariation of two variables.

Criterion variable: in regression, the variable that is to be predicted, the Y variable.

Critical value: in tests of hypotheses, the value of the test statistic that is the cutoff point between the rejection and nonrejection regions.

Cumulative frequency (Cum f): the frequency of observations at or below a given score value.

Cumulative graph: a graph of a cumulative frequency or cumulative percentage distribution.

Cumulative percentage (Cum P): the percentage of observations at or below a given score value.

Data file: a machine- (computer-) readable table of data.

Degrees of freedom (df): the number of observations in a data set that are free to vary.

Degrees of freedom between (df between): in analysis of variance, the degrees of freedom associated with the variation between group means; equal to the number of groups minus one.

Degrees of freedom within (df within): in analysis of variance, the degrees of freedom associated with the variation of observations within each group around the group mean; equal to the total number of observations minus the number of groups.

Dependent variable: in research, the variable affected by the independent variable.

Descriptive statistics: the body of concepts and procedures used for describing a data set; also called *exploratory data analysis.*

Design: in experiments, the plan for manipulating variables and collecting data in a research study.

Directional hypothesis: a hypothesis that an effect, difference, or relationship will be in a particular direction.

Discrete probability distribution: see *Probability distribution.*

Discrete variable: a quantitative variable that takes only certain (usually whole number) values with gaps in between.

Distribution: the set of measurements of a variable in a group of individuals, objects, or events.

Distribution-free test: a test that makes no assumptions concerning the shape(s) of the underlying population(s). Most nonparametric tests are distribution-free.

Effect size: in meta-analysis, a standardized measure of a treatment effect. In correlational research, *r* (adjusted if necessary) is a measure of effect size. In comparisons of means, effect size is the difference between the means divided by the standard deviation and is symbolized *d*.

Empirical sampling distribution: see *Sampling distribution*.

Error of estimate: the difference between an individual's actual criterion score, *Y*, and the value predicted by the regression equation, $\hat{Y}$. Also called *residual*.

Error of inference: in a hypothesis test, either rejecting a true hypothesis (Type I error) or not rejecting a false hypothesis (Type II error).

Error variance: in analysis of variance, the mean square that measures chance variation. The error variances in particular analyses are: for one-way and two-way ANOVA, mean square within; for repeated measures ANOVA, residual mean square.

Estimated standard error of the difference between means (s_{diff}): an estimate, based on sample data, of the actual standard error of the difference between means.

Exact limits: the upper and lower limits of the interval represented by a reported measurement of a continuous variable.

Expected frequency (f_e): in chi square tests, the frequency expected in a category or cell if the hypothesis being tested is true.

Experimental research: comparative research in which one (or more than one) variable is manipulated to determine its effect on another variable. The manipulated variable is called the *independent variable*, and the other variable is called the *dependent variable*.

Explained variance ($S^2_{\hat{y}}$): in regression, the variance of predicted *Y* scores around the mean of *Y*.

***F* distribution:** a continuous probability distribution; the basis of the *F* test in the analysis of variance.

***F* test:** the statistical test in the analysis of variance.

Factorial design: a design containing two (or more) independent variables or factors.

Family of comparisons: in a data set containing three or more conditions or treatments, the set of potential comparisons of interest to the researcher.

Frequency (*f*): The number of observations having a particular value or class of values.

Frequency distribution: a tabular representation of a data set listing the frequency of occurrence of each value (or class of values).

Frequency polygon: a graph of a distribution of continuous quantitative data.

General linear model: the model underlying regression and analysis of variance as well as most other multivariate analyses.

Grand mean (*GM*): in analysis of variance, the mean of the entire data set.

Grouped frequency distribution: a frequency distribution in which two or more adjacent score values are grouped together in each interval.

Heteroscedasticity: in the relation between two variables, the condition of unequal scatter (variation) of *Y* at different values of *X*.

Histogram: a bar graph for displaying a distribution of a continuous quantitative variable.

Homogeneity-of-variance assumption: in the *t* test for the difference between means and analysis of variance, the assumption that all the populations have the same variance.

Homoscedasticity: in the relation between two variables, the condition of equal scatter (variation) of *Y* at all values of *X*.

Hypothesis testing: a method of inference; the process of using sample data to test a hypothesis about some property, characteristic, or parameter of a population.

Independent samples: two or more samples selected in such a way that assignments to one sample do not affect (and are not affected by) assignments to the other(s).

Independent variable: in research, the variable whose effect is under investigation; the manipulated variable in an experiment.

Inferential statistics: the body of concepts and procedures used to draw conclusions about populations from sample data.

Interaction: a situation in which the effect of one variable differs, depending on the level of another variable.

Interquartile range: a measure of variability; the distance between the first and third quartiles, $Q_3 - Q_1$; the range of the middle 50% of observations.

Interval estimation: a method of inference; the process of constructing an interval around a sample statistic within which we are confident, to some degree, that the corresponding population parameter lies.

J-shaped distribution: a distribution having the highest frequency of observations at the highest or lowest score value; the extreme of a skewed distribution.

Level of confidence: the degree of confidence, expressed as a percentage, such as 90% or 95%, that a parameter lies within a confidence interval around the sample estimator.

Level of significance: the standard for rejecting a hypothesis. For example, if the probability of a result, given that the hypothesis is true, is less than .05, the result is significant at the .05 level, and the hypothesis is rejected at the .05 level. Also called *alpha* (α).

Linear transformation: a transformation of a distribution involving addition or subtraction of a constant from each score and/or multiplication or division of each score by a constant.

Main effect: in a factorial design, the effect of one variable or factor, independent of the other(s).

Matching design: a design in which individuals are paired prior to data collection, and one individual within each pair is assigned to one condition and the other individual to the other condition.

Matching variable: the variable on the basis of which individuals are paired in a matching design.

Mean: short for arithmetic mean; a measure of central tendency; the sum of the scores divided by the number of scores. Symbols: M or X (sample mean), μ (population mean).

Mean square between groups (*MS* between): in analysis of variance, the variance of the group means around the grand mean.

Mean square within groups (*MS* within): in analysis of variance, the pooled variance of scores within each group around the group mean.

Median (*Mdn*): a measure of central tendency; the score value below which 50% of observations fall; equal to the 50th percentile (P_{50}) and the second quartile (Q_2).

Meta-analysis: analysis of analyses; methods for combining data over multiple studies to draw conclusion about effects of one or more variables.

Midpoint: the value that is halfway between the upper and lower limits of an interval.

Mode: a measure of central tendency; the most frequently occurring value in a distribution.

Multiple correlation coefficient (*R*): an index of the linear relation between a criterion and the weighted combination of two or more predictors.

Multiple regression: the process of predicting a criterion variable from two or more predictors.

Multiple regression equation: an equation used to predict a criterion variable from two or more predictors.

Multiplication law (of probability): a law stating that, if two (or more) events are independent, the probability that all of them will occur is the product of their separate probabilities.

Negatively skewed distribution: a distribution in which most observations fall nearer to the highest than to the lowest score; when graphed, the longer tail is at the lower (left) end.

Nondirectional hypothesis: a hypothesis that an effect, difference, or relationship may be in either a positive or negative direction.

Nonexperimental comparative research: Research involving comparisons of two or more preexisting groups.

Nonparametric tests: tests about attributes of populations other than specific parameters. See *Distribution-free test*.

Nonrejection region: in a statistical test, the range of values of the test statistic that result in nonrejection of the hypothesis.

Normal curve: a theoretical, symmetrical, bell-shaped distribution.

Null hypothesis (H_o): the hypothesis of no effect, no difference, or no relationship; the statistical hypothesis in most tests.

Observed frequency: the number of individuals, objects, or events falling in a particular category.

Obtained distribution: a distribution of observations.

One-tailed test: a test in which the entire rejection region falls in one tail of the probability distribution; tests of directional hypotheses are one-tailed.

One-way analysis of variance: analysis of variance for studies involving one independent variable varying over two or more independent groups.

Ordinate: the vertical axis of a graph.

Orthogonal comparisons: two or more comparisons that are independent of one another.

Outlier: an unusual observation in a distribution, one that falls relatively far from most of the distribution.

Pairwise comparison: a comparison between two treatment means (rather than two combinations of means).

Parameter: a numerical property of a population or theoretical distribution.

Parametric tests: tests of hypotheses about population parameters; most parametric tests assume that the population(s) is/are normally distributed.

Pearson correlation coefficient: *see* Correlation coefficient.

Percentage: a relative frequency; a proportion multiplied by 100.

Percentile (P_k): The score value below which k percent of observations fall.

Percentile rank: the percentage of observations falling below a given score value.

Phi coefficient (ϕ): a measure of the correlation between two dichotomous variables.

Pilot study: a tryout of the proposed procedure for a research study prior to actual data collection.

Planned comparisons: in data sets containing three or more conditions or treatments, comparisons that were planned in advance of data collection.

Point biserial correlation coefficient (r_{pb}): a measure of the correlation between a dichotomous variable and a continuous variable.

Pooled variance estimate (s_p^2): in the t test for the difference between means, an estimate of the variance of both populations obtained by combining (pooling) data over both samples.

Population: the totality of individuals, objects, or events about which information is sought.

Positively skewed distribution: a distribution in which most observations fall nearer to the lowest than to the highest score; when graphed, the longer tail is at the upper (right) end.

Post hoc tests: in data sets containing three or more conditions or treatments, comparisons selected on the basis of the results of the analysis of variance.

Power: (of a test) the probability of rejecting a false null hypothesis.

Prediction interval: in regression, a confidence interval for an actual criterion score, constructed around the predicted criterion score for a given X.

Predictor variable: in regression, the variable used for making predictions, the X variable.

Probability (p): the likelihood of an event. Empirical definition: the relative frequency of the event over many trials; theoretical definition (assuming that all possible outcomes are equally likely): the proportion of possible outcomes that are examples of the event.

Probability distribution: a distribution of the probabilities of all possible, mutually exclusive outcomes of an event. Probability distributions may be empirical or theoretical. A probability distribution for outcomes of a discrete variable is a *discrete probability distribution*; a probability distribution for outcomes of a continuous variable is a *continuous probability distribution.*

Proportion: a relative frequency; the frequency of observations having a particular value divided by the group size.

Qualitative variable: a variable whose values differ in kind rather than amount.

Quantitative variable: a variable whose values differ in amount.

Quartiles: score values dividing a distribution into four equal quarters. 25% of observations fall below the first quartile (Q_1), 50% below the second quartile (Q_2), and 75% below the third quartile (Q_3).

Random numbers: a set of digits between 0 and 9 in which each digit (1) occurs equally often and (2) follows each digit (including itself) equally often; the sequence of digits is determined by chance.

Random sample: a sample selected in such a way that each member of the population is equally likely to be in the sample.

Randomized groups design: an experimental design in which individuals are randomly assigned to one of two or more groups.

Range: the difference between the highest and lowest scores of a distribution.

Rectangular distribution: a distribution with an approximately equal frequency of observations at all score values.

Regression: the process of predicting one variable, the criterion variable, from one or more predictors.

Regression coefficient (b): the number (weight) by which a value of X is multiplied in a regression equation; the slope of the regression line.

Regression equation: a linear equation used for predicting values of Y from one or more values of X; the equation of the regression line.

Regression line: the straight line that minimizes *errors of estimate* (*residuals*) in predicting a criterion variable from one or more predictors.

Regression toward the mean: the fact that, unless $r = \pm 1$, the predicted Y for any X is closer to the mean of Y (in SD units) than X is to the mean of X.

Rejection region: in a statistical test, the range of values of the test statistic that result in rejection of the hypothesis.

Relative frequency: a frequency expressed relative to the size of the group; *proportions* and *percentages* are relative frequencies.

Repeated measures design: a design in which each individual is measured under two or more conditions or on two or more occasions.

Research hypothesis: the hypothesis that states the nature of the expected effect, difference, or relation. In a statistical test, usually the alternative hypothesis.

Residual: see *Error of estimate*.

Residual mean square: the error variance in repeated measures analysis of variance; a measure of the variation due to the interaction of subjects and treatments.

Residual variance S^2_{res} or s^2_{res}: in regression, the variance of actual Y scores around predicted Y scores.

Robustness: (of a test) the degree to which a test yields relatively accurate conclusions even if underlying assumptions are violated to some degree.

Sample: a subset of a population, usually selected in order to draw conclusions about the population.

Sampling distribution: the distribution of a statistic over many or all possible samples of a certain size from a population. A sampling distribution based on a limited number of random samples in an *empirical sampling distribution*; a sampling distribution based on all possible samples is a *theoretical sampling distribution*.

Sampling error: the difference between a sample statistic and the actual value of the population parameter.

Sampling with replacement: sampling from a population in such a way that, after each selection of an individual for the sample, the individual is returned to the population before the next selection.

Sampling without replacement: sampling from a population in such a way that individuals selected for the sample are not replaced in the population before the next individual is selected.

Scatter diagram: a graphical representation of the relationship between two quantitative variables. Also called *scatter plot*.

Scheffé test: a post hoc test for the family of all possible comparisons.

Score limits: in a grouped frequency distribution, the lowest and highest scores in a particular interval.

Semi-interquartile range (Q): a measure of variability; half the distance between the first and third quartiles; also called the *quartile deviation.*

Significant result or difference: in a statistical test, a result that has such a small probability of occurring if the hypothesis is true that the hypothesis is rejected; a result or difference that would be unlikely to occur by chance alone.

Spearman rank-order correlation coefficient (r_s): a measure of the correlation between paired ranks.

Standard deviation (*SD*): the square root of the variance; a measure of variability particularly important in dealing with normal or near-normal distributions. Symbols: σ (population *SD*), *S* or *s* (sample *SD*).

Standard error of the difference between means: the standard deviation of the sampling distribution of the difference between means. Symbols: σ_{diff} (actual) or s_{diff} (estimated).

Standard error of estimate (S_{est} or s_{est}): in regression, the standard deviation of actual *Y* scores around predicted *Y* scores; the square root of the *residual variance.*

Standard error of the mean: the standard deviation of a sampling distribution of the mean. Symbols: σ_M (actual) or s_M (estimated).

Standard normal curve (or distribution): the normal curve with mean = 0 and standard deviation = 1. Any normal distribution can be converted into a standard normal distribution by converting score values into *z* scores.

Standard score: (*z* score) a score expressed in terms of its standard deviation distance from the mean.

Statistic: a numerical property of a sample.

Statistics: the body of concepts and procedures for organizing, analyzing, and interpreting data.

Stem-and-leaf plot: a tabular/graphical representation of a distribution of quantitative data. Also called *stem plot.*

Studentized range statistic (*q*): the test statistic in the Tukey "honestly significant difference" test.

Subhypothesis: a null hypothesis for a particular comparison.

Sum of squares (*SS*): a sum of squared deviations from the mean. Used in calculating variances or, in analysis of variance, mean squares.

Sum of squares between groups (*SS* between): in analysis of variance, the sum of squared deviations of the group means from the grand mean.

Sum of squares within groups (*SS* within): in analysis of variance, the pooled sum of the squared deviations of scores within each group around the group mean.

Summary table: a table displaying the partitioning of variation and the results of an analysis of variance.

Symmetrical distribution: a distribution whose right and left halves, when graphed, are approximate mirror images.

***t* distribution:** a continuous probability distribution; used in *t* tests.

***t* tests:** tests of hypotheses about the mean of a single population or about the difference between the means of two populations.

Test statistic: in a statistical test, the statistic whose probability distribution describes the probability of various outcomes. In the chi square test, for example, the test statistic is χ^2.

Theoretical distribution: a distribution generated by a mathematical formula or by logical reasoning.

Theoretical sampling distribution: see *Sampling distribution.*

Total degrees of freedom (*df* total): in analysis of variance, the degrees of freedom associated with the total variation of all the observations in the data set; equal to the total number of observations minus one.

Total sum of squares (*SS* total): in analysis of variance, the sum of squared deviations of scores from the grand mean.

Treatment variable: the independent variable in an experiment. Each level of the treatment variable is a *treatment.*

Tukey honestly significant difference (HSD) test: a post hoc test for the family of pairwise comparisons.

Two-tailed test: a test in which the rejection region is evenly divided between the two tails; tests of nondirectional hypotheses are two-tailed.

Two-way analysis of variance: analysis of variance for a factorial design with two independent variables (factors).

Type I error: rejection of a true hypothesis.

Type I error rate: the probability of making at least one Type I error over two or more tests on a set of data.

Type II error: nonrejection of a false hypothesis.

Unbiased estimator: an estimator of a parameter whose mean value, over all possible samples, is equal to the parameter.

Ungrouped frequency distribution: a tabular representation of a data set that lists the frequency of occurrence of each value of the variable.

Variable: any property of individuals, objects, or events that can take more than one value.

Variance: the mean of the squared deviations from the mean, found by dividing the sum of squares (SS) by N, or, when estimating the population variance, by $N - 1$. Symbols: σ^2 (population variance), S^2 or s^2 (sample variance).

Weighted mean: the overall mean of two or more samples.

z score: see *Standard score*.

Appendix A: Arithmetic and Algebra Self-Evaluation Quiz

The quiz is divided into 11 parts. Take the whole quiz. Then compare your answers with the key at the end of the appendix. If you have trouble with any part, read the corresponding part in Appendix B. Then take the posttest at the end of Appendix B. Use your calculator whenever you find it useful in this quiz.

Part 1: Rounding

Using your calculator, you will often obtain answers with many more digits than you need to report. It is a good idea to carry extra digits during the course of your calculations. But you usually need to round your final answer.

Round the following numbers to two decimal places:

1. 163.4839704
2. 6.72876431
3. 42.8967
4. 0.003
5. 0.008
6. 0.1655
7. 0.165
8. 0.155

Part 2: Converting Fractions Into Decimals

Any fraction can be expressed in decimal form. Since statistical results are usually presented in decimal form, and since decimal operations are simpler on a calculator than are computations with fractions, it is usually a good idea to convert fractions to decimals before performing further calculations.

Express the following fractions as decimals. If necessary, round your answer to 4 decimal places:

1. $\frac{1}{2}$
2. $\frac{2}{3}$
3. $\frac{2}{1678}$
4. $\frac{98}{99}$
5. $\frac{131}{20}$

Part 3: Addition and Subtraction

In some statistical calculations, it is necessary to add or subtract sets of numbers, some of which are positive and some negative.

Add the following:

1. (-5) + (-1) =
2. 3 + (-6) =
3. $\begin{array}{r} 2 \\ -5 \\ 6 \\ \underline{-4} \end{array}$
4. $\begin{array}{r} -2 \\ -4 \\ 5 \\ \underline{10} \end{array}$
5. $\begin{array}{r} -3 \\ 5 \\ 2 \\ \underline{-4} \end{array}$

Perform the indicated subtractions:

6. $10 - 14 =$

7. $-8 - (+4) =$

8. $-8 - (-12) =$

9. $6 - (-3) =$

Part 4: Multiplication and Division

It is sometimes necessary to multiply or divide positive and negative numbers. Do the indicated operations and, if necessary, round the answer to two decimal places.

1. $(13)(0) =$
2. $4(-3) =$
3. $(-6)(-2) =$
4. $\frac{-6}{2} =$
5. $\frac{3}{1} =$
6. $\frac{1}{3} =$
7. $\frac{-6}{-2} =$
8. $\frac{-6.3}{21} =$
9. $\frac{0}{67} =$
10. $\left(\frac{-2}{3}\right)\left(\frac{5}{7}\right) =$
11. $(-.5)\left(\frac{2}{3}\right) =$

Part 5: Proportions and Percentages

Occasionally you will need to convert frequencies, proportions, and percentages into one another. Do the following:

1. Express 30 out of 200 as a proportion; as a percentage.

2. Express 30 out of 50 as a proportion; as a percentage.

3. Express 30 out of 500 as a proportion; as a percentage.

Convert the following proportions into percentages:

4. .03

5. .837

6. .0084

Convert the following percentages into proportions:

7. 83%

8. 4%

9. 100%

10. 0.5%

Answer the following questions:

11. How much is 25% of 40?

12. How much is 43% of 200?

13. If the proportion of male newborns is .52, then how many of 500 newborns are expected to be male?

Part 6: Squares and Square Roots

Statistical formulas often require the calculation of squares or square roots of numbers. Use your calculator when necessary. Except in question 3, round answers to two decimal places.

1. $5^2 =$
2. $(1.2)^2 =$
3. $(0.004)^2 =$ (*Do not round*)
4. $(-5)^2 =$
5. $(5 + 2)^2 =$
6. $(2 - 5)^2 =$
7. $[(2)(3)]^2 =$
8. $\frac{(4)^2}{2}$
9. $\left(\frac{4}{2}\right)^2 =$
10. $\sqrt{16} =$
11. $\sqrt{160} =$
12. $\sqrt{4+9} =$

13. $\sqrt{(4)(9)}$ =

14. $\sqrt{\frac{25}{16}}$ =

15. $\sqrt{3}\sqrt{6}$ =

16. $\frac{\sqrt{6}}{\sqrt{3}}$ =

Part 7: Combining Arithmetic Operations

Statistical formulas usually require a combination or sequence of two or more operations. Do the indicated operations. If necessary, round the answer to two decimal places.

1. $(5)(6)^2$

2. $\frac{60-50}{12}=$

3. $\frac{160-(10)^2}{12}=$

4. $160-\frac{(10)^2}{12}=$

5. $\frac{(10-15.5)^2}{15.5}=$

6. $13.5+\frac{1}{3}(14.5-13.5)=$

7. 50 + (3)(10) =

8. 50 + (−3)(10) =

9. $\sqrt{\frac{(6)^2}{10}+\frac{(4)^2}{12}}=$

10. $\sqrt{[10(964)-(40)^2][10(350)-(25)^2]}=$

Part 8: Absolute Values

Sometimes it is necessary to ignore the plus or minus sign attached to a number and to consider just its absolute value. Answer the following questions:

1. $|21|$ =

2. $|-21|$ =

3. $|-0.83|$ =

4. $|5-10|$ =

If you are told, "Reject H if $|t|$ is greater than 2.33," tell whether you should reject H in each of the following cases:

5. t = 1.92

6. t = 4.10

7. t = −1.92

8. t = −4.10

Part 9: Inequalities

In order to talk about quantities falling above, below, or between certain values, we use inequality notation.

Read the following:

1. $x > 3$

2. $p < .05$

3. $.01 < p < .05$

4. $y \leq 25$

Place the proper sign, > or <.

5. 5 __ 3

6. 3 __ 5

7. −3 __ −5

8. −5 __ −3

9. 5 __ −3

10. −5 __ 3

11. $|-3|$ __ $|-5|$

12. $|-5|$ __ $|3|$

13. Express the following, using < and >: "z falls between 2 and 7."

You are told, "Reject H if $z < -1.96$." Should you reject H in each of the following?

14. $z = -2.33$

15. $z = -1.10$

16. $z = 1.10$

17. $z = 2.33$

Part 10: Using Formulas

Statistical procedures often require you to plug numbers into formulas and solve. In the following, round your answers to three decimal places.

1. Find Y if $N = 7$ and $X = 5$:

$$Y = \frac{NX^2 - X}{(NX)^2}$$

2. Using the same formula, find Y if $N = 3$ and $X = -2$.

3. If $A = 2$, $B = 3$, and $C = 5$, find Y:

$$Y = \frac{(B-A)^2(C+2)}{\sqrt{2C^2 - AB}}$$

Part 11: Manipulating Simple Equations

As you proceed through this book, you will sometimes encounter simple derivations or transformations involving algebraic manipulations of formulas. To follow the steps in these transformations, you need to understand rules for manipulating equations. Use whatever rules are needed to solve for X in each of the following:

1. $Y = BX + 3$

2. $Y = X - B$

3. $Y = \dfrac{X}{B}$

4. $Y = \dfrac{B+2}{X}$

5. $Y = \dfrac{B - X^2}{B}$

6. $Y = \sqrt{\dfrac{B+2}{X}}$

END OF PRETEST

Check your answers to the pretest by using the key on the following pages. If you find that you need to review procedures related to any part, refer to the instructional material in Appendix B.

Answers to Self-Evaluation Quiz

Part 1

1. 163.48
2. 6.73
3. 42.90
4. 0.00
5. 0.01
6. .17
7. .16 or .17
8. .16 or .15

Part 2

1. .5
2. .6667
3. 0.0012
4. .9899
5. 6.55

Part 3

1. −6
2. −3
3. −1
4. 9
5. 0
6. −4
7. −12
8. 4
9. 9

Part 4

1. 0
2. −12
3. 12
4. −3
5. 3
6. 0.33
7. 3
8. −0.3
9. 0
10. −.48
11. −.33

Part 5

1. .15, 15%
2. .6, 60%
3. .06, 6%
4. 3%
5. 83.7%
6. .84%
7. .83
8. .04
9. 1.00
10. .005
11. 10
12. 86
13. 260

Part 6

1. 25
2. 1.44
3. .000016
4. 25
5. 49
6. 9
7. 36
8. 8
9. 4
10. 4
11. 12.65
12. 3.61
13. 6
14. 1.25
15. 4.24
16. 1.41

Part 7

1. 1.180
2. 0.83
3. 5
4. 151.67
5. 1.95
6. 13.83
7. 80
8. 20
9. 2.22
10. 4807.81

Part 8

1. 21
2. 21
3. 0.83
4. 5
5. No
6. Yes
7. No
8. Yes

Part 9

1. X is greater than 3
2. p is less than .05
3. p is between .01 and .05
4. y is less than or equal to 25
5. $>$
6. $<$
7. $>$
8. $<$
9. $>$
10. $<$
11. $<$
12. $>$
13. $2 < z < 7$
14. Yes
15. No
16. No
17. No

Part 10

1. 0.139
2. 0.389
3. 1.055

Part 11

1. $X = \frac{Y-3}{B}$
2. $X = Y + B$
3. $X = BY$
4. $X = \frac{B+2}{Y}$
5. $X = \sqrt{B - BY}$ or $X = \sqrt{B(1-Y)}$
6. $X = \frac{B+2}{Y^2}$

Appendix B: Review of Arithmetic and Algebra

This appendix provides a brief review of the concepts and procedures tested in Appendix A. It is not intended as a substitute for high school and college courses in general mathematics and algebra, but you should find it helpful if your skills have gotten rusty.

Part 1: Rounding

When you buy an item at the store for $9.99, you are likely to say that it cost you "about $10.00" (though the store would prefer that you say "just a little over $9.00"). You have rounded off the price to the nearest dollar. Rounding is often necessary in reporting the results of a statistical computation. Usually you need to round to a certain number of decimal places.

Whenever you want to round a number, look at the first digit to be dropped and use the following rules:

1. If the first digit to be dropped is *less* than 5, just drop it and any digits that follow.

   Example: Rounded to two decimal places, 163.4839704 is 163.48.

2. If the first digit to be dropped is *greater* than 5, increase the last digit to be retained by one.

   Example: In rounding 6.72876431 to two decimal places, the first digit to be dropped is 8, which is greater than 5. Therefore we increase the last digit to be retained (2) by one, and the answer is 6.73.

   If the first digit to be dropped is greater than 5 and the last digit to be retained is a 9, adding one to 9 means changing the "9" to "0" and adding one to the *previous* digit.

   Example: 42.8967, rounded to two decimal places, is 42.90.

3. If the first digit to be dropped is *equal* to 5, look at the digits following the 5. If there are any additional digits greater than zero, increase the last digit to be retained by one.

   Example: 0.1655, rounded, is 0.17; 0.165001, rounded, is 0.17.

   If the first digit to be dropped is equal to 5 and there are no additional digits greater than zero, we can equally well justify leaving the last digit to be retained as is or increasing it by one. The following rule may be used in such cases: If the last digit to be retained is *even* (0, 2, 4, 6, 8), leave it as it is; if it is *odd* (1, 3, 5, 7, 9), increase it by one.

   Example: Rounded to two decimal places, 0.165 is 0.16; 0.155 is 0.16.

Part 2: Converting Fractions Into Decimals

Any fraction can be converted into a decimal number simply by dividing the numerator (the number above the line) by the denominator (the number below the line).

Thus ½ is 1 divided by 2, or 0.5. ⅔ is 2 divided by 3, or 0.66666 . . . ; rounded to four decimal places it is 0.6667. Similarly, $\frac{98}{99}$ = 0.989898 . . . or, rounded to four decimal places, 0.9899; $\frac{131}{20} = 6.55$; and $\frac{2}{1678} = 0.0012$.

Incidentally, a common careless error when dividing a small number by a much larger one is reporting too many or too few zeros after the decimal point in the answer. Thus, $\frac{2}{1678}$ is 0.0012, as we have seen. It is *not* 0.012 or 0.00012. When dividing a small number by a much larger one, always be sure to include the proper number of zeros after the decimal point.

Part 3: Addition and Subtraction

When you first learned to count, you always started from zero (or one) and went up from there. You were counting only *positive* numbers, that is, numbers greater than zero. However, numbers can also be less than zero. These are called *negative* numbers, as shown on the number line below.

−10 −9 −8 −7 −6 −5 −4 −3 −2 −1 0 +1 +2 +3 +4 +5 +6 +7 +8 +9 +10

Negative Numbers | Positive Numbers

When we work with positive numbers we usually leave off the + sign, so that "4" means "+4," "0.87" means "+0.87," and so on. But when we deal with negative numbers, it is very important to write the − sign. Many mistakes in statistical computations result from simply forgetting the minus sign in front of negative numbers.

Adding Positive and Negative Numbers

Two simple rules:

1. When adding two or more numbers having the *same* sign, add the numbers and attach the same sign to the sum.

   Example: (+3) + (+2) or 3 + 2 = +5 or 5
   (−5) + (−1) = −6
   (−3) + (−2) + (−6) = −11

2. When adding two numbers having different signs, subtract the smaller from the larger (ignoring signs); then attach the sign of the *larger* number to the sum.

   Example: +3 + (-2) = + 1
   3 + (−6) = −3
   −2 + 6 = 4
   −2 + 1 = −1
   $2X + (-4X) = -2X$

When adding three or more numbers, some of which are positive and some negative,

a. Add all the positive numbers using Rule 1.
b. Add all the negative numbers using Rule 1.
c. Add the positive sum and the negative sum using Rule 2.

Example:

2	a. 2 + 6 = 8
−5	b. (−5) + (−4) = −9
6	c. 8 + (−9) = −1
−4	

−2	a. 5 + 10 = 15
−4	b. (−2) + (−4) = −6
5	c. (15) + (−6) = 9
10	

−3	a. 5 + 2 = 7
5	b. (−3) + (−4) = −7
2	c. 7 + (−7) = 0
−4	

Subtraction of Positive and Negative Numbers

When subtracting signed numbers, one simple rule always works: First, change the sign of the number that you are subtracting; then *add* the numbers using the rules discussed in the previous section.

Example: 10 − 14 = 10 − (+14) = 10 + (−14) = −4
−8 − (+4) = −8 + (−4) = −12
−8 − (−12) = −8 + (+12) = −8 + 12 = 4
6 − (−3) = 6 + (+3) = 6 + 3 = 9

Part 4: Multiplication and Division

There are several ways to symbolize the operation of multiplication. Thus "4 times 3" may be written as

4 × 3   4 · 3   4(3)   (4)(3)

All of them mean "multiply 4 by 3." The last two are the most commonly used in statistics. Similarly, $3X$ means "multiply X by 3," and XY means "multiply X times Y."

Multiplication by Zero and One

The product of zero times any number is always zero. The product of a number times one equals the number.

EXAMPLE:
$13\,(0) = 0$
$(-10)(0) = 0$
$(12)(-16)(54)(0)(81)(-10) = 0$
$(0)(10{,}561{,}981) = 0$
$(-10)^5(0) = 0$
$OY^2 = 0$
$1(13) = 13$
$1(-13) = -13$
$1\,Y^2 = Y^2$

Multiplying Positive and Negative Numbers

A single rule always applies: Count how many negative numbers there are. If there are no negative numbers or if there is an even number of negative numbers (2, 4, 6, etc.), the product is positive. If there is an odd number of negative numbers, the product is negative.

EXAMPLE: $4(-3)$ [one negative number] $= -12$
$(-6)(-2)$ [two negative numbers] $= 12$

EXAMPLE: $(3)(-5)(-2)(2)(4) = +240$
↑ ↑
[two negative numbers]

$(3)(-5)(-2)(-2)(4) = -240$
↑ ↑ ↑
[three negative numbers]

Note that if just two numbers are being multiplied, the product is positive when the two numbers have the same sign; the product is negative when they have different signs.

EXAMPLE:

$(4)(5)$	$= 20$	Plus times plus = plus
$(4)(-5)$	$= -20$	Plus times minus = minus
$(-4)(5)$	$= -20$	Minus times plus = minus
$(-4)(-5)$	$= 20$	Minus times minus = plus

Dividing Positive and Negative Numbers

In arithmetic there are three ways to symbolize the operation of division:

$6 \div 3 \qquad 3\overline{)6} \qquad \dfrac{6}{3}$

Only the last of these is used in statistics. As you can see, any fraction represents the operation of division; the numerator is divided by the denominator, never the reverse. In statistics, we rarely leave a fraction in fractional form. Instead, we perform the indicated division and get a whole or decimal number.

In division, when the numerator is greater than the denominator, the quotient (answer) is always greater than one; when the numerator is less than the denominator, the quotient is less than one. Any number divided by itself equals one.

EXAMPLE:

$\frac{7}{3} = 2.3333\cdots = 2.33$, rounded to 2 decimal places

$\frac{3}{7} = 0.428 = 0.43$, rounded to 2 decimal places

$\frac{7}{7} = 1 \qquad \frac{-3}{-3} = 1$

The rules for dividing positive and negative numbers are the same as for multiplication: If both numbers have the same sign (positive or negative), the quotient is positive; if the two numbers have different signs, the quotient is negative.

EXAMPLE: $\frac{6}{2} = 3 \qquad \frac{-6}{-2} = 3$

$\frac{6}{-2} = -3 \qquad \frac{-6.3}{21} = -0.3 \qquad \frac{-3}{3} = -1$

One final note about division: Zero divided *by* any number is always zero; but zero cannot be divided *into* any number. Division by zero is impossible, and no answer can be given.

EXAMPLE: $\frac{0}{67} = 0 \qquad \frac{0}{-10{,}000} = 0 \qquad \frac{0}{0.0001} = 0$

$\frac{67}{0}$ is indeterminate

Multiplying Fractions

There are two ways to multiply fractions. One way is to convert the fractions into decimal numbers by doing the indicated division, then multiply.

EXAMPLE: $\left(\frac{-2}{3}\right)\left(\frac{5}{7}\right) = (-.667)(.714) = -0.476$

However, fractions can also be multiplied directly, yielding a fraction as a product. To do this, multiply the numerators of the fractions to get the numerator of the product; multiply the denominators to get the denominator of the product.

EXAMPLE: $\left(\frac{-2}{3}\right)\left(\frac{5}{7}\right) = \frac{-10}{21} = -0.476$

To multiply a whole number or decimal number by a fraction, multiply the number by the numerator of the fraction; then divide by the denominator.

EXAMPLE: $.5\left(\frac{2}{3}\right) = \frac{(.5)(2)}{3} = \frac{1}{3} = .33$

Part 5: Proportions and Percentages

In statistics we often divide a set of data into subsets, then express the size or *frequency* of each subset as a proportion or percentage of the whole. Thus, in a color preference study, if 50 of 100 people prefer blue, 25 prefer red, and 25 prefer yellow, we may say that half (or .5 or 50%) chose blue, one-fourth (or .25 or 25%) chose red, and one-fourth (or .25 or 25%) chose yellow. Proportions and percentages are sometimes called *relative frequencies* because they express frequencies relative to the size of the entire group.

A proportion is always a number between zero and one. It may be expressed as a fraction (e.g., ¼) or a decimal (e.g., .25); decimal notation is more common in statistics. When a set of data has been divided into subsets and the frequency of each subset converted into a proportion, the sum of all the proportions equals one. Thus, in the previous example

Choice	Frequency	Proportion
Blue	50	.50
Red	25	.25
Yellow	25	.25
Sum	100	1.00

A frequency is converted into a proportion by dividing the frequency by the size of the entire data set. Thus, the proportion of blue choices is $\frac{50}{100} = .50$, the proportion of red choices is $\frac{25}{100} = .25$, and the proportion of yellow choices is $\frac{25}{100} = .25$.

EXAMPLE: A frequency of 30 in a data set of 200 yields the proportion

$$\frac{30}{200} = .15$$

30 out of 50, expressed as a proportion, is

$$\frac{30}{50} = .6$$

30 out of 500, expressed as a proportion, is

$$\frac{30}{500} = .06$$

A percentage is usually a number between zero and 100.† Any proportion can be converted into a percentage by multiplying the proportion by 100. This is equivalent to moving the decimal point two places to the right.

EXAMPLE:

Proportion	*Percentage*
.5	.5 × 100 = 50%
.25	.25 × 100 = 25%
.15	.15 × 100 = 15%
.6	.6 × 100 = 60%
.06	.06 × 100 = 6%
.837	.837 × 100 = 83.7%
.0084	.0084 × 100 = 0.84%

When the percentages of all the subsets in a set of data are added, the sum is always 100%. In the color preference example, 50% + 25% + 25% = 100%.

To convert a percentage to a proportion, divide the percentage by 100; that is, move the decimal point two places to the left.

†Percentages can be greater than 100, but such percentages are not encountered in statistics.

EXAMPLE:

Percentage	*Proportion*
83%	$83 \div 100 = .83$
4%	$4 \div 100 = .04$
100%	$100 \div 100 = 1.00$
0.5%	$.5 \div 100 = .005$

When we are asked questions like, "How much is 25% of 40?" or "How many of 500 babies are expected to be boys if the proportion of boys is .52?" we need to convert proportions or percentages into frequencies. Because a proportion is obtained by dividing a frequency by total set size, a frequency may be obtained by multiplying a proportion times total set size. Thus, if the proportion of baby boys is .52, and there are 500 babies altogether, the expected frequency of the subset boys is

$$(.52)(500) = 260$$

To convert a percentage into a frequency, first convert the percentage to a proportion, then multiply by total set size, as noted earlier.

EXAMPLE: How much is 25% of 40?
a. 25% is a proportion equal to .25
b. Frequency = $(.25)(40) = 10$
How much is 43% of 200?
a. 43% is a proportion equal to .43
b. Frequency = $(.43)(200) = 86$

Part 6: Squares and Square Roots

You will usually compute squares and square roots on your calculator. However, some definitions and rules are necessary.

Computing Squares of Numbers

To square a number means to multiply it by itself. When a number is to be squared, this is indicated by placing a small superscript 2 above and to the right of the number.

EXAMPLE: 5^2 is read "5 squared." It means, multiply 5 by itself. Thus
$5^2 = (5)(5) = 25$
$(-5)^2 = (-5)(-5) = 25$
(*The square of a negative number is a positive number.*)
$(1.2)^2 = (1.2)(1.2) = 1.44$
$(0.004)^2 = (0.004)(0.004) = 0.000016$
(*Be careful to write the correct number of zeros after the decimal point.*)

Note that the square of any number, positive or negative, is always a positive number.

When an *addition* or *subtraction* operation is included within parentheses, and the entire quantity is to be squared, the addition or subtraction must be performed *first* and the result is then squared.

EXAMPLE: $(5 + 2)^2 = 7^2 = 49$; *not* $5^2 + 2^2$, which is $25 + 4$ or 29
$(2 - 5)^2 = (-3)^2 = 9$; *not* $2^2 - 5^2 = 4 - 25 = -21$

However, when a *multiplication* or *division* operation is included within parentheses, and the entire quantity is squared, it is possible to get the answer in either of two ways: (1) first multiply or divide, then square the result, or (2) first square each number, then multiply or divide.

EXAMPLE: $[(2)(3)]^2 = [6]^2 = 36$; or
$[(2)(3)]^2 = (2)^2(3)^2 = (4)(9) = 36$

$$\left(\frac{4}{2}\right)^2 = (2)^2 = 4; \quad \text{or}$$

$$\left(\frac{4}{2}\right)^2 = \frac{(4)^2}{(2)^2} = \frac{16}{4} = 4$$

Remember: You can compute the squares first when the operation in the parentheses is multiplication or division. However, when the operation is addition or subtraction, you must add or subtract first, then square the result.

Computing Square Roots

To find the square root of a number means to find the number that, when multiplied by itself, results in the given number. The symbol $\sqrt{\ }$ is used to indicate "square root of." Thus $\sqrt{25} = 5$ because $(5)(5)$ or $5^2 = 25$.

EXAMPLE: $\sqrt{16} = 4$ because $4^2 = 16$

$\sqrt{9} = 3$

$\sqrt{.09} = 3$

$\sqrt{160} = 12.65$

Check: $12.65 \times 12.65 = 160.02$, which is the same as 160 except for a small difference due to rounding the square root.

Strictly speaking, any positive number has both a positive and a negative square root because a negative number multiplied by itself yields a positive product. Thus, $\sqrt{25}$ is -5 as well as 5 because $(-5)^2 = 25$. However, in statistical computations it is always the positive root that is required.

Note that for our purposes square roots of negative numbers are indeterminate since no number multiplied by itself can yield a negative product. If in the course of a statistical computation you get a result such as $\sqrt{-38.71}$, go back over your computations for you have surely made a mistake.

If an *addition* or *subtraction* operation is indicated under a square root sign, always perform the addition or subtraction first, then find the square root of the sum or difference.

EXAMPLE:

$$\sqrt{4+9} = \sqrt{13} = 3.61; \textit{ not } \sqrt{4} + \sqrt{9} = 2 + 3 = 5$$

$$\sqrt{10-6} = \sqrt{4} = 2$$

If a *multiplication* or *division* operation is indicated under a square root sign, you may *either* multiply or divide first then take the square root; *or* take the square roots first then multiply or divide.

EXAMPLE:

$$\sqrt{(4)(9)} = \sqrt{36} = 6; \text{ or } \sqrt{(4)(9)} = \sqrt{4}\sqrt{9} = (2)(3) = 6$$

$$\sqrt{\frac{25}{16}} = \sqrt{1.5625} = 1.25; \text{ or } \sqrt{\frac{25}{16}} = \frac{\sqrt{25}}{\sqrt{16}} = \frac{5}{4} = 1.25$$

$$\sqrt{3}\sqrt{6} = (1.732)(2.449) = 4.243; \text{ or } \sqrt{3}\sqrt{6} = \sqrt{18} = 4.243$$

$$\frac{\sqrt{6}}{\sqrt{3}} = \frac{2.449}{1.732} = 1.41; \text{ or } \frac{\sqrt{6}}{\sqrt{3}} = \sqrt{\frac{6}{3}} = \sqrt{2} = 1.41$$

Part 7: Combining Arithmetic Operations

Several rules pertain to combining arithmetic operations:

1. When a series of operations is indicated, none of which is included within parentheses, squaring is performed first, then multiplication and/or division, and finally addition and/or subtraction.

EXAMPLE: $5(6)^2 = 5(36) = 180$

$50 + (3)(10) = 50 + 30 = 80$

$50 + (-3)(10) = 50 + (-30)$

$= 50 - 30 = 20$

2. Any operations within parentheses or brackets must be completed first.

EXAMPLE:

$$13.5 + \frac{1}{3}(14.5 - 13.5) = 13.5 + \frac{1}{3}(1)$$

$$= 13.5 + \frac{1}{3} = 13.5 + .33 = 13.83$$

If there are parentheses within brackets, work from the inside out. That is, first complete the operations in the inner parentheses, then those within the brackets.

EXAMPLE:

$$\sqrt{[10(964) - (40)^2][10(350) - (25)^2]}$$

$$= \sqrt{[9640 - 1600][3500 - 625]}$$

$$= \sqrt{[8040][2875]} = \sqrt{23{,}115{,}000} = 4807.8061$$

3. If the numerator or denominator of a fraction, or both of them, require a sequence of operations, work out the numerator and denominator separately, then divide the numerator by the denominator.

EXAMPLE:

$$\frac{60-50}{12} = \frac{10}{12} = 0.83$$

$$\frac{(10-15.5)^2}{15.5} = \frac{(-5.5)^2}{15.5} = \frac{30.25}{15.5} = 1.95$$

$$\sqrt{\frac{(6)^2}{10}+\frac{(4)^2}{12}} = \sqrt{\frac{36}{10}+\frac{16}{12}} = \sqrt{3.6+1.333}$$
$$= \sqrt{4.9333} = 2.22$$

A word of caution: Be careful to distinguish between cases like the following when performing sequential operations:

$$\frac{160-(10)^2}{12} = \frac{160-100}{12} = \frac{60}{12} = 5$$

but $160 - \frac{(10)^2}{12} = 160 - \frac{100}{12} = 160 - 8.33 = 151.67$

Part 8: Absolute Values

The absolute value of any quantity is its numerical value, disregarding its sign. It is indicated by vertical lines placed before and after the quantity. Thus, $|21|$ means "the absolute value of 21," and it is equal to 21. $|-21|$ also equals 21, $|-0.83| = 0.83$, and $|5-10| = |-5| = 5$.

If you are told "Reject H if $|t|$ is greater than 2.33," you should reject H for any t whose absolute value (disregarding sign) is larger than 2.33. If $t = 1.92$, its absolute value is 1.92, which is less than 2.33. Therefore, you should not reject H. However, for $t = 4.10$, $|t| = 4.10$, which is greater than 2.33; therefore, you should reject H. If $t = -1.92$, $|t| = |-1.92| = 1.92$, which is less (not greater) than 2.33; therefore, do not reject H. But if $t = -4.10$, $|t| = |-4.10| = 4.10$; because this is greater than 2.33, you should reject H.

Part 9: Inequalities

The symbol $>$ means "is greater than," and $<$ means "is less than." "$\geq$" is read "greater than or equal to," and "$\leq$" means "less than or equal to." For example, to say "Y is greater than 5," we write $Y > 5$; to say, "A is less than B," we write $A < B$. (Obviously, $A < B$ is the same as $B > A$.) We read $X > 3$ as "X is greater than 3," and $p < .05$ as "p is less than .05." The expression $.01 < p < .05$ means "p is greater than .01 but less than .05"; in other words, p is between .01 and .05. Similarly, "Z falls between 2 and 7" is expressed $2 < Z < 7$. If we want to say, "X is greater than or equal to 6," we write $X \geq 6$. Similarly, $p \leq .10$ means "p is less than or equal to .10."

Most people have no trouble expressing inequalities for positive numbers. After all, everyone knows that $5 > 3$ and $2 < 4$. Sometimes, though, negative numbers present a problem. An easy way to deal with positive and negative numbers is to consider the numbers as lying on a number line, as illustrated in Part 3 of this appendix. *On a number line, any number is greater than all numbers to its left and less than all numbers to its right.*

Example: On the number line:
5 is to the right of -3, so $5 > -3$.
-3 is to the right of -5, so $-3 > -5$, and $-5 < -3$.
-5 is to the left of 3, so $-5 < 3$, and $3 > -5$.

Remember that the absolute value of a number is its value disregarding its sign. Therefore, to decide whether $|-3|$ is greater than or less than $|-5|$, we compare the values 3 and 5. Since $3 < 5$, therefore, $|-3| < |-5|$. Similarly, $|-5|$ (which equals 5) > 3.

Part 10: Using Formulas

A formula expresses a set of numerical operations in the form of symbols so that various numbers may be substituted for the symbols as needed. For example, you once learned that the area of a rectangle can be found by the formula $A = lw$, where l is the length and w is the width of the rectangle. If you are told that $l = 6$ and $w = 4$, you can *substitute* and compute as follows:

$$A = (6)(4) = 24$$

For another rectangle with $l = 2$ and $w = .5$, $A = (2)(.5) = 1$. In other words, to use a formula, first substitute the appropriate quantity for each symbol in the formula, then perform the indicated operations using the rules given in Part 7.

Example:

1. Find Y if $N = 7$ and $X = 5$.

$$Y = \frac{NX^2 - X}{(NX)^2} = \frac{(7)(5)^2-(5)}{[(7)(5)]^2} = \frac{7(25)-5}{(35)^2}$$
$$= \frac{175-5}{1225} = \frac{170}{1225} = 0.139$$

Note that NX^2 says: First square X, then multiply by N. $(NX)^2$ says: First multiply N by X, then square.

2. Using the same formula, find Y if $N = 3$ and $X = -2$.

$$Y = \frac{(3)(-2)^2 - (-2)}{[(3)(-2)]^2} = \frac{(3)(4) + (+2)}{(-6)^2}$$
$$= \frac{12+2}{36} = \frac{14}{36} = 0.389$$

3. If $A = 2$, $B = 3$, and $C = 5$, find Y.

$$Y = \frac{(B-A)^2(C+2)}{\sqrt{2C^2 - AB}}$$

Substituting

$$Y = \frac{(3-2)^2(5+2)}{\sqrt{2(5)^2 - (2)(3)}} = \frac{(1)^2(7)}{\sqrt{2(25) - 6}} = \frac{(1)(7)}{\sqrt{50-6}}$$
$$= \frac{7}{\sqrt{44}} = \frac{7}{6.6332} = 1.055$$

Part 11: Manipulating Simple Equations

An equation has an expression on each side of an = sign, indicating that the two expressions are equal to one another. For example, $X = 4$ is an equation, as are $3X = 8 + 2X$, $\frac{X-3}{5} = \frac{Y+7}{2}$, and $X^2 - 2X + 6 = 0$.

If the same operation is performed on *both* sides of an equation, the two sides remain equal. For example, starting with the equation $X = 4$, we can

Add 5 to each side:	$X + 5 = 4 + 5$
	$X + 5 = 9$
Add A to each side:	$X + A = 4 + A$
Subtract 8 from each side:	$X - 8 = 4 - 8$
	$X - 8 = -4$
Subtract k from each side:	$X - k = 4 - k$
Multiply each side by -2:	$-2X = (-2)(4)$
	$-2X = -8$
Multiply each side by C:	$CX = 4C$
Divide each side by 3:	$\frac{X}{3} = \frac{4}{3}$
	$\frac{X}{3} = 1.33$
Divide each side by n:	$\frac{X}{n} = \frac{4}{n}$
Square each side:	$X^2 = 4^2$
	$X^2 = 16$

Often, we want to solve an equation for a particular term within it, such as X or Y. Suppose, for example, that we have the equation $Y - 3B = 2X + 4$, and we want to solve the equation for Y. This means that we want to isolate Y on one side of the equation, with everything else on the other side. To accomplish this, we perform one or more of the operations just illustrated. Of course we must be careful that any operation performed on one side of the equal sign is performed on the other side as well. Let us solve the equation $Y - 3B = 2X + 4$ for Y. To isolate Y on the left of the equal sign, we

Add $3B$ to each side: $Y - 3B + 3B = 2X + 4 + 3B$
Since $(-3B + 3B) = 0$: $Y = 2X + 4 + 3B$

Let us solve the same equation for X:

We start with	$Y - 3B = 2X + 4$
Subtract 4 from each side	$Y - 3B - 4 = 2X + 4 - 4$
Since $(4 - 4) = 0$:	$Y - 3B - 4 = 2X$
Divide both sides by 2:	$\frac{Y - 3B - 4}{2} = \frac{2X}{2}$
Since $\frac{2X}{2} = \left(\frac{2}{2}\right)X = X$:	$\frac{Y - 3B - 4}{2} = X$

In the following examples, we solve for X:

1.

We start with:	$Y = BX + 3$
Subtract 3 from both sides:	$Y - 3 = BX + 3 - 3$
Since $(+3-3) = 0$:	$Y - 3 = BX$
Divide both sides by B:	$\frac{Y-3}{B} = \frac{BX}{B}$
Since $\frac{BX}{B} = X$:	$\frac{Y-3}{B} = X$

2.

We start with:	$Y = X - B$
Add B to both sides:	$Y + B = X - B + B$
Since $(-B+B) = 0$:	$Y + B = X$

3. We start with: $Y = \frac{X}{B}$

Multiply both sides by B: $BY = \frac{BX}{B}$

$BY = X$

4. We start with: $Y = \frac{B+2}{X}$

Multiply both sides by X: $XY = \frac{X(B+2)}{X}$

Since $\frac{X(B+2)}{X} = \left(\frac{X}{X}\right)(B+2) = B+2$: $XY = B + 2$

Divide both sides by Y: $\frac{XY}{Y} = \frac{B+2}{Y}$

Since $\frac{XY}{Y} = X\left(\frac{Y}{Y}\right) = X$:

$X = \frac{B+2}{Y}$

5. We start with: $Y = \frac{B-X^2}{B}$

Multiply both sides by B: $BY = \frac{B(B-X^2)}{B}$

$BY = \left(\frac{B}{B}\right)(B-X^2)$

Since $\frac{B}{B} = 1$: $BY = B - X^2$

Add X^2 to each side: $X^2 + BY = B - X^2 + X^2$

Since $(-X^2 + X^2) = 0$: $X^2 + BY = B$

Subtract BY from each side: $X^2 + BY - BY = B - BY$

Since $(+BY - BY) = 0$: $X^2 = B - BY$

Take the square root of each side: $X = \sqrt{B - BY}$

6. We start with: $Y = \sqrt{\frac{B+2}{X}}$

To get rid of the $\sqrt{\ }$ square both sides: $Y^2 = \frac{B+2}{X}$

Multiple both sides by X: $XY^2 = \frac{X(B+2)}{X}$

Since $\frac{X}{X} = 1$: $XY^2 = B + 2$

Divide both sides by Y^2: $\frac{XY^2}{Y^2} = \frac{B+2}{Y^2}$

Since $\frac{Y^2}{Y^2} = 1$: $X = \frac{B+2}{Y^2}$

This completes the instructional component of Appendix B. Now take the posttest that follows.

Self–Evaluation Posttest

Part 1: Rounding

Round to three decimal places:

1. 0.8477604
2. 953.17449
3. 8.0450032
4. 16.9997
5. 1061.27651
6. 0.0754
7. 0.00013
8. 0.00076

Part 2: Converting Fractions Into Decimals

Convert the following fractions into decimals. If necessary, round your answer to four decimal places.

1. $\frac{3}{5}$
2. $\frac{5}{6}$
3. $\frac{1}{2000}$
4. $\frac{49}{99}$
5. $\frac{53}{27}$

Part 3: Addition and Subtraction

Add the following:

1. $(-5) + (-2) =$
2. $-7 + 4 =$
3. $2 + (-5) + 3 =$
4. $\begin{array}{r} -5 \\ 9 \\ 7 \\ \underline{-4} \end{array}$
5. $\begin{array}{r} 3 \\ -1 \\ -3 \\ \underline{-2} \end{array}$

Perform the indicated subtractions:

6. $5 - 10 =$
7. $-5 - (+7) =$
8. $-2 - (-3) =$
9. $2 - (-2) =$

Part 4: Multiplication and Division

1. $(12)(4)(0)(10) =$
2. $3(-1) =$
3. $(-3)(-1) =$
4. $\frac{-10}{5} =$
5. $\frac{-10}{-5} =$
6. $\frac{10}{-5} =$
7. $\frac{0}{72} =$
8. $\left(\frac{1}{2}\right)\left(\frac{-3}{5}\right) =$
9. $2\left(\frac{-3}{5}\right) =$

Part 5: Proportions and Percentages

1. Express 25 out of 4000 as a proportion; as a percentage.
2. Express 25 out of 40 as a proportion; as a percentage.

Convert the following proportions into percentages:

3. .645

4. .01

5. .001

Convert the following percentages into proportions:

6. 25%

7. 7%

8. 99.9%

9. $\frac{3}{5}\%$

10. How much is 27% of 30?

11. How much is 98% of 300?

12. How many of 900 college students are extroverts if the proportion of extroverts is .08?

Part 6: Squares and Square Roots

1. $9^2 =$

2. $(0.6)^2 =$

3. $(0.001)^2 =$

4. $(-9)^2 =$

5. $(3 + 2)^2 =$

6. $(1 - 7)^2 =$

7. $[(4)(2)]^2 =$

8. $\left(\frac{12}{4}\right)^2 =$

9. $\sqrt{9} =$

10. $\sqrt{90} =$

11. $\sqrt{3+6} =$

12. $\sqrt{(2)(8)} =$

13. $\sqrt{\frac{9}{16}} =$

14. $\frac{\sqrt{8}}{\sqrt{2}} =$

Part 7: Combining Arithmetic Operations

1. $52 - \frac{(12)^2}{5} =$

2. $\frac{(7-5)^2}{5} =$

3. $18 - \frac{2}{5}(16 - 14) =$

4. $25 + (2)(10) =$

5. $25 + (-2)(10) =$

6. $\sqrt{\frac{(10)^2}{5} + \frac{(3)^2}{7}} =$

7. $12(850) - (100)^2 =$

Part 8: Absolute Values

1. $|32.9| =$

2. $|-0.71| =$

3. $|17 - 13| =$

4. $|13 - 17| =$

Rule: Reject H if $|z|$ is greater than 1.96. Tell whether you should reject H in each of the following cases:

5. $z = 2.50$

6. $z = 1.50$

7. $z = -1.50$

8. $z = -2.50$

Part 9: Inequalities

Write the following in English:

1. $Z \leq -2$

2. $p < .025$

3. $.001 < p < .005$

4. $F \geq 2.96$

Place the proper sign, > or <:

5. 7 ___ 4

6. 4 ___ 7

7. −4 ___ −7

8. |−7| ___ |−4|

9. 7 ___ −4

10. |−7| ___ |4|

11. Express the following using < and/or >: "Y falls between −3 and +4."

If you are told, "Reject H if $z < -2.58$," decide whether you would reject H in each of the following:

12. $z = 5.13$

13. $z = 1.47$

14. $z = -1.47$

15. $z = -5.13$

Part 10: Using Formulas

1. Find X if $n = 6$ and $Y = 10$:

$$X = \frac{(nY)^2 - (Y+2)}{nY^2}$$

2. If $k = 3$, $n = 5$, and $B = 2$, find X:

$$X = \frac{kB^2}{n}$$

3. If $k = 3$, $n = 5$, and $B = 2$, find X:

$$X = \frac{nkB - kB^2}{\sqrt{nB^2 + kB}}$$

Part 11: Manipulating Simple Equations

Solve for X.

1. $N = \dfrac{A+2}{X}$

2. $Z = \dfrac{X}{3}$

3. $Y = kX - 2$

4. $2Y = X + 3B$

5. $Y = \sqrt{X+6}$

END OF POSTTEST

Check your answers on the following pages.

Answers to Self–Evaluation Posttest

Part 1

1. 0.848
2. 953.174
3. 8.045
4. 17.000
5. 1061.277
6. 0.075
7. 0.000
8. 0.001

Part 2

1. .6
2. .8333
3. .0005
4. .4949
5. 1.9630

Part 3

1. −7
2. −3
3. 0
4. 7
5. −3
6. −5
7. −12
8. 1
9. 4

Part 4

1. 0
2. −3
3. 3
4. −2
5. 2
6. −2
7. 0
8. −.3
9. −1.2

Part 5

1. .006; 0.6%
2. .625; 62.5%
3. 64.5%
4. 1%
5. .1%
6. .25
7. .07
8. .999
9. .006
10. 8.1
11. 294
12. 72

Part 6

1. 81
2. .36
3. .000001
4. 81
5. 25
6. 36
7. 64
8. 9
9. 3
10. 9.49
11. 3
12. 4
13. .75
14. 2

Part 7

1. 23.2
2. 0.8
3. 17.2
4. 45
5. 5
6. 4.614
7. 200

Part 8

1. 32.9
2. 0.71
3. 4
4. 4
5. Yes
6. No
7. No
8. Yes

Part 9

1. Z is less than or equal to -2
2. p is less than .025
3. p is between .001 and .005
4. F is greater than or equal to 2.96
5. $>$
6. $<$
7. $>$
8. $>$
9. $>$
10. $>$
11. $-3 < y < 4$
12. No
13. No
14. No
15. Yes

Part 10

1. 5.98
2. 2.4
3. 3.53

Part 11

1. $X = \frac{A+2}{N}$
2. $X = 3Z$
3. $X = \frac{Y+2}{k}$
4. $X = 2Y - 3B$
5. $X = Y^2 - 6$

Appendix C: Answers to Within-Chapter Exercises

Chapter 1

1–1 Computers cannot select the appropriate procedure or interpret the results.

1–2

1. a. quantitative; continuous
   b. qualitative
   c. quantitative; discrete
   d. quantitative; discrete
   e. qualitative
   f. quantitative; continuous
2. a. classification into categories
   b. measurement of amount
   c. ranking
   d. classification into categories
   e. ranking

1–3

1. a. difference between two populations
   b. status of one variable in one population
   c. difference between three populations
   d. difference between two population (video vs. no video)
   e. relation between two variables
   f. status of one variable in one population
   g. difference between two populations
2. a. nonexperimental comparative study
   c. experiment: independent variable: level of background noise
      dependent variable: perception of tones
   d. experiment: independent variable: whether videos are used
      dependent variable: performance in the course
   g. experiment: independent variable: presence or absence of lab
      dependent variable: performance in statistics

Chapter 2

2–1

1, 2

Major*	f	p**
Sociology	2	8
Psychology	9	36
Music	2	8
English	1	4
Secondary education	1	4
Philosophy	1	4
Chemistry	1	4
Business	3	12
Painting	1	4
Elementary education	1	4
Biology	1	4
Physics	2	8

*Order in which majors are listed is arbitrary.

**For each major, $P = \frac{f}{25} \times 100$.

3. The exact appearance of your two graphs will differ, depending on the order of majors displayed on the abscissa. Check to be sure that you have labeled both axes and that all bars are the same width with equal space between them. The two bar graphs have different units on the ordinate but are otherwise identical.

2–2

1. 450 + 375 + 675 = 1500
2. 375 + 80 + 3000 = 3455
3. You must divide the number in each cell by the row sum.

	Liberal Arts	*Business*	*Preprofessional*
College A	30	25	45
College B	60	10	30
College C	25	60	15

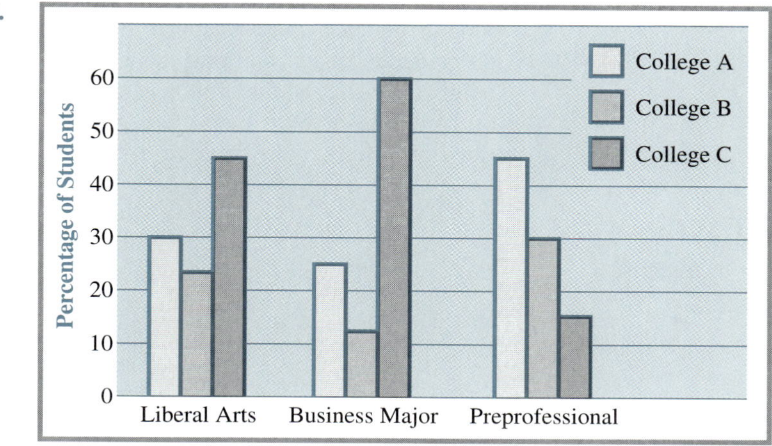

2–3

1.

(A)[a]	(A) Divided by (B)	Class Interval Size
c. 110	7.3	7 (or 8)
d. 37	4.6	5 (or 4)
e. 55	4.6	5 (or 4)

[a]Range = Highest score − Lowest score.

2. (1)

Score	*f*	Cum *f*	Cum *P*
24	1	20	100
23	1	19	95
22		18	90
21	1	18	90
20		17	85
19		17	85
18	1	17	85
17	1	16	80
16	1	15	75
15		14	70
14	2	14	70
13		12	60
12		12	60
11		12	60
10	2	12	60
9	2	10	50
8	2	8	40
7	1	6	30
6	2	5	25
5	1	3	15
4	1	2	10
3	1	1	5

(2) Range = 24 − 3 = 21.
21/10 = 2.1. Use class interval size = 2.
Highest interval may be 23–24 or 24–25.
Lowest interval may be 3–4 or 2–3.
Exact appearance of distribution depends on choice of intervals.

(3) 21/7 = 3. Use class interval size of 3. Exact appearance of distribution depends on choice of intervals.

b. Answers will vary.

c. Gained: A smoother distribution.
Lost: Exact values of individual scores

2–4

Scores in Interval	Class Interval Size	Exact Limits	Midpoint
b. 30, 31, 32, 33, 34, 35, 36, 37, 38, 39	10	29.5–39.5	34.5
c. 30, 31	2	29.5–31.5	30.5
d. 30, 31, 32	3	29.5–32.5	31.0
e. 30, 31, 32, 33, 34, 35, 36, 37, 38, 39, 40, 41, 42, 43, 44, 45, 46, 47, 48, 49	20	29.5–49.5	39.5
f. 30	1	29.5–30.5	30.0

2–5 For cumulative frequencies and proportions, see the answer to Exercise 2–3, number 2.

1. 15
2. 70
3. 9.5
4. 16.5

2–6

Graphs will vary, depending on the method of grouping. Check to make sure you have (1) labeled both axes, (2) started from a frequency of zero on the ordinate, (3) drawn each bar to a height equal to the frequency in the interval. All bars should be the same width. Values on the abscissa should go from lowest on the left to highest on the right. The height of the figure should be about two-thirds to three-quarters the width.

2–7

Graphs will vary, depending on the method of grouping. Check to make sure you have plotted a point for each interval above the *midpoint* of the interval, that you have connected all the points in order, and that you have closed the figure properly at both ends. Otherwise, use the same guidelines as in Exercise 2–6.

2–8

0	34
0	56678899
1	0044
1	678
2	134

2–9

A		B
32	0	01134
9886	0	5667789
4311	1	1224
99876655	1	5666689
4443220	2	00334
88655	2	78

Method B seems to do a better job (fewer defective widgets).

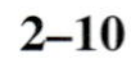

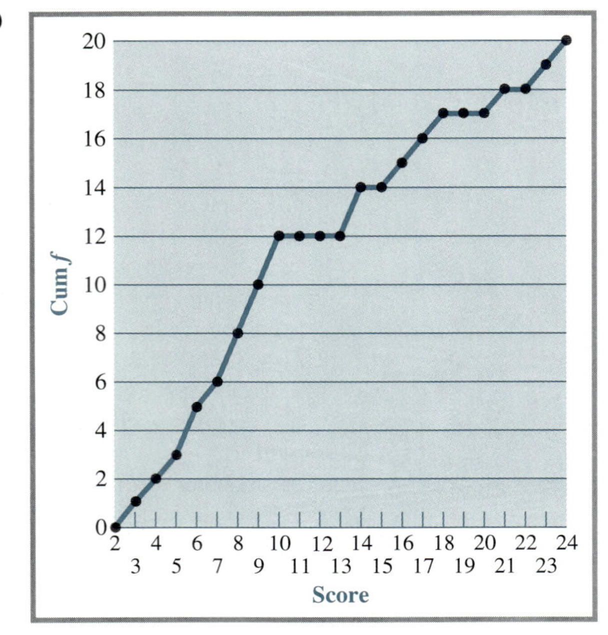

2–11

1. a. symmetrical
   b. positively skewed
   c. bimodal (scores pile up at scores of 2 and 4)
2. a. .10 + .45 = .55
   b. .45 + .40 = .85
   c. .50 − .10 = .40
   d. 8.5

2–12

1. Frequencies and cumulative frequencies are shown in the answer to Exercise 2–3, number 2(1). For each score, Midpoint Cum $f = ½ f$ + next lower score's Cum f, and percentile rank = $\frac{\text{Midpoint Cum} f}{N} \times 100.$

Score	Midpoint Cum *f*	Percentile Rank
24	19.5	97.5
23	18.5	92.5
22	18.0	90.0
21	17.5	87.5
20	17.0	85.0
19	17.0	85.0
18	16.5	82.5
17	15.5	77.5
16	14.5	72.5
15	14.0	70.0
14	13.0	65.0
13	12.0	60.0
12	12.0	60.0
11	12.0	60.0
10	11.0	55.0
9	9.0	45.0
8	7.0	35.0
7	5.5	27.5
6	4.0	20.0
5	2.5	12.5
4	1.5	7.5
3	0.5	2.5

2.

Score	Percentile Rank
5	12.5
10	55.0
15	70.0
20	85.0

Percentile ranks are not equidistant because the distribution is positively skewed. Because most observations fall at the low scores, percentile ranks change most rapidly there (e.g., between scores of 5 and 10).

2–13

Rating	*f*	Cum *f*
7	1	25
6		24
5	2	24
4	4	22
3	7	18
2	8	11
1	3	3
	25	

1. Number of observations below P_{10} = $\frac{(10)(25)}{100} = 2.5$. The next higher Cum f, 3, is that for a rating of 1. P_{10} is in the interval 0.5–1.5.

$$p_{10} = 0.5 + \left(\frac{25-0}{3}\right) = 0.5 + .83 = 1.33$$

Number of observations below P_{90} = $\frac{(90)(25)}{100} = 22.5$.

The next higher Cum f, 24, is that for a rating of 5.
P_{90} is in the interval 4.5–5.5.

$$p_{90} = 4.5 + \left(\frac{22.5-22}{2}\right) = 4.5 + .25 = 4.75$$

2. The distribution is positively skewed.
3. Number of observations below Q_1 (P_{25}) = $\frac{(25)(25)}{100} = 6.25$.. The next higher Cum f, 8, is that for a rating of 2. Q_1 is in the interval 1.5–2.5.

$$Q_1 = 1.5 + \left(\frac{6.25-3}{8}\right) = 1.5 + .4 = 1.9$$

Number of observations below Q_2 (P_{50}) = $\frac{(50)(25)}{100} = 12.5$.

The next higher Cum f, 18, is that for a rating of 3.
Q_2 is in the interval 2.5–3.5.

$$Q_2 = 2.5 + \left(\frac{12.5-11}{7}\right) = 2.5 + .2 = 2.7$$

Number of observations below Q_3 (P_{75}) = $\frac{(75)(25)}{100} = 18.75$. The next higher Cum f, 22, is that for a rating of 4. Q_3 is in the interval 3.5–4.5.

$$Q_3 = 3.5 + \left(\frac{18.75 - 18}{4}\right) = 3.5 + .2 = 3.7$$

$Q_2 - Q_1 = 2.7 - 1.9 = 0.8$; $Q_3 - Q_2 = 3.7 - 2.7 = 1.0$

$Q_3 - Q_2 > Q_2 - Q_1$; distribution is positively skewed.

4. P_{50} = 2nd quartile = 50th percentile = median
   P_{25} = 25th percentile = first quartile
   Q_1 = 25th percentile = first quartile
   Q_3 = 3rd quartile = 75th percentile
   P_{75} = 3rd quartile = 75th percentile
   Q_2 = 2nd quartile = 50th percentile = median

2–14

1.

(The plot can be drawn vertically rather than horizontally. If so, the Ratings scale has 7 at the top and 1 at the bottom.)

The positive skew is indicated by the longer upper whisker and by the greater distance between the median and Q_3 than between the median and Q_1.

2. The frequency polygons are best for displaying the shapes of the distributions, but with a loss of individual scores. Even more details are lost in the box plots, though they, like the frequency polygons, facilitate comparison of the two distributions. The stemplots retain all details, but are less visually compelling than the other graphs.

Chapter 3

3–1

1. Distributions (1) and (3)
2. Distributions (2) and (3)
3. Distribution (4)

3–2

1. a. Mode = 7.
   Scores in order: 7, 7, 7, 8, 9 $N = 5$
   ↑
   Mdn = 7

   b. Modes = 12 and 15
   Scores in order:
   10, 11, 12, 12, 13, 15, 15, 16 $N = 8$
   ↑
   Mdn = 12.5

   c. No mode
   Scores in order:
   8, 9, 15, 18, 19, 21, 24 $N = 7$
   ↑
   Mdn = 18

   d. Mode = 10
   Scores in order:
   6, 8, 8, 10, 10, 10, 11, 15 $N = 8$
   ↑
   Mdn = 10

2.

```
0 | 3 4                $N$ = 20
0 | 5 6 6 7 8 8 9 9 ↑  Mdn is halfway between
1 | 0 0 4 4        Mdn 9 and 10 = 9.5
1 | 6 7 8
2 | 1 3 4
```

3–3

1. a. $N = 7$.

$$\text{Mean} = \frac{8+6+5+9+12+3+2}{7} = \frac{45}{7} = 6.4$$

   b. $N = 4$. $\text{Mean} = \frac{2+5+1+8}{4} = \frac{16}{4} = 4.0$

   c. $N = 5$.

$$\text{Mean} = \frac{3.6+9.4+6.7+7.1+6.5}{5} = \frac{33.3}{5} = 6.66$$

2. $N_1 = 20$, $M_1 = 26.9$, $N_2 = 35$, $M_2 = 20.2$

$$\text{Weighted mean} = \frac{(20)(26.9)+(35)(20.2)}{20+35} = \frac{1245}{55} = 22.6$$

3. a. Mean will increase
   $\Sigma X = 550 - 35 + 45 = 560$

$$M = \frac{560}{20} = 28.0$$

   b. Mean will decrease
   $\Sigma X = 550 - 30 + 20 = 540$

$$M = \frac{540}{20} = 27.0$$

c. Mean will decrease
$\Sigma X = 550 - 50 = 500$; $N = 20 - 1 = 19$

$$M = \frac{500}{19} = 26.3$$

d. Mean will increase
$\Sigma X = 550 - 15 = 535$; $N = 20 - 1 = 19$

$$M = \frac{535}{19} = 28.2$$

e. Mean will increase
$\Sigma X = 550 + 40 = 590$; $N = 20 + 1 = 21$

$$M = \frac{590}{21} = 28.1$$

f. Mean will decrease
$\Sigma X = 550 + 25 = 575$; $N = 20 + 1 = 21$

$$M = \frac{575}{21} = 27.4$$

3–4

1. $M > Mdn >$ mode; distribution is positively skewed.
2. a. Scores in order: 1.4, 2.0, 2.0, 2.0, 2.4, 2.7, 3.1, 3.5, 3.8, 5.2, 6.7, 8.0, 8.5, 9.0, 9.5
   ↑
   Mdn
   Mode = 2.0; median = 3.5;
   $$\text{mean} = \frac{1.4 + 2.0 + \cdots + 9.5}{15} = \frac{69.8}{15} = 4.7$$
   b. median
   c. mode
   d. mean
   e. All three are measures of central tendency. However, because the distribution is very skewed, they differ markedly.
3. a. Distribution is positively skewed. Prediction: $M > Mdn >$ mode.

$$M = \frac{2(20) + 3(19) + 7(18) + 8(17) + 4(16) + 15}{25} =$$

$$\frac{438}{25} = 17.52$$

Mdn is between 16.5 and 17.5 (exact value = 17.4)
Mode = 17

b. Distribution is positively skewed. Prediction: $M > Mdn >$ mode.

$$M = \frac{20 + 2(19) + 3(18) + 5(17) + 10(16) + 4(15)}{25} =$$

$$\frac{417}{25} = 16.7$$

Mdn is between 15.5 and 16.5 (exact value = 16.35)
Mode = 16

c. Distribution is negatively skewed. Prediction: $M < Mdn <$ mode.

$$M = \frac{6(20) + 10(19) + 5(18) + 2(17) + 16 + 15}{25} =$$

$$\frac{465}{25} = 18.6$$

Mdn is between 18.5 and 19.5 (exact value = 18.85)
Mode = 19

4. There is usually a small number of extremely high-priced houses in the region. These high outliers increase the mean to a value well above the price of most houses.

3–5

1. a. Scores in order: 95, 97, 105, 107 109, 110, 110, 110, 113, 113, 115, 117 (N = 12)
   ↑
   Mdn = 110
   Range = 117 – 95 = 22
   Scores below median: 95, 97, 105, 107, 109, 110
   ↑
   $Q_1 = 106$
   Scores above median: 110, 110, 113, 113, 115, 117
   ↑
   $Q_3 = 113$
   Interquartile range (IQR) = 113 – 106 = 7

95 96 97 98 99 100 101 102 103 104 105 106 107 108 109 110 111 112 113 114 115 116 117

b. Scores in order: 2, 3, 3, 3, 4,
5, 6, 6, 6, 6, 7 (N = 11)
↑
Mdn = 5
Range = 7 − 2 = 5
Scores below median: 2, 3, 3, 3, 4
↑
$Q_1 = 3$
Scores above median: 6, 6, 6, 6, 7
↑
$Q_3 = 6$

IQR = 6 − 3 = 3

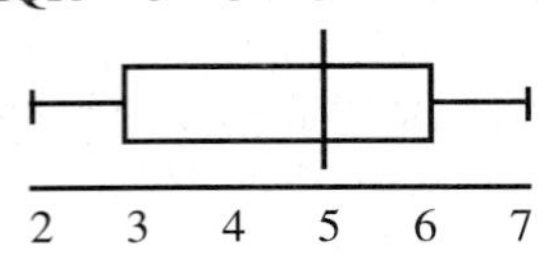

c. Scores in order: 0.6, 0.7, 0.9, 1.0,
1.1, 1.2, 1.2, 1.3, 2.3 $N = 9$
↑
Mdn = 1.1
Range = 2.3 − 0.6 = 1.7
Scores below median: 0.6, 0.7, 0.9, 1.0
↑
$Q_1 = 0.8$
Scores above median: 1.2, 1.2, 1.3, 2.3
↑
$Q_3 = 1.25$

.6 .7 .8 .9 1.0 1.1 1.2 1.3 1.4 1.5 1.6 1.7 1.8 1.9 2.0 2.1 2.2 2.3

d. Scores in order: 24, 26, 27,
28, 30, 32, 33, 35 $N = 8$
↑
Mdn = 29
Range = 35 − 24 = 11
Scores below median: 24, 26, 27, 28
↑
$Q_1 = 26.5$
Scores above median: 30, 32, 33, 35
↑
$Q_3 = 32.5$

IQR = 32.5 − 26.5 = 6.0

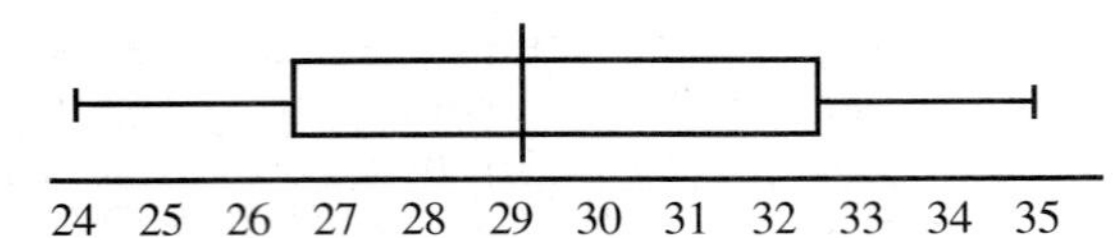

2. a. Set C. The score 2.3 is an outlier.
b. Scores in order: 0.6, 0.7, 0.9,
1.0, 1.1, 1.2, 1.2, 1.3 $N = 8$
↑
Mdn = 1.05
Range = 1.3 − 0.6 = 0.7
Scores below median: 0.6, 0.7, 0.9, 1.0
↑
$Q_1 = 0.8$
Scores above median: 1.1, 1.2, 1.2, 1.3
↑
$Q_3 = 1.2$

IQR = 1.2 − 0.8 = 0.4

Removing the outlier reduces the range by more than half, from 1.7 to 0.7.

Removing the outlier has little effect on IQR, which is reduced from 0.45 to 0.4.

3. Distribution F: *Mdn* = 6
$Q_1 = 4$, $Q_3 = 8$ IQR = 8 − 4 = 4
Distribution G: *Mdn* = 6
$Q_1 = 5$, $Q_3 = 7$ IQR = 7 − 5 = 2

4. a. *Mdn* = 61
b. Mode = 58
c. Range = 94 − 40 = 54
d. $Q_1 = 55$, $Q_3 = 68$; IQR = 68 − 55 = 13;

$$Q = \frac{13}{2} = 6.5$$

e. $Q_2 - Q_1 = 61 - 55 = 6$; $Q_3 - Q_2 = 68 - 61 = 7$; $Q_3 - Q_2 > Q_2 - Q_1$; distribution is positively skewed.

3–6

1. $SS = 640$, $N = 10$.

$$\text{Variance} = \frac{SS}{N} = \frac{640}{10} = 64;$$

$$SD = \sqrt{\text{variance}} = \sqrt{64} = 8$$

2. Sum of squares (SS) = N(variance) = 6(45) = 270

3. Variance = $(SD)^2 = 27.04$; $SS = 100\ (27.04) = 2704$

4. a.

X	$X - M$	$(X - M)^2$
5	−2	4
6	−1	1
8	1	1
9	2	4
28	0	10

$$M = \frac{28}{4} = 7$$

$$\text{Var} = \frac{10}{4} = 2.5$$

$$SD = \sqrt{2.5} = 1.58$$

b. Mean will decrease; Var and *SD* will increase.

X	$X - M$	$(X - M)^2$
2	−4	16
5	−1	1
6	0	0
8	2	4
9	3	9
30	0	30

$$M = \frac{30}{5} = 6$$

$$\text{Var} = \frac{30}{5} = 6$$

$$SD = \sqrt{6} = 2.45$$

c. Mean will increase; Var and *SD* will increase.

X	$X - M$	$(X - M)^2$
5	−3	9
6	−2	4
8	0	0
9	1	1
12	4	16
40	0	30

$$M = \frac{40}{5} = 8$$

$$\text{Var} = \frac{30}{5} = 6$$

$$SD = \sqrt{6} = 2.45$$

3–7

1. a.

X	X^2
2	4
4	16
8	64
6	36
5	25
7	49
8	64
40	258

$$\Sigma X = 40 \quad \Sigma X^2 = 258 \quad N = 7$$

$$SS = 258 - \frac{(40)^2}{7} = 258 - \frac{1600}{7}$$

$$= 258 - 228.57 = 29.43$$

$$\text{Var} = \frac{29.43}{7} = 4.204$$

$$SD = \sqrt{4.204} = 2.05$$

Range = 8 − 2 = 6
SD/Range = 2.05/6 = .34;
SD is about 1/3 range

b.

X	X^2
22	484
14	196
23	529
20	400
18	324
23	529
29	841
26	676
175	3979

$$\Sigma X = 175 \quad \Sigma X^2 = 3979 \quad N = 8$$

$$SS = 3979 - \frac{(175)^2}{8} = 3979 - \frac{30{,}625}{8}$$

$$= 3979 - 3828.125 = 150.875$$

$$\text{Var} = \frac{150.875}{8} = 18.86$$

$$SD = \sqrt{18.86} = 4.34$$

Range = 29 − 14 = 15
SD/Range = 4.34/15 = .29;
SD is about 3/10 range

2. The sum of squares is a sum of squared quantities. Because each square is positive, the sum is positive. The variance is *SS/N*. Because both *SS* and *N* are positive, the variance must be positive.

3–8

1. a. New mean = 60 − 13 = 47
   New *SD* = 12; new Var = $(12)^2$ = 144 (Var and *SD* do not change)
   b. New mean = 4(60) = 240
   New *SD* = 4(12) = 48; new Var = $(48)^2$ = 2304
   c. New mean = 60 + 7 = 67
   New *SD* = 12; new Var = 144 (Var and *SD* do not change)
   d. New mean = 60/10 = 6
   New *SD* = 12/10 = 1.2; new Var = $(1.2)^2$ = 1.44
   e. New mean = $\frac{60-60}{12} = 0$
   New *SD* = 12/12 = 1; new Var = 1^2 = 1

2. a.

X	$X - M$	$(X - M)^2$
2	−1	1
1	−2	4
5	2	4
3	0	0
4	1	1
15		10

$$M = \frac{15}{5} = 3$$

$$\text{Var} = \frac{10}{5} = 2$$

$$SD = \sqrt{2} = 1.414$$

b. New mean = 2(3) + 3 = 9
New *SD* = 2(1.414) = 2.83; new Var = $2^2(2)$ = 8

c.

New X	X − M	$(X - M)^2$
7	−2	4
5	−4	16
13	4	16
9	0	0
11	2	4
45		40

$M = \frac{45}{5} = 9$

$\text{Var} = \frac{40}{5} = 8$

$SD = \sqrt{8} = 2.83$

d.

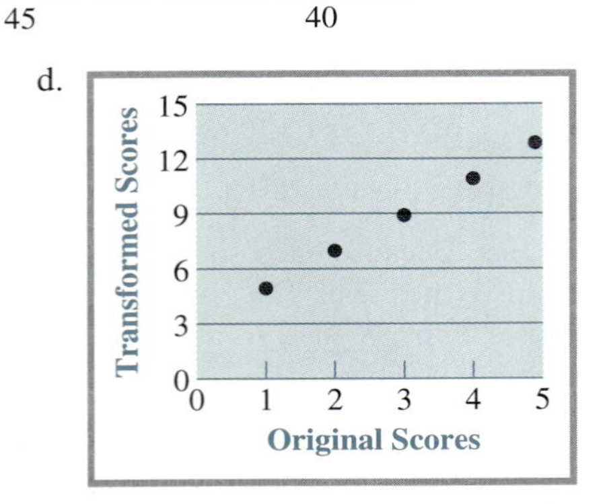

c. $z = \frac{27-25}{7} = 0.29$. A score of 27 is .29 *SD* above the mean.

2. a. *z* = 2; IQ score = 100 + 2(15) = 100 + 30 = 130.
b. *z* = 2/5 = .4; IQ score = 100 + .4(15) = 100 + 6 = 106.
c. *z* = −1.33; IQ score = 100 + (−1.33)(15) = 100 − 20 = 80.

3. a.

X	X − M	$(X - M)^2$
4	−1	1
8	3	9
5	0	0
2	−3	9
6	1	1
25		20

$M = \frac{25}{5} = 5$

$\text{Var} = \frac{20}{5} = 4$

$SD = \sqrt{4} = 2$

b. For each score, $z = \frac{X-5}{2}$.

c.

X	z	(z–Mean of z)	(z–Mean of z)^2
4	−0.5	−0.5	0.25
8	1.5	1.5	2.25
5	0	0	0
2	−1.5	−1.5	2.25
6	0.5	0.5	0.25
	Σz = 0		5.00

$\text{Mean of } z = \frac{\Sigma z}{5} = \frac{0}{5} = 0$

$\text{Var of } z = \frac{5}{5} = 1$

$SD \text{ of } z = \sqrt{1} = 1$

3–9

1. a. $z = \frac{10-25}{7} = -2.14$. A score of 10 is 2.14 *SD*s below the mean.

b. $z = \frac{25-25}{7} = 0$. A score of 25 is at the mean.

Chapter 4

4–1

1.

01 Amy	06 Alan	11 Kevin	16 Bob	21 William
02 Catherine	07 Richard	12 Josephine	17 Clark	22 George
03 John	08 Suzy	13 Barry	18 Wendy	23 Mark
04 Clare	09 Leonard	14 Zoe	19 Carol	24 Jean
05 Joseph	10 Brian	15 Tim	20 Mary	25 Leah

(*Note:* It is also possible to number across rows rather than down columns. In that case, Amy is 01, Alan is 02, etc.)
Numbers selected: 05, 10, 02, 20, 25
Children selected (based on the numbers in the table): Joseph, Brian, Catherine, Mary, Leah

2. a. random
b. not random

3. a. 1.8% is a parameter; 2.5% is a statistic
b. 45% is a parameter; 48% is a statistic
c. 68.5 inches is a parameter; 68.7 inches is a statistic

4–2

1. a. population distribution
   b. sampling distribution of the proportion
   c. sample distribution
2. In order for a sample of size 5 to have mean = 10, all five members must have a score of 10. This is very unlikely since only 17 out of the 100 members of the population have a score of 10. A mean of 10 in a sample of size 20 is impossible. A sample of size 20 must include at least three individuals with scores less than 20, thus reducing the sample mean to a number less than 10.
3. Answers will vary.

4–3

1. p (fish is under 10 inches) = 1 − .1 − .5 = .4
   p (fish is 10 or more inches) = .1 + .5 = .6
2. 15 members of the population have scores of 22 or less. The size of the population is 38. p (score $\leq$ 22) = 15/38 = .39.
3. Of the 52 cards in a standard deck, 4 are kings. p (king) = 4/52 = 1/13. 4 cards are aces. p (ace) = 4/52 = 1/13. Since 13 cards are spades, 39 are not spades; p (not spade) = 39/52 = 3/4 = .75.
4. Event C is most likely to occur; event B is least likely to occur.

4–4

1. a. $p\,(\text{junior}) = \dfrac{350}{1000} = .175;\ p\,(\text{senior}) = \dfrac{400}{2000} = .2$
   p (junior or senior) = .175 + .2 = .375
   b. $p\,(\text{female}) = .5;\ p\,(\text{freshman}) = \dfrac{700}{2000} = .35$
   p (*female freshman*) = .5.35 = .175
2. If Samuel has no preference, p (right turn on one trial) = .5, and p (6 right turns) = .5 × .5 × .5 × .5 × .5 × .5 = .015625. Because this is a very small probability, it is likely that Samuel has a right-turning preference.
3. p (smoker) = 1000/5000 = .2; p (nonsmoker) = 4000/5000 = .8
   p (member) = 1500/5000 = .3; p (nonmember) = 3500/5000 = .7
   p (member and smoker) = .3 × .2 = .06; expected frequency = 5000 (.06) = 300
   p (member and nonsmoker) = .3 × .8 = .24; expected frequency = 5000 (.24) = 1200
   p (nonmember and smoker) = .7 × .2 = .14; expected frequency = 5000 (.14) = 700
   p (nonmember and nonsmoker) = .7 × .8 = .56; expected frequency = 5000 (.56) = 2800

Table of expected frequencies:

	Smoker	Nonsmoker
Member	300	1200
Nonmember	700	2800

4–5

1. a. p (man) = 4/10 = .4
   b. p (man) = 4/10 = .4
   c. p (man) = 4/8 = .5
   d. Not justified, because probabilities with and without replacement are quite different.
2. a. p (man) = 2000/5000 = .4
   b. p (man) = 2000/5000 = .4
   c. p (man) = 2000/4998 = .40016
   d. Justified, because probabilities with and without replacement are almost the same.

4–6

1. a.

HHHH	HTHH	THHH	TTHH
HHHT	HTHT	THHT	TTHT
HHTH	HTTH	THTH	TTTH
HHTT	HTTT	THTT	TTTT

   b.

Number of Heads	p
4	1/16 = .0625
3	4/16 = .2500
2	6/16 = .3750
1	4/16 = .2500
0	1/16 = .0625

   c. p (3 or 4 heads) = p (3 heads) + p (4 heads) = .25 + .0625 = .3125
2. Let H = right turn, T = left turn
   p (3 right turns) = p (3 heads) = .25
   p (4 right turns) = p (4 heads) = .0625
   p (3 or 4 right turns) = .25 + .0625 =.3125
3. a. p (mode = 4, 5, or 6) = p (4) + p (5) + p (6) = .20 + .13 + .09 = .42

b. p (sample mode = 5) = .24; p (sample mode $\neq$ 5) = 1 − .24 = .76

4–7

1. p (5 tails if coin is fair) = .0312. Because this probability is quite small, I conclude that the coin is biased. (*Note:* If, in your opinion, an event with probability .0312 is not very unusual, you may conclude that the coin could be fair.)
2. p (fewer than 3) = p (0 or 1 or 2) = .0010 + .0098 + .0439 = .0547. p (fewer than 2 or more than 8) = p (0 or 1 or 9 or 10) = .0010 + .0098 + .0098 + .0010 = .0216.

4–8

1. a. p ($95 < X < 100$) = .20
   b. p ($X > 110$) = .05
   c. p ($95 < X < 105$) = .20 + .35 = .55
   d. p ($X < 95$) = .05 + .25 = .30
   e. p ($X < 80$ or $X > 110$) = p ($X < 80$) + p ($X > 110$) = .05 + .05 = .10
2. $X = 105$

Chapter 5

5–1

To test H, you need to collect data (make observations). Examples of relevant kinds of observations: comments of students, grades on quizzes and tests, pass/fail rates. For example, comments like "This course isn't as hard as I expected" and high test scores might lead you to reject H, whereas comments like "I just can't learn this stuff" and high failing rates might lead you not to reject it. Answers to the rest of the exercise will vary.

5–2

1. b and c could be hypotheses in statistical tests.
2. b

5–3

1. Less unusual than the PMPI sample
2. The sample's χ^2 is smaller than that at PMPI.
3.

	Thought	Affective	Character	Neuroses	Other
f_o	70	60	35	20	15
f_e	80	60	30	20	10
$f_o - f_e$	−10	0	5	0	5
$(f_o - f_e)^2$	100	0	25	0	25
$\frac{(f_o - f_e)^2}{f_e}$	1.25	0	0.83	0	2.50

$$\chi^2 = \Sigma\left[\frac{(f_o - f_e)^2}{f_e}\right] = 1.25 + 0 + 0.83 + 0 + 2.50 = 4.58$$

5–4

1. a. With 4 df, p ($\chi^2 > 10.73$) is between .05 and .025. ($p < .05$).
   b. Yes.
   c. No, because $p > .01$.
   d. With 5 df, p ($\chi^2 > 10.73$) is between .10 and .05. ($p > .05$).
2. a. $\chi^2 = 23.21$
   b. $\chi^2 = 15.09$
   c. $\chi^2 = 26.12$
3. a. $p < .05$
   b. $p > .01$

5–5

H: 60% pass, 20% fail, 20% withdraw.
In a sample of size 50, expected frequencies are

Pass:	50(.6) = 30
Fail:	50(.2) = 10
Withdraw:	50(.2) = 10

	Pass	Fail	Withdraw	
f_o	45	1	4	
f_e	30	10	10	
$f_o - f_e$	15	−9	−6	
$(f_o - f_e)^2$	225	81	36	
$\frac{(f_o - f_e)^2}{f_e}$	7.5	8.1	3.6	$\chi^2 = 7.5 + 8.1 + 3.6 = 19.2$

$df = 3 - 1 = 2$

With 2 df, $p\ (\chi^2 > 19.2) < .005$.
Since $p < .05$, reject H. The hypothesis does not fit the data.

5–6

H_o: 25% of students plan to major in each field. Since $N = 100$, f_e in each major $= 100(.25) = 25$.

	Humanities	Natural Sciences	Social Sciences	Education	
f_o	32	19	27	22	
f_e	25	25	25	25	
$f_o - f_e$	7	−6	2	−3	
$(f_o - f_e)^2$	49	36	4	9	
$\frac{(f_o - f_e)^2}{f_e}$	1.96	1.44	0.16	0.36	$\chi^2 = 3.92$

$df = 4 - 1 = 3$

With 3 df, $p\ (\chi^2 > 3.92) > .25$.

Since $p > .05$, do not reject H_o. The dean cannot conclude that some majors are chosen more often than others.

5–7

1.

		Major: Hum	Sci	SS
Sex	M	2	3	6
	F	5	3	9

2. H_o: There is no difference between the majors of males or females
   H_o: Sex and major are unrelated

5–8

1. a. $df = (3–1)(4–1) = 2 \times 3 = 6$.
   b. Answers will vary.
2. H_o: Type of prior information and rated comprehension are not related.

   Table of expected frequencies with row and column sums:

	Low	Moderate	High	
Relevant	8.8	9.8	11.4	30
Irrelevant	8.2	9.2	10.6	28
	17	19	22	58 = N

(In each cell, $f_e = \frac{\text{(row sum)(column sum)}}{N}$. For example, in the cell Relevant-Low

$$f_e = \frac{(30)(17)}{58} = 8.8.)$$

$$\chi^2 = \frac{(2-8.8)^2}{8.8} + \frac{(11-9.8)^2}{9.8} + \frac{(17-11.4)^2}{11.4} + \frac{(15-8.2)}{8.2} + \frac{(8-9.2)^2}{9.2} + \frac{(5-10.6)^2}{10.6}$$

$$= 5.25 + 0.15 + 2.75 + 5.64 + 0.16 + 2.96$$

$$= 16.91$$

$$df = (3-1)(2-1) = 2 \times 1 = 2$$

With 2 df, $p(\chi^2 > 16.91) < .0005$.
Because $p < .05$, reject H_o. Conclusion: Type of prior information and rated comprehension are related. Rated comprehension is higher with relevant information.

3. No. To be significant at the .01 level, p must be less than .01. A result that is not significant at the .05 level must have $p > .05$. p cannot be $< .01$ yet $> .05$.

5–9

1. The observations are not independent, because children who solved both puzzles are entered in two cells. The χ^2 test of independence cannot be used.
2. Expected frequencies:

	"Murder"	"20/20"	"Designing"	"Murphy"	"48 Hr."	"Columbo"
Urban	2.8	1.2	4.0	4.8	3.2	4.0
Rural	4.2	1.8	6.0	7.2	4.8	6.0

A chi square test should not be conducted because most of the expected frequencies are less than 5.

5–10

1. a. Type II error
   b. No error of inference
   c. Type I error
2. H_o: 50% of voters will vote for each candidate.

	Smith	**Jones**
f_o	226	174
f_e	200	200
$f_o - f_e$	26	−26
$(f_o - f_e)^2$	676	676
$\frac{(f_o - f_e)^2}{f_e}$	3.38	3.38

$\chi^2 = 3.38 + 3.38 = 6.76$

With 1 *df*, $p(\chi^2 > 6.76)$ is between .01 and .05 ($p < .05$).
Reject H_o. Conclusion: More than half the voters plan to vote for Smith.
If the true population percentage is 50%, a sample percentage of 56.5% is more unusual in a sample of size 400 than in a sample of size 200. (*Remember:* As sample size increases, sample statistics vary less and tend to be closer to the population parameter.)

5–11

1. $p = .0470$ is less than .05; reject H_o.
   $p = .0635$ is greater than .05; do not reject H_o.
2. .00365 is less than .05, .01, and .005, but greater than .001; the result is significant at the .05, .01, and .005 levels but not at the .001 level.

Chapter 6

6–1

1. $\mu = 40.3$, $\sigma^2 = 64$, $\sigma = 8$, $M = 36.4$, Var $= (6.81)^2 = 46.38$, $S = 6.81$.
2.

X	**X − M**	**(X − M)²**
6	0	0
5	−1	1
8	2	4
9	3	9
6	0	0
4	−2	4
4	−2	4
42		22
ΣX		$\Sigma(X - M)^2$

Mean $(M) = \frac{42}{7} = 6$
Sum of squares $(SS) = \Sigma(X - M)^2 = 22$
(*Note*: *SS* can also be calculated by the formula $\Sigma X^2 - \frac{(\Sigma X)^2}{N} = 274 - \frac{(42)^2}{7} = 22.$)
Variance (Var) $= \frac{SS}{N} = \frac{22}{7} = 3.14$
$SD\ (S) = \sqrt{3.14} = 1.77$

6–2

1. σ_M^2 is less for samples of size 3 than size 2 because, as a sample size increases, the variability of sample means decreases.
2. a. $\mu = \frac{2+8+14}{3} = \frac{24}{3} = 8$

$$\sigma^2 = \frac{(2-8)^2 + (8-8)^2 + (14-8)^2}{3} = \frac{72}{3} = 24$$

$\sigma = \sqrt{24} = 4.90$

   b. All possible samples of size 2:

Sample	Sample Mean (M)	$M - \mu$	$(M - \mu)^2$
2, 2	2	−6	36
2, 8	5	−3	9
2, 14	8	0	0
8, 2	5	−3	9
8, 8	8	0	0
8, 14	11	3	9
14, 2	8	0	0
14, 8	11	3	9
14, 14	14	6	36
Sum	72		108

c.

Sample Mean	Number of Samples
2	1
5	2
8	3
11	2
14	1

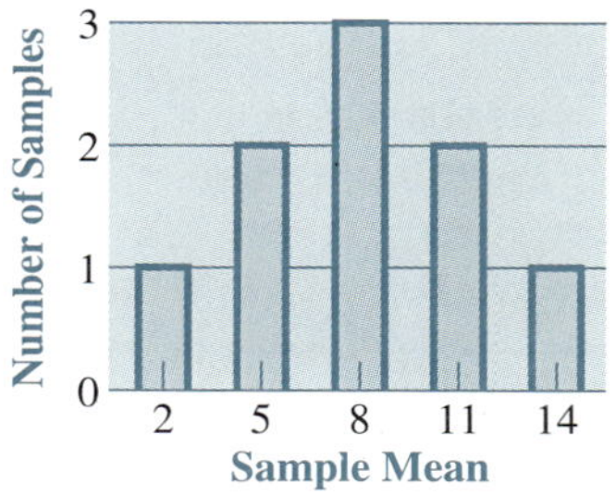

d. $\Sigma M = 72$

$\mu_M = 72/9 = 8 = \mu$

e. $\Sigma (M - \mu)^2 = 108$

$$\sigma_M^2 = \frac{108}{9} = 12$$

$$\sigma_M = \sqrt{12} = 3.464$$

$$\sigma_M^2 = \frac{\sigma^2}{N} = \frac{24}{2} = 12$$

3. M = the mean typing speed of the sample of 30 secretaries.

μ = the mean typing speed of all 250 secretaries.

Variance of typing speed of all the secretaries is σ^2.

Variance of sample is S^2.

The distribution of sample means is the (theoretical) sampling distribution of the mean (for samples of size 30).

The mean of the distribution is μ_M.

The *SD* of the distribution is σ_M and is called the standard error of the mean.

6–3

1. a. Sampling distribution of the mean

b. Normal (because the population is normal)

c. $\mu_M = 30$

d. Standard error of the mean, σ_M

e. $\sigma_M = 5/\sqrt{9} = 1.67$

f. Answer to part e would change. $\sigma_M = 5/\sqrt{16} = 1.25$

g. Answer to part b would change.

2. The sampling distribution of the mean for samples of size 25 is more nearly normal.

6–4

1.

X	$X - M$	$(X - M)^2$
6	0	0
3	−3	9
4	−2	4
6	0	0
5	−1	1
8	2	4
5	−1	1
9	3	9
8	2	4
54		32

$N = 9$

$$M = \frac{54}{9} = 6$$

$SS = 32$

Estimate of population variance $= s^2 = \frac{SS}{N-1} = \frac{32}{8} = 4$

Estimate of population $SD = s = \sqrt{4} = 2$

2.

Sample	s^2*	s	Sample	s^2	s
3, 3	0	0	7, 3	8	2.828
3, 5	2	1.414	7, 5	2	1.414
3, 7	8	2.828	7, 7	0	0
3, 9	18	4.243	7, 9	2	1.414
5, 3	2	1.414	9, 3	18	4.243
5, 5	0	0	9, 5	8	2.828
5, 7	2	1.414	9, 7	2	1.414
5, 9	8	2.828	9, 9	0	0

*See Table 6–3.

$\sum s = 28.282$; mean of $s = \dfrac{28.282}{16} = 1.768 \neq 2.236$

6–5

Answers will vary.

Chapter 7

7–1

1.

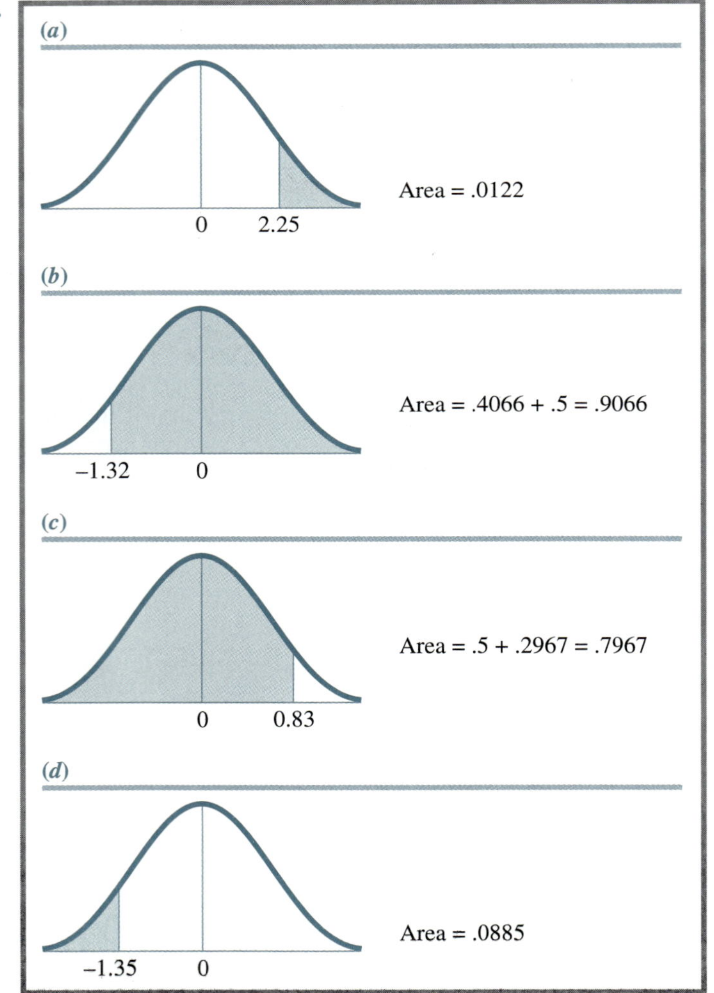

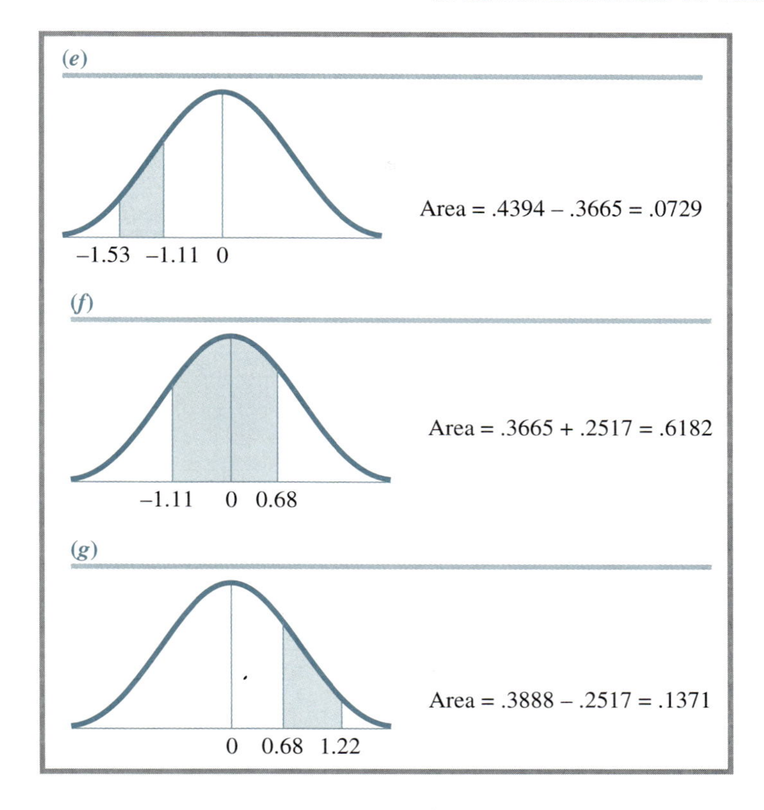

2.

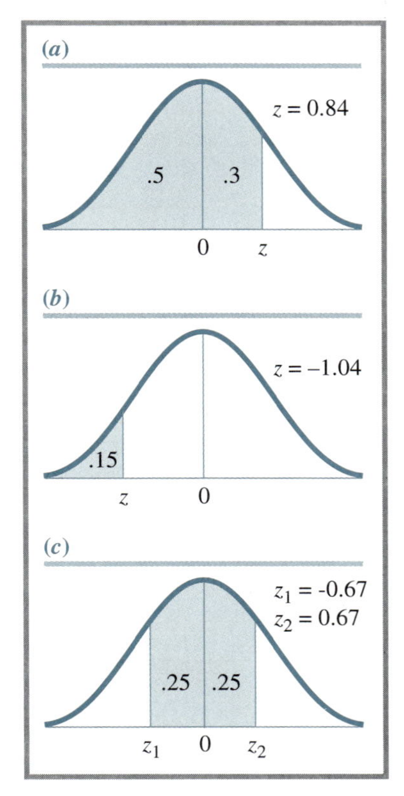

3. For all $z > 4$, the area between the mean and z is almost .5, and the area beyond z is almost zero.
4. Area between mean and $z = 1$ is .3413. Area between $z = -1$ and $z = 1$ is 2(.3413) = .6826, or 68%. Area between mean and $z = 2$ is .4772. Area between $z = -2$ and $z = 2$ is 2 (.4772) = .9544, or 95.5%. Area between mean and $z = 3$ is .4987. Area between $z = -3$ and $z = 3$ is 2(.4987) = .9974, or almost 100%.

7–2

1. a.

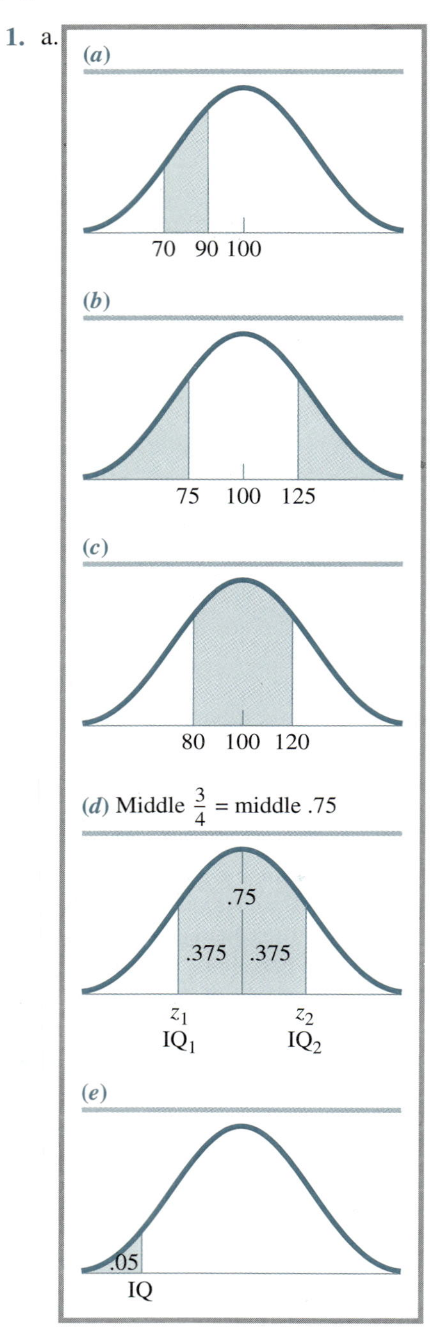

(a)

$$z_{70} = \frac{70-100}{15} = -2$$

$$z_{90} = \frac{90-100}{15} = -0.67$$

Area between $z = -2$ and $z = -0.67$ is $.4772 - .2486 = .2286 = 22.86\%$.

(b)

$$z_{75} = \frac{75-100}{15} = -1.67$$

$p\ (z < -1.67) = .0475$

$$z_{125} = \frac{125-100}{15} = 1.67$$

$p\ (z > 1.67) = .0475$

$p\ (z < -1.67 \text{ or } z > 1.67) = .0475 + .0475 = .095$

(c)

$$z_{80} = \frac{80-100}{15} = -1.33$$

$$z_{120} = \frac{120-100}{15} = 1.33$$

Area between $z = -1.33$ and $z = 1.33 =$ $.4082 + .4082 = .8164$.

Expected frequency $= 1000\ (.8164) = 816.4$.

(d)

$z_1 = -1.15$, $z_2 = 1.15$
$IQ_1 = 100 + (-1.15)(15) = 82.75$
$IQ_2 = 100 + (1.15)(15) = 117.25$

(e)

$z = -1.645$
$IQ = 100 + (-1.645)(15) = 75.325$

2. Converting the scores to z scores will not result in a standard normal curve but a negatively skewed distribution of z scores. The table can be used only if the z scores are normally distributed.

7–3

1. a. Exact $p(3 \text{ or fewer}) = p(3) + p(2) + p(1) + p(0) = .1172 + .0439 + .0098 + .0010 = .1719$

b. Mean = 10(.5) = 5

$SD = \sqrt{10(.5)(.5)} = 1.58$

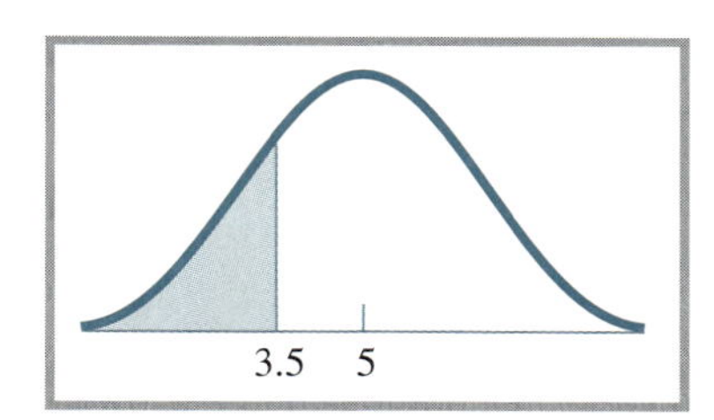

$p(3 \text{ or fewer}) = p(X < 3.5)$

$$z = \frac{3.5 - 5}{1.58} = -0.95$$

Approximate $p =$ area below z of $-0.95 = .1711$

2. Exact value of p (X = 8 or more) = .2816 + .1877 + .0563 = .5256

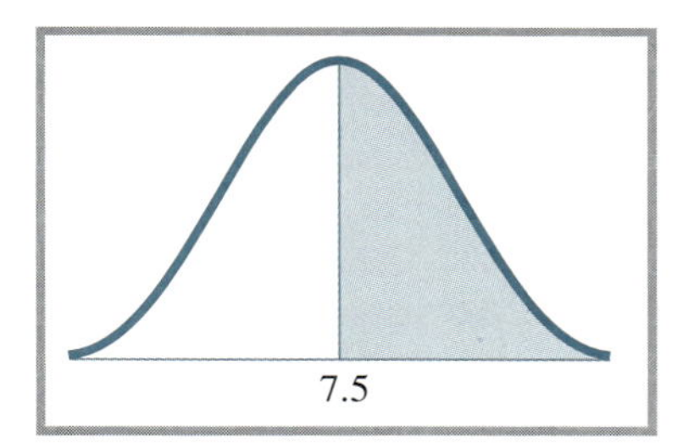

Normal approximation :

mean = 10(.75) = 7.5

$SD = \sqrt{10(.75)(.25)} = 1.37$

Approximate $p(8 \text{ or more}) =$

$p(X > 7.5) = .5000$

The normal approximation is less accurate when p_E = .75 than when p_E = .5 (for the same N).

7–4

1.

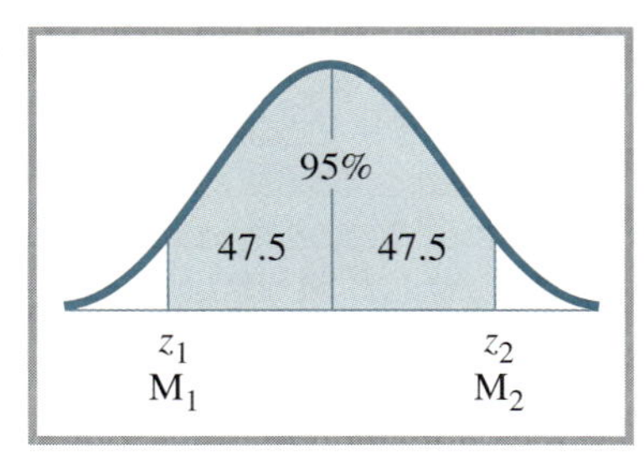

$z_1 = -1.96$, $z_2 = 1.96$ as in Example 7–12

$$\sigma_M = \frac{15}{\sqrt{25}} = 3$$

$M_1 = 100 + (-1.96)(3) = 94.12$

$M_2 = 100 + (1.96)(3) = 105.88$

The middle 95% of sample means fall within a smaller range around the population mean because, as sample size increases, the variability of sample means decreases.

2. μ = 10, σ = 2, N = 20

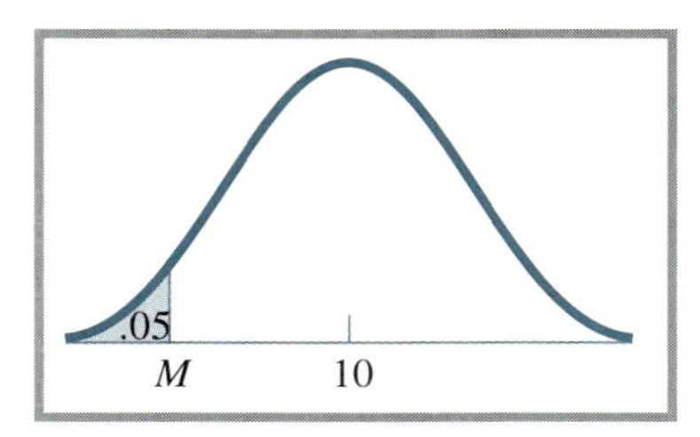

$z = -1.645$

$$\sigma_M = \frac{2}{\sqrt{20}} = 0.447$$

$M = 10 + (-1.645)(0.447) = 9.26$

3. μ = 50, σ = 10

a.

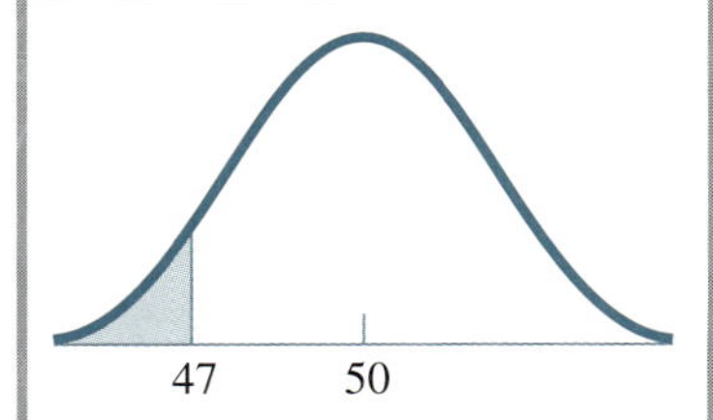

$$\sigma_M = \frac{10}{\sqrt{10}} = 3.16$$

$$z = \frac{47 - 50}{3.16} = -0.95$$

$p(M < 47) =$ area below z of $-0.95 = .1711$

b. p (M < 47) will be smaller.

$$\sigma_M = \frac{10}{\sqrt{50}} = 1.414; \; z = \frac{47 - 50}{1.414} = -2.12$$

p (M < 47) = area below z of –2.12 = .0170

c. $$z = \frac{47 - 50}{10} = -0.30$$

p (X < 47) = area below z of –0.30 = .3821

7–5

1.

a. Upper one-tailed test
H_o: $\mu \leq 58$
H_A: $\mu > 58$

b. Two-tailed test
H_o: $\mu = 75$
H_A: $\mu \neq 75$

c. Lower one-tailed test
H_o: $\mu \geq 32$
H_A: $\mu < 32$

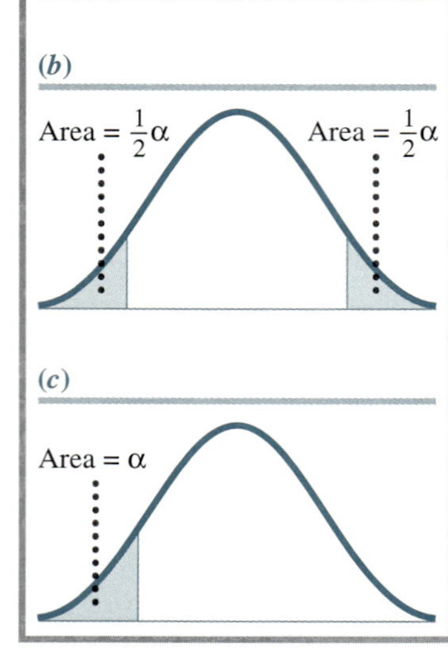

2. a. p ($z > 1.82$) = .0344. Since $p < .05$, reject H_o.
b. p ($z < -1.82$ or $z > 1.82$) = 2(.0344) = .0688. Because $p > .05$, do not reject H_o.
c. Same as part b.
d. p ($z < -1.82$) = .0344. Since $p > .01$, do not reject H_o.

7–6

1. a. Two-tailed test
H_o: $\mu = 10$
H_A: $\mu \neq 10$
Assume $\sigma = 5$.

$$M = \frac{6+9+1+6}{4} = 5.5$$

$$\sigma_M = 5/\sqrt{4} = 2.5$$

$$z = \frac{5.5-10}{2.5} = -1.8$$

p ($z < -1.8$ or $z > 1.8$) = 2(.0359) = .0718. Because $p > .05$, do not reject H_o. The researcher cannot conclude that his population differs from the population with $\mu = 10$, $\sigma = 5$.

b.
$$\sigma_M = \frac{5}{\sqrt{25}} = 1$$

$$z = \frac{5.5-10}{1} = -4.5$$

p ($z < -4.5$ or $z > 4.5$) is almost zero. Because $p < .05$, reject H_o. Conclude that the researcher's population differs from the known population.

c. A sample mean this far from the hypothesized population mean, 10, is much more unusual in a sample of size 25 than a sample of size 4.

2. Upper one-tailed test
H_o: $\mu \leq 20$
H_A: $\mu > 20$
Assume $\sigma = 8$.

$$\sigma_M = \frac{8}{\sqrt{16}} = 2$$

$$z = \frac{24.6-20}{2} = 2.30$$

p ($z > 2.30$) = .0107. Since $p > .01$, do not reject H_o. The personality theorist cannot conclude (at the .01 level) that boys from large families have a higher mean.

3. Lower one-tailed test
H_o: $\mu \geq 8.5$ days
H_A: $\mu < 8.5$ days

Assume $\sigma = 2.4$ days.

$$\sigma_M = \frac{2.4}{\sqrt{50}} = .34$$

$$z = \frac{7.6-8.5}{.34} = -2.65$$

p ($z < -2.65$) = .0040. Since $p < .05$, reject H_o. Conclude that the new drug reduces treatment time.

Chapter 8

8–1

1. s is an estimate of the population standard deviation, σ. s_M is an estimate of the standard error of the mean, σ_M. (Recall that the standard error of the mean is the standard deviation of a distribution of sample means.)
2. $\Sigma X = 23 \qquad \Sigma X^2 = 73 \qquad N = 10$

$$SS = 73 - \frac{(23)^2}{10} = 20.1$$

$$s = \sqrt{\frac{20.1}{9}} = 1.494$$

$$s_M = \frac{s}{\sqrt{N}} = \frac{1.494}{\sqrt{10}} = .473$$

$$\text{In one step: } s_M = \sqrt{\frac{73 - \frac{(23)^2}{10}}{10(9)}} = .473$$

3. $$s_M = \frac{12.4}{\sqrt{25}} = 2.48$$

8–2

1. a. Because $\dfrac{M - \mu}{s_M}$ is not normally distributed; it is distributed as t.
   b. The area above t = 1.20 is greater than .1151.
2. Less in error because, as N increases, the t distribution approaches the normal distribution.

8–3

1. With 7 df, $p(t > 2.02)$ is between .05 and .025 ($p < .05$). Reject H_o at .05 level.
2. With 12 df, $p\ (t < -2.02)$ is between .05 and .025 ($p < .05$). Reject H_o at .05 level.
3. With 14 df, minimum $|t|$ needed to reject H_o with α = .001 (two-tailed test) is 4.140.

8–4

1. a. Hypothesis is nondirectional.
   H_o: $\mu = 50$
   H_A: $\mu \neq 50$
   Rejection region lies in both tails.
   b. Hypothesis is directional.
   H_o: $\mu \leq 54$
   H_A: $\mu > 54$
   Rejection region lies in the upper tail.
   c. Hypothesis is directional.
   H_o: $\mu \geq 25$ sessions
   H_A: $\mu < 25$ sessions
   Rejection region lies in the lower tail.

2. $$s_M = \frac{6.2}{\sqrt{12}} = 1.79$$

$$t = \frac{15.3 - 25}{1.79} = -5.42$$

   df = 11. With 11 df, $p(t < -5.42) < .0005$. Since $p < .05$, reject H_o. Conclusion: The new therapy reduces mean treatment time.

3. H_o: $\mu = 10$
   H_A: $\mu \neq 10$
   $\Sigma X = 41$, $\Sigma X^2 = 351$, $N = 5$

$$M = \frac{41}{5} = 8.2$$

$$s_M = \sqrt{\frac{351 - \frac{(41)^2}{5}}{5(4)}} = 0.86$$

$$t = \frac{8.2 - 10}{0.86} = -2.09$$

   df = 4. With 4 df (two-tailed test), p is between .20 and .10 ($p > .01$). The sample mean does not differ significantly from 10 at the .01 level.

8–5

1. $\Sigma X = 64 \qquad \Sigma X^2 = 844 \qquad N = 5$

$$M = \frac{64}{5} = 12.8$$

$$s_M = \sqrt{\frac{844 - \frac{(64)^2}{5}}{5(4)}} = 1.114$$

$t_{.10/2}$ for 4 $df = 2.132$

90% confidence interval (CI) = 12.8 ± (2.132)(1.114) = 10.42 to 15.18

2. His interpretation implies that μ varies. Sometimes it is between 17.3 and 21.5 and sometimes it is not. This is incorrect because μ is fixed; sample means (and confidence intervals) vary. Correct interpretation: Although the 95% confidence interval varies over samples, the intervals in 95% of samples will include the value of μ.

8–6

1. a. $s_M = \frac{9.83}{\sqrt{25}} = 1.966$

   $t_{.01/2}$ for 24 $df = 2.797$
   99% CI = 78.4 ± (2.797)(1.966) = 72.90 to 83.90

   b. 95% interval will be narrower
   $t_{.05/2}$ for 24 $df = 2.064$
   95% CI = 78.4 ± (2.064)(1.966) = 74.34 to 82.46

   c. If N = 16, CI will be wider
   $s_M = \frac{9.83}{\sqrt{16}} = 2.4575$

   $t_{.01/2}$ for 15 $df = 2.947$
   99% CI = 78.4 ± (2.947)(2.4575) = 71.16 to 85.64

   d. Reject (1) H_o: μ = 70 and (4) H_o: μ = 85, because 70 and 85 fall outside the 99% confidence interval.
2. It does not differ significantly from −10 at the .05 level, because −10 is inside the interval. It does differ significantly from −15 at the .05 level, because −15 is outside the interval.
3. Yes, 25 is within the 99% confidence interval for μ.
4. No, 60 is not within the 95% confidence interval for μ.

8–7

1. $P = \frac{62}{150} \times 100 = 41.3\%$

   $s_P = \sqrt{\frac{41.3(58.7)}{150}} = 4.02\%$

   $z_{.01/2} = 2.575$
   99% CI = 41.3 ± 2.575 (4.02) = 30.95% to 51.65%
2. $z_{.05/2} = 1.96$
   P = 56.5%

   $$N = \left(\frac{1.96}{5}\right)^2 (56.5)(43.5) = 377.7$$

   A sample size of 378 is needed.

8–8

1. a. Yes, Type II error
   b. Yes, Type I error
2. α
3. a. decrease
   b. not affect
4. a. (3) greater than .40 (as α increases, β decreases)
   b. (2) less than .40 (β is smaller with a one-tailed test)
5. A Type II error is more likely if true μ = 15, because 15 is more different from 20.
6. A Type II error is more likely if population SD = 7, because, as population variability increases, β increases.
7. Answers will vary.

8–9

1. Answers will vary
2. a. Test A (smaller population variance)
   b. Test A (larger α)
   c. Test B (larger difference between μ_H and true μ)
   d. Test A (larger N)
   e. Test A (one-tailed test)
3. Remains the same
4. A Type II error is more likely if true μ = 25
5. Increase N, increase α, use a one-tailed test if appropriate

Chapter 9

9–1

1. a. positive correlation
   b. no correlation
   c. negative correlation
   d. negative correlation
   e. no correlation
   f. positive correlation
   g. positive correlation
2. Students who turned their exams in early tended to score lower than students who took longer to finish.
3. Age and fear of monsters co-vary negatively.
4. Answers will vary.

9–2

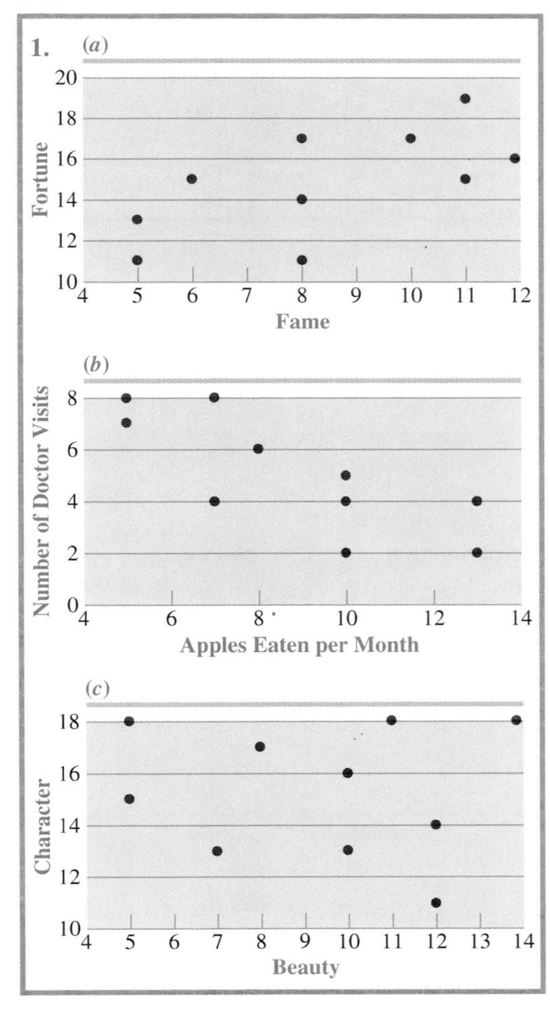

(*Note:* In each scatter diagram, the variable assigned to a given axis is arbitrary.)

2. A—positive, moderate
   B—negative, moderate to strong
   C—no relation
3. B is strongest; C is weakest.

9–3

1. If $r = -.60$, the variables are related (inversely). A correlation of + .40 is weaker than a correlation of –.60. A correlation of 1.20 is impossible because r is always between 0 and ± 1.

2. a. (1) $r = \frac{-50}{(8)(10)} = \frac{-50}{80} = -.625$

   (2) $r = \frac{65}{(8)(10)} = \frac{65}{80} = .8125$

   (3) $r = \frac{5}{(8)(10)} = \frac{5}{80} = .0625$

   b. If $r = +1$, $\text{CoVar}_{xy} = +80$.

3. a.

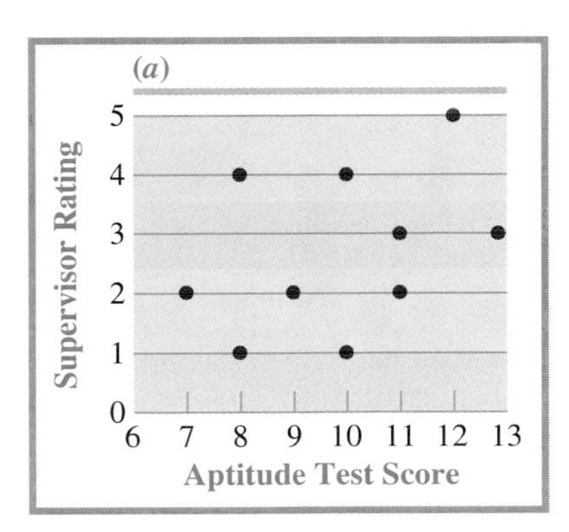

The correlation is positive but not strong.

   b. No effect; the correlation remains the same.

c.

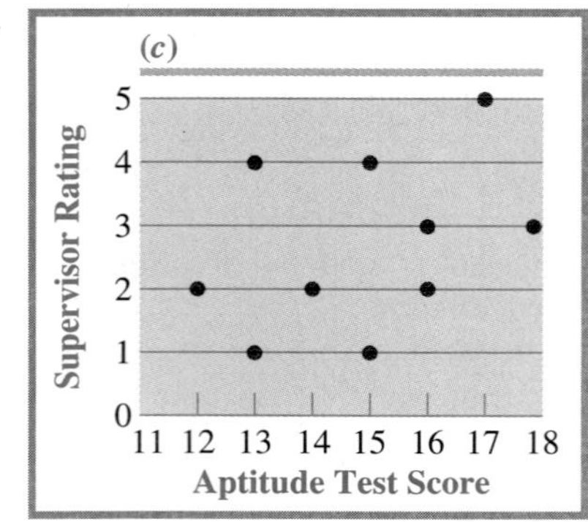

The two graphs are identical. Adding 5 points to each employee's aptitude test score did not change the correlation between test scores and supervisor ratings.

9–4

1.

Child	Self–Esteem X	Reading Ability Y	XY	X^2	Y^2
1	4	13	52	16	169
2	6	10	60	36	100
3	7	16	112	49	256
4	8	13	104	64	169
5	10	17	170	100	289
6	11	12	132	121	144
7	13	14	182	169	196
8	13	17	221	169	289
Sum	72	112	1033	724	1612

$$r = \frac{8(1033)-(72)(112)}{\sqrt{[8(724)-(72)^2][8(1612)-(112)^2]}} = \frac{8264-8064}{\sqrt{(5792-5184)(12,896-12,544)}} = \frac{200}{\sqrt{(608)(352)}} = \frac{200}{\sqrt{214,016}} = \frac{200}{462.62} = .43$$

2.

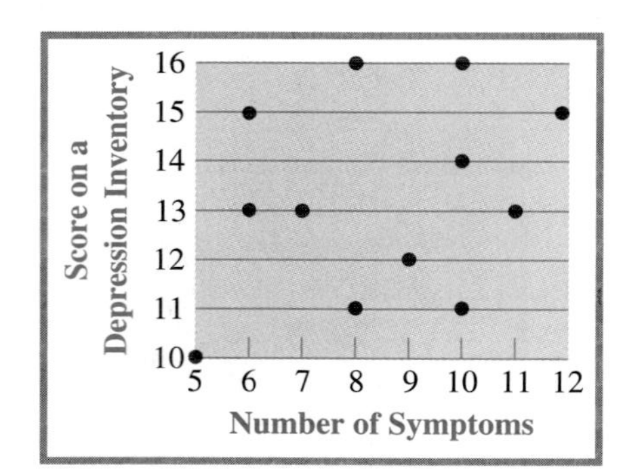

The correlation is positive but weak.

Patient	Number of Symptoms (X)	Score on a Depression Inventory (Y)	XY	X^2	Y^2
1	10	14	140	100	196
2	8	16	128	64	256
3	8	11	88	64	121
4	5	10	50	25	100
5	12	15	180	144	225
6	6	15	90	36	225

Continued

Patient	Number of Symptoms (X)	Score on a Depression Inventory (Y)	XY	X^2	Y^2
7	7	13	91	49	169
8	10	11	110	100	121
9	9	12	108	81	144
10	11	13	143	121	169
11	10	16	160	100	256
12	6	13	78	36	169
102	159	1366	920	2151	

$$r = \frac{12(1366) - (102)(159)}{\sqrt{[12(920) - (102)^2][12(2151) - (159)^2]}} = \frac{16,392 - 16,218}{\sqrt{(11,040 - 10,404)(25,812 - 25,281)}}$$

$$= \frac{174}{\sqrt{(636)(531)}} = \frac{174}{\sqrt{337,716}} = \frac{174}{581.13} = .299$$

9–5

1. a.

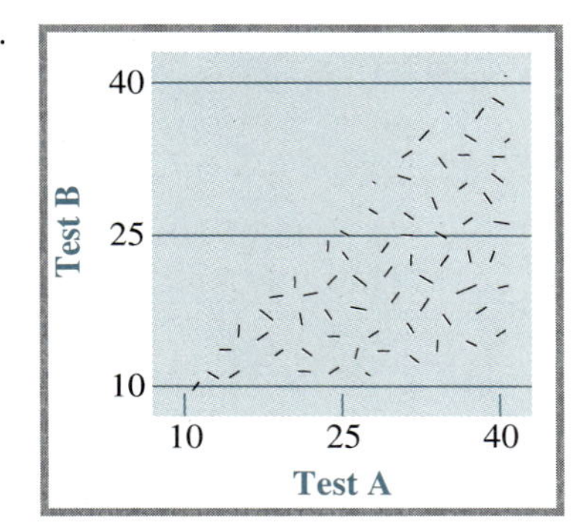

   b. r measures the average correlation between Tests A and B. However, the correlation is strong at low scores on A and B but weak at high scores on A and B.
2. r is higher in the group of high school freshmen because they vary more widely in manual dexterity and eye-hand coordination than master carpenters, all of whom are probably high on both variables.
3. Joe Blow's group had a wide range of heights and weights (from the smallest 6-year-old to the biggest 13-year-old). The range of height and weight is smaller in a group of fourth graders who are all close to the same age.
4. Answers will vary.

9–6

1. The argument is invalid; a strong negative correlation means that the variables are related but does not mean that one causes the other.
2. a. No
   b. (1) Aggressiveness may cause a preference for violent TV:
      aggressiveness → watching violent TV.
      (2) Some other variable (e.g., amount of parent supervision) may affect both aggressiveness and TV watching, causing them to co-vary:

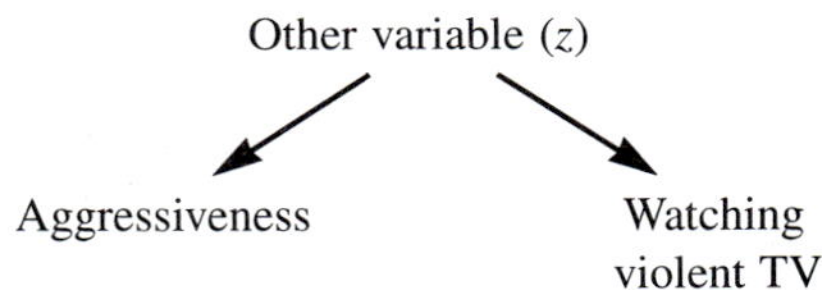

   c. $(.55)^2$ = .3025; 3025 (about 30%) of aggressiveness variance is accounted for by number of violent TV programs. 1 – .3025 = .6975 (about 70%) is not accounted for. The coefficient of determination is .3025; the coefficient of nondetermination is .6975.
3. The actual "strengths" of the correlations are $r^2_{AB} = .16$, $r^2_{CD} = 64$. The correlation between C and D is four times as strong as that between A and B.
4. $r^2 = .30$

   $r = \sqrt{.30} = .55$ or $-.55$

9–7

A different conclusion is reached:

With $N = 50$, $t = \dfrac{.32\sqrt{48}}{\sqrt{1-(-.32)^2}} = -2.34$

With 48 df, $p(t < -2.34)$ is between .02 and .01 ($p < .05$). The correlation is signifcantly lower than zero at the .05 level.

9–8

1. a. $df = 6$. To be signifcantly greater than zero at the .05 level, r must be at least .622. The correlation between self-esteem and reading ability is not significantly greater than zero at the .05 level.
   b. $df = 14$; $N = 16$.
2. a. Low SES women are more likely to be pro-choice.
   b. A two-tailed test is called for.
   c. H_o: $\rho = 0$
      H_A: $\rho \neq 0$
      $N = 30$, $r = -.42$
      With $df = 28$, $p(|r| > .42)$ is between .05 and .02.
   d. Yes, SES is significantly related to attitudes toward abortion at the .05 level because $p < .05$.
3. a. $df = 45$; $N = 47$
   b. $df = 35$; $N = 37$
   c. $df = 45$; $N = 47$

9–9

1. Table 9–11: $N = 5$, $\Sigma X = 15$, $\Sigma Y = 15$, $\Sigma XY = 52$, $\Sigma X^2 = 55$, $\Sigma Y^2 = 55$

$$r = \frac{5(52)-(15)(15)}{\sqrt{[5(55)-(15)^2][5(55)-(15)^2}} = \frac{35}{50} = .70$$

Table 9–12: $N = 10$, $\Sigma X = 6$, $\Sigma Y = 195$, $\Sigma XY = 123$, $\Sigma X^2 = 6$, $\Sigma Y^2 = 3941$

$$r = \frac{10(123)-(6)(195)}{\sqrt{[10(6)-(6)^2][10(3941)-(195)^2}}$$

$$= \frac{60}{\sqrt{(24)(1385)}} = \frac{60}{182.3} = .33$$

Table 9–13: $N = 10$, $\Sigma X = 6$, $\Sigma Y = 5$, $\Sigma XY = 4$, $\Sigma X^2 = 6$, $\Sigma Y^2 = 5$

$$r = \frac{10(4)-(6)(5)}{\sqrt{[10(6)-(6)^2][10(5)-(5)]^2}} = \frac{10}{\sqrt{(24)(25)}}$$

$$= \frac{10}{24.5} = .41$$

2. No, this item is not a good item. The negative correlation shows that students with low scores on the exam answered the item correctly more often than students with high scores on the exam.
3. $a = 23$, $b = 7$, $c = 12$, $d = 8$, $N = 50$

$$\phi = \frac{(23)(8)-(7)(12)}{\sqrt{(23+7)(12+8)(23+12)(7+8)}}$$

$$= \frac{100}{561.25} = .18$$

$\chi^2 = 50(.18)^2 = 1.62$
$df = 1$. With 1 df, $p(\chi^2 > 1.62)$ is between .25 and .10 ($p > .05$). ϕ is not significantly different from zero at the .05 level.

Chapter 10

10–1

1. The predictor is years of teaching experience; the criterion is burnout.
2. a. Scores on the diagnostic test = X; severity of emotional disturbance = Y.
   b. $a = 4.2$, $b = 12$
   c. $\hat{Y} = 4.2 + 12(10) = -4.2 + 120 = 115.8$
3. a. X = anxiety, Y = assertiveness
   b. $a = 16.3$, $b = 1.5$
   c. $\hat{Y} = 16.3 - 1.5(5) = 16.3 - 7.5 = 8.8$
4. $\hat{Y} = 14.92 - .07X$
5. $\hat{Y} = -71.2 + 10.3X$
6. a. $\hat{Y} = 6.59 + 4.40(2) = 6.59 + 8.80 = 15.39$
   b. $\hat{Y} = 6.59 + 4.40(0) = 6.59$
   c. $\hat{Y} = 6.59 + 4.40(5) = 6.59 + 22 = 28.59$
   d. $\hat{Y} = 6.59 + 4.40(4) = 6.59 + 17.6 = 24.19$
   e. Tom, Ted, and Mary

10–2

1. a. X = aptitude test scores, Y = final exam scores
   $M_x = 31.5$, $S_x = 6.8$, $M_y = 71.3$, $S_y = 12.5$, $r = .55$

$$b = .55\left(\frac{12.5}{6.8}\right) = 1.011$$

$a = 71.3 - (1.011)(31.5) = 39.45$

Regression equation: $\hat{Y} = 39.45 + 1.01X$

b. $\hat{Y} = 39.45 + 1.01(60) = 100.05$

2.

Child	Reading (X)	Math (Y)	XY	X^2
1	7	10	70	49
2	4	4	16	16
3	3	5	15	9
4	6	12	72	36
5	8	9	72	64
	28	40	245	174

a.

$$b = \frac{5(245)-(28)(40)}{5(174)-(28)^2} = \frac{105}{86} = 1.22$$

$$a = \frac{40-(1.22)(28)}{5} = \frac{5.84}{5} = 1.17$$

Regression equation: $\hat{Y} = 1.17 + 1.22X$

b. $\hat{Y} = 1.17 + 1.22(5) = 7.27$

10–3

1. No, this is not correct. A different equation must be calculated from the data to predict reading from social studies.
2. a. Less than 40 because Josh is above the mean on X and the correlation between X and Y is negative.
   b. Since Buffy scored at the mean on X, her predicted score on Y is the mean of $Y = 40$.
   c. 40. If $r = 0$, the predicted Y for everyone is the mean of Y.
3. a. Less than 125 (closer to 100), because of regression toward the mean.
   b. Greater than 80 (closer to 100), because of regression toward the mean.

10–4

1. a.

Student	Anxiety (X)	Extraversion (Y)	$\hat{Y} = 7 - X$	$Y - \hat{Y}$	$(Y - \hat{Y})^2$
1	4	4	3	1	1
2	2	6	5	1	1
3	4	1	3	−2	4
4	3	2	4	−2	4
5	5	3	2	1	1
6	1	5	6	−1	1
7	2	7	5	2	4
					16 $\Sigma(Y - \hat{Y})^2$

$$S^2_{\text{res}} = \frac{16}{7} = 2.2857$$

$$S_{\text{est}} = \sqrt{2.2857} = 1.51$$

b. S^2_{res} is less than the total variance of the extraversion scores, because the correlation between anxiety and extraversion is not equal to zero.

Student	Extraversion (Y)	$Y - M_Y$	$(Y - M_Y)^2$
1	4	0	0
2	6	2	4
3	1	−3	9
4	2	−2	4
5	3	−1	1
6	5	1	1
7	7	3	9
	28		28

$$M_y = \frac{28}{7} = 4$$

$$S^2_y = \frac{28}{7} = 4, \text{ which is greater than } S^2_{\text{res}} = 2.2857$$

2. a. For X = 12, $\hat{Y} = 10 + .75(12) = 19$
   68% of individuals with $X = 12$ have Y between $19 \pm 3 = 16$ and 22
   b. For $X = 8$, $\hat{Y} = 10 + .75(8) = 16$
   About 95% of individuals with $X = 8$ have Y between $16 \pm 2(3) = 10$ and 22

10–5

1. a. X = aptitude test, Y = secretarial skills
   $r = .30$, $S_y = 7.5$

$$S_{\text{est}} = 7.5\sqrt{1-(.30)^2} = 7.5\sqrt{1-.09}$$
$$= 7.5\sqrt{.91} = 7.5(.954) = 7.15$$

b. *SD* of secretarial skills for secretaries with aptitude test score = 25 is 7.15.

c. $s_{\hat{y}}^2$ = explained variance

$s_{\hat{y}}^2 = s_y^2 - s_{\text{res}}^2 = (7.5)^2 - (7.15)^2 = 56.25 - 51.12 = 5.13$

d. The standard error of estimate will decrease

$$S_{\text{est}} = 7.5\sqrt{1-(.60)^2} = 6.0$$

2. a. X = SAT, Y = GPA

$M_x = 425$, $S_x = 75$, $M_y = 2.20$, $S_y = 0.50$, $r = .60$

$$b = .60\left(\frac{0.50}{75}\right) = .004$$

$a = 2.20 - (.004)(425) = 0.5$

Regression equation: $\hat{Y} = 0.5 + .004X$

b. $$S_{\text{res}}^2 = S_y^2(1-r^2) = (0.50)^2[1-(.60)^2]$$
$$= 0.16$$
$$S_{\text{est}} = \sqrt{0.16} = 0.4$$

c. For SAT = 350, predicted GPA ($\hat{Y}$) = $0.5 + .004(350) = 1.9$

The middle 95% of freshmen with SAT = 350 will have GPA between $1.9 \pm 2(0.4) = 1.9 \pm .8 = 1.1$ and 2.7.

10–6

1. The 99% prediction interval will be wider. $t_{\alpha/2}$ for $\alpha = .01$ and 24 *df* is 2.797.

$$99\%\ \text{PI} = \hat{Y} \pm (2.797)(6.57)\sqrt{1.0385 + \frac{(X_i - 2.46)^2}{72.46}}$$
$$= \hat{Y} \pm 18.376\sqrt{1.0385 + \frac{(X_i - 2.46)^2}{72.46}}$$

Carl: $X_i = 5$, $\hat{Y} = 28.59$

$$99\%\ \text{PI} = 28.59 \pm 18.376\sqrt{1.0385 + \frac{(5-2.46)^2}{72.46}}$$
$$= 28.59 \pm 19.513$$
$$= 9.077 \text{ to } 48.103$$

Cindy: $X_i = 3$, $\hat{Y} = 19.79$

$$99\%\ \text{PI} = 19.79 \pm 18.376\sqrt{1.0385 + \frac{(3-2.46)^2}{72.46}}$$
$$= 19.79 \pm 18.763$$
$$= 1.027 \text{ to } 38.553$$

2. a. Mr. Smith's prediction interval will be wider, because his score is farther from the mean.

b. $df = 23$. $t_{\alpha/2}$ for 23 *df* and $\alpha = .05$ is 2.069.

$$\text{PI} = \hat{Y} \pm (2.069)(2.57)\sqrt{1 + \frac{1}{25} + \frac{(X_i - 35.4)^2}{1350}}$$
$$= \hat{Y} \pm 5.317\sqrt{1.04 + \frac{(X_i - 35.4)^2}{1350}}$$

Mr. Smith: $\hat{Y} = 2.29 + .15(15) = 4.54$

$$95\%\ \text{PI} = 4.54 \pm 5.317\sqrt{1.04 + \frac{(15-35.4)^2}{1350}}$$
$$= 4.54 \pm 6.174$$
$$= -1.634 \text{ to } 10.714$$

(*Note:* Because a negative number of treatment sessions is impossible, the interval can be stated as zero to 10.714.)

Ms. Jones: $\hat{Y} = 2.29 + .15(40) = 8.29$

$$95\%\ \text{PI} = 8.29 \pm 5.317\sqrt{1.04 + \frac{(40-35.4)^2}{1350}}$$
$$= 8.29 \pm 5.463$$
$$= 2.827 \text{ to } 13.753$$

10–7

1. a. (1) Joe: $\hat{Y} = -200 + 2.7(50) + 4.5(25) + 0.6(8) = 52.3$

(2) Max: $\hat{Y} = -200 + 2.7(30) + 4.5(40) + 0.6(5) = 64$

b. Max has a higher job knowledge score. Because job knowledge has the largest weight, his predicted productivity is higher.

2. b. Reading ability plus mechanical comprehension will yield better predictions. Because there is little overlap between the two uncorrelated predictors, together they predict more of the pilot training variance.

3. a. $r^2 = (.64)^2 = .4096$ or about 41%.
   b. Since N is large, no correction is needed.
   $S_{est} = 7\sqrt{1-(.64)^2} = 5.38$
   c. $t_{\alpha/2}$ for $\alpha = .05$ and 498 df is about 1.961. 95% PI = 21 ± (1.961)(5.38) = 21 ± 10.55 = 10.45 to 31.55.

Chapter 11

11–1

1. (1) Experiment because the variable method of teaching is manipulated.
   (2) Correlated samples because of matching.
2. (1) Experiment because the training variable is manipulated.
   (2) Independent samples because workers are randomly assigned to conditions.
3. (1) Not an experiment because the independent variable, level of math anxiety, is not manipulated.
   (2) Independent samples because high-anxiety and low-anxiety students are not matched.
4. (1) Experiment because type of stimulus presented is manipulated.
   (2) Correlated samples because each subject receives both stimuli.
5. (1) Experiment because the variable information is manipulated.
   (2) Independent samples because tossing a coin is a method of random assignment.

11–2

1. a. Sampling distribution of the difference between means.
   b. Normal because both populations are normal.
   c. Mean = $\mu_1 - \mu_2 = 50 - 30 = 20$.
   d. Standard error of the difference between means, σ_{diff}.
   e. $\sigma_{diff} = \sqrt{\frac{\sigma_1^2}{n_1} + \frac{\sigma_2^2}{n_2}} = \sqrt{\frac{(10)^2}{16} + \frac{(8)^2}{16}} = 3.20$
   f. The standard deviation would decrease because, as n_1 and n_2 increase, the variability of the difference between means decreases.
   g. $\sigma_{diff} = \sqrt{\frac{(10)^2}{100} + \frac{(8)^2}{100}} = 1.28$
2. a. The answers to parts e and g would change because the formula for σ_{diff} of correlated samples includes an additional factor.
   b. You need to know $\rho_{M_1M_2}$, the correlation between the sample means.
3. e. $\sigma_{diff} = \sqrt{\frac{(10)^2}{16} + \frac{(8)^2}{16} - 2(.60)\left(\frac{10}{\sqrt{16}}\right)\left(\frac{8}{\sqrt{16}}\right)}$
   $= 2.06$
   g. $\sigma_{diff} = \sqrt{\frac{(10)^2}{100} + \frac{(8)^2}{100} - 2(.60)\left(\frac{10}{\sqrt{100}}\right)\left(\frac{8}{\sqrt{100}}\right)}$
   $= 0.82$
4. a. The distribution based on samples of size 40 and 35 is more nearly normal.
   b. The distribution based on samples of size 40 and 35 has a smaller standard deviation.

11–3

1. Lower one-tailed test.
   H_o: $\mu_1 - \mu_2 \geq 0$
   H_A: $\mu_1 - \mu_2 < 0$
2. Two-tailed test.
   H_o: $\mu_1 - \mu_2 = 0$
   H_A: $\mu_1 - \mu_2 \neq 0$
3. Upper one-tailed test.
   H_o: $\mu_1 - \mu_2 \leq 0$
   H_A: $\mu_1 - \mu_2 > 0$

11–4

1. Let first-borns be Group 1. Upper one-tailed test.
   H_o: $\mu_1 - \mu_2 \leq 0$
   H_A: $\mu_1 - \mu_2 > 0$
   $n_1 = 7$, $\Sigma X_1 = 58$, $\Sigma X_1^2 = 518$
   $n_2 = 10$, $\Sigma X_2 = 64$, $\Sigma X_2^2 = 464$

$$M_1 = \frac{58}{7} = 8.29,\ M_2 = \frac{64}{10} = 6.4$$

$$SS_1 = 518 = \frac{(58)^2}{7} = 37.429$$

$$SS_2 = 464 = \frac{(64)^2}{10} = 54.4$$

$$s_{\text{diff}} = \sqrt{\frac{37.429 + 54.4}{7 + 10 - 2}\left(\frac{1}{7} + \frac{1}{10}\right)} = 1.219$$

$$t = \frac{8.29 - 6.4}{1.219} = 1.55$$

$df = 15$. With 15 df, $p(t > 1.55)$ is between .10 and .05 ($p > .05$).
First-borns did not cry significantly more, on the average, than later-borns at the .05 level.

2. Two-tailed test.
H_o: $\mu_1 - \mu_2 = 0$
H_A: $\mu_1 - \mu_2 \neq 0$
$SS_1 = (11)(8.51)^2 = 796.6211$
$SS_2 = (16)(7.6)^2 = 924.16$

$$s_{\text{diff}} = \sqrt{\frac{796.6211 + 924.16}{12 + 17 - 2}\left(\frac{1}{12} + \frac{1}{17}\right)} = 3.01$$

$$t = \frac{12 - 17}{3.01} = -1.66$$

$df = 27$. With 27 df, p for $|t| = 1.66$ is between .20 and .10 ($p > .05$). The means do not differ significantly at the .05 level.

11–5

1. H_o: $\mu_1 - \mu_2 \leq 0$
H_A: $\mu_1 - \mu_2 > 0$
$SS_1 = (4)(10.4)^2 = 432.64$
$SS_2 = (22)(2.7)^2 = 160.38$

$$s_{\text{diff}} = \sqrt{\frac{432.64 + 160.38}{5 + 23 - 2}\left(\frac{1}{5} + \frac{1}{23}\right)} = 2.357$$

$$t = \frac{30.6 - 22.1}{2.357} = 3.606$$

$df = 26$. With 26 df, p ($t > 3.606$) is between .0025 and .0005 ($p < .05$). The difference is significant at the .05 level. Conclude that the mean slant of left-handed students' writing is greater.

2. a. Pooled-variance test is appropriate because sample sizes are large and nearly equal.
   b. Pooled-variance test is appropriate because the sample standard deviations are very similar.
   c. Nonpooled procedure is appropriate because the standard deviations are very different and the sample sizes are small and quite unequal.

11–6

1. Let posttest = Condition 1 and pretest = Condition 2, so that $X_1 - X_2$ = posttest − pretest = gain score. Research hypothesis: Mean population gain score is greater than zero. Upper one-tailed test.
H_o: $\mu_1 - \mu_2 \leq 0$
H_A: $\mu_1 - \mu_2 > 0$
(*Note:* If pretest = X_1 and posttest = X_2, H_o and H_A are
H_o: $\mu_1 - \mu_2 \geq 0$
H_A: $\mu_1 - \mu_2 < 0$

Student	Posttest (X_1)	Pretest (X_2)	Gain ($X_1 - X_2 = D$)	D^2
1	7	4	3	9
2	7	8	−1	1
3	12	6	6	36
4	14	8	6	36
5	8	5	3	9
6	12	10	2	4
7	6	5	1	1
8	13	7	6	36
	79	53	26	132

$$M_1 = \frac{79}{8} = 9.875$$

$$M_2 = \frac{53}{8} = 6.625$$

$$s_{\text{diff}} = \sqrt{\frac{132 - \frac{(26)^2}{8}}{8(7)}} = 0.921$$

$$t = \frac{9.875 - 6.625}{0.921} = 3.529$$

$df = 7$. With 7 df, p ($t > 3.529$) is between .005 and .0025 ($p < .005$). Scores were significantly higher after instruction than before at the .005 level.

2. Two-tailed test
   H_o: $\mu_1 - \mu_2 = 0$
   H_A: $\mu_1 - \mu_2 \neq 0$

$$s_{diff} = \sqrt{\frac{1}{20}[(7.45)^2 + (7.75)^2 - 2(.45)(7.45)(7.75)]}$$
$$= 1.783$$

$$t = \frac{38.8 - 35.2}{1.783} = 2.109$$

   With 19 df, $p(|t| > 2.019)$ is between .10 and .05 ($p > .05$). The difference between the methods is not significant at the .05 level.

3. a. H_o: $\mu_1 - \mu_2 \leq 0$
      H_A: $\mu_1 - \mu_2 > 0$
      $\alpha = .05$
      $SS_1 = (24)(10.5)^2 = 2646$
      $SS_2 = (24)(11.8)^2 = 3341.76$

$$s_{diff} = \sqrt{\frac{2646 + 3341.76}{25 + 25 - 2}\left(\frac{1}{25} + \frac{1}{25}\right)} = 3.159$$

$$t = \frac{65.9 - 61.5}{3.159} = 1.393$$

      With 48 df, $p(t > 1.393)$ is between .10 and .05 ($p > .05$).
      If the data were from independent samples, the difference between the means would not be significant at the .05 level.
   b. The positive correlation between the paired scores of the matched samples resulted in a smaller value of s_{diff} and a larger t when the correlated-samples method was used.

11–7

1. a. $M_1 = 17.25$, $M_2 = 14.75$, $n_1 = n_2 = 8$, $s_{diff} = 1.009$.
      $df = 14$. With 14 df, $t_{.05/2} = 2.145$
      95% confidence interval (CI)
      $= (17.25–14.75) \pm (2.145)(1.009)$
      $= 2.5 \pm 2.164 = 0.336$ to 4.664
      Because the interval does not include zero, we reject H_o: $\mu_1 - \mu_2 = 0$ and reach the same conclusion as the test.
   b. With 14 df, $t_{.01/2} = 2.977$
      99% CI $= (17.25 - 14.75) \pm (2.977)(1.009) = 2.5 \pm 3.004 = -0.504$ to 5.504. Zero is in the interval. The difference between the means is not significant at the .01 level.
2. a. Yes, the difference is significant at the .01 level.
   b. Only c is possible (because only c includes zero in the interval).

11–8

1. a. Test A because the difference between μ_1 and μ_2 is larger.
   b. Test B because sample sizes are larger.
   c. Test A because α is larger.
   d. Test A because it is a one-tailed test.
   e. Test A because the correlation reduces s_{diff} and increases t.
   f. Test A because the population variance is smaller.
2. White noise present: $\Sigma X_1 = 54$, $\Sigma X_1^2 = 304$, $n_1 = 10$
   White noise absent: $\Sigma X_2 = 74$, $\Sigma X_2^2 = 580$, $n_2 = 10$

$$SS_1 = 304 - \frac{(54)^2}{10} = 12.4$$

$$SS_2 = 580 - \frac{(74)^2}{10} = 32.4$$

$$s_{diff} = \sqrt{\frac{12.4 + 32.4}{10 + 10 - 2}\left(\frac{1}{10} + \frac{1}{10}\right)} = .706.$$ This

   is smaller than the actual s_{diff} (.760). The use of repeated measures did not reduce s_{diff} in Tom's study.
3. a. $df = 10 + 10 - 2 = 18$

$$r_{pb}^2 = \frac{(-7.83)^2}{(-7.83)^2 + 18} = .77.$$ 77% of the dependent variable's variance is accounted for by the independent variable.

   b. $df = 100 + 100 - 2 = 198$

$$r_{pb}^2 = \frac{(-7.83)^2}{(-7.83)^2 + 198} = .24.$$. 24% of the dependent variable's variance is accounted for by the independent variable.

   c. The effect is more impressive with the smaller sample size.

Chapter 12

12–1

1. Answers will vary.
2. Correct statements of H_A: a, e.
3. If H_o is true, *MS* between is expected to be = MS within; if H_o is false, *MS* between is expected to be > *MS* within.

12–2

$n_1 = n_2 = n_3 = n_4 = 5$; $N_T = 20$
$\Sigma X_1 = 20, \Sigma X_2 = 25, \Sigma X_3 = 40, \Sigma X_4 = 15, \Sigma X_T = 100$
$M_1 = 20/5 = 4$; $M_2 = 25/5 = 5$; $M_3 = 40/5 = 8$; $M_4 = 15/5 = 3$; $GM = 100/20 = 5$

Sample 1 ($M_1 = 4$)

X_1	$X_1 - GM$	$(X_1 - GM)^2$	$X_1 - M_1$	$(X_1 - M_1)^2$
2	−3	9	−2	4
6	1	1	2	4
5	0	0	1	1
1	−4	16	−3	9
6	1	1	2	4
Sum		27		22

Sample 2 ($M_2 = 5$)

X_2	$X_2 - GM$	$(X_2 - GM)^2$	$X_x - M_1$	$(X_2 - M_2)^2$
5	0	0	0	0
4	−1	1	−1	1
7	2	4	2	4
2	−3	9	−3	9
7	2	4	2	4
Sum		18		18

Sample 3 ($M_3 = 8$)

X_3	$X_3 - GM$	$(X_3 - GM)^2$	$X_3 - M_3$	$(X_3 - M_3)^2$
9	4	16	1	1
8	3	9	0	0
12	7	49	4	16
6	1	1	−2	4
5	0	0	−3	9
Sum		75		30

Sample 4 ($M_4 = 3$)

X_4	$X_4 - GM$	$(X_4 - GM)^2$	$X_4 - M_4$	$(X_4 - M_4)^2$
3	−2	4	0	0
5	0	0	2	4
1	−4	16	−2	4
4	−1	1	1	1
2	−3	9	−1	1
Sum		30		10

SS total $= 27 + 18 + 75 + 30 = 150$
SS within $= 22 + 18 + 30 + 10 = 80$
SS between $= 5(4-5)^2 + 5(5-5)^2 + 5(8-5)^2 + 5(3-5)^2 = 70$

12–3

1. a. $N = 20$; $k = 4$
   df within $= 20 - 4 = 16$
   df between $= 4 - 1 = 3$
   df within + *df* between $= 16 + 3 = 19 =$ *df* total
   b. *SS* within = 80 (see answer to Exercise 12–2)

   $$MS \text{ within} = \frac{80}{16} = 5$$

   SS within = 70 (see answer to Exercise 12–2)

   $$MS \text{ between} = \frac{70}{3} = 23.33$$

2. a. $k = 5$, $N_T = 12 + 10 + 12 + 8 + 8 = 50$
   df between $= 5 - 1 = 4$
   df within $= 50 - 5 = 45$
   df total $= 50 - 1 = 49$
   b. $N_T = 35$, $k = 7$
   df between $= 7 - 1 = 6$
   df within $= 35 - 7 = 28$
   df total $= 35 - 1 = 34$

12–4

1. Answers will vary.
2. H_o; $\mu_1 = \mu_2 = \mu_3 = \mu_4$
   H_A: Not all μ are equal.
   $\alpha = .05$
   MS between = 23.33; *MS* within = 5 (see answer to Exercise 12–3, number 1b)

   $$F = \frac{23.33}{5} = 4.67$$

   With (3, 16) *df*, $p(F > 4.67)$ is between .05 and .01 ($p < .05$). The variation among the sample means is greater than would be expected by chance (at the .05 level).

12–5

Source of Variation	SS	df	MS	F	p
Between	698.2	3	232.733	4.69	.05
Within	793.6	16	49.6		
Total	1491.8	19			

Four groups were compared.
With 20 subjects in all, there were 20/4 = 5 subjects in each group.

12–6

1. $n_1 = 8$ $\sum X_1 = 72$ $\sum X_1^2 = 700$ $n_2 = 8$ $\sum X_2 = 88$ $\sum X_2^2 = 1018$

   $n_3 = 8$ $\sum X_3 = 80$ $\sum X_3^2 = 894$ $N_T = 24$ $\sum X_T = 240$ $\sum X_T^2 = 700$

$$SS \text{ total} = 2612 - \frac{(240)^2}{24} = 212$$

$$SS \text{ between} = \frac{(72)^2}{8} + \frac{(88)^2}{8} + \frac{(80)^2}{8} - \frac{(240)^2}{24} = 16$$

$$SS_1 = 700 - \frac{(72)^2}{8} = 52$$

$$SS_2 = 1018 - \frac{(88)^2}{8} = 50$$

$$SS_3 = 894 - \frac{(80)^2}{8} = 94$$

$$SS \text{ within} = 52 + 50 + 94 = 196$$

2. $n_1 = 5$ $\sum X_1 = 20$ $\sum X_1^2 = 102$ $n_2 = 5$ $\sum X_2 = 25$ $\sum X_2^2 = 143$

   $n_3 = 5$ $\sum X_3 = 40$ $\sum X_3^2 = 350$ $n_4 = 5$ $\sum X_4 = 15$ $\sum X_4^2 = 55$ $N_T = 20$ $\sum X_T = 100$ $\sum X_T^2 = 650$

$$SS \text{ total} = 650 - \frac{(100)^2}{20} = 150$$

$$SS \text{ between} = \frac{(20)^2}{5} + \frac{(25)^2}{5} + \frac{(40)^2}{5} + \frac{(15)^2}{5} - \frac{(100)^2}{20} = 70$$

$$SS_1 = 102 - \frac{(20)^2}{5} = 22$$

$$SS_2 = 143 - \frac{(25)^2}{5} = 18$$

$$SS_3 = 350 - \frac{(40)^2}{5} = 30$$

$$SS_4 = 55 - \frac{(15)^2}{5} = 10$$

SS within = 22 + 18 + 30 + 10 = 80

Summary table:

Source of Variation	SS	df	MS	F	p
Between	70	3	23.33	4.67	.05
Within	80	16	5		
Total	150	19			

Yes, reject H_o.
The variation among the sample means is significant at the .05 level.

12–7

$n_1 = 7$ $\sum X_1 = 76$ $\sum X_1^2 = 852$ $M_1 = 10.86$

$n_2 = 8$ $\sum X_2 = 96$ $\sum X_2^2 = 1238$ $M_2 = 12$

$n_3 = 7$ $\sum X_3 = 81$ $\sum X_3^2 = 1085$ $M_3 = 11.57$

$n_4 = 6$ $\sum X_4 = 34$ $\sum X_4^2 = 246$ $M_4 = 5.67$

$N_T = 28$ $\sum X_T = 287$ $\sum X_T^2 = 3421$

$$SS \text{ total} = 3421 - \frac{(287)^2}{28} = 479.25$$

$$SS \text{ between} = \frac{(76)^2}{7} + \frac{(96)^2}{8} + \frac{(81)^2}{7} + \frac{(34)^2}{6} - \frac{(287)^2}{28} = 165.35$$

$$SS \text{ within} = \left[852 - \frac{(76)^2}{7}\right] + \left[1238 - \frac{(96)^2}{8}\right] + \left[1085 - \frac{(81)^2}{7}\right] + \left[246 - \frac{(34)^2}{6}\right] = 313.90$$

Source of Variation	SS	df	MS	F	p
Major (between)	165.33	3	55.12	4.21	.05
Within	313.90	24	13.08		
Total	479.25	27			

The variation among the sample means is significant at the .05 level. Conclusion: Students with different college majors respond differently to academic frustration. Judging from the group means, it appears that business majors have a lower mean level of frustration than English, psychology, and mathematics majors.

12–8

1. $n_1 = n_2 = 5,\ \Sigma X_1 = 41,\ \Sigma X_1^2 = 347,$

$$\Sigma X_2 = 24,\ \Sigma X_2^2 = 128;\ M_1 = \frac{41}{5} = 8.2;$$

$$M_2 = \frac{24}{5} = 4.8$$

$$SS_1 = 347 - \frac{(41)^2}{5} = 10.8$$

$$SS_2 = 128 - \frac{(24)^2}{5} = 12.8$$

$$s_{\text{diff}} = \sqrt{\frac{10.8 + 12.8}{5 + 5 - 2}\left(\frac{1}{5} + \frac{1}{5}\right)}$$

$$= \sqrt{2.95(.4)} = 1.086$$

$$\uparrow$$
$$s_p^2$$

$$t = \frac{8.2 - 4.8}{1.086} = 3.131$$

With 8 *df*, $p(|t| > 3.131)$ is between .02 and .01 ($p < .02$). The difference between the means is significant at the .05 level.

2. $N_T = 10,\ \Sigma X_T = 65,\ \Sigma X_T^2 = 475$

$$SS\ \text{total} = 475 - \frac{(65)^2}{10} = 52.5$$

$$SS\ \text{between} = \frac{(41)^2}{5} + \frac{(24)^2}{5} - \frac{(65)^2}{10} = 28.9$$

Summary table and remaining calculations:

Source of Variation	*SS*	*df*	*MS*	*F*	*p*
Between	28.9	1	28.9	9.797	.05
Within	23.6	8	2.95		
Total	52.5	9			

The difference between the means is significant at the .05 level, as in the t test.
df of t test = 8 = *df* within of ANOVA
s_p^2 of t test = 2.95 = *MS* within of ANOVA
t^2 of t test = $(3.131)^2$ = 9.803 = F of ANOVA (the difference is due to rounding).

12–9

(a) ANOVA A because sample sizes are larger.
(b) ANOVA A because α is larger.
(c) ANOVA B because the population variance is smaller.
(d) ANOVA B because the population means vary more.

12–10

1. $$\hat{\omega}^2 = \frac{165.345 - (3)(13.079)}{479.25 + 13.079} = .416,\ \text{or } 41.6\%$$

2. John's result is more impressive.
3. Amanda Low had a larger sample size.

12–11

1.

	Weights				
Comparison	M_1	M_2	M_3	M_4	M_5
a. $M_3 - M_4$	0	0	1	−1	0
b. M_1 and M_4 vs. M_5	½	0	0	½	−1
or:	1	0	0	1	−2
c. M_1 vs. (M_1, M_3, M_4, M_5)	−¼	1	−¼	−¼	−¼
or:	−1	4	−1	−1	−1

2. Comparison a vs. b: (0)(1) + (0)(0) + (1)(0) + (−1)(1) + (0)(−2) = 1
Comparisons a and b are not orthogonal.
Comparison a vs. c: (0)(−1) + (0)(4) + (1)(−1) + (−1)(−1) + (0)(−1) = 0
Comparisons a and c are orthogonal.
Comparison b vs. c: (1)(−1) + (0)(4) + (0)(−1) + (1)(−1) + (−2)(−1) = 0
Comparisons b and c are orthogonal.

3. a.

	Meaningful				
	Verbal		Nonverbal	Rote	
Comparison	1	2	3	4	5
(1)	2	2	2	−3	−3
(2)	1	1	−2	0	0
(3)	0	0	0	1	−1
(4)	1	−1	0	0	0

b. (1) vs. (2): (2)(1) + (2)(1) + (2)(−2) + (−3)(0) + (−3)(0) = 0
(1) and (2) are orthogonal.
(1) vs. (3): (2)(0) + (2)(0) + (2)(0) + (−3)(1) + (−3)(−1) = 0
(1) and (3) are orthogonal.
(1) vs. (4): (2)(1) + (2)(−1) + (2)(0) + (−3)(0) + (−3)(0) = 0
(1) and (4) are orthogonal.
(2) vs. (3): (1)(0) + (1)(0) + (−2)(0) + 0(1) + 0(−1) = 0
(2) and (3) are orthogonal.
(2) vs. (4): (1)(1) + (1)(−1) + (−2)(0) + (0)(0) + (0)(0) = 0
(2) and (4) are orthogonal.
(3) vs. (4): (0)(1) + (0)(−1) + (0)(0) + (1)(0) + (−1)(0) = 0
(3) and (4) are orthogonal.

12–12

1.

	Experimental		Control	
Comparison	**1**	**2**	**1**	**2**
(a)	1	1	−1	−1
(b)	1	−1	0	0
(c)	0	0	1	−1

(a) vs. (b): (1)(1) + (1)(−1) + (−1)(0) + (−1)(0) = 0: orthogonal
(a) vs. (c): (1)(0) + (1)(0) + (−1)(1) + (−1)(−1) = 0: orthogonal
(b) vs. (c): (1)(0) + (−1)(0) + (0)(1) + (0)(−1) = 0: orthogonal

2. (a) C = (1)(24) + (1)(18) + (−1)(12) + (−1)(15) = 15

$$F = \frac{(15)^2}{30\left[\frac{(1)^2}{5} + \frac{(1)^2}{5} + \frac{(-1)^2}{5} + \frac{(-1)^2}{5}\right]} = 9.375$$

df within = 20 − 4 = 16
With $df = (1, 16)$, $p(F > 9.375) < .01$

(b) C = (1)(24) + (−1)(18) + (0)(12) + (0)(15) = 6

$$F = \frac{(6)^2}{30\left[\frac{(1)^2}{5} + \frac{(-1)^2}{5} + \frac{(0)^2}{5} + \frac{(0)^2}{5}\right]} = 3.00$$

With $df = (1, 16)$, $p(F > 3.00) > .05$

(c) C = (0)(24) + (0)(18) + (1)(12) + (−1)(15) = −3

$$F = \frac{(-3)^2}{30\left[\frac{(0)^2}{5} + \frac{(0)^2}{5} + \frac{(1)^2}{5} + \frac{(-1)^2}{5}\right]} = 0.75,\ p > .05$$

Only the difference between the combined experimental groups vs. the combined control groups is significant at the .05 level.

3. With 3 tests, α for each test = $\frac{.05}{3} = .017$

12–13

1.

	Group		
	1	**2**	**3**
Mean	**7**	**11**	**12**
Remaining Comparisons		**Weights**	
1 vs. 3	1	0	−1
2 vs. 3	0	1	−1
2 vs. 1 and 3	−1	2	−1
3 vs. 1 and 2	−1	−1	2

Comparisons:
1 vs. 3: C = (1)(7) + (0)(11) + (−1)(12) = −5

$$F = \frac{(-5)^2}{9.14\left[\frac{(1)^2}{8} + \frac{(0)^2}{8} + \frac{(-1)^2}{8}\right]} = 10.94$$

Because $F < 11.56$, difference between M_1 and M_3 is not significant
2 vs. 3: C = (0)(7) + (1)(11) + (−1)(12) = −1

$$F = \frac{(-1)^2}{9.14\left[\frac{(0)^2}{8} + \frac{(1)^2}{8} + \frac{(-1)^2}{8}\right]} = 0.44$$

Difference between M_2 and M_3 is not significant.
2 vs. 1 and 3: C = (−1)(7) + (2)(11) + (−1)(12) = 3

$$F = \frac{(-3)^2}{9.14\left[\frac{(-1)^2}{8} + \frac{(2)^2}{8} + \frac{(-1)^2}{8}\right]} = 1.31$$

Difference between 2 vs. 1 and 3 is not significant.
3 vs. 1 and 2: C = (−1)(7) + (−1)(11) + (2)(12) = 6

$$F = \frac{(6)^2}{9.14\left[\frac{(-1)^2}{8} + \frac{(-1)^2}{8} + \frac{(2)^2}{8}\right]} = 5.25$$

Difference between 3 vs. 1 and 2 is not significant.

2. a. $k - 1 = 4 = df$ between; df within = 25 − 5 = 20
   Critical value of F for (4, 20) df and α = .01 is 4.43
   Critical value for Scheffé test = 4(4.43) = 17.72
   b. The calculated F for at least one comparison will be greater than 17.72.
   c.

Comparison	Weights 1	2	3	4	5
(1) Group 2 vs. 4	0	1	0	−1	0
(2) Group 2 vs. 1 and 3	−1	2	−1	0	0

Comparison (1): $C = (0)(16) + (1)(12) + (0)(18) + (-1)(25) + (0)(23) = -13$

$$F = \frac{(-13)^2}{17\left[\frac{(0)^2}{6} + \frac{(1)^2}{4} + \frac{(0)^2}{5} + \frac{(-1)^2}{5} + \frac{(0)^2}{5}\right]} = 22.09$$

The difference between the Group 2 and Group 4 means is significant at the .01 level.

Comparison (2): $C = (-1)(16) + (2)(12) + (-1)(18) + (0)(25) + (0)(23) = -10$

$$F = \frac{(-10)^2}{17\left[\frac{(-1)^2}{6} + \frac{(2)^2}{4} + \frac{(-1)^2}{5} + \frac{(0)^2}{5} + \frac{(0)^2}{5}\right]} = 4.30$$

The difference between Group 2 and the combination of Groups 1 and 3 is not significant at the .01 level.

12–14

Table of differences between means:

		Mean 15.8	17.2	19.0	19.4
	15.8 (Group 3)	0	1.4	3.2	3.6
	17.2 (Group 1)		0	1.8	2.2
Mean	19.0 (Group 4)			0	0.4
	19.4 (Group 2)				0

$k = 4$, df within = 40 − 4 = 36, $\alpha = .01$
$q_{crit} = 4.80$ (using df within = 30)

$$\text{HSD} = 4.80\sqrt{\frac{3.06}{10}} = 2.66$$

Significant differences: Group 2 vs. Group 3, Group 3 vs. Group 4.

Chapter 13

13–1

1. a. There are 9 combinations:
   A_1B_1 High dose of Prozac
   A_1B_2 Medium dose of Prozac
   A_1B_3 Low dose of Prozac
   A_2B_1 High dose of Paxil
   A_2B_2 Medium dose of Paxil
   A_2B_3 Low dose of Paxil
   A_3B_1 High dose of Xanax
   A_3B_2 Medium dose of Xanax
   A_3B_3 Low dose of Xanax
   b. 9 × 10 = 90 rats
2. The effect of heredity is different in different environments; or the effect of environment depends upon one's heredity.

13–2

1. Effect of A: H_o: $\mu_{A1} = \mu_{A2} = \mu_{A3}$
   (The effects of Prozac, Paxil, and Xanax are the same.)
   H_A: Not all μ_A are equal.
   (At least one drug has an effect different from the others.)
   Effect of B: H_o: $\mu_{B1} = \mu_{B2} = \mu_{B3}$
   (The effect is the same at all dosage levels.)
   H_A: Not all μ_B are equal
   (At least one dosage level has an effect different from the others.)
   Interaction: H_o: No population interaction between drug and dosage level. (The effect of dosage is the same for all three drugs,

and drug differences are the same at all dosage levels.)

H_A: Drug and dosage level interact. (The effect of dosage level depends on the drug, and differences between the drugs differ at different dosages.)

2. $df_A = 3 - 1 = 2$
$df_B = 3 - 1 = 2$
$df_{AB} = (2)(2) = 4$
df within $= 90 - 9 = 81$

13–3

$\Sigma X_T = 7 + 18 + 20 + 21 = 66, \quad N_T = 3 \times 4 = 12$

$$SS\text{ total} = 490 - \frac{(66)^2}{12} = 127$$

$$SS\text{ between} = \frac{(7)^2}{3} + \frac{(20)^2}{3} + \frac{(18)^2}{3} + \frac{(21)^2}{3} - \frac{(66)^2}{12} = 41.667$$

$$SS\text{ within} = 127 - 41.667 = 85.333$$

$$\Sigma A_1 = 7 + 20 = 27; \qquad \Sigma A_2 = 18 + 21 = 39$$

$$SS_A = \frac{(27)^2}{6} + \frac{(39)^2}{6} - \frac{(66)^2}{12} = 12$$

$$\Sigma B_1 = 7 + 18 = 25; \qquad \Sigma B_2 = 20 + 21 = 41$$

$$SS_B = \frac{(25)^2}{6} + \frac{(41)^2}{6} - \frac{(66)^2}{12} = 21.333$$

$$SS_{AB} = 41.667 - 12 - 21.333 = 8.334$$

Source of Variation	SS	df	MS	F
A	12	1	12	1.12
B	21.333	1	21.333	2.00
A × B	8.334	1	8.334	0.78
Within	85.333	8	10.667	
Total	127	11		

All three F's are evaluated with (1, 8) df. None of the effects is significant at the .05 level.

13–4

Answers will vary. Examples of possible answers are given next.

1.

	Instruction	No Instruction
Initially Low Scores	25	5
Initially High Scores	25	5

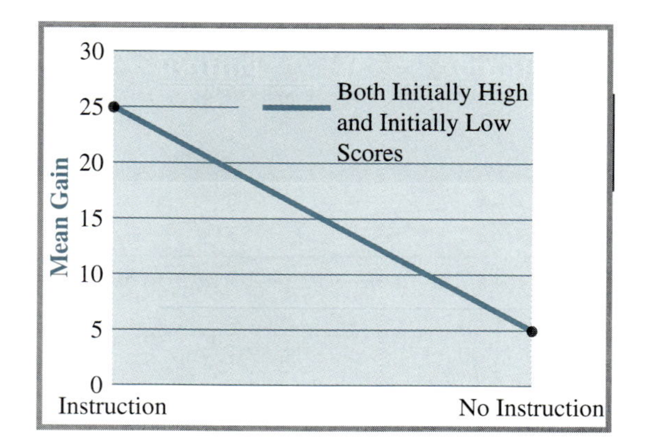

2.

	Instruction	No Instruction
Initially Low Scores	30	0
Initially High Scores	20	10

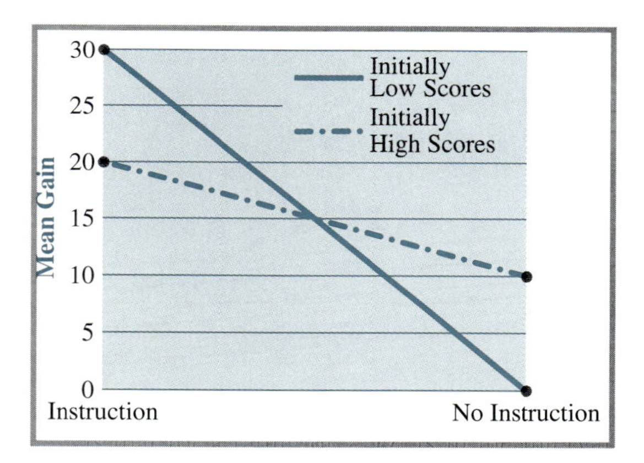

13–5

1.

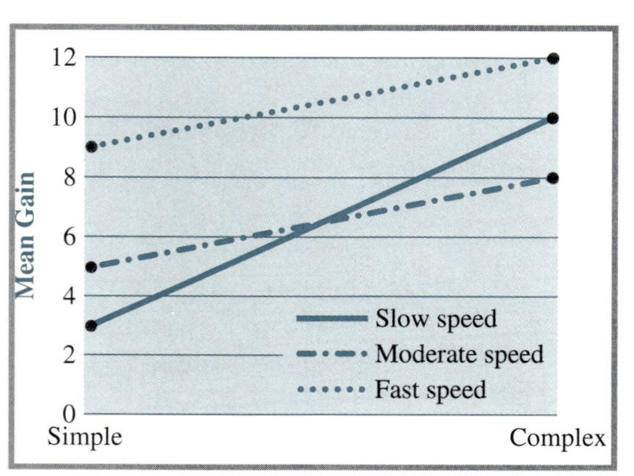

Most people prefer the graph with speed on the abscissa because it includes only two lines and appears less cluttered.

2. a.

Source of Variation	SS	df	MS	F	p
Type of reward	6.58	2	3.29	6.66	.01
Frequency of reward	0.84	2	0.42	0.85	n.s.
Type x frequency	6.76	4	1.69	3.42	.05
Within	17.80	36	0.494		
Total	31.98	44			

b.

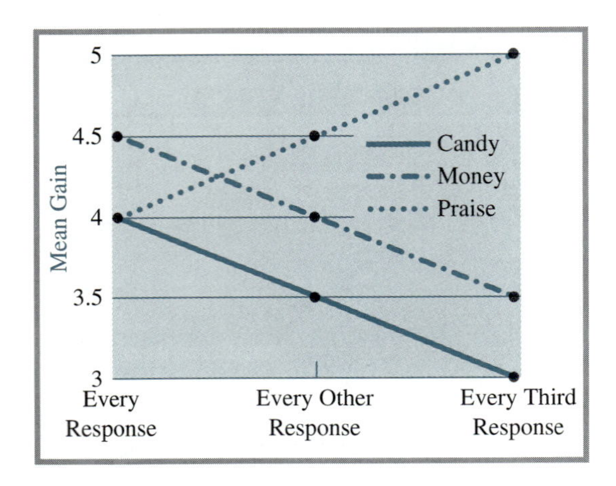

The effect of type of reward is significant at the .01 level, with praise resulting in the most sharing behavior, and candy the least. There was no main effect of frequency of reward. However, the interaction was significant at the .05 level. The effect of praise increased as the frequency of reward decreased, but the effect of money and candy decreased as the frequency of reward decreased.

13–6

$$\text{Complexity: } \hat{\omega}^2 = \frac{92.04-(2-1)(9.04)}{359.96+9.04} = .27$$

$$\text{Speed: } \hat{\omega}^2 = \frac{80.08-(3-1)(9.04)}{359.96+9.04} = .17$$

13–7

1. Type of reward:

$$\text{Type of reward: } \hat{\omega}^2 = \frac{6.58-(3-1)(.494)}{31.98+.494} = .17$$

$$\text{Interaction: } \hat{\omega}^2 = \frac{6.76-(4)(.494)}{31.98+.494} = .15$$

2.

	Money	Candy	Praise
n	15	15	15
Mean	3.93	3.53	4.47
(a) Tangible vs. intangible	1	1	−2
(b) Money vs. candy	1	−1	0

(a) $C = (1)(3.93) + (1)(3.53) + (-2)(4.47) = -1.48$

$$F = \frac{(-1.48)^2}{.494\left[\frac{(1)^2}{15}+\frac{(1)^2}{15}+\frac{(-2)^2}{15}\right]} = 11.09$$

With (1, 36) *df*, $p\ (F > 11.09) < .01$
The difference between tangible and intangible reward is significant at the .01 level. Intangible reward resulted in a higher level of sharing behavior.

(b) $C = (1)(3.93) + (-1)(3.53) + (0)(4.47) = 0.4$

$$F = \frac{(0.4)^2}{.494\left[\frac{(1)^2}{15}+\frac{(-1)^2}{15}+\frac{(0)^2}{15}\right]} = 2.43$$

With (1, 36) *df*, $p\ (F > 2.43) > .01$
The difference between money and candy is not significant at the .01 level.

3. With the Scheffé test, the comparison F's are the same, but the critical value of F is different. Critical value of F at $\alpha = .01$ for (2, 36) *df* is (about) 5.4. Critical value for Scheffé test is 2(5.4) = 10.8. Since F for tangible vs. intangible exceeds 10.58, it is significant at the .01 level. The difference between money and candy is not significant.

4. Table of differences between means:

	Candy	Money	Praise
Candy (3.53)	0	0.4	0.94
Money (3.93)		0	0.54
Praise (4.47)			0

For $k = 3$, df within $= 36$, and $\alpha = .01$, $q_{crit} =$ (about) 4.45

$$\text{HSD} = 4.45\sqrt{\frac{.494}{15}} = 0.81$$

Only the difference between candy and praise is significant at the .01 level.

13–8

1. df total $= (20)(5) - 1 = 99$
   df treatments $= 5 - 1 = 4$
   df subjects $= 20 - 1 = 19$
   df residual $= (4)(19) = 76$
2. The SS within of the two-way ANOVA is based on variability within each cell. In the repeated measures design, there is only one data point within each cell and, therefore, no variability within.

13–9

1. $\Sigma X_T = 82$
   $\Sigma X_T^2 = 442$

Subject sums:
$\Sigma S_1 = 3 + 8 + 2 + 4 = 17$
$\Sigma S_2 = 5 + 7 + 3 + 6 = 21$
$\Sigma S_3 = 2 + 6 + 0 + 2 = 10$
$\Sigma S_4 = 4 + 9 + 1 + 5 = 19$
$\Sigma S_5 = 5 + 5 + 2 + 3 = 15$

Treatment sums:
$\Sigma T_1 = 3 + 5 + 2 + 4 + 5 = 19$
$\Sigma T_2 = 8 + 7 + 6 + 9 + 5 = 35$
$\Sigma T_3 = 2 + 3 + 0 + 1 + 2 = 8$
$\Sigma T_4 = 4 + 6 + 2 + 5 + 3 = 20$

$$SS\text{ total} = 442 - \frac{(82)^2}{20} = 105.8$$

$$SS\text{ subjects} = \frac{(17)^2}{4} + \frac{(21)^2}{4} + \frac{(10)^2}{4} + \frac{(19)^2}{4} + \frac{(15)^2}{4} - \frac{(82)^2}{20} = 17.8$$

$$SS\text{ treatments} = \frac{(19)^2}{5} + \frac{(35)^2}{5} + \frac{(8)^2}{5} + \frac{(20)^2}{5} - \frac{(82)^2}{20} = 73.8$$

$$SS\text{ residual} = 105.8 - 17.8 - 73.8 = 14.2$$

Source of Variation	SS	df	MS	F	p
Subjects	17.8	4			
Treatments	73.8	3	24.6	20.85	.01
Residual	14.2	12	1.18		
Total	105.5	19			

Performance differed significantly over the four puzzles at the .01 level.

2.

Source of Variation	SS	df	MS	F	p
Subjects	7,200	24			
Treatments	4,800	5	960	4.40	.01
Residual	26,160	120	218		
Total	38,160	149			

The variation among the treatments is significant at the .01 level.

13–10

1. SS total is the same as for the original data = 692.17
   SS treatments is also the same = 84.27

$$SS\text{ subjects} = \frac{(37)^2}{3} + \frac{(44)^2}{3} + \frac{(23)^2}{3} + \frac{(19)^2}{3} + \frac{(24)^2}{3} + \frac{(24)^2}{3} + \frac{(29)^2}{3} + \frac{(41)^2}{3} + \frac{(16)^2}{3} + \frac{(18)^2}{3} - \frac{(275)^2}{30} = 295.50$$

SS residual $= 692.17 - 84.27 - 295.5 = 312.40$

Source of Variation	SS	df	MS	F	p
Workers	295.50	9			
Noise levels	84.27	2	42.135	2.43	n.s.
Residual	312.40	18	17.356		
Total	692.17	29			

The variation among noise levels is not significant at the .05 level. Random ordering of scores within treatments reduced the covariation between treatments to a chance level. This reduced the variation among workers and increased the residual. When *MS* treatments (the same in both analyses) was divided by a larger *MS* residual, the resulting value of *F* was reduced below the level required for significance at the .05 level.

2. In the *t* test for correlated samples, s_{diff} (the "error" term) = $\sqrt{\frac{1}{n}(s_1^2 + s_2^2 - 2r_{12}s_1s_2)}$.

   Subtracting $2r_{12}s_1s_2$ removes the covariation between treatments.

13–11

	Low	Moderate	High
Means	10.9	9.7	6.9
n	10	10	10
Weights	1	1	−2

MS residual = 11.244, *df* residual = 18

(a) Planned comparison: $C = (1)(10.9) + (1)(9.7) + (-2)(6.9) = 6.8$

$$F = \frac{(6.8)^2}{11.244\left[\frac{(1)^2}{10} + \frac{(1)^2}{10} + \frac{(-2)^2}{10}\right]} = 6.85$$

With (1, 9) *df*, $p(F > 6.85) < .05$. The difference is significant at the .05 level. Productivity was significantly higher under low or moderate noise levels than under a high noise level.

(b) The appropriate post hoc test is the Scheffé test. *F* is the same, but the critical value at the .05 level = 2(3.55) = 7.1. Since *F* is less than the critical value, the difference is not significant at the .05 level.

Chapter 14

14–1

Counseled		Uncounseled	
Score	Rank	Score	Rank
2	2.5	8	7
5	5	16	9
7	6	4	4
0	1	18	10
2	2.5	12	8
Sum	17		38

Using the sum of ranks in the counseled group ($T = 17$):

$$U = (5)(5) + \frac{(5)(6)}{2} - 17 = 23$$

or using the sum of ranks in the uncounseled group ($T = 38$):

$$U = (5)(5) + \frac{(5)(6)}{2} - 38 = 2$$

Because both values of *U* lie outside the tabled critical values of *U* for $\alpha = .05$ (3 and 22), the difference between the ranks of the counseled and uncounseled patients is significant. Counseled patients reported significantly less pain than uncounseled patients.

14–2

H_o: Ratings did not increase after watching the film.
H_A: Ratings increased.
If gain scores are calculated (after-film minus before-film), an upper one-tailed test is required.

Student	Before	After	After − Before D	|D|	Rank of |D|	Signed Rank
1	0	5	5	5	6	6
2	3	4	1	1	1	1
3	1	5	4	4	5	5
4	3	9	6	6	7	7
5	2	4	2	2	3	3
6	5	5	0			
7	7	9	2	2	3	3
8	4	6	2	2	3	3

Sum of positive signed ranks = 28
Sum of negative signed ranks = 0; $W = 0$

With $N = 7$, $p(W < 0) = .01$ in an upper one-tailed test. The ratings increased significantly at the .01 level.

14–3

3-Year-Olds		4-Year-Olds		5-Year-Olds		6-Year-Olds	
Successes	**Rank**	**Successes**	**Rank**	**Successes**	**Rank**	**Successes**	**Rank**
10	12	8	8.5	16	17.5	19	20
0	1	1	2.5	6	6.5	9	10
4	5	15	15.5	13	14	18	19
15	15.5	1	2.5	3	4	10	12
8	8.5	16	17.5	6	6.5	10	12
Sum (T)	42.0		46.5		48.5		73.0
T^2	1764		2162.25		2352.25		5329
$\frac{T^2}{n}$	352.8		432.45		470.45		1065.8

$$\sum^{k} \frac{T^2}{n} = 352.8 + 432.5 + 470.45 + 1065.8 = 2321.5$$

$$H = \left[\frac{12}{20(21)}\right](2321.5) - 3(21) = 3.33$$

$df = 4 - 1 = 3$

With $df = 3$, $p(\chi^2 > 3.33)$ is greater than .25. Manual dexterity does not differ significantly as a function of age over the years tested.

14–4

Ranks within each row:

	Level of Complexity				
Infant	**1**	**2**	**3**	**4**	**5**
1	1	2	3	5	4
2	1	2	3	4	5
3	1	2	3	5	4
4	1.5	1.5	4	5	3
5	1	2.5	2.5	4	5
6	3	1.5	5	1.5	4
7	1	2	3.5	3.5	5
8	2	4	1	3	5
9	1.5	3.5	3.5	1.5	5
10	1.5	4.5	3	4.5	1.5
11	3.5	3.5	2	5	1
12	1	2.5	2.5	4	5
Sum (T)	19	31.5	36	46	47.5
T^2	361	992.25	1296	2116	2256.25

$$\sum^{K} T^2 = 361 + 992.25 + 1296 + 2116 + 2256.25 = 7021.5$$

$$\chi_r^2 = \frac{12}{(5)(12)(6)}(7021.5) - 3(12)(6) = 18.05$$

With $df = 4$, $p(\chi^2 > 18.05)$ is less than .005. Length of gaze increased significantly as complexity increased.

14–5

1. The Wilcoxon test is appropriate.

Subject	Present	Absent	D	$\lvert D \rvert$	Rank of $\lvert D \rvert$	Signed Rank
1	6	9	–3	3	6	–6
2	4	6	–2	2	3	–3
3	6	5	1	1	1	1
4	7	10	–3	3	6	–6
5	4	10	–6	6	9	–9
6	5	8	–3	3	6	–6
7	5	7	–2	2	3	–3
8	7	5	2	2	3	3
9	6	6	0			
10	4	8	–4	4	8	–8

Sum of positive signed ranks = 1 + 3 = 4
Sum of negative signed ranks = 6 + 3 + 6 + 9 + 6 + 3 + 8 = 41
$W = 4$
With $N = 9$ and a one-tailed test, $p(W \leq 4)$ is between .025 and .01. The t test yielded a value of t with a probability between .025 and .01. In this case, the results of the two tests are the same.

2. a. Parametric test. Though sample sizes are small, the assumptions of population normality and equality of variances and covariances are met.
   b. Nonparametric test because the populations are skewed and sample sizes are small.
   c. Nonparametric test because of the presence of extreme outliers in small samples.

Chapter 15

15–1

1. a. Normal curve (z) test or t test of H_o: $\mu = 75$ (depending on whether the population SD is known).
   b. Confidence interval for the difference between means (of males and females).
   c. One-way analysis of variance comparing the mean dependency of the three groups.
   d. Chi square test of goodness of fit, because preferences are categorical and only one population is involved.
   e. Chi square test of independence to compare distributions of preferences of younger and older consumers.
   f. t test for the difference between means of correlated samples (pretest mean vs. posttest mean of a single group of students).
   g. Correlation coefficient between pretest and posttest scores.
   h. Regression analysis.

15–2

1. The numerator is actually $r - \rho_H$, where ρ_H (the hypothesized value of ρ) is zero. The numerator is the obtained difference between r and ρ_H, and the denominator, $\sqrt{\frac{1-r^2}{N-2}}$, is a measure of the difference expected by chance.
2. Standard error of the correlation coefficient.

15–3

Dr. Smith first answers question (1) by calculating the correlation (r) between the two variables. Then he answers (2) by conducting a test of the significance of r. Dr. Jones first answers question (2) by conducting a t test for the difference between means. Then he answers (1) by calculating r^2_{pb} based on t and degrees of freedom.

15–4

1. The critical value of r for any alpha (the minimum r that is significant at level alpha) decreases as sample size increases. The number of males in Nestor's study was only 32, but the entire group included 59 students
2. Answers will vary.
3. Answers will vary.

15–5

1. Study 1: $d = \frac{24-30}{3} = -2$
   The means differ by two standard deviations.
   Study 2: $d = \frac{95-84}{15} = 0.73$
   The means differ by about three-fourths of a standard deviation.
   Study 3: $d = \frac{58-47}{3} = 3.67$
   The means differ by three-and-two-thirds standard deviations.
2. The gifted children have a restricted range of scores on one or both measures as compared with the public school students. In order to compare two or more correlation coefficients, the groups must be equally heterogeneous.

Appendix D: Tables

Table D–1 Critical Values of Chi Square (χ^2)

Each entry in the table is the value of χ^2 such that, in the chi square distribution with the given degrees of freedom (*df*), the probability of χ^2 greater than the tabled value is equal to the column heading.

	Proportion of Area in the Upper Tail						
df	.25	.15	.10	.05	.025	.01	.005
1	1.32	2.07	2.71	3.84	5.02	6.63	7.88
2	2.77	3.79	4.61	5.99	7.38	9.21	10.60
3	4.11	5.32	6.25	7.81	9.35	11.34	12.84
4	5.39	6.74	7.78	9.49	11.14	13.28	14.86
5	6.63	8.12	9.24	11.07	12.83	15.09	16.75
6	7.84	9.45	10.64	12.59	14.45	16.81	18.55
7	9.04	10.75	12.02	14.07	16.01	18.48	20.28
8	10.22	12.03	13.36	15.51	17.53	20.09	21.95
9	11.39	13.29	14.68	16.92	19.02	21.67	23.59
10	12.55	14.53	15.99	18.31	20.48	23.21	25.19
11	13.70	15.77	17.28	19.68	21.92	24.72	26.76
12	14.85	16.99	18.55	21.03	23.34	26.22	28.30
13	15.98	18.20	19.81	22.36	24.74	27.69	29.82
14	17.12	19.41	21.06	23.68	26.12	29.14	31.32
15	18.25	20.60	22.31	25.00	27.49	30.58	32.80
16	19.37	21.79	23.54	26.30	28.85	32.00	34.27
17	20.49	22.98	24.77	27.59	30.19	33.41	35.72
18	21.60	24.16	25.99	28.87	31.53	34.81	37.16
19	22.72	25.33	27.20	30.14	32.85	36.19	38.58
20	23.83	26.50	28.41	31.41	34.17	37.57	40.00
25	29.34	32.28	34.38	37.65	40.65	44.31	46.93
30	34.80	37.99	40.26	43.77	46.98	50.89	53.67
40	45.62	49.24	51.81	55.76	59.34	63.69	66.77
60	66.98	71.34	74.40	79.08	83.30	88.38	91.95
100	109.1	114.7	118.5	124.3	129.6	135.8	140.2

Table D–2 **Proportions of Area in the Standard Normal Curve**

(A) z	(B) Area Between Mean and z	(C) Area Beyond z	(A) z	(B) Area Between Mean and z	(C) Area Beyond z	(A) z	(B) Area Between Mean and z	(C) Area Beyond z
0.00	.0000	.5000	0.40	.1554	.3446	0.80	.2881	.2119
0.01	.0040	.4960	0.41	.1591	.3409	0.81	.2910	.2090
0.02	.0080	.4920	0.42	.1628	.3372	0.82	.2939	.2061
0.03	.0120	.4880	0.43	.1664	.3336	0.83	.2967	.2033
0.04	.0160	.4840	0.44	.1700	.3300	0.84	.2995	.2005
0.05	.0199	.4801	0.45	.1736	.3264	0.85	.3023	.1977
0.06	.0239	.4761	0.46	.1772	.3228	0.86	.3051	.1949
0.07	.0279	.4721	0.47	.1808	.3192	0.87	.3078	.1922
0.08	.0319	.4681	0.48	.1844	.3156	0.88	.3106	.1894
0.09	.0359	.4641	0.49	.1879	.3121	0.89	.3133	.1867
0.10	.0398	.4602	0.50	.1915	.3085	0.90	.3159	.1841
0.11	.0438	.4562	0.51	.1950	.3050	0.91	.3186	.1814
0.12	.0478	.4522	0.52	.1985	.3015	0.92	.3212	.1788
0.13	.0517	.4483	0.53	.2019	.2981	0.93	.3238	.1762
0.14	.0557	.4443	0.54	.2054	.2946	0.94	.3264	.1736
0.15	.0596	.4404	0.55	.2088	.2912	0.95	.3289	.1711
0.16	.0636	.4364	0.56	.2123	.2877	0.96	.3315	.1685
0.17	.0675	.4325	0.57	.2157	.2843	0.97	.3340	.1660
0.18	.0714	.4286	0.58	.2190	.2810	0.98	.3365	.1635
0.19	.0753	.4247	0.59	.2224	.2776	0.99	.3389	.1611
0.20	.0793	.4207	0.60	.2257	.2743	1.00	.3413	.1587
0.21	.0832	.4168	0.61	.2291	.2709	1.01	.3438	.1562
0.22	.0871	.4129	0.62	.2324	.2676	1.02	.3461	.1539
0.23	.0910	.4090	0.63	.2357	.2643	1.03	.3485	.1515
0.24	.0948	.4052	0.64	.2389	.2611	1.04	.3508	.1492
0.25	.0987	.4013	0.65	.2422	.2578	1.05	3531	.1469
0.26	.1026	.3974	0.66	.2454	.2546	1.06	.3554	.1446
0.27	.1064	.3936	0.67	.2486	.2514	1.07	.3577	.1423
0.28	.1103	.3897	0.68	.2517	.2483	1.08	.3599	.1401
0.29	.1141	.3859	0.69	.2549	.2451	1.09	.3621	.1379
0.30	.1179	.3821	0.70	.2580	2420	1.10	.3643	.1357
0.31	.1217	.3783	0.71	.2611	.2389	1.11	.3665	.1335
0.32	.1255	.3745	0.72	.2642	.2358	1.12	.3686	.1314
0.33	.1293	.3707	0.73	.2673	.2327	1.13	.3708	.1292
0.34	.1331	.3669	0.74	.2704	.2296	1.14	.3729	.1271
0.35	..1368	.3632	0.75	.2734	.2266	1.15	.3749	.1251
0.36	.1406	.3594	0.76	.2764	.2236	1.16	.3770	.1230
0.37	.1443	.3557	0.77	.2794	.2206	1.17	.3790	.1210
0.38	.1480	.3520	0.78	.2823	.2177	1.18	.3810	.1190
0.39	.1517	.3483	0.79	.2852	.2148	1.19	.3830	.1170

Table D–2 Continued

(A) z	(B) Area Between Mean and z	(C) Area Beyond z	(A) z	(B) Area Between Mean and z	(C) Area Beyond z	(A) z	(B) Area Between Mean and z	(C) Area Beyond z
1.20	.3849	.1151	1.60	.4452	.0548	2.00	.4772	.0228
1.21	.3869	.1131	1.61	.4463	.0537	2.01	.4778	.0222
1.22	.3888	.1112	1.62	.4474	.0526	2.02	.4783	.0217
1.23	.3907	.1093	1.63	.4484	.0516	2.03	.4788	.0212
1.24	.3925	.1075	1.64	.4495	.0505	2.04	.4793	.0207
1.25	.3944	.1056	1.65	.4505	.0495	2.05	.4798	.0202
1.26	.3962	.1038	1.66	.4515	.0485	2.06	.4803	.0197
1.27	.3980	.1020	1.67	.4525	.0475	2.07	.4808	.0192
1.28	.3997	.1003	1.68	.4535	.0465	2.08	.4812	.0188
1.29	.4015	.0985	1.69	.4545	.0455	2.09	.4817	.0183
1.30	.4032	.0968	1.70	.4554	.0446	2.10	.4821	.0179
1.31	.4049	.0951	1.71	.4564	.0436	2.11	.4826	.0174
1.32	.4066	.0934	1.72	.4573	.0427	2.12	.4830	.0170
1.33	.4082	.0918	1.73	.4582	.0418	2.13	.4834	.0166
1.34	.4099	.0901	1.74	.4591	.0409	2.14	.4838	.0162
1.35	.4115	.0885	1.75	.4599	.0401	2.15	.4842	.0158
1.36	.4131	.0869	1.76	.4608	.0392	2.16	.4846	.0154
1.37	.4147	.0853	1.77	.4616	.0384	2.17	.4850	.0150
1.38	.4162	.0838	1.78	.4625	.0375	2.18	.4854	.0146
1.39	.4177	.0823	1.79	.4633	.0367	2.19	.4857	.0143
1.40	.4192	.0808	1.80	.4641	.0359	2.20	.4861	.0139
1.41	.4207	.0793	1.81	.4649	.0351	2.21	.4864	.0136
1.42	.4222	.0778	1.82	.4656	.0344	2.22	.4868	.0132
1.43	.4236	.0764	1.83	.4664	.0336	2.23	.4871	.0129
1.44	.4251	.0749	1.84	.4671	.0329	2.24	.4875	.0125
1.45	.4265	.0735	1.85	.4678	.0322	2.25	.4878	.0122
1.46	.4279	.0721	1.86	.4686	.0314	2.26	.4881	.0119
1.47	.4292	.0708	1.87	.4693	.0307	2.27	.4884	.0116
1.48	.4306	.0694	1.88	.4699	.0301	2.28	.4887	.0113
1.49	.4319	.0681	1.89	.4706	.0294	2.29	.4890	.0110
1.50	.4332	.0668	1.90	.4713	.0287	2.30	.4893	.0107
1.51	.4345	.0655	1.91	.4719	.0281	2.31	.4896	.0104
1.52	.4357	.0643	1.92	.4726	.0274	2.32	.4898	.0102
1.53	.4370	.0630	1.93	.4732	.0268	2.33	.4901	.0099
1.54	.4382	.0618	1.94	.4738	.0262	2.34	.4904	.0096
1.55	.4394	.0606	1.95	.4744	.0256	2.35	.4906	.0094
1.56	.4406	.0594	1.96	.4750	.0250	2.36	.4909	.0091
1.57	.4418	.0582	1.97	.4756	.0244	2.37	.4911	.0089
1.58	.4429	.0571	1.98	.4761	.0239	2.38	.4913	.0087
1.59	.4441	.0559	1.99	.4767	.0233	2.39	.4916	.0084

Table D–2 **Continued**

(A) z	(B) Area Between Mean and z	(C) Area Beyond z
2.40	.4918	.0082
2.41	.4920	.0080
2.42	.4922	.0078
2.43	.4925	.0075
2.44	.4927	.0073
2.45	.4929	.0071
2.46	.4931	.0069
2.47	.4932	.0068
2.48	.4934	.0066
2.49	.4936	.0064
2.50	.4938	.0062
2.51	.4940	.0060
2.52	.4941	.0059
2.53	.4943	.0057
2.54	.4945	.0055
2.55	.4946	.0054
2.56	.4948	.0052
2.57	.4949	.0051
2.58	.4951	.0049
2.59	.4952	.0048
2.60	.4953	.0047
2.61	.4955	.0045
2.62	.4956	.0044
2.63	.4957	.0043
2.64	.4959	.0041
2.65	.4960	.0040
2.66	.4961	.0039
2.67	.4962	.0038
2.68	.4963	.0037
2.69	.4964	.0036
2.70	.4965	.0035
2.71	.4966	.0034
2.72	.4967	.0033
2.73	.4968	.0032
2.74	.4969	.0031

(A) z	(B) Area Between Mean and z	(C) Area Beyond z
2.75	.4970	.0030
2.76	.4971	.0029
2.77	.4972	.0028
2.78	.4973	.0027
2.79	.4974	.0026
2.80	.4974	.0026
2.81	.4975	.0025
2.82	.4976	.0224
2.83	.4977	.0023
2.84	.4977	.0023
2.85	.4978	.0022
2.86	.4979	.0021
2.87	.4979	.0021
2.88	.4980	.0020
2.89	.4981	.0019
2.90	.4981	.0019
2.91	.4982	.0018
2.92	.4982	.0018
2.93	.4983	.0017
2.94	.4984	.0016
2.95	.4984	.0016
2.96	.4985	.0015
2.97	.4985	.0015
2.98	.4986	.0014
2.99	.4986	.0014
3.00	.4987	.0013
3.01	.4987	.0013
3.02	.4987	.0013
3.03	.4988	.0012
3.04	.4988	.0012

(A) z	(B) Area Between Mean and z	(C) Area Beyond z
3.05	.4989	.0011
3.06	.4989	.0011
3.07	.4989	.0011
3.08	.4990	.0010
3.09	.4990	.0010
3.10	.4990	.0010
3.11	.4991	.0009
3.12	.4991	.0009
3.13	.4991	.0009
3.14	.4992	.0008
3.15	.4992	.0008
3.16	.4992	.0008
3.17	.4992	.0008
3.18	.4993	.0007
3.19	.4993	.0007
3.20	.4993	.0007
3.21	.4993	.0007
3.22	.4994	.0006
3.23	.4994	.0006
3.24	.4994	.0006
3.30	.4995	.0005
3.40	.4997	.0003
3.50	.4998	.0002
3.60	.4998	.0002
3.70	.4999	.0001
3.80	.49993	.00007
3.90	.49995	.00005
4.00	.49997	.00003

Table D–3 Critical Values of t

	Proportion of Area Cut Off by the Tabled t (or by $-t$) in One Tail						
	.10	.05	.025	.01	.005	.0025	.0005
	Proportion of Area Cut Off by $|t|$ in Both Tails						
df	**.20**	**.10**	**.05**	**.02**	**.01**	**.005**	**.001**
1	3.078	6.314	12.71	31.82	63.66	127.3	636.6
2	1.886	2.920	4.303	6.965	9.925	14.09	31.60
3	1.638	2.353	3.182	4.541	5.841	7.453	12.92
4	1.533	2.132	2.776	3.747	4.604	5.598	8.610
5	1.476	2.015	2.571	3.365	4.032	4.773	6.869
6	1.440	1.943	2.447	3.143	3.707	4.317	5.959
7	1.415	1.895	2.365	2.998	3.499	4.029	5.408
8	1.397	1.860	2.306	2.896	3.355	3.833	5.041
9	1.383	1.833	2.262	2.821	3.250	3.690	4.781
10	1.372	1.812	2.228	2.764	3.169	3.581	4.587
11	1.363	1.796	2.201	2.718	3.106	3.497	4.437
12	1.356	1.782	2.179	2.681	3.055	3.428	4.318
13	1.350	1.771	2.160	2.650	3.012	3.372	4.221
14	1.345	1.761	2.145	2.624	2.977	3.326	4.140
15	1.341	1.753	2.131	2.602	2.947	3.286	4.073
16	1.337	1.746	2.120	2.583	2.921	3.252	4.015
17	1.333	1.740	2.110	2.567	2.898	3.222	3.965
18	1.330	1.734	2.101	2.552	2.878	3.197	3.922
19	1.328	1.729	2.093	2.539	2.861	3.174	3.883
20	1.325	1.725	2.086	2.528	2.845	3.153	3.850
21	1.323	1.721	2.080	2.518	2.831	3.135	3.819
22	1.321	1.717	2.074	2.508	2.819	3.119	3.792
23	1.319	1.714	2.069	2.500	2.807	3.104	3.768
24	1.318	1.711	2.064	2.492	2.797	3.091	3.745
25	1.316	1.708	2.060	2.485	2.787	3.078	3.725
26	1.315	1.706	2.056	2.479	2.779	3.067	3.707
27	1.314	1.703	2.052	2.473	2.771	3.057	3.690
28	1.313	1.701	2.048	2.467	2.763	3.047	3.674
29	1.311	1.699	2.045	2.462	2.756	3.038	3.659
30	1.310	1.697	2.042	2.457	2.750	3.030	3.646
40	1.303	1.684	2.021	2.423	2.704	2.971	3.551
50	1.299	1.676	2.009	2.403	2.678	2.937	3.496
60	1.296	1.671	2.000	2.390	2.660	2.915	3.460
80	1.292	1.664	1.990	2.374	2.639	2.887	3.416
100	1.290	1.660	1.984	2.364	2.626	2.871	3.390
1000	1.282	1.646	1.962	2.330	2.581	2.813	3.300
∞*	1.282	1.645	1.960	2.326	2.576	2.807	3.291

*Critical values of t at $df = \infty$ are equal to critical values of z in the standard normal curve.

Table D–4 **Critical Values of the Pearson Correlation Coefficient (r)**

	Level of Significance for One-Tailed Test					
	.10	.05	.025	.01	.005	.0005
	Level of Significance for Two-Tailed Test					
df	.20	.10	.05	.02	.01	.001
1	.951	.988	.997	.9995	.9999	1.000
2	.800	.900	.950	.980	.990	.999
3	.687	.805	.878	.934	.959	.991
4	.608	.729	.811	.882	.917	.974
5	.551	.669	.755	.833	.875	.951
6	.507	.622	.707	.789	.834	.925
7	.472	.582	.666	.750	.798	.898
8	.443	.549	.632	.716	.765	.872
9	.419	.521	.602	.685	.735	.847
10	.398	.497	.576	.658	.708	.823
11	.380	.476	.553	.634	.684	.801
12	.365	.457	.532	.612	.661	.780
13	.351	.441	.514	.592	.641	.760
14	.338	.426	.497	.574	.623	.742
15	.327	.412	.482	.558	.606	.725
16	.317	.400	.468	.543	.590	.708
17	.308	.389	.456	.529	.575	.693
18	.299	.378	.444	.516	.561	.688
19	.291	.369	.433	.503	.549	.665
20	.284	.360	.423	.492	.537	.652
21	.277	.352	.413	.482	.526	.640
22	.271	.344	.404	.472	.515	.629
23	.265	.337	.396	.462	.505	.618
24	.260	.330	.388	.453	.496	.607
25	.255	.323	.381	.445	.487	.597
26	.250	.317	.374	.437	.479	.588
27	.245	.311	.367	.430	.471	.579
28	.241	.306	.361	.423	.463	.570
29	.237	.301	.355	.416	.456	.562
30	.233	.296	.349	.409	.449	.554
35	.216	.275	.325	.381	.418	.519
40	.202	.257	.304	.358	.393	.490
60	.165	.211	.250	.295	.325	.408
80	.143	.183	.217	.257	.283	.357
100	.128	.164	.195	.230	.254	.321

Table D–5 Critical Values of F for $\alpha = .05$ (Lightface) and $\alpha = .01$ (Boldface)

Degrees of Freedom for Numerator Mean Square (columns); Degrees of Freedom for Denominator Mean Square (rows)

	1	2	3	4	5	6	7	8	9	10	11	12	14	16	20
1	161	200	216	225	230	234	237	239	241	242	243	244	245	246	248
	4052	**4999**	**5403**	**5625**	**5764**	**5859**	**5928**	**5981**	**6022**	**6056**	**6082**	**6106**	**6142**	**6169**	**6208**
2	18.51	19.00	19.16	19.25	19.30	19.33	19.36	19.37	19.38	19.39	19.40	19.41	19.42	19.43	19.44
	98.49	**99.00**	**99.17**	**99.25**	**99.30**	**99.33**	**99.34**	**99.36**	**99.38**	**99.40**	**99.41**	**99.42**	**99.43**	**99.44**	**99.45**
3	10.13	9.55	9.28	9.12	9.01	8.94	8.88	8.84	8.81	8.78	8.76	8.74	8.71	8.69	8.66
	34.12	**30.82**	**29.46**	**28.71**	**28.24**	**27.91**	**27.67**	**27.49**	**27.34**	**27.23**	**27.13**	**27.05**	**26.92**	**26.83**	**26.69**
4	7.71	6.94	6.59	6.39	6.26	6.16	6.09	6.04	6.00	5.96	5.93	5.91	5.87	5.84	5.80
	21.20	**18.00**	**16.69**	**15.98**	**15.52**	**15.21**	**14.98**	**14.80**	**14.66**	**14.54**	**14.45**	**14.37**	**14.24**	**14.15**	**14.02**
5	6.61	5.79	5.41	5.19	5.05	4.95	4.88	4.82	4.78	4.74	4.70	4.68	4.64	4.60	4.56
	16.26	**13.27**	**12.06**	**11.39**	**10.97**	**10.67**	**10.45**	**10.27**	**10.15**	**10.05**	**9.96**	**9.89**	**9.77**	**9.68**	**9.55**
6	5.99	5.14	4.76	4.53	4.39	4.28	4.21	4.15	4.10	4.06	4.03	4.00	3.96	3.92	3.87
	13.74	**10.92**	**9.78**	**9.15**	**8.75**	**8.47**	**8.26**	**8.10**	**7.98**	**7.87**	**7.79**	**7.72**	**7.60**	**7.52**	**7.39**
7	5.59	4.47	4.35	4.12	3.97	3.87	3.79	3.73	3.68	3.63	3.60	3.57	3.52	3.49	3.44
	12.25	**9.55**	**8.45**	**7.85**	**7.46**	**7.19**	**7.00**	**6.84**	**6.71**	**6.62**	**6.54**	**6.47**	**6.35**	**6.27**	**6.15**
8	5.32	4.46	4.07	3.84	3.69	3.58	3.50	3.44	3.39	3.34	3.31	3.28	3.23	3.20	3.15
	11.26	**8.65**	**7.59**	**7.01**	**6.63**	**6.37**	**6.19**	**6.03**	**5.91**	**5.82**	**5.74**	**5.67**	**5.56**	**5.48**	**5.36**
9	5.12	4.26	3.86	3.63	3.48	3.37	3.29	3.23	3.18	3.13	3.10	3.07	3.02	2.98	2.93
	10.56	**8.02**	**6.99**	**6.42**	**6.06**	**5.80**	**5.62**	**5.47**	**5.35**	**5.26**	**5.18**	**5.11**	**5.00**	**4.92**	**4.80**
10	4.96	4.10	3.71	3.48	3.33	3.22	3.14	3.07	3.02	2.97	2.94	2.91	2.86	2.82	2.77
	10.04	**7.56**	**6.55**	**5.99**	**5.64**	**5.39**	**5.21**	**5.06**	**4.95**	**4.85**	**4.78**	**4.71**	**4.60**	**4.52**	**4.41**
11	4.84	3.98	3.59	3.36	3.20	3.09	3.01	2.95	2.90	2.86	2.82	2.79	2.74	2.70	2.65
	9.65	**7.20**	**6.22**	**5.67**	**5.32**	**5.07**	**4.88**	**4.74**	**4.63**	**4.54**	**4.46**	**4.40**	**4.29**	**4.21**	**4.10**
12	4.75	3.88	3.49	3.26	3.11	3.00	2.92	2.85	2.80	2.76	2.72	2.69	2.64	2.60	2.54
	9.33	**6.93**	**5.95**	**5.41**	**5.06**	**4.82**	**4.65**	**4.50**	**4.39**	**4.30**	**4.22**	**4.16**	**4.05**	**3.98**	**3.86**
13	4.67	3.80	3.41	3.18	3.02	2.92	2.84	2.77	2.72	2.67	2.63	2.60	2.55	2.51	2.46
	9.07	**6.70**	**5.74**	**5.20**	**4.86**	**4.62**	**4.44**	**4.30**	**4.19**	**4.10**	**4.02**	**3.96**	**3.85**	**3.78**	**3.67**
14	4.60	3.74	3.34	3.11	2.96	2.85	2.77	2.70	2.65	2.60	2.56	2.53	2.48	2.44	2.39
	8.86	**6.51**	**5.56**	**5.03**	**4.69**	**4.46**	**4.28**	**4.14**	**4.03**	**3.94**	**3.86**	**3.80**	**3.70**	**3.62**	**3.51**
15	4.54	3.68	3.29	3.06	2.90	2.79	2.70	2.64	2.59	2.55	2.51	2.48	2.43	2.39	2.33
	8.68	**6.36**	**5.42**	**4.89**	**4.56**	**4.32**	**4.14**	**4.00**	**3.89**	**3.80**	**3.73**	**3.67**	**3.56**	**3.48**	**3.36**
16	4.49	3.63	3.24	3.01	2.85	2.74	2.66	2.59	2.54	2.49	2.45	2.42	2.37	2.33	2.28
	8.53	**6.23**	**5.29**	**4.77**	**4.44**	**4.20**	**4.03**	**3.89**	**3.78**	**3.69**	**3.61**	**3.55**	**3.45**	**3.37**	**3.25**
17	4.45	3.59	3.20	2.96	2.81	2.70	2.62	2.55	2.50	2.45	2.41	2.38	2.33	2.29	2.23
	8.40	**6.11**	**5.18**	**4.67**	**4.34**	**4.10**	**3.93**	**3.79**	**3.68**	**3.59**	**3.52**	**3.45**	**3.35**	**3.27**	**3.16**
18	4.41	3.55	3.16	2.93	2.77	2.66	2.58	2.51	2.46	2.41	2.37	2.34	2.29	2.25	2.19
	8.28	**6.01**	**5.09**	**4.58**	**4.25**	**4.01**	**3.85**	**3.71**	**3.60**	**3.51**	**3.44**	**3.37**	**3.27**	**3.19**	**3.07**
19	4.38	3.52	3.13	2.90	2.74	2.63	2.55	2.48	2.43	2.38	2.34	2.31	2.26	2.21	2.15
	8.18	**5.93**	**5.01**	**4.50**	**4.17**	**3.94**	**3.77**	**3.63**	**3.52**	**3.43**	**3.36**	**3.30**	**3.19**	**3.12**	**3.00**
20	4.35	3.49	3.10	2.87	2.71	2.60	2.52	2.45	2.40	2.35	2.31	2.28	2.23	2.18	2.12
	8.10	**5.85**	**4.94**	**4.43**	**4.10**	**3.87**	**3.71**	**3.56**	**3.45**	**3.37**	**3.30**	**3.23**	**3.13**	**3.05**	**2.94**
21	4.32	3.47	3.07	2.84	2.68	2.57	2.49	2.42	2.37	2.32	2.28	2.25	2.20	2.15	2.09
	8.02	**5.78**	**4.87**	**4.37**	**4.04**	**3.81**	**3.65**	**3.51**	**3.40**	**3.31**	**3.24**	**3.17**	**3.07**	**2.99**	**2.88**

Table D–5 Continued

Degrees of Freedom for Denominator Mean Square	Degrees of Freedom for Numerator Mean Square														
	1	2	3	4	5	6	7	8	9	10	11	12	14	16	20
22	4.30	3.44	3.05	2.82	2.66	2.55	2.47	2.40	2.35	2.30	2.26	2.23	2.18	2.13	2.07
	7.94	**5.72**	**4.82**	**4.31**	**3.99**	**3.76**	**3.59**	**3.45**	**3.35**	**3.26**	**3.18**	**3.12**	**3.02**	**2.94**	**2.83**
23	4.28	3.42	3.03	2.80	2.64	2.53	2.45	2.38	2.32	2.28	2.24	2.20	2.14	2.10	2.04
	7.88	**5.66**	**4.76**	**4.26**	**3.94**	**3.71**	**3.54**	**3.41**	**3.30**	**3.21**	**3.14**	**3.07**	**2.97**	**2.89**	**2.78**
24	4.26	3.40	3.01	2.78	2.62	2.51	2.43	2.36	2.30	2.26	2.22	2.18	2.13	2.09	2.02
	7.82	**5.61**	**4.72**	**4.22**	**3.90**	**3.67**	**3.50**	**3.36**	**3.25**	**3.17**	**3.09**	**3.03**	**2.93**	**2.85**	**2.74**
25	4.24	3.38	2.99	2.76	2.60	2.49	2.41	2.34	2.28	2.24	2.20	2.16	2.11	2.06	2.00
	7.77	**5.57**	**4.68**	**4.18**	**3.86**	**3.63**	**3.46**	**3.32**	**3.21**	**3.13**	**3.05**	**2.99**	**2.89**	**2.81**	**2.70**
26	4.22	3.37	2.98	2.74	2.59	2.47	2.39	2.32	2.27	2.22	2.18	2.15	2.10	2.05	1.99
	7.72	**5.53**	**4.64**	**4.14**	**3.82**	**3.59**	**3.42**	**3.29**	**3.17**	**3.09**	**3.02**	**2.96**	**2.86**	**2.77**	**2.66**
27	4.21	3.35	2.96	2.73	2.57	2.46	2.37	2.30	2.25	2.20	2.16	2.13	2.08	2.03	1.97
	7.68	**5.49**	**4.60**	**4.11**	**3.79**	**3.56**	**3.39**	**3.26**	**3.14**	**3.06**	**2.98**	**2.93**	**2.83**	**2.74**	**2.63**
28	4.20	3.34	2.95	2.71	2.56	2.44	2.36	2.29	2.24	2.19	2.15	2.12	2.06	2.02	1.96
	7.64	**5.45**	**4.57**	**4.07**	**3.76**	**3.53**	**3.36**	**3.23**	**3.11**	**3.03**	**2.95**	**2.90**	**2.80**	**2.71**	**2.60**
29	4.18	3.33	2.93	2.70	2.54	2.43	2.35	2.28	2.22	2.18	2.14	2.10	2.05	2.00	1.94
	7.60	**5.42**	**4.54**	**4.04**	**3.73**	**3.50**	**3.33**	**3.20**	**3.08**	**3.00**	**2.92**	**2.87**	**2.77**	**2.68**	**2.57**
30	4.17	3.32	2.92	2.69	2.53	2.42	2.34	2.27	2.21	2.16	2.12	2.09	2.04	1.99	1.93
	7.56	**5.39**	**4.51**	**4.02**	**3.70**	**3.47**	**3.30**	**3.17**	**3.06**	**2.98**	**2.90**	**2.84**	**2.74**	**2.66**	**2.55**
32	4.15	3.30	2.90	2.67	2.51	2.40	2.32	2.25	2.19	2.14	2.10	2.07	2.02	1.97	1.91
	7.50	**5.34**	**4.46**	**3.97**	**3.66**	**3.42**	**3.25**	**3.12**	**3.01**	**2.94**	**2.86**	**2.80**	**2.70**	**2.62**	**2.51**
34	4.13	3.28	2.88	2.65	2.49	2.38	2.30	2.23	2.17	2.12	2.08	2.05	2.00	1.95	1.89
	7.44	**5.29**	**4.42**	**3.93**	**3.61**	**3.38**	**3.21**	**3.08**	**2.97**	**2.89**	**2.82**	**2.76**	**2.66**	**2.58**	**2.47**
36	4.11	3.26	2.86	2.63	2.48	2.36	2.28	2.21	2.15	2.10	2.06	2.03	1.98	1.93	1.87
	7.39	**5.25**	**4.38**	**3.89**	**3.58**	**3.35**	**3.18**	**3.04**	**2.94**	**2.86**	**2.78**	**2.72**	**2.62**	**2.54**	**2.43**
38	4.10	3.25	2.85	2.62	2.46	2.35	2.26	2.19	2.14	2.09	2.05	2.02	1.96	1.92	1.85
	7.35	**5.21**	**4.34**	**3.86**	**3.54**	**3.32**	**3.15**	**3.02**	**2.91**	**2.82**	**2.75**	**2.69**	**2.59**	**2.51**	**2.40**
40	4.08	3.23	2.84	2.61	2.45	2.34	2.25	2.18	2.12	2.07	2.04	2.00	1.95	1.90	1.84
	7.31	**5.18**	**4.31**	**3.83**	**3.51**	**3.29**	**3.12**	**2.99**	**2.88**	**2.80**	**2.73**	**2.66**	**2.56**	**2.49**	**2.37**
42	4.07	3.22	2.83	2.59	2.44	2.32	2.24	2.17	2.11	2.06	2.02	1.99	1.94	1.89	1.82
	7.27	**5.15**	**4.29**	**3.80**	**3.49**	**3.26**	**3.10**	**2.96**	**2.86**	**2.77**	**2.70**	**2.64**	**2.54**	**2.46**	**2.35**
44	4.06	3.21	2.82	2.58	2.43	2.31	2.23	2.16	2.10	2.05	2.01	1.98	1.92	1.88	1.81
	7.24	**5.12**	**4.26**	**3.78**	**3.46**	**3.24**	**3.07**	**2.94**	**2.84**	**2.75**	**2.68**	**2.62**	**2.52**	**2.44**	**2.32**
46	4.05	3.20	2.81	2.57	2.42	2.30	2.22	2.14	2.09	2.04	2.00	1.97	1.91	1.87	1.80
	7.21	**5.10**	**4.24**	**3.76**	**3.44**	**3.22**	**3.05**	**2.92**	**2.82**	**2.73**	**2.66**	**2.60**	**2.50**	**2.42**	**2.30**
48	4.04	3.19	2.80	2.56	2.41	2.30	2.21	2.14	2.08	2.03	1.99	1.96	1.90	1.86	1.79
	7.19	**5.08**	**4.22**	**3.74**	**3.42**	**3.20**	**3.04**	**2.90**	**2.80**	**2.71**	**2.64**	**2.58**	**2.48**	**2.40**	**2.28**
50	4.03	3.18	2.79	2.56	2.40	2.29	2.20	2.13	2.07	2.02	1.98	1.95	1.90	1.85	1.78
	7.17	**5.06**	**4.20**	**3.72**	**3.41**	**3.18**	**3.02**	**2.88**	**2.78**	**2.70**	**2.62**	**2.56**	**2.46**	**2.39**	**2.26**
55	4.02	3.17	2.78	2.54	2.38	2.27	2.18	2.11	2.05	2.00	1.97	1.93	1.88	1.83	1.76
	7.12	**5.01**	**4.16**	**3.68**	**3.37**	**3.15**	**2.98**	**2.85**	**2.75**	**2.66**	**2.59**	**2.53**	**2.43**	**2.35**	**2.23**
60	4.00	3.15	2.76	2.52	2.37	2.25	2.17	2.10	2.04	1.99	1.95	1.92	1.86	1.81	1.75
	7.08	**4.98**	**4.13**	**3.65**	**3.34**	**3.12**	**2.95**	**2.82**	**2.72**	**2.63**	**2.56**	**2.50**	**2.40**	**2.32**	**2.20**

Table D–5 Continued

Degrees of Freedom for Denominator Mean Square	Degrees of Freedom for Numerator Mean Square: 1	2	3	4	5	6	7	8	9	10	11	12	14	16	20
65	3.99	3.14	2.75	2.51	2.36	2.24	2.15	2.08	2.02	1.98	1.94	1.90	1.85	1.80	1.73
	7.04	**4.95**	**4.10**	**3.62**	**3.31**	**3.09**	**2.93**	**2.79**	**2.70**	**2.61**	**2.54**	**2.47**	**2.37**	**2.30**	**2.18**
70	3.98	3.13	2.74	2.50	2.35	2.23	2.14	2.07	2.01	1.97	1.93	1.89	1.84	1.79	1.72
	7.01	**4.92**	**4.08**	**3.60**	**3.29**	**3.07**	**2.91**	**2.77**	**2.67**	**2.59**	**2.51**	**2.45**	**2.35**	**2.28**	**2.15**
80	3.96	3.11	2.72	2.48	2.33	2.21	2.12	2.05	1.99	1.95	1.91	1.88	1.82	1.77	1.70
	6.96	**4.88**	**4.04**	**3.56**	**3.25**	**3.04**	**2.87**	**2.74**	**2.64**	**2.55**	**2.48**	**2.41**	**2.32**	**2.24**	**2.11**
100	3.94	3.09	2.70	2.46	2.30	2.19	2.10	2.03	1.97	1.92	1.88	1.85	1.79	1.75	1.68
	6.90	**4.82**	**3.98**	**3.51**	**3.20**	**2.99**	**2.82**	**2.69**	**2.59**	**2.51**	**2.43**	**2.36**	**2.26**	**2.19**	**2.06**
125	3.92	3.07	2.68	2.44	2.29	2.17	2.08	2.01	1.95	1.90	1.86	1.83	1.77	1.72	1.65
	6.84	**4.78**	**3.94**	**3.47**	**3.17**	**2.95**	**2.79**	**2.65**	**2.56**	**2.47**	**2.40**	**2.33**	**2.23**	**2.15**	**2.03**
150	3.91	3.06	2.67	2.43	2.27	2.16	2.07	2.00	1.94	1.89	1.85	1.82	1.76	1.71	1.64
	6.81	**4.75**	**3.91**	**3.44**	**3.14**	**2.92**	**2.76**	**2.62**	**2.53**	**2.44**	**2.37**	**2.30**	**2.20**	**2.12**	**2.00**
200	3.89	3.04	2.65	2.41	2.26	2.14	2.05	1.98	1.92	1.87	1.83	1.80	1.74	1.69	1.62
	6.76	**4.71**	**3.88**	**3.41**	**3.11**	**2.90**	**2.73**	**2.60**	**2.50**	**2.41**	**2.34**	**2.28**	**2.17**	**2.09**	**1.97**
400	3.86	3.02	2.62	2.39	2.23	2.12	2.03	1.96	1.90	1.85	1.81	1.78	1.72	1.67	1.60
	6.70	**4.66**	**3.83**	**3.36**	**3.06**	**2.85**	**2.69**	**2.55**	**2.46**	**2.37**	**2.29**	**2.23**	**2.12**	**2.04**	**1.92**
1000	3.85	3.00	2.61	2.38	2.22	2.10	2.02	1.95	1.89	1.84	1.80	1.76	1.70	1.65	1.58
	6.66	**4.62**	**3.80**	**3.34**	**3.04**	**2.82**	**2.66**	**2.53**	**2.43**	**2.34**	**2.26**	**2.20**	**2.09**	**2.01**	**1.89**
∞	3.84	2.99	2.60	2.37	2.21	2.09	2.01	1.94	1.88	1.83	1.79	1.75	1.69	1.64	1.57
	6.64	**4.60**	**3.78**	**3.32**	**3.02**	**2.80**	**2.64**	**2.51**	**2.41**	**2.32**	**2.24**	**2.18**	**2.07**	**1.99**	**1.87**

Table D–6 **Critical Values of q for α = .05 (Lightface) and α= .01 (Boldface) for Tukey's Honestly Significant Difference Test**

Degrees of Freedom for Error Mean Square	k = Number of Treatments										
	2	3	4	5	6	7	8	9	10	11	12
5	3.64	4.60	5.22	5.67	6.03	6.33	6.58	6.80	6.99	7.17	7.32
	5.70	**6.98**	**7.80**	**8.42**	**8.91**	**9.32**	**9.67**	**9.97**	**10.24**	**10.48**	**10.70**
6	3.46	4.34	4.90	5.30	5.63	5.90	6.12	6.32	6.49	6.65	6.79
	5.24	**6.33**	**7.03**	**7.56**	**7.97**	**8.32**	**8.61**	**8.87**	**9.10**	**9.30**	**9.48**
7	3.34	4.16	4.68	5.06	5.36	5.61	5.82	6.00	6.16	6.30	6.43
	4.95	**5.92**	**6.54**	**7.01**	**7.37**	**7.68**	**7.94**	**8.17**	**8.37**	**8.55**	**8.71**
8	3.26	4.04	4.53	4.89	5.17	5.40	5.60	5.77	5.92	6.05	6.18
	4.75	**5.64**	**6.20**	**6.62**	**6.96**	**7.24**	**7.47**	**7.68**	**7.86**	**8.03**	**8.18**
9	3.20	3.95	4.41	4.76	5.02	5.24	5.43	5.59	5.74	5.87	5.98
	4.60	**5.43**	**5.96**	**6.35**	**6.66**	**6.91**	**7.13**	**7.33**	**7.49**	**7.65**	**7.78**
10	3.15	3.88	4.33	4.65	4.91	5.12	5.30	5.46	5.60	5.72	5.83
	4.48	**5.27**	**5.77**	**6.14**	**6.43**	**6.67**	**6.87**	**7.05**	**7.21**	**7.36**	**7.49**
11	3.11	3.82	4.26	4.57	4.82	5.03	5.20	5.35	5.49	5.61	5.71
	4.39	**5.15**	**5.62**	**5.97**	**6.25**	**6.48**	**6.67**	**6.84**	**6.99**	**7.13**	**7.25**
12	3.08	3.77	4.20	4.51	4.75	4.95	5.12	5.27	5.39	5.51	5.61
	4.32	**5.05**	**5.50**	**4.84**	**6.10**	**6.32**	**6.51**	**6.67**	**6.81**	**6.94**	**7.06**
13	3.06	3.73	4.15	4.45	4.69	4.88	5.05	5.19	5.32	5.43	5.53
	4.26	**4.96**	**5.40**	**5.73**	**5.98**	**6.19**	**6.37**	**6.53**	**6.67**	**6.79**	**6.90**
14	3.03	3.70	4.11	4.41	4.64	4.83	4.99	5.13	5.25	5.36	5.46
	4.21	**4.89**	**5.32**	**5.63**	**5.88**	**6.08**	**6.26**	**6.41**	**6.54**	**6.66**	**6.77**
15	3.01	3.67	4.08	4.37	4.59	4.78	4.94	5.08	5.20	5.31	5.40
	4.17	**4.84**	**5.25**	**5.56**	**5.80**	**5.99**	**6.16**	**6.31**	**6.44**	**6.55**	**6.66**
16	3.00	3.65	4.05	4.33	4.56	4.74	4.90	5.03	5.15	5.26	5.35
	4.13	**4.79**	**5.19**	**5.49**	**5.72**	**5.92**	**6.08**	**6.22**	**6.35**	**6.46**	**6.56**
17	2.98	3.63	4.02	4.30	4.52	4.70	4.86	4.99	5.11	5.21	5.31
	4.10	**4.74**	**5.14**	**5.43**	**5.66**	**5.85**	**6.01**	**6.15**	**6.27**	**6.38**	**6.48**
18	2.97	3.61	4.00	4.28	4.49	4.67	4.82	4.96	5.07	5.17	5.27
	4.07	**4.70**	**5.09**	**5.38**	**5.60**	**5.79**	**5.94**	**6.08**	**6.20**	**6.31**	**6.41**
19	2.96	3.59	3.98	4.25	4.47	4.65	4.79	4.92	5.04	5.14	5.23
	4.05	**4.67**	**5.05**	**5.33**	**5.55**	**5.73**	**5.89**	**6.02**	**6.14**	**6.25**	**6.34**
20	2.95	3.58	3.96	4.23	4.45	4.62	4.77	4.90	5.01	5.11	5.20
	4.02	**4.64**	**5.02**	**5.29**	**5.51**	**5.69**	**5.84**	**5.97**	**6.09**	**6.19**	**6.28**
24	2.92	3.53	3.90	4.17	4.37	4.54	4.68	4.81	4.92	5.01	5.10
	3.96	**4.55**	**4.91**	**5.17**	**5.37**	**5.54**	**5.69**	**5.81**	**5.92**	**6.02**	**6.11**
30	2.89	3.49	3.85	4.10	4.30	4.46	4.60	4.72	4.82	4.92	5.00
	3.89	**4.45**	**4.80**	**5.05**	**5.24**	**5.40**	**5.54**	**5.65**	**5.76**	**5.85**	**5.93**
40	2.86	3.44	3.79	4.04	4.23	4.39	4.52	4.63	4.73	4.82	4.90
	3.82	**4.37**	**4.70**	**4.93**	**5.11**	**5.26**	**5.39**	**5.50**	**5.60**	**5.69**	**5.76**
60	3.83	3.40	3.74	3.98	4.16	4.31	4.44	4.55	4.65	4.73	4.81
	3.76	**4.28**	**4.59**	**4.82**	**4.99**	**5.13**	**5.25**	**5.36**	**5.45**	**5.53**	**5.60**
120	2.80	3.36	3.68	3.92	4.10	4.24	4.36	4.47	4.56	4.64	4.71
	3.70	**4.20**	**4.50**	**4.71**	**4.87**	**5.01**	**5.12**	**5.21**	**5.30**	**5.37**	**5.44**
∞	2.77	3.31	3.63	3.86	4.03	4.17	4.29	4.39	4.47	4.55	4.62
	3.64	**4.12**	**4.40**	**4.60**	**4.76**	**4.88**	**4.99**	**5.08**	**5.16**	**5.23**	**5.29**

Table D–7 Critical Values of U in the Mann–Whitney U Test for $\alpha = .05$ (Two-Tailed) and $\alpha = .025$ (One-Tailed)

	N_A																		
N_B	2	3	4	5	6	7	8	9	10	11	12	13	14	15	16	17	18	19	20
2							1	1	1	1	2	2	2	2	2	3	3	3	3
							15	17	19	21	22	24	26	28	30	31	33	35	37
3				1	2	2	3	3	3	4	5	5	6	6	7	7	8	8	9
				14	16	19	21	24	26	29	31	34	36	39	41	44	46	49	51
4			1	2	3	4	5	5	6	7	8	9	10	11	12	12	13	14	14
			15	18	21	24	27	31	34	37	40	43	46	49	52	56	59	62	66
5		1	2	3	4	6	7	8	9	10	12	13	14	15	16	18	19	20	21
		14	18	22	26	29	33	37	41	45	48	52	56	60	64	67	71	75	79
6		2	3	4	6	7	9	11	12	14	15	17	18	20	22	23	25	26	28
		16	21	26	30	35	39	43	48	52	57	61	66	70	74	79	83	88	92
7		2	4	6	7	9	11	13	15	17	19	21	23	25	27	29	31	33	35
		19	24	29	35	40	45	50	55	60	65	70	75	80	85	90	95	100	105
8	1	3	5	7	9	11	14	16	18	20	23	25	27	30	32	35	37	39	42
	15	21	27	33	39	45	50	56	62	68	73	79	85	90	96	101	107	113	118
9	1	3	5	8	11	13	16	18	21	24	27	29	32	35	38	40	43	46	49
	17	24	31	37	43	50	56	63	69	75	81	88	94	100	106	113	119	125	131
10	1	4	6	9	12	15	18	21	24	27	30	34	37	40	43	46	49	53	56
	19	26	34	41	48	55	62	69	76	83	90	96	103	110	117	124	131	137	144
11	1	4	7	10	14	17	20	24	27	31	34	38	41	45	48	52	56	59	63
	21	29	37	45	52	60	68	75	83	90	98	105	113	120	128	135	142	150	157
12	2	5	8	12	15	19	23	27	30	34	38	42	46	50	54	58	62	66	70
	22	31	40	48	57	65	73	81	90	98	106	114	122	130	138	146	154	162	170
13	2	5	9	13	17	21	25	29	34	38	42	46	51	55	60	64	68	73	77
	24	34	43	52	61	70	79	88	96	105	114	123	131	140	148	157	166	174	183
14	2	6	10	14	18	23	27	32	37	41	46	51	56	60	65	68	75	79	84
	26	36	46	56	66	75	85	94	103	113	122	131	140	150	159	170	177	187	196
15	2	6	11	15	20	25	30	35	40	45	50	55	60	65	71	76	81	86	91
	28	39	49	60	70	80	90	100	110	120	130	140	150	160	169	179	189	199	209
16	2	7	12	16	22	27	32	38	43	48	54	60	65	71	76	82	87	93	99
	30	41	52	64	74	85	96	106	117	128	138	148	159	169	180	190	201	211	221
17	3	7	12	18	23	29	35	40	46	52	58	64	68	76	82	88	94	100	106
	31	44	56	67	79	90	101	113	124	135	146	157	170	179	190	201	212	223	234
18	3	8	13	19	25	31	37	43	49	56	62	68	75	81	87	94	100	107	113
	33	46	59	71	83	95	107	119	131	142	154	166	177	189	201	212	224	235	247
19	3	8	14	20	26	33	39	46	53	59	66	73	79	86	93	100	107	114	120
	35	49	62	75	88	100	113	125	137	150	162	174	187	199	211	223	235	247	260
20	3	9	14	21	28	35	42	49	56	63	70	77	84	91	99	106	113	120	128
	37	51	66	79	92	105	118	131	144	157	170	183	196	209	221	234	247	260	272

Table D–8 **Critical Values of W in the Wilcoxon Matched-Pairs Signed-Ranks Test**

	Level of Significance for One-Tailed Test					Level of Significance for One-Tailed Test			
	.05	.025	.01	.005		.05	.025	.01	.005
	Level of Significance for Two-Tailed Test					Level of Significance for Two-Tailed Test			
n	.10	.05	.02	.01	*n*	.10	.05	.02	.01
5	0	—	—	—	28	130	116	101	91
6	2	0	—	—	29	140	126	110	100
7	3	2	0	—	30	151	137	120	109
8	5	3	1	0	31	163	147	130	118
9	8	5	3	1	32	175	159	140	128
10	10	8	5	3	33	187	170	151	138
11	13	10	7	5	34	200	182	162	148
12	17	13	9	7	35	213	195	173	159
13	21	17	12	9	36	227	208	185	171
14	25	21	15	12	37	241	221	198	182
15	30	25	19	15	38	256	235	211	194
16	35	29	23	19	39	271	249	224	207
17	41	34	27	23	40	286	264	238	220
18	47	40	32	27	41	302	279	252	233
19	53	46	37	32	42	319	294	266	247
20	60	52	43	37	43	336	310	281	261
21	67	58	49	42	44	353	327	296	276
22	75	65	55	48	45	371	343	312	291
23	83	73	62	54	46	389	361	328	307
24	91	81	69	61	47	407	378	345	322
25	110	89	76	68	48	426	396	362	339
26	110	98	84	75	49	446	415	379	355
27	119	107	92	83	50	466	434	397	373

To be significant, the obtained W must be *equal to* or *less than* the critical value.

Index

H

I

J

K

L

M

N

O

P

Q

R

S

U

V

W

Y

Z